An Algebraic Introduction to K-Theory

An Algebraic Introduction to K-Theory

This book is an introduction to K-theory and a text in algebra. These two roles are entirely compatible. On the one hand, nothing more than the basic algebra of groups, rings, and modules is needed to explain the classical algebraic K-theory. On the other hand, K-theory is a natural organizing principle for the standard topics of a second course in algebra, and these topics are presented carefully here, with plenty of excercises at the end of each short section. The reader will not only learn algebraic K-theory, but also Dedekind domains, classical groups, semisimple rings, character theory, quadratic forms, tensor products, localization, completion, tensor algebras, symmetric algebras, central simple algebras, and Brauer groups.

The presentation is self-contained, with all the necessary background and proofs, and is divided into short sections with exercises to reinforce the ideas and suggest further lines of inquiry. The prerequisites are minimal: just a first semester of algebra (including Galois theory and modules over a principal ideal domain). No experience with homological algebra, analysis, geometry, number theory, or topology is assumed. The author has successfully used this text to teach algebra to first-year graduate students. Selected topics can be used to construct a variety of one-semester courses; coverage of the entire text requires a full year.

Bruce A. Magurn is Professor of Mathematics at Miami University in Oxford, Ohio, where he has taught for fifteen years. He edited the AMS volume *Reviews in K-Theory, 1940-84*. This book originated from courses taught by the author at Miami University, the University of Oklahoma, and the University of Padua.

ENCYCLOPEDIA OF MATHEMATICS AND ITS APPLICATIONS

An Algebraic Introduction to K-Theory

BRUCE A. MAGURN
Miami University

CAMBRIDGE
UNIVERSITY PRESS

CAMBRIDGE UNIVERSITY PRESS
Cambridge, New York, Melbourne, Madrid, Cape Town, Singapore,
São Paulo, Delhi, Dubai, Tokyo

Cambridge University Press
The Edinburgh Building, Cambridge CB2 8RU, UK

Published in the United States of America by Cambridge University Press, New York

www.cambridge.org
Information on this title: www.cambridge.org/9780521106580

First published 2002
This digitally printed version (with corrections) 2009

A catalogue record for this publication is available from the British Library

Library of Congress Cataloguing in Publication data

Magurn, Bruce A.
  An algebraic introduction to K-theory / Bruce A. Magurn.
    p. cm – (Encyclopedia of mathematics and its applications ; v. 87)
  Includes bibliographical references and index.
  ISBN 0-521-80078-1
  1. K-theory. I. Title. II. Series.

  QA612.33 .M34 2002
  512′.55–dc21   2001043552

ISBN 978-0-521-80078-5 Hardback
ISBN 978-0-521-10658-0 Paperback

Contents

Preface

This book is intended as an introduction to algebraic K-theory that can serve as a second-semester course in algebra. A first algebra course develops the basic structures of groups, rings, and modules. But a reader of research literature in algebra soon encounters a second level of structures, such as class groups, Burnside rings, representation rings, Witt rings, and Brauer groups. These are groups, rings, or modules whose elements are themselves isomorphism classes of groups, rings, or modules. Each of these second-level structures is a variation of a universal construction developed by A. Grothendieck in 1958. Given a category $\mathcal{C}$ of objects, Grothendieck found a natural way to construct an abelian group $K(\mathcal{C})$ of the isomorphism classes of those objects. By Grothendieck's own account, his letter K probably stood for *Klasse*, the German word for class. This is the source of the K in K-theory.

Algebraic K-theory is the study of groups of classes of algebraic objects. It focuses on a sequence of abelian groups $K_n(R)$ associated to each ring R. The first of these is $K_0(R)$, Grothendieck's group $K(\mathcal{C})$, where $\mathcal{C}$ is a certain category of R-modules. It is used to create a sort of dimension for R-modules that lack a basis. The group $K_1(R)$ consists of the row-equivalence classes of invertible matrices over R and is used to study determinants, and $K_2(R)$ measures the fine details of row-reduction of matrices over R.

Many deep problems in algebra have been solved through algebraic K-theory, such as the normal integral basis problem in number fields, the zero-divisor conjecture for integral group rings of solvable groups, the classification of normal subgroups of linear groups, and the description of the Brauer group of a field in terms of cyclic algebras. But, beyond this, K-theory has brought algebraic techniques to bear in the solution of important problems in topology, geometry, number theory, and functional analysis.

Unfortunately, the currently available introductions to algebraic K-theory expect a great deal of the reader: Some background in algebraic topology, algebraic geometry, homological algebra, or functional analysis is taken as a prerequisite. This is due in part to the important role played by K-theory in disciplines outside of algebra, and it is partly because the "higher" K-groups $K_3(R)$, $K_4(R)$, ... are defined and studied by using algebraic geometry and topology.

But, at an introductory level, the groups $K_n(R)$, for $n \leq 2$, and the higher

Milnor K-groups $K_n^M(F)$ of a field F are accessible purely through algebra. To K-theorists, this algebraic part has come to be known as "classical" algebraic K-theory. And, judging by a census of the research literature, much of the power of K-theory for applications lies in this classical realm.

In this text, I have assumed no prerequisite beyond undergraduate mathematics and a single semester of algebra, including Galois theory and the structure of modules over a principal ideal domain, as might be taught from N. Jacobson's *Basic Algebra I*, S. Lang's *Algebra*, S. MacLane and G. Birkhoff's *Algebra*, or T. Hungerford's *Algebra*. I have included self-contained treatments of standard first-year graduate algebra topics: tensor products, categories and functors, Dedekind domains, the Jacobson radical, semisimple rings, character theory of groups, Krull dimension, projective resolutions, quadratic forms, central simple algebras, and symmetric and exterior algebras. The blend of K-theory with these topics motivates and enhances their exposition.

By including these algebra topics with the K-theory, I also hope to reach the mathematically sophisticated reader, who may have heard that K-theory is useful but inaccessible. Even if your algebra is rusty, you can read this book. The necessary background is here, with proofs.

This book provides a broad coverage of classical algebraic K-theory, with complete proofs. The groups $K_0(R)$, $K_1(R)$, and $K_2(R)$ are developed, along with the computational techniques of devissage, resolution, localization, Morita invariance, preservation of products, and the relative, Mayer-Vietoris, and localization exact sequences. The Krull dimension of a ring is shown to lead, through stable range conditions, to results on direct sum cancellation of modules and the injective and surjective stability theorems for $K_1(R)$. These, in turn, lead to a proof of the Bass-Kubota presentation of relative SK_1 and an outline of the solution of the congruence subgroup problem for the general linear group over a ring of arithmetic type. The higher Milnor K-theory is presented in detail, accompanied by a sketch of its connection to the Witt ring. Then Keune's proof of Matsumoto's presentation of $K_2(R)$ is included. The final part develops norm residue symbols and the relevance of K_2 to reciprocity laws and the Brauer group.

Inevitably, this text reflects my own interests and experience in teaching algebra and K-theory over the last 20 years. It incorporates lecture notes I have presented at the University of Oklahoma, Miami University, and the University of Padua, Italy. The categories considered here are explicit categories of modules, rather than subcategories of arbitrary abelian categories; this entails some loss of generality but seems less intimidating to the student who is new to K-theory. Unlike a treatment aimed at the K-theory of schemes, there is more emphasis here on algebraic than on transcendental extensions of a field, and extensive attention to examples over noncommutative rings R.

The text is organized broadly into five parts, and these are divided into chapters (1, 2, ...) and sections (2A, 2B,...). The chapter numbers are consecutive from the beginning of the book to the end. Numbered items, such as theorems, definitions, or displayed equations, are labeled by the chapter in which they appear, a decimal point, and the number of the item within the chapter. For example, (3.27) is the 27th numbered item in Chapter 3. At the end of each section is a list of several exercises, designed to illustrate with examples or point the way to further study.

The text has enough material for a year of study, but a variety of single-semester courses can be created from selected chapters. I recommend that the introductory Chapters 0 on categories, 1 on free modules, and 2 on projective modules, be summarized in a total of three or four lectures. Material not included in these lectures can be introduced, as needed, later in the course.

A short course on the interaction between rings and modules might include Chapters 3–6, §§7A–B, 8A–C, and 14A–D. This would cover Grothendieck groups, direct sum cancellation, Krull dimension, Dedekind domains, semisimple rings, the Jacobson radical, tensor products, and the construction of algebras from a module.

The connections of K-theory to number theory would be introduced by following §§3A–C, 4A, 5C, and Chapters 6 and 7 by one of two variants: Chapters 9, 10, and 11 for Mennicke symbols and the congruence subgroup problem or Chapter 12, §14H, and Chapter 15 for $K_2(R)$, norm residue and Hilbert symbols, and reciprocity laws.

A course on linear representations of finite groups would follow Chapter 3, §4A, and Chapters 5, 6, 8, and 12 by a sketch of the results in Chapter 16. This would include matrix representations of a finite group over fields of arbitrary characteristic; basic character theory; the structure of group algebras; the representation ring and Burnside ring of a group; and the Brauer group of a field.

I envision two possible short courses in algebraic K-theory itself: A module approach, including $K_0(R)$, $K_1(R)$, $K_2(R)$, devissage, resolution, Morita invariance, stability and exact sequences, would be Chapter 3, §§4A–B, 5A, 5C, and Chapters 6, 9, and 12–14. A linear group emphasis (mainly K_1 and K_2) would follow a brief treatment of Chapters 3 and 5 by Chapter 4, 6(mainly §§6D, 6E), and Chapters 9–14. (A truly minimal coverage of K_0, K_1, and K_2 and their connections would be §§3A–C, 4A, 5A, 5C, 6A–C, Chapter 9, §11A, Chapters 12 and 13, and a sketch of §14H.)

This book would never have seen the light of day without the generous assistance of several people. I am grateful to Keith Dennis, Reinhard Laubenbacher, Stephen Mitchell, and Giovanni Zacher for many illuminating conversations, helping me decide the level and general content of the book. My thanks also go to the members of the Miami University algebra seminar: Dennis Davenport, Tom Farmer, Chuck Holmes, Heather Hulett, and Mark deSaint-Rat, whose patient attention to detail and constructive suggestions led to improvements in a substantial portion of the text. I especially wish to acknowledge the help of Reza Akhtar, David Leep, and Katherine Magurn, who read chapters and gave valuable advice. My students in several graduate classes over the last few years have been very helpful as they studied parts of the text. For artfully translating my handwriting into TeX, I thank Jean Cavalieri. Her skill and cheerful attitude made our working partnership enjoyable and efficient. Miami office staff Bonita Porter and Cindie Johnson also did a nice job of typing a few chapters. My colleague Dennis Burke assisted us with the TeX software; he was unfailingly generous with his time and willing to answer my questions on the spot. Paddy Dowling and Steve Wright also contributed some expert software help. I appreciate the encouragement and material support of two successive Chairs of the Miami Mathematics and Statistics Department, Dave Kullman and Mark Smith. Most of all, I am grateful to Katherine, whose insight and support have sustained me throughout.

0
Preliminaries

Algebraic K-theory can be understood as a natural outgrowth of the attempt to generalize certain theorems in the linear algebra of vector spaces over a field to the wider context of modules over a ring. We assume the reader has been exposed to the fundamentals of module theory, including submodules, quotient modules, and the basic isomorphism theorems. To establish notation and terminology, we review some definitions in module theory. Then we present an introduction to the language of categories and functors.

By an **additive abelian group** we mean an abelian group A with operation denoted by "$+$," identity denoted by "0_A," and the inverse of an element $x \in A$ by "$-x$." By a **ring** we mean an associative ring with identity — that is, an additive abelian group R with a multiplication $R \times R \to R$, $(r, s) \mapsto rs$, satisfying

$$
\begin{aligned}
r(s + t) &= rs + rt \, , \\
(r + s)t &= rt + st \, , \\
(rs)t &= r(st) \, ,
\end{aligned}
$$

for all $r, s, t \in R$ and having an element $1 = 1_R \in R$ with

$$
1r = r = r1
$$

for all $r \in R$.

Suppose R is a ring. A **left R-module** is an additive abelian group M together with a function $R \times M \to M$, $(r, m) \mapsto rm$, satisfying

$$
\begin{aligned}
r(m + n) &= rm + rn \\
(r + s)m &= rm + sm \\
(rs)m &= r(sm) \\
1m &= m
\end{aligned}
$$

for all $r, s \in R$ and all $m, n \in M$. An **R-linear map** $f : M \to N$, between left R-modules M and N, is a homomorphism of additive groups that also satisfies

$$
f(rm) = rf(m)
$$

for each $r \in R$ and $m \in M$.

A **right** R-**module** is an additive abelian group M together with a function $M \times R \to M$, $(m, r) \mapsto mr$, satisfying

$$
\begin{aligned}
(m + n)r &= mr + nr \\
m(r + s) &= mr + ms \\
m(sr) &= (ms)r \\
m1 &= m
\end{aligned}
$$

for all $r, s \in R$ and $m, n \in M$. An R-**linear map** $f : M \to N$, between right R-modules M and N, is a homomorphism of additive groups satisfying

$$
f(mr) = f(m)r
$$

for each $f \in R$ and $m \in M$.

In either left or right modules we refer to elements of R as **scalars**, elements of M as **vectors**, and the maps $R \times M \to M$ or $M \times R \to M$ as **scalar multiplication**. The set of all R-linear maps from M to N is denoted by $\mathrm{Hom}_R(M, N)$. An R-linear map $f : M \to M$, from a module M to itself, is called an **endomorphism** of M, and $\mathrm{Hom}_R(M, M)$ is denoted by $\mathrm{End}_R(M)$. Our preference will be to work with left R-modules — that is, to write scalars on the left side of vectors. So when we refer to an R-module with no mention of left or right, we mean a left R-module.

Sometimes a vector multiplication is defined from $M \times M$ to M. If R is a commutative ring, an R-**algebra** A is an R-module that is also a ring and satisfies $r(ab) = (ra)b = a(rb)$, whenever $r \in R$ and $a, b \in A$. For example, if R is a subring of a ring A, and $ra = ar$ for all $r \in R$ and $a \in A$, then A is an R-algebra with scalar multiplication $R \times A \to A$ restricting the ring multiplication $A \times A \to A$.

Whether we study groups, rings, modules, or any other type of mathematical structure, it is often useful to consider the functions which preserve that structure. For instance, the study of groups includes an examination of group homomorphisms. For a comprehensive view of groups, one might consider the class of all groups, together with all group homomorphisms. Imagine a vast diagram of dots connected by arrows, with each group represented by a dot and each group homomorphism by an arrow. What you are picturing is the "category of groups."

(0.1) Definition. A **category** $\mathcal{C}$ consists of a class **Obj** $\mathcal{C}$ of **objects**, and for each pair of objects A, B a set **Hom**(A, B) of **arrows** from A to B, and for each triple of objects A, B, C a function:

$$
\mathrm{Hom}(A, B) \times \mathrm{Hom}(B, C) \to \mathrm{Hom}(A, C)
$$

called the **composite**. The composite of a pair (f, g) is denoted by $g \circ f$. For $\mathcal{C}$ to be a category, the following three axioms must hold:

(i) $\text{Hom}(A, B) \cap \text{Hom}(C, D) = \emptyset$ unless $A = C$ and $B = D$.

(ii) For each object A, there is an arrow $i_A \in \text{Hom}(A, A)$ for which

$$j \circ i_A = j \quad \text{and} \quad i_A \circ k = k$$

whenever $j \in \text{Hom}(A, B)$ and $k \in \text{Hom}(C, A)$ for objects B, C.

(iii) If $f \in \text{Hom}(A, B), g \in \text{Hom}(B, C)$, and $h \in \text{Hom}(C, D)$, then

$$(h \circ g) \circ f = h \circ (g \circ f).$$

In a category $\mathcal{C}$, an arrow $f \in \text{Hom}(A, B)$ is said to have **domain** A and **codomain** B, and this is implied by the expressions:

$$f : A \to B \quad \text{or} \quad A \xrightarrow{f} B.$$

Axiom (i) just says the domain and codomain of f are uniquely determined by f. The arrow i_A in axiom (ii) is called the **identity** arrow on A; it is uniquely determined by the composition in $\mathcal{C}$, since if i and i' are two arrows from A to A with the property described in (ii), then $i = i \circ i' = i'$.

The set of arrows $\text{Hom}(A, B)$ is sometimes denoted by $\textbf{Hom}_{\mathcal{C}}(A, B)$ to emphasize that they are arrows in the category $\mathcal{C}$. We shall often write $\textbf{End}(A)$ or $\textbf{End}_{\mathcal{C}}(A)$ to denote the set $\text{Hom}_{\mathcal{C}}(A, A)$ of arrows from A to itself.

(0.2) Examples. Here are the names, objects, and arrows of some of the categories considered in this book:

𝒮et: sets; functions.

𝒢roup: groups; group homomorphisms.

𝒜b: abelian groups; group homomorphisms between them.

ℛing: rings (associative with 1); ring homomorphisms (preserving 1).

𝒞ℛing: commutative rings (associative with 1); ring homomorphisms (preserving 1) between them.

𝒯op: topological spaces; continuous maps.

𝗠etric: metric spaces; isometries.

𝒫ower(S): subsets of the particular set S; inclusion maps $(i : A \to B,$ $i(a) = a)$ between them.

For each ring R we have the categories:

R-𝗠od: left R-modules; R-linear maps between them.

𝗠od-R: right R-modules; R-linear maps between them.

For each commutative ring R we have the category:

R-𝒜lg: R-algebras; R-linear ring homomorphisms between them.

For the advantage of brevity, when $\mathcal{C}$ is a category, we shall sometimes write $A \in \mathcal{C}$ instead of $A \in \text{Obj } \mathcal{C}$ to indicate that A is an object of $\mathcal{C}$. For instance,

$A \in \mathcal{CRing}$ means A is a commutative ring, and $A \in R\text{-}\mathcal{Mod}$ means A is a left R-module. This is really an abuse of notation, since the category $\mathcal{C}$ is not the same as the class Obj $\mathcal{C}$; and we will use this shortcut only where it does not suggest any ambiguity.

Each of the preceding examples are **concrete** categories, meaning that the objects are sets (possibly with additional structure), the arrows are (structure-preserving) functions between those sets, the identity arrows are identity functions ($i(x) = x$), and the composition is the composition of functions ($(g \circ f)(x) = g(f(x))$). But many useful categories do not fit this description. Here are some non-concrete categories:

(0.3) Examples.

(i) The objects and arrows of a category need not represent anything. For example, we can construct a category with two objects A and B, and four arrows i, j, f, and g:

$$i \ \ \mathrel{\reflectbox{$\circlearrowleft$}} A \underset{g}{\overset{f}{\rightleftarrows}} B \ \circlearrowright \ \ j \ .$$

Since the codomain of f equals the domain of g, there must be a composite $g \circ f$ from the domain A of f to the codomain A of g. Since i is the only arrow from A to A, we must have $g \circ f = i$. In this way, the scarcity of arrows forces the composites to be defined by the table:

$\circ$	i	j	f	g
i	i			g
j		j	f	
f	f			j
g		g	i	

Each of the sets $\mathrm{Hom}(A, A)$, $\mathrm{Hom}(A, B)$, $\mathrm{Hom}(B, A)$, and $\mathrm{Hom}(B, B)$ has only one element. So the category axioms (ii) and (iii) hold automatically, since both sides of each equation belong to the same set $\mathrm{Hom}(X, Y)$.

More generally, any diagram with the following properties has exactly one law of composition making it into a category:

(a) For each object X there is an arrow from X to X.
(b) If there is an arrow from X to Y and an arrow from Y to Z, there is an arrow from X to Z.
(c) For each pair of objects X and Y (possibly equal) there is at most one arrow from X to Y.

(ii) Suppose A is a set with a partial order $\prec$. That is, A has a binary relation $\prec$ that is reflexive (for all $x \in A$, $x \prec x$), antisymmetric ($x \prec y$ and $y \prec x$ imply $x = y$), and transitive ($x \prec y$ and $y \prec z$ imply $x \prec z$). View the elements of A as objects, and draw one arrow from x to y if $x \prec y$, and no arrow from x to y if $x \not\prec y$. Then properties (a), (b), and (c) of the preceding example hold; so this diagram is a category. We use the term **poset** to refer to a partially ordered set, or to its associated category.

(iii) If M is a set with a binary operation $\circ$ that is associative with an identity element $e \in M$, then M is called a **monoid**. For each object A in any category $\mathcal{C}$, the set $\mathrm{End}_{\mathcal{C}}(A)$ is a monoid under composition of arrows. Every monoid $(M, \circ)$ can be obtained in this way: Just create a category with one object, called A, and with the elements of M regarded as the arrows from A to A; and define the composition of these arrows by using the operation $\circ$ in M.

(iv) Suppose R is a ring. There is a category $\mathcal{M}\mathrm{at}(R)$ whose objects are the positive integers 1,2,3,..., and in which $\mathrm{Hom}(m, n)$ is the set of $m \times n$ matrices

$$A = (a_{ij}) = \begin{bmatrix} a_{11} & a_{12} & \cdots & a_{1n} \\ a_{21} & a_{22} & \cdots & a_{2n} \\ \vdots & \vdots & & \vdots \\ a_{m1} & a_{m2} & \cdots & a_{mn} \end{bmatrix}$$

with entries a_{ij} in R. The composition is matrix multiplication: $B \circ A = AB$. Explicitly, if $A = (a_{ij})$ is an $m \times n$ matrix and $B = (b_{ij})$ is an $n \times p$ matrix, their product $AB = (c_{ij})$ is the $m \times p$ matrix with ij-entry

$$c_{ij} = \sum_{k=1}^{n} a_{ik}b_{kj} .$$

From this formula one can show that the multiplication of matrices is associative, and there is an identity $I_n = (\delta_{ij}) \in \mathrm{Hom}(n, n)$ for each n, where:

$$\delta_{ij} = \begin{cases} 1 \text{ if } i = j \\ 0 \text{ if } i \neq j . \end{cases}$$

For details, see (1.26) below.

(0.4) Definition. A category $\mathcal{D}$ is called a **subcategory** of a category $\mathcal{C}$ if every object of $\mathcal{D}$ is an object of $\mathcal{C}$, and for each pair $A, B \in \mathrm{Obj}\,\mathcal{D}$, $\mathrm{Hom}_{\mathcal{D}}(A, B) \subseteq \mathrm{Hom}_{\mathcal{C}}(A, B)$, and composites and identities in $\mathcal{D}$ agree with those in $\mathcal{C}$. In case $\mathrm{Hom}_{\mathcal{D}}(A, B) = \mathrm{Hom}_{\mathcal{C}}(A, B)$, for each pair $A, B \in \mathrm{Obj}\,\mathcal{D}$, the subcategory $\mathcal{D}$ is called **full**.

Note that if $\mathcal{C}$ is a category, every subclass $\mathcal{D} \subset \mathrm{Obj}\,\mathcal{C}$ is the class of objects of a full subcategory $\mathcal{D}$ of $\mathcal{C}$. In particular, $\mathcal{A}b$ is a full subcategory of $\mathcal{G}\mathrm{roup}$, and $\mathcal{CR}\mathrm{ing}$ is a full subcategory of $\mathcal{R}\mathrm{ing}$. However, if a set S has at least two elements, the subcategory $\mathcal{P}\mathrm{ower}(S)$ of $\mathcal{S}\mathrm{et}$ is not full.

Is $\mathfrak{Ring}$ a subcategory of $\mathcal{Ab}$? After all, every ring $(R, +, \cdot)$ is an additive abelian group $(R, +)$ if we forget its multiplication, and every ring homomorphism is a homomorphism between additive groups. The answer is *no:* $\mathfrak{Ring}$ is *not* a subcategory of $\mathcal{Ab}$! For the condition in the definition of subcategory, "every object of $\mathcal{D}$ is an object of $\mathcal{C}$," is to be taken quite literally: A ring has two operations and an abelian group has only one. So $(R, +, \cdot)$ is not the same as $(R, +)$.

(0.5) Definition. In each category $\mathcal{C}$ an arrow $f \in \text{Hom}(A, B)$ is called an **isomorphism** if there is an arrow $g \in \text{Hom}(B, A)$ with

$$f \circ g = i_B \text{ and } g \circ f = i_A .$$

Such an arrow g is called an **inverse** to the arrow f. An isomorphism in $\text{End}(A)$ is called an **automorphism** of A, and the set of automorphisms of A is denoted by $\mathbf{Aut}(A)$ or $\mathbf{Aut}_{\mathcal{C}}(A)$.

In the categories $\mathfrak{Set}$, $\mathfrak{Group}$, $\mathcal{Ab}$, $\mathfrak{Ring}$, $\mathcal{CRing}$, $R\text{-}\mathfrak{Mod}$, $\mathfrak{Mod}\text{-}R$, and $\mathfrak{Metric}$, an arrow is an isomorphism if and only if it is bijective. In $\mathfrak{Top}$, an isomorphism is the same as a homeomorphism; but not every bijective continuous map is a homeomorphism. (Consider the identity function on X with two different topologies, the first finer than the second.)

In $\mathfrak{Mat}(R)$, an isomorphism in $\text{Hom}(n, n)$ is the same as an invertible $n \times n$ matrix over R. There exist rings R for which $\mathfrak{Mat}(R)$ includes an isomorphism in $\text{Hom}(m, n)$ even though $m \neq n$. Such an isomorphism would be an $m \times n$ matrix X over R for which there is an $n \times m$ matrix Y over R with

$$XY = i_m \text{ and } YX = i_n ,$$

where i_m and i_n are identity matrices of different sizes. But such rings R are a little hard to come by. For an example, see (1.37).

(0.6) Definition. An object X in a category $\mathcal{C}$ is called **initial** if, for each object $A \in \text{Obj } \mathcal{C}$, there is one and only one arrow in $\text{Hom}(X, A)$. An object Y in $\mathcal{C}$ is called **terminal** if, for each object $A \in \text{Obj } \mathcal{C}$, there is one and only one arrow in $\text{Hom}(A, Y)$.

(0.7) Proposition. *If objects X and Y of a category $\mathcal{C}$ are both initial or both terminal, then there is one and only one arrow $f : X \to Y$, and f is an isomorphism in $\mathcal{C}$.*

Proof. The existence and uniqueness of f is immediate since X is initial or Y is terminal. Likewise there is a unique arrow $g : Y \to X$, and the identity arrows are the only arrows in $\text{End}(X)$ and $\text{End}(Y)$. So, for lack of options, $g \circ f = i_X$ and $f \circ g = i_Y$. ∎

(0.8) Examples.

(i) In 𝔖et the empty set $\emptyset$ is initial: $\mathrm{Hom}(\emptyset, A)$ has one member $f : \emptyset \to A$ with empty graph. Each set $\{x\}$ with only one member is terminal.

(ii) In 𝔊roup, any trivial group $\{e\}$ is both an initial and a terminal object.

(iii) In ℜing, $\mathbb{Z}$ is an initial object, and the trivial ring $\{0\}$ is a terminal object. The same is true in 𝒞ℜing.

(iv) For each scalar ring R, the trivial R-module $\{0\}$ is both initial and terminal in R-𝔐od.

(v) If R is a ring with at least two elements, 𝔐at(R) has no initial object and no terminal object, since each set $\mathrm{Hom}(m, n)$ has more than one element.

Once categories are viewed as algebraic entities, it is natural to ask what might be meant by a "homomorphism" from one category to another. A function is an arrow in 𝔖et; its domain and codomain are sets. As a generalization, define a **metafunction** from a class A to a class B to be a procedure that assigns to each member $a \in A$ a unique member $f(a) \in B$. Then a homomorphism between categories should be a metafunction that takes objects to objects, arrows to arrows, and respects domains, codomains, composites, and identities.

(0.9) Definition. A **covariant functor** $F : \mathcal{C} \to \mathcal{D}$ from a category $\mathcal{C}$ to a category $\mathcal{D}$ is a metafunction that assigns to each object A of $\mathcal{C}$ an object $F(A)$ of $\mathcal{D}$, and to each arrow $f : A \to B$ in $\mathcal{C}$ an arrow $F(f) : F(A) \to F(B)$ in $\mathcal{D}$, so that

(i) $F(i_A) = i_{F(A)}$, and

(ii) $F(g \circ f) = F(g) \circ F(f)$,

whenever $A \in \mathrm{Obj}\,\mathcal{C}$ and (f, g) are composable arrows in $\mathcal{C}$.

There is also an arrow-reversing version:

(0.10) Definition. A **contravariant functor** $F : \mathcal{C} \to \mathcal{D}$ is a metafunction that assigns to each object A in $\mathcal{C}$ an object $F(A)$ in $\mathcal{D}$, and to each arrow $f : A \to B$ in $\mathcal{C}$ an arrow $F(f) : F(B) \to F(A)$ in $\mathcal{D}$, so that

(i) $F(i_A) = i_{F(A)}$, and

(ii) $F(g \circ f) = F(f) \circ F(g)$,

whenever $A \in \mathrm{Obj}\,\mathcal{C}$ and (f, g) are composable arrows in $\mathcal{C}$.

For both covariant and contravariant functors F, the condition (ii) amounts to saying that if F is applied to every object and arrow in a commutative triangle in $\mathcal{C}$, one obtains a commutative triangle in $\mathcal{D}$ (but with its arrows reversed in the contravariant case).

(0.11) Examples.

  (i) If $\mathcal{C}$ is a subcategory of $\mathcal{D}$, the **inclusion functor** $F : \mathcal{C} \to \mathcal{D}$ is defined by $F(A) = A$ and $F(f) = f$ for each object A and arrow f in $\mathcal{C}$. The inclusion functor $F : \mathcal{C} \to \mathcal{C}$ is called the **identity functor** $i_{\mathcal{C}}$.

  (ii) If $\mathcal{C}$ is a concrete category, such as $\mathcal{G}$roup or $\mathcal{R}$ing, the **forgetful functor** $F : \mathcal{C} \to \mathcal{S}$et is defined by taking $F(A)$ to be the underlying set of elements of A and $F(f)$ to be f as a function. There are also functors involving a partial loss of memory, such as the functor $F : \mathcal{R}$ing $\to \mathcal{A}$b that forgets multiplication: $F(R) = R$ as an additive group and $F(f) = f$ as an additive group homomorphism.

  (iii) If n is a positive integer, there is a functor

$$M_n : \mathcal{R}\text{ing} \ \to \ \mathcal{R}\text{ing}$$

where $M_n(R)$ is the ring of $n \times n$ matrices with entries in R; and if $f : R \to S$ is a ring homomorphism, $M_n(f) : M_n(R) \to M_n(S)$ is the ring homomorphism defined by

$$M_n(f) \begin{bmatrix} a_{11} & \cdots & a_{1n} \\ \vdots & & \vdots \\ a_{n1} & \cdots & a_{nn} \end{bmatrix} = \begin{bmatrix} f(a_{11}) & \cdots & f(a_{1n}) \\ \vdots & & \vdots \\ f(a_{n1}) & \cdots & f(a_{nn}) \end{bmatrix}.$$

  (iv) If R is a ring, the elements of R with (two-sided) multiplicative inverses in R are called **units**. The set R^* of units in R is a group under multiplication. There is a functor

$$U : \mathcal{R}\text{ing} \ \to \ \mathcal{G}\text{roup}$$

defined by $U(R) = R^*$ and $U(f) = f$ restricted to units. This works because every ring homomorphism f takes units to units.

  (v) If $F : \mathcal{C} \to \mathcal{D}$ and $G : \mathcal{D} \to \mathcal{E}$ are functors, there is a composite functor $G \circ F : \mathcal{C} \to \mathcal{E}$ defined by

$$G \circ F(A) \ = \ G(F(A)) , \quad G \circ F(f) \ = \ G(F(f)) ,$$

for objects A and arrows f in $\mathcal{C}$. If F and G are both covariant or both contravariant, then $G \circ F$ is covariant; if one of G, F is covariant and the other is contravariant, then $G \circ F$ is contravariant.

  (vi) The composite of $M_n : \mathcal{R}$ing $\to \mathcal{R}$ing, followed by $U : \mathcal{R}$ing $\to \mathcal{G}$roup, is the functor

$$GL_n = U \circ M_n : \mathcal{R}\text{ing} \ \to \ \mathcal{G}\text{roup} ,$$

which takes each ring R to the group $GL_n(R)$ of $n \times n$ invertible matrices with entries in R, and takes each ring homomorphism f to the group homomorphism that applies f to each entry. The GL stands for "general linear" group.

(vii) In the functor definitions (0.9) and (0.10), axioms (i) and (ii) hold automatically if $\mathcal{D}$ is a poset category, for in a poset category an arrow is uniquely determined by its domain and codomain. Also, the class of objects in a poset category is a set. So a functor from a poset $(A, \prec)$ to a poset $(B, \prec)$ is just a function $F : A \to B$ that preserves order:

$$x \prec y \;\Rightarrow\; F(x) \prec F(y)$$

if F is covariant, or reverses order:

$$x \prec y \;\Rightarrow\; F(y) \prec F(x)$$

if F is contravariant.

If $F : \mathcal{C} \to \mathcal{D}$ is a functor and there is a functor $G : \mathcal{D} \to \mathcal{C}$ for which

$$F \circ G \;=\; i_{\mathcal{D}} \quad \text{and} \quad G \circ F \;=\; i_{\mathcal{C}} \,,$$

then G is called an **inverse** to F, and F is called an **isomorphism** of categories if F is covariant and an **anti-isomorphism** of categories if F is contravariant. If there is an isomorphism (respectively, anti-isomorphism) of categories from $\mathcal{C}$ to $\mathcal{D}$, we say $\mathcal{C}$ and $\mathcal{D}$ are **isomorphic** (respectively, **anti-isomorphic**).

The correspondence theorems of algebra provide either isomorphisms or anti-isomorphisms of poset categories: For instance, if A is a group with a normal subgroup N, the poset of subgroups of A containing N is isomorphic to the poset of subgroups of A/N; the isomorphism takes each H to H/N, and its inverse takes each subgroup K of A/N to its union $\cup K$. As a contravariant example, if $F \subseteq E$ is a finite-degree Galois field extension, the poset of intermediate fields is anti-isomorphic to the poset of subgroups of $\mathrm{Aut}(E/F)$; the anti-isomorphism takes each K to $\mathrm{Aut}(E/K)$, and its inverse takes each subgroup H of $\mathrm{Aut}(E/F)$ to the fixed field E^H.

For each ring R, right R-modules are, in a sense, mirror images of left R-modules. Their definitions are parallel, and so are the theorems that apply to them. But it is also possible to pass through the looking glass: If R is a ring, define its **opposite ring** R^{op} to have the same elements as R and the same addition as in R, but to have a multiplication $\cdot$ defined by $r \cdot s = sr$ (the right side multiplied in R).

(0.12) Proposition. *There are isomorphisms of categories:*

$$R\text{-}\mathcal{M}o\eth \;\cong\; \mathcal{M}o\eth\text{-}R^{op} \,,$$
$$R^{op}\text{-}\mathcal{M}o\eth \;\cong\; \mathcal{M}o\eth\text{-}R \,.$$

Proof. If M is a left R-module with scalar multiplication $R \times M \to M$, $(r, m) \mapsto r * m$, then the additive group of M can also be made into a right R^{op}-module via a scalar multiplication $M \times R^{op} \to M$, $(m, r) \mapsto m \# r$, defined by $m \# r = r * m$:

$$
\begin{aligned}
(m + n) \# r &= r * (m + n) = (r * m) + (r * n) = (m \# r) + (n \# r) , \\
m \#(r + s) &= (r + s) * m = (r * m) + (s * m) = (m \# r) + (m \# s) , \\
m \#(s \cdot r) &= (rs) * m = r * (s * m) = (m \# s) \# r , \\
m \# 1 &= 1 * m = m .
\end{aligned}
$$

Each R-linear map $f : M \to N$ between left R-modules is also an R^{op}-linear map between right R^{op}-modules:

$$
f(m \# r) = f(r * m) = r * f(m) = f(m) \# r .
$$

So there is a functor F from R-$\mathcal{Mod}$ to $\mathcal{Mod}$-R^{op} with $F((M, *)) = (M, \#)$ and $F(f) = f$.

Similarly, there is a functor G from $\mathcal{Mod}$-R^{op} to R-$\mathcal{Mod}$ with $G((M, \#)) = (M, *)$, where $r * m$ is defined to be $m \# r$, and $G(f) = f$. Since F and G are inverses, F is an isomorphism. Since $(R^{op})^{op} = R$, the second isomorphism follows from the first. ∎

So, for the study of properties held in common by all module categories, it suffices to consider only a category R-$\mathcal{Mod}$ of left R-modules. Of course, if R is commutative ($R^{op} = R$), then we can regard R-$\mathcal{Mod}$ and $\mathcal{Mod}$-R as the same category. Every additive abelian group is a $\mathbb{Z}$-module in exactly one way, and every homomorphism between abelian groups is $\mathbb{Z}$-linear. So $\mathbb{Z}$-$\mathcal{Mod} = \mathcal{Mod}$-$\mathbb{Z} = \mathcal{Ab}$, and $\mathbb{Z}$-$\mathcal{Alg} = \mathcal{Ring}$.

Sometimes the values of two functors from $\mathcal{C}$ to $\mathcal{D}$ are closely related in $\mathcal{D}$. For instance, GL_n and U are functors from $\mathcal{CRing}$ to $\mathcal{Group}$, and the determinant connects them: For each commutative ring R, the determinant is a group homomorphism $\det : GL_n(R) \to U(R)$. And if $f : R \to S$ is a ring homomorphism, the square

$$
\begin{array}{ccc}
GL_n(R) & \xrightarrow{\det} & U(R) \\
\scriptstyle{GL_n(f)} \big\downarrow & & \big\downarrow \scriptstyle{U(f)} \\
GL_n(S) & \xrightarrow[\det]{} & U(S)
\end{array}
$$

commutes in $\mathcal{Group}$; so det is compatible with a change of rings.

(0.13) Definition. If F and G are covariant functors from $\mathcal{C}$ to $\mathcal{D}$, a **natural transformation** $\tau : F \to G$ is a metafunction that assigns to each object X of $\mathcal{C}$ an arrow

$$
\tau_X : F(X) \to G(X)
$$

in $\mathcal{D}$, so that for each arrow $f : X \to Y$ in $\mathcal{C}$ the square

$$
\begin{array}{ccc}
F(X) & \xrightarrow{\tau_X} & G(X) \\
{\scriptstyle F(f)}\Big\downarrow & & \Big\downarrow{\scriptstyle G(f)} \\
F(Y) & \xrightarrow[\tau_Y]{} & G(Y)
\end{array}
$$

commutes in $\mathcal{D}$.

For each covariant functor $F : \mathcal{C} \to \mathcal{D}$ there is an "identity" natural transformation $i : F \to F$ with $i_X = i_{F(X)}$ for each $X \in \mathcal{C}$. If F, G, and H are covariant functors from $\mathcal{C}$ to $\mathcal{D}$, and there are natural transformations $\tau : F \to G$ and $\tau' : G \to H$, there is a composite natural transformation $\tau' \circ \tau : F \to H$, with $(\tau' \circ \tau)_X = \tau'_X \circ \tau_X$. So for each pair of categories $\mathcal{C}, \mathcal{D}$, the covariant functors $\mathcal{C} \to \mathcal{D}$ and the natural transformations between them behave like the objects and arrows of a category. A **natural isomorphism** $\tau : F \to G$ is any natural transformation $\tau : F \to G$ with an "inverse" natural transformation $\tau' : G \to F$ so that $\tau \circ \tau'$ and $\tau' \circ \tau$ are identities on G and F, respectively.

Note: If τ is a natural transformation and each τ_X is an isomorphism (in $\mathcal{D}$), then the inverse arrows τ_X^{-1} constitute a natural transformation inverse to τ. *So τ is a natural isomorphism if and only if each τ_X is an isomorphism.*

There is a strictly analogous theory of natural transformations between contravariant functors.

The importance of functors was first realized in connection with algebraic topology. Two topological spaces X and Y are isomorphic in the category $\mathfrak{Top}$ if and only if they are homeomorphic, that is, if and only if there are continuous maps $f : X \to Y$ and $g : Y \to X$ with $f \circ g = i_Y$ and $g \circ f = i_X$. If there is such a homeomorphism $f : X \to Y$, then, from the topological point of view, X and Y are the same space. In order to determine whether or not two spaces are homeomorphic, functors are created from $\mathfrak{Top}$ to $\mathfrak{Group}$: For each non-negative integer n there is a **singular homology** functor H_n and a **homotopy** functor π_n. These are created, roughly speaking, to detect holes of various dimensions in each topological space. These functors send homeomorphic spaces to isomorphic groups; so if one of these functors sends X and Y to non-isomorphic groups, then X is not homeomorphic to Y. This is a general property of functors:

(0.14) Lemma. *If $F : \mathcal{C} \to \mathcal{D}$ is a functor, and f is an isomorphism in $\mathcal{C}$, then $F(f)$ is an isomorphism in $\mathcal{D}$.*

Proof. Suppose F is a covariant functor, and $f : A \to B$ has inverse $g : B \to A$ in $\mathcal{C}$. Then $F(f) : F(A) \to F(B)$ has inverse $F(g) : F(B) \to F(A)$ in $\mathcal{D}$, since

$$
F(f) \circ F(g) = F(f \circ g) = F(i_B) = i_{F(B)} , \quad \text{and}
$$

$$F(g) \circ F(f) \;=\; F(g \circ f) \;=\; F(i_A) \;=\; i_{F(A)} \;.$$

The contravariant case is similar   ∎

Note that non-isomorphic objects of $\mathcal{C}$ can be sent by a functor $F : \mathcal{C} \to \mathcal{D}$ to isomorphic objects of $\mathcal{D}$. For instance, $\mathbb{Z}$ and $\mathbb{Z}/3\mathbb{Z}$ have isomorphic groups of units. So it is sometimes necessary to apply a variety of different functors: $\mathcal{C} \to \mathcal{D}$, in order to distinguish between nonisomorphic objects in $\mathcal{C}$.

Algebraic K-theory is the study of a sequence of covariant functors

$$K_n : \mathfrak{Ring} \;\to\; \mathcal{Ab} \qquad (n \in \mathbb{Z})$$

that arise from linear algebra. If R is a ring, $K_0(R)$ is related to the dimension of "vector spaces" over R, $K_1(R)$ is similarly related to the determinant of matrices over R, and $K_2(R)$ arises from the study of elementary row operations on matrices over R. The "higher" K-groups $K_3(R)$, $K_4(R), \dots$ are more mysterious. They are homotopy groups $\pi_2(X_R), \pi_3(X_R), \dots$ of a certain topological space X_R associated with the ring R, and are studied with the tools of algebraic topology — in particular, homotopy theory. When R is a number field (finite-degree field extension of $\mathbb{Q}$), these higher groups $K_n(R)$ contain deep number-theoretic information related to zeta functions.

In this book, we confine our attention primarily to K_0, K_1, and K_2, with brief excursions involving K_n, $n \leq 0$, and an algebraically constructed sequence of functors K_n^M, $n \geq 3$, which are related to the higher K-groups. In the literature, K_0, K_1, and K_2 are often called "classical" K-groups, even though the whole subject emerged in the second half of the 20th century, and even though, at its outset, some of the creators of K-theory thought of K_n for $n = 0$, 1, 2 as part of an infinite sequence ($n \in \mathbb{Z}$) of K-groups that were not yet realized.

0. Exercises

1. Verify that the functor M_n defined in (0.11) (iii) is a functor. This includes checking that if $f : R \to S$ is a ring homomorphism, then $M_n(f) : M_n(R) \to M_n(S)$ is a ring homomorphism.

2. If $\mathcal{C}$ is a category with an object C, show there are

   (i) a covariant functor $F : \mathcal{C} \to \mathcal{Set}$ with $F(A) = \mathrm{Hom}_{\mathcal{C}}(C, A)$ for each $A \in \mathrm{Obj}\ \mathcal{C}$, and
   (ii) a contravariant functor $G : \mathcal{C} \to \mathcal{Set}$ with $G(A) = \mathrm{Hom}_{\mathcal{C}}(A, C)$ for each $A \in \mathrm{Obj}\ \mathcal{C}$.

3. Prove that there is no isomorphism of categories from $\mathfrak{Group}$ to $\mathfrak{Ring}$. *Hint:* Compare initial and terminal objects.

4. A **small** category is a category $\mathcal{C}$ in which Obj $\mathcal{C}$ is a set. There is a class of all small categories. Show that there is a category $\mathcal{C}at$ whose objects are the small categories and whose arrows are the functors between them.

5. Suppose G is a group with identity element denoted by e. A **G-set** is a set X together with an operation $G \times X \to X, (g, x) \mapsto g \bullet x$, satisfying the two axioms:

(i) $(gh) \bullet x = g \bullet (h \bullet x)$, and

(ii) $e \bullet x = x$,

for all $g, h \in G$ and all $x \in X$. If X and Y are G-sets, a **G-map** $f : X \to Y$ is any function f from X to Y satisfying $f(g \bullet x) = g \bullet f(x)$ for all $g \in G$, $x \in X$. Prove that the G-sets and G-maps form a concrete category G-$\mathcal{S}et$. Also prove that an arrow in G-$\mathcal{S}et$ is an isomorphism if and only if it is bijective.

6. Verify that a category $\mathcal{C}$ is a poset if and only if Obj $\mathcal{C}$ is a set and, for each pair $x, y \in$ Obj $\mathcal{C}$,

$$\text{Hom}_{\mathcal{C}}(x, y) \ \cup \ \text{Hom}_{\mathcal{C}}(y, x)$$

has at most one element.

7. Prove that an arrow f in a category $\mathcal{C}$ can have at most one inverse arrow in $\mathcal{C}$.

8. Suppose, in a category $\mathcal{C}$, that $\text{Hom}(A, A) = \{i_A\}$ and $\text{Hom}(B, B) = \{i_B\}$. If $\text{Hom}(A, B)$ contains at least one arrow, prove $\text{Hom}(B, A)$ contains at most one arrow.

9. If $\mathcal{C}$ is a category and A is an object of $\mathcal{C}$, prove the set Aut(A), of all automorphisms of A in $\mathcal{C}$, is a group under $\circ$. Also prove every group is obtained in this way. (See (0.3) (iii).)

10. Suppose F is a field, V is an F-vector space, and B is a finite basis of V. Consider the category $\mathcal{D}$ in which an object is a function $f : B \to W$ from B into (the underlying set of) an F-vector space W, and an arrow from $f_1 : B \to W_1$ to $f_2 : B \to W_2$ is the same as an F-linear transformation $h : W_1 \to W_2$ with $h \circ f_1 = f_2$. Prove that the inclusion map $i : B \to V$ is an initial object in $\mathcal{D}$.

PART I

Groups of Modules: K_0

Vector spaces over a field F are isomorphic if and only if they have equal dimension. If $c(V)$ denotes the isomorphism class of an F-vector space V, then sending $c(V)$ to $\dim(V)$ defines a one-to-one correspondence from the collection $\mathfrak{I}$ of isomorphism classes of finitely spanned F-vector spaces to the set $\mathbb{N}$ of non-negative integers. Through this correspondence, the addition in $\mathbb{N}$ imposes an addition on $\mathfrak{I}$, with an identity element $c(\{0\})$ and sharing the associative and commutative properties of $+$ in $\mathbb{N}$.

There is also a way to add F-vector spaces V and W by forming their direct sum $V \oplus W$. Since $\dim(V \oplus W) = \dim(V) + \dim(W)$, the addition in $\mathfrak{I}$ described above is $c(V) + c(W) = c(V \oplus W)$. Of course the abelian monoid $\mathbb{N}$ can be completed to the abelian group $\mathbb{Z}$ of all integers by including the differences $m - n$ of non-negative integers. So $\mathfrak{I}$ can be enlarged to an abelian group $K_0(F)$, isomorphic to $\mathbb{Z}$, consisting of all differences $c(V) - c(W)$.

When we replace the scalar field F by an arbitrary ring R, the R-modules need not have bases — so dimension disappears. But if we restrict our attention to an appropriate class of R-modules (called projective modules), the abelian group $K_0(R)$ remains intact as the ghost of departed dimensions. First developed in 1957–58 by Grothendieck and Serre, $K_0(R)$ is now understood to be one of a sequence $K_n(R)$ $(n \in \mathbb{Z})$ of closely related abelian groups called the algebraic K-theory of R. If R is not a field, $K_0(R)$ need not be isomorphic to $\mathbb{Z}$. Its structure reflects the various generalizations of the dimension of vector spaces to "ranks" of R-modules.

1

Free Modules

1A. Bases

In this section we consider the description of an R-module in terms of generators and defining relations, and we take a close look at those R-modules that are free of defining relations.

(1.1) Definitions. Suppose R is a ring and S is a subset of an R-module M. An R-**linear combination** of elements of S is any element $r_1 s_1 + \cdots + r_n s_n$, where n is a positive integer, $r_1, \ldots, r_n \in R$ and $s_1, \ldots, s_n \in S$. Denote by $\langle S \rangle$ the set of all R-linear combinations of elements of S (together with 0_M, which must be included separately if $S = \emptyset$). One can also think of $\langle S \rangle$ as the intersection of all submodules of M containing S. Say S **generates** (or **spans**) M if $\langle S \rangle = M$.

An R-**linear relation** on S is any true equation:

$$r_1 s_1 + \cdots + r_n s_n = 0_M$$

in M, where n is a positive integer, $s_1, \ldots, s_n$ are n different elements of S, and $r_1, \ldots, r_n$ are nonzero elements of R. The set S is called R-**linearly dependent** if there exists an R-linear relation on S, or it is called R-**linearly independent** if there is no R-linear relation on S. An R-**basis** of M is an R-linearly independent set $B \subseteq M$ that generates M. An R-module M is called **free** if it has an R-basis.

When it does not cause confusion about the choice of scalars, we will sometimes drop the prefix "R-" in the terms defined above.

(1.2) Examples of free modules.

(i) Over every ring R, the zero module $\{0\}$ is free with the empty set $\emptyset$ as basis. We often denote the zero R-module by R^0.

(ii) If R is a nontrivial ring, the polynomial ring $R[x]$ is a free R-module with infinite basis $\{1,\ x,\ x^2, \dots\}$.

(iii) If R is a nontrivial ring and n is a positive integer, the n-fold direct sum

$$R^n = R \oplus \cdots \oplus R$$

is a free R-module with basis $\{e_1, \dots, e_n\}$ where

$$e_i = (0, \dots, 0,\ 1,\ 0, \dots, 0)$$

has 1 in the i-coordinate and 0's elsewhere: For $e_1, \dots, e_n$ span R^n because

$$(r_1, \dots, r_n) = r_1 e_1 + \cdots + r_n e_n \ ;$$

and an R-linear relation among $e_1, \dots, e_n$ is impossible because

$$(r_1, \dots, r_n) = (0, \dots, 0)$$

implies $r_i = 0$ for each i.

(iv) Suppose R is a nontrivial ring and S is a nonempty set. A function $f : S \to R$ is said to have **finite support** if $f(s) = 0_R$ for all but finitely many $s \in S$. Let $\oplus_S R$ denote the set of all functions $f : S \to R$ with finite support. Under pointwise sum and scalar multiplication, $\oplus_S R$ is an R-module. If $s \in S$, let $\widehat{s} : S \to R$ denote the characteristic function of s, defined by

$$\widehat{s}(t) = \begin{cases} 1_R & \text{if } t = s\,, \\ 0_R & \text{if } t \neq s\,. \end{cases}$$

Then the set $\widehat{S} = \{\widehat{s} : s \in S\}$ is a basis of $\oplus_S R$: For if $f \in \oplus_S R$, then

$$f = \sum_{s \in S} f(s)\widehat{s} \ ;$$

and an R-linear combination

$$\sum_{s \in S} f(s)\widehat{s} \quad (= f)$$

is the zero map ($=$ the zero element of $\oplus_S R$) if and only if every $f(s) = 0_R$. Note that if $S = \{s_1, \dots, s_n\}$ has n elements, there is an isomorphism:

$$\begin{array}{ccc} \oplus_S R & \cong & R^n \\ f & \mapsto & (f(s_1), \dots, f(s_n)) \end{array}$$

under which the basis $\widehat{s}_1, \dots, \widehat{s}_n$ corresponds to the basis $e_1, \dots, e_n$.

By the next theorem, every free R-module is isomorphic to one of the form $\oplus_S R$.

(1.3) Theorem. *Suppose R is a ring and B is a nonempty subset of an R-module M. The following are equivalent:*

   (i) *B is an R-basis of M.*
   (ii) *The R-linear map $\phi : \oplus_B R \to M$, defined by*

$$\phi(f) \;=\; \sum_{b \in B} f(b)b \;,$$

     is an isomorphism.
  (iii) *Each $m \in M$ has an expression*

$$m \;=\; \sum_{b \in B} f(b)b$$

    for one and only one $f \in \oplus_B R$.

Proof. Surjectivity of ϕ amounts to B generating M, and to existence of the expression in (iii). Injectivity of ϕ amounts to uniqueness of the expression in (iii), and to $\ker(\phi) =$ the zero map, and hence to the linear independence of the set B. ■

A basis of a free R-module M is characterized by the role it plays in constructing R-linear maps, defined on M:

(1.4) Proposition. *Suppose M is an R-module and B is a nonempty subset of M. The following are equivalent:*

   (i) *B is a basis of M.*
   (ii) *Each function $\phi : B \to N$, from B into an R-module N, has one and only one extension to an R-linear map $\widehat{\phi} : M \to N$.*

Proof. Suppose M and N are R-modules, B is a basis of M, and $\phi : B \to N$ is a function. If $\widehat{\phi} : M \to N$ is an R-linear map extending ϕ, then

(1.5)
$$\widehat{\phi}\left(\sum_{b \in B} f(b)b \right) \;=\; \sum_{b \in B} f(b)\phi(b)$$

for each $f \in \oplus_B R$, proving there is at most one R-linear extension $\widehat{\phi}$ of ϕ. The formula (1.5) defines an R-linear map $\widehat{\phi} : M \to N$. If $b_0 \in B$, using $f = \widehat{b_0}$ (the characteristic function of b_0) in (1.5) shows $\widehat{\phi}(b_0) = \phi(b_0)$; so $\widehat{\phi}$ extends ϕ.

Conversely, if (ii) holds, the function $\phi : B \rightarrow \oplus_B R$, taking b to $\widehat{b}$ for each $b \in B$, has an R-linear extension $\widehat{\phi} : M \rightarrow \oplus_B R$. Applying $\widehat{\phi}$ to the equation

$$\sum_{b \in B} f(b)b = 0_M$$

yields $f = 0 \in \oplus_B R$; so there is no R-linear relation on B. Both the zero map and the canonical surjective map from M to $M/\langle B \rangle$ take each $b \in B$ to $0 + \langle B \rangle$. If (ii) holds, these maps are equal; so $M/\langle B \rangle = 0$ and $M = \langle B \rangle$. Thus (ii) implies B is a basis of M.   ∎

(1.6) Note. If M, N, B, ϕ, and $\widehat{\phi}$ are as in (1.4) (ii), then the R-linear extension $\widehat{\phi} : M \rightarrow N$ is surjective if and only if $\phi(B)$ generates N, and is injective if and only if ϕ is injective and $\phi(B)$ is linearly independent.

(1.7) Corollary.

 (i) If $\psi : M \rightarrow N$ is an isomorphism of R-modules and M is free with basis B, then N is free with basis $\psi(B)$.

 (ii) If M is an R-module and n is a positive integer, then M has an n-element basis if and only if $M \cong R^n$.

 (iii) If M is an R-module and n is a positive integer, there is a bijection from the set of isomorphisms $R^n \cong M$ to the set of ordered n-element bases of M, given by

$$\psi \mapsto (\psi(e_1), \ldots, \psi(e_n)) .$$

Proof. Assertion (i) follows from (1.6) if we regard ψ as an R-linear extension of its restriction to B.

Assertion (ii) is an immediate consequence of (iii). For (iii), recall that $e_1, \ldots, e_n$ is a basis of R^n. So by (1.6), if $\psi : R^n \rightarrow M$ is an isomorphism, then $(\psi(e_1), \ldots, \psi(e_n))$ is an ordered basis of M. If $\theta : R^n \rightarrow M$ is also an isomorphism with $\theta(e_i) = \psi(e_i)$ for each i, then

$$\theta \left(\sum_{i=1}^n r_i e_i \right) = \sum_{i=1}^n r_i \theta(e_i) = \sum_{i=1}^n r_i \psi(e_i) = \psi \left(\sum_{i=1}^n r_i e_i \right);$$

so $\theta = \psi$. If $(b_1, \ldots, b_n)$ is an ordered n-element basis of M, then by (1.5) and (1.6) there is an isomorphism $\psi : R^n \rightarrow M$ with $\psi(e_i) = b_i$ for each i.   ∎

Part (ii) of (1.7) characterizes free modules with finite bases and makes it easy to construct a module that is not free. For instance, $\mathbb{Z}^n$ is infinite if $n > 0$; so every nontrivial finite abelian group is a $\mathbb{Z}$-module without a basis.

(1.8) Proposition. *If R is a nontrivial ring, every set S is a basis of a free R-module $F_R(S)$.*

Proof. If S is empty, we can take $F_R(S) = \{0_R\}$. Suppose S is not empty. By Example (1.2) (iv), $\oplus_S R$ is a free R-module with basis $\widehat{S}$. Since R is nontrivial, $1_R \neq 0_R$; so the function $S \to \widehat{S}$, taking s to $\widehat{s}$ for each $s \in S$, is bijective and we can use it to replace $\widehat{S}$ by S in $\oplus_S R$: There is no overlap of S with $\oplus_S R$, since the set-theoretic axiom of regularity implies no function can be a member of its own domain (see Appendix A). So if $T = \oplus_S R - \widehat{S}$, then $S \cap T = \emptyset$, and there is a bijection $h : S \cup T \to \oplus_S R$ defined by:

$$h(x) = \begin{cases} x & \text{if} \quad x \in T , \\ \widehat{x} & \text{if} \quad x \in S . \end{cases}$$

Then $S \cup T$ is an R-module under operations

$$\begin{aligned} x + y &= h^{-1}(h(x) + h(y)) , \\ rx &= h^{-1}(rh(x)) , \end{aligned}$$

for $x, y \in S \cup T$ and $r \in R$. Take $F_R(S)$ to be $S \cup T$ with this R-module structure. Then h is an R-linear isomorphism from $F_R(S)$ to $\oplus_S R$. So h^{-1} is also an R-linear isomorphism, and by (1.7) (i), $F_R(S)$ is a free R-module with basis $h^{-1}(\widehat{S}) = S$. ∎

(1.9) Definition. For any nontrivial ring R and any set S, **the free R-module based on** S is the R-module $\boldsymbol{F_R(S)}$ constructed in the preceding proof.

(1.10) Corollary. *Every R-module M is R-linearly isomorphic to a quotient of a free R-module. For each positive integer n, every R-module generated by a set of n elements is R-linearly isomorphic to a quotient of R^n.*

Proof. If R is trivial, $M = \{0\}$, which is already free. If R is nontrivial, and $M = \langle S \rangle$ for nonempty S, the inclusion $S \to M$ extends to a surjective R-linear map $f : F_R(S) \to M$ with some kernel K, and this induces an R-linear isomorphism

$$\overline{f} : F_R(S)/K \cong M .$$

If $M = \langle v_1, \cdots, v_n \rangle$, there is a surjective R-linear map $f : R^n \to M$ with $f(e_i) = v_i$ for each i. If $K = ker(f)$, there is an induced isomorphism

$$\overline{f} : R^n/K \cong M .$$ ∎

(1.11) Definition. An R-module M is said to have a **presentation** $(S : D)$ with **generators** S and **defining relators** D if there is an R-linear isomorphism $F/K \cong M$ where F is a free R-module with basis S and K is an R-submodule of F spanned by D.

Technically, the "generators" S in a presentation of M need not be a subset of M, and the composite $F \to F/K \cong M$ need not be injective on S. But if we do have $S \subseteq M$ and $\langle S \rangle = M$, the proof of (1.10) shows that M has a presentation $(S : D)$. So presentations provide a means of describing every R-module.

On the other hand, if R is a nontrivial ring, S is any set, and D is any set of R-linear combinations of elements of S, there is an R-module $F_R(S)/\langle D \rangle$ with presentation $(S : D)$. So presentations provide a tool for creating R-modules that are "made to order." To work in $M = F_R(S)/\langle D \rangle$, choose an abbreviated notation for cosets, such as $\overline{x} = x + \langle D \rangle$. Then the set

$$\overline{S} \ = \ \{\overline{s} : s \in S\}$$

is a spanning set for M. An equation

(1.12) $r_1 \overline{s}_1 + \cdots + r_n \overline{s}_n \ = \ \overline{0}$

$(r_i \in R, \ s_i \in S)$ is true in M if and only if

$$r_1 s_1 + \cdots + r_n s_n \ \in \ \langle D \rangle \ .$$

Those equations (1.12) with

$$r_1 s_1 + \cdots + r_n s_n \ \in \ D$$

are called the **defining relations** for the presentation $(S : D)$. Every true equation (1.12) in M is an R-linear combination of the defining relations — so it is a consequence of the defining relations.

(1.13) Definitions. An R-module is **cyclic** if it is generated by one element. If M is an R-module and $m \in M$, the cyclic submodule generated by m,

$$\langle \{m\} \rangle \ = \ \{rm : r \in R\} \, ,$$

is denoted by Rm. An R-module M is a **finitely generated** (or **f.g.**) R-module if it is generated by a finite set $\{m_1, \ldots, m_n\}$, or equivalently, if M is the sum of finitely many cyclic submodules:

$$M \ = \ Rm_1 + \cdots + Rm_n \ .$$

In a vector space over a field, any two bases have equal size. In a module over a ring, the best we can do for now is

(1.14) Proposition. *If M is a free R-module that is not finitely generated, any two bases of M have the same cardinality. If M is a f.g. free R-module, every basis of M is finite.*

Proof. Suppose B is an R-basis of M and M is generated by a set T. For each $t \in T$ there is a finite subset $B(t)$ of B with $t \in \langle B(t) \rangle$. Since T generates M, the union

$$U = \bigcup_{t \in T} B(t)$$

also generates M. Any element of $B - U$ would be in $M = \langle U \rangle$; so it would be a term in an R-linear relation on B. Therefore $B = U$.

If M is not finitely generated, T is infinite. Since each $B(t)$ is contained in a countable set, an exercise in cardinal arithmetic shows card $(U) \leq$ card (T). So if B and T are both bases of M, card $(B) =$ card (T).

If M is generated by a finite set T, then $B = U$ is finite as well. $\blacksquare$

As with modules, rings can be presented by generators and relations; but the relators are built from the generators by ring addition and multiplication. Such presentations provide rings with made-to-order properties.

A **monoid** is a set N with a binary operation $\cdot$ that is associative and has an identity element 1_N. As a first step toward "free rings," consider how one might make a given monoid $(N, \cdot)$ into part of the multiplicative monoid of a ring R. Such a ring R must include all finite length sums of elements from N and their negatives. The product of two such sums would be computed by the distributive laws, the rules $(-a)b = a(-b) = -(ab)$, and the operation in N. Collecting like terms, each sum can be written as a $\mathbb{Z}$-linear combination of elements of N. If $f, g : N \to \mathbb{Z}$ are functions with finite support, the multiplication would satisfy:

$$(1.15) \qquad \left(\sum_{\nu \in N} f(\nu)\nu \right) \left(\sum_{\nu \in N} g(\nu)\nu \right) = \sum_{\nu \in N} \left(\sum_{\sigma \cdot \tau = \nu} f(\sigma)g(\tau) \right) \nu \, .$$

(1.16) Definition. Suppose $(N, \cdot)$ is a monoid. The **monoid ring** $\mathbb{Z}[N]$ is the free $\mathbb{Z}$-module $F_{\mathbb{Z}}(N)$ based on N, with multiplication given by (1.15). More generally, suppose R is a commutative ring. The **monoid ring** $R[N]$ is the free R-module $F_R(N)$ based on N, with multiplication (1.15).

It is straightforward to verify (if somewhat notationally challenging) that $R[N]$ is an R-algebra whose multiplication restricts to the monoid operation in N, and whose multiplicative identity is the identity 1_N of the monoid N.

The monoid ring is a "free" construction in the sense that it has a universal mapping property. If $(M, \cdot)$ and $(N, \cdot)$ are monoids, a **monoid homomorphism** from M to N is any function $\phi : M \to N$ satisfying $\phi(x \cdot y) = \phi(x) \cdot \phi(y)$ for all $x, y \in M$, and $\phi(1_M) = 1_N$.

(1.17) Proposition. *Suppose $(N, \cdot)$ is a monoid and R is a commutative ring. The inclusion $N \to R[N]$ is a monoid homomorphism into $(R[N], \cdot)$. Each monoid homomorphism $\phi : N \to A$, from N into the multiplicative monoid of an R-algebra A, has one and only one extension to an R-linear ring homomorphism $\widehat{\phi} : R[N] \to A$.*

Proof. Such an extension must satisfy

$$\widehat{\phi}\left(\sum_{\nu \in N} f(\nu)\nu\right) \;=\; \sum_{\nu \in N} f(\nu)\widehat{\phi}(\nu) \;=\; \sum_{\nu \in N} f(\nu)\phi(\nu) \,,$$

and this formula defines an R-linear ring homomorphism $\widehat{\phi}$ extending ϕ.   ■

(1.18) Definition. Suppose S is a set. The **free monoid based on S** is the set $\mathbf{Mon}(S)$ of all strings $s_1 s_2 \ldots s_m$, where $m \geq 0$ and each $s_i \in S$. Two strings are multiplied by concatenation:

$$(s_1 \cdots s_m)(s'_1 \cdots s'_n) \;=\; s_1 \cdots s_m s'_1 \cdots s'_n \,.$$

The empty string (with $m = 0$) is denoted by 1, since it serves as the identity element of $\mathrm{Mon}(S)$.

This construction is also universal:

(1.19) Proposition. *Each function $\psi : S \to N$, from a set S into a monoid N, has one and only one extension to a monoid homomorphism $\widehat{\psi} : \mathrm{Mon}(S) \to N$.*

Proof. Such an extension must satisfy

$$\widehat{\psi}(s_1 \cdots s_m) \;=\; \widehat{\psi}(s_1) \cdots \widehat{\psi}(s_m) \;=\; \psi(s_1) \cdots \psi(s_m) \,,$$

and this formula defines a monoid homomorphism $\widehat{\psi}$ extending ψ.   ■

(1.20) Definition. If S is any set, the **free ring based on S** is the monoid ring $\mathbb{Z}[\mathrm{Mon}(S)]$ of the free monoid based on S.

(1.21) Corollary. *Each function $\theta : S \to R$, from a set S into a ring R, has one and only one extension to a ring homomorphism $\widehat{\theta} : \mathbb{Z}[Mon(S)] \to R$.*

Proof. By (1.17) and (1.19) it suffices to note that each ring homomorphism from $\mathbb{Z}[\mathrm{Mon}(S)]$ to R extending θ extends a monoid homomorphism on $\mathrm{Mon}(S)$ that extends θ.   ■

(1.22) Definition. A ring R is said to have a **presentation** $(S : D)$ **as a ring**, with **generators** S and **defining relators** D, if there is a ring isomorphism $F/I \cong R$, where F is the free ring based on S and I is an ideal of F generated by D.

Suppose S is any set and D is any set of $\mathbb{Z}$-linear combinations of strings from $\text{Mon}(S)$. If I denotes the ideal of $F = \mathbb{Z}[\text{Mon}(S)]$ generated by D, then the quotient $R = F/I$ is a ring with presentation $(S : D)$. For each $x \in F$, denote $x + I$ by $\overline{x}$. Then R consists of the $\mathbb{Z}$-linear combinations of products $\overline{s_1 \cdots s_m} = \overline{s}_1 \cdots \overline{s}_m$ with $s_i \in S, m \geq 0$. If $\sigma_1, \ldots, \sigma_n$ are strings from $\text{Mon}(S)$, an equation

$$r_1 \overline{\sigma}_1 + \cdots + r_n \overline{\sigma}_n = \overline{0}$$

is true in R if and only if

$$r_1 \sigma_1 + \cdots + r_n \sigma_n \in I$$

and is called a **defining relation** for $(S : D)$ if

$$r_1 \sigma_1 + \cdots + r_n \sigma_n \in D .$$

Every true equation in R is a consequence of the defining relations, since D generates I as an ideal.

1A. Exercises

1. Suppose M is an R-module and $S \subseteq T \subseteq M$. If T is linearly independent, prove S is linearly independent. If S spans M, prove T spans M.

2. Prove the additive group $(\mathbb{Q}, +)$ of rational numbers is a $\mathbb{Z}$-module without a basis.

3. In the $\mathbb{Z}$-module $\mathbb{Z}$, find

  (i) a maximal linearly independent subset that is not a basis;
  (ii) a minimal generating set that is not a basis.

4. If D is a division ring, and M is a D-module, prove

  (i) every maximal linearly independent subset of M is a basis of M; and
  (ii) every minimal generating set of M is a basis of M.

5. Suppose R is a ring with q elements, and M is an R-module having a basis with n elements (q, n positive integers). Prove, without using matrices, that $\text{Hom}_R(M, M)$ has $(q^n)^n$ elements. *Hint:* Use (1.4).

6. Suppose an R-module M has an infinite basis B. Prove there is an injective R-linear map $f : M \to M$ that is not surjective, and a surjective R-linear map $g : M \to M$ that is not injective.

7. Give a description by generators and defining relations of the ring $\mathbb{Z}[i]$ of Gaussian integers.

8. Prove that the free ring based on a set $S = \{x\}$ with one element is a commutative integral domain with a non-principal ideal, but that the free ring based on a set $S = \{x, y\}$ with two elements is not even commutative.

9. Show that the field $\mathbb{Q}$ of rational numbers is not finitely generated as a ring.

1B. Matrix Representations

Here we consider the matrix description of R-linear maps between f.g. free R-modules M and N, and the connections between addition and composition of linear maps on the one hand, and addition and multiplication of matrices on the other. This extends the standard linear algebra over a field, and works over an arbitrary ring of scalars.

For each ring R, the category R-$\mathcal{Mod}$ has an additive structure:

(1.23) Proposition. *In R-$\mathcal{Mod}$ each set $\mathrm{Hom}_R(M, N)$ is an additive abelian group. Composition of arrows is distributive over this addition.*

Proof. If $f, g \in \mathrm{Hom}_R(M, N)$, define their sum $f + g : M \to N$ pointwise:
$$(f + g)(m) = f(m) + g(m)$$
for all $m \in M$. Then $f + g$ is R-linear. The zero map $0 : M \to N$ is an identity for this addition, and for each $f \in \mathrm{Hom}_R(M, N)$ there is an additive inverse $-f \in \mathrm{Hom}_R(M, N)$ defined by
$$(-f)(m) = -(f(m))$$
for all $m \in M$. If
$$Q \xrightarrow{\ g\ } M \underset{f_2}{\overset{f_1}{\rightrightarrows}} N \xrightarrow{\ h\ } P$$
is a diagram in R-$\mathcal{Mod}$, then the linearity of h and the definition of pointwise sum imply the distributivity
$$h \circ (f_1 + f_2) = (h \circ f_1) + (h \circ f_2), \quad \text{and}$$
$$(f_1 + f_2) \circ g = (f_1 \circ g) + (f_2 \circ g),$$
of composition over sum. ∎

(1.24) Corollary. *In R-Mod each* $\text{End}_R(M)$ *is a ring under pointwise sum and composition.* ∎

These facts (1.23) and (1.24) remain true when R-Mod is replaced by any of its full subcategories. In particular, consider the full subcategories:

$$\mathcal{Row}(R) \;\subseteq\; \mathcal{F}(R) \;\subseteq\; \mathcal{M}(R)\,,$$

where

$$
\begin{aligned}
\text{Obj } \mathcal{M}(R) &= \text{ all f.g. } R\text{-modules,} \\
\text{Obj } \mathcal{F}(R) &= \text{ all f.g. free } R\text{-modules,} \\
\text{Obj } \mathcal{Row}(R) &= \text{ all } R^n \text{ with } n \geq 1.
\end{aligned}
$$

Properties (1.23) and (1.24), restricted to $\mathcal{Row}(R)$, become the basic facts of matrix arithmetic. Recall the terminology of matrices:

(1.25) Definitions. If R is a ring and m, n are positive integers, denote by $R^{m \times n}$ the set of all $m \times n$ **matrices**

$$
A = (a_{ij}) =
\begin{bmatrix}
a_{11} & a_{12} & \cdots & a_{1n} \\
a_{21} & a_{22} & \cdots & a_{2n} \\
\vdots & \vdots & & \vdots \\
a_{m1} & a_{m2} & \cdots & a_{mn}
\end{bmatrix}
$$

with i,j-entry $a_{ij} \in R$ for $1 \leq i \leq m$ and $1 \leq j \leq n$. On $R^{m \times n}$, **matrix addition** is defined by

$$(a_{ij}) + (b_{ij}) \;=\; (a_{ij} + b_{ij})\,.$$

If m, n, and p are positive integers, **matrix multiplication** $R^{m \times n} \times R^{n \times p} \to R^{m \times p}$ is defined by

$$(a_{ij})(b_{ij}) \;=\; (c_{ij})\,,$$

where

$$c_{ij} \;=\; \sum_{k=1}^{n} a_{ik} b_{kj}\,.$$

The **zero matrix** $0_{m \times n} \in R^{m \times n}$ has every entry equal to 0_R. The **identity matrix** $I_m \in R^{m \times m}$ is

$$
(\delta_{ij}) =
\begin{bmatrix}
1 & 0 & \cdots & 0 \\
0 & 1 & \cdots & 0 \\
\vdots & \vdots & \ddots & \vdots \\
0 & 0 & \cdots & 1
\end{bmatrix},
$$

where

$$\delta_{ij} = \begin{cases} 1 & \text{if} \quad i = j \,, \\ 0 & \text{if} \quad i \neq j \,. \end{cases}$$

A **scalar matrix** is a matrix

$$rI_m = (r\delta_{ij}) \,,$$

where $r \in R$. A matrix $A = (a_{ij}) \in R^{m \times n}$ can be multiplied by a scalar $r \in R$ by

$$rA = (rI_m)A = (ra_{ij}) \,,$$
$$Ar = A(rI_n) = (a_{ij}r) \,.$$

As in Example (0.11) (iii), we also denote $R^{n \times n}$ by $M_n(R)$. By a notational change we can regard each element $(r_1, \ldots, r_n)$ of R^n as a matrix $[r_1 \cdots r_n] \in R^{1 \times n}$. Then the standard basis $e_1, \ldots, e_n$ of R^n becomes the list of rows of I_n.

(1.26) Proposition. *Suppose R is a ring and m, n, p, q are positive integers.*

(i) *$R^{m \times n}$ is an abelian group under matrix addition, with identity $0_{m \times n}$.*

(ii) *If $A \in R^{m \times n}$, $I_m A = A = A I_n$.*

(iii) *If $A \in R^{m \times n}$, $B \in R^{n \times p}$ and $C \in R^{p \times q}$; then $(AB)C = A(BC)$.*

(iv) *If $A \in R^{m \times n}$, B and $C \in R^{n \times p}$ and $D \in R^{p \times q}$; then $A(B + C) = AB + AC$ and $(B + C)D = BD + CD$.*

(v) *$M_n(R)$ is a ring.*

(vi) *There is a category $\mathfrak{Mat}(R)$ in which the objects are the positive integers, $\mathrm{Hom}(m, n) = R^{m \times n}$, and for $A \in R^{m \times n}$ and $B \in R^{n \times p}$, $B \circ A = AB$.*

(vii) *When we identify R^m with $R^{1 \times m}$, each R-linear map $f : R^m \to R^n$ is right multiplication by one and only one matrix $A \in R^{m \times n}$. The i-row of A is $f(e_i)$ for $i = 1, \ldots, m$. Taking f to A defines an isomorphism of additive groups:*

$$\mathrm{Hom}_R(R^m, R^n) \cong R^{m \times n} \,,$$

an isomorphism of rings

$$\mathrm{End}_R(R^n) \cong (M_n(R))^{op} \,,$$

and an isomorphism of categories:

$$\mathfrak{Row}(R) \cong \mathfrak{Mat}(R) \,.$$

Proof. Two matrices in $R^{m \times n}$ are equal if their i, j-entries agree for each i and j. So the additive abelian group axioms for $R^{m \times n}$ follow from those for R, proving (i). Comparing corresponding entries in (ii),

$$\sum_k \delta_{ik} a_{kj} \;=\; a_{ij} \;=\; \sum_k a_{ik} \delta_{kj} \;.$$

Applying the same method in (iii),

$$\sum_s \left(\sum_t a_{it} b_{ts} \right) c_{sj}$$

$$= \sum_{s,t} (a_{it} b_{ts} c_{sj})$$

$$= \sum_t a_{it} \left(\sum_s b_{ts} c_{sj} \right) .$$

For (iv),

$$\sum_k a_{ik} (b_{kj} + c_{kj}) \;=\; \sum_k (a_{ik} b_{kj} + a_{ik} c_{kj})$$

$$= \sum_k a_{ik} b_{kj} \;+\; \sum_k a_{ik} c_{kj} \;,$$

and similarly for the second equation. By (i)–(iv), $M_n(R)$ is a ring under matrix sum and product, proving (v). By (ii) and (iii), $\mathfrak{Mat}(R)$ is a category, proving (vi).

For $f \in \mathrm{Hom}_R(R^m, R^n)$ and $A \in R^{m \times n}$,

$$f \left(\sum_i r_i e_i \right) \;=\; (r_1, \ldots, r_m) A \;,$$

for all $r_1, \ldots, r_m \in R$, if and only if

$$f(e_i) \;=\; e_i A \;=\; i\text{-row of } A \;,$$

for $1 \le i \le m$. If these conditions are true, denote A by $F(f)$ and f by $G(A)$. Then F and G define mutually inverse bijections

$$\mathrm{Hom}_R(R^m, R^n) \underset{G}{\overset{F}{\rightleftarrows}} R^{m \times n} \;.$$

By (iv), $G(A+B) = G(A)+G(B)$; so G and F are additive group isomorphisms. If $m = n$, (iii) and (ii) imply $G(AB) = G(B) \circ G(A)$ and $G(I_m) = $ identity on R^m; so G and F are ring isomorphisms. If $A \in R^{m \times n}$ and $B \in R^{n \times p}$,

$$G(B) \circ G(A) \;=\; G(AB) \;=\; G(B \circ A)$$

by (iii) and the definition of composition in $\mathfrak{Mat}(R)$. If $f = G(A)$ and $g = G(B)$, then

$$F(g \circ f) \;=\; F(G(B \circ A)) \;=\; B \circ A \;=\; F(g) \circ F(f) \,.$$

So F and G are inverse functors. ∎

The functor F in the preceding proof is known as the **matrix representation** of R-linear maps $R^m \to R^n$. Arrows in $\mathcal{F}(R)$ can also be represented by matrices with respect to chosen bases of their domain and codomain. Suppose $f : M \to N$ is an R-linear map, where M has a basis $v_1, \ldots, v_m$ corresponding to an isomorphism $\alpha : M \cong R^m$, and N has a basis $w_1, \ldots, w_n$ corresponding to an isomorphism $\beta : N \cong R^n$. For a matrix $A \in R^{m \times n}$, the following two conditions are equivalent:

(i) $f\left(\sum_i r_i v_i\right) = \sum_i s_i w_i$, where $[r_1 \cdots r_m]A = [s_1 \cdots s_n]$;

(ii) $A = (a_{ij})$, where, for $1 \le i \le m$, $f(v_i) = a_{i1} w_1 + \cdots + a_{in} w_n$.

When these conditions hold, we say A **represents** f **over the bases** $v_1, \ldots, v_n$ of M and $w_1, \ldots, w_n$ of N, and write

$$A \;=\; Mat_\alpha^\beta(f) \,.$$

By condition (i), f is represented by A over the chosen bases if and only if the square

$$
\begin{array}{ccc}
M & \xrightarrow{\;f\;} & N \\
\alpha \downarrow & & \downarrow \beta \\
R^m & \xrightarrow[\cdot A]{} & R^n
\end{array}
$$

commutes, where $\cdot A$ denotes right multiplication by A. So

$$Mat_\alpha^\beta : \operatorname{Hom}_R(M, N) \;\to\; R^{m \times n}$$

is an additive group isomorphism, since it is the composite of isomorphisms

$$\operatorname{Hom}_R(M, N) \;\cong\; \operatorname{Hom}_R(R^m, R^n)$$
$$f \;\mapsto\; \beta \circ f \circ \alpha^{-1}$$

and

$$\operatorname{Hom}_R(R^m, R^n) \;\cong\; R^{m \times n}$$
$$\cdot A \;\mapsto\; A$$

from (1.26) (vii).

If $\alpha : M \to R^m$, $\beta : N \to R^n$ and $\gamma : P \to R^p$ are isomorphisms, and the left and right squares in

$$
\begin{array}{ccccc}
M & \xrightarrow{\ f\ } & N & \xrightarrow{\ g\ } & P \\
\downarrow{\scriptstyle \alpha} & & \downarrow{\scriptstyle \beta} & & \downarrow{\scriptstyle \gamma} \\
R^m & \xrightarrow[\ \cdot A\ ]{} & R^n & \xrightarrow[\ \cdot B\ ]{} & R^p
\end{array}
$$

commute, then the perimeter rectangle commutes. So

$$
Mat_\alpha^\gamma(g \circ f) \;=\; Mat_\alpha^\beta(f) Mat_\beta^\gamma(g) \; ;
$$

and $\alpha \circ i_M = (\cdot I_m) \circ \alpha = \alpha$; so

$$
Mat_\alpha^\alpha(i_M) \;=\; I_m \; .
$$

Combined with the additivity proved above, these equations show

$$
Mat_\alpha^\alpha : \operatorname{End}_R(M) \;\to\; (M_m(R))^{op}
$$

is an isomorphism of rings.

(1.27) Note. In our discussion of modules there is little reason to favor left R-modules over right R-modules, and the reader would be well advised to consider the right R-module version of each definition and theorem encountered. Usually, translating between them involves nothing more than writing scalars on the other side. However, in the matrix representations of R-linear maps, some additional adjustments are needed.

For right R-modules, it is best to regard n-tuples in R^n as *column* vectors in $R^{n \times 1}$. Then each R-linear map $R^m \to R^n$ is *left* multiplication by a unique matrix $F(f) \in R^{n \times m}$, and the j-column of $F(f)$ is $f(e_j)$ for $1 \le j \le m$. This F is a bijection

$$
\operatorname{Hom}_R(R^m,\ R^n) \;\to\; R^{n \times m}
$$

satisfying $F(f + g) = F(f) + F(g)$, $F(i_m) = I_m$, and for R-linear maps

$$
R^m \xrightarrow{\ f\ } R^n \xrightarrow{\ g\ } R^p \; ,
$$

$F(g \circ f) = F(g)F(f)$. In particular, F is a ring isomorphism:

(1.28) $$ \operatorname{End}_R(R^n) \;\cong\; M_n(R) \; . $$

If M and N have ordered bases $v_1, \ldots, v_m$ and $w_1, \ldots, w_n$ with associated R-linear isomorphisms $\alpha : M \to R^m$, $\beta : N \to R^n$, then each R-linear map $f : M \to N$ is represented by the matrix $Mat_\alpha^\beta(f)$ whose j-*column* is the $w_1, \ldots, w_n$-coordinates of $f(v_j)$. Then left multiplication by $Mat_\alpha^\beta(f)$ takes the column of $v_1, \ldots, v_m$-coordinates of v to the column of $w_1, \ldots, w_n$-coordinates of $f(v)$. And Mat_α^β is a bijection

$$\mathrm{Hom}_R(M, N) \ \to \ R^{n \times m}$$

satisfying $Mat_\alpha^\beta(f + g) = Mat_\alpha^\beta(f) + Mat_\alpha^\beta(g), Mat_\alpha^\beta(i_M) = I_m$, and for R-linear maps

$$M \xrightarrow{\ f\ } N \xrightarrow{\ g\ } P$$

and an isomorphism $\gamma : P \cong R^p$, $Mat_\alpha^\gamma(g \circ f) = Mat_\beta^\gamma(g)Mat_\alpha^\beta(f)$. In particular, Mat_α^α is a ring isomorphism:

(1.29) $\mathrm{End}_R(M) \ \cong \ M_m(R)$.

The fact that the opposite ring is not needed in (1.28) and (1.29) makes right R-modules preferable in some contexts. In case R is a *commutative* ring, the transpose is a ring isomorphism

$$M_m(R) \ \cong \ M_m(R)^{op}$$

and is a bijection $R^{m \times n} \to R^{n \times m}$ carrying $Mat_\alpha^\beta(f)$ (left module version) to $Mat_\alpha^\beta(f)$ (right module version). For more details, see Exercise 4.

The proof of parts of (1.26) generalizes to other types of matrices:

(1.30) Example. Suppose R is a ring and m, n are positive integers. If M is an additive abelian group, let $M^{m \times n}$ denote the set of $m \times n$ matrices (m_{ij}) with entries from M. By the proof of (1.26) (i), $M^{m \times n}$ is an additive abelian group under entrywise addition. If $M \in R\text{-}\mathfrak{Mod}$, then the proof of (1.26) (ii–iv) shows that $M^{m \times n}$ is a left $M_m(R)$-module by a scalar multiplication

$$(r_{ij}) \cdot (m_{ij}) \ = \ (m'_{ij}) ,$$

where

$$m'_{ij} \ = \ \sum_k r_{ik} m_{kj} .$$

In the same way, if $M \in \mathfrak{Mod}\text{-}R$, then $M^{m \times n}$ is a right $M_n(R)$-module.

(1.31) Example. An R-linear map between finite direct sums of R-modules can be written as left multiplication of columns by a matrix of maps. For $1 \le i \le t$, $1 \le j \le s$, suppose M_j and N_i are additive abelian groups, and

$$f_{ij} \ \in \ \mathrm{Hom}_{\mathbb{Z}}(M_j, N_i) .$$

The $t \times s$ matrix (f_{ij}) denotes the $\mathbb{Z}$-linear map

$$M_1 \oplus \cdots \oplus M_s \ \to \ N_1 \oplus \cdots \oplus N_t$$

defined by

$$(f_{ij}) \begin{bmatrix} m_1 \\ \vdots \\ m_s \end{bmatrix} = \begin{bmatrix} f_{11}(m_1) & +\cdots+ & f_{1s}(m_s) \\ & \vdots & \\ f_{t1}(m_1) & +\cdots+ & f_{ts}(m_s) \end{bmatrix},$$

which is a kind of matrix multiplication. If the M_j and N_i are all in R-Mod (or all in Mod-R), and the f_{ij} are R-linear, then (f_{ij}) is R-linear.

The sum of maps $(f_{ij}) + (f'_{ij})$ is the matrix sum $(f_{ij} + f'_{ij})$. The composite of matrix maps $(f_{ij}) \circ (g_{ij})$ is the matrix product (h_{ij}), where

$$h_{ij} = \sum_k f_{ik} \circ g_{kj} \,.$$

If 1 denotes an identity map and 0 denotes a zero map, and if A^τ denotes the transpose of a matrix A, then

$$e_i : M_1 \oplus \cdots \oplus M_s \rightarrow M_i$$

is projection to the i-coordinate, and

$$e_i^\tau : M_i \rightarrow M_1 \oplus \cdots \oplus M_s$$

is insertion in the i-coordinate.

For each

$$f \in \mathrm{Hom}_R(M_1 \oplus \cdots \oplus M_s, \ N_1 \oplus \cdots \oplus N_t)$$

there is a unique matrix of maps (f_{ij}) with $f = (f_{ij})$: For if $f(m) = n$, where $m = (m_1, \ldots, m_s)$ and $n = (n_1, \ldots, n_t)$, then

$$n_i = e_i(f(m)) = e_i\left(f\left(\sum_j e_j^\tau(m_j)\right)\right)$$

$$= \sum_j (e_i \circ f \circ e_j^\tau)(m_j) \,.$$

Defining f_{ij} to be $e_i \circ f \circ e_j^\tau$, we have $(f_{ij})m^\tau = n^\tau$, and $f = (f_{ij})$. On the other hand, if $f = (f_{ij})$, then the maps f_{ij} are uniquely determined by f:

$$f_{ij} = e_i(f_{ij})e_j^\tau = e_i \circ f \circ e_j^\tau \,.$$

(1.32) Example. Suppose the summands of $M = M_1 \oplus \cdots \oplus M_s$ and $N = N_1 \oplus \cdots \oplus N_t$ are right R-modules of column vectors:

$$M_i = R^{c(i) \times 1}, \quad N_i = R^{b(i) \times 1},$$

so that each R-linear map

$$f_{ij} \in \mathrm{Hom}_R(M_j, N_i)$$

is left multiplication by a matrix

$$A_{ij} \in R^{b(i) \times c(j)} .$$

As in (1.31), the R-linear map $(f_{ij}) : M \to N$ is left multiplication of a column of columns by a $t \times s$ matrix $A = (A_{ij})$ whose entries A_{ij} are matrices. If also $L = L_1 \oplus \cdots \oplus L_r$, where

$$L_i = R^{d(i) \times 1} ,$$

and $(g_{ij}) : L \to M$ is left multiplication by a matrix $B = (B_{ij})$, where

$$B_{ij} \in R^{c(i) \times d(j)} ,$$

then $(f_{ij}) \circ (g_{ij}) : L \to N$ is left multiplication by the product

$$(A_{ij})(B_{ij}) = (C_{ij}) ,$$

where

$$C_{ij} = \sum_k A_{ik} B_{kj}$$

is a matrix sum of matrix products. This multiplication is called "block multiplication of partitioned matrices."

In more detail, a matrix $A = (A_{ij})$

$$= \begin{bmatrix} A_{11} & \cdots & A_{1s} \\ \vdots & & \vdots \\ A_{t1} & \cdots & A_{ts} \end{bmatrix} \begin{matrix} \} & b(1) & \text{rows} \\ & \vdots & \\ \} & b(t) & \text{rows} \end{matrix}$$

$$\underbrace{\phantom{A_{11}}}_{c(1)\mathrm{col's}} \quad \underbrace{\phantom{A_{1s}}}_{c(s)\mathrm{col's}}$$

of matrices $A_{ij} \in R^{b(i) \times c(j)}$ is said to be a $\boldsymbol{b} \times \boldsymbol{c}$ matrix, where

$$\boldsymbol{b} = (b(1), \ldots, b(t)) ,$$
$$\boldsymbol{c} = (c(1), \ldots, c(s)) .$$

The set of all $\boldsymbol{b} \times \boldsymbol{c}$ matrices over R is denoted by

$$R^{\boldsymbol{b} \times \boldsymbol{c}} .$$

Block addition $(A_{ij}) + (A'_{ij}) = (A_{ij} + A'_{ij})$ makes $R^{\boldsymbol{b} \times \boldsymbol{c}}$ into an additive abelian group. If $\boldsymbol{b}, \boldsymbol{c},$ and $\boldsymbol{d}$ are rows of positive integers, **block multiplication**

$$R^{\boldsymbol{b} \times \boldsymbol{c}} \times R^{\boldsymbol{c} \times \boldsymbol{d}} \to R^{\boldsymbol{b} \times \boldsymbol{d}}$$

is defined by

$$(A_{ij})(B_{ij}) \;=\; \left(\sum_k A_{ik} B_{kj} \right).$$

If $A = (A_{ij}) \in R^{b \times c}$ and b is the sum of the coordinates in $\boldsymbol{b}$ and c is the sum of the coordinates in $\boldsymbol{c}$, then erasing the matrix brackets in each A_{ij} transforms A into a $b \times c$ matrix **erase**(A) with entries in R. Then erase: $R^{b \times c} \to R^{b \times c}$ turns out to preserve sums and products; so block addition and block multiplication agree with ordinary matrix addition and multiplication over R:

(1.33) Proposition. *If R is a ring; $\boldsymbol{b}, \boldsymbol{c}$, and $\boldsymbol{d}$ are rows of positive integers; and $A, A' \in R^{b \times c}$ and $B \in R^{c \times d}$ are partitioned matrices, then*

$$erase(A + A') \;=\; erase(A) \;+\; erase(A') \,,$$

and

$$erase(AB) \;=\; erase(A) \; erase(B) \,.$$

Proof. The first equation is immediate, since block addition $A + A'$ is entrywise addition $(A_{ij}) + (A'_{ij}) = (A_{ij} + A'_{ij})$ and $A_{ij} + A'_{ij}$ is entrywise addition over R. In the block product $(A_{ij})(B_{ij}) = (C_{ij})$, the i', j'-entry of C_{ij} is the sum over k of the i', j'-entries of each $A_{ik} B_{kj}$; so it is the sum over k of the i'-row of A_{ik} times the j'-column of B_{kj}. This coincides with the i'-row of the i-row of A times the j'-column of the j-column of B. So the m, n-entries of erase(AB) and erase(A) erase(B) agree, where

$$m \;=\; \sum_{u < i} b(u) + i' \,,$$

$$n \;=\; \sum_{v < j} d(v) + j' \,,$$

and these m, n cover all entries. ∎

(1.34) Definition. Suppose R is a ring and $f : M_1 \to N_1$, $g : M_2 \to N_2$ are R-linear maps. Then $\boldsymbol{f \oplus g}$ is the R-linear map

$$\begin{bmatrix} f & 0 \\ 0 & g \end{bmatrix} \;:\; M_1 \oplus M_2 \;\to\; N_1 \oplus N_2 \,.$$

For $m_i \in M_i$,

$$(f \oplus g)(m_1, m_2) \;=\; (f(m_1), g(m_2)) \,.$$

Notice that the composite of two such diagonal maps is

$$(h \oplus k) \circ (f \oplus g) \;=\; (h \circ f) \oplus (k \circ g) \,,$$

and the identity on $M_1 \oplus M_2$ is $i_1 \oplus i_2$, where i_j is the identity map on M_j. Therefore, *if f and g are isomorphisms, then $f \oplus g$ is an isomorphism.*

If M_i, N_i are right R-modules of column vectors, then f, g and the zero maps are replaced by matrices:

(1.35) Definition. Suppose R is a ring, $A \in R^{s \times t}$, and $B \in R^{u \times v}$. The **direct sum of matrices** A and B is the matrix:

$$A \oplus B = \begin{bmatrix} A & 0 \\ 0 & B \end{bmatrix} \in R^{(s+u) \times (t+v)} \,.$$

For matrices $A, B, C,$ and D over R, matrix direct sum has the properties:

(i) $(A \oplus B) \oplus C \;=\; A \oplus (B \oplus C) \,,$

(ii) $(A \oplus B)^{\tau} \;=\; A^{\tau} \oplus B^{\tau} \,,$

(iii) $(A \oplus B)(C \oplus D) \;=\; AC \oplus BD \,,$

(iv) $I_m \oplus I_n \;=\; I_{m+n} \,,$

where x^{τ} means the transpose of x and, in (iii), where the products AC and BD are defined. From (iii) and (iv) we get:

(v) If A and B are invertible, so is $A \oplus B$, with $(A \oplus B)^{-1} \;=\; A^{-1} \oplus B^{-1}$.

1B. Exercises

1. Suppose R is a ring. If $1 \leq i \leq s$ and $1 \leq j \leq t$ are positive integers, denote by ϵ_{ij} the matrix in $R^{s \times t}$ with i, j-entry 1_R and all other entries 0_R. Prove:

(i) $R^{s \times t}$ is a free right and left R-module with basis $\{\epsilon_{ij} : 1 \leq i \leq s, 1 \leq j \leq t\}$.

(ii) For $r \in R$, $1 \leq i \leq s$ and $1 \leq j \leq t$, $\; r\epsilon_{ij} = \epsilon_{ij}r$.

(iii) If $\epsilon_{ij} \in R^{s \times t}$ and $\epsilon_{k\ell} \in R^{t \times u}$, then

$$\epsilon_{ij}\epsilon_{k\ell} = \begin{cases} 0_{s \times u} & \text{if } \; j \neq k \\ \epsilon_{i\ell} & \text{if } \; j = k. \end{cases}$$

The matrices ϵ_{ij} are known as **matrix units**. Writing (a_{ij}) as $\sum_{i,j} a_{ij}\epsilon_{ij}$, one can recover the formulas for matrix addition and multiplication from (i), (ii), and (iii) above.

2. Suppose R is a nontrivial ring, so that $0_R \neq 1_R$. Prove that, for $n \geq 2$, the multiplication in $M_n(R)$ is not commutative, and there exist matrices $a, b \in M_n(R)$ with $a \neq 0_{n\times n}$, $b \neq 0_{n\times n}$, and $ab = 0_{n\times n}$. *Hint:* Use exercise 1.

3. The field $\mathbb{C}$ of complex numbers is a vector space over the field $\mathbb{R}$ of real numbers, with basis $1, i$. Let $\alpha : \mathbb{C} \to \mathbb{R}^2$ denote the $\mathbb{R}$-linear isomorphism

$$\alpha(a + bi) \;=\; (a, b)$$

corresponding to this basis. Let $f : \mathbb{C} \to \mathbb{C}$ denote complex conjugation and $g : \mathbb{C} \to \mathbb{C}$ denote multiplication by i. Regarding $\mathbb{C}$ as a left $\mathbb{R}$-module, verify that

$$Mat^\alpha_\alpha(g \circ f) \;=\; Mat^\alpha_\alpha(f) Mat^\alpha_\alpha(g)$$
$$\neq\; Mat^\alpha_\alpha(g) Mat^\alpha_\alpha(f) .$$

Now compute the corresponding matrices with $\mathbb{C}$ regarded as a right $\mathbb{R}$-module to verify

$$Mat^\alpha_\alpha(g \circ f) \;=\; Mat^\alpha_\alpha(g) Mat^\alpha_\alpha(f)$$

in this context.

4. Suppose R is a commutative ring and let $\tau : R^{m\times n} \to R^{n\times m}$ denote the transpose: $\tau((a_{ij})) = (b_{ij})$, where $b_{ij} = a_{ji}$ for each pair i, j.

  (i) Prove τ is a bijection with $\tau(a+b) = \tau(a)+\tau(b)$ and $\tau(ab) = \tau(b)\tau(a)$.
  (ii) If M is a left R-module with a basis $v_1, \ldots, v_m$, define a scalar multiplication $M \times R \to M$, $(m, r) \mapsto m * r$, by $m * r = rm$. Show M is a right R-module via $*$, and $v_1, \ldots, v_m$ is a basis of M as a right R-module.
  (iii) If $f : M \to N$ is an R-linear map between left R-modules M and N, show f is also R-linear when M and N are regarded as right R-modules using $*$ as in part (ii).
  (iv) If $\alpha : M \to R^m$ and $\beta : N \to R^n$ are R-linear isomorphisms, $f : M \to N$ is an R-linear map; $A = Mat^\beta_\alpha(f)$ when M, N are considered left R-modules; and $B = Mat^\beta_\alpha(f)$ when M, N are considered right R-modules, prove $\tau(A) = B$.

5. Suppose $M \in R\text{-}\mathcal{M}\mathfrak{od}$ has two bases $v_1, \ldots, v_m$ and $w_1, \ldots, w_m$.

  (i) In the terminology of Example (1.30), show there is one and only one matrix A over R with

$$\begin{bmatrix} v_1 \\ \vdots \\ v_m \end{bmatrix} = A \begin{bmatrix} w_1 \\ \vdots \\ w_m \end{bmatrix},$$

namely, $A = Mat_\alpha^\beta(i_M)$, where α and β are the isomorphisms $M \cong$
R^m associated to these bases. This matrix A is called the **change of**
basis matrix from $v_1, \ldots, v_m$-coordinates to $w_1, \ldots, w_m$-coordinates.

(ii) If $v_1, \ldots, v_m$ is fixed and $w_1, \ldots, w_m$ varies through all m-element
bases of M, show A varies through all of $GL_m(R)$.

6. Suppose A and B are matrices over R. Prove that if $A \oplus B$ is invertible,
then A and B are invertible.

1C. Absence of Dimension

In the linear algebra of vector spaces over a field, every vector space has a basis,
and any two bases of the same vector space have the same number of elements.
So one can define the **dimension** of a vector space V to be the cardinality of
a basis of V.

For a module M over a ring R, the existence of a basis is not guaranteed,
and in some cases a free R-module can have bases of two different sizes. We
first discuss the rings R whose free modules have unique dimension; then we
characterize the rings R over which every module is free.

(1.36) Definitions. A ring R has **invariant basis number** (or **IBN**) if, in
each free R-module M, each pair of bases of M have equal cardinality. By
(1.14) and (1.7), R has IBN if and only if the following condition holds:

"If m, n are positive integers and $R^m \cong R^n$ as R-modules,
then $m = n$."

If R has IBN, the **free rank** of a free R-module M is the number of elements
in a basis of M.

(1.37) Example of a ring without IBN. Suppose R is a nontrivial ring.
Let $\mathbb{P}$ denote the set of positive integers. Each function $f : \mathbb{P} \times \mathbb{P} \to R$ defines
a "$\mathbb{P} \times \mathbb{P}$ matrix" with infinitely many rows and columns:

$$a = \begin{bmatrix} a_{11} & a_{12} & \cdots \\ a_{21} & a_{22} & \cdots \\ \vdots & \vdots & \end{bmatrix},$$

where $a_{ij} = f(i, j)$. A $\mathbb{P} \times \mathbb{P}$ matrix over R is called **column-finite** if each of
its columns has only finitely many nonzero entries. Denote by A the set of all
column-finite $\mathbb{P} \times \mathbb{P}$ matrices over R.

Ordinary matrix addition and matrix multiplication make A into a ring, and make the set of columns of members of A into an A-module. Let e_j denote the j-column of the identity:

$$1_A = \begin{bmatrix} 1 & 0 & \cdots \\ 0 & 1 & \cdots \\ \vdots & \vdots & \ddots \end{bmatrix}.$$

For each $a \in A$, the j-column of a is ae_j. When A is regarded as an A-module, the vector addition and scalar multiplication are computed columnwise:

$$(a+b)e_j = ae_j + be_j \, ,$$

$$(ab)e_j = a(be_j) \, .$$

For each $a \in A$, define $a', a'' \in A$ so that the columns of a' are the odd-numbered columns of a and the columns of a'' are the even-numbered columns of a. Then the function $f : A \to A \oplus A$, $f(a) = (a', a'')$, is an A-linear isomorphism, with an inverse map that shuffles a' and a'' together to form a.

On the one hand, A has A-basis 1_A. On the other hand, by (1.7) (i), A has a two-element basis:

$$f^{-1}(1_A, 0_A) = \begin{bmatrix} 1 & 0 & 0 & 0 & \cdots \\ 0 & 0 & 1 & 0 & \cdots \\ \vdots & \vdots & \vdots & \vdots & \end{bmatrix},$$

$$f^{-1}(0_A, 1_A) = \begin{bmatrix} 0 & 1 & 0 & 0 & \cdots \\ 0 & 0 & 0 & 1 & \cdots \\ \vdots & \vdots & \vdots & \vdots & \end{bmatrix}.$$

One can as easily find an A-basis of A with any finite number n of basis elements.

(1.38) Note. Suppose M is a free right R-module with a countably infinite basis $\{b_1, b_2, \dots \}$. Just as in (1.29), there is a ring isomorphism from $\operatorname{End}_R(M)$ to the ring A of column-finite $\mathbb{P} \times \mathbb{P}$ matrices over R.

It may have struck the reader that the preceding example is somewhat contrived. This is necessary, as "most" rings do have IBN:

(1.39) Theorem.

   (i) *If a ring R has IBN, then the opposite ring R^{op} has IBN.*
   (ii) *If a ring R has IBN, then $M_n(R)$ has IBN for each positive integer n.*
   (iii) *If $f : R \to S$ is a ring homomorphism and S has IBN, then R has IBN.*
   (iv) *Every nontrivial left noetherian ring has IBN.*

Proof. By (1.26) (vii), the existence of an R-linear isomorphism $R^m \cong R^n$ is equivalent to the existence of matrices $a \in R^{m \times n}$ and $b \in R^{n \times m}$ with $ab = I_m$ and $ba = I_n$. So assertion (i) follows from the fact that the transpose of a matrix product ab is the product over R^{op} of b-transpose times a-transpose.

For assertion (ii) use block multiplication of matrices (see (1.32) and (1.33)): If $a \in (M_s(R))^{m \times n}$ and $b \in (M_s(R))^{n \times m}$, with $ab = I_{ms}$ and $ba = I_{ns}$, then by IBN for R, $ms = ns$; so $m = n$.

For (iii), if we apply f to each entry of a and b, we get matrices $a' \in S^{m \times n}$ and $b' \in S^{n \times m}$, with $a'b' = I_m$ and $b'a' = I_n$. So if S has IBN, then $m = n$.

To prove (iv), suppose R is left noetherian and there are positive integers $m < n$ and matrices $a \in R^{m \times n}$ and $b \in R^{n \times m}$ with $ab = I_m$ and $ba = I_n$. Adjoining zero rows below a and zero columns to the right of b, we create $n \times n$ matrices:

$$a' = \begin{bmatrix} a \\ 0 \end{bmatrix} , \quad b' = [b \ 0] ,$$

with $b'a' = ba = I_n$. Right multiplication by a' defines an R-linear map $g : R^n \to R^n$, which is surjective because $v = vb'a'$ for each $v \in R^n$. By (B.12) in Appendix B, since R is left noetherian, g is also injective. But this can't be, since the last row e_n of I_n is not 0, but $g(e_n) = e_n a' = 0$. ∎

From Theorem (1.39) we see that a ring R has IBN if there is a ring homomorphism from R into a left or right noetherian ring. For instance, every commutative ring has IBN, since it has a maximal ideal, and therefore has a quotient ring that is a field. Then every subring of a matrix ring $M_n(R)$, over a commutative ring R, has IBN. This includes rings such as

$$\begin{bmatrix} \mathbb{Z} & \mathbb{Q} \\ 0 & \mathbb{Z} \end{bmatrix} \subseteq M_2(\mathbb{Q}) ,$$

which are neither left nor right noetherian (see Appendix B, Example (B.1)).

Next, consider the *existence* of dimension, which is problematic over a wide variety of rings. Even the ring $\mathbb{Z}$ of integers has modules with no basis:

(1.40) Examples. Suppose M is an R-module. An R-**torsion element** of M is any $m \in M - \{0\}$ for which there is an $r \in R - \{0\}$ with $rm = 0$. The R-module M is R-**torsion free** if M has no R-torsion elements. The R-torsion elements of R (as an R-module) are the **zero-divisors** in R. A nontrivial ring R is called a **domain** if R has no zero-divisors (i.e., if R is R-torsion free).

Now suppose R is a domain and M is a free R-module, with basis B. If $m \in M - \{0\}$, then $m = r_1 b_1 + \cdots + r_n b_n$ for some $n \in \mathbb{P}$, distinct $b_1, \ldots, b_n \in B$, and nonzero $r_1, \ldots, r_n \in R$. If $r \in R$ and $rm = 0$, then by linear independence of B, $r r_1 = \cdots = r r_n = 0$; so $r = 0$. Thus, if R is a domain, every free R-module is R-torsion free.

If M is an abelian group, a $\mathbb{Z}$-torsion element of M is the same as a nontrivial element of finite order. So each abelian group with a nontrivial element of finite order is a $\mathbb{Z}$-module without a basis.

The next theorem illustrates one of the links between the internal structure of a ring and the shared properties of its modules.

(1.41) Definition. An R-module M is called **simple** if $M \neq \{0\}$ and the only submodules of M are $\{0\}$ and M.

(1.42) Theorem. *Suppose R is a nontrivial ring. The following are equivalent:*

(i) *Every R-module is free.*
(ii) *Every f.g. R-module is free.*
(iii) *Every cyclic R-module is free.*
(iv) *Some simple R-module is free.*
(v) *R is a division ring.*

Proof. The implications (i) $\Rightarrow$ (ii) $\Rightarrow$ (iii) are purely formal. Applying Zorn's Lemma to the set of left ideals of R that do not contain 1_R, we find that R has a maximal left ideal J. If N is an R-submodule of R/J, the union of the cosets belonging to N is a left ideal of R containing J. So R/J is a simple R-module. Since $R/J = R(1 + J)$ is cyclic, (iii) implies (iv).

Assume (iv), and let M denote a free simple R-module with basis B. Since $M \neq 0$, B has a nonzero element b. Since M is simple, $Rb = M$; so $B = \{b\}$. Since $\{b\}$ is linearly independent, the R-linear map $R \to Rb$, $r \mapsto rb$, is an isomorphism. So R is a simple R-module.

If $a \in R$ and $a \neq 0_R$, then $Ra = R$ (since R is simple). So for some $a' \in R$, $a'a = 1_R$. Since $0_R a = 0_R \neq 1_R$, $a' \neq 0_R$. By the same argument with a' in place of a, for some $a'' \in R$, $a''a' = 1_R$. Then $a'' = a''(a'a) = (a''a')a = a$; so a is a unit in R, proving assertion (v).

Assume (v) and suppose M is an R-module. If $M = 0$, M is free with basis $\emptyset$. Suppose $M \neq 0$. Each nonzero element $m \in M$ forms a linearly independent set $\{m\}$, since $rm = 0_R$ and $r \neq 0_R$ implies $m = r^{-1}rm = r^{-1}0_M = 0_M$.

Let $\mathcal{L}$ denote the set of all linearly independent subsets of M, partially ordered by containment. If C is a totally ordered nonempty subset of $\mathcal{L}$, the union U of the members of C is a member of $\mathcal{L}$, since a linear relation on U would have only finitely many terms, and so would be a linear relation on some member of C. By Zorn's Lemma, $\mathcal{L}$ has a maximal element L. If $m \in L$, then of course $m \in \langle L \rangle$. On the other hand, if $m \in M$ and $m \notin L$, then $L \cup \{m\}$ is linearly dependent. So there is a linear relation:

$$r_1 \lambda_1 + \cdots + r_{n-1} \lambda_{n-1} + r_n m = 0_M$$

with $r_1, \ldots, r_n \in R - \{0\}$ and $\lambda_1, \ldots, \lambda_{n-1} \in L$. Since $\{m\}$ is linearly independent, $n > 1$. Then

$$m = (-r_n^{-1}r_1)\lambda_1 + \cdots + (-r_n^{-1}r_{n-1})\lambda_{n-1}$$

belongs to $\langle L \rangle$. Thus $M = \langle L \rangle$ and L is a basis of M, proving (i). ∎

This theorem tells us that, if R is not $\{0\}$ and is not a division ring, then some f.g. R-modules are just too small to be free. In the next chapter we investigate fragments of free R-modules called "projective" modules.

1C. Exercises

1. Suppose M is a free left or right R-module with an infinite (not necessarily countable) basis B. Prove the ring $A = \mathrm{End}_R(M)$ does not have IBN. *Hint:* Show

$$A \cong \mathrm{Hom}_R(M \oplus M, M) \cong A \oplus A$$

as A-modules.

2. If h and k are positive integers, a ring R is said to be of **type** (h, k) if the following are equivalent:

   (i) $R^m \cong R^n$ as R-modules;
   (ii) either $m = n$, or else $m, n > h$ and $m \equiv n \pmod{k}$.

Prove each nontrivial ring without IBN has type (h, k) for some positive integers h, k. (For examples of rings of each type (h, k), see Cohn [66].)

3. Prove directly that commutative nontrivial rings have IBN by applying the determinant in the proof of (1.39) (iv).

4. Suppose S is a ring with a left noetherian subring R, and S is finitely generated as a left R-module (where the scalar multiplication $R \times S \to S$ is the ring multiplication in S). Prove S has IBN.

5. Prove that if R is a nontrivial ring, then R is a domain if and only if every principal left ideal of R is a free R-module.

6. If $R = \{0\}$ is the trivial ring, prove every R-module is free, with bases of two different sizes.

2
Projective Modules

2A. Direct Summands

In this section we review some basic facts about direct sums and exact sequences. The notion of direct summand, in (2.9)–(2.11) below, is essential for the understanding of projective modules, to be introduced in §2B as a natural generalization of free modules. Assume R is a ring and L, M, and N are R-modules.

(2.1) Definitions.

    (i) M is the (**external**) **direct sum** $L \oplus N$ if M is the cartesian product $L \times N = \{(x, y) : x \in L, \ y \in N\}$ with addition

$$(x_1, y_1) + (x_2, y_2) \ = \ (x_1 + x_2, y_1 + y_2)$$

       and scalar multiplication

$$r(x, y) \ = \ (rx, ry) \ .$$

    (ii) M is the **internal direct sum** $L \overset{\bullet}{\oplus} N$ if L and N are submodules of M, $L + N = M$, and $L \cap N = \{0_M\}$.

    (iii) In any ring S, an **idempotent** is an element $e \in S$ with $ee = e$. So an idempotent in $\mathrm{End}_R(M)$ is any R-linear map $e : M \to M$ with $e \circ e = e$.

(2.2) Proposition. *Suppose L and N are submodules of M. The following are equivalent:*

    (i) $M = L \overset{\bullet}{\oplus} N$.

    (ii) *Each $m \in M$ has one and only one expression $m = x + y$ with $x \in L$ and $y \in N$.*

    (iii) *There is an idempotent $e \in \mathrm{End}_R(M)$ with image L and kernel N.*

Proof. Assume (i). Since $M = L + N$, each $m \in M$ has an expression $m = x + y$ with $x \in L$ and $y \in N$. If also $m = x' + y'$ with $x' \in L$ and $y' \in N$, then $x - x' = y' - y \in L \cap N = \{0_M\}$; so $x = x'$ and $y = y'$, proving (ii).

Assume (ii). Then there is an R-linear map e from M to M defined by $e(x + y) = x$ if $x \in L$, $y \in N$. Then $(e \circ e)(x + y) = e(x) = e(x + 0) = x = e(x + y)$; so e is idempotent, with image L and kernel N.

Assume (iii). For each $m \in M$, $m = e(m) + m - e(m)$ with $e(m) \in L$ and $m - e(m) \in \ker(e) = N$. If $x \in L \cap N$, then $x = e(y)$ for some $y \in M$, and $0_M = e(x) = e(e(y)) = e(y) = x$. So $M = L \overset{\bullet}{\oplus} N$. ∎

External and internal direct sums are closely related:

(2.3) Proposition. *Suppose L, M and N are R-modules. Then $M \cong L \oplus N$ if and only if $M = L' \overset{\bullet}{\oplus} N'$ for submodules L', N' of M with $L' \cong L$ and $N' \cong N$.*

Proof. If $f : L \oplus N \to M$ is an isomorphism, then $L \cong L \oplus 0 \cong f(L \oplus 0)$, $N \cong 0 \oplus N \cong f(0 \oplus N)$, and

$$M = f(L \oplus 0) \overset{\bullet}{\oplus} f(0 \oplus N) .$$

Conversely, if $\alpha : L \to L'$, $\beta : N \to N'$ are isomorphisms and $M = L' \overset{\bullet}{\oplus} N'$, then

$$f : L \oplus N \to M$$
$$(x, y) \mapsto \alpha(x) + \beta(y)$$

is R-linear, and is bijective by (2.2) (ii). ∎

(2.4) Definitions. A sequence of R-modules and R-linear maps

$$\cdots \longrightarrow L \overset{f}{\longrightarrow} M \overset{g}{\longrightarrow} N \longrightarrow \cdots$$

is **exact** at M if $\operatorname{im}(f) = \ker(g)$. The whole sequence is **exact** if it is exact at each module between two consecutive maps of the sequence. A **short exact sequence** is an exact sequence

$$0 \longrightarrow L \overset{f}{\longrightarrow} M \overset{g}{\longrightarrow} N \longrightarrow 0$$

where 0 denotes the zero R-module $\{0\}$.

Note that a sequence of R-modules and R-linear maps

$$0 \longrightarrow L \overset{f}{\longrightarrow} M \overset{g}{\longrightarrow} N \longrightarrow 0$$

is exact if and only if f is injective, g is surjective, $g \circ f$ is the zero map, and $g(m) = 0$ implies $m = f(x)$ for some $x \in L$.

In the ring $\mathrm{End}_R(M)$, let i_M denote the identity under composition (so $i_M(m) = m$).

(2.5) Proposition. *Suppose R is a ring and $L, M,$ and N are R-modules.*

(i) *If* $0 \longrightarrow L \overset{f}{\longrightarrow} M \overset{g}{\longrightarrow} N \longrightarrow 0$ *is an R-linear short exact sequence and $h : N \to M$ is an R-linear map with $g \circ h = i_N$, then there is an R-linear map $k : M \to L$ with $k \circ f = i_L$ and $(f \circ k) + (h \circ g) = i_M$.*

(ii) *If* $0 \longrightarrow L \overset{f}{\longrightarrow} M \overset{g}{\longrightarrow} N \longrightarrow 0$ *is an R-linear short exact sequence and $k : M \to L$ is an R-linear map with $k \circ f = i_L$, then there is an R-linear map $h : N \to M$ with $g \circ h = i_N$ and $(f \circ k) + (h \circ g) = i_M$.*

(iii) *If*

$$L \underset{k}{\overset{f}{\rightleftarrows}} M \underset{h}{\overset{g}{\rightleftarrows}} N$$

are R-linear maps with $k \circ f = i_L$, $(f \circ k) + (h \circ g) = i_M$ and $g \circ h = i_N$, then

$$0 \longrightarrow L \overset{f}{\longrightarrow} M \overset{g}{\longrightarrow} N \longrightarrow 0 \,,$$

$$0 \longleftarrow L \overset{k}{\longleftarrow} M \overset{h}{\longleftarrow} N \longleftarrow 0 \,,$$

are short exact sequences,

$$M = f(L) \overset{\bullet}{\oplus} h(N)$$

$$= \ker(g) \overset{\bullet}{\oplus} \ker(k) \,,$$

and the maps

$$\psi : M \to L \oplus N \,,$$
$$x \mapsto (k(x), g(x))$$

$$\theta : L \oplus N \to M \,,$$
$$(x, y) \mapsto f(x) + h(y)$$

are mutually inverse R-linear isomorphisms.

Proof. Under the hypotheses in (i), $h \circ g$ is an idempotent in $\mathrm{End}_R(M)$, with kernel $\ker(g) = f(L)$ and image $h(g(M)) = h(N)$. By (2.2), $M = f(L) \overset{\bullet}{\oplus} h(N)$. So, by injectivity of f and h, each element of M has an expression $f(x) + h(y)$ for unique $x \in L$ and $y \in N$. Define $k : M \to L$ by $k(f(x) + h(y)) = x$. A direct check shows k is R-linear. For each $x \in L$, $k(f(x)) = k(f(x) + h(0)) = x$; so $k \circ f = i_L$. For each $x \in L$ and $y \in N$,

$$
\begin{aligned}
(f \circ k)(f(x) + h(y)) &= f(x) \quad \text{and} \\
(h \circ g)(f(x) + h(y)) &= h(0 + i_N(y)) = h(y) .
\end{aligned}
$$

So $(f \circ k) + (h \circ g) = i_M$.

Under the hypotheses in (ii), $f \circ k$ is an idempotent in $\mathrm{End}_R(M)$ with kernel $\ker(k)$ and image $f(k(M)) = f(L) = \ker(g)$. By (2.2), $M = \ker(g) \overset{\bullet}{\oplus} \ker(k)$. So the intersection of these kernels is $\{0_M\}$, and g restricts to an isomorphism $\ker(k) \to N$. Following the (R-linear) inverse of this isomorphism by inclusion $\ker(k) \to M$, defines an R-linear map $h : N \to M$ with $g \circ h = i_N$. If $x \in \ker(g)$ and $y \in \ker(k)$, then $x = f(z)$ for some $z \in L$, and $h(g(y)) = y$; so

$$
\begin{aligned}
(f \circ k)(x + y) &= f((k \circ f)(z) + 0) = f(z) = x , \quad \text{and} \\
(h \circ g)(x + y) &= h(0 + g(y)) = y .
\end{aligned}
$$

So $(f \circ k) + (h \circ g) = i_M$.

Under the conditions in (iii), if $m \in \ker(g)$, then

$$
\begin{aligned}
m - f(k(m)) &= (i_M - f \circ k)(m) \\
&= (h \circ g)(m) \\
&= h(0) = 0 ;
\end{aligned}
$$

so $m \in f(L)$. Also

$$
\begin{aligned}
g \circ f &= g \circ h \circ g \circ f \\
&= g \circ (i_M - (f \circ k)) \circ f \\
&= (g \circ f) - (g \circ f \circ k \circ f) \\
&= (g \circ f) - (g \circ f) = 0 ;
\end{aligned}
$$

so $\ker(g) = \mathrm{im}(f)$. Since $k \circ f = i_L$, $\ker(f) = 0$. Since $g \circ h = i_N$, $\mathrm{im}(g) = N$. So the sequence

$$
0 \longrightarrow L \overset{f}{\longrightarrow} M \overset{g}{\longrightarrow} N \longrightarrow 0
$$

is exact. By a similar argument

$$
0 \longleftarrow L \overset{k}{\longleftarrow} M \overset{h}{\longleftarrow} N \longleftarrow 0
$$

is exact.

Just as in the proofs of (i) and (ii),

$$
M = f(L) \overset{\bullet}{\oplus} h(N) = \ker(g) \overset{\bullet}{\oplus} \ker(k) .
$$

In the notation of Example (1.31),

$$\psi = \begin{bmatrix} k \\ g \end{bmatrix} \quad \text{and} \quad \theta = [f \ h] \ ,$$

which are R-linear. By the exactness above and the hypotheses in (iii), matrix multiplication gives

$$\psi\theta = \begin{bmatrix} 1 & 0 \\ 0 & 1 \end{bmatrix} \quad \text{and} \quad \theta\psi = 1 \ . \qquad \blacksquare$$

(2.6) Definition. An R-linear short exact sequence

$$0 \longrightarrow L \xrightarrow{f} M \xrightarrow{g} N \longrightarrow 0$$

splits (or **is split**) if there is an R-linear map $h : N \to M$ with $g \circ h = i_N$.

(2.7) Proposition. *There is an R-linear isomorphism $M \cong L \oplus N$ if and only if there is a split R-linear short exact sequence:*

$$0 \longrightarrow L \xrightarrow{f} M \xrightarrow{g} N \longrightarrow 0 \ .$$

Proof. The "if" is immediate from (2.5) (i) and (iii). For the "only if," suppose $h : M \to L \oplus N$ is an isomorphism. There are R-linear maps

$$L \underset{\pi_1}{\overset{i_1}{\rightleftarrows}} L \oplus N \underset{\pi_2}{\overset{i_2}{\leftrightarrows}} N$$

defined by

$$i_1(x) = (x,0) \ , \quad i_2(y) = (0,y) \ ,$$
$$\pi_1(x,y) = x \ , \quad \pi_2(x,y) = y \ .$$

These maps satisfy $\pi_1 \circ i_1 = i_L$, $(i_1 \circ \pi_1) + (i_2 \circ \pi_2) = i_{L \oplus N}$ and $\pi_2 \circ i_2 = i_N$. Then the sequence

$$0 \longrightarrow L \xrightarrow{h^{-1} \circ i_1} M \xrightarrow{\pi_2 \circ h} N \longrightarrow 0$$

is exact, and it splits because

$$(\pi_2 \circ h) \circ (h^{-1} \circ i_2) = i_N \ . \qquad \blacksquare$$

(2.8) Examples. There are $\mathbb{Z}$-linear short exact sequences:

$$0 \longrightarrow \mathbb{Z} \xrightarrow{f} \mathbb{Z} \xrightarrow{g} \mathbb{Z}/2\mathbb{Z} \longrightarrow 0$$

$$0 \longrightarrow \mathbb{Z} \xrightarrow{i} \mathbb{Z} \oplus \mathbb{Z}/2\mathbb{Z} \xrightarrow{\pi} \mathbb{Z}/2\mathbb{Z} \longrightarrow 0 \ ,$$

where $f(x) = 2x$, $g(x) = x + 2\mathbb{Z}$, $i(x) = (x,0)$, $\pi((x,y)) = y$. The middle terms $\mathbb{Z}$ and $\mathbb{Z} \oplus \mathbb{Z}/2\mathbb{Z}$ are not isomorphic since $\mathbb{Z}$ has no elements of additive order 2. So, by (2.7), the first sequence is not split. Since g is surjective, there are functions $h : \mathbb{Z}/2\mathbb{Z} \to \mathbb{Z}$ with $g \circ h = i_{\mathbb{Z}/2\mathbb{Z}}$, but none of them are $\mathbb{Z}$-linear.

However, by (2.7), any two split short exact sequences

$$
\begin{array}{ccccccccc}
0 & \longrightarrow & L & \longrightarrow & M_1 & \longrightarrow & N & \longrightarrow & 0 \\
  &   & \| &   &   &   & \| &   &   \\
0 & \longrightarrow & L & \longrightarrow & M_2 & \longrightarrow & N & \longrightarrow & 0
\end{array}
$$

must have $M_1 \cong L \oplus N \cong M_2$.

Consider again internal direct sums.

(2.9) Definition. If L is a submodule of M, a **complement** of L in M is any submodule N of M with $M = L \overset{\bullet}{\oplus} N$. Call L a **direct summand** of M if L has a complement in M.

The complement of a direct summand need not be unique:

(2.10) Example. In the real vector space $\mathbb{R}^2$, a line V through the origin is a direct summand of $\mathbb{R}^2$, and any other line through the origin is a complement of V.

By (2.2), L is a direct summand of M if and only if $L = e(M)$ for an idempotent $e \in \mathrm{End}_R(M)$. Since $M = \mathrm{im}(e) \overset{\bullet}{\oplus} \ker(e)$, and e acts as the identity map on $\mathrm{im}(e)$ and the zero map on $\ker(e)$, each idempotent e is uniquely determined by its image and kernel. So for each direct summand L of M, there is a bijection between the set of idempotents $e \in \mathrm{End}_R(M)$ with $e(M) = L$, and the set of complements to L in M.

(2.11) Proposition.

   (i) *For R-modules M and N, N is isomorphic to a direct summand of M if and only if there are R-linear maps $h : N \to M$ and $g : M \to N$ with $g \circ h = i_N$.*

   (ii) *If $g : M \to N$ is a surjective R-linear map, there is a bijection $h \mapsto h(N)$ from the set of R-linear right inverses of g to the set of complements to $\ker(g)$ in M.*

   (iii) *If $h : N \to M$ is an injective R-linear map, there is a bijection $g \mapsto \ker(g)$ from the set of R-linear left inverses of h to the set of complements to $h(N)$ in M.*

Proof. If R-linear maps $h : N \to M$ and $g : M \to N$ satisfy $g \circ h = i_N$, there is a split short exact sequence

$$0 \longrightarrow \ker(g) \overset{\subseteq}{\longrightarrow} M \underset{h}{\overset{g}{\rightleftarrows}} N \longrightarrow 0 .$$

So, by (2.5), $M = \ker(g) \overset{\bullet}{\oplus} h(N)$, where $N \cong h(N)$.

Conversely, if $M = L \overset{\bullet}{\oplus} N'$, where $N \cong N'$, then $M \cong L \oplus N$ by (2.3). So by (2.7) there is a split short exact sequence

$$0 \longrightarrow L \overset{f}{\longrightarrow} M \underset{h}{\overset{g}{\rightleftarrows}} N \longrightarrow 0 ,$$

proving (i).

Suppose g is an R-linear surjection $M \to N$. If h is an R-linear right inverse to g, $h(N)$ is a complement to $\ker(g)$ as in (i). Suppose h' is also an R-linear right inverse to g with $h(N) = h'(N)$. For each $n \in N, h(n) - h'(n)$ belongs to $\ker(g) \cap h(N) = \{0\}$; so $h = h'$. If N' is any complement to $\ker(g)$ in M, $N' = h(N)$, where h is the composite:

$$N \cong M/\ker(g) \cong N' \subseteq M ,$$

proving (ii).

Suppose h is an R-linear injection $N \to M$. If g is an R-linear left inverse to h, $\ker(g)$ is a complement to $h(N)$ as in (i). Suppose g' is also an R-linear left inverse to h and $\ker(g') = \ker(g)$. If $m \in M$, $m = a+b$ with $a \in \ker(g) = \ker(g')$ and $b = h(n)$ for some $n \in N$. Then $g'(m) = n = g(m)$; so $g' = g$. If L is any complement to $h(N)$ in M, each $m \in M$ has a unique expression $a + b$ with $a \in L$ and $b \in h(N)$; the composite of $M \to h(N)$, $a + b \mapsto b$, followed by $h(N) \cong N$ is an R-linear map g, left inverse to h, with $\ker(g) = L$. ∎

2A. Exercises

1. Suppose M is an R-module.

    (i) If $e \in \mathrm{End}_R(M)$ is idempotent, prove $1 - e$ is idempotent (where 1 denotes i_M), $\mathrm{im}(e) = \ker(1 - e), \ker(e) = \mathrm{im}(1 - e)$, and

$$M = e(M) \overset{\bullet}{\oplus} (1 - e)(M) .$$

    (ii) If $M = L \overset{\bullet}{\oplus} N$, prove that there is one and only one idempotent $e \in \mathrm{End}_R(M)$ with $e(M) = L$ and $(1 - e)(M) = N$.

2. Suppose L, M and N are R-modules and

$$0 \longrightarrow L \xrightarrow{f} M \xrightarrow{g} N \longrightarrow 0$$

is a short exact sequence of R-linear maps. Let K denote the kernel of g. Prove this sequence is split if and only if there is a submodule N' of M consisting of one member of each coset of K in M. Then use this criterion to prove

$$0 \longrightarrow \mathbb{Z}/2\mathbb{Z} \longrightarrow \mathbb{Z}/4\mathbb{Z} \longrightarrow \mathbb{Z}/2\mathbb{Z} \longrightarrow 0$$
$$\overline{a} \longmapsto \overline{2a}, \ \overline{a} \longmapsto \overline{a}$$

is not split, but

$$0 \longrightarrow \mathbb{Z}/2\mathbb{Z} \longrightarrow \mathbb{Z}/6\mathbb{Z} \longrightarrow \mathbb{Z}/3\mathbb{Z} \longrightarrow 0$$
$$\overline{a} \longmapsto \overline{3a}, \ \overline{a} \longmapsto \overline{a}$$

is split.

3. Suppose L, M, N, L', M', and N' are R-modules;

$$0 \longrightarrow L \xrightarrow{f} M \xrightarrow{g} N \longrightarrow 0$$
$$0 \longrightarrow L' \xrightarrow{f'} M' \xrightarrow{g'} N' \longrightarrow 0$$

are short exact sequences of R-linear maps; and there are isomorphisms $u : L \to L'$ and $v : N \to N'$. If the two sequences split, prove there is an isomorphism $w : M \to M'$ making the diagram

$$
\begin{array}{ccccccccc}
0 & \longrightarrow & L & \xrightarrow{f} & M & \xrightarrow{g} & N & \longrightarrow & 0 \\
& & \downarrow{\scriptstyle u} & & \downarrow{\scriptstyle w} & & \downarrow{\scriptstyle v} & & \\
0 & \longrightarrow & L' & \xrightarrow{f'} & M' & \xrightarrow{g'} & N' & \longrightarrow & 0
\end{array}
$$

commute.

4. **(Snake Lemma).** Suppose

$$
\begin{array}{ccccccccc}
0 & \longrightarrow & L & \xrightarrow{f} & M & \xrightarrow{g} & N & \longrightarrow & 0 \\
& & \downarrow{\scriptstyle u} & & \downarrow{\scriptstyle w} & & \downarrow{\scriptstyle v} & & \\
0 & \longrightarrow & L' & \xrightarrow{f'} & M' & \xrightarrow{g'} & N' & \longrightarrow & 0
\end{array}
$$

is a commutative diagram of R-modules and R-linear maps, and that the two rows are exact. Prove there is an exact sequence of R-linear maps

$$0 \longrightarrow \ker(u) \longrightarrow \ker(w) \longrightarrow \ker(v)$$
$$\xrightarrow{\partial} \operatorname{coker}(u) \longrightarrow \operatorname{coker}(w) \longrightarrow \operatorname{coker}(v) \longrightarrow 0 \ .$$

How much of this sequence remains exact if the first sequence is not exact at L? Or if the second sequence is not exact at N'? (*Hint:* The map ∂ is defined by snaking through the diagram.)

5. Suppose L, M, and N are R-modules and

$$0 \longrightarrow L \xrightarrow{f} M \xrightarrow{g} N \longrightarrow 0$$

is a short exact sequence of R-linear maps. Prove

    (i)   If L is finite with b elements and N is finite with c elements, then M is finite with bc elements.

    (ii) If $r, s \in R$ and $rL = 0$ and $sN = 0$, then $rsM = 0$.

   (iii) If $L \cong R^s$ and $N \cong R^t$, then $M \cong R^{s+t}$.

   (iv) If L and N are finitely generated, then M is finitely generated.

For what rings R is the converse of (iv) true?

6. The distinction between equal and isomorphic submodules is important: Prove that, in an R-module M, two complements of the same submodule L must be isomorphic, but complements of isomorphic submodules need not be isomorphic. (For examples of this second phenomenon, look in Chapter 4.)

2B. Summands of Free Modules

In algebraic K-theory, the most fundamental type of module is a "projective" module. Here we define projective modules, and look them over from a variety of perspectives.

(2.12) Definition. An R-module is called **projective** if it is a direct summand of a free R-module.

The term "projective" is appropriate here, since each rank 2 idempotent $e \in \mathrm{End}_R(\mathbb{R}^3)$ is a geometric projection, carrying all points of $\mathbb{R}^3$, in a direction parallel to the line kernel(e), to a "screen," which is the plane $e(\mathbb{R}^3)$.

(2.13) Examples.

    (i) Each free R-module M is projective: $M = M \overset{\bullet}{\oplus} \{0_M\}$.

   (ii) If $R = \mathbb{Z}/6\mathbb{Z}$, not every projective R-module is free: Every f.g. free R-module has 6^n elements for some $n \geq 0$. But the free R-module R is the internal direct sum $\{\overline{0}, \overline{2}, \overline{4}\} \overset{\bullet}{\oplus} \{\overline{0}, \overline{3}\}$; so the summands $\{\overline{0}, \overline{2}, \overline{4}\}$ and $\{\overline{0}, \overline{3}\}$ are non-free projective R-modules.

(2.14) Proposition. *If R is a ring and $M \cong P$ are isomorphic R-modules, and if P is projective, then M is projective.*

Proof. There is a free R-module F with P as a direct summand. There is a set S with $M \cap S = \emptyset$ and with a bijection $f : S \to F - P$. Let $g : M \to P$ be an isomorphism. Then $h : M \cup S \to F$, defined by

$$h(x) = \begin{cases} f(x) & \text{if } x \in S , \\ g(x) & \text{if } x \in M , \end{cases}$$

is bijective. Under the operations:

$$\begin{aligned} x + y &= h^{-1}(h(x) + h(y)) , \\ rx &= h^{-1}(rh(x)) , \end{aligned}$$

$M \cup S$ becomes an R-module with M as a submodule, and then h becomes an R-linear isomorphism. By (1.7) (i), $M \cup S$ is a free R-module. By (2.2), there is an idempotent $e \in \text{End}_R(F)$ with $e(F) = P$. Then $e' = h^{-1} \circ e \circ h$ is an idempotent in the ring $\text{Hom}_R(M \cup S, M \cup S)$ and $e'(M \cup S) = M$. So M is a direct summand of $M \cup S$, and hence is projective. ∎

(2.15) Corollary. *An R-module P is projective if and only if there is a free R-module F and there are R-linear maps*

$$F \underset{h}{\overset{g}{\rightleftarrows}} P$$

with $g \circ h = i_P$.

Proof. By (2.11), the latter condition is equivalent to P being isomorphic to a direct summand of F. ∎

(2.16) Corollary. *An R-module P is projective if and only if there is an R-module Q for which $P \oplus Q$ is free.*

Proof. If $P \oplus Q$ is free, its direct summand $P \oplus \{0\}$ is projective; since $P \cong P \oplus \{0\}$, P is projective.

If P is projective, there is a free R-module F and a submodule Q of F with $F = P \overset{\bullet}{\oplus} Q$. By (2.3), $F \cong P \oplus Q$. So by (1.7) (i), $P \oplus Q$ is free. ∎

Certain useful mapping properties of free modules actually characterize the wider class of projective modules:

(2.17) Proposition. *For an R-module P, the following are equivalent:*

   (i)   *P is projective.*

  (ii)   *For each diagram of R-linear maps*

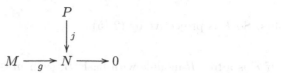

$$P$$
$$\downarrow j$$
$$M \xrightarrow{\;g\;} N \longrightarrow 0$$

      with exact row, there is an R-linear map $h : P \to M$ with $g \circ h = j$.

 (iii)   *Every surjective R-linear map $g : M \to P$ has an R-linear right inverse (i.e., R-linear $h : P \to M$ with $g \circ h = i_P$).*

 (iv)   *Every short exact sequence of R-linear maps:*

$$0 \longrightarrow L \xrightarrow{\;f\;} M \xrightarrow{\;g\;} P \longrightarrow 0$$

      splits.

Proof. We first prove (ii) for free modules. Suppose F is a free R-module with basis B, and

$$F$$
$$\downarrow j$$
$$M \xrightarrow{\;g\;} N \longrightarrow 0$$

are R-linear maps with exact row (meaning g is surjective). For each $b \in B$ choose $b' \in M$ with $g(b') = j(b)$. The function $B \to M$, $b \mapsto b'$, extends to an R-linear map $h : F \to M$ with $g(h(b)) = j(b)$ for each $b \in B$. Since $F = \langle B \rangle$, $g \circ h = j$.

Now assume (i), and choose an R-module Q for which $P \oplus Q$ is free. Suppose

$$P \oplus Q \underset{i}{\overset{\pi}{\rightleftarrows}} P$$
$$\downarrow j$$
$$M \xrightarrow{\;g\;} N \longrightarrow 0$$

are R-linear maps with g surjective, and where $\pi((x,y)) = x$, $i(x) = (x,0)$. Since $P \oplus Q$ is free, there is an R-linear map $k : P \oplus Q \to M$ with $g \circ k = j \circ \pi$. If $h = k \circ i$, then $g \circ h = g \circ k \circ i = j \circ \pi \circ i = j$, proving (ii).

Assertion (iii) follows from (ii) by taking $N = P$ and $j = i_P$, and (iv) is immediate from (iii)

Assume (iv). If F is a free R-module based on a generating set S of P, the inclusion $S \to P$ extends to a surjective R-linear map $g : F \to P$. By (iv), the exact sequence

$$0 \longrightarrow \ker(g) \overset{\subseteq}{\longrightarrow} F \overset{g}{\longrightarrow} P \longrightarrow 0$$

splits. So P is projective by (2.15).   ∎

If F is a free R-module with basis B, each $m \in F$ has a unique expression

$$m = \sum_{b \in B} c(b, m)b ,$$

where $c(b, m) \in R$ and $c(b, m) = 0$ for all but finitely many $b \in B$. For each $b \in B$, **projection to the b-coordinate**

$$b^* : F \;\to\; R$$
$$m \;\mapsto\; c(b, m)$$

is R-linear. So we have a function

$$(\;\;)^* : B \;\to\; \mathrm{Hom}_R(F, R)$$
$$b \;\mapsto\; b^* .$$

(2.18) Definition. A **projective basis** of an R-module M is any function

$$(\;\;)^* : S \;\to\; \mathrm{Hom}_R(M, R)$$
$$s \;\mapsto\; s^* ,$$

where $S \subseteq M$, and where, for each $m \in M$,

(i) $s^*(m) = 0$ for all but finitely many $s \in S$, and

(ii) $m = \displaystyle\sum_{s \in S} s^*(m)s$.

(2.19) Note. By (ii) in this definition, S generates M.

(2.20) Proposition. *An R-module P is projective if and only if P has a projective basis. If P is projective, every generating set S of P is the domain of a projective basis.*

Proof. Suppose $P = \langle S \rangle$. Let $F = F_R(S)$ denote the free R-module based on S. The inclusion $S \to P$ extends to a surjective R-linear map $g : F \to P$.

If P has a projective basis

$$(\ )^* : S \ \to \ \text{Hom}_R(P, R) \, ,$$

there is an R-linear map $h : P \to F$ defined by

$$h(p) \ = \ \sum_{s \in S} s^*(p)s \, ,$$

and $g \circ h = i_P$. So P is projective by (2.15).

Conversely, if P is projective, then by (2.15) there is an R-linear map $h : P \to F$ with $g \circ h = i_P$. For each $s \in S$, projection to the s-coordinate is an R-linear map $s^* : F \to R$; so $s^* \circ h$ is an R-linear map $P \to R$. For each $p \in P$, $s^*(h(p)) = 0$ for all but finitely many $s \in S$. And

$$h(p) \ = \ \sum_{s \in S} s^*(h(p))s$$

in F; so, applying g,

$$p \ = \ \sum_{s \in S} (s^* \circ h)(p)s$$

in P. Thus $s \mapsto s^* \circ h$ defines a projective basis of P. ■

Finitely generated projective R-modules have a particularly simple and elegant description:

(2.21) Proposition. *Suppose P is an R-module and n is a positive integer. The following are equivalent:*

(i) *P is projective and is generated by n elements.*
(ii) *P is isomorphic to a direct summand of R^n.*
(iii) *P is isomorphic to the R-module generated by the rows of an idempotent matrix in $M_n(R)$.*
(iv) *There is an R-module Q with $P \oplus Q \cong R^n$.*

Proof. Assume, as in (i), that P is projective and $P = \langle p_1, \ldots, p_n \rangle$. Then the R-linear map $f : R^n \to P$, with $f(e_i) = p_i$ for each i, is surjective. By (2.17) (iii), there is an R-linear map $g : P \to R^n$ with $f \circ g = i_P$. So, by (2.11), (ii) holds.

Suppose M is a direct summand of R^n. By (2.2) (iii), there is an idempotent $e \in \text{End}_R(R^n)$ with $e(R^n) = M$. By (1.26), e is right multiplication by an idempotent matrix $E \in M_n(R)$; so $R^n E = M$. But $R^n E$ is the R-submodule of R^n generated by the rows of E, so (ii) implies (iii).

Assume, as in (iii), that $E \in M_n(R)$ is idempotent and $P \cong R^n E$. Right multiplication by E is an idempotent element e of $\text{Hom}_R(R^n, R^n)$, and $e(R^n) =$

$R^n E$. By (2.2), $R^n = R^n E \overset{\bullet}{\oplus} \ker(e)$. So by (2.3), $R^n \cong P \oplus \ker(e)$. This proves assertion (iv).

Assume (iv). Then P is projective by (2.16). If $\pi : P \oplus Q \to P$ is the projection to the first coordinate and $f : R^n \to P \oplus Q$ is an isomorphism, then $\pi \circ f : R^n \to P$ is a surjective R-linear map. So P is generated by $\pi(f(e_1)), \ldots, \pi(f(e_n))$, proving (i).   ■

Because of their mapping properties (2.17), f.g. projective modules are usually regarded as the natural generalization to R-modules of finite-dimensional vector spaces over a field. In Part II we see that they arise naturally in connection with unique factorization in number theory, and with matrix representations of finite groups.

2B. Exercises

1. If R is a ring with IBN, prove

$$\begin{bmatrix} R & 0 \\ R & 0 \end{bmatrix} \ \left(= \left\{ \begin{bmatrix} a & 0 \\ b & 0 \end{bmatrix} : a, b \in R \right\} \right)$$

is a projective but not a free $M_2(R)$-module. It *is* free if $R \cong R^2$.

2. If R is a domain, prove every projective R-module is torsion free. If R is a principal ideal (commutative) domain, prove every torsion free f.g. R-module is projective.

3. Prove $\mathbb{Q}$ is not a projective $\mathbb{Z}$-module.

4. Suppose P is an R-module. Prove P is projective if and only if, for every short exact sequence in R-$\mathfrak{Mod}$:

$$0 \longrightarrow L \overset{f}{\longrightarrow} M \overset{g}{\longrightarrow} N \longrightarrow 0 \ ,$$

the sequence

$$0 \to \mathrm{Hom}_R(P, L) \overset{f \circ (\ )}{\longrightarrow} \mathrm{Hom}_R(P, M) \overset{g \circ (\ )}{\longrightarrow} \mathrm{Hom}_R(P, N) \to 0$$

is exact as a sequence in $\mathbb{Z}$-$\mathfrak{Mod}$.

5. Show that the f.g. projective R-modules form the smallest (under $\subseteq$) class of R-modules that is closed under isomorphisms, direct sums, and direct summands, and includes R.

3

Grothendieck Groups

In this chapter we carry out the generalization of dimension outlined in the introduction to Part I. In §3A we develop abelian monoids of R-modules under direct sum. For this section, the reader may benefit from a review of sets and classes, as is provided in Appendix A. As examples, we include monoids of similarity classes of matrices, and monoids of isometry classes of bilinear forms. Then in §3B we give the universal construction of an abelian group from a semigroup. Section 3C contains the construction by A. Grothendieck of the algebraic K-group $K_0(R)$ of f.g. projective R-modules, and the corresponding group $G_0(R)$ of f.g. R-modules, and §3D is devoted to the relation between $K_0(R)$ and $G_0(R)$.

3A. Semigroups of Isomorphism Classes

Suppose R is a ring, and M, N, and P are R-modules. Let 0 denote the zero R-module. The direct sum of R-modules has some properties resembling familiar algebraic axioms; there are R-linear isomorphisms:

$$(M \oplus N) \oplus P \cong M \oplus (N \oplus P) ,$$
$$((m, n), p) \leftrightarrow (m, (n, p)) ;$$

$$M \oplus 0 \cong M \cong 0 \oplus M ,$$
$$(m, 0) \leftrightarrow m \leftrightarrow (0, m) ;$$

$$M \oplus N \cong N \oplus M ,$$
$$(m, n) \leftrightarrow (n, m) .$$

To replace these isomorphisms by equality, we can work with isomorphism classes:

(3.1) Definitions. Objects A and B of a category $\mathcal{C}$ are **isomorphic** (and we write $A \cong B$) if there is an isomorphism $f \in \mathrm{Hom}(A, B)$. The **isomorphism class** of $A \in \mathrm{Obj}\ \mathcal{C}$ is the class $\mathrm{cl}(A)$ of all objects $B \in \mathrm{Obj}\ \mathcal{C}$ with $A \cong B$.

From the definition of an isomorphism as an invertible arrow (0.5), it follows that "$\cong$" is an equivalence relation on the class $\mathrm{Obj}\ \mathcal{C}$: $A \cong A$, $A \cong B$ implies $B \cong A$, and $A \cong B \cong C$ implies $A \cong C$. So each object of $\mathcal{C}$ belongs to one and only one isomorphism class, $A \in \mathrm{cl}(A)$, and $\mathrm{cl}(A) = \mathrm{cl}(B)$ if and only if $A \cong B$.

If $f : M \to M'$ and $g : N \to N'$ are isomorphisms in R-$\mathfrak{Mod}$, so is $f \oplus g : M \oplus N \to M' \oplus N'$. So we can define the sum of two isomorphism classes by

$$\mathrm{cl}(M) + \mathrm{cl}(N) = \mathrm{cl}(M \oplus N) .$$

Now it appears that we can declare the collection of isomorphism classes of R-modules to be an abelian monoid under this addition. However, a monoid, like a group, is defined to be a *set* with a binary operation, while the collection of isomorphism classes of all R-modules is not a set — in fact, it is not even a class!

There are two set-theoretic problems here. The first is that there are too many isomorphism classes. Specifically, no set of R-modules is big enough to include a member of every isomorphism class of R-modules (at least if $R \neq \{0\}$). For details, see Exercise 1.

The second problem is that isomorphism classes of R-modules are too large. Even the isomorphism class of the zero R-module is too big to be a member of any class. For details, see Exercise 2.

To avoid these problems we must work with a restricted subcategory $\mathcal{C}$ of R-$\mathfrak{Mod}$:

(3.2) Definitions. A category $\mathcal{C}$ is **represented by a set** S if S is a set, $S \subseteq \mathrm{Obj}\ \mathcal{C}$, and S meets each isomorphism class in $\mathcal{C}$. A category $\mathcal{C}$ is **modest** if $\mathcal{C}$ is represented by a set. If $\mathcal{C}$ is represented by the set S, the **restricted isomorphism class** of $A \in \mathrm{Obj}\ \mathcal{C}$ is the set

$$\mathbf{c}(A) = \{B \in S : A \cong B\} = S \cap \mathrm{cl}(A) ,$$

which is an isomorphism class within the set S.

Note that $A \in \mathrm{c}(A)$ if and only if $A \in S$; but for $A, B \in \mathrm{Obj}\ \mathcal{C}$,

$$\mathrm{c}(A) = \mathrm{c}(B) \iff A \cong B \iff \mathrm{cl}(A) = \mathrm{cl}(B) .$$

So intersection with S defines a one-to-one correspondence from the collection of isomorphism classes in $\mathcal{C}$ to the *set* of restricted isomorphism classes in $\mathcal{C}$ (= the set of isomorphism classes within S).

Technically, the restricted isomorphism classes depend on the choice of set S representing $\mathcal{C}$, but we shall avoid constructions that depend, in any important way, on the choice of S (see Exercise 4). Also, we suppress the term "restricted";

if $\mathcal{C}$ is a modest category, the **set of isomorphism classes in** $\mathcal{C}$ means the set of restricted isomorphism classes with respect to some set S representing $\mathcal{C}$.

Having dealt with the requirements of set theory, we can now say that, if R is a ring and $\mathcal{C}$ is a modest full subcategory of R-$\mathcal{M}o\mathfrak{d}$ with Obj $\mathcal{C}$ closed under $\oplus$, the set $\mathfrak{I}(\mathcal{C})$ of isomorphism classes in $\mathcal{C}$ is an abelian semigroup under

$$c(M) + c(N) = c(M \oplus N) .$$

If the zero R-module 0 is in Obj $\mathcal{C}$ as well, then $\mathfrak{I}(\mathcal{C})$ is an abelian monoid with identity element $c(0)$.

The same arguments provide a more general construction of a semigroup from a modest category:

(3.3) Definitions. A **binary operation on a category** $\mathcal{C}$ is a procedure $*$ that constructs, for each pair of objects X and Y, a unique object $X * Y$, so that for all $X, X', Y, Y' \in$ Obj $\mathcal{C}$, $X \cong X'$ and $Y \cong Y'$ implies $X * Y \cong X' * Y'$. The operation $*$ is **associative** if

$$(X * Y) * Z \cong X * (Y * Z)$$

for all $X, Y, Z \in$ Obj $\mathcal{C}$; **commutative** if

$$X * Y \cong Y * X$$

for all $X, Y \in$ Obj $\mathcal{C}$; and **has identity object** E if $E \in$ Obj $\mathcal{C}$ and

$$X * E \cong X \cong E * X$$

for all $X \in$ Obj $\mathcal{C}$.

(3.4) Proposition. *If there is an associative binary operation $*$ on a modest category $\mathcal{C}$, then the set $\mathfrak{I}(\mathcal{C})$, of isomorphism classes in $\mathcal{C}$, is a semigroup under*

$$c(x) + c(y) = c(x * y) .$$

This semigroup is abelian (i.e., commutative) if $*$ is commutative, and is a monoid with identity $c(E)$ if $*$ has identity object E in $\mathcal{C}$. ∎

(3.5) Examples: Monoids of boundedly generated modules. The full subcategory $\mathcal{M}(R)$ of R-$\mathcal{M}o\mathfrak{d}$, consisting of f.g. R-modules, is modest, represented by the set of quotient modules R^n/L $(n > 0)$, according to (1.10). If M and $N \in \mathcal{M}(R)$ and M is generated by X, while N is generated by Y, then $M \oplus N$ is generated by $(X \times \{0\}) \cup (\{0\} \times Y)$. So Obj $\mathcal{M}(R)$ is closed under $\oplus$. Also, 0 is finitely generated. So $\mathfrak{I}(\mathcal{M}(R))$ is an abelian monoid under $\oplus$.

If T is any infinite set, let $T\mathfrak{gen}(R)$ denote the full subcategory of R-$\mathcal{M}o\mathfrak{d}$ whose objects are generated by sets $S \neq \emptyset$ of cardinality less than or equal to that of T. Each object of $T\mathfrak{gen}(R)$ is a homomorphic image of $F_R(T)$ under the R-linear extension of a surjection $T \to S$. So $T\mathfrak{gen}(R)$ is modest, represented by the quotients of $F_R(T)$. Also, $0 \in T\mathfrak{gen}(R)$; and since T is infinite, Obj $T\mathfrak{gen}(R)$ is closed under $\oplus$. So there is an abelian monoid $\mathfrak{I}(T\mathfrak{gen}(R))$ under $\oplus$. For instance, if T is countable, $T\mathfrak{gen}(R)$ is the category of **countably generated** R-modules.

(3.6) Example: The monoid generated by R. The full subcategory $\mathfrak{F}(R)$ of R-$\mathcal{Mod}$, consisting of all f.g. free R-modules, is modest, since it is represented by the set of R-modules R^n with $n \geq 0$. If $M \cong R^m$ and $N \cong R^n$, then $M \oplus N \cong R^{m+n}$; so Obj $\mathfrak{F}(R)$ is closed under $\oplus$. Thus $\mathfrak{I}(\mathfrak{F}(R))$ is an abelian monoid under $\oplus$. Its structure is considered in Exercise 3.

(3.7) Example: The monoid of f.g. projectives. The binary operation $\oplus$ on $\mathcal{M}(R)$ restricts to a binary operation on the category $\mathfrak{P}(R)$ of f.g. projective R-modules: For if $P, Q \in \mathfrak{P}(R)$, there are $P', Q' \in R$-$\mathcal{Mod}$ and positive integers m, n, with $P \oplus P' \cong R^m$ and $Q \oplus Q' \cong R^n$. Then

$$(P \oplus Q) \oplus (P' \oplus Q') \cong (P \oplus P') \oplus (Q \oplus Q')$$
$$\cong R^m \oplus R^n \cong R^{m+n} ;$$

so $P \oplus Q \in \mathfrak{P}(R)$. Also, $0 \in \mathfrak{P}(R)$. Thus $\oplus$ is an associative, commutative binary operation on $\mathfrak{P}(R)$ with identity object 0.

The category $\mathfrak{P}(R)$ is modest, represented by the set of row spaces $R^n E$ of idempotent matrices $E \in M_n(R)$, for $n \geq 1$. So $\mathfrak{I}(\mathfrak{P}(R))$ is an abelian monoid under $c(P) + c(Q) = c(P \oplus Q)$.

Using the idempotent matrices over R, one can describe this monoid in more concrete terms: If $X, Y \in M_s(R), X$ is **similar** to Y if $Y = CXC^{-1}$ for some $C \in GL_s(R)$. Similarity is an equivalence relation on the square matrices over R. More generally, if $X \in M_m(R)$ and $Y \in M_n(R)$, then X is **stably similar** to Y if

$$X \oplus 0_{s-m} \qquad \text{and} \qquad Y \oplus 0_{s-n}$$

are similar in $M_s(R)$ for some $s \geq \max\{m, n\}$. The word "stably" is used here because

$$C(X \oplus 0_{s-m})C^{-1} = Y \oplus 0_{s-n}$$

implies

$$(C \oplus I_{t-s})(X \oplus 0_{t-m})(C \oplus I_{t-s})^{-1} = Y \oplus 0_{t-n}$$

for all $t \geq s$. Stable similarity is also an equivalence relation on the square matrices over R.

(3.8) Lemma. *Idempotents $E \in M_m(R)$ and $F \in M_n(R)$ have isomorphic row spaces $R^m E \cong R^n F$ if and only if they are stably similar.*

Proof (Rosenberg [94, (1.2.1)]). If $C \in GL_s(R)$ with $C(E \oplus 0)C^{-1} = F \oplus 0$, then $R^s C = R^s$, and $\cdot C^{-1}$ restricts to the second isomorphism in the sequence:

$$R^m E \cong R^s(E \oplus 0) \cong R^s(E \oplus 0)C^{-1}$$
$$= R^s C(E \oplus 0)C^{-1} = R^s(F \oplus 0) \cong R^n F.$$

Conversely, suppose

$$R^m E \xrightleftharpoons[\beta]{\alpha} R^n F$$

are mutually inverse R-linear isomorphisms. The composite of R-linear isomorphisms

$$R^{m+n} \cong R^m \oplus R^n$$
$$= (R^m E \overset{\bullet}{\oplus} R^m(1-E)) \oplus (R^n F \overset{\bullet}{\oplus} R^n(1-F))$$
$$\cong R^m E \oplus R^m(1-E) \oplus R^n F \oplus R^n(1-F)$$

induces an isomorphism of endomorphism rings. The R-linear endomorphisms of the right side include

$$X = \begin{bmatrix} 1 & 0 & 0 & 0 \\ 0 & 0 & 0 & 0 \\ 0 & 0 & 0 & 0 \\ 0 & 0 & 0 & 0 \end{bmatrix}, \quad Y = \begin{bmatrix} 0 & 0 & 0 & 0 \\ 0 & 0 & 0 & 0 \\ 0 & 0 & 1 & 0 \\ 0 & 0 & 0 & 0 \end{bmatrix},$$

$$\text{and} \quad Z = \begin{bmatrix} 0 & 0 & \beta & 0 \\ 0 & 1 & 0 & 0 \\ \alpha & 0 & 0 & 0 \\ 0 & 0 & 0 & 1 \end{bmatrix},$$

which satisfy $Z^2 = 1$ and $ZXZ = Y$. In $\text{End}_R(R^{m+n})$, X and Y correspond to $\cdot(E \oplus 0)$ and $\cdot(0 \oplus F)$, and Z corresponds to $\cdot C$ for some matrix C. So $C^2 = I_{m+n}$ and $C(E \oplus 0)C^{-1} = 0 \oplus F$. If

$$D = \begin{bmatrix} 0 & I_n \\ I_m & 0 \end{bmatrix},$$

then $D(0 \oplus F)D^{-1} = F \oplus 0$. So E is stably similar to F. $\blacksquare$

To summarize, if S is the set of stable similarity classes of idempotent matrices over R, and $ss(E)$ denotes the stable similarity class of E, then $c(R^m E) \mapsto ss(E)$ defines a bijection

$$\mathcal{I}(\mathcal{P}(R)) \to S.$$

This becomes a monoid isomorphism if we add in S by the rule

$$ss(E) \ + \ ss(F) \ = \ ss(E \oplus F) \,,$$

since $R^m E \oplus R^n F \cong R^{m+n}(E \oplus F)$.

(3.9) Example: A monoid of endomorphisms. Suppose $R[x]$ is the ring of polynomials in one indeterminate x over a ring R, and $\mathcal{C}$ is a modest full subcategory of R-$\mathcal{M}od$ that is closed under $\oplus$ and includes the zero module 0. (For instance, $\mathcal{C}$ could be $\mathcal{M}(R)$, $\mathcal{P}(R)$, or $\mathcal{F}(R)$.) Denote by **end** $\mathcal{C}$ the full subcategory of $R[x]$-$\mathcal{M}od$ whose objects become objects of $\mathcal{C}$ when scalars are restricted to R. Each scalar multiplication $R[x] \times M \to M$ is determined by its restriction to $R \times M \to M$ and the R-linear endomorphism $x \cdot (\ ) : M \to M$. Each endomorphism $f : M \to M$ in $\mathcal{C}$ determines an object (M, f) of end $\mathcal{C}$, which is M, with scalar multiplication $\cdot : R[x] \times M \to M$, extending its scalar multiplication from R so that $x \cdot m = f(m)$. In this way the objects of end $\mathcal{C}$ are in bijective correspondence with the endomorphisms in $\mathcal{C}$.

An $R[x]$-linear map $\alpha : (M, f) \to (N, g)$ in end $\mathcal{C}$ is the same as an R-linear map $\alpha : M \to N$ in $\mathcal{C}$ for which $\alpha(x \cdot m) = x \cdot \alpha(m)$ for all $m \in M$ — that is, for which the square

$$
\begin{array}{ccc}
M & \xrightarrow{\ f\ } & M \\
\alpha \downarrow & & \downarrow \alpha \\
N & \xrightarrow[\ g\ ]{} & N
\end{array}
$$

commutes. Since end $\mathcal{C}$ is full in $R[x]$-$\mathcal{M}od$, an arrow α in end $\mathcal{C}$ is an isomorphism in end $\mathcal{C}$ if and only if it is bijective.

The external direct sum $\oplus$ is an associative, commutative binary operation on $R[x]$-$\mathcal{M}od$, and end $\mathcal{C}$ is closed under $\oplus$:

$$(M, f) \oplus (N, g) \ = \ (M \oplus N, f \oplus g) \,.$$

So $\oplus$ restricts to an associative, commutative binary operation on end $\mathcal{C}$, with identity object $(0, 0)$.

If $\beta : M \to N$ is an isomorphism in $\mathcal{C}$ and $f : M \to M$ is an endomorphism in $\mathcal{C}$, then $\beta \circ f \circ \beta^{-1} : N \to N$ is an endomorphism in $\mathcal{C}$ and β is an isomorphism

$$(M, f) \ \cong \ (N, \beta \circ f \circ \beta^{-1})$$

in end $\mathcal{C}$. So if S is a set of objects representing the isomorphism classes in $\mathcal{C}$, the set of endomorphisms of objects in S represents the isomorphism classes in end $\mathcal{C}$. So end $\mathcal{C}$ is modest, and $\mathcal{I}(\text{end } \mathcal{C})$ is an abelian monoid under $\oplus$.

When $\mathcal{C} = \mathcal{F}(R)$, this monoid has a concrete description in terms of matrices. There is a natural set of objects in end $\mathcal{F}(R)$ that meets every isomorphism class: If $A \in M_n(R)$, then $(R^n, \cdot A)$ is an object of end $\mathcal{F}(R)$. Suppose (M, f) is any object of end $\mathcal{F}(R)$. There is an isomorphism $\alpha : (M, f) \to (R^n, \cdot A)$ if and only if there is an R-linear isomorphism $\alpha : M \to R^n$ for which the square

$$
\begin{array}{ccc}
M & \xrightarrow{\ f\ } & M \\
\alpha \downarrow & & \downarrow \alpha \\
R^n & \xrightarrow[\cdot A]{} & R^n
\end{array}
$$

commutes — that is, if and only if A represents f over some R-basis of M ($A = Mat_\alpha^\alpha(f)$). Every f.g. free R-module M has a finite basis, so every endomorphism in $\mathcal{F}(R)$ is represented by some matrix; even the zero map $R^0 \to R^0$ can be regarded as right multiplication by the "empty matrix" $\emptyset$. Thus end $\mathcal{F}(R)$ is represented by the endomorphisms $(R^n, \cdot A)$.

Now *assume R has IBN.* If $(R^m, \cdot A) \cong (R^n, \cdot B)$ for matrices $A \in M_m(R)$ and $B \in M_n(R)$, then $R^m \cong R^n$; so $m = n$. Right multiplication by a matrix $C \in GL_n(R)$ is an isomorphism from $(R^n, \cdot B)$ to $(R^n, \cdot A)$ if and only if

$$
\begin{array}{ccc}
R^n & \xrightarrow{\ \cdot B\ } & R^n \\
\cdot C \downarrow & & \downarrow \cdot C \\
R^n & \xrightarrow[\cdot A]{} & R^n
\end{array}
$$

commutes, which is to say, if and only if $B = CAC^{-1}$. So $(R^m, \cdot A) \cong (R^n, \cdot B)$ if and only if $m = n$ and A and B are similar. Denote the similarity class of a matrix $A \in M_n(R)$, with $n \geq 1$, by

$$
s(A) = \{CAC^{-1} : C \in GL_n(R)\},
$$

and take the similarity class of $\emptyset \in M_0(R)$ to be $s(\emptyset) = \{\emptyset\}$. Let $\mathbf{sim}(R)$ denote the set of similarity classes of square matrices (including $\emptyset$) over R. The next proposition is now evident.

(3.10) Proposition. *Suppose R is a ring with IBN.*

(i) *The matrices representing an endomorphism (M, f) in $\mathcal{F}(R)$ over various bases of M form a similarity class.*

(ii) *Endomorphisms (M, f) and (N, g) are isomorphic in end $\mathcal{F}(R)$ if and only if they are represented by the same similarity class of matrices.*

(iii) *Matrix representation defines a bijection*

$$
\mathcal{I}(\text{end } \mathcal{F}(R)) \to \text{sim}(R)
$$

taking $c(R^n, \cdot A)$ to $s(A)$ for each $A \in M_n(R)$, $n \geq 0$. ∎

The direct sum on $R[x]$-$\mathfrak{Mod}$ has a simple description when applied to the objects $(R^m, \cdot A)$:

$$(R^m, \cdot A) \oplus (R^n, \cdot B) = (R^{m+n}, \cdot(A \oplus B)) ,$$

where

$$A \oplus B = \begin{bmatrix} A & 0 \\ 0 & B \end{bmatrix} .$$

So $\mathrm{sim}(R)$ becomes an abelian monoid, and the map in (3.10) (iii) becomes a monoid isomorphism, when similarity classes are added by the rule:

$$s(A) + s(B) = s(A \oplus B) .$$

Here, identity is $s(\emptyset) = \{\emptyset\}$, representing $(R^0, 0)$, and we use the convention that $\emptyset \oplus A = A = A \oplus \emptyset$ for each matrix A.

(3.11) Example: Monoids of bilinear forms. Suppose R is a commutative ring. A **bilinear form** on an R-module M is a function $b : M \times M \to R$ for which the maps $M \to R$, given by $v \mapsto b(v, w)$ and $v \mapsto b(w, v)$, are R-linear for each $w \in M$. A **bilinear module** over R is a pair (M, b), where M is an R-module and b is a bilinear form on M. The bilinear module (M, b) and the form b are called **symmetric** if $b(v, w) = b(w, v)$ for all $v, w \in M$. A **homomorphism of bilinear modules** $f : (M, b) \to (M', b')$ over R is an R-linear map $f : M \to M'$ with

$$b'(f(v), f(w)) = b(v, w)$$

for all $v, w \in M$.

Suppose $\mathcal{C}$ is a modest full subcategory of R-$\mathfrak{Mod}$, closed under $\oplus$, and including the zero module 0. Denote by **bil** $\mathcal{C}$ the category of symmetric bilinear modules (M, b) with $M \in \mathcal{C}$, and with arrows the homomorphisms of bilinear modules (M, b) between them. An arrow in bil $\mathcal{C}$ is an isomorphism if and only if it is bijective, in which case it is called an **isometry**.

The **orthogonal sum** of bilinear modules (M, b) and (M', b') over R is the bilinear module

$$(M, b) \perp (M', b') = (M \oplus M', b \perp b') ,$$

where $b \perp b'$ is defined by

$$(b \perp b') ((v, v'), (w, w')) = b(v, w) + b'(v', w') .$$

Evidently $b \perp b'$ is symmetric if b and b' are symmetric. The isomorphisms showing that $\oplus$ is a well-defined associative commutative binary operation on R-$\mathfrak{Mod}$, with identity 0, become isometries showing $\perp$ is a well-defined associative, commutative binary operation on bil $\mathcal{C}$ with identity object $(0, 0)$.

If $\beta : M \to M'$ is an isomorphism in $\mathcal{C}$ and $b : M \times M \to R$ is a symmetric bilinear form on M, then $b' = b \circ (\beta^{-1} \oplus \beta^{-1})$ is a symmetric bilinear form on M', and β is an isometry from (M,b) to (M',b'). So if S is a set of objects representing the isomorphism classes in $\mathcal{C}$, the set of symmetric bilinear modules (M,b) with $M \in S$ represents the isomorphism classes in bil $\mathcal{C}$. So bil $\mathcal{C}$ is modest, and $\mathcal{I}(\text{bil } \mathcal{C})$ is an abelian monoid under $\perp$.

As in (3.9), when $\mathcal{C} = \mathcal{F}(R)$, this monoid has a matrix description. There is a natural set of objects in bil $\mathcal{F}(R)$ that meets every isometry class: Denote the transpose of a matrix X by X^τ. If $A \in M_n(R)$, there is a bilinear form $b_A : R^n \times R^n \to R$ given by

$$b_A(v, w) = vAw^\tau .$$

The form b_A is symmetric if and only if A is symmetric ($A^\tau = A$). Suppose (M,b) is any object of bil $\mathcal{F}(R)$, and $v_1, ..., v_n$ is a basis of M. For a matrix $A \in M_n(R)$, the following are equivalent:

(i) $\quad A = (b(v_i, v_j)) = \begin{bmatrix} b(v_1, v_1) & \cdots & b(v_1, v_n) \\ \vdots & & \vdots \\ b(v_n, v_1) & \cdots & b(v_n, v_n) \end{bmatrix},$

(ii) $\quad b\left(\sum r_i v_i , \sum s_i v_i \right) = \sum_{i,j} r_i b(v_i, v_j) s_j$

$$= [r_1, \cdots r_n] \, A \begin{bmatrix} s_1 \\ \vdots \\ s_n \end{bmatrix},$$

for all scalars $r_i, s_i \in R$. When these conditions hold, the matrix A is said to **represent** b over the basis $v_1, \dots, v_n$. Since b is symmetric, each matrix representing b is symmetric.

Condition (ii) says that, if A represents b, then the coordinate mapping

$$\sum r_i v_i \mapsto [r_1 \cdots r_n]$$

is an isometry from (M,b) to (R^n, b_A). Conversely, if there is an isometry

$$f : (R^n, b_A) \to (M,b)$$

for some symmetric $A \in M_n(R)$, then $f(e_1), \dots, f(e_n)$ is a basis of M, and

$$b(f(e_i), f(e_j)) = b_A(e_i, e_j) = e_i A e_j^\tau ;$$

so $A = (b(f(e_i), f(e_j)))$ represents b over this basis. Therefore, A represents b if and only if $(M,b) \cong (R^n, b_A)$.

Every f.g. free R-module M has a basis; so every (M, b) in bil $\mathcal{F}(R)$ is represented by a symmetric matrix. This is even true of $(0, 0)$ if we say it is represented by the empty matrix $\emptyset$ in $M_0(R)$. So bil $\mathcal{F}(R)$ is represented by the spaces (R^n, b_A).

If $(R^n, b_A) \cong (R^m, b_B)$ for symmetric matrices $A \in M_n(R)$ and $B \in M_m(R)$, then $R^n \cong R^m$; since R is commutative, and so has IBN, $m = n$. Right multiplication by a matrix $C \in GL_n(R)$ is an isometry from (R^n, b_B) to (R^n, b_A) if and only if $vCAC^\tau w^\tau = vBw^\tau$ for all $v, w \in R^n$ — that is, if and only if $CAC^\tau = B$. Matrices $A, B \in M_n(R)$ are **congruent** if $CAC^\tau = B$ for some $C \in GL_n(R)$. So (R^n, b_A) and (R^m, b_B) are isometric if and only if $m = n$ and A is congruent to B.

Congruence is an equivalence relation on the set of square matrices over R; and if A is symmetric, so is every matrix that is congruent to A. Denote the congruence class of a matrix $A \in M_n(R)$, with $n \geq 1$, by

$$\langle A \rangle \ = \ \{ CAC^\tau : C \in GL_n(R) \} \, ,$$

and take the congruence class of $\emptyset \in M_0(R)$ to be $\langle \emptyset \rangle = \{\emptyset\}$. Let **cong**$(R)$ denote the set of congruence classes of symmetric matrices (including $\emptyset$) over R and refer to the objects of bil $\mathcal{F}(R)$ as **symmetric bilinear spaces**. We have proved:

(3.12) Proposition. *Suppose R is a commutative ring.*

   (i)   *The matrices representing a symmetric bilinear space (M, b) over R form a congruence class of symmetric matrices.*

  (ii) *Symmetric bilinear spaces (M, b) and (M', b') are isometric if and only if they are represented by the same congruence class of matrices.*

 (iii) *Matrix representation defines a bijection*

$$\mathfrak{I}(\text{bil } \mathcal{F}(R)) \ \to \ \text{cong}(R)$$

taking $c(R^n, b_A)$ to $\langle A \rangle$, for each symmetric $A \in M_n(R)$, $n \geq 0$. ∎

The orthogonal sum on bil $\mathcal{F}(R)$ has a simple description when applied to the objects (R^m, b_A):

$$(R^m, b_A) \perp (R^n, b_B) \ = \ (R^{m+n}, b_{A \oplus B}) \, .$$

So cong(R) becomes an abelian monoid, and the map in (3.12) (iii) becomes a monoid isomorphism, when congruence classes are added by the rule

$$\langle A \rangle \perp \langle B \rangle \ = \ \langle A \oplus B \rangle \, .$$

The identity is $\langle \emptyset \rangle = \{\emptyset\}$, representing $(R^0, 0)$.

When R is a field F in which $2 \neq 0$, there is a simpler representation of forms: A list of vectors $v_1, \dots, v_n$ in a symmetric bilinear space (M, b) over F is **orthogonal** if $b(v_i, v_j) = 0$ whenever $i \neq j$.

(3.13) Theorem. *Suppose F is a field with characteristic $\neq 2$. Every nonzero symmetric bilinear space (M, b) over F has an orthogonal basis. Equivalently, every symmetric matrix $A \in M_n(F)$ ($n \geq 1$) is congruent to a diagonal matrix:*

$$\mathbf{diag}(a_1, \ldots, a_n) = \begin{bmatrix} a_1 & 0 & \cdots & 0 \\ 0 & a_2 & \cdots & 0 \\ \vdots & \vdots & \ddots & \vdots \\ 0 & 0 & \cdots & a_n \end{bmatrix} .$$

Proof. Work by induction on $n = \dim_F(M)$, the case $n = 1$ being trivial. If $b = 0$, every basis of M is orthogonal. Assume $b \neq 0$. Since

$$b(v, w) = \tfrac{1}{2}(b(v + w, v + w) - b(v, v) - b(w, w)) ,$$

there is a vector $v \in M$ with $b(v, v) = d \neq 0$. Then $b(v, -) : M \to F$ is F-linear and surjective; so its kernel $v^\perp$ has dimension $n-1$. By the induction hypothesis, $v^\perp$ has an orthogonal basis $v_1, \ldots, v_{n-1}$ with respect to the restriction of b to $v^\perp \times v^\perp$. Then $v_1, \ldots, v_{n-1}, v$ is an orthogonal basis of M. ∎

If $D = \mathbf{diag}(a_1, \ldots, a_n)$, it is usual to denote the congruence class of D by

$$\langle D \rangle = \langle a_1, \ldots, a_n \rangle .$$

Then the orthogonal sum becomes

$$\langle a_1, \ldots, a_m \rangle \perp \langle a_{m+1}, \ldots, a_{m+n} \rangle = \langle a_1, \ldots, a_{m+n} \rangle .$$

Suppose R is a commutative ring. The **dual** of an R-module M is the R-module

$$M^* = \mathrm{Hom}_R(M, R)$$

under the scalar multiplication defined by $(r \cdot f)(m) = r(f(m))$. The **dual** of an R-linear map $f : M \to N$ is the R-linear map $f^* : N^* \to M^*$ given by $f^* = (-) \circ f$. The dual defines a contravariant functor $()^*$ from R-$\mathfrak{Mod}$ to itself.

A bilinear module (M, b) over R, and the bilinear form b, are **nonsingular** if the R-linear map $d_b : M \to M^*$, $m \mapsto b(m, -)$, is an isomorphism. In other words, b is nonsingular if $b(m, m') = 0$ for all $m' \in M$ implies $m = 0$, and every $f \in M^*$ is $b(m, -)$ for some $m \in M$

Suppose (M, b) and (N, c) are bilinear modules over R. An R-linear map $f : M \to N$ is a homomorphism of bilinear modules if and only if the square

$$\begin{array}{ccc} M & \xrightarrow{\;d_b\;} & M^* \\ {\scriptstyle f}\downarrow & & \uparrow{\scriptstyle f^*} \\ N & \xrightarrow{\;d_c\;} & N^* \end{array}$$

commutes. If f is an isometry, then f and f^* are isomorphisms; so (M, b) is nonsingular if and only if (N, c) is nonsingular.

For R-modules M and N, there is an isomorphism

$$\alpha : (M \oplus N)^* \to M^* \oplus N^*$$

taking each f to $(f \circ i_1, f \circ i_2)$, where i_1 and i_2 are insertions of M and N into the first and second coordinates in $M \oplus N$. (The inverse takes (f, g) to $(f \circ \pi_1) + (g \circ \pi_2)$, where π_1 and π_2 are the coordinate projections.) If (M, b) and (N, c) are bilinear modules over R, then

$$\alpha \circ d_{b \perp c} = d_b \oplus d_c .$$

So $(M, b) \perp (N, c)$ is nonsingular if both (M, b) and (N, c) are nonsingular.

With $\mathcal{C}$ defined as in (3.11), it follows that the orthogonal sum $\perp$ is an associative commutative binary operation on the full subcategory bil* $\mathcal{C}$ of the category bil $\mathcal{C}$, whose objects are the nonsingular symmetric bilinear modules (M, b) with $M \in \mathcal{C}$. And $\mathcal{I}(\text{bil}^* \, \mathcal{C})$ is a submonoid of $\mathcal{I}(\text{bil } \mathcal{C})$ under $\perp$.

If $\mathcal{C} = \mathcal{F}(R)$, the matrix description of $\mathcal{I}(\text{bil } \mathcal{C})$ restricts to a matrix description of $\mathcal{I}(\text{bil}^* \, \mathcal{C})$: If (M, b) is a bilinear module over R with an R-basis $m_1, \ldots, m_n$, the dual module M^* has a **dual R-basis** $m_1^*, \ldots, m_n^*$, where m_i^* projects each vector to its m_i-coefficient. The matrix $A = (b(m_i, m_j))$ representing (M, b) is also the matrix representing $d_b : M \to M^*$ with respect to these bases. So (M, b) is nonsingular if and only if A is invertible. (This applies even to the nonsingular module $(R^0, 0)$, if we regard the empty matrix $\emptyset \in M_0(R)$ as invertible.) So the isomorphism of monoids

$$\mathcal{I}(\text{bil } \mathcal{F}(R)) \cong \text{cong}(R)$$
$$c(R^n, b_A) \mapsto \langle A \rangle$$

restricts to an isomorphism of monoids

$$\mathcal{I}(\text{bil}^* \, \mathcal{F}(R)) \cong \text{cong}^*(R) ,$$

where **cong**$^*(R)$ is the abelian submonoid of $\text{cong}(R)$ consisting of congruence classes $\langle A \rangle$ of *invertible* symmetric matrices A over R.

When $R = F$ is a field of characteristic $\neq 2$, the elements of $\text{cong}^*(F)$ are $\langle \emptyset \rangle$ and the congruence classes $\langle a_1, \ldots, a_n \rangle$, where $a_1, \ldots, a_n$ are nonzero elements of F. As in $\text{cong}(F)$, these are added by concatenation.

3A. Exercises

1. Suppose R is a nontrivial ring. Prove there is no set S of R-modules that includes a member of every isomorphism class of R-modules. *Hint:* If such a

set S exists, then for each $M \in S$, the power set $P(M)$ is a set, and so is the union

$$U = \bigcup_{M \in S} P(M) \,.$$

Prove that for every set X there is an injective map $X \to U$. This is a contradiction, since there is no injective map $P(U) \to U$.

2. Suppose R is a ring and 0 is the zero R-module. Prove $\mathrm{cl}(0)$ cannot be a member of any class. *Hint:* From Appendix A, each member of a class is a set. But if $\mathrm{cl}(0)$ is a set, then by the axiom of replacement (see Appendix A), there is a set T of all sets having one member. Then the union $\cup T$ is the set of all sets, which does not exist.

3. Determine the structure of the monoid $\mathcal{I}(\mathcal{F}(R))$ in each of the cases:

(i) R has IBN,

(ii) $R = \{0\}$,

(iii) R is of type (h, k) as in §1C, Exercise 2.

4. Suppose $\mathcal{C}$ is a modest category, represented by sets S and T. For each $X \in \mathrm{Obj}\ \mathcal{C}$ let

$$c_S(X) = \{Y \in S : X \cong Y\} \,,$$
$$c_T(X) = \{Y \in T : X \cong Y\} \,.$$

Let $\mathcal{I}_S(\mathcal{C})$ (resp. $\mathcal{I}_T(\mathcal{C})$) denote the set of restricted isomorphism classes with respect to S (resp. with respect to T). If $*$ is an associative binary operation on $\mathcal{C}$, prove there is a semigroup isomorphism $\mathcal{I}_S(\mathcal{C}) \to \mathcal{I}_T(\mathcal{C})$, defined by $c_S(X) \mapsto c_T(X)$. (A homomorphism of semigroups $f : (A, *) \to (B, \#)$ is a function $f : A \to B$ satisfying

$$f(a * a') = f(a) \# f(a')$$

for all $a, a' \in A$.)

5. Prove that a full subcategory of a modest category is modest. For each ring R, prove the modest category $\mathcal{M}(R)$ has a subcategory that is not modest.

6. This exercise illustrates the value of assuming F has characteristic $\neq 2$ in (3.11). Suppose F is any field and V is a finite-dimensional F-vector space. A **quadratic form** on V is a function $q : V \to F$ for which

$$q(av) = a^2 q(v)$$

for all $a \in F$ and $v \in V$, and for which the function

$$B_q : V \times V \to F \,,$$

given by

$$B_q(v, w) = q(v + w) - q(v) - q(w) \,,$$

is bilinear. Let $\mathcal{B}(V)$ denote the set of symmetric bilinear forms $V \times V \to F$, and let $\mathcal{Q}(V)$ denote the set of quadratic forms $V \to F$. If $b \in \mathcal{B}(V)$, the **length function** associated to b is the function $|\;\;|_b : V \to F$ defined by $|v|_b = b(v,v)$.

(i) Show $\mathcal{B}(V)$ and $\mathcal{Q}(V)$ are F-vector spaces (under pointwise operations), and the map

$$\lambda : \mathcal{B}(V) \;\to\; \mathcal{Q}(V)$$
$$b \;\mapsto\; |\;\;|_b$$

is F-linear.

(ii) If $V = F^n$, prove the image of λ is the subspace of all polynomial functions in n-variables $x_1, \ldots, x_n$ of the form

$$\sum_{i \leq j} a_{ij} x_i x_j \;,$$

where $a_{ij} \in 2F$ for $i < j$.

(iii) Prove that the set Q of all polynomial functions $F^n \to F$ of the form

$$\sum_{i \leq j} a_{ij} x_i x_j \;,$$

where $a_{ij} \in F$, is a subspace of $\mathcal{Q}(F^n)$, and that

$$\sum_{i \leq j} a_{ij} x_i x_j \;=\; \sum_{i \leq j} b_{ij} x_i x_j$$

as functions $F^n \to F$ if and only if $a_{ij} = b_{ij}$ for each pair $i \leq j$.

(iv) If F has characteristic 2, prove the image of $\lambda : \mathcal{B}(F^n) \to \mathcal{Q}(F^n)$ is a proper subset of Q.

(v) If F has characteristic other than 2, and V is a finite-dimensional F-vector space, prove $\lambda : \mathcal{B}(V) \to \mathcal{Q}(V)$ is bijective, with inverse taking q to $(1/2)B_q$.

A primary motivation for studying symmetric bilinear forms over F is their connection to homogeneous polynomials of degree 2 over F, and this connection is most complete when $2 \neq 0$ in F.

7. Suppose R is a commutative ring and $R[x, x^{-1}]$ is the Laurent polynomial ring, which is the free R-module based on $\{\ldots, \; x^{-2}, \; x^{-1}, \; 1, \; x, \; x^2, \ldots\}$, with polynomial-like multiplication:

$$\left(\sum_{i \in \mathbb{Z}} a_i x^i \right) \left(\sum_{i \in \mathbb{Z}} b_i x^i \right) \;=\; \sum_{k \in \mathbb{Z}} \left(\sum_{i+j=k} a_i b_j \right) x^k \;.$$

Show the full subcategory **aut** $\mathcal{F}(R)$ of $R[x, x^{-1}]$-$\mathfrak{Mod}$, consisting of $R[x, x^{-1}]$-modules that are finitely generated and free as R-modules, is modest. Give a concrete description of the monoid $\mathfrak{I}(\text{aut } \mathcal{F}(R))$ under $\oplus$ in terms of matrices (as in Example (3.9)).

8. For R a ring, prove directly that

$$s(A) + s(B) \; = \; s(A \oplus B)$$

defines a binary operation making $\text{sim}(R)$ into an abelian monoid. If R is commutative, prove directly that

$$\langle A \rangle \perp \langle B \rangle \; = \; \langle A \oplus B \rangle$$

defines a binary operation making $\text{cong}(R)$ into an abelian monoid.

9. If R is a ring and π is a member of the symmetric group S_n $(n > 0)$, prove the diagonal matrices

$$\text{diag}(a_1, \ldots, a_n) \; \text{ and }$$
$$\text{diag}(a_{\pi(1)}, \ldots, a_{\pi(n)})$$

are both similar and congruent in $M_n(R)$.

3B. Semigroups to Groups

The advantages gained by working with groups instead of semigroups are no less than the advantages offered by the use of negative numbers. There is a universal construction of an abelian group from a semigroup, which generalizes the construction of $\mathbb{Z}$ from the semigroup $\mathbb{P}$ of positive integers. The version we present here generalizes a construction of A. Grothendieck (see (3.20) and §3C), and so it is called the "Grothendieck group" of a semigroup by some authors. We shall call it the "group completion" of a semigroup. An abelian semigroup embeds in its group completion when it is cancellative. For examples, we consider Krull-Schmidt cancellation and Witt cancellation.

Recall that a **semigroup** $(S, *)$ is a nonempty set S with an associative binary operation $*$. If $(S, *)$ and $(T, \#)$ are semigroups, a **semigroup homomorphism** $f : (S, *) \to (T, \#)$ is a function $f : S \to T$ with

$$f(s * s') \; = \; f(s) \# f(s')$$

for all $s, s' \in S$.

(3.14) Definition. Suppose $(S, *)$ is a semigroup. The **group completion** of $(S, *)$ is the quotient

$$\bar{\bar{S}} \; = \; F_{\mathbb{Z}}(S)/\langle D \rangle \, ,$$

where $F_{\mathbb{Z}}(S)$ is the free $\mathbb{Z}$-module based on S, and D is the set of elements in $F_{\mathbb{Z}}(S)$ of the form $(x * y) - x - y$, where $x, y \in S$.

If $x \in S$, let $\overline{x}$ denote the coset $x + \langle D \rangle$. Then $\overline{\overline{S}}$ is presented as a $\mathbb{Z}$-module by the generating set $\{\overline{x} : x \in S\}$ and the defining relations $\overline{x} + \overline{y} = \overline{x * y}$ for $x, y \in S$. The map

$$\gamma : S \to \overline{\overline{S}}$$
$$x \mapsto \overline{x}$$

is called the **group completion map** on $(S, *)$. It is the first semigroup homomorphism from $(S, *)$ into an abelian group:

(3.15) Proposition. *If $(S, *)$ is a semigroup, the group completion map $\gamma :$ $S \to \overline{\overline{S}}$ is a semigroup homomorphism. If $g : S \to A$ is also a semigroup homomorphism into an abelian group $(A, +)$, there is one and only one group homomorphism $\overline{g} : \overline{\overline{S}} \to A$ with $\overline{g} \circ \gamma = g$.*

Proof. The relation $\overline{x * y} = \overline{x} + \overline{y}$ says γ is a semigroup homomorphism. The function $g : S \to A$ extends to a $\mathbb{Z}$-linear map $\widehat{g} : F_{\mathbb{Z}}(S) \to A$. For $x, y \in S$,

$$\widehat{g}((x * y) - x - y) \;=\; g(x * y) - g(x) - g(y) \;=\; 0_A .$$

So $\widehat{g}$ induces a group homomorphism $\overline{g} : \overline{\overline{S}} \to A$ with $\overline{g}(\gamma(x)) = \overline{g}(\overline{x}) = \widehat{g}(x) = g(x)$ for each $x \in S$. Any group homomorphism $h : \overline{\overline{S}} \to A$ with $h \circ \gamma = g$ satisfies $h(\overline{x}) = h(\gamma(x)) = g(x) = \overline{g}(\overline{x})$ for each $x \in S$. Since the elements $\overline{x}$ generate $\overline{\overline{S}}$, $h = \overline{g}$. ∎

If $(S, *)$ and $(T, \#)$ are monoids (= semigroups with identity), a **monoid homomorphism** $f : (S, *) \to (T, \#)$ is a semigroup homomorphism taking the identity of S to the identity of T. If $(S, *)$ is a monoid and $(A, +)$ is a group, every semigroup homomorphism $f : (S, *) \to (A, +)$ is also a monoid homomorphism, since the only idempotent in $(A, +)$ is the identity. So Proposition (3.15) remains true when "semigroup" is replaced by "monoid" throughout.

(3.16) Proposition. *Each element of the group completion $\overline{\overline{S}}$ of a semigroup $(S, *)$ has the form $\overline{x} - \overline{y}$ with $x, y \in S$.*

Proof. Each element of $\overline{\overline{S}}$ is a $\mathbb{Z}$-linear combination of cosets:

$$a_1 \overline{x}_1 + \cdots + a_n \overline{x}_n$$

with $a_i \in \mathbb{Z}$, $x_i \in S$. Each coefficient a_i can be made 1 or -1 by separating the term $a_i \overline{x}_i$ into a sum. Adding $0 = \overline{x}_0 - \overline{x}_0$ ($x_0 \in S$) if necessary, we may assume both 1 and -1 occur as coefficients. By commutativity in A and the defining relations in $\bar{\bar{S}}$, this typical element can be written as

$$(\bar{u}_1 + \cdots + \bar{u}_s) - (\bar{v}_1 + \cdots + \bar{v}_t) = \overline{u_1 * \cdots * u_s} - \overline{v_1 * \cdots * v_t}$$

for $u_i, v_i \in S$. ∎

(3.17) Proposition. *Suppose $\bar{\bar{S}}$ is the group completion of an abelian semigroup $(S, *)$. For $x, y, x', y' \in S$,*

(i) $\bar{x} = \bar{y}$ *in* $\bar{\bar{S}}$ *if and only if $x * z = y * z$ for some $z \in S$;*
(ii) $\bar{x} - \bar{y} = \bar{x}' - \bar{y}'$ *in* $\bar{\bar{S}}$ *if and only if $x * y' * z = x' * y * z$ for some $z \in S$.*

Proof. If $x * z = y * z$, then in $\bar{\bar{S}}$, $\bar{x} + \bar{z} = \bar{y} + \bar{z}$; so $\bar{x} = \bar{y}$. For the converse, suppose $\bar{x} = \bar{y}$ in $\bar{\bar{S}}$. Then in $F_{\mathbb{Z}}(S)$, $x - y \in \langle D \rangle$. That is,

$$x - y = \sum_{i=1}^{n} a_i((x_i * y_i) - x_i - y_i) \,,$$

where $a_i = 1$ or -1 and $x_i, y_i \in S$. Bringing terms with negative coefficients to the other side,

$$x + \sum_{a_i = -1} (x_i * y_i) + \sum_{a_i = 1} (x_i + y_i)$$
$$= y + \sum_{a_i = 1} (x_i * y_i) + \sum_{a_i = -1} (x_i + y_i) \,.$$

Since S is a basis of $F_{\mathbb{Z}}(S)$, the terms on one side of the equation are a permutation of those on the other side. Since $(S, *)$ is abelian, it follows that, in S,

$$x * \prod_{a_i = -1} (x_i * y_i) * \prod_{a_i = 1} (x_i * y_i)$$
$$= y * \prod_{a_i = 1} (x_i * y_i) * \prod_{a_i = -1} (x_i * y_i) \,.$$

So $x * z = y * z$, where

$$z = \prod_{i=1}^{n} (x_i * y_i) \,.$$

To prove (ii), note that $\bar{x} - \bar{y} = \bar{x}' - \bar{y}'$ if and only if $\bar{x} + \bar{y}' = \bar{x}' + \bar{y}$, and the latter is equivalent to $\overline{x * y'} = \overline{x' * y}$. Now apply (i). ∎

A semigroup $(S, *)$ is **cancellative** if, for $x, y, z \in S$, $x * z = y * z$ implies $x = y$.

(3.18) Corollary. *The group completion map $S \to \bar{\bar{S}}$ is injective if and only if $(S, *)$ is abelian and cancellative.* ∎

There is one situation in which an abelian monoid is guaranteed to be cancellative and to have an easily described group completion. Suppose an abelian monoid S is written additively ($* = +, 1_S = 0$). A subset $B \subseteq S$ is a **monoid basis** of S if, for each $x \in S$, there is one and only one function

$$f_x : B \to \mathbb{N} = \{0, 1, 2, \dots\}$$

with finite support, for which

$$x = \sum_{b \in B} f_x(b) b .$$

If S has a monoid basis B, then S is called a **free abelian monoid**.

(3.19) Proposition. *If S is a free abelian monoid with monoid basis B, then S is cancellative, the group completion map $\gamma : S \to \bar{\bar{S}}$ is injective, and $\bar{\bar{S}}$ is a free abelian group with basis $\gamma(B)$.*

Proof. Suppose S is a free abelian monoid with monoid basis B. Suppose $x, y, z \in S$ and $x + z = y + z$. For each $b \in B$,

$$f_x(b) + f_z(b) = f_{x+z}(b) = f_{y+z}(b) = f_y(b) + f_z(b) .$$

So $f_x = f_y$ and $x = y$. By (3.18), γ is injective. By (3.16), the additive abelian group $\bar{\bar{S}}$ is generated by the set $\gamma(B)$ of all $\bar{b}$ for $b \in B$. Suppose $\sum_{b \in B} f(\bar{b})\bar{b} = \bar{0}$ for some function $f : \gamma(B) \to \mathbb{Z}$ with finite support. Let B^- denote the set of $b \in B$ with $f(\bar{b}) < 0$ and $B^+ = B - B^-$. Then

$$\sum_{b \in B^+} f(\bar{b})\bar{b} = \sum_{b \in B^-} -f(\bar{b})\bar{b} .$$

Since γ is additive and injective,

$$\sum_{b \in B^+} f(\bar{b})b = \sum_{b \in B^-} -f(\bar{b})b$$

in S. Since B is a monoid basis of S and $B^+ \cap B^-$ is empty, each $f(\bar{b}) = 0$. ∎

Now we can apply these results to the semigroups constructed in §3A:

(3.20) Definition. Suppose $\mathcal{C}$ is a modest category with an associative binary operation $*$. The group completion of the semigroup $(\mathfrak{I}(\mathcal{C}), *)$ is denoted by

$$K_0(\mathcal{C}, *)$$

and is called the **Grothendieck group of $\mathcal{C}$ with respect to** $*$. It is the quotient $\mathbb{Z}$-module

$$F_{\mathbb{Z}}(\mathfrak{I}(\mathcal{C}))/\langle D \rangle ,$$

where D consists of the elements

$$(c(X) * c(Y)) - c(X) - c(Y) = c(X * Y) - c(X) - c(Y) ,$$

for $X, Y \in \mathrm{Obj}\ \mathcal{C}$.

(3.21) Notation. In $K_0(\mathcal{C}, *)$, for each $X \in \mathrm{Obj}\ \mathcal{C}$ let

$$[X] = c(X) + \langle D \rangle$$

denote the coset of $c(X)$. Then $K_0(\mathcal{C}, *)$ is presented as a $\mathbb{Z}$-module by the generators $[X]$, one for each isomorphism class $c(X) \in \mathfrak{I}(\mathcal{C})$, and defining relations

$$[X] + [Y] = [X * Y]$$

for all $X, Y \in \mathrm{Obj}\ \mathcal{C}$.

If $X \cong Y$ in $\mathcal{C}$, then $c(X) = c(Y)$; so $[X] = [Y]$ in $K_0(\mathcal{C}, *)$. From (3.16) and (3.17), we have

(3.22) Corollary. *Suppose $\mathcal{C}$ is a modest category with a commutative associative binary operation $*$. Then in $K_0(\mathcal{C}, *)$*

(i) *each element has the form $[X] - [Y]$ for $X, Y \in \mathrm{Obj}\ \mathcal{C}$;*

(ii) *$[X] = [Y]$ if and only if $X * Z \cong Y * Z$ for some $Z \in \mathrm{Obj}\ \mathcal{C}$;*

(iii) *$[X] - [Y] = [X'] - [Y']$ if and only if $X * Y' * Z \cong X' * Y * Z$ for some $Z \in \mathrm{Obj}\ \mathcal{C}$; and*

(iv) *if 0 is an identity object for $*$ on $\mathcal{C}$, then $[0]$ is the zero element of $K_0(\mathcal{C}, *)$.*

Proof. Part (i) is (3.16). Parts (ii) and (iii) are (3.17) because

$$c(X) * c(Z) = c(Y) * c(Z)$$

in $(\mathfrak{I}(\mathcal{C}), *)$ if and only if

$$c(X * Z) = c(Y * Z) .$$

For part (iv), $[0] + [0] = [0 * 0] = [0]$; now subtract $[0]$.   ∎

Suppose $\mathcal{C}$ and $*$ are as in (3.20). The relation $\sim$ on Obj $\mathcal{C}$, defined by

$$X \sim Y \Leftrightarrow [X] = [Y] \text{ in } K_0(\mathcal{C}, *)$$
$$\Leftrightarrow X * Z \cong Y * Z \text{ for some } Z \in \mathcal{C} ,$$

is coarser than $\cong$ in $\mathcal{C}$, unless $\mathcal{I}(\mathcal{C})$ is cancellative, which is to say, unless

$$X * Z \cong Y * Z \Rightarrow X \cong Y$$

holds in $\mathcal{C}$. The loss of information in passing from $\mathcal{I}(\mathcal{C})$ to the group $K_0(\mathcal{C}, *)$ is measured by the failure of cancellation.

In the categories $\mathcal{C}$, end $\mathcal{C}$, bil $\mathcal{C}$, and bil* $\mathcal{C}$ (for $\mathcal{C} = \mathcal{M}(R)$, $\mathcal{P}(R)$ or $\mathcal{F}(R)$), each endowed with its own version of direct sum in (3.5)–(3.11), cancellation holds over some rings R, but not over others. Cancellation in some of these categories is considered in Exercises 3–5 below. Cancellation in $\mathcal{P}(R)$ is considered at length in Chapter 4 but, for artinian rings R, there is a unique decomposition of f.g. R-modules whose proof parallels that of the unique factorization of positive integers into primes:

(3.23) Definitions. Over any ring R, a nonzero R-module M is **indecomposable** if M is not a direct sum of two nonzero submodules — or, equivalently, if $M \cong M_1 \oplus M_2$ in R-$\mathcal{M}\mathfrak{od}$ implies $M_1 = 0$ or $M_2 = 0$. An R-module M has a **Krull-Schmidt decomposition** if M has a unique expression as a direct sum of indecomposable R-modules — that is, if

$$M \quad \cong \quad M_1 \oplus \cdots \oplus M_n \quad (1 \leq n < \infty)$$

for indecomposable R-modules M_i, and if

$$M_1 \oplus \cdots \oplus M_n \quad \cong \quad N_1 \oplus \cdots \oplus N_r \quad (1 \leq r < \infty) ,$$

for indecomposable R-modules N_i, implies that $r = n$ and there is a permutation σ of $\{1, \ldots, n\}$ with $N_i \cong M_{\sigma(i)}$ for each i.

(3.24) Theorem. *If R is a left artinian ring, each nonzero $M \in \mathcal{M}(R)$ has a Krull-Schmidt decomposition.*

Proof. Since R is left artinian, by (B.9) and (B.21), each f.g. R-module is both noetherian and artinian. Suppose $0 \neq M \in \mathcal{M}(R)$. Let S denote the set of nonzero submodules of M that are not isomorphic to a direct sum of finitely many indecomposable R-modules. If $N \in S$, $N \neq 0$ and N is not indecomposable. So $N = N_1 \overset{\bullet}{\oplus} N_2$, where $N_1, N_2 \neq 0$, and either $N_1 \in S$ or $N_2 \in S$.

So there is a choice function, sending each $N \in S$ to one of its proper direct summands $f(N) \in S$. If $M \in S$, there is a descending chain

$$M \supsetneq f(M) \supsetneq f^2(M) \supsetneq \cdots ,$$

contradicting the fact that M is artinian. So $M \cong M_1 \oplus \cdots \oplus M_n$ for indecomposables M_i, where $1 \le n < \infty$.

To prove uniqueness of this decomposition, we need:

(3.25) Fitting's Lemma. *Suppose R is a ring and A is an artinian, noetherian, indecomposable R-module. In the ring $\mathrm{End}_R(A)$, every element is either a unit or nilpotent; so the sum of nonunits is a nonunit.*

Proof. Suppose $f \in \mathrm{End}_R(A)$. The chains

$$A \supseteq f(A) \supseteq f^2(A) \supseteq \cdots ,$$
$$0 \subseteq \ker(f) \subseteq \ker(f^2) \subseteq \cdots ,$$

have only finitely many steps that are proper containments. So for some integer $n > 0$, $f^{2n}(A) = f^n(A)$ and $\ker(f^{2n}) = \ker(f^n)$. If $a \in A$ and $f^n(f^n(a)) = 0$, then $f^n(a) = 0$; so

$$f^n(A) \ \cap \ \ker(f^n) \ = \ 0 .$$

If $a \in A$, $f^n(a) = f^{2n}(b)$ for some $b \in A$; so $a = f^n(b) + (a - f^n(b))$, proving

$$f^n(A) \ + \ \ker(f^n) \ = \ A .$$

So $A \cong f^n(A) \oplus \ker(f^n)$.

Since A is indecomposable, either $\ker(f^n) = 0$ and $f^n(A) = A$, or $f^n(A) = 0$. Therefore, in the ring $\mathrm{End}_R(A)$, f is either a unit or nilpotent.

Suppose f and g are nonunits of $\mathrm{End}_R(A)$, but $f + g = u$ is a unit. Then $s = fu^{-1}$ and $t = gu^{-1}$ are nonunits with $s + t = 1$. So $s \ne 0$ $(t \ne 1)$, and for some $n \ge 1$, $s^n \ne 0$ while $s^{n+1} = 0$. Also, $t^m = 0$ for some $m \ge 1$. So $s^n = s^{n+1} + s^n t = s^n t$, and $s^n = s^n t = s^n t^2 = \cdots = s^n t^m = 0$, a contradiction. ∎

For any ring R, an R-module A is called **strongly indecomposable** if the nonunits of $\mathrm{End}_R(A)$ are closed under addition. Each strongly indecomposable R-module is indecomposable, since $i_{M \oplus N} = (i_M \oplus 0_N) + (0_M \oplus i_N)$.

If R is left artinian and $M \in \mathcal{M}(R)$, each direct summand of M is also in $\mathcal{M}(R)$, and so it is artinian and noetherian by (B.9) and (B.21). So by Fitting's Lemma, each indecomposable direct summand of M is strongly indecomposable. The proof of (3.24) is completed by the next lemma.

(3.26) Lemma. *Suppose R is a ring, $M \in \mathcal{M}(R)$, and*

$$M_1 \oplus \cdots \oplus M_n \cong M \cong N_1 \oplus \cdots \oplus N_r$$

for indecomposable R-modules M_i and N_i. If each M_i is strongly indecomposable, then $r = n$ and there is a permutation σ of $\{1, \ldots, n\}$ with $N_i \cong M_{\sigma(i)}$ for each i.

Proof. Let $A = M_1$ and $B = M_2 \oplus \cdots \oplus M_n$. Since $M \cong A \oplus B$, there exist R-linear maps

$$A \xrightarrow{\ f_A\ } M \xleftarrow{\ f_B\ } B$$
$$A \xleftarrow{\ g_A\ } \qquad \xrightarrow{\ g_B\ } B$$

with $g_A f_A = i_A$, $g_B f_B = i_B$, and $f_A g_A + f_B g_B = i_M$. Since $M \cong N_1 \oplus \cdots \oplus N_r$, there exist R-linear maps

$$M \xrightarrow{\ g_i\ } N_i$$
$$\xleftarrow{\ f_i\ }$$

for $1 \le i \le r$, with $g_i f_i = i_{N_i}$ and $\sum_i f_i g_i = i_M$. Then

$$i_A \;=\; g_A f_A \;=\; g_A \left(\sum_i f_i g_i \right) f_A \;=\; \sum_i g_A f_i g_i f_A \,,$$

where each $g_A f_i g_i f_A \in \operatorname{End}_R(A)$. Since $i_A = 1 \in \operatorname{End}_R(A)$, and A is strongly indecomposable, some $g_A f_j g_j f_A$ is a unit $\alpha \in \operatorname{End}_R(A)$. Then

$$\beta = g_j f_A : A \;\rightarrow\; N_j$$

has an R-linear left inverse

$$\gamma = \alpha^{-1} g_A f_j : N_j \;\rightarrow\; A \,.$$

So $N_j = \beta(A) \oplus \ker(\gamma)$, and $\beta(A) \cong A \neq 0$. Since N_j is indecomposable, $\ker(\gamma) = 0$ and $\beta(A) = N_j$. Thus

$$\beta : A = M_1 \;\rightarrow\; N_j$$

is an isomorphism.

Let C denote
$$N_1 \oplus \cdots \oplus N_{j-1} \oplus N_{j+1} \oplus \cdots \oplus N_r \,.$$

The isomorphism (3.26), followed by a reordering of summands, is an isomorphism

$$M_1 \oplus B \xrightarrow{\ \begin{bmatrix} \beta & x \\ y & z \end{bmatrix}\ } N_j \oplus C$$

in the matrix notation for maps between direct sums in (1.31). Composing this with the isomorphism

$$N_j \oplus C \xrightarrow{\begin{bmatrix} 1 & 0 \\ -y\beta^{-1} & 1 \end{bmatrix}} N_j \oplus C$$

produces an isomorphism

$$\begin{bmatrix} 1 & 0 \\ -y\beta^{-1} & 1 \end{bmatrix} \begin{bmatrix} \beta & x \\ y & z \end{bmatrix} = \begin{bmatrix} \beta & x \\ 0 & z' \end{bmatrix}$$

from $M_1 \oplus B$ to $N_j \oplus C$. The inverse

$$\begin{bmatrix} \beta & x \\ 0 & z' \end{bmatrix}^{-1} = \begin{bmatrix} \beta^{-1} & x' \\ y' & z'' \end{bmatrix}$$

has $y'\beta = 0$; so $y' = 0$. Then z'' is a two-sided inverse to z'. So $z' : B \to C$ is an isomorphism. That is,

$$M_2 \oplus \cdots \oplus M_n \cong N_1 \oplus \cdots \oplus N_{j-1} \oplus N_{j+1} \oplus \cdots \oplus N_r \ .$$

Repeating the argument, we continue to cancel isomorphic terms from both sides. Since the M_i and N_i are indecomposable, $r = n$. ■ ■

(3.27) Corollary. *If R is a left artinian ring and $\mathcal{C} = \mathcal{M}(R)$ or $\mathcal{P}(R)$, then $\mathcal{I}(\mathcal{C})$ is a cancellative monoid under $\oplus$, and its group completion $K_0(\mathcal{C}, \oplus)$ is free abelian, with one basis element $[P]$ for each isomorphism class $c(P)$ of indecomposable R-modules $P \in \mathcal{C}$. If $R \cong P_1 \oplus \cdots \oplus P_n$ is a decomposition into indecomposable R-modules, every indecomposable $P \in \mathcal{P}(R)$ is isomorphic to one of the P_i. So if $P_1, \ldots, P_m$ are pairwise nonisomorphic, and each P_i with $i > m$ is isomorphic to one of them, then $K_0(\mathcal{P}(R), \oplus)$ is free abelian with basis $[P_1], \ldots, [P_m]$.*

Proof. Since $\mathcal{C}$ is closed under direct summands, Theorem (3.24) implies $\mathcal{I}(\mathcal{C})$ is a free abelian monoid based on isomorphism classes of indecomposable R-modules in $\mathcal{C}$. The rest of the first sentence follows from Proposition (3.19).

Suppose $R \cong P_1 \oplus \cdots \oplus P_n$ is the Krull-Schmidt decomposition of the R-module R into indecomposables, and suppose P is an indecomposable in $\mathcal{P}(R)$. For some $Q \in \mathcal{P}(R)$ and $r \geq 1$,

$$P \oplus Q \cong R^r \cong (P_1 \oplus \cdots \oplus P_n)^r.$$

Since Q has a Krull-Schmidt decomposition, uniqueness of the decomposition for R^r implies $P \cong P_i$ for some i. So, if $P_1, \ldots, P_m$ represent the different isomorphism classes of $P_1, \ldots, P_n$, they represent the isomorphism classes of all indecomposable $P \in \mathcal{P}(R)$. ■

One of the oldest cancellation theorems appeared in a paper of Witt [37] in 1937. It takes place in the category bil $\mathcal{F}(F)$, for F a field:

(3.28) Witt Cancellation Theorem. *Suppose F is a field of characteristic $\neq 2$, and (L, b_1), (M, b_2), and (N, b_3) are finite-dimensional symmetric bilinear modules over F. If*

$$(L, b_1) \perp (N, b_3) \cong (M, b_2) \perp (N, b_3) ,$$

then $(L, b_1) \cong (M, b_2)$.

Proof. An isometry is a linear isomorphism, so $\dim_F(L) = \dim_F(M)$. Call this common dimension n. If $n = 0$, the cancellation is immediate: $(0, 0) \cong (0, 0)$. Suppose $n \geq 1$.

Since F is a field of characteristic $\neq 2$, the desired cancellation can be restated in terms of congruence $\sim$ of diagonal matrices: Suppose D, D' are diagonal matrices in $M_n(F)$ and D'' is any diagonal matrix over F. We must prove

$$D \oplus D'' \sim D' \oplus D'' \;\Rightarrow\; D \sim D' .$$

Since $\mathrm{diag}(a_1, \ldots, a_m) = [a_1] \oplus \cdots \oplus [a_m]$, it suffices to prove this when $D'' = [d]$ for $d \in F$. Write $X \oplus [d]$ as $X \oplus d$, and assume $D \oplus d \sim D' \oplus d$.

Case 1: Suppose $d = 0$. If D or $D' = 0$, then $D = 0 = D'$ and we are done. Permuting diagonal entries by applying $P(-)P^\tau$ for a permutation matrix P, $D \sim D_1 \oplus 0_r$ and $D' \sim D_1' \oplus 0_s$ for invertible matrices D_1 and D_1'. Switching D and D' if necessary, we may assume $r \leq s$. Write $D_1' \oplus 0_s = D_2 \oplus 0_r$. By hypothesis:

$$C(D_2 \oplus 0_{r+1})C^\tau = D_1 \oplus 0_{r+1}$$

for some $C \in GL_{n+1}(F)$. If $B \in M_{n-r}(F)$ is the upper left corner of C, block multiplication of partitioned matrices implies $BD_2B^\tau = D_1$. Since D_1 is invertible, so is B. So $D_2 \sim D_1$, and

$$D' \sim D_2 \oplus 0_r \sim D_1 \oplus 0_r \sim D .$$

Case 2: Suppose $d \neq 0$. On F^{n+1} there are symmetric bilinear forms $b = b_{D \oplus d}$ and $b' = b_{D' \oplus d}$. By hypothesis, there is an isometry

$$\cdot C : (F^{n+1}, b) \rightarrow (F^{n+1}, b') ,$$

where $C \in GL_{n+1}(F)$. So

$$
\begin{aligned}
b'(e_{n+1}C, e_{n+1}C) &= b(e_{n+1}, e_{n+1}) \\
&= d = b'(e_{n+1}, e_{n+1}) .
\end{aligned}
$$

Assume there is an isometry

$$\cdot C' : (F^{n+1}, b') \rightarrow (F^{n+1}, b')$$

with $e_{n+1}CC' = e_{n+1}$. Then $CC' \in GL_{n+1}(F)$, and

$$CC'(D' \oplus d)(CC')^\tau = D \oplus d .$$

Then

$$e_{n+1}^\tau = \frac{1}{d}(D \oplus d)e_{n+1}^\tau$$

$$= \frac{1}{d}CC'(D' \oplus d)(CC')^\tau e_{n+1}^\tau$$

$$= \frac{1}{d}CC'(D' \oplus d)e_{n+1}^\tau$$

$$= CC'e_{n+1}^\tau .$$

Therefore $CC' = C_0 \oplus 1$. Block multiplication shows $C_0 \in GL_n(F)$ and $C_0 D' C_0^\tau = D$, as desired.

To show the last assumption was justified, we prove another theorem of Witt. Suppose (M, b) is a symmetric bilinear module over a commutative ring R. By analogy of b with the euclidean inner product, we call

$$S(d) = \{v \in M : b(v, v) = d\}$$

a **sphere** in (M, b), and we call the group of self-isometries of (M, b) the **orthogonal group** $O(M, b)$.

(3.29) Theorem. *If (M, b) is a finite-dimensional symmetric bilinear module over a field F of characteristic $\neq 2$, and d is a nonzero element of F, then the orthogonal group $O(M, b)$ acts transitively on the sphere $S(d)$.*

Proof (Lam [80]). For $x \in M$ and $\sigma \in O(M, b)$, $b(\sigma(x), \sigma(x)) = b(x, x)$; so $O(M, b)$ does act on $S(d)$. If $y \in M$ with $b(y, y) \neq 0$, the map $\tau_y : M \to M$ defined by

$$\tau_y(v) = v - \frac{2b(y, v)}{b(y, y)}y$$

is a member of $O(M, b)$ called **reflection in the hyperplane orthogonal to** y. (For a justification of this name, do Exercise 8.)

In the cartesian plane, the sum of squares of the diagonals of a parallelogram is the sum of the squares of its sides. This is a simple identity, even valid in (M, b) : For $x, y \in M$,

$$b(x + y, x + y) + b(x - y, x - y) = 2b(x, x) + 2b(y, y) .$$

Suppose $x, y \in S(d)$, with $d \neq 0$; so $b(x,x) = d = b(y,y)$. It follows that $b(x+y, x+y)$ and $b(x-y, x-y)$ cannot both be 0.

If $b(x-y, x-y) \neq 0$,

$$\tau_{x-y}(x) = x - \frac{2b(x-y, x)}{b(x-y, x-y)}(x-y) = y \,,$$

since

$$
\begin{aligned}
b(x - y, x - y) \\
&= b(x-y, x) - b(x, y) + b(y,y) \\
&= b(x-y, x) + b(x,x) - b(y,x) \\
&= 2b(x-y, x) \,.
\end{aligned}
$$

Or, if $b(x+y, x+y) \neq 0$,

$$\tau_{x+y}(x) = \tau_{x-(-y)}(x) = -y \,;$$

so $(-\tau_{x+y})(x) = y$, and $-\tau_{x+y} \in O(M, b)$ as well.   ■ ■

The Witt cancellation described above is by no means the most general cancellation theorem on categories of forms. For generalizations, see Bak [69]; Magurn, van der Kallen, and Vaserstein [88]; and Stafford [90]. But much attention is still given to the groups

$$K_0(\text{bil } \mathcal{F}(F), \perp) \,,$$
$$K_0(\text{bil}^* \ \mathcal{F}(F), \perp) \,,$$

for fields F of characteristic $\neq 2$. In both groups, $[(M, b)] = [(M', b')]$ if and only if $(M, b) \cong (M', b')$; so the group contains the classification of symmetric bilinear spaces up to isometry. The group $K_0(\text{bil}^* \ \mathcal{F}(F), \perp)$ was introduced in the paper by Witt [37] mentioned above and is known in the current literature as the **Witt-Grothendieck group**, $\widehat{W}(F)$, of the field F. It is one of the earliest precursors of algebraic K-theory.

3B. Exercises

1. If $(S, *)$ is a group, prove its group completion $\bar{\bar{S}}$ is its abelianization $S/[S,S]$, where $[S,S]$ is the commutator subgroup of S. *Hint:* Use (3.15), and a similar universal property of $S/[S,S]$.

2. Suppose $(A, +)$ is an abelian group, and S is a nonempty subset of A, closed under $+$. Show S is a semigroup under $+$, and $\bar{\bar{S}}$ is isomorphic to the subgroup of A generated by S.

3. Prove $(\mathcal{F}(R), \oplus)$ has cancellation if and only if R has IBN or $R = \{0\}$. What does this say about $\mathcal{P}(R)$ and $\mathcal{M}(R)$?

4. For idempotent matrices E, F over R, prove their row spaces P, Q have $[P] = [Q]$ in $K_0(\mathcal{P}(R), \oplus)$ if and only if $E \oplus I_n$ is stably similar to $F \oplus I_n$ for some $n \geq 0$.

5. Suppose $\mathcal{C}$ is the full subcategory of $\mathbb{Z}\text{-}\mathcal{M}\mathfrak{o}\mathfrak{d}$ consisting of all finite abelian groups.

   (i) Prove $\mathcal{I}(\mathcal{C})$ is a free abelian monoid based on the isomorphism classes $c(\mathbb{Z}/p^e)$ where p is prime and $e \geq 1$. *Hint:* Use the Structure Theorem for f.g. $\mathbb{Z}$-modules.
   (ii) For $A, B \in \mathcal{C}$, prove $[A] = [B]$ in $K_0(\mathcal{C}, \oplus)$ if and only if $A \cong B$.

6. Recall the definition of end $\mathcal{C}$ from (3.9), and the connection between end $\mathcal{F}(R)$ and sim(R) in (3.10). Suppose K is a field.

   (i) Prove $\mathcal{I}(\text{end } \mathcal{P}(K))$ is a free abelian monoid based on the isomorphism classes $c(K[x]/p(x)^e)$ where $p(x)$ is a monic irreducible in $K[x]$ and $e \geq 1$. *Hint:* Use the Structure Theorem for f.g. $K[x]$-modules.
   (ii) Then show sim(K) is a free abelian monoid based on the similarity classes $s(C)$ where C is the companion matrix of $K[x]/p(x)^e$, where $p(x)$ is monic irreducible and $e \geq 1$. (This gives the "primary" rational canonical form of matrices over K.) *Hint:* $K[x]/p(x)^e$ has K-basis $\overline{1}, \overline{x}, \overline{x}^2, \cdots, \overline{x}^{d-1}$, where d is the degree of $p(x)^e$.
   (iii) For A and B in $M_n(K)$, prove $[K^n, \cdot A] = [K^n, \cdot B]$ in $K_0(\text{end } \mathcal{P}(K), \oplus)$ if and only if A and B are similar.

7. Suppose $\mathbb{R}$ is the field of real numbers and $\mathbb{N}$ is the monoid of nonnegative integers under addition. Prove the map $f : \mathbb{N} \times \mathbb{N} \to \mathcal{I}(\text{bil}^*\mathcal{F}(\mathbb{R}))$, given by $f((a, b)) = a\langle 1 \rangle \perp b\langle -1 \rangle$, is a monoid isomorphism. *Hint:* Show f is surjective, since every positive real number is a square, and injective, since -1 is not a sum of squares in $\mathbb{R}$. Prove the Witt-Grothendieck group $\widehat{W}(\mathbb{R})$, of the field of real numbers $\mathbb{R}$, is a free abelian group based on $\{\langle 1 \rangle, \langle -1 \rangle\}$.

8. In the proof of (3.29), show $\tau_y(y) = -y$, and $\tau_y(v) = v$ if and only if $b(y, v) = 0$. (When $F = \mathbb{R}$ and $b = b_I$, this shows τ_y is reflection in the hyperplane orthogonal to y.)

3C. Grothendieck Groups

In 1957 (see Borel and Serre [58]), Alexander Grothendieck founded the subject of algebraic K-theory by constructing a group from the set of isomorphism classes of certain "sheaves over an algebraic variety," in order to generalize the Riemann-Roch Theorem in algebraic geometry. Within a year or so, papers of

Serre [58], Rim [59], and Swan [59] developed analogous Grothendieck groups for isomorphism classes of various types of modules over a ring. These developments were also inspired by the construction of the ideal class group by Dedekind [93] in 1893.

This Grothendieck group construction applies to any modest category with short exact sequences; it uses these sequences instead of a binary operation on the category. We present the construction here for modest subcategories of R-$\mathcal{M}\mathfrak{od}$ and compute several examples. In (3.35) we arrive at the algebraic K-group $K_0(R)$ and Grothendieck group $G_0(R)$ of a ring. For left artinian rings R, we use techniques from the preceding section to compute $K_0(R)$ and the technique of "devissage" to compute $G_0(R)$. Devissage (3.42) is a major tool in K-theory based on the connection between a module and its composition factors.

(3.30) Definition. Suppose R is a ring and $\mathcal{C}$ is a modest subcategory of R-$\mathcal{M}\mathfrak{od}$. The **Grothendieck group** of $\mathcal{C}$ is the quotient $\mathbb{Z}$-module

$$\boldsymbol{K_0}(\mathcal{C}) \;=\; F_{\mathbb{Z}}(\mathfrak{I}(\mathcal{C}))/\langle E \rangle \;,$$

where E is the set of elements

$$c(M) \;-\; c(N) \;-\; c(L)$$

for which there is a short exact sequence in $\mathcal{C}$:

$$0 \longrightarrow L \longrightarrow M \longrightarrow N \longrightarrow 0 \;.$$

(3.31) Notation. In $K_0(\mathcal{C})$, for each $M \in \mathrm{Obj}\, \mathcal{C}$, let

$$[\boldsymbol{M}] \;=\; c(M) \;+\; \langle E \rangle$$

denote the coset of $c(M)$. Then $K_0(\mathcal{C})$ is presented as a $\mathbb{Z}$-module by generators $[M]$, one for each isomorphism class $c(M)$ in $\mathcal{C}$, and defining relations

$$[M] \;=\; [L] \;+\; [N]$$

for each short exact sequence

$$0 \longrightarrow L \longrightarrow M \longrightarrow N \longrightarrow 0$$

in $\mathcal{C}$.

Of course if $0 \notin \mathrm{Obj}\,\mathcal{C}$, there are no short exact sequences in $\mathcal{C}$, and $K_0(\mathcal{C}) = F_{\mathbb{Z}}(\mathfrak{I}(\mathcal{C}))$. On the other hand, if $0 \in \mathrm{Obj}\,\mathcal{C}$, then there is a short exact sequence

$$0 \longrightarrow 0 \longrightarrow 0 \longrightarrow 0 \longrightarrow 0$$

in $\mathbb{C}$; so $[0] = 0$ in $K_0(\mathbb{C})$. For the most part we shall consider modest full subcategories of $R\text{-}\mathcal{Mod}$ that include 0 as an object.

(3.32) Example. Suppose $\mathbb{C}$ is the full subcategory of $R\text{-}\mathcal{Mod}$ whose objects are the simple R-modules and the trivial R-modules. If

$$0 \longrightarrow L \longrightarrow M \longrightarrow N \longrightarrow 0$$

is a short exact sequence in $\mathbb{C}$, then $L \cong 0$ or $N \cong 0$. So each relator in the definition of $K_0(\mathbb{C})$ is

$$c(M) - c(L) - c(N) = -c(0) .$$

If $S \subseteq \mathcal{I}(\mathbb{C})$ is the set of isomorphism classes of simple R-modules, then the function

$$S \rightarrow K_0(\mathbb{C})$$
$$c(X) \mapsto [X]$$

extends to a $\mathbb{Z}$-linear map

$$F_{\mathbb{Z}}(S) \rightarrow K_0(\mathbb{C}) ,$$

which is surjective since $\mathcal{I}(\mathbb{C}) = S \cup \{c(0)\}$ and injective since $c(0) \notin S$. So $K_0(\mathbb{C})$ *is a free $\mathbb{Z}$-module with one generator $[X]$ for each isomorphism class* $c(X)$ *of simple R-modules.* Note that, in this example, $[X] = [Y]$ in $K_0(\mathbb{C})$ if and only if $X \cong Y$.

In the preceding example, the number of terms required to express an element of $K_0(\mathbb{C})$ can be any positive integer up to the cardinality of $\mathcal{I}(\mathbb{C})$. A simpler expression of elements is available when Obj $\mathbb{C}$ is closed under direct sum:

(3.33) Proposition. *Suppose $\mathbb{C}$ is a modest full subcategory of $R\text{-}\mathcal{Mod}$ for which $0 \in$ Obj $\mathbb{C}$ and Obj $\mathbb{C}$ is closed under $\oplus$. Then there is a surjective group homomorphism*

$$K_0(\mathbb{C}, \oplus) \rightarrow K_0(\mathbb{C})$$
$$[M] \mapsto [M] .$$

So each element of $K_0(\mathbb{C})$ has an expression $[M] - [N]$ for some $M, N \in$ Obj $\mathbb{C}$.

Proof. For any $L, N \in$ Obj $\mathbb{C}$, there is an exact sequence

$$0 \longrightarrow L \longrightarrow L \oplus N \longrightarrow N \longrightarrow 0$$

in $\mathbb{C}$. So the relators E in the definition of $K_0(\mathbb{C})$ include the relators D in the definition (3.19) of $K_0(\mathbb{C}, \oplus)$. Now apply (3.22) (i). ∎

The next example carries this simplification further. If $M_1, M_2, \ldots$ are R-modules, the **countable cartesian product**

$$\prod_{i=1}^{\infty} M_i = M_1 \times M_2 \times \cdots$$

is an R-module under coordinatewise addition and scalar multiplication, and the **countable direct sum**

$$\bigoplus_{i=1}^{\infty} M_i = M_1 \oplus M_2 \oplus \cdots$$

is the R-submodule consisting of those sequences $(m_1, m_2, \ldots)$ with $m_i = 0$ for all but finitely many i.

(3.34) Example. (Swan [68]) Suppose $\mathcal{C}$ is a modest full subcategory of $R\text{-}\mathfrak{Mod}$ with 0 in Obj $\mathcal{C}$, and suppose Obj $\mathcal{C}$ is closed under direct sums and countable direct sums (resp. countable cartesian products). If $M \in$ Obj $\mathcal{C}$ and N denotes

$$\bigoplus_{i=1}^{\infty} M \quad \left(\text{resp.} \prod_{i=1}^{\infty} M\right),$$

then $M \oplus N \cong N$ via

$$(m, (m_1, m_2, \ldots)) \quad \leftrightarrow \quad (m, m_1, m_2, \ldots) .$$

So in $K_0(\mathcal{C}, \oplus)$, hence also in $K_0(\mathcal{C})$,

$$\begin{aligned}[M] &= [M] + [N] - [N] \\ &= [M \oplus N] - [N] = 0 .\end{aligned}$$

This works for every $M \in$ Obj $\mathcal{C}$; so $K_0(\mathcal{C}, \oplus) = K_0(\mathcal{C}) = 0$.

Example (3.34) includes the category of countably generated R-modules, and, for any infinite set T, the category of R-modules generated by sets of cardinality at most that of T.

Examples (3.32) and (3.34) represent opposite extremes. Between them are full subcategories $\mathcal{C}$ of $\mathcal{M}(R)$ for which Obj $\mathcal{C}$ includes 0 and is closed under direct sum, such as

$$\mathcal{F}(R) \subseteq \mathcal{P}(R) \subseteq \mathcal{M}(R) .$$

The Grothendieck groups of the last two categories are useful enough to have special names:

(3.35) Definition. For each ring R, the **Grothendieck group of the ring** R is

$$G_0(R) = K_0(\mathcal{M}(R)) ,$$

and the **zeroth algebraic K-group** of R is

$$K_0(R) \; = \; K_0(\mathcal{P}(R)) \, .$$

Since every short exact sequence in $\mathcal{P}(R)$ splits, $K_0(R) = K_0(\mathcal{P}(R), \oplus)$, because these two groups share the same generators and relations. However, there are rings R with $G_0(R) \neq K_0(\mathcal{M}(R), \oplus)$:

(3.36) Example. The exact sequence of f.g. $\mathbb{Z}$-modules

$$0 \longrightarrow \mathbb{Z} \longrightarrow \mathbb{Z} \longrightarrow \mathbb{Z}/2\mathbb{Z} \longrightarrow 0$$

$$n \longmapsto 2n$$

shows that $[\mathbb{Z}/2\mathbb{Z}] = 0 = [0]$ in $G_0(\mathbb{Z})$. If $G_0(\mathbb{Z})$ were $K_0(\mathcal{M}(\mathbb{Z}), \oplus)$, then (3.22) (ii) would imply the existence of a f.g. $\mathbb{Z}$-module A with

$$\mathbb{Z}/2\mathbb{Z} \oplus A \; \cong \; 0 \oplus A \, ,$$

which violates the structure theorem for finitely generated abelian groups.

If $0 \to L \to M \to N \to 0$ is a short exact sequence in $R\text{-}\mathcal{M}\mathit{od}$ and $L \cong R^a$, $N \cong R^b$ for nonnegative integers a and b, then N is projective, and so $M \cong L \oplus N \cong R^{a+b}$. For this reason, we expect any reasonable notion of the rank of a module to be additive over short exact sequences:

(3.37) Definition. A **generalized rank** on a subcategory $\mathcal{C}$ of $R\text{-}\mathcal{M}\mathit{od}$ is a metafunction

$$f : \mathrm{Obj} \; \mathcal{C} \; \to \; (A, *)$$

from the class $\mathrm{Obj} \; \mathcal{C}$ to an abelian group $(A, *)$, for which

$$f(M) \; = \; f(L) * f(N)$$

whenever there is a short exact sequence $0 \to L \to M \to N \to 0$ in $\mathcal{C}$.

By its construction, $K_0(\mathcal{C})$ is the codomain of the first generalized rank on the category $\mathcal{C}$:

(3.38) Proposition. *Suppose $\mathcal{C}$ is a modest full subcategory of $R\text{-}\mathcal{M}\mathit{od}$ with $0 \in \mathrm{Obj} \; \mathcal{C}$. Then $d : \mathrm{Obj} \; \mathcal{C} \to K_0(\mathcal{C})$, defined by $d(M) = [M]$, is a generalized rank on $\mathcal{C}$. If $r : \mathrm{Obj} \; \mathcal{C} \to A$ is any generalized rank on $\mathcal{C}$, there is one and only one group homomorphism $\overline{r} : K_0(\mathcal{C}) \to A$ with $r = \overline{r} \circ d$.*

Proof. By the defining relations for $K_0(\mathcal{C})$, d is a generalized rank. Suppose $r : \mathrm{Obj} \; \mathcal{C} \to A$ is a generalized rank. From the exact sequence of $0's$, $r(0) = 0_A$. So from exact sequences

$$0 \longrightarrow 0 \longrightarrow M \longrightarrow N \longrightarrow 0 \, ,$$

r is constant on isomorphism classes in $\mathcal{C}$ and defines a function $\mathfrak{I}(\mathcal{C}) \to A$, whose $\mathbb{Z}$-linear extension induces a group homomorphism:

$$\bar{r} : K_0(\mathcal{C}) \to A .$$
$$[M] \mapsto r(M)$$

Then $(\bar{r} \circ d)(M) = \bar{r}([M]) = r(M)$; so $\bar{r} \circ d = r$. If also $r' : K_0(\mathcal{C}) \to A$ is a group homomorphism and $r' \circ d = r$, then, for $M \in \mathrm{Obj}\ \mathcal{C}$,

$$r'([M]) = r(M) = \bar{r}([M]) .$$

Since the elements $[M]$ generate $K_0(\mathcal{C})$, $r' = \bar{r}$.   ∎

This universal property is the natural means of constructing homomorphisms on $K_0(\mathcal{C})$, and can lead to its computation:

(3.39) Examples.

(i) If R is a ring with IBN, the free rank (= size of a basis) is a generalized rank from $\mathcal{F}(R)$ to $(\mathbb{Z}, +)$. By (3.38) there is an induced group homomorphism

$$f : K_0(\mathcal{F}(R)) \to (\mathbb{Z}, +)$$

with $f([M]) = $ free rank (M). Since free rank $(R) = 1$, f is surjective. If $[M] - [N] \in \ker(f)$, then M and N have the same free rank $n \geq 0$. So $M \cong R^n \cong N$ and $[M] - [N] = 0$. Thus f is a group isomorphism.

(ii) Suppose $\mathcal{C}$ is the full subcategory of $\mathbb{Z}$-Mod consisting of the finite abelian groups. If $A \in \mathrm{Obj}\ \mathcal{C}$, let $|A|$ denote the order of A. If

$$0 \longrightarrow A \xrightarrow{f} B \xrightarrow{g} C \longrightarrow 0$$

is a short exact sequence in $\mathcal{C}$, then $A \cong f(A)$ and $C \cong B/f(A)$; so $|B| = |A||C|$. Thus the group order is a generalized rank from $\mathcal{C}$ to $(\mathbb{Q}^+, \cdot)$, the group of positive rationals under multiplication. There are finite abelian groups of every (positive integer) order; so the induced homomorphism

$$f : K_0(\mathcal{C}) \to (\mathbb{Q}^+, \cdot)$$
$$[A] - [B] \mapsto |A|/|B|$$

is surjective.

Suppose $[A] - [B] \in \ker(f)$. Then $|A| = |B|$. If a prime p divides $|A|$, then, by Cauchy's Theorem, A has an element of order p. So there is a short exact sequence

$$0 \longrightarrow \mathbb{Z}/p\mathbb{Z} \longrightarrow A \longrightarrow A' \longrightarrow 0 ,$$

where $|A| = p|A'|$. Then

$$[A] = [\mathbb{Z}/p\mathbb{Z}] + [A']$$

in $K_0(\mathcal{C})$. So by induction on the number of prime factors of the order, $|A| = |B|$ implies $[A] = [B]$ in $K_0(\mathcal{C})$. Then $[A] - [B] = 0$, and f is an isomorphism of groups.

This last example can be approached another way: Every finite abelian group A has a composition series

$$A = A_0 \supset A_1 \supset \cdots \supset A_n = \{0_A\} \, ,$$

where each $A_i/A_{i+1} \cong \mathbb{Z}/p_i\mathbb{Z}$ for some prime p_i. The composition factors A_i/A_{i+1} are thought of as the building blocks from which the group A is constructed, just as $|A|$ is the product of the prime orders $|A_i/A_{i+1}|$. This analogy is especially apt in $K_0(\mathcal{C})$, since

$$
\begin{aligned}
[A] &= [A] - [0] \\
&= \sum_{i=0}^{n-1} ([A_i] - [A_{i+1}]) \\
&= \sum_{i=0}^{n-1} [A_i/A_{i+1}] \, ,
\end{aligned}
$$

by the exact sequences:

$$0 \longrightarrow A_{i+1} \longrightarrow A_i \longrightarrow A_i/A_{i+1} \longrightarrow 0 \, .$$

The reduction to composition factors leads, below, to an alternate computation of $K_0(\mathcal{C})$; the technique is quite general and is called **devissage** (from the French, meaning "to unscrew"):

(3.40) Definition. If $\mathcal{D}$ is a subcategory of R-$\mathfrak{Mod}$ and M is an R-module, a $\mathcal{D}$-**filtration** of M is any descending chain of submodules

$$M = M_0 \supseteq M_1 \supseteq \cdots \supseteq M_n = \{0_M\}$$

of finite length n, with $M_i/M_{i+1} \in \text{Obj } \mathcal{D}$ for $0 \le i \le n-1$.

(3.41) Lemma. *If $\mathcal{C}$ is a modest full subcategory of R-$\mathfrak{Mod}$ and M is an R-module with a $\mathcal{C}$-filtration*

$$M = M_0 \supseteq M_1 \supseteq \cdots \supseteq M_n = \{0_M\}$$

in $\mathcal{C}$, *then*

$$[M] = \sum_{i=0}^{n-1} [M_i/M_{i+1}]$$

in $K_0(\mathcal{C})$.

Proof.

$$
\begin{aligned}
[M] &= [M] - [0] \\
&= \sum_{i=0}^{n-1} ([M_i] - [M_{i+1}]) \\
&= \sum_{i=0}^{n-1} [M_i/M_{i+1}]
\end{aligned}
$$

by the exact sequences

$$0 \longrightarrow M_{i+1} \longrightarrow M_i \longrightarrow M_i/M_{i+1} \longrightarrow 0 . \qquad \blacksquare$$

(3.42) Theorem (Devissage). *Suppose R is a ring and $\mathcal{D} \subset \mathcal{C}$ are modest full subcategories of R-$\mathcal{M}o\eth$. Assume* $\mathrm{Obj}\ \mathcal{D}$ *is closed under submodules and homomorphic images in R-$\mathcal{M}o\eth$. If each object of $\mathcal{C}$ has a $\mathcal{D}$-filtration consisting of objects of $\mathcal{C}$, there is a group isomorphism*

$$\phi : K_0(\mathcal{D}) \rightarrow K_0(\mathcal{C})$$

with $\phi([X]) = [X]$ for each $X \in \mathrm{Obj}\ \mathcal{D}$.

Proof. Suppose $M \in \mathrm{Obj}\ \mathcal{C}$ has $\mathcal{D}$-filtration:

(3.43) $$M = M_0 \supseteq \cdots \supseteq M_n = \{0_M\} .$$

We claim the element

$$\sum_{i=0}^{n-1} [M_0/M_{i+1}] \in K_0(\mathcal{D})$$

depends only on M, and not on the choice of $\mathcal{D}$-filtration:

If N is an R-module and $M_i \supseteq N \supseteq M_{i+1}$, there is an exact sequence

$$0 \longrightarrow \frac{N}{M_i} \longrightarrow \frac{M_{i+1}}{M_i} \longrightarrow \frac{M_{i+1}}{N} \longrightarrow 0 ,$$

which lies in $\mathcal{D}$ by the closure properties assumed for Obj $\mathcal{D}$. So the chain

$$M_0 \supseteq \cdots \supseteq M_i \supseteq N \supseteq M_{i+1} \supseteq \cdots \supseteq M_n$$

is also a $\mathcal{D}$-filtration, yielding the same element of $K_0(\mathcal{D})$. The claim now follows from Schreier's Theorem (see (B.16) in Appendix B), which implies that any two finite length chains $M \supseteq \cdots \supseteq \{0_M\}$ have equivalent refinements.

Thus there is a metafunction $\psi :$ Obj $\mathcal{C} \to K_0(\mathcal{D})$, taking M with $\mathcal{D}$-filtration (3.43) to

$$\psi(M) = \sum_{i=0}^{n-1} [M_i/M_{i+1}]$$

in $K_0(\mathcal{D})$. Suppose

$$0 \longrightarrow L \longrightarrow M \longrightarrow N \longrightarrow 0$$

is a short exact sequence in $\mathcal{C}$. By the Correspondence Theorem for submodules (see (B.7) in Appendix B), a $\mathcal{D}$-filtration of M can be constructed as the image of a $\mathcal{D}$-filtration of L, continued by the preimage of a $\mathcal{D}$-filtration of N. Thus $\psi(M) = \psi(L) + \psi(N)$, and ψ is a generalized rank. So there is an induced group homomorphism $\psi : K_0(\mathcal{C}) \to K_0(\mathcal{D})$ with $\psi([M]) = \psi(M)$ for each $M \in$ Obj $\mathcal{C}$.

Since every short exact sequence in $\mathcal{D}$ is also a short exact sequence in $\mathcal{C}$, there is a generalized rank $\phi :$ Obj $\mathcal{D} \to K_0(\mathcal{C})$ with $\phi(X) = [X]$ for each $X \in$ Obj $\mathcal{D}$. So there is an induced group homomorphism $\phi : K_0(\mathcal{D}) \to K_0(\mathcal{C})$ with $\phi([X]) = [X]$.

For each $M \in$ Obj $\mathcal{C}$ with $\mathcal{D}$-filtration (3.43) in $\mathcal{C}$,

$$[M] = \sum_{i=0}^{n-1} [M_i/M_{i+1}]$$

in $K_0(\mathcal{C})$; so $\phi(\psi([M])) = [M]$. Each $X \in$ Obj $\mathcal{D}$ has $\mathcal{D}$-filtration $X \supseteq \{0_X\}$ in $\mathcal{C}$; so $\psi(\phi([X])) = [X]$. Thus ψ and ϕ are mutually inverse isomorphisms. ∎

If we apply devissage to the example (3.39) (ii) of the category $\mathcal{C}$ of finite abelian groups, we obtain an isomorphism

$$\phi : K_0(\mathcal{D}) \to K_0(\mathcal{C}) ,$$
$$[X] \mapsto [X]$$

where $\mathcal{D}$ is the full subcategory of $\mathbb{Z}$-$\mathcal{M}\mathfrak{od}$ consisting of the cyclic groups of prime order and the trivial groups. As in Example (3.32), $K_0(\mathcal{D})$ is free abelian, with a basis consisting of one element $[\mathbb{Z}/p\mathbb{Z}]$ for each prime p. So $K_0(\mathcal{C})$ has the same description. Applying the isomorphism

$$K_0(\mathcal{C}) \cong (\mathbb{Q}^+, \cdot)$$

induced by the group order, we see that devissage is a generalization of the theorem that $(\mathbb{Q}^+, \cdot)$ is a free abelian group with basis the set of primes — which is equivalent to unique prime factorization in $\mathbb{Z}$, the Fundamental Theorem of Arithmetic!

(3.44) Corollary. *If R is a left artinian ring, then:*

(i)   *For simple R-modules X and Y, $[X] = [Y]$ in $G_0(R)$ if and only if $X \cong Y$.*

(ii) *Every simple R-module is isomorphic to a composition factor of the R-module R. If $X_1, \ldots, X_n$ represent the different isomorphism classes of composition factors of R, then $G_0(R)$ is a free abelian group with basis $[X_1], \ldots, [X_n]$.*

Proof. Since R is left artinian, every f.g. R-module M has a composition series (see (B.20) and (B.21) in Appendix B). By devissage, if $\mathcal{D}$ is the full subcategory in R-$\mathfrak{Mod}$ whose objects are simple or trivial, there is an isomorphism $K_0(\mathcal{D}) \cong G_0(R)$ taking $[X]$ to $[X]$. Now apply the corresponding facts for $K_0(\mathcal{D})$ proved in Example (3.32) to show that, if S is a set of simple R-modules, one from each isomorphism class, then $G_0(R)$ is free abelian with basis $\{[X] : X \in S\}$. Suppose $X \in S$ and $0 \neq x \in X$. Since X is simple, $r \mapsto rx$ is a map f in a short exact sequence

$$0 \longrightarrow K \longrightarrow R \overset{f}{\longrightarrow} X \longrightarrow 0$$

in $\mathcal{M}(R)$. So in $G_0(R)$,

$$[X] + [K] = [R] = \sum_{i=1}^{s} [M_i/M_{i+1}] \, ,$$

where $0 = M_S \subseteq \cdots \subseteq M_1 = R$ is a composition series for R. Each composition factor M_i/M_{i+1} is simple, and so it is isomorphic to some $Y \in S$; so $[M_i/M_{i+1}] = [Y]$. By uniqueness of the expression of $[R]$ in terms of the basis of $G_0(R)$, $[X] = [M_i/M_{i+1}]$ for some i; so $X \cong M_i/M_{i+1}$.

Therefore, if $X_1, \ldots, X_n$ represent the isomorphism classes of composition factors of R, then $\{[X] : X \in S\}$ has the n elements $[X_1], \ldots, [X_n]$.   ∎

For any ring R, the homomorphism $K_0(R) \to G_0(R)$, taking $[P]$ to $[P]$ for each $P \in \mathcal{P}(R)$, is known as the **Cartan map**. From (3.27) and (3.44), if R is left artinian, then $K_0(R)$ is free abelian based on $[P_1], \ldots, [P_m]$, where $P_1, \ldots, P_m$ represent the isomorphism classes of indecomposables in $\mathcal{P}(R)$, and $G_0(R)$ is free abelian based on $[X_1], \ldots, [X_n]$, where $X_1, \ldots, X_n$ represent the isomorphism classes of simple R-modules. By Lemma (3.41),

$$[P_i] = \sum_{j=1}^{n} a_{ij}[X_j]$$

in $G_0(R)$, where a_{ij} is the number of composition factors of P_i isomorphic to X_j. So the Cartan map is represented over these bases as right multiplication by the **Cartan matrix** $(a_{ij}) \in R^{m \times n}$. Allowing for reorderings of the bases, the Cartan matrix is determined up to a permutation of its rows and a permutation of its columns.

3C. Exercises

1. Suppose R is a nontrivial ring and A is the ring of column-finite $\mathbb{P} \times \mathbb{P}$ matrices over R, defined in (1.37). Prove $K_0(A) = 0$. *Hint:* Use an argument similar to the proof of the countability of $\mathbb{Q}$ to construct an A-linear isomorphism $A^\infty \cong A$, where $A^\infty = A \times A \times \cdots$. Then, if $P \in \mathcal{P}(A)$, there exists $Q \in A\text{-}\mathcal{M}o\eth$ with $P^\infty \oplus Q^\infty \cong R^\infty \cong R$. So $P^\infty \in \mathcal{P}(A)$.

2. Suppose F is a field, and R is a ring with F a subring of its center. Suppose R is finite-dimensional as an F-vector space (under multiplication in R). If $M \in \mathcal{M}(R)$, then M is also a finite-dimensional F-vector space. Prove the dimension over F defines a generalized rank

$$\dim_F : \text{Obj } \mathcal{M}(R) \; \to \; (\mathbb{Z}, +) \, .$$

Thus a generalized rank from $\mathcal{M}(R)$ to $\mathbb{Z}$ need not take R to 1.

3. Suppose $\mathcal{C}$ is the full subcategory of $\mathbb{Z}\text{-}\mathcal{M}o\eth$ consisting of the finite cyclic groups. Even though Obj $\mathcal{C}$ is not closed under $\oplus$, prove every element of $K_0(\mathcal{C})$ has the form $[M] - [N]$ for some $M, N \in \text{Obj } \mathcal{C}$. Then prove $K_0(\mathcal{C}) \cong (\mathbb{Q}^+, \cdot)$.

4. Suppose p is a prime number and $\mathcal{C}$ is the full subcategory of $\mathbb{Z}\text{-}\mathcal{M}o\eth$ consisting of finite abelian p-groups. Prove $K_0(\mathcal{C}) \cong (\mathbb{Z}, +)$.

5. Suppose R is a commutative principal ideal domain. Prove $K_0(R) \cong (\mathbb{Z}, +)$ via an isomorphism taking $[R]$ to 1. *Hint:* Apply (3.39) (i).

6. If D is a division ring, prove $G_0(D) = K_0(D) \cong (\mathbb{Z}, +)$ via an isomorphism taking $[D]$ to 1.

The canonical map $K_0(\mathcal{C}, \oplus) \to K_0(\mathcal{C})$ generally loses information when there are short exact sequences in $\mathcal{C}$ that do not split. For $\mathcal{C}$ the category of finite abelian groups, $[A] = [B]$ in $K_0(\mathcal{C}, \oplus)$ if and only if $A \cong B$, by §3B Exercise 5. But $[A] = [B]$ in $K_0(\mathcal{C})$ if and only if A and B have the same order, by (3.39) (ii). Similarly, when K is a field and $\mathcal{C} = \text{end } \mathcal{P}(K) = \text{end } \mathcal{F}(K)$, two matrices $A, B \in M_n(K)$ determine the same element $[K^n, \cdot A] = [K^n, \cdot B]$ of $K_0(\mathcal{C}, \oplus)$ if and only if they are similar, by §3B Exercise 6 (iii). By the next two exercises, $[K^n, \cdot A] = [K^n, \cdot B]$ in $K_0(\mathcal{C})$ if and only if A and B have the same characteristic polynomial:

7. (Based on Kelley and Spanier [68].) An R-module M is **torsion** if, for each $m \in M$, there is a nonzero $r \in R$ with $rm = 0$. Suppose R is a

commutative principal ideal domain, with field of fractions F. Let $\mathcal{T}$ denote the full subcategory of R-$\mathcal{M}\mathit{o}\mathit{d}$ consisting of the f.g. torsion R-modules.

(i) Prove there is a function $\delta : \mathcal{I}(\mathcal{T}) \to F^*/R^*$ taking M to $\det(A)R^*$ whenever M is isomorphic to the cokernel of right multiplication $(\cdot A)$: $R^n \to R^n$ by a matrix $A \in M_n(R)$ with R-linearly independent rows. *Hint:* To show δ independent of the choice of A, use the Smith-Normal form of A and the Structure Theorem for $\mathcal{T}$.

(ii) Prove δ is a generalized rank on $\mathcal{T}$. *Hint:* For $Q \subseteq P \cong R^n$ in R-$\mathcal{M}\mathit{o}\mathit{d}$ with $P/Q \cong M$, show there is an injective endomorphism $\sigma : P \to P$ with image Q; and for any such σ, $\delta(M) = \text{determinant}(\sigma)R^*$. Then prove that if $N \subseteq M$ in $\mathcal{T}$, $\delta(M/N)\delta(N) = \delta(M)$.

(iii) Show the induced homomorphism $\overline{\delta} : K_0(\mathcal{T}) \to F^*/R^*$ is an isomorphism. *Hint:* Use relations arising from short exact sequences

$$0 \longrightarrow \frac{R}{aR} \longrightarrow \frac{R}{abR} \longrightarrow \frac{R}{bR} \longrightarrow 0$$

whenever a and b are nonzero elements of R, and use the fact that $\delta(R/aR) = aR^*$.

8. (Also from Kelley and Spanier [68].) If K is a field, show there is an isomorphism

$$K_0(\text{end } \mathcal{P}(K)) \cong K(x)^*/K^* \, ,$$

where $K(x)$ is the field of fractions of the polynomial ring $K[x]$. In particular, show that, for $A, B \in M_n(K)$, $[K^n, \cdot A] = [K^n, \cdot B]$ in $K_0(\text{end } \mathcal{P}(K))$ if and only if A and B have the same characteristic polynomial. (Note that Almkvist [73] has proved the latter statement when K is replaced by an arbitrary commutative ring!) *Hint:* Show end $\mathcal{P}(K)$ is the category of f.g. torsion $K[x]$-modules, and apply Exercise 7. For the second statement use the "characteristic exact sequence"

$$0 \longrightarrow K[x]^n \overset{f}{\longrightarrow} K[x]^n \overset{g}{\longrightarrow} (K^n, \cdot A) \longrightarrow 0 \, ,$$

where f is right multiplication by $xI_n - A$, and $g(e_i) = e_i$ for $1 \leq i \leq n$. For exactness of this sequence, see Jacobson [85, p. 196].

9. If $\mathcal{C}$ is a modest full subcategory of R-$\mathcal{M}\mathit{o}\mathit{d}$, including 0 and closed under $\oplus$, let **aut** $\mathcal{C}$ denote the full subcategory of end $\mathcal{C}$ consisting of those (M, f) with $f : M \to M$ an R-linear automorphism. When K is a field, prove the inclusion aut $\mathcal{P}(K)$ into end $\mathcal{P}(K)$ induces an injective homomorphism

$$K_0(\text{aut } \mathcal{P}(K)) \to K_0(\text{end } \mathcal{P}(K)) \, ,$$

and determine the image of $K_0(\text{aut } \mathcal{P}(K))$ in $K(x)^*/K^*$ under $\overline{\delta}$ from Exercises 7 and 8. *Hint:* Show δ defines a generalized rank on aut $\mathcal{P}(K)$, inducing an injective homomorphism on $K_0(\text{aut } \mathcal{P}(K))$, factoring through $K_0(\text{end } \mathcal{P}(K))$.

10. If R is a left artinian ring, show each indecomposable $P \in \mathcal{P}(R)$ is isomorphic to Re for some idempotent $e \in R$, which is not expressible as a sum of two nonzero idempotents. *Hint:* By (3.27), $R \cong P_1 \oplus \cdots \oplus P_n$ for indecomposables $P_i \in \mathcal{P}(R)$, and $P \cong P_i$ for some i. Show $R = Q_1 \overset{\bullet}{\oplus} \cdots \overset{\bullet}{\oplus} Q_n$, where the isomorphism takes Q_i to $0 \oplus \cdots \oplus P_i \oplus \cdots \oplus 0$, for each i. Then $1 = e_1 + \cdots + e_n$, where $e_i \in Q_i$ for each i. Show each e_i is idempotent and each $Q_i = Re_i$. Then use the indecomposability of Q_i. The indecomposables Re are known as the **principal indecomposables** in $\mathcal{P}(R)$, since they are principal left ideals of R.

11. If n is an integer greater than 1, find a Cartan matrix of the ring $\mathbb{Z}/n\mathbb{Z}$. *Hint:* Consider the prime factorization of n.

12. If R is a commutative ring and G is an abelian group, the **group ring** RG is the free R-module based on G, made into a commutative ring by a multiplication extending the multiplications in R and G as well as the scalar multiplication. The identity element of G serves as the 1 in RG. Suppose $\mathbb{F}_q$ denotes a finite field with q elements, and C_p denotes a cyclic group $\langle \sigma \rangle$ of order p generated by σ.

(a) If R is the group ring $\mathbb{F}_2 C_2 = \{0, 1, \sigma, 1 + \sigma\}$, find a Cartan matrix of R. *Hint:* Determine a Krull-Schmidt decomposition of R and a composition series of R. Show multiplication by $1 + \sigma$ induces an isomorphism between the composition factors.

(b) Do the same for $\mathbb{F}_2 C_4$ and $\mathbb{F}_3 C_3$. *Hint:* To determine the units and nonunits, take the fourth power in $\mathbb{F}_2 C_4$ and the cube in $\mathbb{F}_3 C_3$.

(c) What can you prove about FG where F is a perfect field of prime characteristic p, and G is an abelian group of order p^r, $r \geq 1$? *Hint:* If M is a simple R-module and $0 \neq m \in M$, then $R \to M$, $r \mapsto rm$, is surjective, with kernel J a maximal left ideal of R; so $R/J \cong M$.

(d) Find a Cartan matrix for $\mathbb{F}_2 C_3$.

3D. Resolutions

When we describe a module M in terms of generators, relations among those generators, relations among those relations, etc., we obtain a "free resolution" of M — a long exact sequence ending with $M \to 0$. This is the beginning of a branch of algebra called "homological algebra." In a Grothendieck group of modules, M is connected to an alternating sum of the terms in its resolution. This connection leads to the Resolution Theorem (3.52) for the comparison of Grothendieck groups.

First, consider Grothendieck groups of modules over a commutative principal ideal domain D. Since D is commutative, it has IBN, and we can speak of the free rank of f.g. free D-modules. On the way to proving the Structure Theorem for f.g. D-modules, one encounters the following lemma:

(3.45) Lemma. *If D is a commutative principal ideal domain and M is a free D-module of free rank m, then every submodule N of M is free, of free rank $n \leq m$.*

Proof. Since free rank is preserved by isomorphisms, it is enough to prove the assertion when $M = D^m$. If $m = 1$, N is a principal ideal Dd; and since D is an integral domain, $d = 0$ or d is linearly independent. So N is free of free rank 0 or 1.

Suppose $m > 1$. An exact sequence

$$0 \longrightarrow D^{m-1} \overset{i}{\longrightarrow} D^m \overset{\pi}{\longrightarrow} D \longrightarrow 0$$

restricts to an exact sequence

$$0 \longrightarrow i^{-1}(N) \longrightarrow N \longrightarrow \pi(N) \longrightarrow 0 \ .$$

By the case $m = 1, \pi(N)$ is free, and hence projective, and the latter sequence splits. So

$$N \cong i^{-1}(N) \oplus \pi(N) \ ,$$

and the lemma follows by induction on m. ∎

One consequence of this lemma is that every f.g. projective D-module is free. So $\mathcal{P}(D) = \mathcal{F}(D)$. As in Example (3.39) (i), the free rank induces an isomorphism:

$$K_0(D) = K_0(\mathcal{F}(D)) \cong \mathbb{Z} \ .$$

A second consequence is that, in a f.g. free D-module F, no spanning set can have fewer elements than a basis: For, suppose

$$F = \langle g_1, \ldots, g_p \rangle \cong D^n \ .$$

There is a short exact sequence

$$0 \longrightarrow \ker(\mathrm{f}) \overset{\subseteq}{\longrightarrow} D^p \overset{f}{\longrightarrow} F \longrightarrow 0$$
$$e_i \longmapsto g_i$$

that splits since F is projective. A splitting map $g : F \to D^p$ embeds F as a submodule $g(F)$ of D^p. So $n \leq p$.

Now consider $G_0(D)$. Suppose M is a f.g. D-module. So there is a finite set $S = \{s_1, \ldots, s_m\}$ and a surjective D-linear map $f : F_D(S) \to M$. By the lemma, $K = \ker(f)$ is free of free rank $n \leq m$. If Δ is a basis of K, M has presentation $(S : \Delta)$ with m generators and n relators. Call such a presentation **efficiently related** for f, since no smaller set of relators could generate K, by the preceding paragraph.

We claim that the number n is independent of the map f — that is, of the choice of generators $f(s_1), \ldots, f(s_m)$ of M. More generally, the difference $m - n$ depends only on M, and we call it the **presentation rank** of M. This claim is a consequence of the exact sequence

$$0 \longrightarrow D^n \longrightarrow D^m \longrightarrow M \longrightarrow 0$$

and of Schanuel's Lemma (3.46), below.

Further, the presentation rank is additive over short exact sequences; this is a consequence of the Horseshoe Lemma, (3.51), below. So it induces an inverse to the group homomorphism $\mathbb{Z} \to G_0(D), 1 \mapsto [D]$, and

$$G_0(D) \cong \mathbb{Z} \cong K_0(D) \,.$$
$$[D] \;\leftrightarrow\; 1 \;\leftrightarrow\; [D]$$

This approach to $G_0(D)$ can be applied to $G_0(R)$ for a much wider class of rings R:

(3.46) Schanuel's Lemma. *Suppose R is a ring and*

$$0 \longrightarrow K \longrightarrow P \longrightarrow M \longrightarrow 0$$
$$0 \longrightarrow L \longrightarrow Q \longrightarrow N \longrightarrow 0$$

are exact sequences in R-$\mathcal{M}\mathfrak{o}\mathfrak{d}$, P and Q are projective, and $M \cong N$. Then

$$K \oplus Q \cong L \oplus P \,.$$

Proof. Since P is projective we claim the given sequences are the rows of a commutative diagram

$$
\begin{array}{ccccccccc}
0 & \longrightarrow & K & \overset{f}{\longrightarrow} & P & \overset{g}{\longrightarrow} & M & \longrightarrow & 0 \\
 & & \downarrow{\scriptstyle c} & & \downarrow{\scriptstyle b} & & \downarrow{\scriptstyle a} & & \\
0 & \longrightarrow & L & \underset{h}{\longrightarrow} & Q & \underset{i}{\longrightarrow} & N & \longrightarrow & 0
\end{array}
$$

in R-$\mathcal{M}\mathfrak{o}\mathfrak{d}$, where a is the isomorphism assumed to exist from M to N: For, the existence of b making the right square commute is a consequence of the projectivity of P, the surjectivity of i, and (2.17). Then b restricts to $\beta : f(K) \to h(L)$ and h restricts to an isomorphism $\eta : L \to h(L)$. So we can define c, to make the left square commute, by $c = \eta^{-1} \circ \beta \circ f$.

Since a is an isomorphism, the sequence

$$0 \longrightarrow K \xrightarrow{\sigma} L \oplus P \xrightarrow{\tau} Q \longrightarrow 0 \, ,$$

with $\sigma(k) = (f(k),\ c(k))$, and $\tau(\ell, p) = b(p) - h(\ell)$, is exact: Injectivity of σ follows from that of f, and $\tau \circ \sigma = 0$ because the left square commutes. That τ is surjective and $\ker(\tau) \subseteq im(\sigma)$ are proved by chasing elements around the diagram.

Since Q is projective, the latter exact sequence splits, and

$$L \oplus P \cong K \oplus Q \, . \qquad\qquad \blacksquare$$

(3.47) Corollary. *Suppose R is a ring and*

$$0 \to K \to P_n \to P_{n-1} \to \cdots \to P_0 \to M \to 0$$
$$0 \to L \to Q_n \to Q_{n-1} \to \cdots \to Q_0 \to N \to 0$$

are exact sequences in R-$\mathfrak{Mod}$, the modules P_i and Q_i are projective, and $M \cong N$. Then

$$K \oplus Q_n \oplus P_{n-1} \oplus Q_{n-2} \oplus \cdots \oplus \begin{cases} P_0 & \text{if } 2 \nmid n \\ Q_0 & \text{if } 2 \mid n \end{cases}$$

$$\cong \ L \oplus P_n \oplus Q_{n-1} \oplus P_{n-2} \oplus \cdots \oplus \begin{cases} Q_0 & \text{if } 2 \nmid n \\ P_0 & \text{if } 2 \mid n \end{cases}$$

and K is projective if and only if L is projective.

Proof. For $n = 0$, this is Schanuel's Lemma. Suppose $n > 0$ and the given maps $P_n \to P_{n-1}$ and $Q_n \to Q_{n-1}$ have images K' and L', respectively. Define

$$C \ = \ P_{n-1} \oplus Q_{n-2} \oplus \cdots \oplus \begin{cases} P_0 & \text{if } 2 \nmid n \\ Q_0 & \text{if } 2 \mid n \, , \end{cases}$$

$$D \ = \ Q_{n-1} \oplus P_{n-2} \oplus \cdots \oplus \begin{cases} Q_0 & \text{if } 2 \nmid n \\ P_0 & \text{if } 2 \mid n \, . \end{cases}$$

Then there are exact sequences:

$$0 \to K' \to P_{n-1} \to \cdots \to P_0 \to M \to 0 \, ,$$
$$0 \to L' \to Q_{n-1} \to \cdots \to Q_0 \to N \to 0 \, ,$$
$$0 \to K \to P_n \oplus D \to K' \oplus D \to 0 \, ,$$
$$0 \to L \to Q_n \oplus C \to L' \oplus C \to 0 \, .$$

If $K' \oplus D \cong L' \oplus C$, then, by Schanuel's Lemma, $K \oplus Q_n \oplus C \cong L \oplus P_n \oplus D$. So the corollary follows by induction on n. $\blacksquare$

(3.48) Definitions. An exact sequence

$$\cdots \to P_n \to P_{n-1} \to \cdots \to P_0 \to M \to 0$$

in $R\text{-}\mathfrak{Mod}$ is called a **resolution** of M. This resolution is **projective** if each P_i is, and **finite** if, for some integer $n \geq 0$, $P_i = 0$ for all $i \geq n$. A ring R is **left regular** if R is left noetherian and every f.g. R-module has a finite projective resolution.

As in Appendix B, (B.9), (B.10), if R is left noetherian, every submodule of a finitely generated R-module is finitely generated.

(3.49) Corollary. *If R is a left regular ring, every f.g. R-module has a finite resolution by f.g. projective R-modules.*

Proof. If M is a f.g. R-module spanned by a finite set S, inclusion $S \to M$ extends to an R-linear map $F_R(S) \to M$, whose kernel K_0 is spanned by a finite set S_0. Inclusion $S_0 \to F_R(S)$ extends to an R-linear map $F_R(S_0) \to F_R(S)$, whose kernel K_1 is spanned by a finite set S_1. Continuing, we produce a resolution of M by f.g. free R-modules:

$$\cdots \to F_m \to F_{m-1} \to \cdots \to F_0 \to M \to 0 \; ,$$

where $F_i = F_R(S_i)$. By Corollary (3.47), the existence of a projective resolution

$$0 \to P_n \to P_{n-1} \to \cdots \to P_0 \to M \to 0$$

implies

$$0 \to K_n \to F_{n-1} \to \cdots \to F_0 \to M \to 0$$

is a resolution of M by f.g. projective R-modules. $\blacksquare$

(3.50) Proposition. *Suppose R is a ring, and an R-module M has two finite resolutions by f.g. projective R-modules:*

$$0 \to P_n \to \cdots \to P_0 \to M \to 0 \; ,$$
$$0 \to Q_m \to \cdots \to Q_0 \to M \to 0 \; .$$

Then, in $K_0(R)$,

$$\sum_{i=0}^{n}(-1)^i[P_i] \;\; = \;\; \sum_{i=0}^{m}(-1)^i[Q_i] \; .$$

Proof.   By using some 0's as f.g. projectives, we can assume $m = n$. Then by Corollary (3.47),

$$\left(\bigoplus_{i \text{ even}} P_i\right) \oplus \left(\bigoplus_{i \text{ odd}} Q_i\right) \cong \left(\bigoplus_{i \text{ odd}} P_i\right) \oplus \left(\bigoplus_{i \text{ even}} Q_i\right).$$

The desired equation in $K_0(R)$ follows by bringing $[P_i]$s to one side and $[Q_i]$s to the other.   ∎

Over any ring R, let $\mathcal{P}_{<\infty}(R)$ denote the full subcategory of $\mathcal{M}(R)$ whose objects have finite resolutions by f.g. projective R-modules. So

$$\mathcal{P}(R) \subseteq \mathcal{P}_{<\infty}(R) \subseteq \mathcal{M}(R) ,$$

the last two being equal when R is left regular. If $\mathcal{C}$ is a full subcategory of $\mathcal{P}_{<\infty}(R)$, (3.50) implies there is a metafunction

$$\chi : \text{Obj } \mathcal{C} \;\to\; K_0(R)$$

defined by

$$\chi(M) \;=\; \sum_{i=0}^{n} (-1)^i \, [P_i]$$

whenever

$$0 \to P_n \to \cdots \to P_0 \to M \to 0$$

is a resolution by f.g. projective R-modules P_i. Often χ is called an **Euler characteristic**. It generalizes the presentation rank for modules over a principal ideal domain.

(3.51) Horseshoe Lemma. *Suppose R is a ring and*

$$0 \longrightarrow L \xrightarrow{\;f\;} M \xrightarrow{\;g\;} N \longrightarrow 0$$

is an exact sequence of R-modules. If there are projective resolutions

$$0 \to P'_n \to \cdots \to P'_0 \xrightarrow{\;\alpha\;} L \to 0$$

$$0 \to P''_n \to \cdots \to P''_0 \xrightarrow{\;\beta\;} N \to 0$$

in R-$\mathcal{M}\mathfrak{o}\mathfrak{d}$, there is also a projective resolution

$$0 \to P'_n \oplus P''_n \to \cdots \to P'_0 \oplus P''_0 \to M \to 0 .$$

So if $L, N \in \mathcal{P}_{<\infty}(R)$, then $M \in \mathcal{P}_{<\infty}(R)$ and $\chi(M) = \chi(L) + \chi(N)$.

Proof.   There are R-linear maps h, k making the diagram

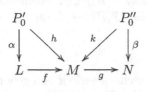

commute: $h = f \circ \alpha$, and k exists because P_0'' is projective. Define

$$\gamma : P_0' \oplus P_0'' \to M$$

by $\gamma(x, y) = h(x) + k(y)$. Then there is a commutative diagram

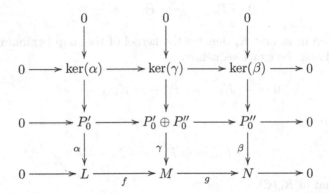

with all three columns and the last two rows exact. Since α and β are surjective, the Snake Lemma (see Exercise 4 in §2A) shows γ is surjective and the row of kernels is exact. Replacing the bottom row by the row of kernels, we can repeat this process to get the desired projective resolution of M. Then, if P_i', P_i'' are finitely generated,

$$\sum_{i=0}^{n}(-1)^n \left[P_i' \bigoplus P_i'' \right] = \sum_{i=0}^{n}(-1)^n [P_i'] + \sum_{i=0}^{n}(-1)^n [P_i'']$$

in $K_0(R)$.   ∎

The inclusion of $\mathcal{P}(R)$ into any larger subcategory $\mathcal{C}$ of $\mathcal{M}(R)$ preserves short exact sequences; so $P \mapsto [P] \in K_0(\mathcal{C})$ defines a generalized rank on $\mathcal{P}(R)$, inducing a group homomorphism

$$c : K_0(R) \to K_0(\mathcal{C}) .$$
$$[P] \mapsto [P]$$

(3.52) Resolution Theorem. *For any ring R, suppose $\mathcal{C}$ is a full subcategory of $\mathcal{P}_{<\infty}(R)$, containing $\mathcal{P}(R)$, and including as objects the kernels of all its arrows. The inclusion $\mathcal{P}(R) \to \mathcal{C}$ induces an isomorphism of groups $c : K_0(R) \to K_0(\mathcal{C})$.*

Proof. By (3.51), the Euler characteristic is a generalized rank on $\mathcal{C}$, inducing a group homomorphism $\overline{\chi} : K_0(\mathcal{C}) \to K_0(R)$ with $\overline{\chi}([M]) = \chi(M)$. Since each f.g. projective R-module P has a resolution

$$0 \longrightarrow P \longrightarrow P \longrightarrow 0 \ ,$$

$\overline{\chi} \circ c = $ the identity on $K_0(R)$. Suppose

$$0 \to P_n \to \cdots \to P_0 \to M \to 0$$

is a resolution in $\mathcal{C}$, and K_i denotes the kernel of the map beginning at P_i for $0 \leq i \leq n$. From the exact sequences

$$0 \longrightarrow K_i \longrightarrow P_i \longrightarrow K_{i-1} \longrightarrow 0$$

for $1 \leq i \leq n$ and

$$0 \longrightarrow K_0 \longrightarrow P_0 \longrightarrow M \longrightarrow 0 \ ,$$

it follows that in $K_0(\mathcal{C})$,

$$\sum_{i=0}^{n} (-1)^i [P_i] \ = \ [M] \pm [K_n]$$
$$= \ [M] \ ;$$

so $c \circ \overline{\chi} = $ the identity on $K_0(\mathcal{C})$. ∎

(3.53) Corollary. *If R is a left regular ring, the Cartan map $K_0(R) \to G_0(R)$, $[P] \mapsto [P]$, is a group isomorphism.* ∎

3D. Exercises

1. Suppose R is a commutative principal ideal domain. The Structure Theorem for $\mathcal{M}(R)$ says each f.g. R-module M is isomorphic to

$$R^n \oplus (R/d_1 R) \oplus \cdots \oplus (R/d_m R),$$

where each $d_{i+1} \in d_i R$. Call n the **torsion free rank** of M. Prove the torsion free rank equals the presentation rank. Deduce from this that the torsion free rank is well defined, constant on isomorphism classes, and additive over short exact sequences in $\mathcal{M}(R)$.

2. Verify the exactness of the sequence

$$0 \longrightarrow K \longrightarrow L \oplus P \longrightarrow Q \longrightarrow 0$$

in the proof of Schanuel's Lemma.

3. Suppose R is a ring, and

$$0 \longrightarrow P_n \longrightarrow \cdots \longrightarrow P_0 \longrightarrow 0$$

is an exact sequence in $\mathcal{P}(R)$. Prove

$$\sum_{i=0}^{n} (-1)^i [P_i] = 0$$

in $K_0(R)$. So for any generalized rank $\rho : \mathrm{Obj}\ \mathcal{P}(R) \to (A, +)$, we have $\Sigma(-1)^i \rho(P_i) = 0$. *Hint:* Use (3.50).

4

Stability for Projective Modules

Finitely generated projective R-modules are "stably isomorphic" if they become equal in $K_0(R)$, and "stably equivalent" if they become congruent in $K_0(R)$ modulo f.g. free modules. A f.g. projective that is stably equivalent to a free module is said to be "stably free." In §4A we describe these relations among modules in terms of the addition of copies of R. Examples in §4B show f.g. projectives need not be stably free, and stably free modules need not be free. Section 4C discusses conditions on $GL_n(R)$ that force sufficiently large stably free modules to be free; and in §4D these conditions are related to the "stable rank" of a ring (which is also used to simplify higher K-groups $K_n(R)$, as we will see in Chapter 10 for $K_1(R)$). The computation of the stable rank is made easier by a discussion of its properties in §4D and by its connection to Krull dimension in §4E.

4A. Adding Copies of R

Sometimes, dissimilar mathematical objects can become closely related when given enough "elbow room" for the connection between them. The sine and cosine functions are apparently unrelated to the exponential function in real variable calculus, but they are all closely connected as functions of a complex variable. And the dot and cross product of vectors in $\mathbb{R}^3$ become two aspects of quaternion multiplication in $\mathbb{R}^4$.

So also for f.g. projective R-modules P and Q, it can happen that $P \not\cong Q$, but that $P \oplus R^n \cong Q \oplus R^n$ for some integer $n > 0$ (and hence for all larger integers). One might be led to this connection by looking at $K_0(R)$, where $[P \oplus R^n] = [Q \oplus R^n]$ implies $[P] = [Q]$.

On the abelian monoid $\mathcal{I} = \mathcal{I}(\mathcal{P}(R))$ under $\oplus$, consider the function "add the isomorphism class $c(R)$":

$$\sigma : \mathcal{I} \to \mathcal{I}$$
$$c(P) \mapsto c(P \oplus R).$$

There is a descending chain

$$\mathcal{I} \supseteq \sigma(\mathcal{I}) \supseteq \sigma^2(\mathcal{I}) \supseteq \cdots ,$$

and σ restricts to a sequence of surjective maps

(4.1) $$\mathcal{I} \rightarrow \sigma(\mathcal{I}) \rightarrow \sigma^2(\mathcal{I}) \rightarrow \cdots .$$

For each integer $m \geq 0$, there is a function

$$s_m : \sigma^m(\mathcal{I}) \rightarrow K_0(R) ,$$
$$c(P \oplus R^m) \mapsto [P]$$

and the diagrams

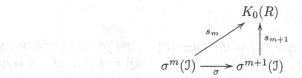

commute. Each of the maps s_m has the same image, which is the image of the group completion map

$$s_0 : \mathcal{I} \rightarrow K_0(R) ,$$

namely, the set

$$\boldsymbol{K_0^+(R)} = \{[P] : P \in \mathcal{P}(R)\} .$$

This set $K_0^+(R)$ is a submonoid of $K_0(R)$, analogous to $\mathbb{N}$ in $\mathbb{Z}$. In a sense (made precise in Exercise 1) $K_0^+(R)$ is a limit of the sequence (4.1).

This sequence (4.1) becomes **stable at** n, meaning

$$\sigma^n(\mathcal{I}) \rightarrow \sigma^{n+1}(\mathcal{I}) \rightarrow \cdots$$

are bijections, if and only if R has n-**cancellation**:

"*For* $P, Q \in \mathcal{P}(R)$, $P \oplus R^{n+1} \cong Q \oplus R^{n+1}$ *implies* $P \oplus R^n \cong Q \oplus R^n$."

In §§4D–E, we shall see that many rings have n-cancellation for some $n \geq 0$; the least such n is a kind of "dimension" of the ring R. In this context, the phrase "for sufficiently large n" is often replaced by "stably" or "in the stable range."

(4.2) **Definition.** If R is a ring, f.g. projective modules P and Q are **stably isomorphic** if, for some integer $n \geq 0$, $P \oplus R^n \cong Q \oplus R^n$ (that is, if $c(P)$ and $c(Q)$ become equal after n iterations of σ). In the same vein, P and Q are **stably equivalent** if, for some integers $m, n \geq 0$, $P \oplus R^m \cong Q \oplus R^n$.

On the class of f.g. projective R-modules, "stably isomorphic" and "stably equivalent" define equivalence relations. Evidently, isomorphic R-modules are stably isomorphic, and stably isomorphic R-modules are stably equivalent.

Within $K_0(R)$, the f.g. free R-modules $F \cong R^n$ $(n \geq 0)$ become the elements $[R^n] = n[R]$ $(n \geq 0)$ inside the cyclic subgroup $\langle[R]\rangle$ generated by $[R]$. The quotient

$$\widetilde{K}_0(R) = K_0(R)/\langle[R]\rangle$$

is known as the **projective class group** of R and can loosely be thought of as "projectives mod frees." For $P \in \mathcal{P}(R)$, let $\overline{[P]}$ denote the coset $[P] + \langle[R]\rangle$ in $\widetilde{K}_0(R)$. The elements of $K_0^+(R)$ correspond to the stable isomorphism classes in $\mathcal{P}(R)$ and the elements of $\widetilde{K}_0(R)$ correspond to the stable equivalence classes in $\mathcal{P}(R)$:

(4.3) Proposition. *Suppose R is a ring.*

(i) *Each element of $K_0(R)$ has the form $[P] - [R^n]$ for some $P \in \mathcal{P}(R)$ and integer $n \geq 0$. For $P, Q \in \mathcal{P}(R), [P] = [Q]$ in $K_0(R)$ if and only if P and Q are stably isomorphic.*

(ii) *Each element of $\widetilde{K}_0(R)$ has the form $\overline{[P]}$ for some $P \in \mathcal{P}(R)$. For $P, Q \in \mathcal{P}(R), \overline{[P]} = \overline{[Q]}$ if and only if P and Q are stably equivalent.*

Proof. By (3.22), each element of $K_0(R) = K_0(\mathcal{P}(R), \oplus)$ has the form $[P] - [Q]$ for $P, Q \in \mathcal{P}(R)$. If $Q \oplus Q' \cong R^n$,

$$[P] - [Q] = [P] + [Q'] - [Q] - [Q']$$
$$= [P \oplus Q'] - [R^n].$$

Also by (3.22), $[P] = [Q]$ if and only if $P \oplus M \cong Q \oplus M$ for some $M \in \mathcal{P}(R)$. Adding a complement of M, this is equivalent to saying $P \oplus R^n \cong Q \oplus R^n$ for some integer $n \geq 0$.

For (ii), each element of $\widetilde{K}_0(R)$ is $\overline{[P]} - \overline{[R^n]} = \overline{[P]}$ for some $P \in \mathcal{P}(R)$. If $\overline{[P]} = \overline{[Q]}$, then

$$[P] - [Q] = n[R], \quad n \in \mathbb{Z}.$$

If $n \geq 0$, this implies $[P] = [Q \oplus R^n]$. If $n < 0$, it implies $[P \oplus R^{-n}] = [Q]$. Either way, P is stably equivalent to Q by (i).

Conversely, if $P \oplus R^m \cong Q \oplus R^n$ for integers $m, n \geq 0$, then $[P] - [Q] = (n - m)[R]$ in $K_0(R)$; so $\overline{[P]} = \overline{[Q]}$. ∎

If the sequence (4.1) is stable at n, then it achieves its limit $K_0^+(R)$, since, for $m \geq n$, each map

$$s_m : \sigma^m(\mathcal{I}) \rightarrow K_0^+(R)$$

is bijective, by (4.3) (i) and m-cancellation.

4A. Exercises

1. Suppose $\mathcal{C}$ is a category and

$$A_0 \xrightarrow{f_0} A_1 \xrightarrow{f_1} A_2 \xrightarrow{f_2} \cdots$$

is a sequence of arrows in $\mathcal{C}$. A **cone** $(B, \{g_i\})$ **from** this sequence is a commutative diagram in $\mathcal{C}$:

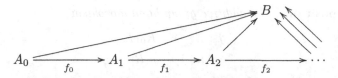

where the arrows going up are $g_i : A_i \to B$. These cones are the objects of a category $\mathcal{D}$ in which an arrow from $(B, \{g_i\})$ to $(B', \{g'_i\})$ is an arrow in $\mathcal{C}$, $h : B \to B'$, with $h \circ g_i = g'_i$ for each i. An initial object in $\mathcal{D}$ is called a **direct limit diagram** for the given sequence, and its apex B is called the **direct limit** : $B = \varinjlim_n A_n$.

If $\mathcal{C} = \mathbf{Set}$, prove the sequence (4.1) has direct limit $\varinjlim_n \sigma^n(\mathfrak{I}) = K_0^+(R)$.

2. Suppose R is a ring. On $\mathcal{P}(R)$, prove:

   (a) "stably equivalent implies stably isomorphic" if and only if $R^n \cong R^{n+1}$ for some $n \geq 0$;

   (b) "stably isomorphic implies isomorphic" if and only if R has 0-cancellation.

3. Suppose D is a division ring, $R = M_2(D)$,

$$P = \begin{bmatrix} D & 0 \\ D & 0 \end{bmatrix} \quad \text{and} \quad Q = \begin{bmatrix} 0 & D \\ 0 & D \end{bmatrix} .$$

Prove:

   (a) P and Q are isomorphic simple projective R-modules, and are maximal left ideals of R.

   (b) If J is a left ideal of R, then $J = \{0\}$, $J = R$, $J = P$, or $J \overset{\bullet}{\oplus} P = R$.

   (c) If M is a cyclic R-module, $M \cong 0$, P, or P^2.

   (d) If M is a f.g. R-module, M is a sum of finitely many copies of P. If this sum is minimal (no summand contains another summand), it is direct.

   (e) R has 0-cancellation (use part (d) and dimension over D). So $[P_1] = [P_2]$ in $K_0(R)$ implies $P_1 \cong P_2$.

   (f) $K_0(R)$ is infinite cyclic, generated by $[P]$, and $\widetilde{K}_0(R) \cong \mathbb{Z}/2\mathbb{Z}$.

4B. Stably Free Modules

The algebraic K-group $K_0(R)$ cannot determine whether all f.g. projective R-modules are free. But it comes close. In this section we see how close.

(4.4) Definition. An R-module P is **stably free** if $P \oplus R^m \cong R^n$ for some integers $m, n \geq 0$.

Note that every stably free R-module belongs to $\mathcal{P}(R)$, and that $P \in \mathcal{P}(R)$ is stably free if and only if P is stably equivalent to 0, that is, if and only if $\overline{[P]} = 0$ in $\widetilde{K}_0(R)$. So $\widetilde{K}_0(R)$ should really be thought of as "projectives mod stably frees."

(4.5) Proposition. *The additive group homomorphism*

$$f : \mathbb{Z} \quad \to \quad K_0(R)$$
$$n \quad \mapsto \quad n[R]$$

is injective if and only if R has IBN, and is surjective if and only if every f.g. projective R-module is stably free.

Proof. The map f fails to be injective if and only if $[R^n] = [0]$ for some integer $n > 0$; the latter happens if and only if $R^{n+m} \cong R^m$ for some integers $n > 0$ and $m \geq 0$. The map f fails to be surjective if and only if $[P] \notin \langle [R] \rangle$ for some $P \in \mathcal{P}(R)$, that is, if and only if some $P \in \mathcal{P}(R)$ is not stably free. ∎

Suppose P is a stably free R-module with $P \oplus R^m \cong R^n$ and $P \oplus R^s \cong R^t$. Then in $K_0(R)$,

$$(n - m)[R] \quad = \quad [P] \quad = \quad (t - s)[R] .$$

If R has IBN, then by (4.5), $n - m = t - s$.

(4.6) Definition. If R is a ring with IBN and P is an R-module with $P \oplus R^m \cong R^n$ for integers $m, n \geq 0$, the **stably free rank** of P is the difference $n - m$.

On f.g. free R-modules, the stably free rank equals the free rank, since $P \cong R^n$ implies $P \oplus R^0 \cong R^n$.

Every f.g. free R-module is stably free, and every stably free R-module is projective. But the reverse implications are not always true; they depend on the particular ring R :

(4.7) Examples of $P \in \mathcal{P}(R)$ that are not stably free.

(i) The ring $R = \mathbb{Z}/6\mathbb{Z}$ is $\{\overline{0}, \overline{3}\} \overset{\bullet}{\oplus} \{\overline{0}, \overline{2}, \overline{4}\}$; but $P = \{\overline{0}, \overline{3}\}$ is not a stably free R-module, since $P \oplus R^m$ has $2(6^m)$ elements, while R^n has 6^n elements.

(ii) If F is a field, the ring $R = M_2(F)$ is

$$\begin{bmatrix} F & 0 \\ F & 0 \end{bmatrix} \overset{\bullet}{\oplus} \begin{bmatrix} 0 & F \\ 0 & F \end{bmatrix} .$$

But

$$P = \begin{bmatrix} F & 0 \\ F & 0 \end{bmatrix}$$

is not stably free, since each R-linear isomorphism is F-linear, and we have $\dim_F(P \oplus R^m) = 2 + 4m$, while $\dim_F(R^n) = 4n$. More generally, suppose R is any ring with IBN. By (1.39), $A = M_2(R)$ also has IBN. If

$$P = \begin{bmatrix} R & 0 \\ R & 0 \end{bmatrix}, \quad Q = \begin{bmatrix} 0 & R \\ 0 & R \end{bmatrix},$$

then $A = P \overset{\bullet}{\oplus} Q$; so $P, Q \in \mathcal{P}(A)$. Switching columns, $P \cong Q$; so $A \cong P \oplus P$. If P were stably free, so that $P \oplus A^m \cong A^n$, we would have $A \oplus A^{2m} \cong A^{2n}$, forcing $2m + 1 = 2n$.

(4.8) Examples of nonfree stably free modules.

(i) (Ojanguren and Sridharan [71]) Recall Hamilton's division ring of quaternions $\mathbb{H}$, which is a vector space over its center $\mathbb{R}$ (the real numbers) with basis $\{1, i, j, ij\}$ and has $i^2 = j^2 = (ij)^2 = -1$, so that $ji = -ij \neq ij$. The polynomial ring $R = \mathbb{H}[x, y] = (\mathbb{H}[x])[y]$ has IBN, since there is a ring homomorphism $R \to \mathbb{H}$ (taking x and y to 0) and $\mathbb{H}$ is left noetherian.

Right multiplication by the matrix

$$\begin{bmatrix} x + j \\ -(y + i) \end{bmatrix}$$

is an R-linear map $f : R^2 \to R$. Since

$$f(y + i, \; x + j) \;=\; ij - ji \;=\; 2ij \in R^*,$$

the map f is surjective.

Let P denote the kernel of f:

$$P \;=\; \{(a, b) \in R^2 : a(x + j) \;=\; b(y + i)\}.$$

Since $y + i$ is not a zero-divisor in the domain R, projection to the first coordinate, $R^2 \to R$, takes P isomorphically onto the left ideal

$$J \;=\; \{a \in R : a(x + j) \in R(y + i)\}$$

of R. Since R is a free R-module, the short exact sequence

$$0 \longrightarrow P \overset{\subseteq}{\longrightarrow} R^2 \overset{f}{\longrightarrow} R \longrightarrow 0$$

splits; so $J \oplus R \cong P \oplus R \cong R^2$.

Assume J is a free R-module. Since R has IBN, this means J has free rank 1; so $J = Rz$ for some nonzero $z \in J$. Since

$$
\begin{aligned}
(y+i)(x-j)(x+j) &= (y+i)(x^2+1) \\
&= (x^2+1)(y+i),
\end{aligned}
$$

the product $(y+i)(x-j)$ belongs to J. So z has y-degree ≤ 1. Every nonzero element of J has y-degree ≥ 1. So $z = cy + d$ for $c, d \in \mathbb{H}[x]$. Since

$$
\begin{aligned}
(y^2+1)(x+j) &= (x+j)(y^2+1) \\
&= (x+j)(y-i)(y+i),
\end{aligned}
$$

we also have $y^2 + 1 \in J$. So

$$
y^2 + 1 = r(cy + d)
$$

for some $r \in R$. Then $r = c'y + d'$ for $c', d' \in \mathbb{H}[x]$ and $c'c = 1$. So c' is a nonzero quaternion, and $c' \in R^*$. Then $J = Rz = Rc'z = R(y+c'd)$. Therefore

$$
(y+i)(x-j) = s(y + c'd),
$$

where $s \in R$. So s has y-degree 0. Comparing y-coefficients, $s = x - j$. Then comparing y-degree 0 terms,

$$
i(x-j) = (x-j)c'd.
$$

So $c'd$ has x-degree 0; and comparing x-coefficients, $c'd = i$. But then $-ij = -ji$, which is false in $\mathbb{H}$. So J is a nonfree stably free left ideal of R.

Replacing $\mathbb{H}$ by $\mathbb{H}[t_1, \ldots, t_{n-2}]$, the same argument constructs a nonfree stably free left ideal in $\mathbb{H}[t_1, \ldots, t_n]$ whenever $n \geq 2$. The authors of this example actually constructed a nonfree stably free left ideal in $D[t_1, \ldots, t_n]$ ($n \geq 2$) for each noncommutative division ring D. For a generalization to twisted polynomial rings, see McConnell and Robson [87, Chapter 11, §2]. At this point it is worth mentioning that if F is a field, all f.g. projective $F[t_1, \ldots, t_n]$- modules are free. In 1958, Serre [58] asked if this were true; and for the next 18 years this question was one of the motivating problems for the development of algebraic K-theory, and was known as the "Serre Conjecture." In 1976 it was affirmed independently by Quillen [76] and Suslin [76]. So a division ring D is commutative if and only if all stably free $D[t_1, t_2]$-modules are free.

(ii) (Kaplansky) Suppose $\mathbb{R}$ is the field of real numbers and

$$
R = \mathbb{R}[X, Y, Z]/(X^2 + Y^2 + Z^2 - 1).
$$

Let $\overline{X}, \overline{Y}, \overline{Z}$ denote the cosets in R of X, Y, Z. Right multiplication by the matrices

$$
\begin{bmatrix} \overline{X} \\ \overline{Y} \\ \overline{Z} \end{bmatrix} \quad \text{and} \quad [\overline{X} \ \ \overline{Y} \ \ \overline{Z}]
$$

defines R-linear maps $g : R^3 \to R$ and $h : R \to R^3$, respectively, with $g \circ h = i_R$. Let P denote $\ker(g)$. There is a split short exact sequence

$$0 \longrightarrow P \overset{\subseteq}{\longrightarrow} R^3 \underset{h}{\overset{g}{\longleftrightarrow}} R \longrightarrow 0 .$$

By (2.5) there is an isomorphism $\theta : P \oplus R \to R^3$ defined by $\theta((p, r)) = p + h(r)$. So P is stably free.

Suppose P is a free R-module. Since R is commutative, it has IBN. So P has free rank 2, and there is an isomorphism $f : R^2 \to P$. So there is an isomorphism:

$$\phi : R^3 \to P \oplus R$$
$$(r, s, t) \mapsto (f(r, s), t) .$$

The composite isomorphism $\theta \circ \phi : R^3 \to R^3$ takes $(0, 0, 1)$ to $h(1) = (\overline{X}, \overline{Y}, \overline{Z})$; so it is right multiplication by an invertible matrix:

$$A = \begin{bmatrix} a_1 & a_2 & a_3 \\ b_1 & b_2 & b_3 \\ \overline{X} & \overline{Y} & \overline{Z} \end{bmatrix} \in GL_3(R) .$$

Then the determinant of A is a unit $u \in R^*$. Multiplying the first row by u^{-1} creates a matrix

$$B = \begin{bmatrix} c_1 & c_2 & c_3 \\ b_1 & b_2 & b_3 \\ \overline{X} & \overline{Y} & \overline{Z} \end{bmatrix} \in GL_3(R)$$

with determinant 1_R.

Let S^2 denote the unit sphere

$$\{(x, y, z) \in \mathbb{R}^3 : x^2 + y^2 + z^2 = 1\} .$$

Let $C(S^2)$ denote the $\mathbb{R}$-algebra of continuous functions $S^2 \to \mathbb{R}$, added, multiplied, and scalar multiplied pointwise. If $\pi_i : S^2 \to \mathbb{R}$ is the projection to the i-coordinate ($i = 1, 2, 3$), there is a unique $\mathbb{R}$-linear ring homomorphism

$$\psi : \mathbb{R}[X, Y, Z] \to C(S^2)$$

taking $X \mapsto \pi_1$, $Y \mapsto \pi_2$, and $Z \mapsto \pi_3$. The kernel of ψ includes $X^2 + Y^2 + Z^2 - 1$; so ψ induces a ring homomorphism

$$\overline{\psi} : R \to C(S^2) .$$

Applying $\overline{\psi}$ to each entry of B produces a matrix

$$C = \begin{bmatrix} \gamma_1 & \gamma_2 & \gamma_3 \\ \beta_1 & \beta_2 & \beta_3 \\ \pi_1 & \pi_2 & \pi_3 \end{bmatrix} \in GL_3(C(S^2))$$

of determinant $1 \in C(S^2)$. Denoting the rows of C by γ, β, and π, the cross product $\beta \times \pi$ is a continuous vector field on S^2, perpendicular to the radial vector field π, and nowhere vanishing because

$$\gamma \cdot (\beta \times \pi) \;=\; \text{determinant}(C)$$

is the constant function to 1.

It is a famous theorem of topology (see Brouwer [12], Milnor [78]) that S^2 has no continuous nonvanishing tangent vector field. So P is a nonfree stably free module over the commutative ring R. For generalizations of this example to other affine rings over arbitrary fields, see §II of Suslin [79]. It is no accident that we chose an example with stably free rank ≥ 2 over a commutative ring — if R is commutative, there are no nonfree stably free R-modules of stably free rank 1 (see §14D, Exercise 6).

(iii) (Cohn [66]) Suppose R is the ring presented (as in (1.22)) by generators x_1, x_2, y_1, y_2 and defining relators $x_1 y_1 - 1$, $x_2 y_2 - 1$, $x_1 y_2$, $x_2 y_1$. Then R is the free $\mathbb{Z}$-module based on the set of all strings

$$s_1 \cdots s_m t_1 \cdots t_n \quad (m, n \geq 0) \;,$$

where each $s_i \in \{y_1, y_2\}$ and each $t_i \in \{x_1, x_2\}$. The empty string $(m = n = 0)$ is the multiplicative identity 1_R. The multiplication in R is determined by the partial multiplication table:

	y_1	y_2
x_1	1_R	0_R
x_2	0_R	1_R

Over this ring we have the matrix equation:

$$\begin{bmatrix} x_1 \\ x_2 \end{bmatrix} \begin{bmatrix} y_1 & y_2 \end{bmatrix} = \begin{bmatrix} 1_R & 0_R \\ 0_R & 1_R \end{bmatrix} \;.$$

Right multiplication by

$$\begin{bmatrix} y_1 & y_2 \end{bmatrix} \quad \text{and} \quad \begin{bmatrix} x_1 \\ x_2 \end{bmatrix}$$

defines R-linear maps $g : R \to R^2$ and $h : R^2 \to R$, respectively, with $g \circ h = $ the identity map on R^2. So there is a split short exact sequence

$$0 \longrightarrow P \overset{\subseteq}{\longrightarrow} R \underset{h}{\overset{g}{\longleftrightarrow}} R^2 \longrightarrow 0 \;,$$

where P is the kernel of g. Thus $P \oplus R^2 \cong R$ and P is stably free.

By using the universal Hattori-Stallings trace function and methods of universal algebra, Cohn proved the remarkable fact that R has IBN. Thus P cannot be a free R-module because P has negative stably free rank! For generalizations of this example, see Cohn [66], Bergman [74], and Lam [76].

(iv) Suppose A is the ring of column-finite $\mathbb{P} \times \mathbb{P}$ matrices over a nontrivial ring R (see (1.37)). Let P denote the set of first columns of matrices in A. Then $P \oplus A \cong A$ as A-modules, by the map bumping each column of $a \in A$ to the right to make room for $p \in P$. So P is stably free. But P has no A-linearly independent elements; so P is not a free A-module.

4B. Exercises

1. Suppose R is a ring and $P \in \mathcal{P}(R)$. Prove P is stably free if and only if P has a finite free resolution (i.e., there is a finite length exact sequence

$$0 \to F_n \to F_{n-1} \to \cdots \to F_0 \to P \to 0$$

in R-$\mathcal{M}\mathfrak{o}\mathfrak{d}$ in which each F_i is f.g. free).

2. Suppose R is a ring without IBN and $f : \mathbb{Z} \to K_0(R)$ is the homomorphism in (4.5). If the first repetition in $R^1, R^2, R^3, \ldots$ is the isomorphism $R^h \cong R^{h+k}$, prove $f(\mathbb{Z}) \cong \mathbb{Z}/k\mathbb{Z}$, and $a[R] = b[R]$ for integers a, b if and only if $a \equiv b \pmod{k}$. So one could define a modular stably free rank with values in $\mathbb{Z}/k\mathbb{Z}$.

3. ("Eilenberg Swindle"). Suppose R is a ring and P is a projective R-module. Prove there is a free R-module F for which $P \oplus F$ is free. Why doesn't this contradict (4.7)? *Hint:* For some R-module Q, $P \oplus Q = F$ is free and not finitely generated. Then $F \cong F^\infty$ (= the infinite cartesian product $F \times F \times \cdots$), and $P \oplus F \cong F$.

4C. When Stably Free Modules Are Free

There are many rings R over which all stably free R-modules are free. Even if R does not have IBN this can happen: Cohn [94] has shown that every nonzero algebra over a field is a subring of a ring R for which every nonzero $P \in \mathcal{P}(R)$ is isomorphic to R; so $R^2 \cong R$ and stably free R-modules are free.

In general, to determine which stably free modules are free, it is useful to classify them by stably free rank; so we focus on rings with IBN. To avoid modules with negative stably free rank, as in Example (4.8) (iii), consider a condition that is slightly stronger than IBN:

(4.9) Definition. A ring R is **weakly n-finite** if $P \oplus R^n \cong R^n$ implies $P \cong 0$; and R is **weakly finite** if it is weakly n-finite for each integer $n \geq 0$.

(4.10) Proposition. *A ring $R \neq \{0\}$ is weakly finite if and only if R has IBN and every R-module with stably free rank ≤ 0 is free.*

Proof. Suppose $P \oplus R^m \oplus R^n \cong R^n$ for $m, n \geq 0$. If R is weakly finite, $P \oplus R^m \cong 0$; so $P \cong 0$ and $m = 0$. Thus R has IBN and an R-module P of stably free rank ≤ 0 must be 0, which is free.

Conversely, if R has IBN and every R-module of stably free rank ≤ 0 is free, then $P \oplus R^n \cong R^n$ implies P is free, of free rank 0; so $P \cong 0$. ∎

Weakly n-finite is a condition that arises naturally in basic linear algebra:

(4.11) Proposition. *Suppose R is a ring and n is an integer ≥ 1. The following are equivalent:*

(i)   *R is weakly n-finite.*
(ii) *If $a, b \in M_n(R)$ and $ab = I_n$, then $ba = I_n$.*
(iii) *Each surjective R-linear map $f : R^n \to R^n$ is also injective.*
(iv) *Each n-element spanning set of R^n is a basis.*

Proof. Suppose $a, b \in M_n(R)$ with $ab = I_n$. From the split short exact sequence

$$0 \longrightarrow P \overset{\subseteq}{\longrightarrow} R^n \underset{\cdot a}{\overset{\cdot b}{\rightleftarrows}} R^n \longrightarrow 0 \, ,$$

$P \oplus R^n \cong R^n$. If R is weakly finite, $P \cong 0$. So $\cdot b$ is an isomorphism with inverse $\cdot a$, and $ba = I_n$, proving (i) implies (ii).

Since R^n is projective, any short exact sequence

$$0 \longrightarrow P \longrightarrow R^n \overset{f}{\longrightarrow} R^n \longrightarrow 0$$

splits. If f is right multiplication by $b \in M_n(R)$ and the splitting map is right multiplication by $a \in M_n(R)$, then $ab = I_n$. If (ii) holds, $ba = I_n$ and f is injective. So (ii) implies (iii).

For equivalence of (iii) and (iv), use the R-linear map f taking $e_1, \ldots, e_n$ to the chosen spanning set of R^n.

Finally, suppose $\alpha : R^n \to P \oplus R^n$ is an isomorphism and $\pi : P \oplus R^n \to R^n$ is the projection to the second summand. The composite $\pi \circ \alpha$ is surjective. Assuming (iii), it is also injective. So π is injective and $P \cong 0$. Thus (iii) implies (i). ∎

Weakly finite rings form a very large class, including most types of rings with IBN: By (4.11) (i) ⇔ (ii), the class of weakly finite rings is closed under subrings, isomorphisms, full matrix rings (R weakly finite implies $M_m(R)$ weakly finite — use block multiplication as in (1.32)), and cartesian products of rings. All commutative rings are weakly finite, by the cofactor formula for the inverse of a matrix. Left noetherian rings satisfy (4.11) (iii) by (B.12), and so they are weakly finite.

Next, consider modules of stably free rank $r \geq 1$. Our treatment follows that of Swan [68, Chapter 12]. Suppose M is an R-module with $M \oplus R^s \cong R^{s+m}$ for integers $m, s > 0$. It appears that we need 0-cancellation (see the discussion preceding (4.2)) to remove all s copies of R and prove $M \cong R^m$. But since the right side is free, we can get by with the condition:

$$P \oplus R \cong R^n \implies P \cong R^{n-1}$$

for all $n > m$.

For any ring R, with or without IBN, consider how an isomorphism $f : P \oplus R \cong R^n$ is reflected inside R^n. The submodule $Q = f(P \oplus \{0\})$ of R^n is isomorphic to P; so $P \cong R^{n-1}$ if and only if $Q \cong R^{n-1}$. The element $v = f((0,1)) \in R^n$ is linearly independent and generates a direct summand:

$$R^n = Q \overset{\bullet}{\oplus} Rv \,.$$

When v is regarded as a row matrix in $R^{1 \times n}$, its properties have a simple matrix version:

(4.12) Definitions. A vector $v = (a_1, \ldots, a_n) \in R^n$ is **right unimodular** if it forms a right invertible row $v \in R^{1 \times n}$ — so there is a column $w \in R^{n \times 1}$ with $vw = 1$. Equivalently, v is right unimodular if and only if $a_1 b_1 + \cdots + a_n b_n = 1$ for some $b_1, \ldots, b_n \in R$ — that is, if and only if

$$a_1 R + \cdots + a_n R = R \,.$$

A vector $w = (b_1, \ldots, b_n) \in R^n$ is **left unimodular** if it forms a left invertible column $w \in R^{n \times 1}$ — so $a_1 b_1 + \cdots + a_n b_n = 1$ for some $a_1, \ldots, a_n \in R$, and

$$R b_1 + \cdots + R b_n = R \,.$$

(4.13) Lemma. *Suppose n is a positive integer. Then $v \in R^n$ is right unimodular if and only if v is (left) R-linearly independent and Rv is a direct summand of R^n.*

Proof. Suppose $v = [a_1 \ \cdots \ a_n]$, w transpose $= [b_1 \ \cdots \ b_n]$, and $vw = 1$. The R-linear map $f : R^n \to Rv$ defined by

$$(r_1, \ldots, r_n) \mapsto \sum_{i=1}^{n} r_i b_i v$$

restricts to the identity map on Rv. So if K is its kernel,

$$R^n = K \overset{\cdot}{\oplus} Rv .$$

If $r \in R$ and $rv = 0$, then $r = r(vw) = (rv)w = 0w = 0$.
  Conversely, suppose

$$R^n = Q \overset{\cdot}{\oplus} Rv ,$$

where $v = (a_1, \dots, a_n)$. For $1 \le i \le n$, there exist $q_i \in Q$ and $b_i \in R$ with

$$e_i = q_i + b_i v .$$

Then

$$v = \sum_{i=1}^{n} a_i e_i = q + \sum_{i=1}^{n} a_i(b_i v) ,$$

where $q \in Q \cap Rv = \{0\}$. So

$$1v = \left(\sum_{i=1}^{n} a_i b_i \right) v .$$

If v is linearly independent, then $1 = \sum a_i b_i$; so v is right unimodular. ∎

(4.14) Proposition. *If R is any ring, $n > 1$, and $v = (a_1, \dots, a_n)$ is a right unimodular element in R^n with $R^n = Q \overset{\cdot}{\oplus} Rv$, the following are equivalent:*

  (i) $Q \cong R^{n-1}$.
  (ii) *v is part of an n-element basis of R^n.*
  (iii) *v is the first row of an invertible matrix in $GL_n(R)$.*
  (iv) *For some $A \in GL_n(R)$, $(a_1, \dots, a_n)A = (1, 0, , \dots, 0)$.*
  (v)   *For some $A \in GL_n(R)$, and some $a_1', \dots, a_{n-1}' \in R$, $(a_1, \dots, a_n)A = (a_1', \dots, a_{n-1}', 0)$.*

Proof. Suppose (i); so Q has a basis $v_1, \dots, v_{n-1}$. Then

$$Rv + Rv_1 + \dots + Rv_{n-1} = Rv + Q = R^n .$$

And $\{v, v_1, \dots, v_{n-1}\}$ is linearly independent since v is, and since $R^n = Rv \overset{\cdot}{\oplus} Q$. This proves (ii).

  Assume as in (ii) that $\{v, v_1, \dots, v_{n-1}\}$ is a basis of R^n. Let B denote the matrix

$$\begin{bmatrix} v \\ v_1 \\ \vdots \\ v_{n-1} \end{bmatrix} \in M_n(R) .$$

Since the rows of the identity matrix I_n are in the span of $v, v_1, \ldots, v_{n-1}$, there is a matrix $C \in M_n(R)$ with $CB = I_n$. Then $BCB = B$. Since the rows of B are linearly independent, B can be canceled from the right. So $BC = I_n$ and $B \in GL_n(R)$, proving (iii).

As in (iii), assume there is a matrix

$$D = \begin{bmatrix} v \\ w_1 \\ \vdots \\ w_{n-1} \end{bmatrix} \in GL_n(R)$$

with inverse E. Inspecting the first row of $DE = I_n$,

$$(a_1, \ldots, a_n)E = (1, 0, \ldots, 0),$$

proving (iv). That (iv) implies (v) is a formality.

Assume (v). Since A^{-1} has a left inverse and can be canceled from the right, its rows form a basis of R^n. Call these rows $e'_1, \ldots, e'_n$. Then

$$\begin{aligned} v &= (a'_1, \ldots, a'_{n-1}, 0)A^{-1} \\ &= a'_1 e'_1 + \cdots + a'_{n-1} e'_{n-1} . \end{aligned}$$

By hypothesis, $vw = 1$ for some $w \in R^{n \times 1}$. So the row $v' = (a'_1, \ldots, a'_{n-1})$ in R^{n-1} is right unimodular:

$$1 = vw = (a'_1, \ldots, a'_{n-1}, 0)A^{-1}w .$$

So Rv' is a direct summand of R^{n-1}. The isomorphism

$$R^{n-1} \rightarrow Re'_1 + \cdots + Re'_{n-1} ,$$

taking e_i to e'_i for each i, takes v' to v. So Rv is a direct summand of $Re'_1 + \cdots + Re'_{n-1}$, having some complement Q'. Then

$$\begin{aligned} R^n &= (Q' \overset{\bullet}{\oplus} Rv) \overset{\bullet}{\oplus} Re'_n \\ &= Rv \overset{\bullet}{\oplus} (Q' \overset{\bullet}{\oplus} Re'_n) . \end{aligned}$$

Thus

$$\begin{aligned} Q &\cong R^n/Rv \cong Q' \overset{\bullet}{\oplus} Re'_n \\ &\cong Q' \oplus R \cong Q' \overset{\bullet}{\oplus} Rv \cong R^{n-1} , \end{aligned}$$

proving (i) . ∎

(4.15) Definition. A ring R has n-**matrix completion** if every right unimodular vector in R^n is the first row of an invertible matrix $A \in GL_n(R)$. A ring has **matrix completion** if it has n-matrix completion for every $n > 1$. The set of right unimodular vectors in R^n is denoted by $Um_r(n, R)$. If $v \in Um_r(n, R)$ and $A \in GL_n(R)$, then $vA \in Um_r(n, R)$, since $vw = 1$ implies $vAA^{-1}w = 1$. So right matrix multiplication defines a right group action of $GL_n(R)$ on $Um_r(n, R)$. (Similarly there is a left group action of $GL_n(R)$ on the set $Um_c(n, R)$ of left unimodular columns in $R^{n \times 1}$.)

For $n > 1$, Proposition (4.14) and the discussion preceding it show R has n-matrix completion if and only if $P \oplus R \cong R^n$ implies $P \cong R^{n-1}$. So, if $m \geq 1$ and R has n-matrix completion for all $n > m$, then all R-modules of stably free rank m are free, as are those of stably free rank $r \geq m$. If R has IBN, the converse follows easily:

(4.16) Corollary. *Suppose R is a ring with IBN and $m \geq 1$. The following are equivalent:*

(i) *Every R-module of stably free rank $r \geq m$ is free.*

(ii) *For each $n > m$, each right unimodular vector v in R^n is part of a basis of R^n.*

(iii) *R has n-matrix completion for each $n > m$.*

(iv) *For each $n > m$, $GL_n(R)$ acts transitively by right multiplication on $Um_r(n, R)$.*

(v) *For each $n > m$ and each right unimodular $v \in R^n$, there exists a matrix $A \in GL_n(R)$ and there exist $a'_1, \ldots, a'_{n-1} \in R$ with*

$$vA \;=\; (a'_1, \ldots, a'_{n-1}, 0) \, . \qquad \blacksquare$$

(4.17) Corollary. *If R is a ring with IBN, then every stably free R-module is free if and only if R is weakly finite and has matrix completion.* $\qquad \blacksquare$

(4.18) Note. Matrix completion means n-matrix completion for all $n > 1$. It would be convenient if n-matrix completion implied $(n + 1)$-matrix completion — for infinitely many conditions could be replaced by one. But that is not the case: Every commutative ring R has 2-matrix completion, since $ax + by = 1$ in R implies

$$\begin{bmatrix} a & b \\ -y & x \end{bmatrix}^{-1} \;=\; \begin{bmatrix} x & -b \\ y & a \end{bmatrix} \, .$$

But in the commutative ring

$$R \;=\; \mathbb{R}[X, Y, Z]/(X^2 + Y^2 + Z^2 - 1)$$

of Example (4.8) (ii), the unimodular vector $(\overline{X}, \overline{Y}, \overline{Z})$ does not complete to a matrix in $GL_3(R)$; so R does not have 3-matrix completion.

4C. Exercises

1. If R is a ring, say R has **WF** if R is weakly finite; say R has **NNP** (no negative projectives) if $P \oplus R^m \cong R^n$ implies $m \le n$.

   (a) For nonzero rings R, prove WF $\Rightarrow$ NNP $\Rightarrow$ IBN.
   (b) Show R has NNP if and only if for $A \in R^{m \times n}$ and $B \in R^{n \times m}$, $AB = I_m$ implies $m \le n$.
   (c) P. M. Cohn's example (4.8) (iii) shows IBN $\not\Rightarrow$ NNP. Prove NNP $\not\Rightarrow$ WF. *Hint:* Suppose R, S are nonzero rings, R has WF, and S does not. Consider $R \times S$.
   (d) Prove $K_0^+(R) = K_0(R)$ if and only if R does not have NNP.
   (e) Prove every stably free R-module is stably isomorphic to a free R-module if and only if R has NNP or does not have IBN.

2. Prove Theorem (1.39) with NNP in place of IBN.

3. Suppose R is a ring with IBN. Show that all stably free right R-modules are free if and only if all stably free left R-modules are free. *Hint:* To prove R^{op} weakly finite implies R weakly finite, use the isomorphism

$$M_n(R)^{op} \;\cong\; M_n(R^{op})$$

given by the transpose. To prove matrix completion for R^{op} implies that for R, note that, if $v \in R^{1 \times n}$, $w \in R^{n \times 1}$, and $vw = 1$, then for some $A \in GL_n(R)$, Aw is the transpose of e_1; then $vA^{-1}Aw = 1$ implies there exists $B \in GL_n(R)$ with $vA^{-1}B = e_1$.

4. Suppose R is a ring and n, m are positive integers with $m \le n$. Prove $A \in R^{m \times n}$ has a right inverse $B \in R^{n \times m}$ if and only if the rows of A are (left) R-linearly independent and span a direct summand of R^n. Then show the following are equivalent:

   (i) $P \oplus R^m \cong R^n \;\Rightarrow\; P \cong R^{n-m}$.
   (ii) Every right invertible matrix $A \in R^{m \times n}$ is the first m rows of some $A' \in GL_n(R)$.

5. Suppose R is a commutative ring, and $(f_1, \ldots, f_n)$ is a unimodular row in $R[t]$, for which the leading coefficients form a unimodular row in R. It is a theorem of Vaserstein (see Lam [78, III. 2]) that, for some $A \in GL_n(R[t])$,

$$(f_1, \ldots, f_n)A \;=\; (f_1(0), \ldots, f_n(0)) \,.$$

Use this to prove that, if $m \geq 1$ and F is a field, all f.g. stably free $F[t_1, \ldots, t_m]$-modules are free. Compare this with Example (4.8) (i).

4D. Stable Rank

In the preceding section, (4.16) (v) showed that a key to matrix completion, and hence to proving stably free R-modules are free, is the ability to shorten unimodular rows. Specifically, if v is a right unimodular vector in R^{n+1}, one needs a matrix $A \in GL_{n+1}(R)$ for which vA has last entry 0.

Sometimes this can be done with a matrix A of a very simple type: For any $r_1, \ldots, r_n \in R$, the matrix

$$(4.19) \qquad E = \begin{bmatrix} 1 & 0 & \cdots & 0 & 0 \\ 0 & 1 & \cdots & 0 & 0 \\ \vdots & \vdots & & \vdots & \vdots \\ 0 & 0 & \cdots & 1 & 0 \\ r_1 & r_2 & \cdots & r_n & 1 \end{bmatrix}$$

is in $GL_{n+1}(R)$, with inverse of the same form with $-r_i$ in place of r_i for each i. (Likewise, the transpose of E is in $GL_{n+1}(R)$.) If

$$v = (a_1, \ldots, a_{n+1}) \in R^{n+1}$$

is right unimodular, so is

$$vE = (a_1 + a_{n+1}r_1, \ldots, a_n + a_{n+1}r_n, a_{n+1}).$$

If $r_1, \ldots, r_n$ can be chosen so the first n coordinates of vE form a right unimodular vector in R^n, then, for some $s_1, \ldots, s_n \in R$,

$$-a_{n+1} = (a_1 + a_{n+1}r_1)s_1 + \cdots + (a_n + a_{n+1}r_n)s_n .$$

In that case,

$$vEE' = (a_1 + a_{n+1}r_1, \ldots, a_n + a_{n+1}r_n, 0) ,$$

where

$$E' = \begin{bmatrix} 1 & 0 & \cdots & 0 & s_1 \\ 0 & 1 & \cdots & 0 & s_2 \\ \vdots & \vdots & \ddots & \vdots & \vdots \\ 0 & 0 & \cdots & 1 & s_n \\ 0 & 0 & \cdots & 0 & 1 \end{bmatrix} .$$

So v has been shortened, in the sense of (4.16) (v).

(4.20) Definitions. A right unimodular vector $(a_1, \ldots, a_{n+1})$ in R^{n+1} **can be shortened** if there exist $r_1, \ldots, r_n \in R$ for which

$$(a_1 + a_{n+1}r_1 \ , \ \ldots \ , \ a_n + a_{n+1}r_n)$$

is right unimodular in R^n. The ring R has n-**shorten** if every right unimodular row in R^{n+1} can be shortened. The **stable rank** of R (abbreviated **sr**(R)) is the least positive integer n for which R has n-shorten; if no such n exists, say $sr(R) = \infty$.

By the preceding discussion and (4.14), n-shorten implies $(n + 1)$-matrix completion. The condition n-shorten is often stronger than $(n + 1)$-matrix completion (see Exercise 1), and it has more convenient properties:

(4.21) Proposition. *For each ring R, n-shorten implies $(n + 1)$-shorten.*

Proof. Suppose $a_i, b_i \in R$ for $1 \le i \le n + 2$, and

$$a_1 b_1 + \cdots + a_{n+2} b_{n+2} \ = \ 1 \ .$$

Isolating the last two terms, $b = a_{n+1} b_{n+1} + a_{n+2} b_{n+2}$, we have

$$a_1 b_1 + \cdots + a_n b_n + b \cdot 1 \ = \ 1 \ .$$

Assuming n-shorten, there exist $c_1, \ldots, c_n \in R$ with

$$(a_1 + bc_1 \ , \ \ldots \ , \ a_n + bc_n)$$

right unimodular in R^n. So

$$v \ = \ (a_1 + bc_1 \ , \ \ldots \ , \ a_n + bc_n \ , \ a_{n+1})$$

is also right unimodular in R^{n+1}. If E is the matrix (4.19) with $r_i = -b_{n+1} c_i$ for each i, then

$$vE \ = \ (a_1 + a_{n+2}(b_{n+2}c_1) \ , \ \ldots \ , \ a_n + a_{n+2}(b_{n+2}c_n), a_{n+1} + a_{n+2}0)$$

is right unimodular in R^{n+1}. ∎

(4.22) Corollary. *If R is a nonzero ring without IBN, then $sr(R) = \infty$.*

Proof. Suppose R has type (h, k), which means $R^h \cong R^{h+k}$ is the first repetition in the list $R^1, R^2, R^3, \ldots$. If $P = R^{h-1}$, then for each integer $m \ge 1$,

$$P \oplus R \ \cong \ R^{h+mk}, \ \text{but}$$
$$P \ \not\cong \ R^{h+mk-1} \ .$$

By (4.14), there is a unimodular row $v \in R^{h+mk}$ that cannot be shortened. By (4.21), n-shorten cannot hold if $n \le h + mk$; and this is the case for every $m \ge 1$. ∎

(4.23) Corollary. *If R is a nonzero ring with finite stable rank, each R-module of stably free rank $r \geq sr(R)$ is free.*

Proof. By (4.22), R has IBN. Since $(n-1)$-shorten holds for all $n > sr(R)$, (4.16) (v) is true with $m = sr(R)$. ∎

To shorten extra-long unimodular rows, one need not alter more than $sr(R)$ coordinates:

(4.24) Skipping Lemma. *Suppose R is a ring, $sr(R) = n < m$, and*
$$(a_1, \ldots, a_{m+1}) \in R^{m+1}$$
is right unimodular. Then $(a_1 + a_{m+1}r_1 , \ldots , a_m + a_{m+1}r_m)$ is right unimodular for some $r_1, \ldots, r_m \in R$ with $r_{n+1} = \cdots = r_m = 0$.

Proof. Permute coordinates to place a_{m+1} just before a_{n+1}. After a sequence of shortenings, obtain a right unimodular row
$$(a_1 + \alpha_1 , \ldots , a_n + \alpha_n , a_{m+1} + \alpha_{m+1})$$
with each α_i in the right ideal $J = a_{n+1}R + \cdots + a_m R$. Shorten again to obtain a right unimodular row $v + \beta$, where
$$v = (a_1 + a_{m+1}r_1 , \ldots , a_n + a_{m+1}r_n)$$
and $\beta \in J^n$. Regard v as a row matrix in $R^{1 \times n}$ and let w denote the row matrix:
$$w = [a_{n+1} \cdots a_m] \in R^{1 \times (m-n)} .$$
The partitioned matrix $[v + \beta \quad w] \in R^{1 \times m}$ is right unimodular, since $v + \beta$ is. Expressing the coordinates of β in terms of $a_{n+1}, \ldots, a_m$, there is a matrix $A \in R^{(m-n) \times n}$ with $\beta = wA$. Then
$$[v \quad w] = [v + \beta \quad w] \begin{bmatrix} I_n & 0 \\ -A & I_{m-n} \end{bmatrix}$$
is also right unimodular. ∎

One could also define the **left stable rank** of a ring R to be the least positive integer n for which R has the condition **left n-shorten**: "For each left unimodular vector $(b_1, \ldots, b_{n+1})$ in R^{n+1}, there exist $r_1, \ldots, r_n \in R$ for which
$$(b_1 + r_1 b_{n+1} , \ldots , b_n + r_n b_{n+1})$$
is left unimodular in R^n." The next proposition of Vaserstein [71] shows left n-shorten is equivalent to n-shorten, and left stable rank equals stable rank:

(4.25) Proposition. *For each ring R, $sr(R) = sr(R^{op})$.*

Proof. Suppose $sr(R) = m$ ($\neq \infty$) and $(b_1, \ldots, b_{m+1})$ is right unimodular in $(R^{op})^{1 \times (m+1)}$. So there exist $a_1, \ldots, a_{m+1} \in R^{op}$ with

$$b_1 a_1 + \cdots + b_{m+1} a_{m+1} = 1 .$$

Then, in R,

$$a_1 b_1 + \cdots + a_m b_m + (a_{m+1} b_{m+1}) \cdot 1 = 1 .$$

Let

$$a = [a_1 \cdots a_m] \in R^{1 \times m} ,$$

$$b = \begin{bmatrix} b_1 \\ \vdots \\ b_m \end{bmatrix} \in R^{m \times 1} .$$

Since R has m-shorten, there exists $v \in R^{1 \times m}$ for which

$$a' = a + a_{m+1} b_{m+1} v$$

is right unimodular in R^m. So, for some $c \in R^{m \times 1}$, $a'c = -a_{m+1}$. Define

$$u = -(I_m - bv)c \in R^{m \times 1} .$$

Working with block matrices

$$\begin{bmatrix} a_{11} & a_{12} & a_{13} \\ a_{21} & a_{22} & a_{23} \\ a_{31} & a_{32} & a_{33} \end{bmatrix}$$

with $a_{11}, a_{31}, a_{13}, a_{33} \in R^{1 \times 1}$, $a_{21}, a_{23} \in R^{m \times 1}$, $a_{12}, a_{32} \in R^{1 \times m}$, and $a_{22} \in R^{m \times m}$, the following are matrices in $GL_{m+2}(R)$:

$$A = \begin{bmatrix} 1 & -a & -a_{m+1} \\ 0 & 1 & 0 \\ 0 & 0 & 1 \end{bmatrix}$$

$$B = \begin{bmatrix} 1 & 0 & 0 \\ b & 1 & 0 \\ b_{m+1} & 0 & 1 \end{bmatrix}$$

$$C = \begin{bmatrix} 1 & 0 & 0 \\ 0 & 1 & 0 \\ 0 & b_{m+1} v & 1 \end{bmatrix}$$

$$D = \begin{bmatrix} 1 & -v & 0 \\ 0 & 1 & 0 \\ 0 & 0 & 1 \end{bmatrix}$$

$$E = \begin{bmatrix} 1 & 0 & 0 \\ 0 & 1 & c \\ 0 & 0 & 1 \end{bmatrix}$$

$$F = \begin{bmatrix} 1 & 0 & 0 \\ 0 & 1 & u \\ 0 & 0 & 1 \end{bmatrix}$$

$$G = \begin{bmatrix} 1 & 0 & 0 \\ 0 & 1 & 0 \\ -b_{m+1} & 0 & 1 \end{bmatrix}.$$

Multiplying these in the order indicated by the parentheses

$$H = (F((((AB)C)D)E))G \,,$$

we obtain the invertible matrix

$$H = \begin{bmatrix} 0 & -a' & 0 \\ b + ub_{m+1} & 1 - bv & 0 \\ 0 & 0 & 1 \end{bmatrix}$$

in $GL_{m+2}(R)$. If the first row of H^{-1} is $(w_0, \ldots, w_{m+1})$, and if

$$u = \begin{bmatrix} u_1 \\ \vdots \\ u_m \end{bmatrix},$$

then

$$\sum_{i=1}^{m} w_i(b_i + u_i b_{m+1}) = 1$$

in R. So

$$\sum_{i=1}^{m} (b_i + b_{m+1} u_i) w_i = 1$$

in R^{op}, proving $(b_1, \ldots, b_{m+1})$ can be shortened in R^{op}. Thus

$$sr(R^{op}) \leq m = sr(R).$$

Since $(R^{op})^{op} = R$, it follows that $sr(R^{op}) = sr(R)$. ∎

The next two theorems show that stable rank leads to more general cancellation results,

$$M \oplus R \;\cong\; N \oplus R \;\Rightarrow\; M \cong N \,,$$

than those already obtained in (4.23), where N was free.

(4.26) Theorem. *Suppose R is a ring and X, Y are R-modules. For each integer $n \geq sr(R)$, $X \oplus R^{n+1} \cong Y \oplus R$ implies $X \oplus R^n \cong Y$.*

Proof. Suppose σ is an isomorphism from $X \oplus R^{n+1}$ to $Y \oplus R$. The R-linear maps

$$\alpha \;=\; [0 \;\; i_R] \circ \sigma : X \oplus R^{n+1} \to R \,,$$

$$\beta \;=\; \sigma^{-1} \circ \begin{bmatrix} 0 \\ i_R \end{bmatrix} : R \to X \oplus R^{n+1} \,,$$

have composite $\alpha \circ \beta = i_R$. Suppose their matrix forms (as in (1.31)) are

$$\alpha \;=\; [\alpha_0 \;\cdots\; \alpha_{n+1}] \,,$$

$$\beta \;=\; \begin{bmatrix} \beta_0 \\ \vdots \\ \beta_{n+1} \end{bmatrix} .$$

Each $f \in \operatorname{Hom}_R(R,R)$ is right multiplication by some $r \in R$. Say $\alpha_0 \circ \beta_0 = (-)r_0$, and for $1 \leq i \leq n+1$, $\alpha_i = (-)a_i$ and $\beta_i = (-)b_i$. Then

$$1 \;=\; (\alpha \circ \beta)(1) \;=\; r_0 + b_1 a_1 + \cdots + b_{n+1} a_{n+1} \,.$$

So $(r_0, a_1, \ldots, a_{n+1})$ is left unimodular. By (4.25) and the Skipping Lemma (4.24), there exist $c_1, \ldots, c_n \in R$ with

$$(r_0 \,,\; a_1 + c_1 a_{n+1} \,,\; \ldots \,,\; a_n + c_n a_{n+1})$$

left unimodular. That is, there exist $d_0, \ldots, d_n \in R$ with

$$(4.27) \qquad\qquad 1 \;=\; d_0 r_0 \;+\; \sum_{i=1}^{n} d_i (a_i + c_i a_{n+1}) \,.$$

Define R-linear maps ρ, γ, τ in the diagram

$$
\begin{array}{ccccc}
X \oplus R^{n+1} & \xrightarrow{\;\rho\;} & X \oplus R^{n+1} & \xrightarrow{\;\sigma\;} & Y \oplus R \\
\gamma \uparrow & & \alpha \downarrow & \nearrow \pi = [0 \;\; i_R] & \\
X \oplus R^n & \xleftarrow{\;\tau\;} & R & &
\end{array}
$$

by

$$\rho(x, r_1, \ldots, r_{n+1}) = \left(x, r_1, \ldots, r_n, r_{n+1} + \sum_{i=1}^{n} r_i c_i\right),$$

$$\gamma(x, r_1, \ldots, r_n) = (x, r_1, \ldots, r_n, 0),$$

$$\tau(r) = (\beta_0(rd_0), rd_1, \ldots, rd_n).$$

Then ρ is an isomorphism and, by (4.27), $\alpha \circ \rho \circ \gamma \circ \tau = i_R$.
So there are split short exact sequences:

$$0 \longrightarrow W \overset{\subseteq}{\longrightarrow} X \oplus R^n \underset{\tau}{\overset{\alpha\rho\gamma}{\rightleftarrows}} R \longrightarrow 0$$

and

$$0 \longrightarrow Y \overset{\left[\begin{smallmatrix}1\\0\end{smallmatrix}\right]}{\longrightarrow} Y \oplus R \underset{\sigma\rho\gamma\tau}{\overset{\pi}{\rightleftarrows}} R \longrightarrow 0.$$

So

$$X \oplus R^n \cong W \oplus R \cong \frac{(X \oplus R^n) \oplus R}{\tau(R) \oplus \{0\}}$$

$$= \frac{X \oplus R^{n+1}}{(\gamma \circ \tau)(R)} \cong \frac{Y \oplus R}{(\sigma \circ \rho \circ \gamma \circ \tau)(R)} \cong Y. \qquad \blacksquare$$

(4.28) Theorem. *If R is a ring with $sr(R) = 1$, and X, Y are R-modules, then $X \oplus R \cong Y \oplus R$ implies $X \cong Y$.*

Proof. First we prove R is weakly 1-finite. In R, suppose $ab = 1$ and $c = 1 - ba$. Then $ba + c1 = 1$; so (b, c) is right unimodular. Since $sr(R) = 1$, there exist $d, e \in R$ with $(b + cd)e = 1$. Since $ac = a - aba = 0$, $a(b + cd) = 1$. So

$$a = a(b + cd)e = e$$

is a unit, as is $b = a^{-1}$.

Now begin, as in the proof of (4.26), with $n = 0$. Then $1 = r_0 + b_1 a_1$. So (b_1, r_0) is right unimodular, and for some $s \in R$,

$$u = b_1 + r_0 s \in R^*.$$

Let $\phi : R \to R$ denote right multiplication by s, and $\psi : R \to R$ denote right multiplication by u^{-1}. Then there are split short exact sequences:

$$0 \longrightarrow X \overset{\left[\begin{smallmatrix}1\\-\phi\alpha_0\end{smallmatrix}\right]}{\longrightarrow} X \oplus R \underset{\beta}{\overset{\psi[\phi\alpha_0 \ \ 1]}{\rightleftarrows}} R \longrightarrow 0$$

$$0 \longrightarrow Y \xrightarrow{\begin{bmatrix} 1 \\ 0 \end{bmatrix}} Y \oplus R \underset{\sigma\beta}{\overset{\pi}{\rightleftarrows}} R \longrightarrow 0 \ .$$

So

$$X \cong \frac{X \oplus R}{\beta(R)} \cong \frac{Y \oplus R}{(\sigma \circ \beta)(R)} \cong Y \ . \qquad \blacksquare$$

(4.29) Corollary. *If R is a ring with $sr(R) = 1$, every stably free R-module is free.*

Proof. For positive integers m and n, suppose $P \oplus R^m \cong R^n$. By (4.28) we may cancel R's. If $m \le n$, it follows that $P \cong R^{n-m}$. If $m > n$, $P \oplus R^{m-n} \cong 0$; so $P \cong 0$. Either way, P is free. $\qquad \blacksquare$

4D. Exercises

1. Prove $\mathbb{Z}$ has 2-matrix completion, but not 1-shorten.

2. If R is a commutative principal ideal domain, prove $sr(R) \le 2$.

3. Use the Skipping Lemma (4.24) to prove that, if I is an ideal of a ring R, then $sr(R/I) \le sr(R)$.

4. If R_1 and R_2 are rings, prove

$$sr(R_1 \times R_2) = \max\{sr(R_1), sr(R_2)\} \ .$$

Then generalize to n $(< \infty)$ direct factors.

5. If a ring R is local (meaning its nonunits form an ideal), prove $sr(R) = 1$.

6. If D is a division ring and n is a positive integer, prove $sr(M_n(D)) = 1$. *Hint:* If $A \in M_n(D)$, its column space, as a right D-module, is $AM_n(D)$.

7. For the ring A of column-finite $\mathbb{P} \times \mathbb{P}$ matrices over a nontrivial ring R (see (1.37)), prove directly that $sr(A) = \infty$. *Hint:* If $A \cong A^n$ is the isomorphism taking a to $(a_1, \ldots, a_m)$, where the columns of a_i are the $i, i+n, i+2n, \ldots$ columns of a, prove that the image of the identity 1_A is a right unimodular vector in A^n that cannot be shortened.

8. The Jacobson radical rad R of a ring R is the intersection of the maximal left ideals of R. By (6.30), rad R is also the intersection of the maximal right ideals of R. If I is an ideal of R contained in rad R, show the stable rank of R/I equals the stable rank of R.

4E. Dimensions of a Ring

The dimension of a f.g. vector space V over a field can be computed as the number of steps r in a longest possible chain of subspaces

$$V_0 \subsetneq V_1 \subsetneq \cdots \subsetneq V_r$$

of V. There is a well-known analog for commutative rings:

(4.30) Definition. If R is a commutative ring, the **Krull dimension Kdim(R)** of R is the number of steps r in a longest chain of prime ideals

$$A_0 \subsetneq A_1 \subsetneq \cdots \subsetneq A_r$$

in R.

Recall that an ideal A of a commutative ring R is **prime** if its complement $R - A$ is nonempty and closed under multiplication — or equivalently, if R/A is an integral domain. So if R is a field, its only prime ideal is $\{0\}$, and Kdim$(R) = 0$. For each integer $n > 1$, Kdim$(\mathbb{Z}/n\mathbb{Z}) = 0$ since if n is composite, its prime ideals are $p\mathbb{Z}/n\mathbb{Z}$ for primes $p|n$.

If R is a principal ideal domain, each nonzero prime ideal of R is a maximal ideal; so, unless R is a field, Kdim$(R) = 1$. If F is a field, the polynomial ring $F[x_1, \ldots, x_n]$, in n indeterminates, has a chain of prime ideals

$$\{0\} \subsetneq \langle x_1 \rangle \subsetneq \langle x_1, x_2 \rangle \subsetneq \cdots \subsetneq \langle x_1, \ldots, x_n \rangle \ .$$

By a well-known theorem of commutative algebra (see Matsumura [89]), no longer chain of primes can exist; so

$$\text{Kdim } (F[x_1, \ldots, x_n]) \ = \ n \ .$$

As we are about to show, the stable rank of a commutative ring is closely related to the Krull dimension. Suppose R is any ring, commutative or not, and $S \subseteq R$. Denote by $\mathbf{J}(S)$ the intersection of R and all maximal right ideals M of R with $S \subseteq M$. Then $J(R) = R$, and $S \subseteq J(S)$. A maximal right ideal of R contains $J(S)$ if and only if it contains S; so $J(J(S)) = J(S)$.

(4.31) Definitions. A vector $(a_1, \ldots, a_{n+1}) \in R^{n+1}$ **can be shortened** if there exist $r_1, \ldots, r_n \in R$ for which

$$J(a_1 + a_{n+1}r_1, \ldots, a_n + a_{n+1}r_n) \ = \ J(a_1, \ldots, a_{n+1}) \ .$$

Note that this equation is equivalent to

$$a_{n+1} \ \in \ J(a_1 + a_{n+1}r_1 \ , \ \ldots \ , \ a_n + a_{n+1}r_n) \ .$$

The ring R has **absolute n-shorten** if every element of R^{n+1} can be shortened. The **absolute stable rank** of R (abbreviated **asr(R)**) is the least positive integer n for which R has absolute n-shorten.

A vector $(a_1, \ldots, a_{n+1})$ is right unimodular if and only if $J(a_1, \ldots, a_{n+1}) = R$. So the phrase "can be shortened" has the same meaning here as it did in Definition (4.20) — only now it can be applied to all vectors in R^{n+1}. If you can shorten every vector in R^{n+1}, you can shorten those that are right unimodular; so absolute n-shorten implies n-shorten, and $sr(R) \leq asr(R)$.

In case R is a *commutative* ring, take $\mathcal{J}$ to be the set of all ideals of R that are intersections of R with a set of maximal ideals of R. Then

$$\mathcal{J} = \{A \lhd R : J(A) = A\} .$$

(4.32) Definition. If R is a commutative ring, the **maximal spectrum dimension $\mathrm{Kdim}_{\mathcal{J}}(R)$** of R is the number of steps r in a longest chain of prime ideals of R in $\mathcal{J}$.

Comparing their definitions, (4.30) and (4.32), it is clear that

$$\mathrm{Kdim}_{\mathcal{J}}(R) \leq \mathrm{Kdim}(R) .$$

Recall that a commutative ring is **noetherian** if its set of ideals has acc (the ascending chain condition) — see Appendix B for details. We shall call a commutative ring $\mathcal{J}$-**noetherian** if $\mathcal{J}$ has acc.

(4.33) Lemma. *If A is a proper ideal of a $\mathcal{J}$-noetherian commutative ring R, then $J(A)$ is an intersection of finitely many prime ideals from $\mathcal{J}$.*

Proof. Denote by E the set of proper ideals in $\mathcal{J}$ that contain A, and by F the set of intersections of finitely many prime ideals in E. Assume the lemma fails for A. Then $J(A) \in E - F$. So $E - F$ is a nonempty subset of $\mathcal{J}$; since $\mathcal{J}$ has acc, $E - F$ has a maximal element M (see (B.3)).

By the definition of F, M is not prime; so there exist $x, y \in R$ with $x, y \notin M$ but $xy \in M$. Define

$$M_1 = J(M \cup \{x\}) , \quad M_2 = J(M \cup \{y\}) .$$

Then $M_1, M_2 \in E$. Since each maximal ideal containing M is prime, it contains x or y; so

$$M = J(M) = M_1 \cap M_2 .$$

Since $x, y \notin M$, $M \subsetneqq M_1$ and $M \subsetneqq M_2$. By the maximality of M in $E - F$, M_1 and M_2 belong to F, as does their intersection M, a contradiction. ∎

(4.34) Theorem. (Estes and Ohm [67], Stein [78]) *If R is a $\mathcal{J}$-noetherian commutative ring,*

$$asr(R) \ \leq \ Kdim_{\mathcal{J}}(R) + 1 .$$

Proof. First we recall a basic fact about prime ideals. If a prime ideal P of a commutative ring R contains an intersection of finitely many ideals $Q_1, \ldots, Q_n$ of R, then $P \supseteq Q_i$ for some i; for otherwise, there would exist $x_i \in Q_i - P$ for each i, with $x_1 \cdots x_n \in P$. We use this twice below.

Suppose R is a $\mathcal{J}$-noetherian commutative ring with $Kdim_{\mathcal{J}}(R) = d$, and suppose $(a_1, \ldots, a_{d+2}) \in R^{d+2}$ cannot be shortened. Then

$$a_{d+2} \notin J(a_1', \ldots, a_{d+1}')$$

for all vectors $(a_1', \ldots, a_{d+1}')$ in

$$(a_1 + a_{d+2}R) \times \cdots \times (a_{d+1} + a_{d+2}R) .$$

Note that, for $1 \leq i \leq d+1$, each list $a_1', \ldots, a_i'$ generates a proper ideal of R, so by Lemma (4.33),

$$J(a_1', \ldots, a_i') \ = \ \bigcap \mathcal{P}$$

for a finite set $\mathcal{P}$ of prime ideals from $\mathcal{J}$. We may assume the set $\mathcal{P}$ is *minimal*, in the sense that no member of $\mathcal{P}$ is contained in another member of $\mathcal{P}$. Set $a_0' = 0$.

Claim. *One can choose $(a_1', \ldots, a_{d+1}')$ in*

$$(a_1 + d_{d+2}R) \times \cdots \times (a_{d+1} + a_{d+2}R)$$

and finite minimal sets $\mathcal{P}_0, \ldots, \mathcal{P}_d$ of prime ideals from $\mathcal{J}$, so that for $0 \leq i \leq d$,

$$J(a_0', a_1', \ldots, a_i') \ = \ \bigcap \mathcal{P}_i$$

and each member of $\mathcal{P}_i$ that does not contain a_{d+2} also does not contain a_{i+1}'.

Proof of Claim. Transform a_{i+1} to a_{i+1}', one i at a time, in the order $i = 0, 1, \ldots, d$ as follows: Suppose $a_0' = 0$, $a_1', \ldots, a_i'$ have been chosen, and

$$J(a_0', \ldots, a_i') \ = \ \bigcap \mathcal{P}_i$$

for a finite minimal set $\mathcal{P}_i$ of prime ideals from $\mathcal{J}$. Let $\mathcal{Q}$ denote the set of all $Q \in \mathcal{P}_i$ with $a_{i+1}, a_{d+2} \notin Q$. If $P \in \mathcal{P}_i$ with $a_{i+1} \in P$ but $a_{d+2} \notin P$, then P is unequal to, and hence does not contain, each $Q \in \mathcal{Q}$. Since P is prime, P also does not contain $\cap \mathfrak{Q}$. So there is some $b \in (\cap \mathfrak{Q}) - P$. Then $a_{i+1} + a_{d+2}b$ is not in P, and is not in any member of $\mathfrak{Q}$.

Repeating the process with $a_{i+1} + a_{d+2}b$ in place of a_{i+1}, and with P in the new $\mathfrak{Q}$, we eventually reach $a'_{i+1} \in a_{i+1} + a_{d+2}R$, which does not belong to any $P \in \mathcal{P}_i$ not containing a_{d+2}. This proves the claim.

Suppose $a'_0, \ldots, a'_{d+1}$ and $\mathcal{P}_0, \ldots, \mathcal{P}_d$ are chosen as in the claim. Since

$$J(a'_1, \ldots, a'_{d+1})$$

is an intersection of prime ideals from $\mathcal{J}$ and does not contain a_{d+2}, one of these prime ideals P_{d+1} does not contain a_{d+2}. Since

$$P_{d+1} \supseteq J(a'_1, \ldots, a'_d) = \bigcap \mathcal{P}_d$$

and P_{d+1} is prime, $P_{d+1} \supseteq P_d$ for some $P_d \in \mathcal{P}_d$. Similarly, we have prime ideals

$$P_{d+1} \supseteq P_d \supseteq \cdots \supseteq P_0$$

with $P_i \in \mathcal{P}_i$ for $0 \le i \le d$. None of them contain a_{d+2}; so by the choices made in the claim, $a'_i \in P_i - P_{i-1}$ for $1 \le i \le d+1$. Thus

$$P_0 \subsetneq P_1 \subsetneq \cdots \subsetneq P_{d+1}$$

is a chain of prime ideals from $\mathcal{J}$ of length $d+1$. Since $\mathrm{Kdim}_{\mathcal{J}}(R) = d$, this cannot happen; so R has absolute $(d+1)$-shorten, and $asr(R) \le d+1$. ∎

To summarize, for a $\mathcal{J}$-noetherian commutative ring R,

$$sr(R) \le asr(R) \le \mathrm{Kdim}_{\mathcal{J}}(R) + 1 \le \mathrm{Kdim}(R) + 1 .$$

These inequalities have immediate consequences for cancellation and stably free modules over commutative rings, via (4.26), (4.28), and (4.29) in the last section.

The dimensions of a commutative ring can be generalized, and so can the corresponding cancellation theorems. In the paper Bass [64A], which introduced much of algebraic K-theory, he proved that

$$sr(A) \le \mathrm{Kdim}_{\mathcal{J}}(R) + 1$$

if A is finitely generated as a module over a subring R of its center. And in Magurn, van der Kallen, and Vaserstein [88], this inequality is improved to

$$asr(A) \le \mathrm{Kdim}_{\mathcal{J}}(R) + 1 .$$

For left (or right) noetherian rings, there is a generalization of Krull dimension due to Rentschler and Gabriel [67], and using this definition, Stafford [90] proved

$$asr(R) \leq 1 + \text{Kdim}(R/J(0))$$

for every left (or right) noetherian ring R.

The cancellation theorems that follow from these inequalities can be sharpened considerably for specific classes of rings R. For examples, see Bass [68], Swan [74], and McConnell and Robson [87]. Such theorems are known as "pre-stability" results for K_0, because they establish earlier stability than is implied by the stable rank.

4E. Exercises

1. Prove absolute n-shorten implies absolute $(n + 1)$-shorten.

2. If R is a ring and every right ideal of R is principal, prove $asr(R) = sr(R)$.

3. If R is a commutative integral domain and $R - R^*$ is a non-zero ideal of R, prove $\text{Kdim}_g(R) = 0$ while $\text{Kdim}(R) > 0$. Show

$$R = \left\{ \frac{a}{b} \in \mathbb{Q} : b \text{ is odd}, a \in \mathbb{Z} \right\}$$

is such a ring.

4. If R is a commutative artinian ring, prove $\text{Kdim}(R) = 0$, so $sr(R) = 1$. *Hint:* If P is a prime ideal of R and $0 \neq x \in R/P$, use the descending chain condition on R/P to prove x is invertible. (This result generalizes to noncommutative rings: see Rowen [91, Examples 3.5.39–3.5.40].)

5

Multiplying Modules

Just as modules are added by the direct sum, so they can be multiplied by the "tensor product." If R is a commutative ring, this arithmetic among modules turns $K_0(R)$ into a commutative ring. In the same way, some other Grothendieck-type groups become rings under a suitable product. In §5A we set forth the required properties of such a product, illustrated in §5B by the Burnside ring. The main focus of this chapter is §5C, where the tensor product is defined and its arithmetic properties established. We include applications to the Grothendieck rings of modules, linear endomorphisms, and bilinear forms. The latter specializes to the Witt ring of a field, and we discuss its use in classifying forms.

5A. Semirings

In §3B, we saw how to complete a semigroup $(S, *)$ to an abelian group $(\bar{\bar{S}}, +)$. A prototype is the completion of the natural numbers $(\mathbb{N}, +)$ to the integers $(\mathbb{Z}, +)$. But $\mathbb{N}$ also has a multiplication, which extends to a multiplication on its completion $\mathbb{Z}$.

(5.1) Definition. A **semiring** $(S, *, \cdot)$ is a set S with two binary operations, $*$ and $\cdot$, for which $(S, *)$ is an abelian monoid with identity 0_S, $(S, \cdot)$ is a monoid with identity 1_S, and $\cdot$ distributes over $*$. The semiring $(S, *, \cdot)$ is **commutative** if $(S, \cdot)$ is abelian.

(5.2) Proposition. *The group completion of a semiring is a ring. The group completion of a commutative semiring is a commutative ring.*

Proof. The group completion of the semiring $(S, *, \cdot)$ is the quotient

$$\bar{\bar{S}} = F_{\mathbb{Z}}(S)/\langle D \rangle,$$

where D is the set of elements

$$(x * y) \; - \; x \; - \; y \; \in \; F_{\mathbb{Z}}(S)$$

for $x, y \in S$. The free $\mathbb{Z}$-module $F_{\mathbb{Z}}(S)$ is a ring (the monoid ring $\mathbb{Z}[S]$ of (1.16)) under a multiplication extending the operation $\cdot$ in S. Since $\cdot$ distributes over $*$, $\langle D \rangle$ is an ideal of $F_{\mathbb{Z}}(S)$; so $\bar{\bar{S}}$ is a ring under addition and multiplication of cosets. A typical element of $\bar{\bar{S}}$ can be written in the form $\bar{x} - \bar{y}$, where $x, y \in S$, $\bar{x} = x + \langle D \rangle$, and $\bar{y} = y + \langle D \rangle$. So the multiplication in $\bar{\bar{S}}$ takes the form

$$(\bar{x} - \bar{y})(\bar{x}' - \bar{y}') \;\; = \;\; \overline{x \cdot x'} \; - \; \overline{x \cdot y'} \; - \; \overline{y \cdot x'} \; + \; \overline{y \cdot y'} \; .$$

If $\cdot$ commutes, so does this multiplication. The additive and multiplicative identities of the ring $\bar{\bar{S}}$ are $\overline{0_S}$ and $\overline{1_S}$. ∎

If $(S, *, \cdot)$ and $(T, *, \cdot)$ are semirings, a **semiring homomorphism** $f : S \to T$ is a function from S to T satisfying

$$f(0_S) \; = \; 0_T \; , \quad f(1_S) \; = \; 1_T \; ,$$
$$f(x * y) \;\; = \;\; f(x) * f(y) \; ,$$
$$\text{and} \quad f(x \cdot y) \;\; = \;\; f(x) \cdot f(y) \; ,$$

for each $x, y \in S$. The group completion $\gamma : S \to \bar{\bar{S}}$ is a semiring homomorphism. If $g : S \to R$ is a semiring homomorphism into a ring $(R, +, \cdot)$, the unique additive group homomorphism $\bar{g} : \bar{\bar{S}} \to R$ with $\bar{g} \circ \gamma = g$ (see (3.15)) is necessarily a ring homomorphism. *So γ is an initial semiring homomorphism from S into a ring.*

(5.3) Examples.

(i) The set of natural numbers $0, 1, 2, \dots$ is a semiring under ordinary sum and product, with completion the ring of integers $\mathbb{Z}$.

(ii) Suppose R is a ring and $\mathfrak{id}(R)$ is the set of all (two-sided) ideals of R. If A and B are ideals of R, so are

$$A + B \;\; = \;\; \{a + b : a \in A, \;\; b \in B\}$$

and

$$AB \;\; = \;\; \left\{ \sum_{i=1}^{n} a_i b_i : a_i \in A, \;\; b_i \in B, \;\; n > 0 \right\} \; .$$

Notice that $A + B$ is contained in every ideal that contains $A \cup B$, and that $AB \subseteq A \cap B$.

If $A, B, C \in \mathfrak{id}(R)$, then $A(B + C)$ contains both AB and AC, and so it also contains $AB + AC$. On the other hand, each element of $A(B + C)$ is a sum of terms $a(b + c) = ab + ac$, for $a \in A$, $b \in B$, and $c \in C$; so $A(B + C) = AB + AC$.

In a similar fashion one can verify that the ideal sum and product satisfy all the properties needed to make $\mathfrak{id}(R)$ a semiring.

The identity elements of this semiring are the zero ideal (for sum) and the whole ring R (for product). Since $A + R = R = B + R$ for all $A, B \in \mathfrak{id}(R)$, the group completion of $\mathfrak{id}(R)$ is the trivial ring $\{0\}$.

In the next section we describe a classical example of a semiring with a very nontrivial completion to a ring.

5A. Exercises

1. Suppose R is a commutative principal ideal domain, with group of units R^*. From each associate class R^*p of irreducibles in R, choose one element. Denote by S the set consisting of 0, the chosen irreducibles, and their (finite length) products. Then the map $f : S \to \mathfrak{id}(R)$, $f(x) = Rx$, is bijective, by unique factorization in R. Via this bijection, show that S is a semiring under greatest common divisors in S and the multiplication in R. (Here we regard 0 as the greatest common divisor of 0 and 0.)

2. Verify, for all rings R, that $\mathfrak{id}(R)$ is a semiring under ideal sum and product.

3. If X is any set, show that the power set (= the set of all subsets) of X is a commutative semiring with $* = \cup$, $\cdot = \cap$ and also with $* = \cap$, $\cdot = \cup$. What are the group completions of these two semirings?

4. Suppose $(S, +, \cdot)$ is a semiring and $(S, +)$ is a free abelian monoid with basis $B = \{b_1, \ldots, b_n\}$. Prove the multiplication in the ring $\bar{\bar{S}}$ is determined by n^3 integers. (See (3.19).)

5B. Burnside Rings

The notion of "group" studied by Lagrange, Cauchy, Abel, and Galois in the early 19th century was that of a closed (under composition) set of bijections on a finite set X. A subgroup of the group $\mathrm{Aut}(X)$ of all bijections $f : X \to X$ under composition is called a **permutation group**. A **permutation representation** of any group G is a group homomorphism $\rho : G \to \mathrm{Aut}(X)$. The image $\rho(G)$ can be considered a permutation group "representing" G.

An equivalent notion is that of a group "action." An **action** of a group G on a set X is a map $\bullet : G \times X \to X$, $(g, x) \mapsto g \bullet x$, satisfying

$$g \bullet (h \bullet x) = (gh) \bullet x, \quad \text{and} \quad e \bullet x = x,$$

for all $g, h \in G$ and $x \in X$, where e is the identity element of G. There is a bijection, between the set of actions $\bullet$ of G on X and the set of permutation representations ρ of G on X, determined by the equation $\rho(g)(x) = g \bullet x$.

Suppose we fix a group G and consider its actions on finite sets. A G-**set** is a finite set X together with an action of G on X. If X and Y are G-sets, a function $f : X \to Y$ is a G-**map** if

$$f(g \bullet x) \;=\; g \bullet f(x)$$

for all $g \in G$ and $x \in X$. As in Chapter 0, Exercise 5, for each group G the G-sets and G-maps form a concrete category G-Set, in which a G-map is an isomorphism if and only if it is bijective. Each G-set X having n elements is isomorphic to a G-set $\{1, 2, \ldots, n\}$; so G-Set is a modest category — it has a *set* of isomorphism classes.

If X and Y are G-sets, so is their **disjoint union**

$$X \overset{d}{\cup} Y \;=\; (X \times \{1\}) \cup (Y \times \{2\})$$

via $g\bullet(x, 1) = (g\bullet x, 1)$ and $g\bullet(y, 2) = (g\bullet y, 2)$. If $X \cap Y = \emptyset$, then $X \overset{d}{\cup} Y \cong X \cup Y$ in G-Set, where the action of G on $X \cup Y$ combines the actions on X and Y. Also, the cartesian product $X \times Y$ is a G-set via $g \bullet (x, y) = (g \bullet x, g \bullet y)$. For G-sets X, Y, and Z, there are isomorphisms:

$$X \overset{d}{\cup} (Y \overset{d}{\cup} Z) \;\cong\; (X \overset{d}{\cup} Y) \overset{d}{\cup} Z$$

$$X \overset{d}{\cup} Y \;\cong\; Y \overset{d}{\cup} X$$

$$\emptyset \overset{d}{\cup} X \;\cong\; X$$

$$X \times (Y \times Z) \;\cong\; (X \times Y) \times Z$$

$$X \times Y \;\cong\; Y \times X$$

$$\{1\} \times X \;\cong\; X$$

$$X \times (Y \overset{d}{\cup} Z) \;\cong\; (X \times Y) \overset{d}{\cup} (X \times Z) \,,$$

where $\emptyset$ and $\{1\}$ are G-sets in the only way possible.

In G-Set, if $X \cong X'$ and $Y \cong Y'$, then

$$X \overset{d}{\cup} Y \;\cong\; X' \overset{d}{\cup} Y' \quad \text{and} \quad X \times Y \;\cong\; X' \times Y' \,.$$

So $\overset{d}{\cup}$ and $\times$ are binary operations on the category G-Set. And if $c(X)$ denotes the isomorphism class of X, the set $\mathfrak{I}(G\text{-Set})$ of isomorphism classes is a commutative semiring under

$$c(X) + c(Y) \;=\; c(X \overset{d}{\cup} Y) \,, \quad \text{and}$$
$$c(X) \cdot c(Y) \;=\; c(X \times Y) \,.$$

In a less category-theoretic form, this semiring was partially described in the 1911 edition of the landmark group theory text, Burnside [55]. Consider some basic facts about group actions. Suppose X is a G-set. There is an equivalence relation on X given by saying x is related to y if there exists $g \in G$ with $g \bullet x = y$. The equivalence classes are called **orbits**; the orbit of x is

$$G \bullet x = \{g \bullet x : g \in G\} .$$

A G-set is **transitive** if it has only one orbit. For instance, if $x \in X$, the orbit $G \bullet x$ is a transitive G-set. As another example, if H is a subgroup of finite index in G, the set G/H of left cosets of H is a transitive G-set under $g \bullet g'H = (gg')H$.

Except for the order of terms, *each element of $\mathfrak{I}(G\text{-}\mathfrak{S}\mathfrak{e}\mathfrak{t})$ has a unique expression as a sum of isomorphism classes of transitive G-sets:* If X has orbits $X_1, \dots, X_n$, then $c(X) = c(X_1) + \dots + c(X_n)$. Suppose $T_1, \dots, T_m, T_1', \dots, T_n'$ are transitive G-sets and

$$c(T_1) + \dots + c(T_m) = c(T_1') + \dots + c(T_n') .$$

Define $S_i = T_i \times \{i\}$, with G-action on the first coordinate. Then $T_i \cong S_i$ for each i, and $S_1, \dots, S_m$ are pairwise disjoint transitive G-sets. So they are the orbits of their union S, and $c(S) = c(T_1) + \dots + c(T_m)$. Construct S' in the same way, with orbits $S_1', \dots, S_n'$ isomorphic to $T_1', \dots, T_n'$. Then $c(S) = c(S')$; so there is an isomorphism $f : S \to S'$. Since f is bijective with $f(G \bullet x) = G \bullet f(x)$, it follows that $n = m$ and for some permutation σ of $\{1, \dots, m\}$, $S_i \cong S_{\sigma(i)}'$ and hence $c(T_i) = c(T_{\sigma(i)}')$ for each i.

There is a standard form for each transitive G-set as well. If X is a G-set and $x \in X$, the **stabilizer** of x is the set

$$G_x = \{g \in G : g \bullet x = x\} .$$

For each $x \in X$, G_x is a subgroup of G. For each $g \in G$, the coset gG_x coincides with the set

$$\{g' \in G : g' \bullet x = g \bullet x\} ;$$

so $g \bullet x \mapsto gG_x$ defines a bijection $G \bullet x \to G/G_x$, which is evidently an isomorphism of G-sets:

$$G \bullet x \cong G/G_x .$$

(This isomorphism underlies the standard counting principle of basic group theory: "The size of the orbit equals the index of the stabilizer.") Thus every transitive G-set is isomorphic to G/H for some subgroup H of G.

For each $x \in X$ and $g \in G$, $G_{g \cdot x} = gG_x g^{-1}$. Any isomorphism of G-sets $f : X \to Y$ preserves stabilizers: $G_x = G_{f(x)}$. Combining these two facts, we can prove that, *if H and K are subgroups of finite index in G, then $G/H \cong G/K$ in G-$\mathfrak{S}\mathfrak{e}\mathfrak{t}$ if and only if H and K are conjugate in G:* Under the action of G

on G/H, the stabilizer of $H = eH$ is $G_H = H$. So if $f : G/H \to G/K$ is an isomorphism with $f(H) = gK$, then

$$H = G_H = G_{f(H)} = G_{gK} = gG_K g^{-1} = gKg^{-1} .$$

Conversely, if $H = gKg^{-1}$, then, since G/K is the orbit of gK,

$$G/K \cong G/G_{gK} = G/gG_K g^{-1} = G/gKg^{-1} = G/H .$$

Assembling these facts, we have proved:

(5.4) Theorem. *Suppose G is a group and $\mathcal{H}$ is a set of subgroups of G, one chosen from each conjugacy class of subgroups of finite index in G. Each element of $\mathfrak{I}(G\text{-}\mathfrak{Set})$ has a unique expression*

$$\sum_{H \in \mathcal{H}} f(H)c(G/H) ,$$

where $f : \mathcal{H} \to \{0, 1, 2, \dots\}$ is a function with finite support. If $c(X)$ has this expression, then $f(H)$ is the number of orbits in X that are isomorphic to G/H. ∎

(5.5) Definition. The **Burnside ring** $\Omega(G)$, of a group G, is the group completion of the semiring $\mathfrak{I}(G\text{-}\mathfrak{Set})$.

For a G-set X, let $[X]$ denote the coset of $c(X)$ in $\Omega(G)$; so the group completion $\gamma : \mathfrak{I}(G\text{-}\mathfrak{Set}) \to \Omega(G)$ takes $c(X)$ to $[X]$. By (5.4), $\mathfrak{I}(G\text{-}\mathfrak{Set})$ is a free abelian monoid based on the isomorphism classes $c(G/H)$ for $H \in \mathcal{H}$. So by (3.19), γ is injective, and when we identify $[X]$ with $c(X)$ we obtain:

(5.6) Corollary. *For each group G, $\Omega(G)$ is, additively, the free $\mathbb{Z}$-module based on the isomorphism classes $c(G/H)$ as H ranges through a set $\mathcal{H}$ of representatives of the conjugacy classes of subgroups of G of finite index.* ∎

The multiplication in $\Omega(G)$ is completely determined by the cartesian products $(G/K) \times (G/L)$ for $K, L \in \mathcal{H}$: The coefficient of (G/H) in the expression (5.4) of this product is the number of orbits isomorphic to G/H— that is, the number of orbits in which the stabilizer of any (hence every) element is conjugate to H.

For instance, suppose G is the dihedral group of order 6 with rotation r of order 3, reflection f of order 2, and $fr = r^2 f$. Then $\Omega(G)$ is the free $\mathbb{Z}$-module with basis

$$\begin{aligned}
1 &= (G/G) = (\{1\}) \\
a &= (G/\{e, r, r^2\}) \\
b &= (G/\{e, f\}) \\
c &= (G/\{e\})
\end{aligned}$$

and multiplication determined by the table:

	1	a	b	c
1	1	a	b	c
a	a	$2a$	c	$2c$
b	b	c	$b+c$	$3c$
c	c	$2c$	$3c$	$6c$

Computation of the Burnside ring $\Omega(G)$ is streamlined by the use of ring homomorphisms $\Omega(G) \to R$. For example, suppose H is a subgroup of G. Each G-set is also an H-set when we restrict scalars to H. Denote by X^H the set of fixed points in X under the action of H. Any isomorphism of G-sets $X \cong Y$ restricts to a bijection $X^H \cong Y^H$. For G-sets X and Y,

$$(X \overset{d}{\cup} Y)^H = X^H \overset{d}{\cup} Y^H \, , \quad \emptyset^H = \emptyset \, ,$$
$$(X \times Y)^H = X^H \times Y^H \, , \quad \{1\}^H = \{1\} \, .$$

So the number $\#(X^H)$ of fixed points under H defines a semiring homomorphism $\mathfrak{I}(G\text{-}set) \to \mathbb{Z}$, extending to a ring homomorphism $\phi_H : \Omega(G) \to \mathbb{Z}$, with $\phi_H([X] - [Y]) = \#(X^H) - \#(Y^H)$.

Now suppose $\mathcal{H} = \{H_i\}_{i \in I}$ is a set of subgroups of G, one from each conjugacy class of subgroups of finite index in G. By (5.6), ϕ_H is determined by its effect on each $[G/H_i]$. If H and K are subgroups of G, write $H \leq_G K$ to mean $g^{-1}Hg \subseteq K$ for some $g \in G$. Also denote the index of H in G by $(G : H)$. Note that

$$(G/K)^H = \{gK : g \in G \, , \ HgK = gK\}$$
$$= \{gK : g \in G \, , \ g^{-1}Hg \subseteq K\} \, .$$

So $\phi_H[G/H] = (N_G(H) : H)$, where $N_G(H)$ is the normalizer of H in G. And $\phi_H[G/K] \neq 0$ if and only if $H \leq_G K$.

(5.7) Theorem. (Burnside) *Suppose G is a finite group and $\mathcal{H} = \{H_1, ..., H_d\}$ is a full set of nonconjugate subgroups of the finite group G. For G-sets X and Y, $X \cong Y$ if and only if $\phi_H[X] = \phi_H[Y]$ for all $H \in \mathcal{H}$.*

Proof. The "only if" is immediate. Suppose $X \not\cong Y$, so $[X] \neq [Y]$. By (5.6),

$$[X] = \sum m_i[G/H_i] \, , \quad [Y] = \sum n_i[G/H_i] \, ,$$

for integers m_i, n_i. Choose a largest $K = H_j \in \mathcal{H}$ with $m_j \neq n_j$. Suppose $\phi_K[X] = \phi_K[Y]$. Then

$$\sum m_i \phi_K[G/H_i] = \sum n_i \phi_K[G/H_i] .$$

For those i with $\phi_K[G/H_i] \neq 0$, $K \leq_G H_i$; so either $i = j$ or H_i is larger than K and $m_i = n_i$. So

$$m_j(N_G(K) : K) = n_j(N_G(K) : K) ,$$

and $m_j = n_j$, a contradiction. So $\phi_K[X] \neq \phi_K[Y]$. ∎

Under the hypotheses of (5.7), it follows that

$$\phi : \Omega(G) \rightarrow \mathbb{Z}^d ,$$
$$\phi(x) = (\phi_{H_1}(x), ..., \phi_{H_d}(x))$$

is an injective ring homomorphism, where the operations in $\mathbb{Z}^d$ are coordinate-wise. So $\Omega(G)$ is isomorphic to the image of ϕ, which is the $\mathbb{Z}$-linear span of $\phi[G/H_1], ..., \phi[G/H_d]$.

5B. Exercises

1. Use the map ϕ to give a complete description of $\Omega(G)$ if G is cyclic of prime order p. Determine the units and idempotents of $\Omega(G)$.

2. If G is a finite group, prove the image of ϕ has finite index as an additive subgroup of $\mathbb{Z}^d$, and the maps ϕ_H for $H \in \{H_1, ..., H_d\}$ are $\mathbb{Z}$-linearly independent. *Hint:* The rank of f.g. $\mathbb{Z}$-modules is additive over short exact sequences, since it is the composite of isomorphisms $G_0(\mathbb{Z}) \cong K_0(\mathbb{Z}) \cong \mathbb{Z}$ (see §3D, Exercise 1). Consider the sequence

$$0 \longrightarrow \Omega(G) \overset{\phi}{\longrightarrow} \mathbb{Z}^d \longrightarrow \mathbb{Z}^d/\phi(\Omega(G)) \longrightarrow 0$$

and any composite zero sequence

$$\Omega(G) \overset{\phi}{\longrightarrow} \mathbb{Z}^d \longrightarrow \mathbb{Z} .$$

3. If H and K are subgroups of a group G, prove $\phi_H[G/H]$ divides $\phi_K[G/H]$. *Hint:* Show $(G/H)^K$ is an $N_G(H)/H$-set via $gH \bullet xH = xg^{-1}H$, and that the stabilizer of each element under this action is the trivial group $\{eH\}$. Conclude

that $(G/H)^K$ is a union of pairwise disjoint orbits, each of size $(N_G(H) : H) = \phi_H[G/H]$.

4. Suppose G is a group of order p^r (p prime, $r \geq 1$). Prove each proper subgroup H of G is a proper subgroup of its normalizer $N_G(H)$, and each element of $\phi(\Omega(G))$ has all its coordinates congruent modulo p. *Hint:* When H acts on G/H by $h \bullet xH = hxH$, the size of each orbit divides the order of H, so it is 1 or a power of p. Since p also divides the sum $(G : H)$ of the orbit sizes, p divides the number of fixed points $\phi_H[G/H]$. Now apply Exercise 3 to the coordinates of each $\phi[G/H_i], H_i \in \mathcal{H}$.

5. For the groups G considered in Exercise 4, find all idempotents in $\Omega(G)$. If p is also odd, find all units in $\Omega(G)$.

6. For a finite group G with $\mathcal{H} = \{H_1, ..., H_d\}$, the Burnside ring $\Omega(G)$ can be identified with the $\mathbb{Z}$-module $\mathbb{Z}^d$ via the basis $[G/H_1], ..., [G/H_d]$. So the map ϕ becomes right multiplication by the matrix whose i-row is $\phi[G/H_i]$, for $1 \leq i \leq d$. Determine this matrix when G is a cyclic group of order p^r (p prime, $r \geq 1$). Then describe its row-space $\phi(\Omega(G))$, as a subring of $\mathbb{Z}^d$.

7. Suppose G is a finite group with $\mathcal{H} = \{H_1, ..., H_d\}$. Suppose X is a G-set and $1 \leq j \leq d$. In $\Omega(G)$, prove

$$[X] \cdot [G/H_j] \;=\; \sum n_i[G/H_i] \,,$$

where $n_i = 0$ unless $H_i \leq_G H_j$, and where $n_j = \phi_{H_j}[X]$.

8. If G is a finite group, prove every ring homomorphism $f : \Omega(G) \to R$, into an integral domain R, factors as a composite of ring homomorphisms

$$\Omega(G) \xrightarrow{\;\phi_H\;} \mathbb{Z} \longrightarrow R \,,$$

where H is the smallest member of $\mathcal{H}$ with $f[G/H] \neq 0$. Then show that $\Omega(G)$ is an integral domain if and only if $\Omega(G) \cong \mathbb{Z}$. *Hint:* Apply f to the equation in Exercise 7 with $H = H_j$. For the second assertion, take $f : \Omega(G) \to \Omega(G)$ to be the identity map.

5C. Tensor Products of Modules

If F is a field and $\mathcal{I}$ is the set of isomorphism classes of finite-dimensional F-vector spaces, then the dimension over F defines an isomorphism of monoids

$$(\mathcal{I}, \oplus) \;\cong\; (\mathbb{N}, +) \,,$$

where $\mathbb{N}$ is the set of natural numbers $0, 1, 2, \ldots$. But $\mathbb{N}$ is also a semiring under $+, \cdot$; so $\mathcal{I}$ is a semiring via the dimension map. Thus, for finite-dimensional vector spaces V and W, there must exist a finite-dimensional vector space X with $c(V)c(W) = c(X)$ in $\mathfrak{J}$, and $\dim_F X = (\dim_F V)(\dim_F W)$. Is there a natural construction of X from V and W?

A similar question arises from ideals. As in Example (5.3) (ii), the set of ideals $\mathfrak{id}(R)$ of a ring R is a semiring under ideal sum and product. For R-modules, the analog of ideal sum is the direct sum. What is the analog of ideal multiplication for R-modules?

If A and B are ideals of a ring R, a typical element of AB is a sum of terms ab where $a \in A$ and $b \in B$. From the distributive and associative properties in R,

$$
\begin{aligned}
(a + a')b &= ab + a'b , \\
a(b + b') &= ab + ab' , \\
(ar)b &= a(rb) ,
\end{aligned}
$$

whenever $a, a' \in A$, $b, b' \in B$, and $r \in R$. To extend this ideal multiplication to modules, we need a way to multiply elements of one module by elements of another.

(5.8) Definitions. If R is a ring and M is a module over R, write M as $_RM$ if it is a left R-module, and as M_R if it is a right R-module. If R has modules M_R and $_RN$, and if A is an additive abelian group, a function $f : M \times N \to A$ is R-**balanced** if

$$
\begin{aligned}
f((m + m', n)) &= f((m, n)) + f((m', n)) , \\
f((m, n + n')) &= f((m, n)) + f((m, n')) , \\
f((mr, n)) &= f((m, rn)) ,
\end{aligned}
$$

for all $m, m' \in M$, $n, n' \in N$, and $r \in R$.

There is a universal R-balanced map $M_R \times {_RN} \to A$, whose codomain A is the product of modules we are seeking:

(5.9) Definition. If R is a ring, the **tensor product** of modules M_R and $_RN$ over R is the quotient group

$$
M \otimes_R N = F_{\mathbb{Z}}(M \times N)/\langle D \rangle ,
$$

where $F_{\mathbb{Z}}(M \times N)$ is the free $\mathbb{Z}$-module based on $M \times N$, and D is the set of all its elements having any of the three forms:

$$
\begin{aligned}
(m + m', n) &- (m, n) - (m', n) , \\
(m, n + n') &- (m, n) - (m, n') , \\
(mr, n) &- (m, rn) ,
\end{aligned}
$$

for $m, m' \in M$, $n, n' \in N$, and $r \in R$. If $m \in M$ and $n \in N$, the coset $(m, n) + \langle D \rangle$ of a pair (m, n) is denoted by $m \otimes n$, and called a **little tensor**.

By its construction, $M \otimes_R N$ is the additive abelian group with generators the little tensors $m \otimes n$ and defining relations

$$(m + m') \otimes n = (m \otimes n) + (m' \otimes n),$$
$$m \otimes (n + n') = (m \otimes n) + (m \otimes n'),$$
$$mr \otimes n = m \otimes rn,$$

for $m, m' \in M$, $n, n' \in N$, and $r \in R$. So the natural map

$$t : M \times N \rightarrow M \otimes_R N$$
$$(m, n) \mapsto m \otimes n$$

is R-balanced. In fact, it is the initial R-balanced map on $M \times N$:

(5.10) Proposition. *If R is a ring with modules M_R and $_R N$, then, for each R-balanced function $f : M \times N \to A$, there is one and only one additive group homomorphism*

$$\overline{f} : M \otimes_R N \to A \quad with \quad \overline{f}(m \otimes n) = f((m, n))$$

for all $m \in M$ and $n \in N$, that is, one and only one group homomorphism $\overline{f}$ making the triangle

$$M \times N \xrightarrow{\ t\ } M \otimes_R N$$
$$f \searrow \quad \downarrow \overline{f}$$
$$A$$

commute.

Proof. Uniqueness of $\overline{f}$ follows from the generation of $M \otimes_R N$ by little tensors. Since f is R-balanced, its $\mathbb{Z}$-linear extension to $F_{\mathbb{Z}}(M \times N)$ takes D to 0; so it induces a $\mathbb{Z}$-linear map $\overline{f}$ on $M \otimes_R N$ with $\overline{f}(m \otimes n) = f((m, n))$. ∎

By the defining relations for $M \otimes_R N$, for each $m \in M$ and $n \in N$, the maps

$$(-) \otimes n : M \rightarrow M \otimes_R N, \quad \text{and}$$
$$m \otimes (-) : N \rightarrow M \otimes_R N,$$

are additive group homomorphisms. So $0 \otimes n = m \otimes 0 = 0$, and $-(m \otimes n) = (-m) \otimes n = m \otimes (-n)$. Thus every element of $M \otimes_R N$ can be written as a sum of little tensors:

$$\sum_{i=1}^{a} (m_i \otimes n_i) \quad (a \geq 1).$$

The analog for modules of a two-sided ideal is a "bimodule:"

(5.11) Definition. Suppose R and S are rings. An additive abelian group M is an R, S-**bimodule** (written $_RM_S$) if M is a left R-module and a right S-module, and $r(ms) = (rm)s$ for all $r \in R$, $s \in S$, and $m \in M$.

(5.12) Examples of bimodules.

(i) If R is a *commutative* ring, any one-sided R-module $_RM$ or M_R is an R, R-bimodule $_RM_R$ when the same scalar multiplication is used on both sides: $rm = mr$. (For noncommutative rings, this need not work — a scalar multiplication on one side can fail to be a scalar multiplication on the other side because the associative module axiom can fail.)

(ii) One-sided modules are bimodules, for $_RM = {_RM_{\mathbb{Z}}}$ and $M_S = {_{\mathbb{Z}}M_S}$ in exactly one way.

(iii) If $f : R \to T$ and $g : S \to U$ are ring homomorphisms, each T, U-bimodule M is also an R, S-bimodule via

$$r \cdot m = f(r)m \ , \quad m \cdot s = mg(s) \ .$$

This is true, in particular, when f and g are inclusions of subrings $R \subseteq T$, $S \subseteq U$.

(iv) Ring multiplication makes each ring R into an R, R-bimodule. By (iii), if $f : S \to R$ is a ring homomorphism, then R becomes an S, S-bimodule via $s \cdot r = f(s)r$, $r \cdot s = rf(s)$.

(v) If m and n are positive integers, R is a ring, $M = R^{m \times n}$, $A = M_m(R)$, and $B = M_n(R)$, then matrix multiplication makes M a bimodule $_AM_B$. Since R embeds as a subring of $M_t(R)$ via

$$r \mapsto \begin{bmatrix} r & 0 & \cdots & 0 \\ 0 & r & \cdots & 0 \\ \vdots & \vdots & \ddots & \vdots \\ 0 & 0 & \cdots & r \end{bmatrix} ,$$

there are also bimodules $_RM_B$, $_AM_R$, and $_RM_R$.

(5.13) Proposition. *For all rings R, S, and T and bimodules $_RM_S$ and $_SN_T$, their tensor product over S is a bimodule $_R(M \otimes_S N)_T$.*

Proof. If $r \in R$, the function

$$f_r : M \times N \ \to \ M \otimes_S N$$
$$(m, n) \ \mapsto \ (rm \otimes n)$$

is S-balanced. (Note: The proof of $f_r((ms, n)) = f_r((m, sn))$ requires the bimodule property $(rm)s = r(ms)$.) So there is a $\mathbb{Z}$-linear map

$$\overline{f}_r : M \otimes_S N \to M \otimes_S N$$

with $\overline{f}_r(m \otimes n) = (rm) \otimes n$, for $m \in M$ and $n \in N$. The function

$$R \times (M \otimes_S N) \to M \otimes_S N$$
$$(r, x) \mapsto r \cdot x = \overline{f}_r(x)$$

is a scalar multiplication making $M \otimes_S N$ a left R-module, since $\overline{f}_r$ is $\mathbb{Z}$-linear, and

$$
\begin{aligned}
(r + r') \cdot (m \otimes n) &= (rm \otimes n) + (r'm \otimes n) \,, \\
(rr') \cdot (m \otimes n) &= r \cdot (r' \cdot (m \otimes n)) \,, \\
1 \cdot (m \otimes n) &= m \otimes n \,.
\end{aligned}
$$

Under this scalar multiplication,

$$r \cdot \sum_{i=1}^{a} (m_i \otimes n_i) = \sum_{i=1}^{a} (rm_i) \otimes n_i \,.$$

Similarly, $M \otimes_S N$ is a right T-module via

$$\left(\sum_{i=1}^{a} (m_i \otimes n_i) \right) \cdot t = \sum_{i=1}^{a} m_i \otimes (n_i t) \,.$$

To prove $M \otimes_S N$ is an R, T-bimodule, we need only note that

$$
\begin{aligned}
r \cdot \left(\sum (m_i \otimes n_i) \cdot t \right) &= \sum (rm_i) \otimes (n_i t) \\
&= \left(r \cdot \sum (m_i \otimes n_i) \right) \cdot t \,.
\end{aligned}
$$ ∎

(5.14) Definition. If R and S are rings, with bimodules $_R M_S$ and $_R N_S$, a map $f : M \to N$ is **left linear** if it is R-linear and **right linear** if it is S-linear.

Just as with $\oplus$, there are isomorphisms resembling the axioms of arithmetic for $\otimes$:

(5.15) Proposition. *For rings* $R, S, T,$ *and* U *and bimodules* $_RM_S,$ $_SN_T,$ $_SN_T',$ *and* $_TP_U,$ *there are (both left and right) linear isomorphisms:*

(i) $M \otimes_S (N \otimes_T P) \quad \cong \quad (M \otimes_S N) \otimes_T P$,
       $\quad m \otimes (n \otimes p) \qquad \mapsto \qquad (m \otimes n) \otimes p$

(ii) $R \otimes_R M \quad \cong \quad M$, $M \otimes_S S \quad \cong \quad M$,
        $\quad r \otimes m \quad \mapsto \quad rm \qquad m \otimes s \quad \mapsto \quad ms$

(iii) $M \otimes_S (N \oplus N') \quad \cong \quad (M \otimes_S N) \oplus (M \otimes_S N')$,
         $\quad m \otimes (n, n') \qquad \mapsto \qquad (m \otimes n, m \otimes n')$

(iv) $(N \oplus N') \otimes_T P \quad \cong \quad (N \otimes_T P) \oplus (N' \otimes_T P)$.
        $\quad (n, n') \otimes p \qquad \mapsto \qquad (n \otimes p, n' \otimes p)$

Proof. For each $m \in M$, there is a T-balanced map

$$f_m : N \times P \quad \to \quad (M \otimes_S N) \otimes_T P$$
$$\quad (n, p) \quad \mapsto \quad (m \otimes n) \otimes p$$

inducing an additive homomorphism $\overline{f_m}$ on $N \otimes_T P$. Then

$$M \times (N \otimes_T P) \quad \to \quad (M \otimes_S N) \otimes_T P$$
$$\quad (m, x) \quad \mapsto \quad \overline{f_m}(x)$$

is an S-balanced map, inducing an additive homomorphism on $M \otimes_S (N \otimes_T P)$, taking $m \otimes (n \otimes p)$ to $(m \otimes n) \otimes p$. The inverse homomorphism is constructed in the same way. Both maps are left and right linear, proving (i).

By the module axioms, the scalar multiplications $R \times M \to M$ and $M \times S \to M$ are R-balanced and S-balanced, respectively, so they induce additive homomorphisms on $R \otimes_R M$ and $M \otimes_S S$, with inverse maps $m \mapsto 1 \otimes m$ and $m \mapsto m \otimes 1$. These maps are left and right linear, proving (ii).

For (iii), each of the maps

$$f : M \times (N \oplus N') \quad \to \quad (M \otimes_S N) \oplus (M \otimes_S N')$$
$$\quad (m, (n, n')) \quad \mapsto \quad (m \otimes n, m \otimes n')$$

$$g : M \times N \quad \to \quad M \otimes_S (N \oplus N')$$
$$\quad (m, n) \quad \mapsto \quad m \otimes (n, 0)$$

$$h : M \times N' \quad \to \quad M \otimes_S (N \oplus N')$$
$$\quad (m, n') \quad \mapsto \quad m \otimes (0, n')$$

is S-balanced, inducing additive homomorphisms $\overline{f}, \overline{g}$, and $\overline{h}$ on the tensor products over S. Then $\overline{f}$ has inverse map taking (x, y) to $\overline{g}(x) + \overline{h}(y)$. And $\overline{f}$ is left and right linear, proving (iii). The proof of (iv) is similar. ∎

To show $\otimes$ is an operation well defined on isomorphism classes, we must define the tensor product of linear maps. Here the parallel with $\oplus$ is striking: If $f : M \to M'$ and $g : N \to N'$ are R-linear, so is the map

$$f \oplus g : M \oplus N \to M' \oplus N' \, ,$$

defined by $(f \oplus g)(m, n) = (f(m), g(n))$. If also $h : M' \to M''$ and $k : N' \to N''$ are R-linear, then

$$(h \oplus k) \circ (f \oplus g) = (h \circ f) \oplus (k \circ g) \, .$$

The direct sum of identity maps is an identity map:

$$i_M \oplus i_N = i_{M \oplus N} \, .$$

So the direct sum of isomorphisms f, g is an isomorphism $f \oplus g$ with inverse $f^{-1} \oplus g^{-1}$. In §3A we used this to show $\oplus$ is a binary operation on R-$\mathcal{M}\mathrm{o}\eth$.
Now suppose

$$f : {}_R M_S \to {}_R M'_S \, , \quad g : {}_S N_T \to {}_S N'_T$$

are left and right linear maps between bimodules. The map

$$M \times N \to M' \otimes_S N'$$
$$(m, n) \mapsto f(m) \otimes g(n)$$

is S-balanced; so it induces an additive homomorphism

$$f \otimes g : M \otimes_S N \to M' \otimes_S N'$$

taking $m \otimes n$ to $f(m) \otimes g(n)$. Since f is left linear and g is right linear, $f \otimes g$ is left and right linear. If also $h : {}_R M'_S \to {}_R M''_S$ and $k : {}_S N'_T \to {}_S N''_T$ are left and right linear, then

$$(h \otimes k) \circ (f \otimes g) = (h \circ f) \otimes (k \circ g) \, .$$

And the tensor product of identity maps is the identity map:

$$i_M \otimes i_N = i_{M \otimes_S N} \, .$$

So the tensor product of isomorphisms f, g is an isomorphism $f \otimes g$ with inverse $f^{-1} \otimes g^{-1}$.

(5.16) Proposition. *Suppose R is a commutative ring. Then $\otimes_R$ is a commutative, associative binary operation on the category R-$\mathcal{M}\mathrm{o}\mathrm{d}$, with identity R. If $\mathcal{C}$ is a modest full subcategory of R-$\mathcal{M}\mathrm{o}\mathrm{d}$ closed under $\oplus$ and $\otimes_R$ and including the objects 0 and R, the set $\mathcal{I}(\mathcal{C})$ of isomorphism classes in $\mathcal{C}$ is a commutative semiring under the operations:*

$$c(M) * c(N) = c(M \oplus N) ,$$
$$c(M) \cdot c(N) = c(M \otimes_R N) .$$

So its group completion $K_0(\mathcal{C}, \oplus)$ is a commutative ring $\boldsymbol{K_0(\mathcal{C}, \oplus, \otimes_R)}$ in which

$$[M] \cdot [N] = [M \otimes_R N] ,$$

for all $M, N \in \mathcal{C}$.

Proof. Since R is commutative, each left R-module M is an R, R-bimodule with $r \cdot m = m \cdot r$ for each $r \in R$ and $m \in M$. By (5.13), each tensor product $M \otimes_R N$, of R-modules M and N, is also an R, R-bimodule. And R-linear maps are both left and right linear. So $M \cong M'$ and $N \cong N'$ in R-$\mathcal{M}\mathrm{o}\mathrm{d}$, implies $M \otimes_R N \cong M' \otimes_R N'$.

That $\otimes_R$ is associative with identity R follows from (5.15) (i) and (ii). If $M, N \in R$-$\mathcal{M}\mathrm{o}\mathrm{d}$, the map

$$M \times N \rightarrow N \otimes_R M$$
$$(m, n) \mapsto n \otimes m$$

is R-balanced and induces an additive homomorphism on $M \otimes_R N$ that is R-linear, with an inverse map defined in the same way. So $\otimes_R$ is commutative.

That $(\mathcal{I}(\mathcal{C}), *, \cdot)$ is a commutative semiring is now a consequence of the isomorphisms in §3A and (5.15). That $K_0(\mathcal{C}, \oplus)$ is a commutative ring with $[M] \cdot [N] = [M \otimes_R N]$ is immediate from Proposition (5.2) on the completion of semirings. ∎

To apply this proposition to the categories $\mathcal{M}(R)$, $\mathcal{P}(R)$, and $\mathcal{F}(R)$, we first note that tensor product multiplies dimensions, as envisioned at the beginning of this section:

(5.17) Proposition. *Suppose R is a ring and m, n are positive integers. If there are left and right linear isomorphisms of R, R-bimodules $M \cong R^m$ and $N \cong R^n$, there is a left and right linear isomorphism: $M \otimes_R N \cong R^{mn}$.*

Proof. The tensor product of isomorphisms $M \cong R^m$ and $N \cong R^n$ is the first in the sequence of isomorphisms:

$$M \otimes_R N \cong R^m \otimes_R R^n \cong (R^m \otimes_R R)^n$$
$$\cong (R^m)^n \cong R^{mn} .$$

The second and third come from the distributivity of $\otimes_R$ over $\oplus$ in (5.15), and the fourth from the associativity of $\oplus$. ∎

(5.18) Examples: Grothendieck rings of modules.

(i) If R is a commutative ring, then $\mathcal{F}(R)$ is closed under $\otimes_R$ by (5.17). So $K_0(\mathcal{F}(R), \oplus, \otimes_R)$ is a commutative ring.

(ii) If R is a commutative ring, m and n are positive integers, and there are surjective R-linear maps $R^m \to M$ and $R^n \to N$, their tensor product is a surjective R-linear map

$$(R^{mn} \cong)\ R^m \otimes_R R^n \ \to \ M \otimes_R N \ ;$$

so $\mathcal{M}(R)$ is closed under $\otimes_R$. Thus $K_0(\mathcal{M}(R), \oplus, \otimes_R)$ is a commutative ring.

(iii) If R is a commutative ring, m and n are positive integers, $P \oplus P' \cong R^m$, and $Q \oplus Q' \cong R^n$ in R-$\mathfrak{Mod}$, then

$$\begin{aligned} R^{mn} &\cong\ (P \oplus P') \otimes_R (Q \oplus Q') \\ &\cong\ (P \otimes_R Q) \oplus (\text{three terms}) \end{aligned}$$

by distributivity of $\otimes_R$ over $\oplus$. So $\mathcal{P}(R)$ is closed under $\otimes_R$, and $K_0(R) = K_0(\mathcal{P}(R), \oplus, \otimes_R)$ is a commutative ring.

(iv) If R is a commutative regular ring, $G_0(R)$ is a commutative ring via the Cartan isomorphism

$$\begin{aligned} K_0(R) &\cong\ G_0(R)\ , \\ [P] &\mapsto\ [P] \end{aligned}$$

in the Resolution Theorem (3.52). In this case, in $G_0(R)$, $[P][Q] = [P \otimes_R Q]$ for $P, Q \in \mathcal{P}(R)$. In fact, $[M][Q] = [M \otimes_R Q]$ for $M \in \mathcal{M}(R)$ and $Q \in \mathcal{P}(R)$. But it need not be true that $[M][N] = [M \otimes_R N]$ for $M, N \in \mathcal{M}(R)$! For a counterexample, see Example (7.55) below.

(5.19) Example: Grothendieck rings of endomorphisms.
Suppose R is a commutative ring and $\mathcal{C}$ is a modest full subcategory of R-$\mathfrak{Mod}$, closed under both $\oplus$ and $\otimes_R$, and including the objects 0 and R. For instance, $\mathcal{C}$ could be $\mathcal{M}(R)$, $\mathcal{P}(R)$, or $\mathcal{F}(R)$. The category end $\mathcal{C}$ of endomorphisms of objects in $\mathcal{C}$ is defined in (3.9). There, $\mathcal{I}(\text{end } \mathcal{C})$ is shown to be an abelian monoid under $\oplus$. In fact (with the closure under $\otimes_R$ and with $R \in \mathcal{C}$) it is a commutative semiring under

$$\begin{aligned} (M, f) \oplus (N, g) &=\ (M \oplus N, f \oplus g)\ , \\ (M, f) \otimes (N, g) &=\ (M \otimes_R N, f \otimes g)\ , \end{aligned}$$

on end $\mathcal{C}$: Here $M \otimes_R N$ is the $R[x]$-module with the usual action of R,

$$r \cdot (m \otimes n) = (rm \otimes n)\ ,$$

and with diagonal action of x:

$$x \cdot (m \otimes n) \ = \ (xm) \otimes (xn) \ = \ f(m) \otimes g(n) \ .$$

The proof that $\otimes$ is well defined on isomorphism classes in end $\mathcal{C}$, and that $(\mathcal{I}(\text{end } \mathcal{C}), \oplus, \otimes)$ is a commutative semiring, is just the proof of (5.16) together with the verification that each isomorphism commutes with the diagonal action of x. The multiplicative identity is the isomorphism class of (R, f), where f is the constant map to 1.

Denoting the coset of $c(M, f)$ in the group completion by $[M, f]$, multiplication in the commutative ring $K_0(\text{end } \mathcal{C}, \oplus, \otimes)$ is determined by

$$[M, f] \cdot [N, g] \ = \ [M \otimes_R N, \ f \otimes g] \ .$$

When $\mathcal{C} = \mathcal{F}(R)$, we can replace endomorphisms by the similarity classes of their representing matrices. This transforms $\mathcal{I}(\text{end } \mathcal{F}(R))$ into a commutative semiring $\text{sim}(R)$ of similarity classes of square matrices over R (as in (3.10)). Addition in $\text{sim}(R)$ is

$$s(A) + s(B) \ = \ s(A \oplus B) \ = \ s\left(\begin{bmatrix} A & 0 \\ 0 & B \end{bmatrix} \right) \ ,$$

for square matrices A and B over R. Multiplication in $\text{sim}(R)$ also has a simple description:

(5.20) Definition. If $A = (a_{ij}) \in M_m(R)$ and $B = (b_{ij}) \in M_n(R)$, their **Kronecker product** is the matrix $A \otimes B \in M_{mn}(R)$ that has a decomposition into $n \times n$ blocks:

$$A \otimes B \ = \ \begin{bmatrix} a_{11}B & a_{12}B & \cdots & a_{1m}B \\ a_{21}B & a_{22}B & \cdots & a_{2m}B \\ \vdots & \vdots & \ddots & \vdots \\ a_{m1}B & a_{m2}B & \cdots & a_{mm}B \end{bmatrix} \ ,$$

with i, j-block $a_{ij}B$. For integers $1 \leq x \leq m$ and $1 \leq y \leq n$, let

$$x * y \ = \ (x - 1)n \ + \ y \ .$$

Then the i'', j''-entry of $A \otimes B$ is $a_{ij}b_{i'j'}$, where $i'' = i * i'$ and $j'' = j * j'$.

Now suppose M and N are R-modules with bases $v_1, \ldots, v_m$ and $w_1, \ldots, w_n$, respectively. If we take $u_{i*i'} = v_i \otimes w_{i'}$, then $u_1, \ldots, u_{mn}$ is the list

$$\begin{array}{cccc} v_1 \otimes w_1 \,, & v_1 \otimes w_2 & ,\ldots, & v_1 \otimes w_n \,, \\ v_2 \otimes w_1 \,, & v_2 \otimes w_2 & ,\ldots, & v_2 \otimes w_n \,, \\ \multicolumn{4}{c}{\cdots\cdots\cdots\cdots\cdots\cdots\cdots\cdots\cdots\cdots\cdots} \\ v_m \otimes w_1 \,, & v_m \otimes w_2 & ,\ldots, & v_m \otimes w_n \,, \end{array}$$

of little tensors $v_i \otimes w_j$ in dictionary order. The isomorphism $M \otimes_R N \cong R^{mn}$ in (5.17) takes $u_1, \ldots, u_{mn}$ to the standard basis; so $u_1, \ldots, u_{mn}$ is a basis of $M \otimes_R N$ over R.

(5.21) Proposition. *Assume that* $f \in \text{End}(M)$ *is represented over* $v_1, \ldots, v_m$ *by* $A = (a_{ij}) \in M_m(R)$ *and* $g \in \text{End}(N)$ *is represented over* $w_1, \ldots, w_n$ *by* $B = (b_{ij}) \in M_n(R)$. *Then* $f \otimes g$ *is represented over* $u_1, \ldots, u_{mn}$ *by* $A \otimes B$.

Proof. The coefficient of $u_{j*j'} = v_j \otimes w_{j'}$ in

$$(f \otimes g)(u_{i*i'}) = f(v_i) \otimes g(w_{i'})$$

$$= \left(\sum_j a_{ij} v_j \right) \otimes \left(\sum_{j'} b_{i'j'} w_{j'} \right)$$

is $a_{ij} b_{i'j'}$, which is the $i * i', j * j'$-entry of $A \otimes B$. ∎

So the multiplication in $\text{sim}(R)$ is

$$s(A) \cdot s(B) = s(A \otimes B)$$

for square matrices A, B over R. From this it follows that the Kronecker product is a well-defined operation on similarity classes of square matrices over a commutative ring R. This can be proved directly by matrix computations:

The following properties are evident consequences of the definition of $A \otimes B$ and block multiplication of matrices (see (1.32)).

(5.22) Proposition. *If* R *is a commutative ring and* $A, C \in M_m(R)$ *while* $B, D \in M_n(R)$,

(i) $(A \otimes B)(C \otimes D) = (AC \otimes BD)$.

(ii) $I_m \otimes I_n = I_{mn}$.

(iii) *If* A *and* B *are invertible, so is* $A \otimes B$, *with* $(A \otimes B)^{-1} = A^{-1} \otimes B^{-1}$. ∎

Now $A' = PAP^{-1}$ and $B' = QBQ^{-1}$ imply

$$A' \otimes B' = (P \otimes Q)(A \otimes B)(P \otimes Q)^{-1}.$$

The axioms for a commutative semiring can also be proved for $\text{sim}(R)$ under $\oplus, \otimes$, by matrix similarity identities (see Exercise 5).

In the commutative ring $K_0(\text{end } \mathcal{F}(R), \oplus, \otimes)$, each element has an expression $[A] - [B]$ for square matrices A and B, and $[A] = [B]$ if and only if $A \oplus C$ is similar to $B \oplus C$ for some square matrix C.

(5.23) Example: Grothendieck rings of forms. Suppose R and $\mathcal{C}$ are chosen as in (5.19). The category bil $\mathcal{C}$ of symmetric bilinear modules (M, b) with $M \in \mathcal{C}$ is defined in (3.11). Since $\mathcal{C}$ is closed under $\otimes_R$ and $R \in \mathcal{C}$, the set $\mathcal{I}(\text{bil } \mathcal{C})$ of isometry classes in bil $\mathcal{C}$ is a commutative semiring under operations

$$(M, b) \perp (M', b') = (M \oplus M', \ b \perp b'),$$

$$(M, b) \otimes (M', b') = (M \otimes_R M', \ b \otimes b'),$$

in bil $\mathcal{C}$, defined by

$$(b \perp b')((v, v'), (w, w')) = b(v, w) + b'(v', w') ,$$
$$(b \otimes b')(v \otimes v', w \otimes w') = b(v, w)b'(v', w') :$$

In (3.11), $b \perp b'$ is shown to be a symmetric bilinear form on $M \oplus M'$. To do the same for $b \otimes b'$, first define

$$f(v, v') : M \otimes_R M' \rightarrow R$$

as the additive map induced by the R-balanced map on $M \times M'$ taking (w, w') to $b(v, w)b(v', w')$. This $f(v, v')$ is R-linear, belonging to

$$(M \otimes_R M')^* = \text{Hom}_R(M \otimes_R M', R) .$$

Then f is an R-balanced map inducing an additive map

$$\overline{f} : M \otimes_R M' \rightarrow (M \otimes_R M')^* ,$$

which is also R-linear. Now define $(b \otimes b')(x, y)$ to be $\overline{f}(x)(y)$, whenever $x, y \in M \otimes_R M'$. The bilinearity of $b \otimes b'$ is just the linearity of $\overline{f}$ and $\overline{f}(x)$; its symmetry is immediate from its formula on little tensors and the symmetry of b and b'.

That these operations are well defined on isometry classes and make $\mathfrak{I}(\text{bil } \mathcal{C})$ into a commutative semiring is proved by showing that, when the isomorphisms in the proof of (5.16) are applied to bilinear modules, they are isometries. The multiplicative identity is the isometry class of $(R, b_{[1]})$, which is the ring multiplication $R \times R \rightarrow R$. Denoting the coset of $c(M, b)$ in the group completion by $[M, b]$, multiplication in the commutative ring $K_0(\text{bil } \mathcal{C}, \oplus, \otimes)$ is determined by

$$[M, b] \cdot [M', b'] = [M \otimes_R M', b \otimes b'] .$$

When $\mathcal{C} = \mathcal{F}(R)$, we can replace isometry classes by their congruence classes of representing matrices. This transforms $\mathfrak{I}(\text{bil } \mathcal{F}(R))$ into a commutative semiring $\text{cong}(R)$ of all congruence classes of symmetric matrices over R (see (3.11), (3.12)). Addition and multiplication in $\text{cong}(R)$ are

$$\langle A \rangle \perp \langle B \rangle = \langle A \oplus B \rangle ,$$
$$\langle A \rangle \otimes \langle B \rangle = \langle A \otimes B \rangle :$$

The first of these equations was shown in (3.12). For the second, suppose (M, b) and (N, c) are in bil $\mathcal{F}(R)$, so M and N have bases $v_1, \ldots, v_m$ and $w_1, \ldots, w_n$, respectively. Using the notation of the preceding example, $M \otimes_R N$ has basis

$u_1, \ldots, u_{mn}$ where $u_{i*i'} = v_i \otimes w_{i'}$. If $A = (b(v_i, v_j))$ and $B = (b(w_{i'}, w_{j'}))$, then $A \otimes B$ has $i * i', j * j'$-entry

$$
\begin{aligned}
b(v_i, v_j) \, c \, (w_{i'}, w_{j'}) &= (b \otimes c)(v_i \otimes w_{i'}, v_j \otimes w_{j'}) \\
&= (b \otimes c)(u_{i*i'}, u_{j*j'}) \, .
\end{aligned}
$$

So if A represents (M, b) and B represents (N, c), then $A \otimes B$ represents the pair $(M \otimes_R N, \ b \otimes c)$.

This approach shows the Kronecker product to be a well-defined operation on congruence classes of symmetric matrices over a commutative ring R. A direct proof of this is easy: Suppose A^τ denotes the transpose of A. For matrices A, B over a commutative ring, $(AB)^\tau = B^\tau A^\tau$; so the transpose of an invertible matrix is invertible. Also $(P \otimes Q)^\tau = P^\tau \otimes Q^\tau$. So the Kronecker product of symmetric matrices is symmetric; and by (5.22), if $A' = PAP^\tau$ and $B' = QBQ^\tau$, for invertible matrices P and Q, then

$$
A' \otimes B' = (P \otimes Q)(A \otimes B)(P \otimes Q)^\tau \, ,
$$

with $P \otimes Q$ invertible. That $\mathrm{cong}(R)$ is a commutative semiring under $\perp$ and $\otimes$ can also be proved by matrix congruence identities (see Exercise 6).

Every element of the commutative ring $K_0(\mathrm{bil}\,\mathcal{F}(R), \perp, \otimes)$ has an expression $[A] - [B]$, for symmetric matrices A and B over R; and $[A] = [B]$ if and only if $A \oplus C$ is congruent to $B \oplus C$ for some symmetric matrix C.

(5.24) Example: Witt-Grothendieck and Witt rings. Suppose R is a commutative ring. Recall from §3A that bil* $\mathcal{F}(R)$ is the category of *nonsingular* symmetric bilinear modules (M, b) with $M \in \mathcal{F}(R)$, and that $(M, b) \in \mathrm{bil}\,\mathcal{F}(R)$ is nonsingular if and only if its representing matrices are invertible. By (5.21), the Kronecker product of invertible matrices is invertible; so the submonoid $\mathcal{I}(\mathrm{bil}^*\,\mathcal{F}(R))$ of $\mathcal{I}(\mathrm{bil}\,\mathcal{F}(R))$ under $\perp$ is actually a subsemiring under $\perp$ and $\otimes$.

Passing to representing matrices, the set $\mathrm{cong}^*(R)$ of congruence classes of *invertible* symmetric matrices over R is a commutative semiring under

$$
\begin{aligned}
\langle A \rangle \perp \langle B \rangle &= \langle A \oplus B \rangle \, , \\
\langle A \rangle \otimes \langle B \rangle &= \langle A \otimes B \rangle \, .
\end{aligned}
$$

Every element in the commutative ring

$$
\widehat{W}(R) = K_0(\mathcal{I}(\mathrm{bil}^*\,\mathcal{F}(R), \perp, \otimes)
$$

can be written as $[A] - [B]$ for invertible symmetric matrices A and B over R, and $[A] = [B]$ if and only if $A \oplus C$ is congruent to $B \oplus C$ for an invertible symmetric matrix C over R. The operations are determined by $[A] + [B] = [A \oplus B]$ and $[A] \cdot [B] = [A \otimes B]$.

For the rest of this example, take R to be a field F of characteristic $\neq 2$. Then $\widehat{W}(F)$ is known as the **Witt-Grothendieck ring** of F. By Witt-cancellation

(3.28), $[A] = [B]$ in $\widehat{W}(F)$ if and only if $\langle A \rangle = \langle B \rangle$; so each element can be written as a difference of congruence classes $\langle A \rangle - \langle B \rangle$, with $\mathrm{cong}^*(F)$ regarded as a subsemiring of $\widehat{W}(F)$. Also, by (3.13) each invertible symmetric matrix A is congruent to a diagonal matrix $D = \mathrm{diag}\,(a_1, \ldots, a_n)$ with all $a_i \neq 0$. The congruence class $\langle A \rangle = \langle D \rangle$ is commonly written $\langle a_1, \ldots, a_n \rangle$, and represents the isometry class of the form taking

$$((x_1, \ldots, x_n)\,,\,(y_1, \ldots, y_n)) \ \text{ to } \ \sum_i a_i x_i y_i \,.$$

In this notation, the operations in $\widehat{W}(F)$ are determined by:

$$\langle a_1, \ldots, a_m \rangle \perp \langle a_{m+1}, \ldots, a_{m+n} \rangle \ = \ \langle a_1, \ldots, a_{m+n} \rangle \,,$$

$$\langle a_1, \ldots, a_m \rangle \otimes \langle b_1, \ldots, b_n \rangle \ =$$
$$\langle a_1 b_1, \ldots, a_1 b_n, a_2 b_1, \ldots, a_2 b_n, \ldots, a_m b_1, \ldots, a_m b_n \rangle \,,$$

the latter coming from the Kronecker product of diagonal matrices.

A vector v in a bilinear space (M, b) over F is called **isotropic** if $v \neq 0$ but $b(v, v) = 0$. An **isotropic space** is a space (M, b) that has an isotropic vector. In an F-vector space, every nonzero vector is part of a basis. So $(M, b) \in \mathrm{bil}\ \mathcal{F}(F)$ is isotropic if and only if it is represented by a matrix with a zero on the diagonal. Invertible symmetric matrices can have a zero on the diagonal, so nonsingular bilinear spaces (M, b) can be isotropic. The most fundamental example is the **hyperbolic plane**, which is defined to be any bilinear space represented by the matrix

$$\epsilon_{12} + \epsilon_{21} \ = \ \begin{bmatrix} 0 & 1 \\ 1 & 0 \end{bmatrix} .$$

Note: In euclidean n-space $\mathbb{R}^n$, with the standard inner product, $b(v, v)$ is the square of the length of v, and

$$b(v, w) \ = \ \sqrt{b(v, v) b(w, w)} \ \cos \Theta$$

is related to the angle Θ between v and w. Euclidean intuition does not hold up well in a nonsingular isotropic space (M, b), since an isotropic vector v will have "length zero" ($b(v, v) = 0$) but have varying "angles" with some other vectors ($b(v, w) \neq 0$ for some w). In this light, the term *isotropic* vector sounds a little misleading: The dictionary meaning of isotropic is "the same in all directions," but the meaning we use here (following a nearly universal convention) is more like "length zero."

Consider the congruence class of matrices representing a hyperbolic plane:

(5.25) Lemma. *If F is a field of characteristic $\neq 2$ and $a, b, c \in F$ with $a \neq 0$ and $b \neq 0$, then the matrices*

$$\begin{bmatrix} 0 & b \\ b & c \end{bmatrix} \quad \text{and} \quad \begin{bmatrix} a & 0 \\ 0 & -a \end{bmatrix}$$

are congruent.

Proof. The product

$$P = \begin{bmatrix} 1 & 0 \\ 0 & ab^{-1} \end{bmatrix} \begin{bmatrix} 1 & ab^{-1} \\ 0 & 1 \end{bmatrix} \begin{bmatrix} 1 & 0 \\ -(ac + b^2)(2ab)^{-1} & 1 \end{bmatrix}$$

is invertible, and

$$P \begin{bmatrix} 0 & b \\ b & c \end{bmatrix} P^\tau = \begin{bmatrix} a & 0 \\ 0 & -a \end{bmatrix} . \qquad \blacksquare$$

(5.26) Proposition. *A space (M, b) in bil* $\mathcal{F}(F)$ is isotropic if and only if there is a hyperbolic plane (H, b_1) and a nonsingular space (N, b_2) with $(M, b) \cong (H, b_1) \perp (N, b_2)$.*

Proof. If $(M, b) \cong (H, b_1) \perp (N, b_2)$, and v is an isotropic vector in H, then the vector in M corresponding to $(v, 0) \in H \oplus N$ is isotropic.

Conversely, suppose m_1 is an isotropic vector in M. Since M is nonsingular, there is a vector $m_2 \in M$ with $b(m_1, m_2) = r \neq 0$. Then $m_2 \notin Fm_1$; so m_1, m_2 is a basis of a subspace H of M. If b_1 is the restriction of the form b to H, the space (H, b_1) is represented by the matrix

$$\begin{bmatrix} 0 & r \\ r & s \end{bmatrix} , \quad s = b(m_2, m_2) .$$

By Lemma (5.25), (H, b_1) is also represented by

$$\begin{bmatrix} 0 & 1 \\ 1 & 0 \end{bmatrix} \quad \text{and} \quad \begin{bmatrix} 1 & 0 \\ 0 & -1 \end{bmatrix} ,$$

so it is a hyperbolic plane and has a basis v_1, v_2 with $b(v_1, v_1) = 1$, $b(v_2, v_2) = -1$ and $b(v_1, v_2) = 0$. Consider the subspace of M,

$$H^\perp = \{v \in M : \forall \, w \in H , \; b(v, w) = 0\} .$$

Since H is nonsingular, $H \cap H^\perp = \{0\}$. If $m \in M$ and

$$x = b(v_1, m)v_1 - b(v_2, m)v_2 ,$$

then $x \in H$ and $y = m - x \in H^\perp$, the latter because $b(v_1, y) = b(v_2, y) = 0$. So $m = x + y \in H + H^\perp$, proving $M = H \overset{\bullet}{\oplus} H^\perp$. Take b_2 to be the restriction of the form b to $H^\perp$. By (3.13), $H^\perp$ has an orthogonal basis $v_3, \ldots, v_n$ with respect to b_2. Then $v_1, \ldots, v_n$ is a basis of M, and (M, b) is represented by

$$\langle 1, -1, \; a_3, \ldots, a_n \rangle \;\; = \;\; \langle 1, -1 \rangle \perp \langle a_3, \ldots, a_n \rangle \; ,$$

as is $(H, b_1) \perp (H^\perp, b_2)$. ∎

(5.27) Corollary. *Suppose (M, b) is a space in* bil* $\mathcal{F}(F)$. *The following are equivalent*

   (i)   *(M, b) is isotropic.*
  (ii) *(M, b) is represented by $\langle 1, -1, \; a_3, \ldots, a_n \rangle$.*
 (iii) *(M, b) is represented by $\langle a_1, a_2, \ldots, a_n \rangle$ where $a_i + a_j = 0$ for some $i \neq j$.*

Proof. The equivalence of (i) and (ii) is Proposition (5.26). The equivalence of (ii) and (iii) comes from the commutativity of $\widehat{W}(F)$, and from Lemma (5.25), which implies $\langle a, -a \rangle = \langle 1, -1 \rangle$ for all nonzero $a \in F$. ∎

Caution: It is possible that $\langle b_1, \ldots, b_n \rangle = \langle 1, -1, \; a_3, \ldots, a_n \rangle$ even if there is no relation $b_i + b_j = 0$.

Suppose $\mathbb{H}$ is the isometry class of hyperbolic planes over F. For nonzero $a_1, \ldots, a_n \in F$,

$$
\begin{aligned}
\langle a_1, \ldots, a_n \rangle \otimes \langle 1, -1 \rangle \\
= \; \langle a_1, -a_1, \; a_2, -a_2, \ldots, a_n, \; -a_n \rangle \\
= \; \langle a_1, -a_1 \rangle \perp \cdots \perp \langle a_n, -a_n \rangle \\
= \; n \cdot \langle 1, -1 \rangle \; .
\end{aligned}
$$

So the principal ideal $\widehat{W}(F) \cdot \mathbb{H}$ generated by $\mathbb{H}$ is the additive subgroup $\mathbb{Z} \cdot \mathbb{H}$ generated by $\mathbb{H}$. The quotient ring:

$$\boldsymbol{W(F)} \; = \; \widehat{W}(F)/\mathbb{Z} \cdot \mathbb{H} \; \cong \; \mathrm{cong}^*(F)/\mathbb{Z} \cdot \langle 1, -1 \rangle$$

is the **Witt ring** of the field F.

Since isotropic vectors in a nonsingular space belong to hyperbolic plane summands of that space, reducing modulo hyperbolic planes should get rid of isotropic vectors. That is the sense of the following theorem. A space

(M, b) in bil* $\mathcal{F}(F)$ is **anisotropic** if M has no isotropic vectors. If (M, b) is anisotropic, so is every other space in its isometry class; so we may say the isometry class $c(M, b)$ is **anisotropic**. So we also say a congruence class in cong*(F) is **anisotropic** if it represents anisotropic spaces or, equivalently, if it contains no matrix with a zero on the diagonal.

(5.28) Theorem.   *If S is the set of isometry classes of anisotropic spaces in* bil* $\mathcal{F}(F)$, *the canonical map* $\widehat{W}(F) \to W(F)$ *restricts to a bijection* $S \to W(F)$.

Proof.   For injectivity, suppose $x, y \in$ cong*(F) are anisotropic and $x - y = n\langle 1, -1 \rangle$, where $n \in \mathbb{Z}$. Either

$$x \perp r\langle 1, -1 \rangle \;=\; y \, , \quad \text{or} \quad x \;=\; y \perp r\langle 1, -1 \rangle \, ,$$

where r is the natural number $|n|$. Since x and y are anisotropic, $r = 0$ and $x = y$.

For surjectivity, first note that for nonzero $a_1, \ldots, a_n \in F$,

$$\langle a_1, \ldots, a_n \rangle \perp \langle -a_1, \ldots, -a_n \rangle \;=\; n\langle 1, -1 \rangle \, .$$

So, in $W(F)$,

$$-\overline{\langle a_1, \ldots, a_n \rangle} \;=\; \overline{\langle -a_1, \ldots, -a_n \rangle} \, .$$

Then every element x of $W(F)$ can be written as

$$\overline{\langle a_1, \ldots, a_m \rangle} - \overline{\langle b_1, \ldots, b_n \rangle}$$

$$= \overline{\langle a_1, \ldots, a_m \rangle} + \overline{\langle -b_1, \ldots, -b_n \rangle}$$

$$= \overline{\langle a_1, \ldots, a_m, -b_1, \ldots, -b_n \rangle} \, ,$$

the coset of a single congruence class. By repeated use of (5.27),

$$x \;=\; \overline{\langle c_1, \ldots, c_p \rangle \perp r\langle 1, -1 \rangle} \;=\; \overline{\langle c_1, \ldots, c_p \rangle} \, ,$$

where $\langle c_1, \ldots, c_p \rangle$ is anisotropic. ∎

By the preceding theorem, for each isometry class $c(M, b)$ in bil* $\mathcal{F}(F)$, there is a unique anisotropic isometry class $c(N, b')$ and an integer $r \geq 0$ for which

$$c(M, b) \;=\; c(N, b') \perp r\mathbb{H} \, .$$

Call $c(N, b')$ the **anisotropic part** of $c(M, b)$. The bijection in this theorem then imposes a ring structure on S; so the Witt ring $W(F)$ can be described as the set of isometry classes of anisotropic spaces (M, b) in bil* $\mathcal{F}(F)$, with operations:

$$c(M_1, b_1) + c(M_2, b_2) = c(\text{the anisotropic part of } (M_1, b_1) \perp (M_2, b_2)) \ ,$$
$$c(M_1, b_1) \cdot c(M_2, b_2) = c(\text{the anisotropic part of } (M_1, b_1) \otimes (M_2, b_2)) \ .$$

This is essentially the classical definition of $W(F)$ used by Witt [37].

To aid in the classification of bilinear forms, there are certain **invariants** — that is, quantities assigned to each form that are equal for isometric forms. For instance, isometric spaces in bil $\mathcal{F}(F)$ are isomorphic F-vector spaces, and so they have equal **dimension**. In fact, the dimension over F defines a semiring homomorphism $\mathcal{J}(\text{bil}^* \ \mathcal{F}(F)) \to \mathbb{N}$, extending to a ring homomorphism $\widehat{W}(F) \to \mathbb{Z}$, taking

$$\langle a_1, \ldots, a_m \rangle \ - \ \langle b_1, \ldots, b_n \rangle$$

to $m - n$.

Since isometric spaces have isometric anisotropic parts, the number r of hyperbolic planes in the decomposition $c(M, b) = c(N, b') \perp r\mathbb{H}$, where (N, b') is anisotropic, is an invariant

$$r = \frac{\dim(M) - \dim(N)}{2} \ ,$$

called the **Witt index** of (M, b).

For each $n > 0$, the determinant is a surjective group homomorphism

$$\det : GL_n(F) \to F^* \ ,$$

where F^* is the group of units in F under multiplication. Since $\det(A^t) = \det(A)$ for each square matrix A, and since the transpose of an invertible matrix over F is invertible, the determinants of all matrices in a congruence class $\langle B \rangle \in \text{cong}^*(F)$ form a coset $x(F^*)^2$ in the quotient group $F^*/(F^*)^2$, where

$$(F^*)^2 = \{y^2 : y \in F^*\}$$

is the set of squares in F^*. A coset $x(F^*)^2$ is called a **square class** in F. The **discriminant** of a space (M, b) is the square class

$$d(M, b) = \text{the determinants of all matrices representing } (M, b) \ .$$

If (M, b) is represented by $\langle a_1, \ldots, a_n \rangle$, then $d(M, b)$ is the square class represented by the product $a_1 \cdots a_n$. So

$$d((M, b) \perp (M', b')) = d(M, b) \cdot d(M', b') \ .$$

Note: When F is a field of characteristic 2, quadratic and bilinear forms are not in perfect correspondence (see §3A, Exercise 6). Over rings R in which $2 \notin R^*$, there is, besides bil $\mathcal{P}(R)$, a category quad $\mathcal{P}(R)$ based on quadratic forms, and there are related but different Witt-Grothendieck rings, $\widehat{W}(R)$ and $\widehat{W}_q(R)$, constructed from these categories. For an exposition of this theory, see Baeza [78].

5C. Exercises

1. If $f : R \to S$ is a surjective ring homomorphism and M_S, $_SN$ are S-modules (on the indicated sides), they are also R-modules, with scalars acting via f. Prove $M \otimes_S N = M \otimes_R N$. Note that this is equality, not just an isomorphism.

2. If A and B are finite abelian groups of relatively prime orders, prove $A \otimes_{\mathbb{Z}} B = 0$.

3. If p is prime and $r > s$, prove $(\mathbb{Z}/p^r\mathbb{Z}) \otimes_{\mathbb{Z}} (\mathbb{Z}/p^s\mathbb{Z}) \cong \mathbb{Z}/p^s\mathbb{Z}$. *Hint:* Use Exercise 1 and the fact that $\mathbb{Z}/p^s\mathbb{Z}$ is a $\mathbb{Z}/p^r\mathbb{Z}$-module.

4. Use the preceding exercises and the structure theorem for f.g. abelian groups to describe $G \otimes_{\mathbb{Z}} H$ when G and H are f.g. abelian groups.

5. Suppose R is a commutative ring. In §3A, Exercise 8, sim(R) is shown to be an abelian monoid under $\oplus$. Prove directly that sim(R) is also an abelian monoid under $\otimes$ of matrices, and that $\otimes$ distributes over $\oplus$; so sim(R) is a commutative semiring under $\oplus, \otimes$. *Hint:* Show $(A \otimes B) \otimes C = A \otimes (B \otimes C)$, $A \otimes [1] = [1] \otimes A = A$, $A \otimes B$ is similar to $B \otimes A$, and $A \otimes (B \oplus C)$ is similar to $(A \otimes B) \oplus (A \otimes C)$, for square matrices A, B, and C over R.

6. Repeat Exercise 5 for cong(R). *Hint:* The conjugating matrices involved in the commutative and distributive laws are permutation matrices; so they have orthonormal columns, and hence transpose equal to inverse.

6

Change of Rings

The inventory of specific rings is enlarged by several constructions that build new rings out of old ones. These constructions include cartesian products, matrix rings, opposite rings, quotient rings, rings of fractions, and various monoid and polynomial rings. If S is a ring constructed from R, we consider the relation between $K_0(S)$ and $K_0(R)$, and between $G_0(S)$ and $G_0(R)$. The first three constructions are considered for K_0 in §6A, and the first four for G_0 in §6B. Passing to a quotient R/I is a special case of a ring homomorphism; in §6C we show K_0 is a functor and derive some consequences for the ranks of modules. In §6D, we discuss the Jacobson radical, and in §6E we consider localization: the process of adjoining inverses to R. Group rings will be discussed in Chapter 8 and polynomial rings in §13C.

6A. K_0 of Related Rings

Suppose R and S are rings. Since $K_0(R)$ is constructed from the category $\mathcal{P}(R)$, to find a homomorphism $K_0(R) \to K_0(S)$ it is natural to look for a functor $F : \mathcal{P}(R) \to \mathcal{P}(S)$. Since functors preserve isomorphisms, F would determine a function

$$\mathcal{I}(\mathcal{P}(R)) \to \mathcal{I}(\mathcal{P}(S)) \ , \quad c(P) \mapsto c(F(P)).$$

There would be a $\mathbb{Z}$-linear extension of this map to the free $\mathbb{Z}$-module based on these sets. If F should have the additional property of preserving short exact sequences, there would be an induced homomorphism of groups

$$K_0(R) \to K_0(S) \ , \quad [P] \mapsto [F(P)] \ .$$

(6.1) **Definition.** Suppose R and S are rings, $\mathcal{C}$ is a subcategory of $R\text{-}\mathcal{M}\mathfrak{o}\mathfrak{d}$ or $\mathcal{M}\mathfrak{o}\mathfrak{d}\text{-}R$, and $\mathcal{D}$ is a subcategory of $S\text{-}\mathcal{M}\mathfrak{o}\mathfrak{d}$ or $\mathcal{M}\mathfrak{o}\mathfrak{d}\text{-}S$. A functor $F : \mathcal{C} \to \mathcal{D}$ is **exact** if, for each short exact sequence

$$0 \longrightarrow L \overset{f}{\longrightarrow} M \overset{g}{\longrightarrow} N \longrightarrow 0$$

160

in $\mathcal{C}$, the sequence

$$0 \longrightarrow F(L) \xrightarrow{F(f)} F(M) \xrightarrow{F(g)} F(N) \longrightarrow 0$$

if F is covariant, or

$$0 \longleftarrow F(L) \xleftarrow{F(f)} F(M) \xleftarrow{F(g)} F(N) \longleftarrow 0$$

if F is contravariant, is also exact.

A good source of exact functors $\mathcal{P}(R) \to \mathcal{P}(S)$ is those functors that preserve sums of arrows: If $\mathcal{C}$ is a full subcategory of R-$\mathcal{M}\mathrm{o}\mathrm{d}$ or $\mathcal{M}\mathrm{o}\mathrm{d}$-$R$, and $A, B \in \mathcal{C}$, the set $\mathrm{Hom}_R(A, B)$ is an additive abelian group under pointwise addition of maps, $(f + g)(a) = f(a) + g(a)$. The identity of this group is the zero map ($f(a) = 0$ for all $a \in A$).

(6.2) Definition. Suppose R and S are rings, $\mathcal{C}$ is a full subcategory of R-$\mathcal{M}\mathrm{o}\mathrm{d}$ or $\mathcal{M}\mathrm{o}\mathrm{d}$-$R$, and $\mathcal{D}$ is a full subcategory of S-$\mathcal{M}\mathrm{o}\mathrm{d}$ or $\mathcal{M}\mathrm{o}\mathrm{d}$-$S$. A functor $F : \mathcal{C} \to \mathcal{D}$ is **additive** if, for each $A, B \in \mathcal{C}$ and each $f, g \in \mathrm{Hom}_R(A, B)$,

$$F(f + g) = F(f) + F(g) .$$

Of course, an additive functor restricts to group homomorphisms on the hom-sets $\mathrm{Hom}_R(A, B)$; so it takes zero maps to zero maps. An additive functor

$$F : R\text{-}\mathcal{M}\mathrm{o}\mathrm{d} \to S\text{-}\mathcal{M}\mathrm{o}\mathrm{d}$$

also preserves direct sums

$$F(A \oplus B) \cong F(A) \oplus F(B) ;$$

for if

$$0 \longrightarrow L \underset{k}{\overset{f}{\rightleftarrows}} M \underset{h}{\overset{g}{\rightleftarrows}} N \longrightarrow 0$$

is in R-$\mathcal{M}\mathrm{o}\mathrm{d}$ with

$$k \circ f = i_L , \quad g \circ h = i_N , \quad \text{and} \quad (f \circ k) + (h \circ g) = i_M ,$$

then the same relations hold among $F(f), F(g), F(h)$, and $F(k)$. Put another way, *additive functors preserve split short exact sequences*.

(6.3) Proposition. *Suppose F is an additive functor from R-$\mathcal{M}o\eth$ to S-$\mathcal{M}o\eth$ and $F(R) \in \mathcal{P}(S)$. Then F restricts to an exact functor $\mathcal{P}(R) \to \mathcal{P}(S)$, inducing a group homomorphism $f : K_0(R) \to K_0(S)$ with $f([P]) = [F(P)]$ for each $P \in \mathcal{P}(R)$. If R and S are commutative rings, f is a ring homomorphism if and only if $[F(R)] = [S]$ and $[F(P) \otimes_S F(Q)] = [F(P \otimes_R Q)]$ in $K_0(S)$, for all $P, Q \in \mathcal{P}(R)$.*

Proof. The functor F takes $\mathcal{P}(R)$ into $\mathcal{P}(S)$ because, if $P \oplus Q \cong R^n$, then

$$F(P) \oplus F(Q) \cong F(R)^n \in \mathcal{P}(S) .$$

All short exact sequences in $\mathcal{P}(R)$ split, so the restriction of F to $\mathcal{P}(R) \to \mathcal{P}(S)$ is exact. If R and S are commutative, the last assertion follows from

$$1_{K_0(R)} = [R] , \quad 1_{K_0(S)} = [S]$$

and the fact that $K_0(R)$ is additively generated by the elements $[P]$ with $P \in \mathcal{P}(R)$. ∎

(6.4) Examples of additive functors.

(i) Suppose e is a **central idempotent** of a ring R (which means $e \in R$, $ee = e$, and $re = er$ for all $r \in R$). The ideal $Re = eR = eRe$ is a ring with the same operations as R, but with multiplicative identity e. For each R-module M, there is an Re-module eM. If $f : M \to N$ is R-linear, it restricts to an Re-linear map $eM \to eN$. The functor

$$F : R\text{-}\mathcal{M}o\eth \to Re\text{-}\mathcal{M}o\eth$$
$$F(M) = eM$$
$$F(f) = \text{restriction of } f$$

is additive, and $F(R) = eR = Re \in \mathcal{P}(Re)$. So there is a group homomorphism $K_0(R) \to K_0(Re)$ with $[P] \mapsto [eP]$ for each $P \in \mathcal{P}(R)$. If R is a commutative ring, this is a ring homomorphism $K_0(R) \to K_0(Re)$, since $eP \otimes_{Re} eQ$ is isomorphic to $e(P \otimes_R Q)$: For $(ep, eq) \mapsto ep \otimes eq$ defines an Re-balanced map, inducing an Re-linear map

$$f : eP \otimes_{Re} eQ \to P \otimes_R Q ,$$

with image $e(P \otimes_R Q)$. Since $(p, q) \mapsto (ep \otimes eq)$ is an R-balanced map inducing a left inverse to f, f is injective.

(ii) Suppose $\phi : R \to S$ is a ring homomorphism. Each S-module M is also an R-module M_ϕ via $r \cdot m = \phi(r) \cdot m$, and each S-linear map is also R-linear. The functor

$$F : S\text{-}\mathcal{M}o\eth \to R\text{-}\mathcal{M}o\eth$$
$$F(M) = M_\phi$$
$$F(f) = f$$

is additive and is called **restriction of scalars.** *If S is f.g. and projective as an R-module* under $r \cdot s = \phi(r)s$, there is a group homomorphism $\phi' : K_0(S) \to K_0(R)$ with $\phi'([P]) = [P_\phi]$ for each $P \in \mathcal{P}(S)$. In case ϕ is the inclusion of a subfield R into a field S, and $[S : R] = n < \infty$, $\phi'([S]) = [R^n] = n[R]$; so ϕ' becomes multiplication by $n : \mathbb{Z} \to \mathbb{Z}$. Thus, when R and S are commutative rings, the restriction of scalars map $\phi' : K_0(S) \to K_0(R)$ *need not be a ring homomorphism.*

(iii) Suppose R and S are rings and M is an S, R-bimodule. By (5.13), for each $P \in R$-$\mathcal{M}\mathrm{o}\mathfrak{d}$, there is a module $M \otimes_R P \in S$-$\mathcal{M}\mathrm{o}\mathfrak{d}$ with S acting on the left coordinate of each little tensor. For each arrow $f : P \to Q$ in R-$\mathcal{M}\mathrm{o}\mathfrak{d}$,

$$i_M \otimes f : M \otimes_R P \to M \otimes_R Q$$

is an arrow in S-$\mathcal{M}\mathrm{o}\mathfrak{d}$, and there is a functor

$$F : R\text{-}\mathcal{M}\mathrm{o}\mathfrak{d} \to S\text{-}\mathcal{M}\mathrm{o}\mathfrak{d}$$
$$F(P) = M \otimes_R P$$
$$F(f) = i_M \otimes f.$$

Sometimes we denote this functor F by $M \otimes_R (-)$. Since little tensors distribute over sums, F is additive. And $F(R) = M \otimes_R R \cong M$ in S-$\mathcal{M}\mathrm{o}\mathfrak{d}$; so *if we assume M is f.g. and projective as a left S-module,* then there is a group homomorphism $K_0(R) \to K_0(S)$ with $[P] \mapsto [M \otimes_R P]$ for each $P \in \mathcal{P}(R)$.

(iv) Suppose M is an R, S-bimodule. For each $P \in R$-$\mathcal{M}\mathrm{o}\mathfrak{d}$, the additive group $\operatorname{Hom}_R(P, M)$ is a left S^{op}-module with

$$(s \cdot f)(p) = f(p) \cdot s$$

for $s \in S$, $f \in \operatorname{Hom}_R(P, M)$ and $p \in P$. If $\alpha : P_1 \to P_2$ is R-linear, there is an S^{op}-linear map

$$\operatorname{Hom}_R(P_2, M) \to \operatorname{Hom}_R(P_1, M)$$

defined by $f \mapsto f \circ \alpha$. If also $\beta : P_2 \to P_3$ is R-linear, then $f \circ (\beta \circ \alpha) = (f \circ \beta) \circ \alpha$ for each $f \in \operatorname{Hom}_R(P_3, M)$, and $f \circ i_P = f$ for each $f \in \operatorname{Hom}_R(P, M)$. So there is a contravariant functor

$$F : R\text{-}\mathcal{M}\mathrm{o}\mathfrak{d} \to S^{op}\text{-}\mathcal{M}\mathrm{o}\mathfrak{d}$$
$$F(P) = \operatorname{Hom}_R(P, M)$$
$$F(\alpha) = (-) \circ \alpha.$$

Since composition distributes over the pointwise sum, F is additive. And $F(R) = \operatorname{Hom}_R(R, M) \cong M$ as S^{op}-modules via the map $f \mapsto f(1_R)$; so *if we assume M is f.g. and projective as a right S-module* (i.e., left S^{op}-module), there is a group homomorphism $K_0(R) \to K_0(S^{op})$ taking $[P]$ to $[\operatorname{Hom}_R(P, M)]$ for each $P \in \mathcal{P}(R)$.

Next we apply these examples of additive functors to connect K_0 groups of related rings.

(6.5) Theorem. *If $R \cong S$ as rings, then restriction of scalars defines an isomorphism $K_0(S) \cong K_0(R)$ as groups, and also as rings if R and S are commutative.*

Proof. Say $\phi : R \to S$ is a ring isomorphism. Then ϕ is R-linear; so $S_\phi \in \mathcal{P}(R)$, and there is an induced group homomorphism $\phi' : K_0(S) \to K_0(R)$ with $\phi'([P]) = [P_\phi]$ for each $P \in \mathcal{P}(S)$. Likewise, $(\phi^{-1})' : K_0(R) \to K_0(S)$ takes $[Q]$ to $[Q_{\phi^{-1}}]$ for $Q \in \mathcal{P}(R)$. Since $(P_\phi)_{\phi^{-1}} = P_{\phi\phi^{-1}} = P$ and $(Q_{\phi^{-1}})_\phi = Q_{\phi^{-1}\phi} = Q$, ϕ' and $(\phi^{-1})'$ are inverse maps; so ϕ' is a group isomorphism.

If R and S are commutative and $P, Q \in \mathcal{P}(S)$, then there is an R-linear map

$$(P \otimes_S Q)_\phi \;\to\; P_\phi \otimes_R Q_\phi \,,$$
$$p \otimes q \;\mapsto\; p \otimes q$$

with inverse defined the same way using ϕ^{-1}. So ϕ' is a ring isomorphism. ■

The preceding theorem can hardly be a surprise, since a mere renaming of the scalars cannot affect the structure of its category of f.g. projective modules.

As an application of the functors (6.4) (i) and (ii), we now prove that K_0 respects cartesian products. If S and T are rings, their cartesian product $S \times T$ is a ring with coordinatewise sum and product. The elements $e = (1, 0)$ and $f = (0, 1)$ are central idempotents of $S \times T$ with $e + f = 1_{S \times T} = (1, 1)$. On the other hand, if R is a ring with a central idempotent e, then $f = 1 - e$ is also a central idempotent, and $e + f = 1$. Also, $ef = e - ee = 0$. So Re and Rf are rings, and there is a ring isomorphism $R \cong Re \times Rf$, $r \mapsto (re, rf)$.

(6.6) Theorem. *If S and T are rings, there is an isomorphism*

$$K_0(S \times T) \;\cong\; K_0(S) \times K_0(T)$$

of groups, which is an isomorphism of rings when S and T are commutative.

Proof. Take $R = S \times T$, $e = (1, 0)$, and $f = (0, 1)$. By (6.4) (i) there are group homomorphisms

$$\pi_1 : K_0(R) \;\to\; K_0(Re) \,, \qquad \pi_2 : K_0(R) \;\to\; K_0(Rf) \,.$$
$$[P] \;\mapsto\; [eP] \qquad\qquad\qquad [P] \;\mapsto\; [fP]$$

Since $e + f = 1$ and $ef = 0$, $R = Re \overset{\bullet}{\oplus} Rf$ as R-modules; so $Re, Rf \in \mathcal{P}(R)$. So, by (6.4) (ii) applied to the ring homomorphisms $R \to Re$ $(r \mapsto re)$ and $R \to Rf$ $(r \mapsto rf)$, there are group homomorphisms

$$i_1 : K_0(Re) \;\to\; K_0(R) \,, \qquad i_2 : K_0(Rf) \;\to\; K_0(R) \,.$$
$$[P] \;\mapsto\; [P] \qquad\qquad\qquad [Q] \;\mapsto\; [Q]$$

For the latter P and Q, $eP = P$ and $eQ = Q$; so

$$\pi_1 \circ i_1 = i_{K_0(Re)} \quad \text{and} \quad \pi_2 \circ i_2 = i_{K_0(Rf)} .$$

For $M \in R\text{-}\mathfrak{Mod}$, $M = eM \overset{\bullet}{\oplus} fM$. So

$$(i_1 \circ \pi_1) + (i_2 \circ \pi_2) = i_{K_0(R)} .$$

By (2.5), there is an isomorphism

$$K_0(R) \cong K_0(Re) \oplus K_0(Rf)$$
$$x \mapsto (\pi_1(x), \pi_2(x))$$

of groups. Restriction of scalars, applied to ring isomorphisms $S \cong Re$ and $T \cong Rf$, defines isomorphisms of groups:

$$j_1 : K_0(Re) \cong K_0(S) , \quad j_2 : K_0(Rf) \cong K_0(T) .$$

Then sending x to $(j_1 \circ \pi_1(x), j_2 \circ \pi_2(x))$ defines an isomorphism

$$K_0(S \times T) \cong K_0(S) \times K_0(T)$$

of additive groups, which is a ring isomorphism when S and T are commutative. ∎

As an application of (6.4) (iii), we prove the matrix invariance of K_0:

(6.7) Theorem. *Suppose R is a ring and m, n are positive integers. Then*

$$[P] \mapsto [R^{n \times m} \otimes_{M_m(R)} P] \quad \text{and} \quad [Q] \mapsto [R^{m \times n} \otimes_{M_n(R)} Q]$$

define mutually inverse group isomorphisms:

$$K_0(M_m(R)) \rightleftarrows K_0(M_n(R)) .$$

Proof. In $M_n(R)\text{-}\mathfrak{Mod}$, $(R^{n \times 1})^n \cong M_n(R)$ is free of rank 1; so $R^{n \times m} \cong (R^{n \times 1})^m \in \mathcal{P}(M_n(R))$. Similarly, $R^{m \times n} \in \mathcal{P}(M_m(R))$. So the indicated homomorphisms exist. That they are inverse to each other follows from:

(6.8) Lemma. *Matrix multiplication defines a left and right linear isomorphism of $M_m(R), M_m(R)$-bimodules:*

$$R^{m \times n} \otimes_{M_n(R)} R^{n \times m} \cong M_m(R) .$$

Proof of Lemma. Matrix multiplication from $R^{m \times n} \times R^{n \times m}$ to $M_m(R)$ is $M_n(R)$-balanced, by its associativity and distributivity; so it induces an additive homomorphism

$$f : R^{m \times n} \otimes_{M_n(R)} R^{n \times m} \to M_m(R) ,$$

defined on little tensors by $f(A \otimes B) = AB$. Again by associativity of matrix multiplication, f is left and right $M_m(R)$-linear. So its image is an ideal of $M_m(R)$. This ideal includes

$$f(\sum_{i=1}^{m} \epsilon_{i1} \otimes \epsilon_{1i}) = \sum_{i=1}^{m} \epsilon_{ii} = I_m ,$$

where ϵ_{ij} is the matrix whose only nonzero entry is a 1 in the i,j-position. So f is surjective.

Suppose $\sum A_i \otimes B_i \in \ker(f)$, so that $\sum A_i B_i = 0_{m \times m}$. For $\epsilon_{j1} \in R^{m \times n}$

$$
\begin{aligned}
\sum_i A_i \otimes B_i &= \sum_i (A_i \otimes B_i) \sum_j \epsilon_{j1} \epsilon_{1j} \\
&= \sum_{i,j} A_i \otimes B_i \epsilon_{j1} \epsilon_{1j} \\
&= \sum_{i,j} A_i B_i \epsilon_{j1} \otimes \epsilon_{1j} \\
&= \sum_i (A_i B_i) \sum_j \epsilon_{j1} \otimes \epsilon_{1j} \\
&= 0 .
\end{aligned}
$$

■ ■

In particular, taking $m = 1$, $K_0(M_n(R)) \cong K_0(R)$.

Next, we apply (6.4) (iv) to prove $K_0(R) \cong K_0(R^{op})$. Take M to be the R, R-bimodule R. The additive contravariant functor

$$
\begin{aligned}
D : R\text{-}\mathcal{Mod} &\to R^{op}\text{-}\mathcal{Mod} \\
D(P) &= \text{Hom}_R(P, R) \\
D(f) &= (-) \circ f
\end{aligned}
$$

is called the **dual** functor, and $D(P)$ is called the **dual module** to P. We shall also use the notation P^* for $D(P)$ and f^* for $D(f)$.

The double dual DD is an additive covariant functor from R-$\mathfrak{Mod}$ to itself:

$$P^{**} = DD(P) = \mathrm{Hom}_{R^{op}}(\mathrm{Hom}_R(P, R),\ R^{op})$$
$$f^{**} = DD(f) = (-) \circ ((-) \circ f)\ .$$

Let $\#$ denote multiplication in R^{op}. For each $p \in P$, evaluation at p defines a map

$$\tau(p) : P^* \ \rightarrow\ R^{op}\ ,\ f\ \mapsto\ f(p)\ ,$$

which is R^{op}-linear:

$$\tau(p)(r \cdot f) = (r \cdot f)(p) = f(p)r = r\#f(p) = r\#\tau(p)(f)\ .$$

Taking p to $\tau(p)$ defines an R-linear map $\tau : P \rightarrow P^{**}$:

$$\tau(r \cdot p)(f) = f(r \cdot p) = rf(p)$$
$$= f(p)\#r = \tau(p)(f)\#r = (r \cdot \tau(p))(f)\ ,$$

for each $f \in P^*$. An R-module P is **reflexive** if $\tau : P \rightarrow P^{**}$ is an isomorphism.

(6.9) Lemma. *The maps $\tau : P \rightarrow P^{**}$ define a natural transformation from the identity functor on R-$\mathfrak{Mod}$ to the double dual functor.*

Proof. According to the definition (0.13) of natural transformation, we must show that, for each R-linear map $\alpha : P \rightarrow Q$, the square

$$
\begin{array}{ccc}
P & \xrightarrow{\ \alpha\ } & Q \\
{\scriptstyle\tau}\downarrow & & \downarrow{\scriptstyle\tau} \\
P^{**} & \xrightarrow{\ \alpha^{**}\ } & Q^{**}
\end{array}
$$

commutes. If $f \in Q^*$ and $p \in P$,

$$\alpha^{**}(\tau(p))(f) = (\tau(p) \circ \alpha^*)(f) = \tau(p)(f \circ \alpha) = (f \circ \alpha)(p)$$
$$= f(\alpha(p)) = \tau(\alpha(p))(f)\ .$$

So $\tau \circ \alpha = \alpha^{**} \circ \tau$. ∎

The next theorem shows that $K_0(R)$ is the same group, whether defined by using left R-modules or right R-modules:

(6.10) Theorem. *For each ring R, there is a group isomorphism $K_0(R) \cong K_0(R^{op})$ with $[P] \mapsto [P^*]$ for each $P \in \mathcal{P}(R)$.*

Proof. The bimodule $_R R_R$ is a free rank 1 right R-module; so the dual functor takes $\mathcal{P}(R)$ to $\mathcal{P}(R^{op})$ and induces a homomorphism $K_0(R) \to K_0(R^{op})$, $[P] \mapsto [P^*]$. Likewise, the dual induces a homomorphism $K_0(R^{op}) \to K_0(R)$. That these are mutually inverse maps follows from the fact that:

(6.11) Lemma. *Every f.g. projective R-module is reflexive.*

Proof. First we check that R is reflexive. Among the maps in $R^* = \operatorname{Hom}_R(R, R)$ is the identity i_R. If $\tau(r) = 0 \in R^{**}$, then $r = i_R(r) = \tau(r)(i_R) = 0 \in R$; so τ is injective. If $g \in R^{**}$, then for each $f \in R^*$, $f = f(1) \cdot i_R$ and

$$g(f) = f(1) \cdot g(i_R) = g(i_R)f(1)$$
$$= f(g(i_R)1) = f(g(i_R)) = \tau(g(i_R))(f) .$$

So $g = \tau(g(i_R))$, proving τ is surjective.

Next suppose $P, Q \in R\text{-}\mathfrak{Mod}$ and π_1, π_2 are the projections from $P \oplus Q$ to its first and second coordinates, respectively. The diagram

$$
\begin{array}{ccc}
P \oplus Q & \xrightarrow{\quad \tau \quad} & (P \oplus Q)^{**} \\
 & \searrow{\scriptstyle \tau \oplus \tau} & \downarrow{\scriptstyle (\pi_1^{**}, \pi_2^{**})} \\
 & & P^{**} \oplus Q^{**}
\end{array}
$$

commutes: If $p \in P$ and $q \in Q$, then for each $f \in P^*$ and $g \in Q^*$,

$$\pi_1^{**}(\tau((p,q)))(f) = (\tau((p,q)) \circ \pi_1^*)(f) = \tau((p,q))(f \circ \pi_1)$$
$$= (f \circ \pi_1)((p,q)) = f(p) = \tau(p)(f) ,$$

and similarly,

$$\pi_2^{**}(\tau((p,q)))(g) = \tau(q)(g) .$$

By (2.5), (π_1^{**}, π_2^{**}) is an isomorphism. Therefore $P \oplus Q$ is reflexive if and only if both P and Q are reflexive.

So for each $n \geq 0$, R^n is reflexive. If $M \cong N$ in $R\text{-}\mathfrak{Mod}$, there is a commutative square (since τ is natural):

$$
\begin{array}{ccc}
M & \cong & N \\
\downarrow{\scriptstyle \tau} & & \downarrow{\scriptstyle \tau} \\
M^{**} & \cong & N^{**} .
\end{array}
$$

So M is reflexive if and only if N is reflexive. Thus, if $P \oplus Q \cong R^n$, P is reflexive. ∎ ∎

6A. Exercises

1. Suppose R is a commutative ring and $K_0(R)$ is a cyclic group. Prove R has no idempotents other than 0_R and 1_R.

2. Find a noncommutative ring R, with idempotents other than 0_R and 1_R, for which $K_0(R)$ is infinite cyclic.

3. Compute $K_0(\mathbb{Q}G)$ where G is the cyclic group $\langle a \rangle$ of order 2 generated by a, and $\mathbb{Q}G = \mathbb{Q}[G]$ is the group ring (= monoid ring) of G over $\mathbb{Q}$. *Hint:* Look for idempotents.

4. Suppose a ring R is torsion free as an R-module ($ab = 0$ for $a, b \in R$ implies $a = 0$ or $b = 0$). If M is an R-module, show each torsion element $m \in M$ ($am = 0$ for some nonzero $a \in R$) belongs to the kernel of the standard map $M \to M^{**}$. So if M is not torsion free, it is not reflexive.

6B. G_0 of Related Rings

If R and S are rings, any exact functor $F : \mathcal{M}(R) \to \mathcal{M}(S)$ induces an additive group homomorphism $G_0(R) \to G_0(S)$, taking $[M]$ to $[F(M)]$ for each $M \in \mathcal{M}(R)$. Some short exact sequences of R-modules do not split, so additive functors $\mathcal{M}(R) \to \mathcal{M}(S)$ need not be exact. Fortunately, some of them are.

(6.12) Theorem.

   (i)   *If e is a central idempotent of a ring R, there is a group homomorphism $G_0(R) \to G_0(Re)$ taking $[M]$ to $[eM]$ for each $M \in \mathcal{M}(R)$.*

  (ii)   *If $\phi : R \to S$ is a ring homomorphism and $S_\phi \in \mathcal{M}(R)$, then restriction of scalars defines a group homomorphism $\phi' : G_0(S) \to G_0(R)$ with $\phi'([M]) = [M_\phi]$ for each $M \in \mathcal{M}(S)$. If ϕ is an isomorphism, so is ϕ'.*

 (iii)   *If S and T are rings, $G_0(S \times T) \cong G_0(S) \oplus G_0(T)$.*

Proof. Suppose e is a central idempotent of R. The functor $F : R\text{-}\mathcal{M}\mathfrak{o}\mathfrak{d} \to Re\text{-}\mathcal{M}\mathfrak{o}\mathfrak{d}$ with

$$F(M) = eM$$
$$F(f) = \text{restriction of } f$$

is exact: For, if

$$L \xrightarrow{\ f\ } M \xrightarrow{\ g\ } N$$

is exact in $R\text{-}\mathcal{M}\mathfrak{o}\mathfrak{d}$, the composite of

(6.13) $eL \xrightarrow{\ f\ } eM \xrightarrow{\ g\ } eN$

is the restriction of the zero map $g \circ f$, and so it is zero. If $m \in M$ and $g(em) = 0$, there exists $\ell \in L$ with $f(\ell) = em$. Scalar multiplying by e, $f(e\ell) = em$. So (6.13) is exact.

If $M \in \mathcal{M}(R)$ is generated by $m_1, \ldots, m_n$, then eM is generated as an Re-module by $em_1, \ldots, em_n$. So F is an exact functor $\mathcal{M}(R) \to \mathcal{M}(Re)$, proving (i). The scalar multiplications have no effect on the exactness of a sequence of linear maps; so restriction of scalars is exact. Suppose $\phi : R \to S$ is a ring homomorphism and $S_\phi = \Sigma_{i=1}^{t} Rs_i$, where $s_1, \ldots, s_t \in S$. If $M = \Sigma_{j=1}^{n} Sm_j$, then $M = \Sigma\, Rs_i m_j$. So restriction of scalars is an exact functor $\mathcal{M}(S) \to \mathcal{M}(R)$, inducing the homomorphism ϕ' in (ii). Just as with K_0 (see (6.5)), if ϕ is an isomorphism, $(\phi^{-1})' = (\phi')^{-1}$, proving (ii). The proof in (6.6) that $K_0(S \times T) \cong K_0(S) \oplus K_0(T)$ works as well for G_0.   ∎

Suppose $\mathcal{C}$ is a full subcategory of $R\text{-}\mathcal{M}\mathfrak{od}$ or $\mathcal{M}\mathfrak{od}\text{-}R$ and $\mathcal{D}$ is a full subcategory of $S\text{-}\mathcal{M}\mathfrak{od}$ or $\mathcal{M}\mathfrak{od}\text{-}S$. An additive functor $F : \mathcal{C} \to \mathcal{D}$ is **left exact** if it takes each exact sequence

$$0 \longrightarrow L \longrightarrow M \longrightarrow N$$

in $\mathcal{C}$ to an exact sequence in $\mathcal{D}$, or is **right exact** if it takes each exact sequence

$$L \longrightarrow M \longrightarrow N \longrightarrow 0$$

in $\mathcal{C}$ to an exact sequence in $\mathcal{D}$.

(6.14) Proposition. *If P is an S, R-bimodule and Q is an R, S-bimodule, the functors*

$$P \otimes_R (-) : R\text{-}\mathcal{M}\mathfrak{od} \ \to \ S\text{-}\mathcal{M}\mathfrak{od} \quad and$$
$$(-) \otimes_R Q : \mathcal{M}\mathfrak{od}\text{-}R \ \to \ \mathcal{M}\mathfrak{od}\text{-}S$$

are right exact.

Proof. Suppose

$$L \xrightarrow{\ f\ } M \xrightarrow{\ g\ } N \longrightarrow 0$$

is exact in $R\text{-}\mathcal{M}\mathfrak{od}$, and consider the sequence

$$P \otimes_R L \xrightarrow{\ i \otimes f\ } P \otimes_R M \xrightarrow{\ i \otimes g\ } P \otimes_R N \longrightarrow 0 \ .$$

Since g is surjective, $i \otimes g$ is surjective. Since $g \circ f$ is zero, $i \otimes (g \circ f)$ is zero. It remains to prove

$$A = \frac{\ker(i \otimes g)}{\operatorname{im}(i \otimes f)}$$

is trivial. But A is the kernel of the map

$$\gamma : \frac{P \otimes_R M}{\operatorname{im}(i \otimes f)} \to P \otimes_R N$$

induced by $i \otimes g$. We prove γ is injective by constructing a left inverse: If $p \in P$ and $g(m_1) = g(m_2) = n$, then $m_1 - m_2 \in \operatorname{im}(f)$; so

$$(p \otimes m_1) - (p \otimes m_2) \in \operatorname{im}(i \otimes f).$$

Thus there is a well-defined function

$$d : P \times N \to \frac{P \otimes_R M}{\operatorname{im}(i \otimes f)},$$

taking (p, n) to the coset of $p \otimes m$ whenever $g(m) = n$. Since g is R-linear, d is R-balanced, inducing a map δ on $P \otimes_R N$ with $\delta \circ \gamma$ the identity map.

The proof for $(-) \otimes_R Q$ is similar. ∎

As an aside, we obtain a useful tensor identity:

(6.15) Corollary. *Suppose R is a ring with an ideal I and a left R-module M. Denote by IM the set of finite length sums of products im with $i \in I$, $m \in M$. Then IM is a submodule of M, and in R-$\mathfrak{Mod}$,*

$$(R/I) \otimes_R M \cong M/IM.$$

Proof. Tensoring the right R-linear exact sequence

$$I \longrightarrow R \longrightarrow R/I \longrightarrow 0$$

with M yields an exact sequence

$$I \otimes_R M \longrightarrow R \otimes_R M \longrightarrow (R/I) \otimes_R M \longrightarrow 0.$$

The isomorphism $R \otimes_R M \cong M$ carries the image of $I \otimes_R M$ to IM. So there is an isomorphism

$$M/IM \to (R/I) \otimes_R M,$$
$$m + IM \mapsto (1 + I) \otimes m,$$

which is evidently left R-linear. ∎

(6.16) Definition. A module P_R (respectively $_RP$) is **flat** if $P \otimes_R (-)$ (respectively $(-) \otimes_R P$) is exact.

By (6.14), P is flat if and only if tensoring with P preserves injectivity of maps.

(6.17) Example. Suppose $f : \mathbb{Z} \to \mathbb{Z}$ is multiplication by 2. Then f is injective and $\mathbb{Z}$-linear; but

$$i \otimes f : (\mathbb{Z}/2\mathbb{Z}) \otimes_\mathbb{Z} \mathbb{Z} \to (\mathbb{Z}/2\mathbb{Z}) \otimes_\mathbb{Z} \mathbb{Z}$$

is the zero map: $(i \otimes f)(\overline{a} \otimes b) = \overline{a} \otimes 2b = \overline{a}2 \otimes b = \overline{0} \otimes b = 0$. Since $\mathbb{Z}/2\mathbb{Z} \otimes_\mathbb{Z} \mathbb{Z} \cong \mathbb{Z}/2\mathbb{Z} \neq 0, i \otimes f$ is not injective. So $\mathbb{Z}/2\mathbb{Z}$ is not a flat $\mathbb{Z}$-module.

(6.18) Proposition. *Every f.g. projective left or right R-module is flat.*

Proof. We work with a right f.g. projective R-module P; the proof for left modules is similar. If $P \cong P'$ in $\mathcal{M}\mathfrak{od}$-R, the commutative square

$$
\begin{array}{ccc}
P \otimes_R M & \cong & P' \otimes_R M \\
{\scriptstyle i \otimes f}\downarrow & & \downarrow{\scriptstyle i \otimes f} \\
P \otimes_R N & \cong & P' \otimes_R N
\end{array}
$$

shows P' is flat if P is flat. From the commutative square

$$
\begin{array}{ccc}
(P \oplus Q) \otimes_R M & \cong & (P \otimes_R M) \oplus (Q \otimes_R M) \\
{\scriptstyle i \otimes f}\downarrow & & \downarrow{\scriptstyle (i \otimes f) \oplus (i \otimes f)} \\
(P \oplus Q) \otimes_R N & \cong & (P \otimes_R N) \oplus (Q \otimes_R N)
\end{array}
$$

$P \oplus Q$ is flat if and only if both P and Q are flat. From the commutative square

$$
\begin{array}{ccc}
R \otimes_R M & \cong & M \\
{\scriptstyle i \otimes f}\downarrow & & \downarrow{\scriptstyle f} \\
R \otimes_R N & \cong & N
\end{array}
$$

R_R is flat.   ∎

(6.19) Proposition. *Suppose an S, R-bimodule P is finitely generated as a left S-module and flat as a right R-module. Then there is a group homomorphism $G_0(R) \rightarrow G_0(S)$, defined on generators by $[M] \mapsto [P \otimes_R M]$ for all $M \in \mathcal{M}(R)$.*

Proof. If $M \in \mathcal{M}(R)$ and $_SP_R \in \mathcal{M}(S)$, there are positive integers m and n, and surjective homomorphisms $R^n \rightarrow M$ and $S^m \rightarrow P$ in R-$\mathfrak{Mod}$ and S-$\mathfrak{Mod}$, respectively. Then there are left S-linear surjections:

$$ S^{mn} \rightarrow P^n \cong P \otimes_R R^n \rightarrow P \otimes_R M \ . $$

So $P \otimes_R M \in \mathcal{M}(S)$. Since P_R is flat, tensoring with P is exact, inducing the required homomorphism $G_0(R) \rightarrow G_0(S)$.   ∎

Like K_0, G_0 is matrix-invariant:

(6.20) Theorem. *Suppose R is a ring and m, n are positive integers. Then*

$$ [M] \ \mapsto \ [R^{n \times m} \otimes_{M_m(R)} M] \ , \quad and $$
$$ [N] \ \mapsto \ [R^{m \times n} \otimes_{M_n(R)} N] $$

define mutually inverse isomorphisms

$$ G_0(M_m(R)) \ \underset{\longleftarrow}{\overset{\longrightarrow}{\rule{0pt}{0pt}\hspace{2em}}} \ G_0(M_n(R)) \ \ . $$

Proof. The $M_n(R)$, $M_m(R)$-bimodule $R^{n \times m}$ is generated as an $M_n(R)$-module by its finite set of matrix units ϵ_{ij} and is a f.g. projective right $M_m(R)$-module, since it is isomorphic to $(R^{1 \times m})^n$. So tensoring with $R^{n \times m}$ defines a group homomorphism from $G_0(M_m(R))$ to $G_0(M_n(R))$. Similarly, tensoring with $R^{m \times n}$ defines a map going the other way, which is inverse to the first map by Lemma (6.8).   ∎

6B. Exercises

1. For what rings R can you prove $G_0(R) \cong G_0(R^{op})$? To the author's knowledge it is an open question whether $G_0(R) \cong G_0(R^{op})$ for all rings R.

2. If I and J are ideals of a ring R, prove $(R/I) \otimes_R (R/J) \cong R/(I + J)$ as R-modules.

6C. K_0 as a Functor

If $\phi : R \to S$ is a ring homomorphism, S is an S, R-bimodule via $s' \cdot s \cdot r = s' s \phi(r)$. The resulting additive functor

$$S \otimes_R (-) : R\text{-}\mathcal{M}\mathfrak{o}\mathfrak{d} \ \to \ S\text{-}\mathcal{M}\mathfrak{o}\mathfrak{d}$$

is known as **extension of scalars**. Employing Proposition (6.3), since $S \otimes_R R \cong S \in \mathcal{P}(S)$, this functor induces an additive group homomorphism

$$\boldsymbol{K_0}(\phi) : K_0(R) \ \to \ K_0(S) \ ,$$

with $K_0(\phi)([P]) = [S \otimes_R P]$ for each $P \in \mathcal{P}(R)$. In particular,

(6.21) $K_0(\phi)([R]) \ = \ [S] \ .$

(6.22) Theorem. *By extension of scalars, K_0 is a covariant functor from $\mathcal{R}\mathrm{ing}$ to $\mathcal{A}\mathfrak{b}$ and from $\mathcal{C}\mathcal{R}\mathrm{ing}$ to $\mathcal{C}\mathcal{R}\mathrm{ing}$.*

Proof. For $P \in \mathcal{P}(R)$, $R \otimes_R P \cong P$; so $K_0(i_R) = i_{K_0(R)}$. If $\phi : R \to S$ and $\psi : S \to T$ are ring homomorphisms, making S and T into bimodules $_S S_R$ and $_T T_S$, and if T is also regarded as a bimodule $_T T_R$ via the composite map $\psi \circ \phi$, then the isomorphism

$$T \otimes_S S \cong T$$
$$t \otimes s \ \mapsto \ t \cdot s \ = \ t \psi(s)$$

is both left T-linear and right R-linear. So, for each $P \in \mathcal{P}(R)$,

$$T \otimes_S (S \otimes_R P) \cong (T \otimes_S S) \otimes_R P \cong T \otimes_R P \ ,$$

and $K_0(\psi \circ \phi) = K_0(\psi) \circ K_0(\phi)$.

Now suppose R and S are commutative rings and $\phi : R \to S$ is a ring homomorphism. By (6.21), $K_0(\phi)$ takes $1 = [R] \in K_0(R)$ to $1 = [S] \in K_0(S)$. For all $P, Q \in \mathcal{P}(R)$,

$$\begin{aligned}
(S \otimes_R P) \otimes_S (S \otimes_R Q) \ &\cong \ (P \otimes_R S) \otimes_S (S \otimes_R Q) \\
&\cong \ (P \otimes_R (S \otimes_S S)) \otimes_R Q \\
&\cong \ (P \otimes_R S) \otimes_R Q \\
&\cong \ (S \otimes_R P) \otimes_R Q \\
&\cong \ S \otimes_R (P \otimes_R Q) \ .
\end{aligned}$$

So

$$K_0(\phi)(xy) \ = \ K_0(\phi)(x) \ K_0(\phi)(y)$$

if $x = [P]$ and $y = [Q]$. That the same equation holds when $x = [P] - [P']$ and $y = [Q] - [Q']$ is a consequence of the additivity of $K_0(\phi)$. ∎

Recall from (4.5) that the additive homomorphism $f : \mathbb{Z} \to K_0(R)$, $f(n) = n[R]$, is injective if and only if R has IBN, and surjective if and only if every $P \in \mathcal{P}(R)$ is stably free. Its cokernel is the projective class group $\widetilde{K}_0(R) = K_0(R)/\langle[R]\rangle$.

Once we know K_0 is a functor, the map f arises in a canonical way: The ring $\mathbb{Z}$ is the initial object in the category $\mathfrak{Ring}$. That is, for each ring R, there is a unique ring homomorphism $i : \mathbb{Z} \to R$. Its kernel is generated by the characteristic of R, and its image is the intersection of all subrings of R. So one might expect $K_0(i) : K_0(\mathbb{Z}) \to K_0(R)$ to be simply related to the structure of $K_0(R)$, or even to that of the category $\mathcal{P}(R)$. Since $\mathbb{Z}$ has IBN and each $P \in \mathcal{P}(\mathbb{Z})$ is free, the map

$$\mathbb{Z} \to K_0(\mathbb{Z})$$
$$n \mapsto n[\mathbb{Z}]$$

is an isomorphism. For each ring R, there is a commutative triangle

So f and $K_0(i)$ have equal images $\langle[R]\rangle$ and isomorphic kernels. In particular, $K_0(i)$ is injective if and only if R has IBN, and surjective if and only if each $P \in \mathcal{P}(R)$ is stably free.

The cokernel $\widetilde{K}_0(R)$ of both maps is also a functor $\widetilde{K}_0 : \mathfrak{Ring} \to \mathcal{Ab}$. For if $\phi : R \to S$ is a ring homomorphism, there is a commutative diagram with exact rows:

(6.23)
$$
\begin{array}{ccccccccc}
0 & \longrightarrow & \langle[R]\rangle & \longrightarrow & K_0(R) & \longrightarrow & \widetilde{K}_0(R) & \longrightarrow & 0 \\
& & \downarrow{\scriptstyle\alpha} & & \downarrow{\scriptstyle K_0(\phi)} & & \downarrow{\scriptstyle\widetilde{K}_0(\phi)} & & \\
0 & \longrightarrow & \langle[S]\rangle & \longrightarrow & K_0(S) & \longrightarrow & \widetilde{K}_0(S) & \longrightarrow & 0
\end{array}
$$

where $K_0(\phi)$ restricts to a surjective map α by (6.21), and induces the map $\widetilde{K}_0(\phi)$. That $\widetilde{K}_0(-)$ preserves identities and composites follows from the corresponding properties of $K_0(-)$. Commutativity of the right-hand square says the canonical maps $K_0(R) \to \widetilde{K}_0(R)$ define a natural transformation $K_0 \to \widetilde{K}_0$ in the sense of (0.13).

It is common for a ring R to have IBN, but relatively less common for every $P \in \mathcal{P}(R)$ to be stably free. An intermediate condition between injectivity and bijectivity of

$$f : \mathbb{Z} \to K_0(R)$$
$$n \mapsto n[R]$$

is that f have an additive left inverse g (so $g \circ f = i_N$). An additive map $g : K_0(R) \to \mathbb{Z}$ is left inverse to f if and only if $g([R]) = 1$ (or, equivalently, $g([R^n]) = n$ for all $n \geq 0$).

If g is an additive left inverse to f, there is a split exact sequence

$$0 \longrightarrow \mathbb{Z} \underset{g}{\overset{f}{\rightleftarrows}} K_0(R) \longrightarrow \widetilde{K}_0(R) \longrightarrow 0$$

and $K_0(R) \cong \mathbb{Z} \oplus \widetilde{K}_0(R)$. Also, g assigns to each f.g. projective R-module P an integer $r(P) = g([P])$. This $r : \mathrm{Obj}\ \mathcal{P}(R) \to \mathbb{Z}$ is a generalized rank (i.e., additive over short exact sequences), and $r(R^n) = n$ for each $n \geq 0$.

Conversely, by (3.38), each $\mathbb{Z}$-valued generalized rank r on $\mathcal{P}(R)$ induces an additive map $g : K_0(R) \to \mathbb{Z}$; and if $r(R^n) = n$ for all $n \geq 0$, then g is left inverse to f.

(6.24) Definition. If $\mathcal{C}$ is a subcategory of R-$\mathfrak{Mod}$ containing the category $\mathcal{F}(R)$ of f.g. free R-modules, an **extended free rank** on $\mathcal{C}$ is a generalized rank $r : \mathrm{Obj}\ \mathcal{C} \to \mathbb{Z}$ that extends the free rank on $\mathcal{F}(R)$ (i.e., with $r(R^n) = n$ for all $n \geq 0$).

We have proved:

(6.25) Proposition. *The map* $f : \mathbb{Z} \to K_0(R)$, $n \mapsto n[R]$, *has an additive left inverse if and only if there is an extended free rank on* $\mathcal{P}(R)$. ∎

(6.26) Note. Suppose R is a ring with IBN; so f is injective. The condition, $r(P) = g([P])$ for all $P \in \mathcal{P}(R)$, defines a bijective correspondence between the set of additive left inverses g of f and the collection of extended free ranks r on $\mathcal{P}(R)$. By (2.11), there is also a bijection $g \leftrightarrow \ker(g)$ between the additive left inverses of f and the complements C to $\langle [R] \rangle = f(\mathbb{Z})$ in $K_0(R)$. So the extended free ranks r on $\mathcal{P}(R)$ correspond to these complements C by

$$r \longleftrightarrow C = \{[P] - [Q] : P,\ Q \in \mathcal{P}(R),\ r(P) = r(Q)\}.$$

Of course, each of these complements is isomorphic to $\widetilde{K}_0(R)$.

Suppose there is a ring homomorphism $j : R \to S$. Then S is an S, R-bimodule via j. Any extended free rank r on $\mathcal{P}(S)$ induces an extended free rank r' on $\mathcal{P}(R)$, with $r'(P) = r(S \otimes_R P)$: For if $g : K_0(S) \to \mathbb{Z}$ is additive, with $g([S]) = 1$, then the composite

$$K_0(R) \xrightarrow{K_0(j)} K_0(S) \xrightarrow{g} \mathbb{Z}$$

is additive, taking $[R]$ to $g([S]) = 1$ and $[P]$ to $g([S \otimes_R P])$ for $P \in \mathcal{P}(R)$.

If D is a division ring, D has IBN and $\mathcal{P}(D) = \mathcal{F}(D)$; so the homomorphism $f : \mathbb{Z} \to K_0(D)$, $n \mapsto n[D]$, has an inverse induced by the free rank $\dim_D(-)$. So we have:

(6.27) Proposition. *If $j : R \to D$ is a ring homomorphism, making D into a D, R-bimodule D_j, and D is a division ring, there is an extended free rank r_j on $\mathcal{P}(R)$, defined by*

$$r_j(P) = \dim_D(D_j \otimes_R P) \ ;$$

and $K_0(R) = \langle [R] \rangle \overset{\bullet}{\oplus} C$, where $\langle [R] \rangle \cong \mathbb{Z}$ and where

$$C = \{ [P] - [Q] : P, \ Q \in \mathcal{P}(R), r_j(P) = r_j(Q) \} \cong \tilde{K}_0(R) \ . \qquad \blacksquare$$

From (6.27) we see that the class of all rings R for which there is an extended free rank on $\mathcal{P}(R)$ is fairly broad. It includes all commutative rings ($D = R/M$, M a maximal ideal), all division rings, and is closed under subrings, cartesian products with other rings, and pre-images of ring homomorphisms. But it is not closed under the formation of matrix rings:

(6.28) Proposition. *If R is isomorphic to a matrix ring $R' = M_n(S)$ for an IBN ring S, then $[R]$ is a multiple of n in $K_0(R)$. So if $n > 1$, there is no extended free rank on $\mathcal{P}(R)$, and*

$$0 \longrightarrow \langle [R] \rangle \overset{\subseteq}{\longrightarrow} K_0(R) \longrightarrow \tilde{K}_0(R) \longrightarrow 0$$

does not split.

Proof. The isomorphism $K_0(R') \cong K_0(R)$, induced by $R' \cong R$, takes multiples of n to multiples of n, and $[R']$ to $[R]$. So, without loss of generality, we may assume $R = R'$.

If $\{\epsilon_{ij} : 1 \le i, j \le n\}$ is the standard S-basis of R, then in R-$\mathfrak{Mod}$,

$$R = R\epsilon_{11} \overset{\bullet}{\oplus} R\epsilon_{22} \overset{\bullet}{\oplus} \cdots \overset{\bullet}{\oplus} R\epsilon_{nn} \ ;$$

and switching 1 and i columns defines an isomorphism $P = R\epsilon_{11} \cong R\epsilon_{ii}$. So $[R] = n[P]$ in $K_0(R)$. If $n > 1$, then 1 is not a multiple of n in $\mathbb{Z}$.   ∎

6C. Exercises

1. Suppose $\phi : R \to S$ is a ring homomorphism. Use diagram (6.23) and the Snake Lemma to prove:

(i) $K_0(\phi)$ is surjective if and only if $\widetilde{K}_0(\phi)$ is surjective.
(ii) If S has IBN, $K_0(\phi)$ is injective if and only if $\widetilde{K}_0(\phi)$ is injective.
(iii) Every $P \in \mathcal{P}(S)$ is stably free if and only if, for each ring homomorphism $\phi : R \to S$, $K_0(\phi)$ is surjective.
(iv) If R has IBN, every $P \in \mathcal{P}(R)$ is stably free if and only if, for each ring homomorphism $\phi : R \to S$ where S has IBN, $K_0(\phi)$ is injective.

2. Use the fact that K_0 is a functor to prove that, if $\phi : R \to S$ is a ring homomorphism and S has IBN, then R has IBN.

3. Suppose R is the ring $\mathbb{Z} \times \mathbb{Z}$ and $I = 6\mathbb{Z} \times \mathbb{Z}$. If $f : R \to R/I$ is the canonical ring homomorphism, prove $K_0(f)$ is neither injective nor surjective. *Hint:* It factors through $K_0(\mathbb{Z})$.

6D. The Jacobson Radical

Recall from (1.41) that a left or right R-module M is **simple** if $M \neq \{0_M\}$ and M has no submodules aside from M and $\{0_M\}$. An element a of R **annihilates** a left (resp. right) R-module M if $aM = \{0_M\}$ (resp. $Ma = \{0_M\}$). Some elements of a ring are fairly lethal as scalars:

(6.29) Definition. The **Jacobson radical** of a ring R is the set **rad** R of its elements that annihilate every simple left R-module.

It is evident from the definition that rad R is a two-sided ideal of R.

Although rad R is defined in terms of left R-modules, it has left–right symmetry and can be defined without reference to modules:

(6.30) Proposition. *Suppose R is a ring and $x \in R$. The following are equivalent:*

(i) $xM = \{0_M\}$ *for all simple left R-modules M;*
(ii) $Mx = \{0_M\}$ *for all simple right R-modules M;*

(iii) x belongs to every maximal left ideal of R;
(iv) x belongs to every maximal right ideal of R;
(v)   For each $y \in R, 1 + yx$ has a left inverse in R;
(vi)   For each $y \in R, 1 + xy$ has a right inverse in R;
(vii) $1 + RxR \subseteq R^*$.

Proof. Assume (i). If J is a maximal left ideal of R, then R/J is a simple left R-module (as in the proof of (1.42) (iii) $\Rightarrow$ (iv)). So $x + J = x(1 + J) = 0 + J$, and $x \in J$, proving (iii).

Assume (iii). If $y \in R$ and $1 + yx$ does not have a left inverse, then $R(1+yx)$ is a proper left ideal of R. Applying Zorn's Lemma to the proper left ideals containing $R(1 + yx)$, we find that they include a maximal left ideal J of R. Then $x \in J$; so $1 = (1 + yx) - yx \in J$, forcing $J = R$, a contradiction. This proves (v).

We interrupt this proof for a necessary lemma:

(6.31) Lemma. *If R is a ring and $x, y \in R$ with $1 + xy \in R^*$, then $1 + yx \in R^*$ too.*

Proof. If $v = (1 + xy)^{-1}$, then

$$
\begin{aligned}
(1 - yvx)(1 + yx) &= (1 + yx) - yvx(1 + yx) \\
&= 1 + yx - yv(1 + xy)x \\
&= 1 + yx - yx = 1, \text{ and} \\
(1 + yx)(1 - yvx) &= 1 + yx - (1 + yx)yvx \\
&= 1 + yx - y(1 + xy)vx \\
&= 1 + yx - yx = 1. \quad \blacksquare
\end{aligned}
$$

Now assume (v), and suppose $r, s \in R$. There exist $t, u \in R$ with $t(1 + srx) = 1$ and $ut = u(1 - tsrx) = 1$. Then $u = ut(1 + srx) = (1 + srx)$, and hence $1 + srx \in R^*$, with inverse t. By (6.31), $1 + rxs \in R^*$ as well, proving (vii).

If (i) fails, $xM \neq \{0\}$ for some simple R-module M. So there exists $m \in M$ with $xm \neq 0$. Then Rxm is a nonzero submodule of M; so it equals M. In particular, $m \in Rxm$; so $m = yxm$ for some $y \in R$. Then $(1 - yx)m = 0$. Since $m \neq 0$, $1 - yx1 \notin R^*$, and (vii) fails.

Thus (i), (iii), (v), and (vii) are equivalent. Now $1 + RxR \subseteq R^*$ is equivalent to $1 + R^{op}xR^{op} \subseteq (R^{op})^*$, where R^{op} is the opposite ring. And (i), (iii), and (v) for R^{op} are just (ii), (iv), and (vi) for R. $\blacksquare$

A ring element $a \in R$ is **nilpotent** if $a^n = 0$ for some $n \geq 1$.

(6.32) Proposition. *If R is a commutative ring, every nilpotent element of R belongs to rad R.*

Proof. If $a \in R$ but $a \notin \text{rad } R$, then $aM \neq \{0\}$ for some simple R-module M. Since R commutes, aM is a submodule of M; so $aM = M$. Then $a^n M = M \neq \{0\}$ for all $n \geq 1$, and a cannot be nilpotent. ∎

(6.33) Definitions. An ideal or left ideal of R is **radical** if it is contained in rad R. Of course $\{0_R\}$ and rad R are radical ideals of R. If R is commutative, the nilpotent elements of R form a radical ideal of R called the **nilradical** of R, and denoted by $\sqrt{0}$, since it consists of the nth roots of 0_R for all $n \geq 1$. This is the source of the term "radical."

Over any ring R, commutative or not, if M is an R-module and I a left ideal of R, then

$$IM = \left\{ \sum_{i=1}^n x_i y_i : x_i \in I, y_i \in M, n \geq 1 \right\}$$

is also an R-module. If I, J are two left ideals of R, $I(JM) = (IJ)M$. A left ideal I of R is **nilpotent** if $I^n = \{0_R\}$ for some $n \geq 1$.

(6.34) Proposition. *In any ring R, every nilpotent left ideal is a radical ideal. In every left artinian ring R, every radical left ideal is nilpotent.*

Proof. If I is a left ideal of R with $I \nsubseteq \text{rad } R$, there is a simple R-module M with $IM \neq \{0_M\}$. Since IM is a submodule of M, $IM = M$. Then $I^n M = M$ for all $n \geq 1$. So I is not nilpotent.

Suppose R is a left artinian ring and J is a radical left ideal of R. The descending chain $J \supseteq J^2 \supseteq J^3 \supseteq \cdots$ must be eventually constant; so for some positive integer k, $J^k = J^{k+1} = \cdots$. Set B equal to J^k. Assume $B \neq 0$. Consider the set $\mathcal{B}$ of all left ideals L of R with $L \subseteq B$ and $BL \neq 0$. Since $B^2 = B \neq 0$, $B \in \mathcal{B}$; so $\mathcal{B}$ is not empty. Since R is left artinian, $\mathcal{B}$ has a minimal member I. Since $BI \neq 0$, I has an element x with $Bx \neq 0$. Then $Bx \subseteq I$ and $Bx \in \mathcal{B}$; so $Bx = I$. Then $bx = x$ for some $b \in B$. That is, $(1 - b)x = 0$. Since $b \in \text{rad } R$, $1 - b \in R^*$; so $x = 0$, a contradiction. ∎

For a radical ideal I of R, there are surjective ring homomorphisms

$$R \longrightarrow R/I \longrightarrow R/\text{rad } R .$$

As a general principle, there is much to gain and little to lose in passing from R to $R/\text{rad } R$. The gain is a simple description of $R/\text{rad } R$ when it is artinian; we discuss this in Part II, §8C. That little is lost is the theme of the rest of this section.

(6.35) Nakayama's Lemma. *Suppose R is a ring with a left ideal I. The following are equivalent:*

(i) $I \subseteq \text{rad } R$.

(ii) *If M is a f.g. R-module with $IM = M$, then $M = \{0_M\}$.*

(iii) *If a f.g. R-module M has a submodule N with $N + IM = M$, then $N = M$.*

Proof. Assume (i). If M is a f.g. R-module, there is a smallest positive integer t for which M is generated by t elements $m_1, \ldots, m_t$. If $IM = M$, $m_1 = \Sigma \, r_i m_i$ with $r_i \in I$. Since $r_1 \in \text{rad } R$, $1 - r_1 \in R^*$; so if $t > 1$,

$$m_1 \; = \; (1 - r_1)^{-1} \sum_{i=2}^{t} r_i m_i \; ,$$

and M is generated by the $t-1$ elements $m_2, \ldots, m_t$, a contradiction. Therefore $t = 1$ and $m_1 = (1 - r_1)^{-1} 0_M = 0_M$. So $M = Rm_1 = \{0_M\}$, proving (ii).

If N is a submodule of M with $N + IM = M$, then for each $m \in M$ there are $n \in N, x \in IM$ with $m = n + x$. So $m + N = x + n + N = x + N \in I(M/N)$. If M is finitely generated, the cosets of its generators generate M/N. Assuming (ii), $M/N = \{0\}$; so $M = N$, proving (iii).

If $I \not\subseteq \text{rad } R$, there is a simple R-module $M(\neq \{0\})$, finitely generated (by any one nonzero element), with $\{0\} + IM = M$. So (iii) implies (i). ∎

Suppose I is any ideal of a ring R. Via the formula $r \cdot m = (r + I) \cdot m$, an additive abelian group M is an R/I-module if and only if it is an R-module with $IM = \{0\}$. And an additive homomorphism $M \to N$ between R/I-modules is R/I-linear if and only if it is R-linear.

In particular, for each R-module M, the quotient M/IM is an R/I-module. Each R-linear map $f : M \to N$ carries IM into IN; so it induces an R/I-linear map

$$\overline{f} : M/IM \; \to \; N/IN, \quad m + IM \; \mapsto \; f(m) + IN \; .$$

In this way we obtain an additive functor

$$F_I : R\text{-}\mathcal{Mod} \; \to \; R/I\text{-}\mathcal{Mod}$$
$$M \; \mapsto \; M/IM$$
$$f \; \mapsto \; \overline{f} \; .$$

Since F_I is additive, it preserves split short exact sequences and direct sums:

$$\frac{M \oplus N}{I(M \oplus N)} \; \cong \; \frac{M}{IM} \; \oplus \; \frac{N}{IN} \; .$$

By (6.3), since $F_I(R) = R/I$ belongs to $\mathcal{P}(R/I)$, F_I restricts to an exact functor: $\mathcal{P}(R) \to \mathcal{P}(R/I)$.

So F induces a group homomorphism $K_0(R) \to K_0(R/I)$, taking $[P]$ to $[P/IP]$ for each f.g. projective R-module P. This is no different from the usual homomorphism, taking $[P]$ to $[R/I \otimes_R P]$, since, by (6.15), there is an R-linear isomorphism

$$R/I \otimes_R M \;\cong\; M/IM \;,$$
$$\overline{r} \otimes m \;\mapsto\; \overline{rm}$$

for each R-module M. In fact, this isomorphism is a natural transformation from $R/I \otimes (-)$ to F_I, in the sense of (0.13): For if $f : M \to N$ is an R-linear map, the square

$$
\begin{array}{ccc}
R/I \otimes_R M & \longrightarrow & R/I \otimes_R N \\
\downarrow & & \downarrow \\
M/IM & \longrightarrow & N/IN
\end{array}
$$

commutes (because $\overline{r f(m)} = \overline{f(rm)}$ for each $r \in R, m \in M$).

(6.36) Theorem. *If I is a radical ideal of a ring R and P, Q are f.g. projective R-modules with $P/IP \cong Q/IQ$ as R-modules (or equivalently, as R/I-modules), then $P \cong Q$. So the map*

$$\mathfrak{I}(\mathcal{P}(R)) \;\to\; \mathfrak{I}(\mathcal{P}(R/I)) \;,$$

induced on isomorphism classes by F_I, is injective.

Proof. Suppose $P, Q \in \mathcal{P}(R)$ and $\phi : P/IP \to Q/IQ$ is an R-linear isomorphism. Since P is projective, there is an R-linear map $f : P \to Q$ making the diagram (with exact row)

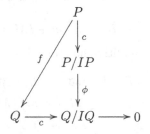

commute, where c denotes canonical maps.

If $q \in Q$, $q + IQ = \phi(p + IP) = f(p) + IQ$ for some $p \in P$. So $Q = f(P) + IQ$. Since Q is finitely generated, Nakayama's Lemma says $Q = f(P)$; so f is surjective. If $K = \ker f$, we have a short exact sequence

$$0 \longrightarrow K \stackrel{i}{\longrightarrow} P \stackrel{f}{\longrightarrow} Q \longrightarrow 0 \;,$$

which splits because Q is projective. Since F_I is additive,

$$0 \longrightarrow \frac{K}{IK} \xrightarrow{\bar{i}} \frac{P}{IP} \xrightarrow{\phi} \frac{Q}{IQ} \longrightarrow 0$$

is exact. Since ϕ is injective, $IK = K$. Because the first sequence splits, K is a homomorphic image of P; so it is finitely generated. By Nakayama's Lemma, $K = 0$ and f is injective. ∎

(6.37) Corollary. *If I is a radical ideal of R, and $c : R \to R/I$ is the canonical map, then $K_0(c) : K_0(R) \to K_0(R/I)$ is injective.*

Proof. If $[P] - [Q]$ is sent to $[P/IP] - [Q/IQ] = 0$, then P/IP and Q/IQ are stably isomorphic over R/I. So there are R/I-linear (hence R-linear) isomorphisms

$$\frac{P \oplus R^n}{I(P \oplus R^n)} \cong \frac{P}{IP} \oplus \left(\frac{R}{I}\right)^n \cong \frac{Q}{IQ} \oplus \left(\frac{R}{I}\right)^n \cong \frac{Q \oplus R^n}{I(Q \oplus R^n)} .$$

By the theorem, $P \oplus R^n \cong Q \oplus R^n$ in $\mathcal{P}(R)$. So $[P] - [Q] = 0$. ∎

(6.38) Corollary. *Suppose I is a radical ideal of a ring R.*

   (i) *If all f.g. projective R/I-modules are stably free, then all f.g. projective R-modules are stably free.*

   (ii) *If all f.g. projective R/I-modules are free, then all f.g. projective R-modules are free.*

Proof. The functor K_0 takes the commutative diagram

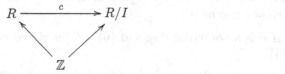

in $\mathfrak{Ring}$ to a commutative diagram

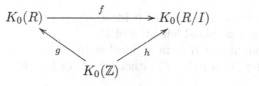

in $\mathcal{Ab}$. Since f is injective, h surjective implies g is surjective. Now apply (4.5).

For (ii), if $P \in \mathcal{P}(R)$ and there are R/I-linear (hence R-linear) isomorphisms

$$P/IP \cong (R/I)^n \cong R^n/I \cdot R^n,$$

then by the preceding theorem, $P \cong R^n$. ∎

Now f.g. projective D-modules are free whenever D is a division ring (see (1.42)). A ring R is **local** if $R/\mathrm{rad}\ R$ is a division ring. Thus:

(6.39) Corollary. *Over a local ring, every f.g. projective module is free.* ∎

6D. Exercises

1. For what rings R can you prove $\mathrm{rad}\ R = \{0\}$? Determine $\mathrm{rad}\ (\mathbb{Z}/n\mathbb{Z})$ when n is a positive integer.

2. Prove every surjective ring homomorphism $f : R \to S$ carries $\mathrm{rad}\ R$ into $\mathrm{rad}\ S$.

3. If I is a radical ideal of a ring R, prove

$$(\mathrm{rad}\ R)/I \;=\; \mathrm{rad}\ (R/I)\,.$$

In particular, $\mathrm{rad}\ (R/\mathrm{rad}\ R) = 0$.

4. If I is a radical ideal of a ring R, prove $r + I \in (R/I)^*$ if and only if $r \in R^*$.

5. If I is a nilpotent ideal of a ring R, prove $r+I$ is idempotent in R/I if and only if $r + I = e + I$ for some idempotent e in R. *Hint:* If $r + I$ is idempotent, $r - r^2 = r(1 - r) \in I$, and so is nilpotent. Say its nth power is 0. Use the binomial theorem on

$$(r + (1 - r))^{2n-1}$$

to write $1 = e + f$, where $ef = fe = 0$, so $e^2 = e$, and where $r^{2n-1} \equiv e \bmod I$. Then show $r \equiv e \bmod I$.

6. If R is a nontrivial ring and $\{0\} \cup R^*$ is closed under addition, show $\mathrm{rad}\ R = \{0\}$.

7. For a ring R, show each of the following conditions is equivalent to R being a local ring:

    (i) R has only one maximal left ideal.
    (ii) $\mathrm{rad}\ R$ is a maximal left ideal of R.
    (iii) The nonunits of R form an ideal of R.
    (iv) If $a + b = 1$ for $a, b \in R$, either $a \in R^*$ or $b \in R^*$.

8. For each of the following rings R, prove every f.g. projective R-module is free.

    (i) $R = \mathbb{Z}[x]/x^n\mathbb{Z}[x]$ for n a positive integer.
    (ii) $R = F[x,y]/y^nF[x,y]$ for n a positive integer and F a field.
    (iii) $R = F[x]/(x^p - 1)F[x]$, where F is a field of prime characteristic p.

9. (From Bass [68, p. 90]) If I is a nilpotent ideal of a ring R, show $M_n(I)$ is a nilpotent ideal of $M_n(R)$ for each $n \geq 1$. Use this to prove that the canonical map $R \to R/I$ induces

$$\mathcal{I}(\mathcal{P}(R)) \cong \mathcal{I}(\mathcal{P}(R/I)) \quad \text{and} \quad K_0(R) \cong K_0(R/I) \,.$$

Hint: As shown in (2.21), $\mathcal{P}(R)$ is represented by the row-spaces $R^n e$ of idempotent matrices $e \in M_n(R)$ for $n \geq 1$. Now use Exercise 5.

10. If R is a commutative ring with exactly $n (< \infty)$ maximal ideals, prove $K_0(R) \cong \mathbb{Z}^m$ for some $m \leq n$. *Hint:* Use (6.37) and the Chinese Remainder Theorem.

11. If R is a commutative ring with exactly $n (< \infty)$ maximal ideals, and if rad R is nilpotent, prove R is isomorphic to a cartesian product of n local rings, so that $K_0(R) \cong \mathbb{Z}^n$. *Hint:* If $(\text{rad } R)^m = 0$, apply the Chinese Remainder Theorem to $\overline{R} = R/(\text{rad } R)^m$. *Note:* The hypotheses of this problem hold for each finite commutative ring.

12. (Dennis and Geller [76]) Suppose A and B are rings and M is an A, B-bimodule. The matrices

$$\begin{bmatrix} a & m \\ 0 & b \end{bmatrix} \quad , \text{ with } \quad a \in A \,, \quad b \in B \,, \quad m \in M \,,$$

form a ring T under the standard matrix operations. Prove $K_0(T)$ is isomorphic to $K_0(A) \oplus K_0(B)$. *Hint:* Sending such a matrix to (a, b) defines a ring homomorphism $T \to A \times B$ with nilpotent kernel. Use Exercise 9. *Note:* An induction argument now proves $K_0(T_n) \cong K_0(R)^n$, where T_n is the ring of $n \times n$ upper triangular matrices over a ring R.

6E. Localization

Linear algebra over a local ring is like linear algebra over a field — by (6.39), every f.g. projective module is free. A commutative ring R is local if and only if R has only one maximal ideal. Any ideal containing a unit must be the whole ring; so if a commutative ring is not local, one might try to enlarge it to a local ring by adjoining inverses of some of its elements, thereby reducing the number of maximal ideals.

For instance, $\mathbb{Z}$ has a maximal ideal $p\mathbb{Z}$ for each prime p. Adjoining the inverses of odd integers, we obtain the local ring

$$R = \left\{ \frac{a}{b} \in \mathbb{Q} : a, b \in \mathbb{Z}, \ b \text{ odd} \right\}$$

with maximal ideal $2R$. In this example, the adjoined inverses $1/b$ come from a preexisting ring $\mathbb{Q}$ containing $\mathbb{Z}$. But $\mathbb{Q}$ itself is a local ring built from $\mathbb{Z}$ by the adjunction of inverses.

The **localization** of a ring A at a subset S, defined in (6.40) below, is the simplest construction of a ring B from A in which the elements of S become units. Even if A is commutative, the ring B may or may not be local; but it is closer to being local than A (see (6.48) below). If A is a noncommutative ring, a localization of A is easiest to construct when the elements to be inverted belong to the **center** of A, which is the subring

$$Z(A) \;=\; \{a \in A : \forall\, b \in A, \;\; ab = ba\}\,.$$

Note that $Z(A)$ is a commutative subring of A, but it need not be the largest commutative subring; for instance, if $\mathbb{H}$ is Hamilton's ring of quaternions, $Z(\mathbb{H}) = \mathbb{R} \subsetneqq \mathbb{C} \subsetneqq \mathbb{H}$.

To motivate the details of the localization of A at S, we recall from §3B the adjunction of inverses to an abelian monoid $(S, +)$ to form its group completion $\overline{\overline{S}}$. The abelian group $(\overline{\overline{S}}, +)$ is the $\mathbb{Z}$-module quotient $F_{\mathbb{Z}}(S)/\langle D \rangle$, where D is the set of all $(x * y) - x - y$ with $x, y \in S$. For each $x \in S$, the coset $x + \langle D \rangle$ is denoted by $\overline{x}$. Every element of $\overline{\overline{S}}$ has the form $\overline{x} - \overline{y}$ for some $x, y \in S$, and $\overline{a} - \overline{b} = \overline{c} - \overline{d}$ if and only if $a * d * e = b * c * e$ for some $e \in S$.

This suggests an alternate construction of the group completion: On $S \times S$ define a relation $\sim$ by

$$(a, b) \sim (c, d) \quad \Leftrightarrow \quad \exists\, e \in S \;\; \text{with} \;\; a * d * e = b * c * e \,.$$

This $\sim$ is an equivalence relation. Let $a - b$ denote the equivalence class of a pair (a, b). Define addition in the set $\mathcal{S}$ of equivalence classes by

$$(a - b) + (c - d) \;=\; (a * c) - (b * d) \,.$$

This $+$ is well-defined, making $\mathcal{S}$ into an additive abelian group. The map $f : \mathcal{S} \to \overline{\overline{S}}$ with $f(a - b) = \overline{a} - \overline{b}$ is an isomorphism of groups. If $(S, *)$ happens to be cancellative, the map $S \to \mathcal{S}$, $x \mapsto x - 0$, is injective, and S may be regarded as a submonoid of $\mathcal{S}$. In that case, $x - y$ takes on the usual meaning of subtraction: $x + (-y)$.

This construction of $\mathcal{S}$ is used to build $\mathbb{Z}$ from the additive monoid $\mathbb{N}$ of natural numbers. Putting it in multiplicative notation (a/b, $\cdot$ in place of $a - b$, $+$), a similar construction yields the field of fractions F of an integral domain R. In this case the multiplicative monoid $R - \{0\}$ is completed to the multiplicative group $F - \{0\}$; but $\sim$ is defined on $R \times (R - \{0\})$, and both $+$ and $\cdot$ are extended from R to F. Here is a general version of **localization**:

(6.40) Definition of $S^{-1}M$, $S^{-1}A$. *For the rest of this section, A is a ring, R is a subring of $Z(A)$, S is a submonoid of $(R, \cdot)$, and M is an A-module.* On $M \times S$ define a relation $\sim$ by:

$$(m_1, s_1) \sim (m_2, s_2) \quad \Leftrightarrow \quad \exists\, s_3 \in S \;\; \text{with} \;\; s_3(s_2 m_1) = s_3(s_1 m_2) \,.$$

In M, the last equation says $s_3(s_2m_1 - s_1m_2) = 0$. The relation $\sim$ is reflexive (since $s_30 = 0$) and symmetric (since $s_3(-m) = -(s_3m)$). Transitivity is slightly tricky — here we use commutativity of S: If

$$s(s_2m_1 - s_1m_2) = s'(s_3m_2 - s_2m_3) = 0,$$

then $ss's_2(s_3m_1 - s_1m_3) = 0$. Denote the equivalence class of a pair (m, s) by m/s and the set of equivalence classes by $S^{-1}M$. The reader is invited to verify that the usual addition of fractions

$$\frac{m_1}{s_1} + \frac{m_2}{s_2} = \frac{s_2m_1 + s_1m_2}{s_1s_2}$$

is a well-defined operation making $S^{-1}M$ an additive abelian group with zero $0/1$. Since S lies in the center of A, the usual multiplication of fractions

$$\frac{a}{s_1} \frac{m}{s_2} = \frac{am}{s_1s_2}$$

makes $S^{-1}A$ into a ring with unit $1/1$ and $S^{-1}M$ into an $S^{-1}A$-module.

Note that $s/s = 1/1$ for all $s \in S$; so cancellation

$$\frac{s_1m}{s_1s} = \frac{m}{s}$$

works as one would expect for fractions. Therefore each finite list of elements in $S^{-1}M$ can be expressed with a common denominator: $m_1/s, m_2/s, \ldots, m_n/s$. In this form, addition is simplified:

$$\frac{m_1}{s} + \frac{m_2}{s} = \frac{m_1 + m_2}{s}.$$

The ring $S^{-1}A$ is the simplest modification of A in which the elements of S become units:

(6.41) Lemma. *The map $\phi : A \to S^{-1}A$, $a \mapsto a/1$, is a ring homomorphism with $\phi(S) \subseteq (S^{-1}A)^*$. For each ring homomorphism $\psi : A \to B$ with $\psi(S) \subseteq B^*$, there is one and only one ring homomorphism $\widehat{\psi} : S^{-1}A \to B$ for which $\widehat{\psi} \circ \phi = \psi$.*

Proof. That ϕ is a ring homomorphism is routine. If $s \in S$, $\phi(s)$ has inverse $1/s$ in $S^{-1}A$. If $\widehat{\psi}$ exists, $\widehat{\psi}(a/s) = \psi(a)\psi(s)^{-1}$. The latter equation defines a function $\widehat{\psi} : S^{-1}A \to B$, because $s_3(s_2a_1 - s_1a_2) = 0$ in A implies $\psi(s_2)\psi(a_1) = \psi(s_1)\psi(a_2)$ in B, so that $\psi(a_1)\psi(s_1)^{-1} = \psi(a_2)\psi(s_2)^{-1}$. Evidently $\widehat{\psi}$ is a ring homomorphism with $\widehat{\psi} \circ \phi = \psi$. ∎

(6.42) Examples.

(i) An ideal $\mathfrak{p}$ of R is prime if and only if its complement $R - \mathfrak{p}$ is a submonoid of $(R, \cdot)$. If $S = R - \mathfrak{p}$, the ring $S^{-1}A$ is denoted by $A_\mathfrak{p}$ and the module $S^{-1}M$ is denoted by $M_\mathfrak{p}$.

(ii) If R is an integral domain, $\{0\}$ is a prime ideal of R, and $R_{\{0\}} = (R - \{0\})^{-1}R$ is the field of fractions of R.

(iii) If $s \in R$, then $S = \{s^n : n \geq 0\}$ is a submonoid of $(R, \cdot)$. The ring $S^{-1}A$ is denoted by $A[1/s]$, since its elements can be expressed as A-linear combinations of $1, 1/s, 1/s^2, \ldots$. But every element actually has the form a/s^n with $a \in A$, $n \geq 0$.

Restricting scalars to the fractions $a/1$, $S^{-1}M$ is an A-module with $a(s/m) = (as)/m$, and the **localization map**

$$\phi_M : M \;\to\; S^{-1}M, \quad m \mapsto m/1 \,,$$

is A-linear. If $T \subseteq R$, we say T **acts through injections** (resp. **bijections**) if, for each $t \in T$, the A-linear map $t \cdot (-) : M \to M$, $m \mapsto tm$, is injective (resp. bijective).

(6.43) Lemma. *The localization map $\phi_M : M \to S^{-1}M$ is injective if and only if S acts through injections on M, and bijective if and only if S acts through bijections on M. The latter is true if $S \subseteq A^*$.*

Proof. When $m, m' \in M$, $m/1 = m'/1$ if and only if $sm = sm'$ for some $s \in S$, proving the claim about injectivity. Assuming ϕ_M and the $s \cdot (-)$ are injective, $m/s = m'/1$ if and only if $sm' = m$, proving the claim about surjectivity. ∎

The condition that S acts through injections is analogous to the condition that the monoid S is cancellative. For $M = A$, S acts through injections if and only if S includes neither zero nor any zero-divisor of A^{op}. When this is so, we can regard A as a subring of $S^{-1}A$ (just as $\mathbb{Z}$ is a subring of $\mathbb{Q}$), and each fraction a/s takes on the usual meaning of a quotient: as^{-1}.

If $f : M \to N$ is an A-linear map, there is a function

$$S^{-1}f : S^{-1}M \;\to\; S^{-1}M, \quad \frac{m}{s} \mapsto \frac{f(m)}{s} \,.$$

It is well-defined because if $m_1/s_1 = m_2/s_2$, there exists $s_3 \in S$ with

$$s_3(s_2 f(m_1) - s_1 f(m_2)) \;=\; f(s_3(s_2 m_1 - s_1 m_2)) \;=\; f(0) \;=\; 0 \,;$$

so $f(m_1)/s_1 = f(m_2)/s_2$. From the A-linearity of f, the $S^{-1}A$-linearity of $S^{-1}f$ follows.

(6.44) Proposition. *With $S^{-1}M$ and $S^{-1}f$ defined as above, $S^{-1}(-)$ is an exact (additive) functor from A-$\mathfrak{Mod}$ to $S^{-1}A$-$\mathfrak{Mod}$.*

Proof. From $S^{-1}f(m/s) = f(m)/s$, it is evident that

$$S^{-1}i_M = i_{S^{-1}M} \quad \text{and} \quad S^{-1}(g \circ f) = S^{-1}g \circ S^{-1}f \,.$$

If $f, g \in \operatorname{Hom}_R(M, N)$, then

$$
\begin{aligned}
S^{-1}(f + g)(\frac{m}{s}) &= \frac{f(m) + g(m)}{s} \\
&= \frac{f(m)}{s} + \frac{g(m)}{s} = (S^{-1}f + S^{-1}g)(\frac{m}{s}) \,.
\end{aligned}
$$

So $S^{-1}(-)$ is an additive functor, and it takes zero maps to zero maps and zero modules to zero modules. Suppose

$$L \xrightarrow{\ f\ } M \xrightarrow{\ g\ } N$$

is exact in A-$\mathfrak{Mod}$. Then $S^{-1}g \circ S^{-1}f = S^{-1}(0\text{-map}) = 0\text{-map}$. If $S^{-1}g(m/s) = 0/1$, then for some $s' \in S$, $g(s'm) = s'g(m) = 0$. So $s'm = f(x)$ for some $x \in L$. Then $m/s = f(x)/s's = S^{-1}f(x/s's)$. $\blacksquare$

If N is an A-submodule of M, inclusion $i : N \to M$ is injective. Since $S^{-1}(-)$ is exact, the $S^{-1}A$-linear map

$$S^{-1}i : S^{-1}N \to S^{-1}M, \quad \frac{n}{s} \mapsto \frac{n}{s} \,,$$

is also injective. So we can identify $S^{-1}N$ with its image, the fractions in $S^{-1}M$ that can be written with numerator in N.

(6.45) Examples.

(i) If $\mathfrak{p}$ is a prime ideal of R, then $R_\mathfrak{p}$ is a local ring, with unique maximal ideal

$$
\begin{aligned}
\mathfrak{p}_\mathfrak{p} &= (R - \mathfrak{p})^{-1}\mathfrak{p} \\
&= \{\frac{r}{s} \ : \ r \in \mathfrak{p}, \ s \in R - \mathfrak{p}\} \,,
\end{aligned}
$$

since $R_\mathfrak{p} - \mathfrak{p}_\mathfrak{p}$ consists of units. If M is an A-module with a nonzero element m, the annihilator of m in R,

$$\operatorname{ann}_R(m) = \{r \in R : rm = 0\} \,,$$

is a proper ideal of R, so it lies in some maximal ideal $\mathfrak{p}$ of R. If $S = R - \mathfrak{p}$, there is no $s \in S$ with $sm = 0$; so $m/1 \neq 0$ in $M_\mathfrak{p}$. This proves that $M = 0$ if *and only if $M_{\mathfrak{p}} = 0$ for all maximal ideals $\mathfrak{p}$ of R. Using exactness of $S^{-1}(-)$, it follows that an A-linear map $f : M \to N$ is injective (resp. surjective) if and only if*

$$(R - \mathfrak{p})^{-1}f = f_{\mathfrak{p}} : M_{\mathfrak{p}} \to N_{\mathfrak{p}}$$

is injective (resp. surjective) for all maximal ideals $\mathfrak{p}$ of R.

(ii) *If $\mathfrak{p}$ is a nonzero prime ideal of an integral domain R with field of fractions F, then $R_{\mathfrak{p}}$ is the subring of F:*

$$R_{\mathfrak{p}} = \{\frac{r}{s} \in F \ : \ r \in R, \ s \in R - \mathfrak{p}\} .$$

More generally, for prime ideals $\mathfrak{p}_1 \subsetneq \mathfrak{p}_2$ of R, $R_{\mathfrak{p}_2}$ is a subring of $R_{\mathfrak{p}_1}$.

(iii) *If R is an integral domain with field of fractions F, then $(R - \{0\})^{-1}A$ is an F-algebra, with F in its center.*

For each A-linear map $f : M \to N$ with kernel K and image I, there is a correspondence (see (B.7) in the Appendix) matching the poset of submodules of M that contain K with the poset of submodules of I. For the A-linear localization map $\phi_M : M \to S^{-1}M$, there is yet another correspondence:

Denote by sub $_AM$ the poset of A-submodules of M, and by sub $_{S^{-1}A}S^{-1}M$ the poset of $S^{-1}A$-submodules of $S^{-1}M$. Within sub $_AM$, let $\mathcal{C}$ denote the subposet of those N with $S(M - N) \subseteq M - N$. If P is an $S^{-1}A$-submodule of $S^{-1}M$, then

$$\mathrm{num}(P) = \text{all numerators of fractions in } P$$
$$= \{m \in M \ : \ \frac{m}{1} \in P\}$$

is an A-submodule of M.

(6.46) Theorem. *The localization functor $S^{-1}(-)$, from the poset sub $_AM$ to the poset sub $_{S^{-1}A}S^{-1}M$, restricts to a poset isomorphism from $\mathcal{C}$ to the poset sub $_{S^{-1}A}S^{-1}M$, with inverse num$(-)$. If $N' \subseteq N$ are in sub $_AM$, the localization map $\phi_N : N \to S^{-1}N$, $n \mapsto n/1$, induces an A-linear map*

$$\overline{\phi}_N : \frac{N}{N'} \to \frac{S^{-1}N}{S^{-1}N'} ,$$

which is injective (resp. bijective) if and only if S acts through injections (resp. bijections) on N/N'. In particular, $\overline{\phi}_N$ is injective if $N' \in \mathcal{C}$.

Proof. To say an A-submodule N of M belongs to $\mathcal{C}$ is to say that, if $s \in S$, $m \in M$ and $sm \in N$, then $m \in N$. Suppose P is a member of sub $_{S^{-1}A}S^{-1}M$. If

$sm \in \text{num}(P)$, then $sm/1 \in P$; so $m/1 \in P$ and $m \in \text{num}(P)$. This shows $\text{num}(P) \in \mathcal{C}$.

Also, $S^{-1}(\text{num}(P)) = \{m/s : m/1 \in P\} = P$. For $N \in \mathcal{C}$, we now show $\text{num}(S^{-1}N) = N$. It is immediate that $\text{num}(S^{-1}N) \supseteq N$. If $m \in \text{num}(S^{-1}N)$, then $m/1 \in S^{-1}N$, and $m/1 = n/s$ for some $n \in N$ and $s \in S$. For some $t \in S$, $tsm = tn \in N$. Since $N \in \mathcal{C}$, $m \in N$.

To prove the second assertion, apply $S^{-1}(-)$ to the exact sequence

$$0 \longrightarrow N' \overset{\subseteq}{\longrightarrow} N \longrightarrow N/N' \longrightarrow 0$$

to obtain an exact sequence

$$0 \longrightarrow S^{-1}N' \overset{\subseteq}{\longrightarrow} S^{-1}N \overset{c}{\longrightarrow} S^{-1}(N/N') \longrightarrow 0 \,,$$

where $c(n/s) = \overline{n}/s$. So c induces an $S^{-1}A$-linear isomorphism

$$\theta : \frac{S^{-1}N}{S^{-1}N'} \;\cong\; S^{-1}\left(\frac{N}{N'}\right)$$

with $\theta(\overline{n/s}) = \overline{n}/s$. Following the localization map $\phi_{N/N'}$, from N/N' to $S^{-1}(N/N')$, by θ^{-1}, defines an A-linear map

$$\overline{\phi}_N : \frac{N}{N'} \to \frac{S^{-1}N}{S^{-1}N'}$$

with $\overline{\phi}_N(\overline{n}) = \overline{n/1}$. Since θ is an isomorphism, $\overline{\phi}_N$ is injective (resp. bijective) if and only if $\phi_{N/N'}$ is. Now apply (6.43). For the last assertion, S acts through injections on N/N' if and only if $S(N - N') \subseteq (N - N')$. ∎

(6.47) Corollary. *If M is a noetherian (resp. artinian) A-module, then $S^{-1}M$ is a noetherian (resp. artinian) $S^{-1}A$-module.* ∎

The **prime spectrum** of a commutative ring R is the topological space **spec**(R) whose points are the prime ideals of R, and whose closed sets are the subsets having the form

$$V(X) = \{Q \in \text{spec}(R) : Q \supseteq X\}$$

for some subset X of R. If $\langle X \rangle$ is the ideal of R generated by X, then $V(X) = V(\langle X \rangle)$.

(6.48) Corollary. *For a commutative ring R, the map* num$(-)$, *from the poset sub $_{S^{-1}R}S^{-1}R$ to the poset sub $_RR$, restricts to a homeomorphism from* spec$(S^{-1}R)$ *onto the subspace Y of* spec(R) *consisting of those prime ideals that do not meet S. The inverse homeomorphism takes each $p \in Y$ to $S^{-1}p$.*

If p is a maximal ideal of R that does not meet S, then for each positive integer n the localization map $\phi : R \to S^{-1}R$, $r \mapsto r/1$, induces an isomorphism of residue rings

$$\overline{\phi} : \frac{R}{p^n} \cong \frac{S^{-1}R}{(S^{-1}p)^n} .$$

Proof. If Q is a prime ideal of R, then $S(R - Q) \subseteq R - Q$ if and only if Q does not meet S. So Y is the set of prime ideals in $\mathcal{C}$.

The correspondence in (6.46) matches prime ideals in $S^{-1}R$ with prime ideals in R that do not meet S: If $Q \in Y$ and $(a/s)(b/t) \in S^{-1}Q$, then $ab \in$ num$(S^{-1}Q) = Q$; so $a \in Q$ or $b \in Q$; then $a/s \in S^{-1}Q$ or $b/t \in S^{-1}Q$. Therefore $S^{-1}Q \in$ spec$(S^{-1}R)$.

In the other direction, if $P \in$ spec$(S^{-1}R)$ and $ab \in$ num(P), then $ab/1 \in P$; so $a/1 \in P$ or $b/1 \in P$; then $a \in$ num(P) or $b \in$ num(P). Thus num$(P) \in Y$.

Next we show that the bijections spec$(S^{-1}R) \rightleftarrows Y$, defined by num$(-)$ and $S^{-1}(-)$, take closed sets to closed sets. Suppose I is an ideal of $S^{-1}R$. Since num$(-)$ preserves containments, num$(V(I))$ is part of $Y \cap V($num$(I))$. If Q belongs to the latter intersection, $Q =$ num$(S^{-1}Q)$; since $S^{-1}(-)$ preserves containments, $S^{-1}Q \supseteq S^{-1}(num(I)) \supseteq I$. So $Q \in$ num$(V(I))$, proving

$$\text{num}(V(I)) = Y \cap V(\text{num}(I)) ,$$

a closed set in Y.

On the other hand, suppose J is an ideal of R. Since $S^{-1}(-)$ preserves containments, $S^{-1}(Y \cap V(J))$ is part of $V(S^{-1}J)$. If $P \in V(S^{-1}J)$, then $P = S^{-1}($num$(P))$. Now num$(P) \in Y$; and since num$(-)$ preserves containments, num$(P) \supseteq$ num$(S^{-1}J) \supseteq J$. So $P \in S^{-1}(Y \cap V(J))$, proving

$$S^{-1}(Y \cap V(J)) = V(S^{-1}J) ,$$

a closed set in spec$(S^{-1}R)$. Finally, if $p \in Y$ is maximal and $s \in S$, then $\overline{s} = s + p^n$ is not in the maximal ideal p/p^n of the local ring R/p^n; so $\overline{s} \in (R/p^n)^*$. The action of R on R/p^n induces an action by R/p^n, with $\overline{r} \cdot x = r \cdot x$. So S acts through bijections on R/p^n. By (6.46), localization induces an R-linear isomorphism $\overline{\phi}$ from R/p^n to $S^{-1}R/S^{-1}(p^n)$, with $\overline{\phi}(\overline{r}) = \overline{r/1}$. Note that $S^{-1}(p^n) = (S^{-1}p)^n$ in $S^{-1}R$, and $\overline{\phi}$ is a ring homomorphism. ∎

(6.49) Definition. An A-module M is S-**torsion** if $S^{-1}M = 0$. Of course, $m/t = 0/1$ if and only if $sm = 0$ in M for some $s \in S$. So a f.g. A-module M is S-torsion if and only if $sM = 0$ for some $s \in S$.

If M is the A-linear span of $m_1, \ldots, m_r$, then $S^{-1}M$ is the $S^{-1}A$-linear span of $m_1/1, \ldots, m_r/1$; so $S^{-1}(-)$ restricts to an exact functor $\mathcal{M}(A) \to \mathcal{M}(S^{-1}A)$. Modulo S-torsion A-modules, $S^{-1}(-)$ is essentially bijective on isomorphism classes:

(6.50) Lemma.

  (i) *For each f.g. $S^{-1}A$-module P there is a f.g. A-module M with $S^{-1}M$ isomorphic to P.*

 (ii) *Suppose M is a f.g. A-module and K is the kernel of $\phi_M : M \to S^{-1}M$. Then $S^{-1}K = 0$, $S^{-1}M \cong S^{-1}(M/K)$, and S acts through injections on M/K.*

(iii) *If S acts through injections on f.g. A-modules M and N, then $S^{-1}M \cong S^{-1}N$ as $S^{-1}A$-modules if and only if there is an A-linear isomorphism $M \cong M'$ where M' is a submodule of N and $S^{-1}(N/M') = 0$.*

Proof. Say P is the $S^{-1}A$-linear span of $m_1, \ldots, m_r$, and M is the A-submodule spanned by $m_1, \ldots, m_r$. Regarding P as an A-module, for each $a \in A$, $s \in S$ and $p \in P$,

$$s \cdot \left(\frac{a}{s} \cdot p\right) = \frac{s}{1} \cdot \frac{a}{s} \cdot p = \frac{a}{1} \cdot p .$$

So in $S^{-1}P$,

$$\frac{\frac{a}{s} \cdot p}{1} = \frac{\frac{a}{1} \cdot p}{s} = \frac{a}{s} \cdot \frac{p}{1} .$$

From this it follows that the A-linear isomorphism $\phi_P : P \to S^{-1}P$ is $S^{-1}A$-linear, and that

$$S^{-1}M = S^{-1}A\frac{m_1}{1} + \cdots + S^{-1}A\frac{m_r}{1} = \phi_P(P) = S^{-1}P ,$$

proving $S^{-1}M \cong P$.

For (ii), if $k \in K$ and $s \in S$, $k/s = (1/s)(k/1) = 0$ in $S^{-1}K$; so $S^{-1}K = 0$. By exactness of $S^{-1}(-)$, $S^{-1}M \cong S^{-1}(M/K)$. If $s \in S$ and $m \in M$, $sm \in K$ implies $m \in K$. So $s \cdot (-)$ is injective on M/K.

For (iii), suppose $\alpha : S^{-1}M \to S^{-1}N$ is an $S^{-1}A$-linear isomorphism and M is the A-linear span of $m_1, \ldots, m_r$. Then $\alpha\phi_M(M)$ is the A-linear span of

$$\alpha\{\frac{m_1}{1}, \ldots, \frac{m_r}{1}\} = \{\frac{n_1}{s}, \ldots, \frac{n_r}{s}\}$$

for some $n_i \in N$, $s \in S$. Since s is a unit in $S^{-1}A$, $s \cdot (-)$ is an A-linear isomorphism from $\alpha\phi_M(M)$ onto the A-linear span of $n_1/1, \ldots, n_r/1$ in $\phi_N(N)$.

Taking M' to be the A-linear span of $n_1, \ldots, n_r$ in N, $M \cong M'$ and $S^{-1}M' = \alpha(S^{-1}M) = S^{-1}N$; so $S^{-1}(N/M') = 0$.

For the converse, suppose β is the A-linear composite $M \cong M' \subseteq N$. By exactness of $S^{-1}(-)$, since $S^{-1}(N/M') = 0$, the map $S^{-1}\beta$ is an $S^{-1}A$-linear isomorphism. $\blacksquare$

If the ring A is left noetherian, the functor $S^{-1}(-)$ from $\mathcal{M}(A)$ to $\mathcal{M}(S^{-1}A)$ is nearly surjective on arrows:

(6.51) Lemma. *Suppose A is a left noetherian ring and M, N are f.g. A-modules. For each $S^{-1}A$-linear map $f : S^{-1}M \to S^{-1}N$, there is an A-linear map $g : M \to N$ and an element $s_0 \in S$ for which the diagram*

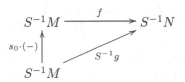

commutes.

Proof. Suppose M is the A-linear span of $m_1, \ldots, m_r$ and

$$f\left\{\frac{m_1}{1}, \ldots, \frac{m_r}{1}\right\} = \left\{\frac{n_1}{s}, \ldots, \frac{n_r}{s}\right\}$$

for $n_i \in N$ and $s \in S$. The A-linear map $v : A^r \to M$, taking each e_i to m_i, is surjective. Since A is left noetherian, the kernel of v is generated by m elements, where $m < \infty$. So there is an exact sequence of A-linear maps

$$A^m \xrightarrow{u} A^r \xrightarrow{v} M \longrightarrow 0.$$

Denote by $\gamma : A^r \to N$ the A-linear map taking each e_i to n_i. In the set of maps $\mathrm{Hom}_{S^{-1}A}(S^{-1}A^r, S^{-1}M)$,

$$\frac{1}{s} \cdot S^{-1}\gamma = f \circ S^{-1}v,$$

since both maps take each $e_i/1$ to n_i/s. So

$$\frac{1}{s} \cdot S^{-1}(\gamma \circ u) = f \circ S^{-1}(v \circ u) = \text{0-map},$$

and $S^{-1}(\gamma \circ u)(S^{-1}A^m) = 0$. That means $(\gamma \circ u)(A^m)$ is S-torsion; so there exists $t \in S$ with $t(\gamma \circ u)(A^m) = 0$. Then $u(A^m)$ is contained in the kernel of the map $t\gamma : A^r \to N$; so there is an induced A-linear map

$$\overline{t\gamma} : \frac{A^r}{u(A^m)} \to N, \quad \overline{w} \mapsto t\gamma(w).$$

Likewise there is an induced isomorphism

$$\overline{v} : \frac{A^r}{u(A^m)} \ \rightarrow \ M, \ \overline{w} \ \mapsto \ v(w) \ .$$

So $g = \overline{t\gamma} \circ \overline{v}^{-1} : M \to N$ is A-linear, and for each i,

$$f(st\frac{m_i}{1}) \ = \ st\frac{n_i}{s} \ = \ \frac{tn_i}{1}$$

$$= \ \frac{g(m_i)}{1} \ = \ S^{-1}g(\frac{m_i}{1}) \ .$$

So $f \circ (st \cdot (-)) = S^{-1}g$.   ∎

Consider the application of G_0 to the ring homomorphism $\phi : A \to S^{-1}A$. The exact functor $S^{-1}(-) : \mathcal{M}(A) \to \mathcal{M}(S^{-1}A)$ induces a group homomorphism

$$G_0(\phi) : G_0(A) \ \rightarrow \ G_0(S^{-1}A)$$

taking $[M]$ to $[S^{-1}M]$ for each $M \in \mathcal{M}(A)$. By (6.50) (i), $G_0(\phi)$ is surjective.

(6.52) Theorem. (Swan [68]) *Suppose A is a left noetherian ring and $\mathcal{T}$ is the full subcategory of A-Mod whose objects are the finitely generated S-torsion A-modules. Inclusion $\mathcal{T} \to \mathcal{M}(A)$ induces the first map in an exact sequence of abelian groups:*

$$K_0(\mathcal{T}) \xrightarrow{\ \psi\ } G_0(A) \xrightarrow{\ G_0(\phi)\ } G_0(S^{-1}A) \longrightarrow 0 \ .$$

Proof. Let T denote the image of ψ, which is the subgroup of $G_0(A)$ generated by those $[M]$ with M a f.g. S-torsion A-module. Evidently T is contained in the kernel of $G_0(\phi)$. For the reverse containment, it suffices to prove the induced homomorphism

$$\frac{G_0(A)}{T} \xrightarrow{\ \overline{G_0(\phi)}\ } G_0(S^{-1}A)$$

has a left inverse, and so it is injective.

Suppose $P \in \mathcal{M}(S^{-1}A)$. By (6.50)(i), there exists $M \in \mathcal{M}(A)$ with $S^{-1}M \cong P$. If also $N \in \mathcal{M}(A)$ and $S^{-1}N \cong P$, there is an $S^{-1}A$-linear isomorphism $f : S^{-1}M \to S^{-1}N$. By (6.51), since A is left noetherian, there is an A-linear map $g : M \to N$ and an $S^{-1}A$-linear isomorphism $h : S^{-1}M \to S^{-1}M$ with $f \circ h = S^{-1}g$. Then $S^{-1}g$ is an isomorphism. Say g has kernel K and cokernel L. Since A is left noetherian, $K, L \in \mathcal{M}(A)$. Applying $S^{-1}(-)$ to the exact sequence

$$0 \longrightarrow K \longrightarrow M \xrightarrow{\ g\ } N \longrightarrow L \longrightarrow 0 \ ,$$

we see that $S^{-1}K = S^{-1}L = 0$. If $Q = g(M)$, then $Q \in \mathcal{M}(A)$, and from the exact sequences

$$0 \longrightarrow K \longrightarrow M \longrightarrow Q \longrightarrow 0$$
$$0 \longrightarrow Q \longrightarrow N \longrightarrow L \longrightarrow 0$$

it follows that in $G_0(A)$, $[M] - [N] = [K] - [L] \in T$ and $[M] + T = [N] + T$. So there is a metafunction

$$\theta : \mathrm{Obj}\, \mathcal{M}(S^{-1}A) \;\to\; G_0(A)/T$$

taking P to $[M] + T$ whenever $S^{-1}M \cong P$. Note that θ is constant on isomorphism classes. Suppose

$$0 \longrightarrow P'' \longrightarrow P \longrightarrow P' \longrightarrow 0$$

is an exact sequence in $\mathcal{M}(S^{-1}A)$. There exist $M, N \in \mathcal{M}(A)$ with $S^{-1}M \cong P$ and $S^{-1}N \cong P'$; so there is an exact sequence in $\mathcal{M}(S^{-1}A)$ that is the bottom row of:

$$
\begin{array}{ccccccccc}
0 & \longrightarrow & S^{-1}M'' & \longrightarrow & S^{-1}M & \overset{S^{-1}j}{\longrightarrow} & S^{-1}M' & \longrightarrow & 0 \;. \\
& & & & \Big\downarrow{h} & \searrow{S^{-1}g} & \Big\downarrow{S^{-1}i} & & \\
0 & \longrightarrow & P'' & \longrightarrow & S^{-1}M & \underset{f}{\longrightarrow} & S^{-1}N & \longrightarrow & 0
\end{array}
$$

To construct the rest of this diagram, use (6.51) to obtain an isomorphism h and an A-linear map $g : M \to N$ so the lower triangle commutes. Factor g as a surjection $j : M \to M'$ to its image M', followed by inclusion $i : M' \to N$; so the upper triangle commutes. Take M'' to be the kernel of j and the top row to be $S^{-1}(-)$ of the A-linear exact sequence:

$$0 \longrightarrow M'' \overset{\subseteq}{\longrightarrow} M \overset{j}{\longrightarrow} M' \longrightarrow 0 \;;$$

so the top row is exact. Since f is surjective, so is $S^{-1}g$; applying $S^{-1}(-)$ to the exact sequence

$$M \overset{g}{\longrightarrow} N \longrightarrow N/M' \longrightarrow 0$$

shows $S^{-1}(N/M') = 0$. So $S^{-1}i$ is an isomorphism.

Since the rows are exact and the vertical maps are isomorphisms, there is an isomorphism $S^{-1}M'' \cong P''$. So $\theta(P) = [M] + T = [M''] + [M'] + T = \theta(P'') + \theta(P')$, and θ induces a homomorphism

$$\overline{\theta} : G_0(S^{-1}A) \;\to\; G_0(A)/T \;.$$

And $\overline{\theta} \circ \overline{G_0(\phi)}([M] + T) = \overline{\theta}([S^{-1}M]) = [M] + T$; so $\overline{\theta}$ is left inverse to $\overline{G_0(\phi)}$. ∎

(6.53) Theorem. (Swan [68]) *If A and R are left noetherian, restriction of scalars defines the first map in an exact sequence of abelian groups:*

$$\bigoplus_{\mathfrak{p} \in \mathcal{M}} G_0(A/\mathfrak{p}A) \longrightarrow G_0(A) \xrightarrow{\ G_0(\phi)\ } G_0(S^{-1}A) \longrightarrow 0 \ ,$$

where $\mathcal{M}$ is the set of prime ideals of R that meet S.

Proof. Each f.g. $A/\mathfrak{p}A$-module is a f.g. S-torsion A-module, since it is annihilated by $\mathfrak{p}$ and $\mathfrak{p}$ meets S. So the composite is zero at $G_0(A)$. For exactness at $G_0(A)$, it suffices to prove:

(6.54) Claim. *Every f.g. A-module M has a filtration $M = M_n \supseteq \cdots \supseteq M_0 = 0$ in $\mathcal{M}(A)$ in which the annihilator of each quotient M_i/M_{i-1} intersects R in a prime ideal $\mathfrak{p}_i$.*

If this is true, and $sM = 0$ for some $s \in S$, then

$$[M] \ = \ \sum [M_i/M_{i-1}]$$

in $G_0(A)$, and s belongs to each $\mathfrak{p}_i$; so each $\mathfrak{p}_i$ meets S, and each M_i/M_{i-1} is a f.g. $A/\mathfrak{p}_i A$-module.

Suppose the claim fails. Since R is noetherian, every nonempty set of ideals of R has a maximal element. Choose $M \in \mathcal{M}(A)$ for which the claim fails, so that

$$\mathfrak{q} \ = \ \mathrm{ann}_R(M) \ = \ \{r \in R : rM = 0\}$$

is maximal. The claim would be true for $M = 0$ or if $\mathrm{ann}_R(M)$ were prime; so $\mathfrak{q}$ is a proper nonprime ideal of R. So there exist $a, b \in R - \mathfrak{q}$ with $ab \in \mathfrak{q}$. Consider the exact sequence:

$$0 \longrightarrow aM \longrightarrow M \longrightarrow M/aM \longrightarrow 0 \ .$$

Since

$$\mathrm{ann}_R(aM) \ \supseteq \ \mathfrak{q} \cup \{b\} \ \text{ and}$$
$$\mathrm{ann}_R(M/aM) \ \supseteq \ \mathfrak{q} \cup \{a\} \ ,$$

the claim holds for aM and M/aM. By the Correspondence Theorem (B.7), in the Appendix, the claim must also be true for M, a contradiction. ∎ ∎

The functor $S^{-1}(-)$ is really just an extension of scalars from A to $S^{-1}A$:

(6.55) Proposition. *There is a natural isomorphism τ from $S^{-1}A \otimes_A (-)$ to $S^{-1}(-)$, given for each A-module M by*

$$\tau_M : S^{-1}A \otimes_A M \;\cong\; S^{-1}M \ .$$
$$\frac{a}{s} \otimes m \;\mapsto\; \frac{am}{s}$$

Proof. Resticting scalars to $a/1$ on the right, $S^{-1}A$ is an $S^{-1}A$, A-bimodule. The map $S^{-1}A \times M \to S^{-1}M$, taking $(a/s, m)$ to am/s, is A-balanced, inducing the left $S^{-1}A$-linear map τ_M. There is a well-defined function $g : S^{-1}M \to S^{-1}A \otimes_A M$, taking m/s to $(1/s) \otimes m$, since $s_3 s_2 m_1 = s_3 s_1 m_2$ implies

$$\frac{1}{s_1} \otimes m_1 \;=\; \frac{1}{s_1 s_2 s_3} \otimes s_3 s_2 m_1 \;=\; \frac{1}{s_1 s_2 s_3} \otimes s_3 s_1 m_2 \;=\; \frac{1}{s_2} \otimes m_2 \ .$$

Since g is additive, it is inverse to τ_M.

For each A-linear map $f : M \to N$, the square

$$
\begin{array}{ccc}
S^{-1}A \otimes_A M & \xrightarrow{\ \tau_M\ } & S^{-1}M \\
{\scriptstyle 1 \otimes f} \downarrow & & \downarrow {\scriptstyle S^{-1}f} \\
S^{-1}A \otimes_A N & \xrightarrow{\ \tau_N\ } & S^{-1}N
\end{array}
$$

commutes; so τ is a natural transformation, as defined in (0.13).   ∎

(6.56) Corollary. *Each localization $S^{-1}A$ is a flat right A-module.*

Proof. If $L \to M \to N$ is exact in A-$\mathfrak{Mod}$, the exactness of the bottom row of the commutative diagram

$$
\begin{array}{ccccc}
S^{-1}A \otimes_A L & \longrightarrow & S^{-1}A \otimes_A M & \longrightarrow & S^{-1}A \otimes_A N \\
{\scriptstyle \tau_L} \downarrow & & {\scriptstyle \tau_M} \downarrow & & \downarrow {\scriptstyle \tau_N} \\
S^{-1}L & \longrightarrow & S^{-1}M & \longrightarrow & S^{-1}N
\end{array}
$$

implies that of the top.   ∎

Now consider K_0. If $\phi : A \to S^{-1}A$ is the localization map, then $K_0(\phi) : K_0(A) \to K_0(S^{-1}A)$ takes $[P]$ to $[S^{-1}A \otimes_A P] = [S^{-1}P]$ for each $P \in \mathcal{P}(A)$. So there is a commutative square

$$
\begin{array}{ccc}
K_0(A) & \xrightarrow{\ K_0(\phi)\ } & K_0(S^{-1}A) \\
{\scriptstyle c} \downarrow & & \downarrow {\scriptstyle c} \\
G_0(A) & \xrightarrow{\ G_0(\phi)\ } & G_0(S^{-1}A)
\end{array}
$$

where the maps c are the Cartan homomorphisms.

If A is left regular, so is $S^{-1}A$: For $S^{-1}(-)$ is additive and carries A to $S^{-1}A$; so it restricts to a functor from $\mathcal{P}(A)$ to $\mathcal{P}(S^{-1}A)$, and $S^{-1}(-)$ is exact, preserving projective resolutions. Finally every f.g. $S^{-1}A$-module is $S^{-1}M$ for a f.g. A-module M.

So if A is left regular, the Cartan maps c are both isomorphisms, and we can replace G_0 by K_0 in all but the first term of the sequences (6.52) and (6.53). Even when A is not left regular, there is a localization sequence for K_0:

(6.57) Notation. Let $\mathcal{T}_1$ denote the full subcategory of $A\text{-}\mathfrak{Mod}$ whose objects are the f.g. S-torsion A-modules M for which there is an A-linear exact sequence

$$0 \longrightarrow P_1 \longrightarrow P_0 \longrightarrow M \longrightarrow 0$$

with $P_1, P_0 \in \mathcal{P}(A)$.

(6.58) Theorem. (Bass [68, p. 494]) *Suppose S contains neither zero nor any zero-divisors of A. There is an exact sequence of group homomorphisms:*

$$K_0(\mathcal{T}_1) \xrightarrow{\ \delta\ } K_0(A) \xrightarrow{\ K_0(\phi)\ } K_0(S^{-1}A) \ .$$

Proof. By (3.50) and (3.51), the Euler characteristic is a generalized rank

$$\chi : \mathrm{Obj}\ \mathcal{T}_1 \ \rightarrow \ K_0(A) \ ,$$

with $\chi(M) = [P_0] - [P_1]$ whenever M has a resolution (6.57). So χ induces a group homomorphism

$$\delta : K_0(\mathcal{T}_1) \ \rightarrow \ K_0(A)$$

taking $[M]$ to $[P_0] - [P_1]$. Then

$$(K_0(\phi) \circ \delta)[M] \ = \ [S^{-1}P_0] - [S^{-1}P_1] \ = \ 0 \ ,$$

since $S^{-1}M = 0$ implies $S^{-1}P_0 \cong S^{-1}P_1$.

On the other hand, suppose $P, Q \in \mathcal{P}(A)$ and

$$[S^{-1}P] - [S^{-1}Q] \ = \ K_0(\phi)([P] - [Q]) \ = \ 0 \ .$$

Then $S^{-1}P$ and $S^{-1}Q$ are stably isomorphic: For some integer $n > 0$,

$$S^{-1}(P \oplus A^n) \cong S^{-1}P \oplus (S^{-1}A)^n \cong S^{-1}Q \oplus (S^{-1}A)^n \cong S^{-1}(Q \oplus A^n) \ .$$

Since S contains neither zero nor any zero-divisors, it acts through injections on the f.g. projective A-modules $M = P \oplus A^n$ and $N = Q \oplus A^n$. By Lemma

(6.50) (iii), there is a submodule N' of M with $N \cong N'$ and $S^{-1}(M/N') = 0$. From the resolution

$$0 \longrightarrow N' \longrightarrow M \longrightarrow M/N' \longrightarrow 0 \, ,$$

we see that

$$\delta[M/N'] \;=\; [M] - [N] \;=\; [P] - [Q] \, . \qquad \blacksquare$$

(6.59) Note. Within the category $\mathcal{T}$ of f.g. S-torsion A-modules, let $\mathcal{T}_{<n}$ denote the full subcategory consisting of those M having $\mathcal{P}(A)$-resolutions

$$0 \to P_i \to \cdots \to P_0 \to M \to 0$$

of length $i < n$. So $\mathcal{T}_1 = \mathcal{T}_{<2}$. With essentially the same proof, Theorem (6.58) remains true if $\mathcal{T}_1$ is replaced by any $\mathcal{T}_{<n}$ for $n \geq 2$, or by

$$\mathcal{T}_{<\infty} \;=\; \bigcup \mathcal{T}_{<n} \, .$$

When A is left regular, $\mathcal{T}_{<\infty} = \mathcal{T}$, and the form of sequence (6.58) using $\mathcal{T}_{<\infty}$ becomes the G_0 sequence (6.52).

6E. Exercises

Retain the notation of A for a ring, R for its center, and S for a submonoid of the monoid $(R, \cdot)$.

1. If $S \subseteq T$ are submonoids of $(R, \cdot)$, prove $T^{-1}A$ is isomorphic to the localization of $S^{-1}A$ at the set $S^{-1}T = \phi(T)$.

2. If $\mathfrak{p}$ is a prime ideal of R and P is a f.g. projective R-module, then $P_{\mathfrak{p}}$ is a f.g. projective module over the local ring $R_{\mathfrak{p}}$; so $P_{\mathfrak{p}} \cong (R_{\mathfrak{p}})^n$ for some $n \geq 0$. The integer n is denoted by $rk_{\mathfrak{p}}(P)$ and called the **local rank** of P at $\mathfrak{p}$. The metafunction $rk_{\mathfrak{p}}$ is just the composite

$$\mathrm{Obj}\ \mathcal{P}(R) \;\to\; K_0(R) \;\to\; K_0(R_{\mathfrak{p}}) \;\cong\; \mathbb{Z} \, ;$$

so it is additive over short exact sequences.

(i) If R is an integral domain, prove $rk_{\mathfrak{p}} = rk_{\mathfrak{q}}$ for each pair of prime ideals $\mathfrak{p}$ and $\mathfrak{q}$. *Hint:* If $S = R - \{0\}$, the localization map $R \to S^{-1}R$ factors through both $R_{\mathfrak{p}}$ and $R_{\mathfrak{q}}$.

(ii) If $R = \mathbb{Q} \times \mathbb{Q}$, find prime ideals $\mathfrak{p}$ and $\mathfrak{q}$ of R with $rk_{\mathfrak{p}} \neq rk_{\mathfrak{q}}$. Illustrate with some $P \in \mathcal{P}(R)$ for which $rk_{\mathfrak{p}}(P) = 2$ and $rk_{\mathfrak{q}}(P) = 3$.

3. If $e \in R$ with $ee = e$, and if $S = \{1, e\}$, show the localization sequence (6.52) becomes a familiar split short exact sequence.

4. Suppose $A = R$ is a principal ideal domain with field of fractions F. Use (6.52) to prove directly that $G_0(\phi) : G_0(R) \to G_0(F)$ is an isomorphism, so that $G_0(R) \cong \mathbb{Z}$. (In case $R = \mathbb{Z}$, this isomorphism is just the rank of f.g. abelian groups.)

5. Suppose $\mathfrak{p}_1, \ldots, \mathfrak{p}_n$ are $n(< \infty)$ different maximal ideals of R. Prove the complement $S = R - \cup \mathfrak{p}_i$ of their union is a submonoid of $(R, \cdot)$. Prove that every ideal of R contained in $\cup \mathfrak{p}_i$ is contained in one of the maximal ideals $\mathfrak{p}_i$. Use this and (6.48) to prove $S^{-1}R$ has exactly n maximal ideals $S^{-1}\mathfrak{p}_1, \ldots, S^{-1}\mathfrak{p}_n$. A commutative ring with only finitely many maximal ideals is called **semilocal**.

6. Prove that a commutative ring R is semilocal if and only if $R/\mathrm{rad}R$ is artinian. (A *non*commutative ring A is *defined* to be **semilocal** if $A/\mathrm{rad}A$ is left artinian, which in this case is equivalent to right artinian — see (8.29).)

7. If M and N are R-modules, prove there is an $S^{-1}R$-linear isomorphism

$$S^{-1}(M \otimes_R N) \;\cong\; S^{-1}M \otimes_{S^{-1}R} S^{-1}N \;,$$

and that the latter equals $S^{-1}M \otimes_R S^{-1}N$. *Hint:* Recall the proof that K_0 takes $\mathcal{C}\mathcal{R}\mathrm{ing}$ to $\mathcal{C}\mathcal{R}\mathrm{ing}$ (see (6.22)).

8. If M and N are R-modules, show there is a homomorphism of $S^{-1}R$-modules

$$\theta : S^{-1}\mathrm{Hom}_R(M, N) \;\to\; \mathrm{Hom}_{S^{-1}R}(S^{-1}M, S^{-1}N) \;,$$

where $\theta(f/s)(m/t) = f(m)/st$. Then show θ is an isomorphism if

(i) $M = R$,
(ii) $M \in \mathcal{F}(R)$,
(iii) $M \in \mathcal{P}(R)$.

Hint: $S^{-1}\mathrm{Hom}_R(-, N)$ and $\mathrm{Hom}_{S^{-1}R}(S^{-1}(-), S^{-1}N)$ are additive contravariant functors.

9. Show $\mathcal{T}_{<n} \subseteq \mathcal{T}_{<n+1} \subseteq \mathcal{T}_{<\infty}$ induce isomorphisms of K_0 groups. *Hint:* Show every object of $\mathcal{T}_{<\infty}$ has a finite $\mathcal{T}_{<2}$-resolution and each $\mathcal{T}_{<n}$ is closed under kernels of its arrows. Then use the Resolution Theorem.

Sources of K_0

Projective modules are of some interest as natural generalizations of free modules, but generalizations become significant when they connect objects already under study. In Chapter 7 we find that the integral domains R whose ideals are f.g. projective R-modules are the rings of primary interest in algebraic number theory. And in Chapter 8 the rings, all of whose modules are projective, turn out to be the gateway to the matrix representations of finite groups.

In connection with these two types of ring are two abelian groups that are precursors of $K_0(R)$. Pinpointing the historical origin of K_0 is like locating the source of a great river; there are many tributaries along the way, and the identity of the true source can be a subjective judgement. I propose that the source of K_0 is the ideal class group, described in quite modern terms by Dedekind in 1893. We consider the class group in Chapter 7. The origin of K_0 of noncommutative rings is perhaps the ring of virtual characters of finite groups. This subject has an equally long history, going back to Frobenius in the 1890s (with earlier roots due to Dirichlet, Dedekind, and Kronecker), but it recognizably enters the stream of K-theory in the 1960 paper of Swan, "Induced Representations and Projective Modules." We develop character theory in Chapter 8.

7

Number Theory

The theory of numbers is, at the same time, one of the simplest and one of the deepest parts of mathematics. It is the search for simple patterns in the addition and multiplication of integers; but the verification of these patterns draws on techniques from algebra, geometry, analysis, and topology. We focus here on algebraic number theory, beginning in §7A with a generalization of the notion of integer to "algebraic integer." In §7B, the ring of algebraic integers in a number field is shown to have unique factorization of its ideals into products of maximal ideals and, equivalently, to have every ideal a projective module. Integral domains with these properties are called Dedekind domains. Section 7C introduces the ideal class group, which measures how far a Dedekind domain is from having unique factorization of elements into primes, or equivalently, from having every ideal a free module. The ideal class group is a 19th-century ancestor of the Grothendieck group, but its calculation is still an active area of research. Section 7D focuses on techniques for factoring ideals, including exercises on the computation of class groups, and §7E relates class groups to the K_0 and G_0 of Dedekind domains.

7A. Algebraic Integers

At its core, number theory is the study of equations and congruences within the ring of integers $\mathbb{Z}$. An effective device for solving such problems is the unique factorization of each nonzero integer as a product of primes:

$$(\pm 1)2^{n(2)}3^{n(3)}5^{n(5)}7^{n(7)} \ldots ,$$

where each exponent $n(p)$ is a nonnegative integer and $n(p) = 0$ for all but finitely many primes p.

(7.1) Examples.

(i) If $\sqrt{2} = a/b$ for integers a and b, then $a^2 = 2b^2$; but the number of 2's in the factorization of the left side is even, while the number of 2's on the right is odd. So $\sqrt{2}$ is irrational.

(ii) Suppose a, b are nonzero integers and a^2 is a factor of b^2. Cutting the exponents in their prime factorizations in half shows a is a factor of b.

(7.2) Definitions. Suppose R is a commutative integral domain. An element of R is **composite** if it is a product of two nonzero nonunits of R. An element of R is **irreducible** if it is not zero, not a unit, and not composite. We say R is **factorial** with respect to T if $T \subseteq R - \{0\}$ and each $r \in R - \{0\}$ can be uniquely written in the form

$$ r = u \prod_{p \in T} p^{n(p)} , $$

where $u \in R^*$ and the exponents $n(p)$ are nonnegative integers that are zero for all but finitely many $p \in T$.

By the uniqueness requirement, no $p \in T$ is a unit of R. So r is a unit if and only if $\sum n(p) = 0$, r is composite if and only if $\sum n(p) > 1$, and r is irreducible if and only if $\sum n(p) = 1$. In particular, the irreducibles are the products up with $u \in R^*$ and $p \in T$.

The set T is not uniquely determined by the factorial ring R : If F is the field of fractions of R, then R^* is a (normal) subgroup of F^*. The set of irreducibles in R is a union of some of the cosets of R^*. The factorial ring R is factorial with respect to T if and only if T consists of exactly one representative from each coset of irreducibles from R.

To solve some problems in $\mathbb{Z}$, somewhat larger factorial domains than $\mathbb{Z}$ can be useful:

(7.3) Example. Each prime number $p \in \mathbb{Z}$ can be written in at most one way as a sum of two squares: For if $p = x^2 + y^2$ in $\mathbb{Z}$, then $p = (x + yi)(x - yi)$ in $\mathbb{Z}[i] = \{a + bi : a, b \in \mathbb{Z}\}$. It happens that $\mathbb{Z}[i]$ is a euclidean ring, and so it is factorial; for T we can take those irreducibles $x + yi$ with $x > 0$ and $y \geq 0$. The complex norm $|\ | : \mathbb{C} \to \mathbb{R}$, defined by $|a + bi| = a^2 + b^2$, restricts to a multiplicative function $N : \mathbb{Z}[i] \to \mathbb{Z}$ that takes zero to zero, units to units, and composites to composites. So

$$ p = (x + yi)(x - yi) = N(x + yi) = N(x - yi) $$

implies both $x + yi$ and $x - yi$ are irreducible in $\mathbb{Z}[i]$. Since $\mathbb{Z}[i]^* = \{\pm 1, \pm i\}$, the irreducible factors of p in $\mathbb{Z}[i]$ are of the form $(\pm x) + (\pm y)i$ or $(\pm y) + (\pm x)i$. So x^2 and y^2 are uniquely determined.

The enlarged domain to use is not always the first one to come to mind:

(7.4) Example. Can each prime number be written in at most one way as a square plus three times a square? If $p = x^2 + 3y^2$ in $\mathbb{Z}$, then $p = (x + y\sqrt{3}i)(x - y\sqrt{3}i)$ in $\mathbb{Z}[\sqrt{3}i] = \{a + b\sqrt{3}i : a, b \in \mathbb{Z}\}$. Again the complex norm restricts to $N : \mathbb{Z}[\sqrt{3}i] \to \mathbb{Z}$, leading to the conclusion that $x + y\sqrt{3}i$ and $x - y\sqrt{3}i$ are irreducible in $\mathbb{Z}[\sqrt{3}i]$. Unfortunately, $\mathbb{Z}[\sqrt{3}i]$ is not factorial:

$$4 = (1 + \sqrt{3}i)(1 - \sqrt{3}i) = (2)(2)$$

while $\mathbb{Z}[\sqrt{3}i]^* = \{\pm 1\}$. So p may have irreducible factors not among $(\pm x) + (\pm y)\sqrt{3}i$. To see that this does not happen, one can work in the larger euclidean domain $\mathbb{Z}[w]$ where $w = (1 + \sqrt{3}i)/2$ and use unique factorization there. (See Exercise 7.)

In the last example, $\mathbb{Z}[\sqrt{3}i]$ and $\mathbb{Z}[w]$ have the same field of fractions. Here is another example in which enlarging an integral domain within its field of fractions yields a factorial domain:

(7.5) Example. In a factorial domain R, if $a^3 = b^2$, then a is a square and b is a cube: This holds because it is true for powers p^n with $p \in T$, since $n \in 2\mathbb{Z} \cap 3\mathbb{Z}$ implies $n \in 6\mathbb{Z}$, and it is true in the group of units, since $u^3 = v^2$ implies $(u^{-1}v)^2 = u$ and $(u^{-1}v)^3 = v$. In the domain

$$\mathbb{Z}[\sqrt{8}] = \{a + b\sqrt{8} : a, b \in \mathbb{Z}\}$$
$$= \{a + 2b\sqrt{2} : a, b \in \mathbb{Z}\},$$

$2^3 = (\sqrt{8})^2$, but 2 is not a square, since $\pm\sqrt{2} \notin \mathbb{Z}[\sqrt{8}]$. Therefore $\mathbb{Z}[\sqrt{8}]$ is not factorial. The field of fractions of $\mathbb{Z}[\sqrt{8}]$ is $\mathbb{Q}(\sqrt{8}) = \mathbb{Q}(\sqrt{2})$, and within it is the larger domain $\mathbb{Z}[\sqrt{2}]$, which is euclidean, and hence factorial.

In the last two examples, what enables $\mathbb{Z}[w]$ and $\mathbb{Z}[\sqrt{2}]$ to be well behaved is that they consist of *all* roots, in their field of fractions, of monic polynomials in $\mathbb{Z}[x]$. To elaborate let's move to a more general setting.

Suppose A is a ring. An A-module M is **faithful** if there is no nonzero scalar $a \in A$ with $aM = 0$. Suppose R is a subring of the center of A. An element $a \in A$ is **integral** over R if there is a monic polynomial $p(x) \in R[x]$ with $p(a) = 0$. The ring A is **integral** over R if every element of A is integral over R.

(7.6) Proposition. *Suppose A is a ring, R is a subring of the center of A, and $a \in A$. The following are equivalent.*

  (i) *The element a is integral over R.*
  (ii) *The ring $R[a]$ is finitely generated as an R-module.*
  (iii) *The ring $R[a]$ has a faithful module M that is finitely generated as an R-module.*

If A is finitely generated as an R-module, A is integral over R.

Proof. Assume

$$a^n + r_{n-1}a^{n-1} + \ldots + r_1 a + r_0 = 0,$$

with each $r_i \in R$. Solving for a^n and repeatedly substituting in a typical element of $R[a]$ shows

$$R[a] = R + Ra + \ldots + Ra^{n-1}.$$

So (i) implies (ii). Since $1 \in R[a]$, $R[a]$ is a faithful $R[a]$-module; therefore (ii) implies (iii).

Assume (iii) and suppose $m_1, \ldots, m_n$ generate M as an R-module. Now M^n (as column vectors) is an $M_n(R[a])$-module via matrix multiplication. There exist $r_{ij} \in R$ with

$$\begin{bmatrix} a & 0 & \ldots & 0 \\ 0 & a & \ldots & 0 \\ \vdots & \vdots & \ddots & \vdots \\ 0 & 0 & \ldots & a \end{bmatrix} \begin{bmatrix} m_1 \\ \vdots \\ m_n \end{bmatrix} = \begin{bmatrix} r_{11} & \ldots & r_{1n} \\ \vdots & & \vdots \\ \vdots & & \vdots \\ r_{n1} & \ldots & r_{nn} \end{bmatrix} \begin{bmatrix} m_1 \\ \vdots \\ m_n \end{bmatrix}.$$

If δ_{ij} is the i,j-entry of the multiplicative identity matrix and N is the matrix with i,j-entry $\delta_{ij}a - r_{ij}$, then

$$N \begin{bmatrix} m_1 \\ \vdots \\ m_n \end{bmatrix} = \begin{bmatrix} 0 \\ \vdots \\ 0 \end{bmatrix}.$$

Left multiply both sides by the adjoint of $N (=$ the transpose of the matrix of cofactors of N) to get

$$\begin{bmatrix} d & 0 & \ldots & 0 \\ 0 & d & \ldots & 0 \\ \vdots & \vdots & \ddots & \vdots \\ 0 & 0 & \ldots & d \end{bmatrix} \begin{bmatrix} m_1 \\ \vdots \\ m_n \end{bmatrix} = \begin{bmatrix} 0 \\ \vdots \\ 0 \end{bmatrix},$$

where d is the determinant of N. So $dM = 0$. Since M is a faithful $R[a]$-module, $d = 0$. But $d = p(a)$, where $p(x)$ is the characteristic polynomial of the matrix (r_{ij}), so it is a monic polynomial in $R[x]$. Therefore (iii) implies (i).

The last assertion follows from the fact that, for each $a \in A$, A is a faithful $R[a]$-module, because $1 \in A$. ∎

Suppose R is a subring of a *commutative* ring E. Say R is **integrally closed** in E if every element of E that is integral over R belongs to R. The **integral closure** of R in E is the set of all elements of E that are integral over R.

(7.7) Proposition. *If E is a commutative ring with a subring R and A is the integral closure of R in E, then A is a subring of E containing R, and A is integrally closed in E.*

Proof. Each $r \in R$ is a root of $x - r$; so $R \subseteq A$. Suppose $a, a' \in A$. Then $R[a]$ is a f.g. R-module, and $R[a, a'] = R[a][a']$ is a f.g. $R[a]$-module. Multiplying generating sets yields a finite generating set of $R[a, a']$ as an R-module. By the last assertion of (7.6), $a + a', -a$, and aa' belong to A; so A is a subring of E. If $b \in E$ is integral over A, b is a root of a monic polynomial $p(x) \in A[x]$ with some coefficients $a_0, \ldots, a_{n-1} \in A$. So b is integral over $R[a_0, \ldots, a_{n-1}]$. We have a chain of subrings $R \subseteq R[a_0] \subseteq \ldots \subseteq R[a_0, \ldots, a_{n-1}, b]$, each f.g. as a module over its predecessor. Multiplying generating sets, the last of these rings is a f.g. R-module; so b is integral over R and belongs to A. ∎

If R is a subring of a field F, we say F is **a field of fractions** of R if every element of F is ab^{-1} for some $a, b \in R, b \neq 0$. Above, **the field of fractions** of R was the field $S^{-1}R$ $(S = R - \{0\})$, with R embedded as a subring. There is little difference here: If F is any field of fractions of R, the inclusion $R \to F$ extends to a ring homomorphism $S^{-1}R \to F$, $a/b \mapsto ab^{-1}$, which is evidently surjective, and is injective because $S^{-1}R$ and F are fields. Since this isomorphism fixes the elements of R, it restricts to an isomorphism between the integral closures of R in $S^{-1}R$ and F. An integral domain R is **integrally closed** if it is integrally closed in any (hence every) field of fractions of R.

(7.8) Proposition. *Every factorial domain is integrally closed.*

Proof. Suppose R is a factorial domain with field of fractions F, and suppose $a, b \in R, b \neq 0$, and a/b is integral over R. Then there exist $r_0, \ldots, r_{n-1} \in R$ with

$$\frac{a^n}{b^n} + r_{n-1}\frac{a^{n-1}}{b^{n-1}} + \ldots + r_1\frac{a}{b} + r_0 = 0 \, .$$

So $a^n = b(-r_{n-1}a^{n-1} - \ldots - r_1ab^{n-2} - r_0b^{n-1})$, and every irreducible factor of b also divides a. Repeatedly canceling the irreducible factors of b from the numerator and denominator of a/b, we eventually reach $b \in R^*$; so $a/b \in R$. ∎

The integral domains $\mathbb{Z}[\sqrt{3}i]$ and $\mathbb{Z}[\sqrt{8}]$ in Examples (7.4) and (7.5) have no chance of being factorial, since they are not integrally closed: The root $w = (1 + \sqrt{3}i)/2$ of $x^2 - x + 1$ is in $\mathbb{Q}(\sqrt{3}i)$, but not in $\mathbb{Z}[\sqrt{3}i]$; the root $\sqrt{2} = \sqrt{8}/2$ of $x^2 - 2$ is in $\mathbb{Q}(\sqrt{8})$ but not in $\mathbb{Z}[\sqrt{8}]$. The larger domains $\mathbb{Z}[w]$ and $\mathbb{Z}[\sqrt{2}]$ are the integral closures of $\mathbb{Z}[\sqrt{3}i]$ and $\mathbb{Z}[\sqrt{8}]$ in their fields of fractions.

The examples (7.4) and (7.5) are just two small illustrations of the algebraic approach to number theory, which focuses on the following types of rings: A finite-degree field extension F of $\mathbb{Q}$ is called an **algebraic number field** (or just **number field**, for short). An element of F that is integral over $\mathbb{Z}$ is an **algebraic integer**. The integral closure of $\mathbb{Z}$ in a number field F is known as the **ring of algebraic integers in** F, denoted briefly by

$$\mathcal{O}_F = \text{alg. int.}(F) \ .$$

The ring of algebraic integers in a number field is integrally closed, but it may or may not be factorial. For instance, if ζ_n is a complex number of multiplicative order n, then alg. int.$(\mathbb{Q}(\zeta_n)) = \mathbb{Z}[\zeta_n]$. (For an elementary proof of this when n is prime, see Theorem 3.5, p. 72, in Stewart and Tall [87]. For the general case, see Theorem 2.6, p. 11, in Washington [97].) It could have been the assumption that $\mathbb{Z}[\zeta_n]$ is factorial which led Fermat to believe he had a proof that $x^n + y^n = z^n$ has no positive integer solutions for $n > 2$. But $\mathbb{Z}[\zeta_p]$ is factorial for a prime p if and only if $p \le 19$. (This was a conjecture of Kummer, finally proved in the 1970s — see Uchida [71], and Masley and Montgomery [76].)

This stumbling block was partly overcome in the last half of the 19th century, when Kummer and his student, Kronecker, developed the notion of "ideal prime numbers," based on the logarithmic functions associated with unique factorization; Dedekind then defined ideals of a ring and proved every nonzero ideal of a ring of algebraic integers $\mathcal{O}_F$ has a unique factorization as a product of maximal ideals (see Dedekind [93]). Factorization of ideals suffices for many problems in number theory. We defer the proof of Dedekind's theorem until the next section, where integral domains with unique ideal factorization are characterized.

7A. Exercises

1. If F is a number field and $a \in F$, show there must be some nonzero $r \in \mathbb{Z}$ for which ra is an algebraic integer.

2. If F is a number field and $a \in F$, prove a is an algebraic integer if and only if its minimal polynomial over $\mathbb{Q}$ lies in $\mathbb{Z}[x]$. *Hint*: Use Gauss' Lemma.

3. If $\mathbb{Q} \subseteq F$ is a degree 2 field extension, prove $F = \mathbb{Q}(\sqrt{d})$ for some square-free integer d. Then prove alg. int.(F) is either $\mathbb{Z}[\sqrt{d}]$ (if $d \not\equiv 1 \bmod 4$) or $\mathbb{Z}[(1 + \sqrt{d})/2]$ (if $d \equiv 1 \bmod 4$). *Hint:* For the first assertion, use Exercise 1 to write F as $\mathbb{Q}(\alpha)$ for an algebraic integer α, and Exercise 2 and the quadratic formula to compute α. For the second assertion, if $\beta \in F - \mathbb{Q}$, its minimal polynomial over $\mathbb{Q}$ is $p(x) = (x - \beta)(x - \sigma(\beta))$, where σ generates $Aut(F/\mathbb{Q})$. If $\beta = (a + b\sqrt{d})/c$ with $a, b, c \in \mathbb{Z}$, find conditions on a, b and c for $p(x) \in \mathbb{Z}[x]$.

4. Suppose $\mathbb{Q} \subseteq F$ is a Galois field extension of finite degree n, with Galois group $Aut(F/\mathbb{Q}) = \{\sigma_1, \ldots, \sigma_n\}$. For $x \in F$, define the **norm** of x to be

$$N_{F/\mathbb{Q}}(x) = \sigma_1(x) \cdots \sigma_n(x) \ .$$

Prove $N_{F/\mathbb{Q}}(x) \in \mathbb{Q}$ and $N_{F/\mathbb{Q}} = N$ defines a multiplicative map $(N(xy) = N(x)N(y))$ from F to $\mathbb{Q}$. Show $N(x)$ is ± 1 times the constant coefficient of the minimal polynomial of x over $\mathbb{Q}$. Prove N restricts to a function $N : \mathcal{O}_F \to \mathbb{Z}$, and this restriction takes zero to zero, units to units, and composites to composites. *Hint:* For the last assertion, note that $N^{-1}(0) = 0$ and $N^{-1}(\pm 1) \cap \mathcal{O}_F = \mathcal{O}_F^*$.

5. If F is a quadratic imaginary number field (i.e., $F = \mathbb{Q}(\sqrt{d})$ where the square-free integer d is negative), prove $\mathcal{O}_F^*$ is finite. *Hint:* In this case $N(x + yi) = x^2 + y^2 = |x + yi|^2$, and hence every element of $\mathcal{O}_F^*$ lies on the unit circle in the complex plane. Use Exercise 3 to describe the distribution of the points in $\mathcal{O}_F$ in the plane.

6. Suppose F is a quadratic imaginary number field (as in Exercise 5) and the distance in the complex plane from each complex number to the nearest number in $\mathcal{O}_F$ is less than 1. Prove $\mathcal{O}_F$ is euclidean, with the size of elements measured by the norm $N(x) = |x|^2$. *Hint:* If $a, b \in \mathcal{O}_F$ with $b \neq 0$, the complex number ab^{-1} is within a distance of 1 of some $q \in \mathcal{O}_F$. Show $N(a - bq) < N(b)$.

7. Suppose p is a prime number and $p = x^2 + 3y^2$ for $x, y \in \mathbb{Z}$. Prove the squares x^2, y^2 are uniquely determined by p. *Hint:* If $w = (1 + \sqrt{3}i)/2$, then $\mathbb{Z}[w] =$alg. int.$\mathbb{Q}(\sqrt{3}i)$ by Exercise 3 and is euclidean (hence factorial) by Exercise 6. As in Exercise 5, $\mathbb{Z}[w]^*$ is cyclic of order 6 generated by w. Using the norm properties in Exercise 4, $x + y\sqrt{3}i$ and $x - y\sqrt{3}i$ are irreducible factors of p in $\mathbb{Z}[w]$. Since p is their product, the only irreducible factors of p in $\mathbb{Z}[w]$ are $w^j(x + y\sqrt{3}i)$ and $w^j(x - y\sqrt{3}i)$ for $j \in \mathbb{Z}$. Among these, show the only ones in $\mathbb{Z}[\sqrt{3}i]$ are $\pm x \pm y\sqrt{3}i$.

8. Suppose R is a subring of the center of a ring A.

   (i) If C is a ring, $A \subseteq B$ are subrings of the center of C, and the extensions $A \subseteq B \subseteq C$ are integral, prove $A \subseteq C$ is integral.

   (ii) If A is the integral closure of R in a commutative ring E, and if B, C are subrings of E with $R \subseteq B \subsetneqq A \subsetneqq C \subseteq E$, prove B is not integrally closed in E and C is not integral over R.

9. If $\mathbb{Q} \subseteq F \subseteq E$ are finite-degree field extensions, prove $\mathcal{O}_E$ is the integral closure in E of $\mathcal{O}_F$.

10. If R is a subring of the center of a ring A and A is integral over R, prove $A^* \cap R = R^*$.

11. If i is the imaginary unit (whose square is -1), prove the ring $\mathbb{Z}[i]$ is euclidean by the method in Exercise 6. Then use the norm as in Exercise 4 to determine the irreducibles in $\mathbb{Z}[i]$ by considering which primes in $\mathbb{Z}$ are sums of two squares in $\mathbb{Z}$.

7B. Dedekind Domains

In this section we see that the integral domains with unique ideal factorization are the integral domains in which every ideal is a projective module. *Through-out, take R to be a commutative integral domain with field of fractions F.* Multiplication in F makes F an R-module. Let **sub** $_RF$ denote the set of nonzero R-submodules of F, and **sub** $_RR$ the set of nonzero R-submodules of R (= the nonzero ideals of R). Under the multiplication

$$IJ = \{\sum_{i=1}^{n} x_i y_i : x_i \in I, y_i \in J, n > 0\}\,,$$

sub $_RF$ is a commutative monoid, with sub $_RR$ as a submonoid. The multiplicative identity is R.

In passing from factorization of elements to factorization of ideals, the relation between R and F is replaced by the relation between sub $_RR$ and sub $_RF$. Although ideal factorization takes place within sub $_RR$, it is useful to consider it within the context of sub $_RF$. To see why one might expect this, consider the groups of units $\mathbb{Z}^*$ and $\mathbb{Q}^*$. The group $\mathbb{Z}^* = \{\pm 1\}$ is too small to contain much information, but unique factorization in $\mathbb{Z}$ amounts to the fact that $\mathbb{Q}^*$ is $\mathbb{Z}^*$ times a free abelian group based on the prime numbers. The group (sub $_RR)^*$ of invertible elements in the monoid sub $_RR$ is trivial, by the absorption property of ideals. But the group (sub $_RF)^*$ of invertible elements in the monoid sub $_RF$ contains information about ideal factorization.

(7.9) Lemma. *If $I \in$ sub $_RF$ has an inverse in sub $_RF$, that inverse is*

$$(R:I) = \{x \in F : xI \subseteq R\}\,.$$

Proof. Whether I has an inverse or not, $(R : I)$ is an R-submodule of F. If I has inverse J, then $J \subseteq (R : I)$. So $R = JI \subseteq (R : I)I \subseteq R$, proving $JI = (R : I)I$. Multiply by J to cancel I. ∎

The notation $(R : I)$ is meant to imitate $R \div I$ or $1/I$ in sub $_RF$. By the lemma, $I \in$ sub $_RF$ is invertible if and only if the ideal $(R : I)I$ is all of R, that is, if and only if $x_1 y_1 + \ldots + x_n y_n = 1$ for some $x_i \in (R : I)$, $y_i \in I$, and $n > 0$.

If $I, J \in$ sub $_RF$, we say J **divides** I, and write $J|I$, to mean that $I = JK$ for some ideal K of R. So "divides" means divides with quotient in sub $_RR$. For ideals of R, "divides" has the expected meaning in the monoid sub $_RR$, but **invertible** means invertible in sub $_RF$.

The R-linear maps between members of sub $_RF$ are just the restrictions of F-linear maps from F to F:

(7.10) Lemma. *Suppose I and J are R-submodules of F. Every R-linear map $f : I \to J$ is multiplication by some $x \in F$.*

Proof. If $I = 0$, any x will do. If $I \neq 0$, the element $x = f(y)/y \in F$ is independent of the choice of nonzero $y \in I$: For if $a, b, s \in R - \{0\}$, then

$$\frac{a}{s}f(\frac{b}{s}) = \frac{1}{s}f(\frac{ab}{s}) = \frac{b}{s}f(\frac{a}{s}) .$$

Then $f(y) = xy$ for all $y \in I$. ∎

Now consider a curious connection between the additive notion of projective module and the multiplication in sub $_RF$:

(7.11) Proposition. *If I is a nonzero R-submodule of F, the following are equivalent:*

   (i) *I is projective,*
   (ii) *I is f.g. projective,*
   (iii) *I is invertible,*
   (iv) *I divides all its submodules.*

Proof. Suppose I is projective. By (2.20), I has a projective basis $(\ )^* : S \to \mathrm{Hom}_R(I, R)$, meaning $S \subseteq I$, and for each $x \in I$, $s^*(x) = 0$ for all but finitely many $s \in S$, and $x = \sum s^*(x)s$. By Lemma (7.10), each s^* is multiplication by some $a_s \in (R : I)$. Since I has a nonzero element x, $a_s = 0$ for all but finitely many $s \in S$. Say T is the set of $s \in S$ with $a_s \neq 0$. Each $x \in I$ has the form

$$x = \sum_{s \in T} a_s x s \qquad (a_s x \in R) ;$$

so I is a f.g. projective R-module, and (i) implies (ii). Taking $x \neq 0$ and multiplying by x^{-1},

$$1 = \sum_{s \in T} a_s s \in (R : I)I ,$$

proving I is invertible. So (ii) implies (iii). Now assume I is invertible, and choose $x_i \in (R : I)$ and $y_i \in I$ with $x_1 y_1 + \ldots + x_n y_n = 1$. Define F-linear maps:

$$F \xrightarrow{\cdot [x_1 \ldots x_n]} F^n \xrightarrow{\begin{bmatrix} y_1 \\ \vdots \\ y_n \end{bmatrix}} F .$$

These restrict to R-linear maps $I \to R^n \to I$, with composite i_I; so I is projective, and (iii) implies (i).

For (iii) implies (iv), assume $R = IK$, where $K \in \text{sub } _RF$. If J is a submodule of I, multiply both sides by J to get $J = I(JK)$, where $JK \subseteq IK = R$; so I divides J.

For the converse, choose a nonzero $x \in I$. Then xR is a submodule of I. If (iv) holds, then $xR = IJ$ for some nonzero ideal J of R. Then $x^{-1}J$ is an inverse to I in sub $_RF$. ∎

To prove principal ideal domains are factorial, an intermediate step is to show that, if p, a, b are elements and p is irreducible, then $p|ab$ implies $p|a$ or $p|b$. The corresponding fact for ideals is true in every commutative ring: Recall that an ideal P of R is **prime** if $P \neq R$ and its complement $R - P$ is closed under multiplication.

(7.12) Lemma. *If $P, I_1, \ldots, I_n$ are ideals of a commutative ring and P is prime, then $P \supseteq I_1 \cdots I_n$ implies $P \supseteq I_i$ for some i.*

Proof. If not, choose $x_i \in I_i - P$ for each i. Then the product $x_1 \cdots x_n$ belongs to $I_1 \cdots I_n$, but not to P, which is a contradiction. ∎

(7.13) Proposition. *If a nonzero ideal I of an integral domain R is projective, it has at most one expression (neglecting the order of factors) as a product of prime ideals of R.*

Proof. Suppose $I = P_1 \cdots P_m = Q_1 \cdots Q_n$ for prime ideals P_i and Q_i of R. Renumbering if necessary, we may assume P_1 is minimal among the P_i. By Lemma (7.12), P_1 contains some Q_j; renumbering the Q_i if necessary, $P_1 \supseteq Q_1$. Also, $Q_1 \supseteq P_i$ for some i. Since P_1 is minimal, $P_i = P_1$; so $P_1 = Q_1$.

Since I is invertible, so is P_1. Multiplying the expressions for I by $(R : P_1)$ leaves $P_2 \cdots P_m = Q_2 \cdots Q_n$. The argument can be repeated, and $n = m$ because the ideals P_i, Q_i and their products are proper subsets of R. ∎

(7.14) Definition. A **Dedekind domain** is a commutative integral domain R satisfying any of the equivalent conditions in the next theorem:

(7.15) Theorem. *If R is a commutative integral domain, the following are equivalent:*

   (i)   *Every ideal of R is projective.*
  (ii) *If I and J are ideals of R, then $I \supseteq J$ if and only if $I|J$.*
 (iii) *Every nonzero ideal of R has unique factorization as a product of maximal ideals.*
 (iv) *The ring R is noetherian, integrally closed, and every nonzero prime ideal of R is maximal.*

Proof. The equivalence of (i) and (ii) is immediate from (7.11). Assume (ii), and suppose I is a nonzero ideal of R. If $I = R$, then I has only the empty factorization – no maximal ideal factors. Suppose $I \neq R$. Then $I \subseteq P_1$ for a maximal ideal P_1 of R. By (ii), $I = P_1 I_1$ for some ideal I_1 of R. If $I_1 \neq R$, we similarly get $I_1 = P_2 I_2$ with P_2 maximal. Continuing as long as $I_i \neq R$, we produce an ascending chain of ideals

$$I = I_0 \subseteq I_1 \subseteq \ldots \quad \text{(where } I_{i-1} = P_i I_i \subseteq I_i \text{)}.$$

Since I is invertible, so is each factor I_i. If some $I_{i-1} = I_i$, multiplying by $(R : I_i)$ would leave $P_i = R$, which is false for a maximal ideal P_i. So the chain is strictly ascending:

$$I_0 \subsetneq I_1 \subsetneq I_2 \subsetneq \ldots \ .$$

The union $\cup_i I_i$ is an ideal of R, so it is finitely generated by (7.11). Each generator occurs in some I_i, so the ascending chain terminates after finitely many steps. Then some $I_n = R$, and I has the factorization $I = P_1 \cdots P_n$, which is unique by (7.13). This proves (iii).

Next we show (iii) implies (i). Assume (iii). Suppose P is a nonzero maximal ideal of R, and choose a nonzero element $x \in P$. By (iii), $xR = P_1 \cdots P_n$ for some maximal ideals P_i. Since xR is invertible (with inverse $(1/x)R$), each P_i is invertible too. By Lemma (7.12), since P is prime and contains xR, $P \supseteq P_i$ for some i; since P_i is maximal, $P = P_i$. So P is invertible. A product of invertible elements of a monoid is invertible; so (iii) now implies every nonzero ideal of R is invertible, proving (i).

Now assume the equivalent conditions (i) and (iii). Suppose $a \in F$ and a is integral over R. Then $R[a]$ is a f.g. R-module. If d is the product of denominators in a set of generators, then $I = dR[a]$ is a nonzero ideal of R with $aI \subseteq I$. By (i), I is invertible; so $aR \subseteq R$ and $a \in R$, proving R is integrally closed. By (7.11), every ideal of R is finitely generated; so R is noetherian. And if P is a nonzero prime ideal of R, Lemma (7.12) implies P contains one of the maximal ideals in its factorization under hypothesis (iii); so P is maximal. Thus (i) and (iii) imply (iv).

To complete the proof, we show (iv) implies (i). The hypothesis that a commutative ring is noetherian implies the **maximum condition**: "Every nonempty set of ideals has a maximal element." (See (B.3) in the Appendix.) We use this twice. The first use yields a lemma of independent interest:

(7.16) Lemma. *In a noetherian commutative ring R, every nonzero ideal contains a product (with ≥ 1 factors) of nonzero prime ideals.*

Proof. If not, the set of nonzero ideals that contain no such product has a maximal element I. Then I is not prime; so $xy \in I$ for some $x, y \in R - I$. But $I + xR$ and $I + yR$ properly contain I, so each contains a product of nonzero prime ideals. Then so does their product

$$(I + xR)(I + yR) \subseteq I + xyR \subseteq I \ ,$$

a contradiction. ∎

Returning to the proof of the theorem, suppose (iv) holds but (i) fails. Then the set of nonzero ideals of R that are not invertible has a maximal element I. Note that I is not principal, since nonzero principal ideals are invertible: $rRr^{-1}R = R$. Choose a nonzero $x \in I$ and a maximal ideal P with $I \subseteq P$. Inside xR, choose a minimal length product $P_1 \cdots P_r$ of nonzero prime (hence maximal) ideals P_i. Here $r > 1$, since we cannot have $P_1 \subseteq xR \subsetneqq I \subseteq P$. By (7.12), $P \supseteq P_i$ for some i; so $P = P_i$. By minimality of r,

$$(P_1 \cdots P_{i-1}P_{i+1} \cdots P_r) \; - \; xR$$

is nonempty; so it has an element a. Then

$$aI \;\subseteq\; aP \;\subseteq\; P_1 \cdots P_r \;\subseteq\; xR \; ;$$

so $(a/x)I$ is an ideal of R.

If F is the field of fractions of R, then $a/x \in F - R$, since $a \notin xR$. Since R is integrally closed, a/x is not integral over R. So $(a/x)I \not\subseteq I$, for otherwise I would be a faithful $R[a/x]$-module that is finitely generated as an R-module. Therefore

$$J \;=\; (a/x)I + I \;=\; ((a/x)R + R)I$$

is a nonzero ideal of R strictly containing I. By the choice of I, J is invertible. But then so is its factor I in sub $_RF$, which is a contradiction. Thus (iv) implies (i). ∎

(7.17) Example. Every commutative principal ideal domain R is Dedekind, since each ideal xR is a free R-module with basis $\{x\}$ if $x \neq 0$, or $\emptyset$ if $x = 0$.

The theorem of Dedekind, that $\mathcal{O}_F$ has unique ideal factorization for every number field F, is a special case of Theorem (7.19) below.

(7.18) Lemma. *Suppose A is an integral domain with field of fractions F and F is a subring of the center of a ring E. If $b \in E$ is algebraic over F, then ab is integral over A for some nonzero $a \in A$.*

Proof. Suppose $p(b) = 0$ for some $p(x) \in F[x] - \{0\}$. Clear denominators to get

$$a_n b^n + a_{n-1}b^{n-1} + \ldots + a_0 \;=\; 0$$

with all $a_i \in A$ and $a_n \neq 0$. Multiply by a_n^{n-1} to show $a_n b$ is integral over A. ∎

(7.19) Theorem. *If A is a Dedekind domain with field of fractions F and $F \subseteq E$ is a finite-degree field extension, and if B is the integral closure of A in E, then B is Dedekind, with field of fractions E. If $F \subseteq E$ is also separable, then B is finitely generated as an A-module.*

Proof. The proof given here follows closely the proof in Jacobson [89, §10.3]. Since $F \subseteq E$ has finite degree, each $e \in E$ is algebraic over F. By (7.18), ae is integral over A for some nonzero $a \in A$. So $ae \in B$ and $e = ae/a$, proving E is a field of fractions for B.

The argument that B is Dedekind resolves into two cases, proved by different methods. Suppose K is the set of all elements of E that are separable over F. Then K is an intermediate field, and $K \subseteq E$ is purely inseparable — meaning no member of $E - K$ is separable over K. Let C denote the integral closure of A in K. An element of E is integral over C if and only if it is integral over A (as in the proof of (7.7)). So B is the integral closure of C in E. If the theorem is proved for the separable and purely inseparable cases, then A Dedekind implies C is Dedekind, which implies B is Dedekind.

Case 1. Suppose $F \subseteq E$ is separable of degree n. So there are exactly n different field embeddings $\sigma_1, \ldots, \sigma_n$, fixing F, from E into a Galois extension L of F containing E. If $\sigma \in \mathrm{Aut}(L/F)$, then $\sigma\sigma_1, \ldots, \sigma\sigma_n$ is a permutation of $\sigma_1, \ldots, \sigma_n$; so for each $e \in E$,

$$Tr_{E/F}(e) = \sigma_1(e) + \ldots + \sigma_n(e)$$

lies in the fixed field F. So the **trace** $Tr_{E/F} : E \to F$ is an F-linear map. By linear independence of characters, $\sigma_1 + \ldots + \sigma_n \neq 0$; so the trace is surjective.

If $b \in B$, $p(b) = 0$ for some monic $p(x) \in A[x]$. Each σ_i fixes the coefficients of $p(x)$; so each $\sigma_i(b)$ is a root of $p(x)$, so is integral over A. Then $Tr_{E/F}(b)$ is an element of F integral over A, and so it lies in A. That is, the trace restricts to an A-linear map $Tr_{E/F} : B \to A$. By Lemma (7.18), E has an F-basis consisting of elements $b_1, \ldots, b_n$ from B. Define an A-linear map $\theta : B \to A^n$ by

$$\theta(x) = (Tr_{E/F}(b_1 x), \ldots, Tr_{E/F}(b_n x)) .$$

If $\theta(x) = (0, \ldots, 0)$, then $Tr_{E/F}(yx) = 0$ for all $y \in E$. Since the trace is not the zero map, that forces $Ex \neq E$; so $x = 0$. Therefore θ is injective, and $B \cong \theta(B) \subseteq A^n$ as A-modules. Since A is noetherian, so is A^n, and $B \cong \theta(B)$ is a f.g. R-module, *proving the final assertion of the theorem.* Consequently, B is noetherian as an A-module, and hence as a B-module. So B is a noetherian ring.

Since B is the integral closure of A in E, B is integrally closed in E by (7.7).

To show the nonzero prime ideals of B are maximal, we prove a more general fact:

(7.20) Lemma. *Suppose $R \subseteq D$ is an integral extension of commutative integral domains, and J is an ideal of D, so that $I = J \cap R$ is an ideal of R.*

   (i)   *If J is prime, I is prime.*
   (ii) *If $J \neq 0$, then $I \neq 0$.*
  (iii) *If I is maximal and J is prime, then J is maximal.*

Proof. If J is prime, then $R - I = R \cap (D - J)$ is closed under multiplication and includes 1; so I is prime.

Suppose d is a nonzero element of J. Let $p(x)$ denote a monic polynomial in $R[x]$ of least degree with $p(d) = 0$. Then $p(x) = xq(x) + p(0)$, where $p(0) \in R$ and $q(x)$ is a monic polynomial in $R[x]$ of lower degree than $p(x)$. So $q(d) \neq 0$. Then $p(0) = -d\, q(d)$ is a nonzero member of I.

Now choose $c \in D - J$. There is a monic polynomial $a(x) \in R[x]$ with $a(c) = 0 \in J$. Let $f(x)$ denote a monic polynomial in $R[x]$ of least degree with $f(c) \in J$. Then $f(x) = xg(x) + f(0)$, where $f(0) \in R$ and $g(x)$ is a monic polynomial in $R[x]$ of smaller degree than $f(x)$. So $g(c) \notin J$. Assuming J is prime, $cg(c) \notin J$. Since $f(c) \in J$, $f(0) = f(c) - cg(c) \in R - J$. If I is maximal in R,

$$1 \in f(0)R + I \subseteq cD + J .$$

So J is maximal in D.   ■

Case 2. Suppose $F \subseteq E$ is purely inseparable. It will suffice to prove every nonzero ideal of B is invertible. If $E = F$, there is nothing to prove. Suppose $e \in E - F$. The minimal polynomial $m(x)$ of e over F has a multiple root, so it has a common factor in $F[x]$ with its formal derivative $m'(x)$. Since $m(x)$ is irreducible in $F[x]$, this forces $m'(x) = 0$; so F has nonzero characteristic p, and $m(x) = n(x^p)$ for some nonzero polynomial $n(x) \in F[x]$ of smaller degree than $m(x)$. So $[F(e^p) : F] < [F(e) : F]$. Inductively, this shows $e^{q(e)} \in F$ for some $q(e) = p^r, r \geq 0$. If $e_1, \ldots, e_n$ is an F-basis of E, let q denote

$$\max \{q(e_i) : 1 \leq i \leq n\} .$$

Since E has characteristic p, the qth power map $x \mapsto x^q$ is a ring homomorphism on E, so it takes E into F.

If $\overline{E}$ is an algebraic closure of E, the qth power map $\theta : \overline{E} \to \overline{E}$ is a ring homomorphism, since $\overline{E}$ has characteristic p, is injective, since $\overline{E}$ is a field, and is surjective, since $\overline{E}$ is algebraically closed. Denote by $F^{1/q}$ the subfield $\theta^{-1}(F)$ of $\overline{E}$, and by $A^{1/q}$ the subring $\theta^{-1}(A)$ of $F^{1/q}$. From the preceding paragraph, $E \subseteq F^{1/q}$. The map θ restricts to a ring isomorphism $A^{1/q} \cong A$; so $A^{1/q}$ is Dedekind. Since A is integrally closed, $\theta(B) \subseteq A$; so $B \subseteq A^{1/q}$.

Suppose I is a nonzero ideal of B. Then $I' = IA^{1/q}$ is a nonzero ideal of $A^{1/q}$. Let J' denote its inverse, $(A^{1/q} : I')$, and let J denote $J' \cap E$. Since $B \subseteq A^{1/q}$, J is a B-submodule of E. It only remains to show $IJ = B$.

On the one hand,

$$IJ \subseteq I'J' \cap E = A^{1/q} \cap E \,,$$

and this intersection is contained in B because $A^{1/q}$ is integral over A. On the other hand,

$$A^{1/q} = I'J' = IA^{1/q}J' = IJ' \,.$$

So $1 = x_1 y_1' + \ldots + x_t y_t'$ for some $x_i \in I, y_i' \in J'$, and $t > 0$. Apply the qth power map θ to get $1 = x_1 y_1 + \ldots + x_t y_t$, where

$$y_i = x_i^{q-1}(y_i')^q \in (I')^{q-1}(J')^q = J' \,.$$

Also, $y_i \in I^{q-1}(J')^q \subseteq BF \subseteq E$. So $y_i \in J' \cap E = J$. This shows $1 \in IJ$ and $IJ = B$. ∎

(7.21) Corollary. *The ring of algebraic integers in a number field is Dedekind, and the number field is its field of fractions.* ∎

Further examples of Dedekind domains arise from localization:

(7.22) Proposition. *Suppose R is an integral domain with field of fractions F, and S is a submonoid of $(R - \{0\}, \cdot)$. Then $S^{-1}R$ can be identified with the subring*

$$\left\{ \frac{r}{s} \in F \,:\, r \in R,\ s \in S \right\}$$

of F containing R. If R is Dedekind, then $S^{-1}R$ is Dedekind.

Proof. Since $S \subseteq F^*$, there is by (6.41) a ring homomorphism $S^{-1}R \to F$, $r/s \mapsto r/s$. If $r/s = 0/1$ in F, then $r = 0$; so $r/s = 0/1$ in $S^{-1}R$. So $S^{-1}R$ is isomorphic to its image in F, and $S^{-1}R$ is an integral domain. By (6.46), every ideal of $S^{-1}R$ is $S^{-1}I$ for some ideal I of R. Since the functor $S^{-1}(-)$ carries $\mathcal{P}(R)$ into $\mathcal{P}(S^{-1}R)$, it follows that $S^{-1}R$ is Dedekind if R is Dedekind. ∎

For a partial converse, we have:

(7.23) Theorem. *A noetherian commutative integral domain R is Dedekind if R_P is Dedekind for each maximal ideal P of R.*

Proof. Suppose R is noetherian and R_P is Dedekind for each maximal ideal P of R. Suppose I is a nonzero ideal of R. It will suffice to prove I is invertible, and hence projective.

The field of fractions F of R is also a field of fractions for R_P. For each $M \in \text{sub }_R F$, identify M_P with its image under $F_P \cong F$; so for $m \in M$ and $s \in R - P$, $m/s = ms^{-1}$.

Since R is noetherian, I is generated as an ideal by a finite set $x_1, \ldots, x_n$, which also generates I_P as an ideal of R_P.

Claim. $(R : I)_P = (R_P : I_P)$.

For, if $c \in (R : I)_P$, then for some $s \in R - P$, $scx_i \in R$ for each i; so each $cx_i \in R_P$, and $c \in (R_P : I_P)$. Conversely, if $cx_i \in R_P$ for each i, there exist $s_i \in R - P$ with $s_i cx_i \in R$ for each i. If $s = \Pi s_i$, then $scx_i \in R$ for each i; so $sc \in (R : I)$ and $c \in (R : I)_P$, proving the claim.

Now, since $0 \neq I \subseteq I_P$ and R_P is Dedekind, the ideal I_P is invertible:

$$1 \in I_P(R_P : I_P) = I_P(R : I)_P .$$

So

$$1 = \sum_{i=1}^{m} \frac{x_i}{s} \frac{y_i}{t} ,$$

where $y_i \in (R : I), x_i \in I$, and $s, t \in R - P$. Then $\sum x_i y_i = st$ does not belong to P, so $I(R : I) \not\subseteq P$. This is true for every maximal ideal P of R; so the ideal $I(R : I)$ equals R, proving I is invertible. ∎

Local Dedekind domains have a pleasantly simple description. A commutative ring is **semilocal** if it has only finitely many maximal ideals.

(7.24) Proposition. *Every semilocal Dedekind domain is a principal ideal domain.*

Proof. Suppose the Dedekind domain R has only n different maximal ideals $P_1, \ldots, P_n$. Since $P_1^2 \neq P_1$, there exists $a \in P_1 - P_1^2$. Since no proper ideal contains (=divides) any two of $P_1^2, P_2, \ldots, P_n$, the Chinese Remainder Theorem supplies $b \in R$ with $b \equiv a (\bmod P_1^2)$ and $b \equiv 1 (\bmod P_i)$ for $2 \leq i \leq n$. Then bR is divisible by P_1 but not by P_1^2, nor by P_i for $2 \leq i \leq n$. So $bR = P_1$, proving P_1 is principal. In the same way, each P_i is principal, as is every product of the P_i. ∎

(7.25) Definitions. A **discrete valuation** on a field F is a surjective function $v : F^* \to \mathbb{Z}$ satisfying

    (i) $v(xy) = v(x) + v(y)$, and

    (ii) $v(x + y) \geq \min\{v(x), v(y)\}$,

whenever $x, y \in F^*$. By property (i), $v(1) = 2v(1) = 2v(-1)$; so $v(-1) = v(1) = 0$ and $v(-x) = v(x)$ for all $x \in F^*$. So the set

$$\mathcal{O}_v = \{x \in F^* : v(x) \geq 0\} \cup \{0\}$$

is a subring of F. We call $\mathcal{O}_v$ the **discrete valuation ring** (or **DVR**) associated with v.

(7.26) Theorem. *Suppose R is not a field. Then R is a local Dedekind domain if and only if R is a discrete valuation ring.*

Proof. Suppose R is a local Dedekind domain with maximal ideal P. By (7.24), $P = R\pi$ for some $\pi \in P$. For each nonzero $r \in R$, there is a unique integer $v(r) \geq 0$ with

$$Rr = P^{v(r)} = R\pi^{v(r)}.$$

Then there is a unique unit $u \in R^*$ with

$$r = u\pi^{v(r)}.$$

So R is factorial, based on one prime π. The map $v : R - \{0\} \to \mathbb{Z}$, defined by this exponent, evidently satisfies properties (7.25) (i) and (ii) since $Rxy = RxRy$, and $R(x + y)$ is contained in, and hence divisible by, $Rx + Ry$. By (i), v is a semigroup homomorphism from $R - \{0\}$ into the abelian group $(\mathbb{Z}, +)$. Since $(F^*, \cdot)$ is the group completion of $(R - \{0\}, \cdot)$, v extends to a group homomorphism $v : F^* \to \mathbb{Z}$, with $v(r/s) = v(r) - v(s)$ for $r, s \in R - \{0\}$. This extension also satisfies (i) and (ii), and is surjective since $v(\pi^n) = n$. So v is a discrete valuation on F. By the factorizations of r and s in terms of π, $r/s \in R$ if and only if $v(r) - v(s) \geq 0$; so R is the discrete valuation ring $\mathcal{O}_v$.

For the converse, supose $R = \mathcal{O}_v$ for a discrete valuation $v : F^* \to \mathbb{Z}$. Since $xy = 1$ implies $v(x) + v(y) = 0$,

$$R^* = \{x \in F^* : v(x) = 0\}.$$

The complement

$$R - R^* = \{x \in F^* : v(x) > 0\} \cup \{0\}$$

is an ideal P of R, so it must be the only maximal ideal of R, proving R is local. Since $v(r) \geq v(s)$ if and only if $r/s \in R$, each ideal of R is either 0 or Rd where d is its nonzero element of least value $v(d)$. So R is a principal ideal domain, and hence it is Dedekind. ∎

Next consider the quotients of a Dedekind domain.

(7.27) Proposition. *Suppose R is (i) the ring of integers in a number field E, (ii) the polynomial ring $F[x]$ over a finite field F, or (iii) a localization of one of the rings (i) or (ii). For each nonzero ideal I of R, the residue ring R/I is finite.*

Proof. In case (i), R is a f.g. $\mathbb{Z}$-module by (7.19), and $I \cap \mathbb{Z}$ contains a nonzero integer by (7.20) (ii). So R/I is a f.g. torsion (hence finite) $\mathbb{Z}$-module.

In case (ii), $I = p(x)F[x]$ for a polynomial $p(x)$ of some degree n. By the division algorithm in $F[x]$, R/I has F-basis $\overline{1}, \overline{x}, \ldots, \overline{x}^{n-1}$, so it has q^n elements, where q is the size of F.

Suppose in either case (i) or (ii) that S is a submonoid of $(R, \cdot)$. By (6.48), each nonzero prime ideal of $S^{-1}R$ is $S^{-1}P$ for a nonzero prime ideal P of R not meeting S, and for each positive integer e, $S^{-1}R/(S^{-1}P)^e \cong R/P^e$. By the Chinese Remainder Theorem, $S^{-1}R/I$ is finite for nonzero ideals I. ∎

(7.28) Proposition. *If R is any Dedekind domain and I is a nonzero ideal of R, every ideal of R/I is principal.*

Proof. If $I = R$, R/I is the zero ring whose only ideal is principal. Suppose $I \neq R$. So

$$I = P_1^{e_1} P_2^{e_2} \cdots P_g^{e_g}$$

for distinct maximal ideals $P_1, \ldots, P_g$ and positive integers $e_1, \ldots, e_g$. By the Chinese Remainder Theorem,

$$R/I \cong (R/P_1^{e_1}) \times \cdots \times (R/P_g^{e_g})$$

as a ring. Taking $P = P_i$ and $e = e_i$, the local Dedekind domain R_P is a principal ideal domain. So every ideal of $R_P/P^e R_P$ is principal. By (6.48), $R/P^e \cong R_P/P^e R_P$. Finally, each ideal of $\Pi(R/P_i^{e_i})$ is principal, since it has the form ΠI_i, where I_i is an ideal of $R/P_i^{e_i}$, for each i. ∎

(7.29) Corollary. *If R is a Dedekind domain, each ideal I of R requires at most two generators.*

Proof. Suppose I has a nonzero element x. The ideal I/Rx of R/Rx is principal, generated by $y + Rx$ for some $y \in I$. For each $z \in I$,

$$z + Rx = (s + Rx)(y + Rx)$$

for some $s \in R$. So $z = rx + sy$ for some $r \in R$, proving $I = Rx + Ry$. ∎

(7.30) Corollary. *If I and J are nonzero ideals of a Dedekind domain R, there is an R-linear isomorphism: $R/I \cong J/IJ$.*

Proof. Since $IJ \neq 0$, the ideal J/IJ of the ring R/IJ is principal, generated by some coset $j + IJ$. So there is a surjective R-linear map

$$f : R/I \;\mapsto\; J/IJ \;,\quad r + I \;\mapsto\; rj + IJ\;.$$

Since $j \in J$, $Rj = KJ$ for some ideal K of R. Taking the union of cosets in J/IJ,

$$J \;=\; Rj + IJ \;=\; KJ + IJ \;=\; (K + I)J\;.$$

Since J is invertible, $R = K + I$. Now if $r + I$ is in the kernel of f, then $rj \in IJ$; so IJ divides $RrRj = RrKJ$. Canceling J again, I divides RrK. So I divides

$$RrI + RrK \;=\; Rr(I + K) \;=\; Rr\;.$$

Then $r \in I$ and $r + I = 0 + I$, proving f injective. ∎

7B. Exercises

1. Give an example of a commutative domain R with nonzero ideals I and J for which $R/I \not\cong J/IJ$. *Hint:* In $R = \mathbb{Z}[\sqrt{-3}]$, take $I = J = R(1 + \sqrt{-3}) + R(2)$. Show $II = 2I$, R/I has only two elements $\overline{0}$ and $\overline{1}$, and I is $\mathbb{Z}$-linearly isomorphic to $\mathbb{Z} \oplus \mathbb{Z}$, so that $I/2I$ has four elements.

2. Prove the converse to Proposition (7.13) is false. *Hint:* In Exercise 1, show $I = \langle 1 + \sqrt{-3}, 2 \rangle$ is a noninvertible maximal ideal that cannot be written (in any other way than I) as a product of prime ideals of $R = \mathbb{Z}[\sqrt{-3}]$.

3. (Jacobson [89]) Suppose R is a commutative domain and every proper ideal of R is a product of prime ideals. Prove R is Dedekind. *Hint:* First show every invertible prime ideal is maximal: Suppose P is an invertible prime ideal that is not maximal; so there exists $a \in R - P$ with $aR + P$ and $a^2R + P$ proper ideals containing P. Then $aR + P = P_1 \cdots P_m$ and $a^2R + P = Q_1 \cdots Q_n$ for prime ideals P_i, Q_i containing P. In R/P, show $\overline{a^2R + P} = (\overline{aR + P})^2$ is invertible and $\overline{P_i}, \overline{Q_i}$ are prime. Conclude that $\overline{Q_1}, \ldots, \overline{Q_n}$ is $\overline{P_1}, \ldots, \overline{P_m}$ listed twice (in some order). So $(aR + P)^2 = a^2R + P$. Now show $P \subseteq aP + P^2 \subseteq P$ and multiply by P^{-1} to get $aR + P = R$, a contradiction. Now that you know each invertible prime is maximal, prove each nonzero prime P is invertible. *Hint:* For nonzero $b \in P$ consider the factorization of $bR\ (\subseteq P)$ into primes.

4. Prove a ring R is a discrete valuation ring if and only if R is a commutative principal ideal domain with exactly one nonzero prime ideal P.

5. If v is a discrete valuation on a field F, prove that for each $x \in F^*$ either $x \in \mathcal{O}_v$ or $x^{-1} \in \mathcal{O}_v$.

6. If v is a discrete valuation on a field F and a is a real number with $0 < a < 1$, prove the function $|\ |_v$ from F to $\mathbb{R}$, defined by $|0|_v = 0$ and $|x|_v = a^{v(x)}$ for $x \neq 0$, has the properties

(i) $|x|_v \geq 0$ (and $=0$ only for $x = 0$),

(ii) $|xy|_v = |x|_v\,|y|_v$,

(iii) $|x + y|_v \leq \max\{|x|_v, |y|_v\} \leq |x|_v + |y|_v$.

So $\rho(x, y) = |x - y|_v$ is a metric on F. Prove that different choices of a always yield the same metric topology on F.

7. Suppose R is a Dedekind domain with field of fractions F. For each maximal ideal P of R, and each $r \in R - \{0\}$, let $v_P(r)$ denote the exponent of P in the ideal factorization of rR. Show $v_P(rs) = v_P(r) + v_P(s)$ for $r, s \in R - \{0\}$. Since $(F^*, \cdot)$ is a group completion of the monoid $(R - \{0\}, \cdot)$, v_P extends to a group homomorphism $v_P : F^* \to \mathbb{Z}$, with $v_P(r/s) = v_P(r) - v_P(s)$. Prove v_P is a discrete valuation on F and $\mathcal{O}_v = R_P$. (This v_P is known as the **P-adic valuation** on F.)

8. Suppose R is a commutative principal ideal domain with field of fractions F. Prove the only discrete valuations v on F with $R \subseteq \mathcal{O}_v$ are the P-adic valuations, for maximal $P \triangleleft R$. *Hint:* The set

$$P_v = \{x \in F : v(x) > 0\} \cup \{0\}$$

is a maximal ideal of $\mathcal{O}_v$. Prove $P_v \cap R = P$ is a maximal ideal of R with $R_P = \mathcal{O}_v$, so that $v = v_P$.

9. Describe all the discrete valuations on $\mathbb{Q}$ and their discrete valuation rings.

7C. Ideal Class Groups

Unique ideal factorization is not as sharp a tool as unique factorization of elements. But the only factorial Dedekind domains are the principal ideal domains:

(7.31) Proposition. *A Dedekind domain is factorial if and only if its ideals are principal.*

Proof. It is a standard fact that principal ideal domains are factorial. (See Exercise 1.) Suppose R is a factorial Dedekind domain, with respect to a set T of irreducibles. Of course the ideals $0 = 0R$ and $R = 1R$ are principal. Suppose P is a nonzero maximal ideal of R, and $0 \neq x \in P$. Since P is prime and x is not a unit, some $p \in T$ in its factorization belongs to P. The factorization of a product $ab \ (\neq 0)$ is obtained from those of a and b by multiplying units and adding exponents; so $p|ab$ implies $p|a$ or $p|b$. So pR is a nonzero prime (hence maximal) ideal, contained in P, and $P = pR$ is principal. Then every product of maximal ideals is principal. ∎

(7.32) Example of a Nonfactorial Dedekind Domain. By §7A, Exercise 3, the ring of algebraic integers in $\mathbb{Q}(\sqrt{-5})$ is

$$\mathbb{Z}[\sqrt{-5}] = \{a + b\sqrt{-5} : a, b \in \mathbb{Z}\} ;$$

so $R = \mathbb{Z}[\sqrt{-5}]$ is Dedekind. In R,

(7.33) $$(1 + \sqrt{-5})(1 - \sqrt{-5}) = 6 = (2)(3) .$$

Let $\overline{x}$ denote the complex conjugate of x. Then

$$(a + b\sqrt{-5}) \overline{(a + b\sqrt{-5})} = a^2 + 5b^2 .$$

Suppose p is any of $1 + \sqrt{-5}$, $1 - \sqrt{-5}$, 2, or 3. If $p = xy$ for nonzero nonunits x and y, then

$$x\overline{x}y\overline{y} = xy\overline{xy} = p\overline{p} = 4, 6, \text{ or } 9 .$$

Since x and y are nonunits, $x\overline{x}$ and $y\overline{y}$ are integers greater than 1. So $x\overline{x} = 2$ or 3. But $a^2 + 5b^2$ cannot be 2 or 3 if $a, b \in \mathbb{Z}$. So p is irreducible.

If $x, y \in R$ and $xy = 1$, then $x\overline{x}y\overline{y} = 1 \cdot \overline{1} = 1$, forcing $x\overline{x} = 1$. But $a^2 + 5b^2 = 1$ for $a, b \in \mathbb{Z}$ only if $a = 1$ or -1 and $b = 0$. So $R^* = \{1, -1\}$. The four irreducibles p belong to different cosets of $\{1, -1\}$. If R were factorial, it would be factorial with respect to a set T including $1 + \sqrt{-5}$, $1 - \sqrt{-5}$, 2, and 3, and equation (7.33) would violate unique factorization.

Of course, unique ideal factorization holds in the Dedekind domain R. If (u, v) denotes the ideal $Ru + Rv$, the following are maximal ideals of R:

$$J_1 = (1 + \sqrt{-5}, 2) , \quad \overline{J}_1 = (1 - \sqrt{-5}, 2) ,$$

$$J_2 = (1 + \sqrt{-5}, 3) , \quad \overline{J}_2 = (1 - \sqrt{-5}, 3) .$$

For instance, J_1 is maximal because it is the kernel of the surjective ring homomorphism $f_1 : R \to \mathbb{Z}/2\mathbb{Z}$ defined by

$$f_1(a + b\sqrt{-5}) = (a - b) + 2\mathbb{Z} .$$

If we replace the factors in (7.33) by the principal ideals they generate, both sides factor further — the left as $J_1 J_2 \overline{J_1} \overline{J_2}$ and the right as $J_1 \overline{J_1} J_2 \overline{J_2}$. Since the factors in (7.33) are irreducible, the ideals $J_1, J_2, \overline{J_1}, \overline{J_2}$ are not principal.

More directly, if J_1 were principal, generated by d, then $d \notin R^*$; so $d\overline{d} > 1$. And $d\overline{d}$ is an integer factor of

$$(1 + \sqrt{-5})\,\overline{(1 + \sqrt{-5})} \;=\; 6 \quad \text{and} \quad 2 \cdot \overline{2} \;=\; 4 \,.$$

So $d\overline{d} = 2$, an impossibility for $d \in R$. Similar direct arguments work for $J_2, \overline{J_1}$, and $\overline{J_2}$.

The **ideal class group** $\mathbf{Cl}(R)$ of a Dedekind domain R is an abelian group designed by Dedekind [93] in the late 1800s to measure the deviation of R from being a principal ideal domain. In its original form, the elements of $Cl(R)$ are the equivalence classes of nonzero ideals of R under the equivalence relation " $\sim$" defined by saying $I \sim J$ if and only if $xI = yJ$ for some nonzero $x, y \in R$. Let (I) denote the equivalence class of I. If $xI = yJ$ and $x'I' = y'J'$, then $xx'II' = yy'JJ'$; so multiplication of ideals defines an operation $(I)(J) = (IJ)$ on $Cl(R)$, which is associative, commutative, and has identity (R). Note that (R) consists of the nonzero principal ideals of R.

This much works for any integral domain R. But since R is Dedekind, every nonzero ideal contains, *and hence divides*, a nonzero principal ideal. So every element of $Cl(R)$ has an inverse, and $Cl(R)$ is an abelian group. Of course $Cl(R)$ is trivial if and only if every ideal of R is principal.

If F is a number field (that is, a finite-degree extension of $\mathbb{Q}$) with ring of algebraic integers $\mathcal{O}_F$, the **class number** of F is the order of the group $Cl(\mathcal{O}_F)$. In §7D, Exercise 13, we see that the class number of a number field is always finite. Along with (7.28) and (7.29), this shows that the rings $\mathcal{O}_F$ are very nearly principal ideal domains.

By Lemma (7.10), ideals I and J of a Dedekind domain R are R-linearly isomorphic if and only if $J = (x/y)I$ for some nonzero $x, y \in R$. So *Dedekind's equivalence classes are just the isomorphism classes* in sub $_R R$. If F is the field of fractions of R, we gain a new perspective on $Cl(R)$ by considering isomorphism classes in sub $_R F$. A **fractional ideal** of R is a set $(x/y)I \subseteq F$, where I is a nonzero ideal of R and x, y are nonzero elements of R. The fractional ideals are just the R-submodules of F that are R-linearly isomorphic to nonzero ideals of R. Note that $J \in$ sub $_R F$ is a fractional ideal of R if and only if its elements have a common denominator — that is, if and only if there is a nonzero $d \in R$ with $dJ \subseteq R$: For, in that case, dJ is an ideal of R and $J = (1/d)dJ$. By contrast, an ordinary ideal $I \in$ sub $_R R$ is sometimes called an **integral ideal** of R.

An R-submodule of F that is isomorphic to R itself is called a **principal fractional ideal** of R, since it equals $(x/y)R$ for some nonzero $x, y \in R$.

(7.34) Proposition. *Suppose R is a Dedekind domain with field of fractions F, and $I \in$ sub $_RF$. The following are equivalent:*

- (i) *I is projective,*
- (ii) *I is invertible,*
- (iii) *I is finitely generated,*
- (iv) *I is a fractional ideal of R.*

Also, the following are equivalent:

- (v) *I is a free R-module,*
- (vi) *I is a principal fractional ideal of R.*

Proof. The equivalence of (i) and (ii) is part of (7.11), as is the assertion that (i) implies (iii). If I is finitely generated, a common denominator for its generators is a common denominator for I; so (iii) implies (iv). Each fractional ideal is isomorphic to an integral ideal, which is projective; so (iv) implies (i).

Each principal fractional ideal is isomorphic to R, so it is a free R-module. Since R-linearly independent elements of F are F-linearly independent, any nonzero free R-submodule of F has free rank 1, and so it is a principal fractional ideal $(x/y)R$. ∎

From this proposition it follows that the set $\mathbf{I}(R)$ of fractional ideals of a Dedekind domain R is the multiplicative abelian group (sub $_RF)^*$. The set $\mathbf{PI}(R)$ of principal fractional ideals of R is a subgroup of $I(R)$. Two elements I, J of sub $_RF$ are isomorphic if and only if $J = (x/y)I = (x/y)RI$ for nonzero $x, y \in R$. So the quotient group $I(R)/PI(R)$ consists of the R-linear isomorphism classes in sub $_RF$ that contain integral ideals of R. The map

$$Cl(R) \ \mapsto \ I(R)/PI(R) \, ,$$

enlarging each isomorphism class in sub $_RR$ to an isomorphism class in sub $_RF$, is evidently a group isomorphism. So we can regard $Cl(R)$ as the quotient "projectives modulo frees" in sub $_RF$, and there is an exact sequence of abelian groups:

$$1 \longrightarrow R^* \overset{\subseteq}{\longrightarrow} F^* \longrightarrow I(R) \longrightarrow Cl(R) \longrightarrow 0 \, .$$

$$\frac{x}{y} \longmapsto \frac{x}{y}R$$

(7.35) Corollary. *If R is a Dedekind domain, $I(R) = (sub\ _RF)^*$ is the free abelian group based on the nonzero maximal ideals of R.*

Proof. If I is a fractional ideal of R, it has a common denominator d; so $dRI = dI \in sub\ _RR$, and $I = (dR)^{-1}dI$. From ideal factorizations of dR and dI, I is in the (multiplicative) $\mathbb{Z}$-linear span of the nonzero maximal ideals of R. If

$$P_1^{e_1} \cdots P_n^{e_n} = R$$

for distinct nonzero maximal ideals $P_1, \ldots, P_n$ and arbitrary integers e_i, then

$$\prod_{e_i > 0} P_i^{e_i} = \prod_{e_j < 0} P_j^{-e_j}\,,$$

and no P_i appears on both sides. By unique ideal factorization, $e_i = \cdots = e_n = 0$; so the nonzero maximal ideals of R are $\mathbb{Z}$-linearly independent. ∎

7C. Exercises

1. Suppose R is a commutative principal ideal domain. Prove R is factorial. *Hint:* For the set T, choose one generator of each maximal ideal of R. Now use unique factorization of ideals in the Dedekind domain R.

2. Suppose R is any commutative domain, with field of fractions F. Verify that Dedekind's equivalence classes on sub $_RR$ form an abelian monoid $M(R)$ under ideal multiplication, with identity (R) consisting of the nonzero principal ideals of R. Define $\mathbf{Cl}(R)$ to be the group $M(R)^*$ of invertible elements of $M(R)$. Suppose I is a nonzero ideal of R.

   (i) Prove $(I) \in Cl(R)$ if and only if I is a f.g. projective R-module. *Hint:* Use (7.11) to show I is invertible if and only if it divides a nonzero principal ideal.
   (ii) Prove $(I) = (R)$ if and only if I is a free R-module.
   (iii) Prove $(I) = (J)$ in $M(R)$ if and only if $I \cong J$ as R-modules.

3. Suppose R, F are as in Exercise 2. If R is not Dedekind, it must have an ideal that is not invertible. We can produce such an ideal under various conditions:

   (i) If R is not noetherian, it has a nonfinitely generated ideal J that is not invertible by (7.11).
   (ii) If a nonzero prime ideal P of R is not maximal, there is an ideal I with $P \subsetneq I \neq R$. Prove either I or P fails to be invertible. *Hint:* If I is invertible, then $I|P$ by (7.11). Say $P = IJ$, where $J \lhd R$. Then $J \subseteq P$ (since P is prime). Conclude $P = IP$, but $R \neq I$.

(iii) If R is not integrally closed, there exist nonzero $a, b \in R$ with $a/b \notin R$ but

$$\left(\frac{a}{b}\right)^n \in I = R\left(\frac{a}{b}\right)^{n-1} + \cdots + R\left(\frac{a}{b}\right) + R \,.$$

Show I and the integral ideal $b^n I$ are not invertible. *Hint:* Show $I^2 = I$ but $I \neq R$.

4. In $F = \mathbb{Q}(\sqrt{-11})$ the ring of algebraic integers is $\mathbb{O}_F = \mathbb{Z}[(1 + \sqrt{-11})/2]$ by §7A, Exercise 3. Consider the subring $R = \mathbb{Z}[\sqrt{-11}]$.

(i) Prove the ideal $I = R(1 + \sqrt{-11}) + R(2)$ is not invertible – so it is a f.g. torsionfree R-module that is not projective. *Hint:* Use Exercise 3 (iii).

(ii) Prove the ideal $J = R(1+\sqrt{-11}) + R(3)$ is invertible but not principal – so it is a f.g. projective R-module that is not free. *Hint:* In $\mathbb{O}_F$, $(1 + \sqrt{-11})/2$ is a greatest common divisor of $1 + \sqrt{-11}$ and 3, while $\mathbb{O}_F^* = \{1, -1\}$. Also, $(1 - \sqrt{-11})/3 \in J^{-1}$, and $1 \in J^{-1}$.

(iii) Calculate the order of (J) in $Cl(R)$.

5. If R is a Dedekind domain, prove $I(R) = (\text{sub }_R F)^*$ is isomorphic to the group completion of the monoid sub $_R R$. *Hint:* Use (7.35) and (3.19).

Is $(\text{sub }_R F)^*$ always a group completion of sub $_R R$ when R is a commutative domain with field of fractions F?

6. Assume F is a number field, and so has finite class number h by §7D, Exercise 13. Prove there is a field extension $F \subseteq L$ with $[L : F] \leq h$ for which each ideal J of $\mathbb{O}_F$ generates a principal ideal $J\mathbb{O}_L = \alpha\mathbb{O}_L$ of $\mathbb{O}_L$. *Hint:* Decompose $Cl(\mathbb{O}_F)$ as a product of cyclic groups $C_1 \times \cdots \times C_n$ and choose a generator (J_i) of C_i for each i, where $J_i \lhd \mathbb{O}_F$. If n_i is the order of C_i, there exists $a_i \in \mathbb{O}_F$ with $J_i^{n_i} = a_i\mathbb{O}_F$. Take L to be F with an n_i-root of a_i adjoined for each i.

7. Under the hypothesis in Exercise 6, suppose n is the least positive integer for which J^n is principal for every ideal J of $\mathbb{O}_F$. Prove there is an ideal I of $\mathbb{O}_F$ for which $I, I^2, \ldots, I^{n-1}$ are not principal.

8. Suppose R is a Dedekind domain and I, J are nonzero ideals of R. Prove there is an ideal K of R for which $J + K = R$ and IK is principal. *Hint:* Write I and J as products of the same maximal ideals $P_1, \ldots, P_n$, so

$$I = P_1^{e_1} \cdots P_n^{e_n} \,.$$

Use the Chinese Remainder Theorem to produce $\alpha \in R$ with

$$\alpha R = P_1^{e_1} \cdots P_n^{e_n} Q \,,$$

where Q is not divisible by any P_i. Show $\alpha R + IJ = I$. Since $\alpha R \subseteq I$, $\alpha R = IK$ for some $K \lhd R$. Now show $K + J = R$.

9. Suppose R is a Dedekind domain and $x, y \in Cl(R)$. Prove there are ideals J, K of R with $J \in x$, $K \in y$, and $J + K = R$. *Hint:* Use Exercise 8.

7D. Extensions and Norms

As indicated by the examples in §7A, number-theoretic problems in $\mathbb{Z}$ can sometimes be solved by working in a larger ring B of algebraic integers. In this setting, it is helpful to know how primes in $\mathbb{Z}$ factor in B, or how pB factors into maximal ideals in B, for each prime $p \in \mathbb{Z}$. In this section we adopt the general version of this setting, used in Theorem (7.19): *Assume A is a Dedekind domain with field of fractions F, assume $F \subseteq E$ is a finite-degree field extension, and take B to be the integral closure of A in E.* By (7.19), B is Dedekind, with field of fractions E. *Assume also that A is not a field, so $A \neq F$.*

Since A is integrally closed, there are elements of F not integral over A; so $B \neq E$ and B is not a field either. In this setting, we shall explore the connections between ideals of A and ideals of B.

If J is an ideal of B, then $J' = J \cap A$ is an ideal of A, and we say J **lies above** J'. By Lemma (7.20), each maximal ideal of B lies above a maximal ideal of A. If I is an ideal of A, the ideal of B generated by I is IB, the sum of products ib with $i \in I$ and $b \in B$. We say IB is **extended from** I.

If p is a maximal ideal of A, then $p \neq 0$ since A is not a field. So pB is a nonzero ideal of B.

(7.36) Lemma. *If p is a maximal ideal of A, then pB is a proper ideal of B.*

Proof. By (7.22), if we take $S = A - p$, we can identify $A_p = S^{-1}A$ and $B_p = S^{-1}B$ as Dedekind subrings of E, and we have inclusions

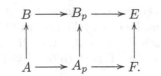

Since B is integral over A, B_p is integral over A_p. Since B_p is integrally closed in its field of fractions E, it follows that B_p is the integral closure of A_p in E. So A_p and B_p are just like A and B.

By (7.20), if J is a maximal ideal of B_p, then J lies above a maximal ideal of A_p. But A_p is local, with unique maximal ideal containing p. So $J \supseteq p$. Then $J \supseteq pB$, proving $pB \neq B$. ∎

Now B is Dedekind; so there exist maximal ideals $P_1, \ldots, P_r$ of B and positive integers $e_1, \ldots, e_r$ with

$$pB = P_1^{e_1} \cdots P_r^{e_r}.$$

Assuming no two of $P_1, \ldots, P_r$ are equal, the exponent

$$e_i = e(P_i/p)$$

is the **ramification index** of P_i over p.

If J is a proper ideal of B, then by maximality of p in A, the following are equivalent:

 (i) $J \supseteq p$,
 (ii) $J \cap A = p$,
 (iii) $J \supseteq pB$,
 (iv) $J | pB$.

So the maximal ideals of B lying above p are the $P_1, \ldots, P_r$ appearing in the factorization of pB. For each maximal ideal P of B lying above p, inclusion $A \to B$ carries p into P, and so it induces a ring homomorphism $A/p \to B/P$. Through this homomorphism, B/P is a vector space over A/p. Quotient rings R/I have also been called **residue rings** and the cosets $r + I$ **residue classes**. So A/p and B/P are known as **residue fields**, and the dimension

$$f_i = f(P_i/p) = [B/P_i : A/p]$$

is called the **residue degree** of P_i over p.

(7.37) Theorem. *In addition to the above conditions on A, B and p, if B is finitely generated as an A-module, then each residue degree f_i is finite, and*

$$\sum_{i=1}^{r} e_i f_i = [E : F] .$$

Proof. By (7.24), the local Dedekind domain A_p is a principal ideal domain. The A-linear generators of B are also A_p-linear generators of B_p; so B_p is a f.g. torsion free A_p-module. By the Structure Theorem for f.g. modules over a principal ideal domain, there is an A_p-linear isomorphism $B_p \cong (A_p)^n$ for some $n > 0$.

Let T denote the multiplicative monoid $A_p - \{0\}$. Identifying x/t with xt^{-1} in E as usual, $T^{-1}A_p = F$, and by (7.18), $T^{-1}B_p = E$. Since $T^{-1}(-)$ is an additive functor, there is an F-linear isomorphism $E \cong F^n$. So $n = [E : F]$. Also

$$\frac{B_p}{pB_p} \cong \frac{A_p^n}{pA_p^n} \cong \left(\frac{A_p}{pA_p}\right)^n$$

as A_p/pA_p-modules. There is a commutative square of ring homomorphisms

$$
\begin{array}{ccc}
B/pB & \xrightarrow{\ \cong\ } & B_p/pB_p \\[4pt]
\big\uparrow & & \big\uparrow \\[4pt]
A/pA & \xrightarrow{\ \cong\ } & A_p/pA_p
\end{array}
$$

where the horizontal isomorphisms are induced by localization as in (6.48) and the verticals are induced by inclusions. Through these maps, the rings on the top row are vector spaces over the fields on the bottom row. Any A_p/pA_p-basis of B_p/pB_p is also an A/pA-basis, and the top isomorphism is A/pA-linear. So

$$[B/pB : A/pA] = n .$$

By the Chinese Remainder Theorem,

$$B/pB \cong B/P_1^{e_1} \oplus \cdots \oplus B/P_r^{e_r}$$

as A/p-algebras. In the filtration

$$B/P_i^{e_i} \supseteq P_i/P_i^{e_i} \supseteq P_i^2/P_i^{e_i} \supseteq \cdots \supseteq 0 ,$$

each quotient of successive terms has an A/p-linear isomorphism to $P_i^m/P_i^{m+1} \cong B/P_i$ by (7.30). So in $K_0(A/p)$,

$$[B/pB] = \sum_{i=1}^r [B/P_i^{e_i}] = \sum_{i=1}^r e_i[B/P_i] .$$

Applying dimension over $A/p : K_0(A/p) \to \mathbb{Z}$,

$$n = \sum_{i=1}^r e_i f_i . \qquad\blacksquare$$

To obtain natural homomorphisms between the groups $I(A)$ and $I(B)$ of fractional ideals, we shall express them in terms of f.g "torsion" modules. Recall some definitions: Suppose R is a commutative integral domain. A f.g. R-module M is **torsion** if its annihilator

$$ann_R(M) = \{r \in R : rM = 0\}$$

is nonzero. The annihilator of any R-module M is an ideal of R. An R-module M is **simple** if $M \neq 0$ and its only submodules are 0 and M.

(7.38) Lemma. *Suppose R is a commutative integral domain. There is a bijection $J \mapsto c(R/J)$, $ann_R(M) \leftarrow c(M)$, between the set of maximal ideals of R and the set of isomorphism classes of simple R-modules.*

Proof. For each maximal ideal J of R, the quotient R/J is a simple R-module, by the submodule correspondence theorem. If M is any simple R-module, it has a nonzero element m and its submodule Rm must be M. So the R-linear map $R \to M$, $r \mapsto rm$, is surjective with kernel $J = ann_R(M)$. Then $R/J \cong M$ and

R/J is simple, forcing J to be a maximal ideal of R. Isomorphic R-modules share the same annihilator, and $ann_R(R/J) = J$. ■

The ideals of B containing a given ideal of A generate a subgroup of $I(B)$ with a Grothendieck group description: Suppose R is a Dedekind domain and $X \subseteq R$. Denote by $I(R, X)$ the subgroup of $I(R)$ generated by the maximal ideals of R that contain X. So $I(R, X)$ is free abelian, based on the set of maximal ideals dividing the ideal $\langle X \rangle$ generated by X. Let $\mathcal{T}_X(R)$ denote the full subcategory of R-$\mathcal{M}o\mathfrak{d}$ consisting of the f.g. torsion R-modules with annihilators containing X.

(7.39) Proposition. *If R is a Dedekind domain and J is an ideal of R, there is a group isomorphism*

$$I(R, J) \cong K_0(\mathcal{T}_J(R))$$

taking J' to $[R/J']$ for each ideal J' of R containing J.

Proof. In R-$\mathcal{M}o\mathfrak{d}$, let $\mathcal{D}$ denote the full subcategory of simple or zero R-modules with annihilator containing J. By (3.32) and (7.38), $K_0(\mathcal{D})$ is a free abelian group based on the set of $[R/P]$, one for each maximal ideal P of R containing J. Since $I(R, J)$ is free abelian based on the maximal ideals of R containing J, there is a group isomorphism

$$f : I(R, J) \rightarrow K_0(\mathcal{D})$$

taking P to $[R/P]$ for each maximal ideal P containing J.

If $M \in \mathcal{T}_J(R)$, then $\mathcal{A} = ann_R(M)$ is nonzero and M is a f.g. $R/\mathcal{A}$-module. The ring $R/\mathcal{A}$ has only finitely many ideals, corresponding to the ideals of R dividing $\mathcal{A}$. So $R/\mathcal{A}$ is a noetherian, artinian ring. By (B.20), in Appendix B, M has a composition series in $R/\mathcal{A}$-$\mathcal{M}o\mathfrak{d}$. This filtration also serves in $\mathcal{T}_J(R)$, with composition factors in $\mathcal{D}$. By Devissage (Theorem (3.42)), there is an isomorphism

$$g : K_0(\mathcal{D}) \rightarrow K_0(\mathcal{T}_J(R)) \quad , \quad [R/P] \mapsto [R/P] .$$

Then $g \circ f$ is an isomorphism from $I(R, J)$ to $K_0(\mathcal{T}_J(R))$, taking P to $[R/P]$ when P is a maximal ideal containing J.

If J_1 and J_2 are nonzero ideals of R with $J_1 J_2$ containing J, there is an exact sequence

$$0 \longrightarrow \frac{R}{J_1} \overset{\alpha}{\longrightarrow} \frac{R}{J_1 J_2} \overset{\beta}{\longrightarrow} \frac{R}{J_2} \longrightarrow 0$$

in $\mathcal{T}_J(R)$, where $\beta(r + J_1 J_2) = r + J_2$ and α is the isomorphism $R/J_1 \cong J_2/J_1 J_2$ of (7.30), followed by inclusion in $R/J_1 J_2$. So

$$[R/J_1] + [R/J_2] = [R/J_1 J_2]$$

in $K_0(\mathfrak{T}_J(R))$. If $P_1, \ldots, P_n$ are maximal ideals, $e_1, \ldots, e_n$ are positive integers, and $J \subseteq P_1^{e_1} \cdots P_n^{e_n}$, then

$$g \circ f(P_1^{e_1} \cdots P_n^{e_n}) = \sum_{i=1}^{n} e_i[R/P_i] = [R/P_1^{e_1} \cdots P_n^{e_n}] . \qquad \blacksquare$$

(7.40) Note. If J is a nonzero ideal of R, then $\mathfrak{T}_J(R)$ coincides with the category $\mathcal{M}(R/J)$ of f.g. R/J-modules and

$$I(R, J) \cong G_0(R/J) \cong \mathbb{Z}^r ,$$

where r is the number of different maximal ideal factors of J. On the other hand, for $J = 0$, $\mathfrak{T}_0(R) = \mathfrak{T}(R)$ is the category of f.g. torsion R-modules and

$$I(R) \cong K_0(\mathfrak{T}(R)) .$$

(7.41) Note. By the exact sequence in the proof of (7.39),

$$[R/J_1 \oplus R/J_2] = [R/J_1 J_2]$$

in $K_0(\mathfrak{T}(R))$ for any nonzero ideals J_1 and J_2 of R.

The extension and restriction of scalars $\mathfrak{T}(A) \rightleftarrows \mathfrak{T}(B)$ induce homomorphisms between K_0-groups, and associated homomorphisms $I(A) \rightleftarrows I(B)$ between groups of fractional ideals: First, the extension of scalars

$$B \otimes_A (-) : \mathcal{M}(A) \to \mathcal{M}(B)$$

carries $\mathfrak{T}(A)$ into $\mathfrak{T}(B)$, since $ann_B(B \otimes_A M) \supseteq ann_A(M)$. Suppose J is an ideal of A, generating the ideal JB of B. By (6.15), there are A-linear isomorphisms

$$B/JB \cong (A/J) \otimes_A B \cong B \otimes_A (A/J)$$

$$b + JB \mapsto (1 + J) \otimes b \mapsto b \otimes (1 + J) ,$$

which are evidently B-linear as well. So the induced homomorphism from $K_0(\mathfrak{T}(A))$ to $K_0(\mathfrak{T}(B))$ takes $[A/J]$ to $[B/JB]$ for each nonzero ideal J of A. And there is an **extension** homomorphism

$$e_{B/A} : I(A) \mapsto I(B) ,$$

taking J to JB if $J \lhd A$.

Now consider restriction of scalars $\rho : B\text{-}\mathcal{M}o\eth \to A\text{-}\mathcal{M}o\eth$. *Assume B is finitely generated as an A-module.* Then ρ carries $\mathcal{M}(B)$ into $\mathcal{M}(A)$. For each B-module M, $ann_A(M) = A \cap ann_B(M)$. By (7.20), a nonzero ideal of B must lie above a nonzero ideal of A; so ρ carries $\mathcal{T}(B)$ into $\mathcal{T}(A)$. If P is a maximal ideal of B, $p = P \cap A$, and

$$f = f(P/p) = [B/P : A/p]$$

is the residue degree, then the induced homomorphism from $K_0(\mathcal{T}(B))$ to $K_0(\mathcal{T}(A))$ takes $[B/P]$ to $[B/P] = f \cdot [A/p] = [A/p^f]$. There is a corrsponding homomorphism from $I(B)$ to $I(A)$:

(7.42) Definition. If B is finitely generated as an A-module, the **ideal norm** $n_{B/A} : I(B) \to I(A)$ is the group homomorphism with

$$n_{B/A}(P) = p^f$$

whenever P is a maximal ideal of B lying above p in A, with residue degree f.

(7.43) Proposition. *For each nonzero ideal J of B, J divides $e_{B/A} \circ n_{B/A}(J)$. So only finitely many ideals of B have the same norm as J.*

Proof. For $1 \le i \le r$, suppose P_i is a maximal ideal of B, $p_i = P_i \cap A$, $f_i = f(P_i/p_i)$, and $n_i \ge 0$. Then

$$n_{B/A}\Big(\prod_i P_i^{n_i}\Big) = \prod_i p_i^{f_i n_i} \subseteq \prod_i P_i^{n_i} ;$$

so $J \supseteq n_{B/A}(J)B = e_{B/A} \circ n_{B/A}(J)$. ∎

(7.44) Proposition. *If $[E : F] = n$ and B is a f.g. A-module, then the composite $n_{B/A} \circ e_{B/A} : I(A) \to I(A)$ is the nth power map.*

Proof. It is enough to prove this on each maximal ideal p of A. By (7.37), if

$$pB = P_1^{e_1} \cdots P_r^{e_r}$$

is the maximal ideal factorization of pB in B, and f_i is the residue degree $f(P_i/p)$, then

$$n_{B/A}(pB) = p^{\Sigma e_i f_i} = p^n . $$ ∎

(7.45) Proposition. *Taking $F = \mathbb{Q}$ and $A = \mathbb{Z}$, if J is a nonzero ideal of $B = alg.\ int.\ (E)$, then $n_{B/\mathbb{Z}}(J) = m\mathbb{Z}$, where m is the number of elements in B/J.*

Proof. After restriction of scalars, $B/J \in \mathcal{T}(\mathbb{Z})$ is isomorphic to $\mathbb{Z}/d_1\mathbb{Z} \oplus \cdots \oplus \mathbb{Z}/d_n\mathbb{Z}$ for positive integers d_i with $d_1 \cdots d_n = m$. By (7.41), $[B/J] = [\mathbb{Z}/m\mathbb{Z}]$ in $K_0(\mathcal{T}(\mathbb{Z}))$. ∎

Recall that the norm $N_{E/F} : E^* \to F^*$ is defined so that $N_{E/F}(x)$ is the determinant of the F-linear map $E \to E$, $e \mapsto xe$. The ideal norm, applied to principal fractional ideals, coincides with this field norm:

(7.46) Proposition. *If $x \in E^*$, then $n_{B/A}(xB) = N_{E/F}(x)A$.*

Proof. Suppose first that A is a principal ideal domain and $x \in B$. Since B is a f.g. torsion free A-module, B has a finite A-basis, which then serves as an F-basis of E. Say $n = [E : F]$. Multiplication by x is an injective A-linear map from B to B, represented over this basis by a matrix $M \in M_n(A)$. Since A is a principal ideal domain, there are invertible matrices $Q, Q' \in GL_n(A)$ for which $QMQ' = N = diag(d_1, \ldots, d_n)$. So there is a commutative diagram in $\mathcal{M}(A)$ with exact rows:

$$
\begin{array}{ccccccccc}
0 & \longrightarrow & A^n & \xrightarrow{\cdot M} & A^n & \longrightarrow & B/xB & \longrightarrow & 0 \\
& & \downarrow{\cdot Q^{-1}} & & \downarrow{\cdot Q'} & & & & \\
0 & \longrightarrow & A^n & \xrightarrow{\cdot N} & A^n & \longrightarrow & (A/d_1A) \oplus \cdots \oplus (A/d_nA) & \longrightarrow & 0
\end{array}
$$

and

$$N_{E/F}(x)A = det(M)A = det(N)A = d_1 \cdots d_n A .$$

Since $\cdot Q^{-1}$ and $\cdot Q'$ are isomorphisms,

$$B/xB \cong (A/d_1A) \oplus \cdots \oplus (A/d_nA) .$$

So in $K_0(\mathcal{T}(A))$,

$$[B/xB] = [A/d_1 \cdots d_nA] ,$$

and $n_{B/A}(xB) = N_{E/F}(x)A$.

Now suppose A is Dedekind, but not necessarily a principal ideal domain, p is a maximal ideal of A, and $S = A - p$. The localization functor $S^{-1}(-)$ defines group homomorphisms $I(A) \to I(A_p)$ and $I(B) \to I(B_p)$ by $M \mapsto S^{-1}M$. The first of these takes maximal ideals other than p (hence meeting S) to A_p, and takes p to the unique maximal ideal $pA_p = S^{-1}p$ of A_p. The second takes maximal ideals of B not lying above p to B_p and, by (6.46) and (6.48), restricts to a bijection from the set of maximal ideals lying above p to the set of maximal ideals in B_p, all of which lie above pA_p.

For each maximal ideal P of B lying above p, there is a commutative square of ring homomorphisms

$$
\begin{array}{ccc}
B/P & \overset{\cong}{\longrightarrow} & B_p/PB_p \\
\uparrow & & \uparrow \\
A/p & \overset{\cong}{\longrightarrow} & A_p/pA_p
\end{array}
$$

induced by inclusions — the top and bottom are isomorphisms as in (6.48). Through these maps, the top fields are vector spaces over the bottom fields. Any A_p/pA_p-basis of B_p/PB_p is also an A/p-basis, and the top map is A/p-linear. So

$$f(PB_p/pA_p) \;=\; f(P/p).$$

It follows that the square

$$
\begin{array}{ccc}
I(B) & \overset{S^{-1}(-)}{\longrightarrow} & I(B_p) \\
{\scriptstyle n_{B/A}}\downarrow & & \downarrow{\scriptstyle n_{B_p/A_p}} \\
I(A) & \overset{S^{-1}(-)}{\longrightarrow} & I(A_p)
\end{array}
$$

commutes, since both composites take a maximal ideal P of B to A_p if P does not lie over p, or to $p^f A_p$ ($f = f(P/p)$) if P lies over p.

Now A_p is a local Dedekind domain; so it is a principal ideal domain. Therefore, for $x \in B$,

$$
S^{-1}n_{B/A}(xB) \;=\; n_{B_p/A_p}(S^{-1}(xB)) \;=\; n_{B_p/A_p}(xB_p)
$$
$$
=\; N_{E/F}(x)A_p \;=\; S^{-1}(N_{E/F}(x)A) \; .
$$

So p has the same exponent in the factorizations of $n_{B/A}(xB)$ and $N_{E/F}(x)A$. This is true for each maximal ideal p of A; so $n_{B/A}(xB) = N_{E/F}(x)A$.

Finally, if $x = bc^{-1}$ where $b, c \in B - \{0\}$, then

$$
n_{B/A}(xB) \;=\; n_{B/A}(bB)n_{B/A}(cB)^{-1}
$$
$$
=\; N_{E/F}(b)A \; N_{E/F}(c)^{-1}A
$$
$$
=\; N_{E/F}(x)A \; . \hspace{2cm} \blacksquare
$$

Next we use the torsion module description of $I(R, J)$ when $J \neq 0$ to obtain a well-known method of Kummer for the explicit factorization of extended ideals. This method applies when (for A, B defined as above) $B = A[\theta]$ for some $\theta \in B$. Let $p(x)$ denote the (monic) minimal polynomial of θ over F. By Gauss' Lemma, since θ is integral over A, $p(x) \in A[x]$. So if $p(x)$ has degree n, then $1, x, \ldots, x^{n-1}$ is an A-basis of B; in particular, B is f.g. as an A-module. The problem at hand is to factor pB when p is a maximal ideal of A.

Let k denote the field A/p, and for $a(x) \in A[x]$ let $\overline{a}(x)$ denote the image of $a(x)$ under the ring homomorphism $A[x] \to k[x]$, reducing coefficients mod p.

(7.47) Kummer's Theorem. *Suppose $B = A[\theta]$ and $p(x)$ is the minimal polynomial of θ over F. There is an isomorphism of groups,*

$$I(k[x], \overline{p}(x)) \cong I(B, pB) ,$$

carrying maximal ideals to maximal ideals, and taking $\overline{a}(x)k[x]$ to $a(\theta)B + pB$ for each $a(x) \in A[x]$ with $\overline{a}(x)$ dividing $\overline{p}(x)$. In particular, if

$$\overline{p}(x) = \overline{p}_1(x)^{e_1} \cdots \overline{p}_r(x)^{e_r} \quad (e_i > 0) ,$$

where $p_i(x) \in A[x]$ and $\overline{p}_1(x), \ldots, \overline{p}_r(x)$ are distinct monic irreducibles in $k[x]$, then

$$pB = P_1^{e_1} \cdots P_r^{e_r}$$

for distinct maximal ideals $P_1, \ldots, P_r$ of B, where

$$P_i = p_i(\theta)B + pB$$

for each i. Further, the residue degree $f_i = f(P_i/p)$ is the degree of the polynomial $\overline{p}_i(x)$.

Proof. Suppose $a(x) \in A[x]$ and $\overline{a}(x)$ divides $\overline{p}(x)$ in $k[x]$. Reduction of coefficients mod p, followed by reduction mod $\overline{a}(x)$, defines a ring homomorphism from $A[x]$ onto $k[x]/\overline{a}(x)k[x]$ with kernel generated by $a(x)$ and p, and containing $p(x)$. So there is a commutative square of ring homomorphisms:

$$
\begin{array}{ccc}
\dfrac{A[x]}{p[x]A[x]} & \xrightarrow{\;\cong\;} & A[\theta] = B \\[2ex]
\downarrow & & \downarrow \\[2ex]
\dfrac{A[x]}{a(x)A[x] + pA[x]} & \xrightarrow{\;\cong\;} & \dfrac{k[x]}{\overline{a}(x)k[x]} \ ,
\end{array}
$$

where the top isomorphism is induced by evaluation at θ and the right map is induced by the left. The left map is surjective, with kernel generated by $\overline{a}(x)$ and $\overline{p}$; so the right map is surjective with kernel $a(\theta)B + pB$, inducing a ring isomorphism

$$\phi : \dfrac{B}{a(\theta)B + pB} \cong \dfrac{k[x]}{\overline{a}(x)k[x]} \ ,$$

taking $\overline{f(\theta)}$ to $\overline{f}(x)$ for $f(x) \in A[x]$.

Let $\mathcal{C}$ denote the full subcategory of $A[x]$-$\mathcal{M}\mathfrak{od}$ consisting of those f.g. $A[x]$-modules with annihilator containing $p(x)$ and p. When given the usual module action of a ring on its homomorphic images, the domain and codomain of ϕ are cyclic $A[x]$-modules in $\mathcal{C}$; and ϕ is $A[x]$-linear, so it is an isomorphism in $\mathcal{C}$. Therefore,

$$\left[\frac{k[x]}{\overline{a}(x)k[x]}\right] = \left[\frac{B}{a(\theta)B + pB}\right]$$

in $K_0(\mathcal{C})$, if $\overline{a}(x)$ divides $\overline{p}(x)$.

The categories of torsion modules:

$$\mathcal{C}_1 = T_{\overline{p}(x)k[x]}(k[x]) ,$$
$$\mathcal{C}_2 = T_{pB}(B) ,$$

are essentially the same as $\mathcal{C}$: Restriction of scalars to $A[x]$ defines exact isomorphisms of categories $\mathcal{C}_1 \to \mathcal{C} \leftarrow \mathcal{C}_2$ inducing isomorphisms of K_0-groups taking $[M] \to [M] \leftarrow [M]$. By (7.39) there is a composite isomorphism

$$I(k[x], \overline{p}(x)) \cong K_0(\mathcal{C}_1) \cong K_0(\mathcal{C}) \cong K_0(\mathcal{C}_2) \cong I(B, pB) ,$$

taking $\overline{a}(x)k[x]$ to $a(\theta)B + pB$ whenever $a(x) \in A[x]$ and $\overline{a}(x)$ divides $\overline{p}(x)$. In particular, it takes $\overline{p}(x)$ to pB. If $\overline{a}(x)$ is an irreducible factor $\overline{p_i}(x)$ of $\overline{p}(x)$, then the rings connected by ϕ are fields and $P_i = a(\theta)B + pB$ is a maximal ideal of B. Since ϕ is k-linear, comparing dimensions over $k = A/p$ yields

$$\text{degree } \overline{p_i}(x) = f(P_i/p) . \qquad \blacksquare$$

7D. Exercises

1. Suppose A is a Dedekind domain with field of fractions F, and $F \subseteq E \subseteq K$ are finite-degree field extensions, with B and C the integral closures of A in E and K, respectively. Suppose Q is a maximal ideal of C lying above P in B and p in A. Prove

$$e(Q/p) = e(Q/P)e(P/p) ,$$
$$f(Q/p) = f(Q/P)f(P/p) , \text{ and}$$
$$n_{C/A} = n_{B/A} \circ n_{C/B} .$$

2. In Stewart and Tall [87, Theorem 3.5, p. 72] it is shown that, if p is a positive prime in $\mathbb{Z}$ and $\zeta_p = e^{2\pi i/p}$, then alg. int. $(\mathbb{Q}(\zeta_p)) = \mathbb{Z}[\zeta_p]$. Now ζ_p is a root of

$$x^p - 1 = (x - 1)(x^{p-1} + x^{p-2} + \cdots + x + 1)$$

other than 1, and the second factor is irreducible in $\mathbb{Q}[x]$ (substitute $x + 1$ for x and apply Eisenstein's Criterion). So the second factor is the minimal polynomial of ζ_p over $\mathbb{Q}$. Use Kummer's Theorem to prove there is only one maximal ideal P of $\mathbb{Z}[\zeta_p]$ lying above $p\mathbb{Z}$, and $f(P/p\mathbb{Z}) = 1$. Find generators of the ideal P.

3. If E is a quadratic number field ($[E : \mathbb{Q}] = 2$) and $R = alg.\ int.\ (E)$, prove that, for each prime $p \in \mathbb{Z}$, the maximal ideal factorization of pR is one of:

(i) pR (p is **inert**),
(ii) P^2 (p **ramifies**),
(iii) $P_1 P_2$ with $P_1 \neq P_2$ (p **splits**).

Then prove pR factors (i.e., p splits or ramifies) if and only if $p\mathbb{Z}$ is the norm of an ideal of R. *Hint:* Use Theorem (7.37).

4. Suppose E is a quadratic imaginary number field $\mathbb{Q}(\sqrt{-\delta})$, where δ is a square-free positive integer. By §7A, Exercise 3, $R = alg.\ int.\ (E)$ is $\mathbb{Z}[\sqrt{-\delta}]$ if $\delta \equiv 1$ or $2 \pmod 4$ and $\mathbb{Z}[(1 + \sqrt{-\delta})/2]$ if $\delta \equiv 3 \pmod 4$. Prove that a positive prime $p \in \mathbb{Z}$ is not the norm $N_{E/\mathbb{Q}}(x)$ of any element $x \in R$ if $\delta > 4p$. *Hint:* Here $N_{E/\mathbb{Q}}(x)$ is the square of the complex absolute value $|x|$ and so is the square of the distance from x to 0 in the complex plane. Locate the points of R in the complex plane.

5. Under the hypotheses in Exercise 4, show $2\mathbb{Z}$ is the norm of an ideal of R, but not the norm of a principal ideal of R, if and only if $\delta = 5$, $\delta = 6$, or $\delta > 8$ and $\delta \not\equiv 3 \pmod 8$. So, under these conditions on R, 2 belongs to a nonprincipal ideal of R, and $Cl(R) \neq 0$. *Hint:* Use Kummer's Theorem, Proposition (7.46), and Exercises 3 and 4.

6. As in Exercise 5, show either 2 or 3 belongs to a nonprincipal ideal of $R = alg.\ int.\ (\mathbb{Q}(\sqrt{-\delta}))$ if and only if $\delta = 5, 6, 10$ or $\delta > 12$ and $\delta \not\equiv 19 \pmod{24}$.

7. If E is a quadratic number field and $R = alg.\ int.\ (E)$, Exercise 3 of §7A says $E = \mathbb{Q}(\sqrt{d})$ for a square-free integer d and $R = \mathbb{Z}[\theta]$, where $\theta = \sqrt{d}$ if $d \not\equiv 1 \pmod 4$ or $\theta = (1 + \sqrt{d})/2$ if $d \equiv 1 \pmod 4$. Say the minimal polynomial of θ over $\mathbb{Q}$ is $x^2 + bx + c$, with discriminant $D = b^2 - 4c$. For each *odd* prime $p \in \mathbb{Z}$, use Kummer's Theorem to prove

(i) p is inert if and only if $\overline{D}$ is not a square in $\mathbb{Z}/p\mathbb{Z}$;
(ii) p ramifies if and only if $p \mid D$;
(iii) p splits if and only if $\overline{D}$ is a nonzero square in $\mathbb{Z}/p\mathbb{Z}$.

What are the conditions on $x^2 + bx + c$ for 2 to be inert? Ramify? Split?

8. In a euclidean domain B, $b_1 B + b_2 B = dB$, where d is a greatest common divisor of b_1 and b_2, and d can be computed by Euclid's algorithm. Use this

and Kummer's Theorem to factor 78 as a unit times a product of irreducibles in $\mathbb{Z}[i]$.

9. Suppose p is a positive odd prime in $\mathbb{Z}$. Since $\mathbb{Z}/p\mathbb{Z}^*$ is cyclic of order $p-1$, $-\overline{1}$ is a square in $\mathbb{Z}/p\mathbb{Z}$ if and only if $4|(p-1)$. Prove p ramifies or splits in $\mathbb{Z}[i]$ if and only if $4|(p-1)$. Show p is a sum of two squares in $\mathbb{Z}$ if and only if $4|(p-1)$. *Hint:* For the last, show that p is a product of two nonunits in $\mathbb{Z}[i]$ if and only if $p = (a+bi)(a-bi)$ for some $a, b \in \mathbb{Z}$.

10. Use torsion modules to prove $S^{-1} : I(R) \to I(S^{-1}R)$ is a group homomorphism when R is a Dedekind domain and S is a submonoid of $(R - \{0\}, \cdot)$.

11. If F is a number field, prove $\mathcal{O}_F$ has a finite $\mathbb{Z}$-basis, and every $\mathbb{Z}$-basis of $\mathcal{O}_F$ is a $\mathbb{Q}$-basis of F. *Hint:* Use (7.19), (3.45), and then (7.18).

12. **Number-valued ideal norm.** Suppose F is a number field and $R = \mathcal{O}_F$. Show there is a group homomorphism

$$|| \; || : I(R) \to (\mathbb{Q}^+, \cdot)$$

with the properties:

   (i)   If J is a nonzero ideal of R, then $||J||$ is the index of J as an additive subgroup of R.

   (ii) If $x \in F^*$, $||xR|| = |N_{F/\mathbb{Q}}(x)|$.

  (iii) For each $y \in \mathbb{Q}^+$, the set of all $J \in I(R)$ with $||J|| = y$ is finite.

Hint: Follow $n_{R/\mathbb{Z}}$ by a group isomorphism from $I(\mathbb{Z})$ to $\mathbb{Q}^+$. Although elementary to construct, this isomorphism can be obtained as the composite $I(\mathbb{Z}) \cong K_0(\mathcal{T}(\mathbb{Z})) \cong \mathbb{Q}^+$, the last step induced by the order of each finite abelian group, as in (3.39) (ii).

13. **Finiteness of the ideal class group of a number field.** Suppose F is a number field and $R = \mathcal{O}_F$ has $\mathbb{Z}$-basis $b_1, \ldots, b_n$ as in Exercise 11. Let $\sigma_1, \ldots, \sigma_n$ denote the field homomorphisms $\sigma_i : F \to \mathbb{C}$, and

$$\beta = \max\{|\sigma_i(b_j)| : 1 \leq i \leq n, \ 1 \leq j \leq n\}.$$

Call the set

$$\{\sum_{i=1}^{n} a_i b_i : a_i \in \mathbb{Z}, \ 0 \leq a_i \leq K\}$$

a K**-box.** Suppose $0 \neq J \lhd R$.

   (i)   Show that, if $K \geq ||J||^{1/n}$, then the K-box has more elements than R/J, and so it has two elements r_1, r_2 with $r_1 - r_2 \in J$.

   (ii) Show that if $K \leq ||J||^{1/n}$, then for any two elements r_1, r_2 in the K-box,

$$|N_{F/\mathbb{Q}}(r_1 - r_2)| \leq (\beta n)^n ||J||.$$

Hint: From (14.61),

$$N_{F/\mathbb{Q}}(x) = \prod_{i=1}^{n} \sigma_i(x) .$$

(iii) Show the ideal class of J^{-1} has a member $I \lhd R$ with $||I|| \le (\beta n)^n$. *Hint:* Choose r_1, r_2 in the $||J||^{1/n}$-box with $r = r_1 - r_2 \in J$. Then $rR \subseteq J$; so $rR = IJ$ for some ideal I of R.

(iv) Show every ideal class in $Cl(R)$ includes the inverse of an integral ideal $J \lhd R$.

(v) Use Exercise 12 (iii) to prove $Cl(R)$ is a finite group.

14. Compute $Cl(\mathbb{Z}[\sqrt{-5}])$. *Hint:* By Exercise 13 (iii), (iv), each ideal class of $\mathcal{O}_F$ includes an integral ideal I with $||I|| \le (\beta n)^n$. For $F = \mathbb{Q}(\sqrt{-5})$, $R = \mathcal{O}_F = \mathbb{Z}[\sqrt{-5}]$ has $\mathbb{Z}$-basis $1, \sqrt{-5}$. So $n = 2$ and $\beta = \sqrt{5}$. Then $Cl(R)$ is generated by the (P) where P lies above the $p\mathbb{Z}$ for p prime and $0 < p < 20$. Use Kummer's Theorem to find these maximal ideals. If p remains prime in R, the maximal ideal pR is principal. The maximal ideal above $5\mathbb{Z}$ is $\sqrt{-5}R$, which is also principal. If p splits, so that $pR = PP'$, then $(P') = (P)^{-1}$. If $||I||$ cannot be written as $a^2 + 5b^2 = N_{F/\mathbb{Q}}(a + b\sqrt{-5})$ for $a, b \in \mathbb{Z}$, then I is not principal; so $(I) \ne 1$. In this way, obtain three maximal ideals P_1, P_2, P_3 whose classes generate $Cl(R)$. If an integral ideal I is contained in a principal ideal rR and $||I|| = ||rR|| = N_{\mathbb{Q}/F}(r)$, then $I = rR$ and $(I) = 1$. Use this to show $P_1^2 \cong P_2^2 \cong P_3^2 \cong P_1 P_2 \cong P_1 P_3 \cong R$.

15. Suppose A is a Dedekind domain (but not a field) with field of fractions F, assume $F \subseteq E$ is a finite degree Galois field extension with Galois group G, and take B to be the integral closure of A in E. Suppose p is a maximal ideal of A. Prove:

(i) For all $\sigma \in G$, $\sigma(B) = B$.

(ii) If P is a maximal ideal of B lying above p, so is $\sigma(P)$.

(iii) The action of G on the maximal ideals of B lying above p is transitive.

(iv) If P_1 and P_2 are maximal ideals of B lying above p, then $e(P_1/p) = e(P_2/p)$ and $f(P_1/p) = f(P_2/p)$.

Hint: For (iii), suppose P_2 differs from each $\sigma(P_1)$, and use the Chinese Remainder Theorem to produce $x \in P_2$ with $x \equiv 1 \mod \sigma(P_1)$ for all $\sigma \in G$. Show $N_{E/F}(x) \in F \cap P_2 = p \subseteq P_1$, even though none of the factors of $N_{E/F}(x) = \prod_{\sigma \in G} \sigma(x)$ belongs to P_1.

7E. K_0 and G_0 of Dedekind Domains

Suppose R is a commutative integral domain, $S = R - \{0\}$, and $F = S^{-1}R$ is its field of fractions. The composite of group homomorphisms

$$r : G_0(R) \to G_0(F) \cong \mathbb{Z}$$

defines a generalized rank of f.g. R-modules M by

$$rk(M) = r([M]) = dim_F(S^{-1}M) .$$

This rank extends the free rank on $\mathcal{F}(R)$, since

    (i) $rk(M \oplus N) = rk(M) + rk(N)$,
    (ii) $rk(R) = 1$, $rk(0) = 0$, and
    (iii) $rk(M) = rk(N)$ if $M \cong N$.

Since R is a domain, S acts through injections on each ideal J of R; so the localization map $J \to S^{-1}J$ is injective. If $J \neq 0$, then $S^{-1}J$ is a nonzero F-submodule of $F(= S^{-1}R)$; so $S^{-1}J = F$. This proves

    (iv) $rk(J) = 1$ if $0 \neq J \lhd R$.

By (iii), $rk(-)$ is an isomorphism invariant on $\mathcal{M}(R)$. When R is Dedekind, another isomorphism invariant with values in the ideal class group $Cl(R)$ was discovered in 1911 by Steinitz ([11] and [12]). Steinitz used these two invariants to classify all f.g. torsion free R-modules. Here is his classification, recast in modern language:

(7.48) Steinitz' Theorem. *Suppose R is a Dedekind domain and $M \in \mathcal{M}(R)$. The following are equivalent:*

    (i)   *M is torsion free,*
    (ii) *M is projective,*
    (iii) *$M \cong J_1 \oplus \cdots \oplus J_m$ for ideals J_i of R,*
    (iv) *$M \cong R^{m-1} \oplus J$, where $m \geq 1$ and J is an ideal of R.*

For nonzero ideals J_i, J_i' of R,

$$J_1 \oplus \cdots \oplus J_m \cong J_1' \oplus \cdots \oplus J_n'$$

if and only if $m = n$ and $J_1 \cdots J_m \cong J_1' \cdots J_n'$. So $R^{m-1} \oplus J \cong R^{n-1} \oplus J'$ for nonzero ideals J, J' if and only if $m = n$ and $J \cong J'$. In particular, the monoid $\mathcal{I}(\mathcal{P}(R))$ under $\oplus$ is cancellative.

Proof. To show (i) implies (iii), work by induction on $rk(M)$. Suppose $M \in \mathcal{M}(R)$ and M is torsion free. Then S acts through injections on M, and the localization map $M \to S^{-1}M$ is injective. Regard it as an inclusion by identifying $m/1$ with m.

If $rk(M) = 0$, then $M \subseteq S^{-1}M = 0$ and $M = 0$. Assume $rk(M) = n > 0$ and every torsion free f.g. R-module of rank $n-1$ is isomorphic to a direct sum of finitely many ideals of R. A generating set of M as an R-module is also a spanning set of the F-vector space $S^{-1}M$; so it contains an F-basis $b_1, \ldots, b_n$. Then

$$I = \{x \in F : xb_1 \in M\}$$

is an R-submodule of F, containing R; so $I \neq 0$. Multiplication by b_1 is an R-linear isomorphism

$$I \cong Ib_1 = M \cap Fb_1 \subseteq M .$$

Since $M \in \mathcal{M}(R)$ and R is noetherian, $I \in \mathcal{M}(R)$. Clearing denominators, $I \cong J \lhd R$. So there is an R-linear exact sequence

$$0 \to J \to M \to \frac{M}{M \cap Fb_1} \to 0 .$$

Evidently $M_1 = M/(M \cap Fb_1)$ is torsion free. Since rk is additive over short exact sequences,

$$rk(M_1) = rk(M) - rk(J) = n - 1 .$$

So, by the induction hypothesis,

$$M_1 \cong J_1 \oplus \cdots \oplus J_{n-1} \quad (= 0 \text{ if } n = 1)$$

for nonzero ideals J_i of R. Then M_1 is projective; so

$$M \cong M_1 \oplus J \cong J_1 \oplus \cdots \oplus J_{n-1} \oplus J .$$

That (iii) implies (ii) is immediate since R is Dedekind, so its ideals are projective. That (ii) implies (i) follows from the embedding of each projective module into a free module and the fact that R is a domain.

The equivalence of (iii) and (iv) follows from:

(7.49) Lemma. *If I and J are nonzero ideals of a Dedekind domain R, then $I \oplus J \cong R \oplus (IJ)$ as R-modules.*

Proof. By (7.30) there is an R-linear isomorphism $R/J \cong I/IJ$. Now apply Schanuel's Lemma (3.46) to the projective resolutions:

$$0 \to J \to R \to R/J \to 0$$

$$0 \to IJ \to I \to I/IJ \to 0 . \qquad \blacksquare$$

Next we establish the isomorphism invariants. Suppose $P = J_1 \oplus \cdots \oplus J_m$ for nonzero ideals J_i of R. The composite

$$P \to S^{-1}P \cong S^{-1}J_1 \oplus \cdots \oplus S^{-1}J_m = F^m$$

is just the inclusion of P in F^m. If also $Q = J_1' \oplus \cdots \oplus J_n'$ for nonzero ideals J_i', any R-linear map $f : P \to Q$ extends to $S^{-1}f : S^{-1}P \to S^{-1}Q$, and so to an F-linear map $F^m \to F^n$. So f is right multiplication by a matrix $A = (a_{ij}) \in F^{m \times n}$. Choosing nonzero $r_i \in J_i$ for each i, P contains the F-basis $r_1 e_1, \ldots, r_m e_m$ of F^m; so the matrix A is uniquely determined by f.

(7.50) Definition. For $I, J \in$ sub $_R F$, $(J : I) = \{x \in F : Ix \subseteq J\}$. Since $PA \subseteq Q$, each entry a_{ij} belongs to $(J'_j : J_i)$.

Now specialize to the case $P = R^{m-1} \oplus J$, $Q = R^{n-1} \oplus J'$, where J and J' are nonzero ideals of R. Assume $f : P \to Q$ is an R-linear isomorphism. Then $m = rk(P) = rk(Q) = n$. As above, f is right multiplication by an $(m - 1, 1)$ by $(m - 1, 1)$ partitioned matrix

$$A = \begin{bmatrix} B & c \\ d & e \end{bmatrix} \in M_m(F) \,,$$

where the entries of B are in $(R : R) = R$, the entries of c are in $(J' : R)$, the entries of d are in $(R : J)$, and $e \in (J' : J)$. Expanding the determinant of A along the last row, the first $m - 1$ terms belong to

$$(J' : R)R(R : J) \subseteq (J' : J) \,,$$

and the last term is in $R(J' : J) \subseteq (J' : J)$. So $\det(A)J \subseteq J'$.

Since f is an isomorphism, so is $S^{-1}f$; so A is an invertible matrix. Multiplying by A^{-1} restricts to $f^{-1} : Q \to P$. Just as for A, $\det(A^{-1})J' \subseteq J$. So multiplication by $\det(A)$ is an isomorphism $J \cong J'$.

More generally, if $J_1 \oplus \cdots \oplus J_m$ is isomorphic to $J'_1 \oplus \cdots \oplus J'_n$ for nonzero ideals J_i, J'_i, then by Lemma (7.49)

$$R^{m-1} \oplus (J_1 \cdots J_m) \cong R^{n-1} \oplus (J'_1 \cdots J'_n) \,;$$

so $m = n$ and $J_1 \cdots J_m \cong J'_1 \cdots J'_n$.

For cancellation, suppose $P, Q \in \mathcal{P}(R)$ and $R \oplus P \cong R \oplus Q$. Taking ranks, if $P \cong 0$, then $Q \cong 0$. On the other hand, if P, Q are nonzero, then $P \cong R^{m-1} \oplus J$ and $Q \cong R^{n-1} \oplus J'$ for nonzero ideals J, J'. Since $R^m \oplus J \cong R^n \oplus J'$, we must have $m = n$ and $J \cong J'$; so $R^{m-1} \oplus J \cong R^{n-1} \oplus J'$. This shows we can cancel R. If instead $P' \oplus P \cong P' \oplus Q$, for $P' \in \mathcal{P}(R)$, add a complement to P' and then cancel R's, one at a time. ∎

The number and product of nonzero ideal direct summands also define invariants on K_0 of a Dedekind domain: For each commutative domain R with field of fractions F, the embedding of R in F and the dimension over F induce *ring* homomorphisms

$$K_0(R) \to K_0(F) \cong \mathbb{Z}$$

by (6.22) and (5.17), and their composite

$$r : K_0(R) \to \mathbb{Z}$$

takes $[P] - [Q]$ to $rk(P) - rk(Q)$ for $P, Q \in \mathcal{P}(R)$. Not only is r surjective, but there is a ring homomorphism

$$i : \mathbb{Z} \ \to \ K_0(R) \, , \quad n \ \mapsto \ n[R]$$

with $r \circ i = i_{\mathbb{Z}}$. Since r is a ring homomorphism, the rank $rk(-)$ is multiplicative on f.g. projective modules over a commutative domain R:

$$rk(P \otimes_R Q) \ = \ rk(P) \, rk(Q) \, .$$

When P and Q are projective ideals of R, their tensor product coincides with their product as ideals:

(7.51) Lemma. *If I and J are ideals of a commutative ring R with I projective, then multiplication induces an R-linear isomorphism $I \otimes_R J \cong IJ$.*

Proof. By (6.18), tensoring with the projective ideal I is an exact functor on R-$\mathfrak{Mod}$. Since inclusion $J \to R$ is injective, the composite

$$I \otimes_R J \ \to \ I \otimes_R R \ \cong \ I$$

is injective, R-linear, and has image IJ.   ∎

In particular, for ideals I and J of a Dedekind domain R, $[I][J] = [IJ]$ in $K_0(R)$.

For the second invariant, consider the projective class group $K_0(R)/\langle[R]\rangle = \widetilde{K}_0(R)$, and the description of its elements in (4.3). It is called the projective class group because of the following connection to the ideal class group:

(7.52) Proposition. *If R is a Dedekind domain, there is a group isomorphism*

$$f : Cl(R) \ \to \ \widetilde{K}_0(R) \, ,$$

taking the isomorphism class $c(J)$ to the coset $\overline{[J]} = [J] + \langle[R]\rangle$.

Proof. By the first half of Steinitz' Theorem, each member of $\widetilde{K}_0(R)$ is $\overline{[J]}$ for some nonzero ideal J of R:

$$\overline{[R^{m-1} \oplus J]} \ = \ \overline{[J]} \, , \quad \overline{[0]} \ = \ \overline{[R]} \, .$$

By the second half of that theorem, nonzero ideals are stably equivalent if and only if they are isomorphic. So the map f is bijective. For nonzero ideals I and J,

$$f(c(I)c(J)) \ = \ f(c(IJ)) \ = \ \overline{[IJ]} \ = \ \overline{[R \oplus IJ]} \ = \ \overline{[I \oplus J]} \ = \ \overline{[I]} + \overline{[J]}$$
$$= \ f(c(I)) + f(c(J)) \, . \qquad \blacksquare$$

The canonical map and f^{-1} are group homomorphisms

$$K_0(R) \ \rightarrow \ \widetilde{K}_0(R) \ \cong \ Cl(R) \ ,$$

and their composite

$$s : K_0(R) \ \rightarrow \ Cl(R)$$

takes $[J_1 \oplus \cdots \oplus J_m]$ to $c(J_1 \cdots J_m)$ for nonzero ideals $J_1, \ldots, J_m$. Not only is s surjective, but there is a map

$$j : Cl(R) \ \rightarrow \ K_0(R)$$
$$c(J) \ \mapsto \ [J] - [R]$$

with $s \circ j = i_{Cl(R)}$, and j is a group homomorphism:

$$j(c(I)c(J)) \ = \ j(c(IJ)) \ = \ [IJ] - [R] = \ [R \oplus IJ] - 2[R] \ = \ [I \oplus J] - 2[R]$$
$$= \ ([I] - [R]) + ([J] - [R]) = \ j(c(I)) + j(c(J)) \ .$$

From the standard short exact sequence (for R with IBN)

$$0 \longrightarrow \mathbb{Z} \xrightarrow{\ i\ } K_0(R) \longrightarrow \widetilde{K}_0(R) \longrightarrow 0$$

we derive the split short exact sequence

$$0 \longrightarrow \mathbb{Z} \underset{r}{\overset{i}{\rightleftarrows}} K_0(R) \underset{j}{\overset{s}{\rightleftarrows}} Cl(R) \longrightarrow 0 \ ,$$

with $r \circ i = 1$, $s \circ j = 1$ and $(i \circ r) + (j \circ s) = 1$, where 1 denotes identity maps.

(7.53) Theorem. *If R is a Dedekind domain, i embeds $\mathbb{Z}$ as the smallest subring $\langle [R] \rangle$ of $K_0(R)$, j embeds $Cl(R)$ as an ideal*

$$C \ = \ \{ [J] - [R] : 0 \neq J \triangleleft R \}$$

of square zero in $K_0(R)$, and, additively, $K_0(R) \ = \ \langle [R] \rangle \overset{\bullet}{\oplus} C$. So the invariants r and s define an isomorphism

$$(r, s) : K_0(R) \ \cong \ \mathbb{Z} \oplus Cl(R)$$

of rings when multiplication in $\mathbb{Z} \oplus Cl(R)$ is defined by

$$(m, c(I)) \cdot (n, c(J)) \ = \ (mn, \ c(I^n J^m)) \ .$$

Proof. The ring homomorphism i is injective because $r \circ i = 1$. The image of the ring homomorphism from $\mathbb{Z}$ into any ring A is the additive subgroup generated by 1_A; so it is the smallest subring of A. The group homomorphism j is injective since $s \circ j = 1$. By the general properties of split exact sequences (2.5), the sequence of splitting maps

$$(7.54) \qquad 0 \longleftarrow \mathbb{Z} \xleftarrow{\ r\ } K_0(R) \xleftarrow{\ j\ } Cl(R) \longleftarrow 0$$

is exact. So $C = \text{image }(j) = \ker (r)$. Since r is a ring homomorphism, C is an ideal of $K_0(R)$. And $C^2 = 0$ since, by (7.49) and (7.51),

$$([I] - [R])([J] - [R]) \ = \ [IJ] - [I] - [J] + [R] \ = \ [IJ \oplus R] - [I \oplus J] \ = \ 0 \ .$$

Again by (2.5),

$$K_0(R) \ = \ \text{image }(i) \ \overset{\bullet}{\oplus} \ \text{image }(j)$$

$$= \ \langle [R] \rangle \ \overset{\bullet}{\oplus} \ C \ ,$$

and (r,s) defines a group isomorphism $K_0(R) \cong \mathbb{Z} \oplus Cl(R)$. The multiplication given on $\mathbb{Z} \oplus Cl(R)$ is obtained through (r,s) from the multiplication in $\langle [R] \rangle \ \overset{\bullet}{\oplus} \ C$. ∎

Now consider $G_0(R)$ for a Dedekind domain R. Since R is a (left) noetherian domain, submodules of f.g. torsion free R-modules are f.g. torsion free. So by Steinitz' Theorem, Obj $\mathcal{P}(R)$ is closed under submodules. (A ring with this property is called **hereditary**.) Therefore each f.g. R-module M has a short projective resolution

$$0 \longrightarrow P \xrightarrow{\ \subseteq\ } R^n \longrightarrow M \longrightarrow 0 \ ;$$

so R is regular, in the sense of (3.48). By the Resolution Theorem (3.53), the Cartan map

$$c : K_0(R) \ \to \ G_0(R) \ , \quad [P] \ \mapsto \ [P]$$

and the Euler characteristic

$$\chi : G_0(R) \ \to \ K_0(R) \ , \quad [M] \ \mapsto \ [R^n] - [P]$$

are mutually inverse group isomorphisms.

As remarked in (5.18) (iv), there is a ring multiplication on $G_0(R)$ matching that in $K_0(R)$, but not necessarily compatible with the tensor product on $\mathcal{M}(R)$:

(7.55) Example. Suppose R is a Dedekind domain with a nonprincipal ideal J. Then $R \not\cong J$; so $[R] \neq [J]$ in $K_0(R)$. By injectivity of the Cartan map, $[R] \neq [J]$ in $G_0(R)$ too. By (7.29), $J = aR + bR$ for some $a, b \in R$. Then

$$(R/aR) \ \otimes_R \ (R/bR) \ \cong \ \frac{R/bR}{a(R/bR)} \ = \ \frac{R/bR}{J/bR} \ \cong \ R/J \ .$$

So in $G_0(R)$,

$$[(R/aR) \otimes_R (R/bR)] = [R/J] = [R] - [J] \neq 0 .$$

But the exact sequences

$$0 \longrightarrow R \overset{a\cdot}{\longrightarrow} R \longrightarrow R/aR \longrightarrow 0 ,$$

$$0 \longrightarrow R \overset{b\cdot}{\longrightarrow} R \longrightarrow R/bR \longrightarrow 0 ,$$

show $[R/aR] [R/bR] = 0 \cdot 0 = 0$ in $G_0(R)$.

Note: The exact sequence (7.54) can also be derived from the localization exact sequence of Swan (6.52):

$$K_0(\mathcal{T}(R)) \longrightarrow G_0(R) \longrightarrow G_0(F) \longrightarrow 0 .$$

First replace $G_0(F) \cong K_0(F)$ with $\mathbb{Z}$ (via rank). Then use the isomorphism (7.39) from $I(R)$ to $K_0(\mathcal{T}(R))$, taking each ideal J of R to $[R/J]$. If J is principal, then $R \cong J$ and in $G_0(R)$

$$[R/J] = [R] - [J] = 0 .$$

So there is an induced homomorphism $k : Cl(R) \to G_0(R)$ taking $c(J)$ to $[R/J]$ and an exact sequence

$$Cl(R) \overset{k}{\longrightarrow} G_0(R) \overset{r}{\longrightarrow} \mathbb{Z} \longrightarrow 0 .$$

Follow k by the Euler characteristic, and precede r by the Cartan map to get the exact sequence

$$Cl(R) \overset{-j}{\longrightarrow} K_0(R) \overset{r}{\longrightarrow} \mathbb{Z} \longrightarrow 0 .$$

Then $-j : c(J) \mapsto [R] - [J]$ can be replaced by j, and both are injective by cancellation in $\mathcal{P}(R)$.

For a Dedekind domain R, the ideal class group embeds both as an additive and as a multiplicative subgroup of the ring $K_0(R)$. The map $j : Cl(R) \to K_0(R)$, taking $c(J)$ to $[J] - [R]$, is additive. But there is a more natural multiplicative homomorphism $j' : Cl(R) \to (K_0(R))^*$, taking $c(J)$ to $[J]$, since

$$[J_1 J_2] = [J_1 \otimes_R J_2] = [J_1][J_2]$$

by (7.51). And both j and j' are injective because the ideal class invariant s is left inverse to both of them: $s \circ j = s \circ j' = i_{Cl(R)}$.

(7.56) Proposition. *If R is a Dedekind domain,*

$$(K_0(R))^* = \{\pm[J] : 0 \neq J \triangleleft R\} \cong \{\pm 1\} \times Cl(R) .$$

Proof. Every element of $K_0(R)$ has the form $[J] - n[R]$ for $0 \neq J \triangleleft R$ and $n \in \mathbb{Z}$, the rank of this element being $1 - n$. The ring homomorphism r takes $(K_0(R))^*$ to $\mathbb{Z}^* = \{\pm 1\}$. So each unit in $K_0(R)$ has the form $[J] - 0[R] = [J]$ or

$$
\begin{aligned}
[J] - 2[R] &= [J] - [R \oplus R] \\
&= [J] - [J \oplus J^{-1}] \\
&= -[J^{-1}] = -[J']
\end{aligned}
$$

for nonzero ideals J, J' of R. The required isomorphism takes x to $(r(x), s(x))$. ∎

Note: The invariant s has the peculiar property that for nonzero ideals I and J,

$$s([I] + [J]) = s([I] \cdot [J]) = s([I])s([J]) .$$

7E. Exercises

1. Show Lemma (7.51) becomes false if we delete the words "with I projective." *Hint:* If f is a field and $R = F[x, y]$, then F is an R-module through the ring homomorphism $R \to F$ taking x and y to 0. Suppose $I = J = Rx + Ry$. Define $f : I \to F$ to take each polynomial to its constant coefficient of x, and $g : J \to F$ to pick out the constant coefficient of y. Show f and g are R-linear, and that $f \otimes g : I \otimes_R J \to F \otimes_R F = F \otimes_F F$, followed by multiplication $F \otimes_F F \to F$, distinguishes $x \otimes y$ from $y \otimes x$. So multiplication $I \otimes_R J \to IJ$ cannot be injective.

2. Suppose R is any ring, D is a division ring, and $j : R \to D$ is a ring homomorphism, making D a D, R-bimodule D_j. Then, by (6.27), there is an extended free rank $r_j : \mathrm{Obj}\ \mathcal{P}(R) \to \mathbb{Z}$ with

$$r_j(P) = \dim_D(D_j \otimes_R P).$$

If R is a commutative domain, prove $r_j(P) = rk(P)$ for every $P \in \mathcal{P}(R)$. *Hint:* The image of j must be a domain; so the kernel is a prime ideal M of R. Then j takes $S = R - M$ into D^*, and so factors as a composite of ring homomorphisms

$R \to S^{-1}R \to D$. Similarly, the localization map $i : R \to F$ $(= (R - \{0\})^{-1}R)$ factors as $R \to S^{-1}R \to F$. The induced diagram

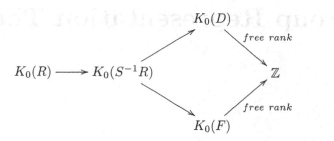

commutes, since $S^{-1}R = R_M$ is local, so that $\mathcal{P}(R_M) = \mathcal{F}(R_M)$.

3. Find a commutative ring R and two homomorphisms of rings $R \to D_1$, $R \to D_2$ from R into division rings D_1, D_2, leading to two *different* extended free ranks j_1, j_2 : Obj $\mathcal{P}(R) \to \mathbb{Z}$. *Hint:* Consider the ring $\mathbb{Z}/6\mathbb{Z} \cong (\mathbb{Z}/2\mathbb{Z}) \times (\mathbb{Z}/3\mathbb{Z})$, and the f.g. projective $\mathbb{Z}/6\mathbb{Z}$-module $P = \mathbb{Z}/2\mathbb{Z}$.

4. Suppose R is a Dedekind domain. Prove the kernel of the map $K_0(R) \to \mathbb{Z}$, taking $[P]$ to $rk(P)$, is the Jacobson radical of the commutative ring $K_0(R)$. Use (6.38) to prove every f.g. projective $K_0(R)$-module is free, so the ring homomorphism $\mathbb{Z} \to K_0(K_0(R))$ is an isomorphism. **Note:** The same conclusions are true for an arbitrary commutative domain R, since $\widetilde{K}_0(R)$ is both the nilradical and the Jacobson radical of $K_0(R)$, as shown by Swan [68, Corollary 10.7]. In Chen [94] it is proved that, for a commutative ring R, the following are equivalent:

    (i) If $xx = x$ for $x \in R$, then $x = 0$ or 1,
    (ii) $\mathcal{P}(K_0(R)) = \mathcal{F}(K_0(R))$,
    (iii) $K_0(K_0(R)) \cong \mathbb{Z}$.

8

Group Representation Theory

For novices in algebra as well as experienced researchers, a reliance on concrete examples serves as a valuable guide to the abstract theory. Groups of matrices can serve this purpose for group theory. The representations of a group G as a group of matrices over R are shown in §8A to correspond to modules M over the group ring RG that have a finite R-basis. The Grothendieck ring of these modules is called the "representation ring" of G over R. When G is finite and R is a field F, §8B compares this representation ring to $K_0(FG)$ and $G_0(FG)$, which are free abelian groups of the same free rank, based on idecomposable and simple modules, respectively. Maschke's Theorem (8.16) shows all three rings are the same when the order of G is not 0 in F. Section 8C characterises semisimple rings $\mathcal{A}$ as rings with $\mathcal{M}(\mathcal{A}) = \mathcal{P}(\mathcal{A})$, proves the Wedderburn-Artin Structure Theorem (8.28) for such rings, and discusses "splitting fields" of a group G, over which the order of G is the sum of the squares of the number of rows in each simple matrix representation of G. In §8D the trace of a matrix representation over a characteristic 0 field F defines a "character" of G, a function from G into F. Operations in F make the differences of characters into a "character ring" isomorphic to the representation ring $K_0(FG)$. Examples are provided to show how "orthogonality relations" among characters can lead to the discovery of all F-characters of G.

8A. Linear Representations

In this section we discuss the ways to represent a group as a multiplicative group of matrices. Assume throughout that R is a commutative ring and M is an R-module. The group $\mathrm{Aut}_R(M)$, of R-linear automorphisms of M under composition, can be thought of as a group of motions in the space M. When $R = \mathbb{R}$ and $M \cong \mathbb{R}^n$ for $n \leq 3$, we can even visualize these motions. Although $\mathrm{Aut}_R(M)$ generally has a wide variety of subgroups, they can be studied through the techniques of R-linear algebra.

(8.1) Definitions. An **R-linear representation** of a group G on M is a group homomorphism $\rho : G \to \text{Aut}_R(M)$. A **matrix representation** of G of **degree** n over R is a group homomorphism $\mu : G \to GL_n(R)$. If ρ or μ is injective, it is called **faithful**, and we say it represents G **faithfully** as a subgroup of $\text{Aut}_R(M)$ or $GL_n(R)$, repectively.

For each positive integer n, there is a group isomorphism from $GL_n(R)$ to $\text{Aut}_R(R^{n \times 1})$, taking each matrix A to left multiplication by A. Composition with this isomorphism defines a bijection from the set of matrix representations $\mu : G \to GL_n(R)$ to the set of R-linear representations $\rho : G \to \text{Aut}_R(R^{n \times 1})$; the R-linear representation corresponding to μ is

$$\mu(-)\cdot \; : \; G \;\to\; \text{Aut}_R(R^{n \times 1}) \, ,$$

taking $g \in G$ to left multiplication by $\mu(g)$. For computations, we shall want to apply the inverse bijection, replacing each R-linear representation ρ of G as a group of motions in $R^{n \times 1}$ by the more concrete μ, representing G as a group of matrices.

More generally, suppose M is an R-module with a finite R-basis $b_1, \ldots, b_n$ and associated "coordinate map"

$$\alpha : M \;\cong\; R^{n \times 1} \, , \qquad \sum r_i b_i \;\mapsto\; \begin{bmatrix} r_1 \\ \vdots \\ r_n \end{bmatrix} .$$

Then, as in (1.26)–(1.29), there is a group isomorphism

$$\text{Mat}_\alpha^\alpha \; : \; \text{Aut}_R(M) \;\cong\; GL_n(R)$$

taking f to A, where the square of R-linear maps

$$
\begin{array}{ccc}
M & \xrightarrow{\;f\;} & M \\
\alpha \downarrow & & \downarrow \alpha \\
R^{n \times 1} & \xrightarrow[\;A\cdot\;]{} & R^{n \times 1}
\end{array}
$$

commutes. If we identify R^n with the column matrices $R^{n \times 1}$, tracing e_j around this square shows that the j-column of $A = \text{Mat}_\alpha^\alpha(f)$ is

$$A e_j \;=\; \alpha \circ f \circ \alpha^{-1}(e_j) \;=\; \alpha(f(b_j)) \, ,$$

the column of $b_1, \ldots, b_n$-coordinates of $f(b_j)$. Then each R-linear representation $\rho : G \to \text{Aut}_R(M)$ and choice of basis $b_1, \ldots, b_n$ of M yields a matrix representation

$$\rho^\alpha \;=\; \text{Mat}_\alpha^\alpha \circ \rho \; : \; G \;\to\; GL_n(R) \, .$$

The j-column of $\rho^\alpha(g)$ is the column of $b_1, \dots, b_n$-coordinates of $\rho(g)(b_j)$.

(8.2) Example. The dihedral group D_4 of order 8 has generators a and b, defining relations $a^4 = 1$, $b^2 = 1$, $ba = a^{-1}b$, and elements $1, a, a^2, a^3, b, ab, a^2b$, a^3b. The standard $\mathbb{R}$-linear representation ρ of D_4 on $\mathbb{R}^2$, as symmetries of a square, takes a to the counter-clockwise rotation $\rho(a)$ by $\pi/2$ radians about the origin, and takes b to the reflection $\rho(b)$ through the x-axis.

If we use the standard $\mathbb{R}$-basis e_1, e_2 of $\mathbb{R}^2$, with coordinate map α,

$$\rho(a)(e_1) \;=\; e_2 \;=\; 0e_1 + 1e_2 \;, \quad \text{and}$$

$$\rho(a)(e_2) \;=\; -e_1 \;=\; (-1)e_1 + 0e_2 \;; \quad \text{so}$$

$$\rho^\alpha(a) \;=\; \begin{bmatrix} 0 & -1 \\ 1 & 0 \end{bmatrix}.$$

Also,

$$\rho(b)(e_1) \;=\; e_1 \;=\; 1e_1 + 0e_2 \;, \quad \text{and}$$

$$\rho(b)(e_2) \;=\; -e_2 \;=\; 0e_1 + (-1)e_2 \;; \quad \text{so}$$

$$p^\alpha(b) \;=\; \begin{bmatrix} 1 & 0 \\ 0 & -1 \end{bmatrix}.$$

Since ρ^α is a group homomorphism, its value on all elements of D_4 is now determined:

$$\rho^\alpha(a^i b^j) \;=\; \begin{bmatrix} 0 & -1 \\ 1 & 0 \end{bmatrix}^i \begin{bmatrix} 1 & 0 \\ 0 & -1 \end{bmatrix}^j.$$

If we use a different basis of $\mathbb{R}^2$, say $b_1 = (1,1)$, $b_2 = (-2,2)$ with coordinate map β, then

$$\rho(a)(b_1) \;=\; (-1,1) \;=\; 0b_1 + (1/2)b_2 \;,$$

$$\rho(a)(b_2) \;=\; (-2,-2) \;=\; (-2)b_1 + 0b_2 \;,$$

$$\rho(b)(b_1) \;=\; (1,-1) \;=\; 0b_1 + (-1/2)b_2 \;,$$

$$\rho(b)(b_2) \;=\; (-2,-2) \;=\; (-2)b_1 + 0b_2 \;; \quad \text{so}$$

$$\rho^\beta(a) \;=\; \begin{bmatrix} 0 & -2 \\ 1/2 & 0 \end{bmatrix}, \; \rho^\beta(b) \;=\; \begin{bmatrix} 0 & -2 \\ -1/2 & 0 \end{bmatrix}$$

$$\text{and} \;\; \rho^\beta(a^i b^j) \;=\; \begin{bmatrix} 0 & -2 \\ 1/2 & 0 \end{bmatrix}^i \begin{bmatrix} 0 & -2 \\ -1/2 & 0 \end{bmatrix}^j.$$

If two R-bases of M give rise to coordinate isomorphisms α and β, there is a matrix $C \in GL_n(R)$ for which

$$\begin{array}{ccc} M & \xrightarrow{\;i_M\;} & M \\ {\scriptstyle \alpha}\big\downarrow & & \big\downarrow{\scriptstyle \beta} \\ R^{n\times 1} & \xrightarrow[\;C.\;]{} & R^{n\times 1} \end{array}$$

commutes. In §1B, the matrix C is called the **change of basis** matrix and is denoted by $\mathrm{Mat}_\alpha^\beta(i_M)$. If ρ is an R-linear representation of G on M, then for each $g \in G$, the diagram

$$
\begin{array}{ccccccc}
M & \xrightarrow{\ i_M\ } & M & \xrightarrow{\ \rho(g)\ } & M & \xrightarrow{\ i_M\ } & M \\
{\scriptstyle \beta}\downarrow & & {\scriptstyle \alpha}\downarrow & & {\scriptstyle \alpha}\downarrow & & {\scriptstyle \beta}\downarrow \\
R^{n\times 1} & \xrightarrow[\ C^{-1}.\ ]{} & R^{n\times 1} & \xrightarrow[\ \rho^\alpha(g).\ ]{} & R^{n\times 1} & \xrightarrow[\ C.\ ]{} & R^{n\times 1}
\end{array}
$$

commutes; so $\rho^\beta(g) = C\rho^\alpha(g)C^{-1}$ for all $g \in G$. That is, ρ^β and ρ^α are "similar" in the following sense:

(8.3) Definitions. If μ and μ' are matrix representations of G over R, say that μ is **similar** to μ' if they have the same degree n and there is a matrix $C \in GL_n(R)$ for which

$$\mu'(g) = C\mu(g)C^{-1}$$

for all $g \in G$. Similarity is an equivalence relation on the set of matrix representations of G over R. A **similarity class** is an equivalence class under similarity. If μ is a matrix representation of G over R of degree n, and $C \in GL_n(R)$, then μ followed by the inner automorphism $C(-)C^{-1}$ of $GL_n(R)$ is also a matrix representation of G. So the similarity class of μ is

$$s(\mu) = \{C\mu(-)C^{-1} : C \in GL_n(R)\} .$$

Denote the set of similarity classes of matrix representations of G over R by $\mathbf{sim}_G(R)$.

Since every invertible matrix $C \in GL_n(R)$ is the change of basis matrix $\mathrm{Mat}_\alpha^\beta(i_M)$ for some coordinate map β, the matrix representations obtained from an R-linear representation ρ of G on $M \in \mathcal{F}(R)$, by choosing various bases, form an entire similarity class. We reach this conclusion in another way in (8.8) below.

To begin a systematic enumeration of the R-linear representations of G, we look at them another way, as modules over the group ring RG : Recall that **RG** is the free R-module based on the elements of G, made into an R-algebra by the multiplication

$$\left(\sum_{g\in G} r_g g\right)\left(\sum_{g\in G} s_g g\right) = \sum_{g\in G}\left(\sum_{hk=g} r_h s_k\right)g$$

(see (1.16)). This complicated looking multiplication is just that calculated by the distributive law, the rule

$$(r_h h)(s_k k) = (r_h s_k)(hk) ,$$

and the module axioms used to collect terms; so it is determined by the multiplications $r_h s_k$ in R and hk in G. When using the group ring RG, it is customary to identify each $r \in R$ with $r \cdot 1_G \in RG$; the map $R \to R \cdot 1_G$, $r \mapsto r \cdot 1_G$, is a ring isomorphism (injective by linear independence of 1_G). This identification poses no problems as long as $R \cap G$ is empty to begin with. If it happens that G meets R, one first makes them disjoint by renaming the elements of G. After the identification, R and G share only the element $1_R = 1_R \cdot 1_G = 1_G$, and R becomes a subring of RG with $rg = gr$ for all $r \in R$ and $g \in G$. The scalar multiplication $R \times RG \to RG$ then simply coincides with the ring multiplication in RG.

Now $\mathrm{Aut}_R(M)$ is just the group of units in $\mathrm{End}_R(M)$, which is an R-algebra with sum, product, and scalar multiplication defined by

$$
\begin{aligned}
(f_1 + f_2)(m) &= f_1(m) + f_2(m) \, , \\
(f_1 \circ f_2)(m) &= f_1(f_2(m)) \, , \\
(rf)(m) &= r(f(m))
\end{aligned}
$$

for $f, f_1, f_2 \in \mathrm{End}_R(M)$, $m \in M$, and $r \in R$. By (1.17), each group homomorphism

$$
\rho : \; G \to \mathrm{Aut}_R(M) \;=\; (\mathrm{End}_R(M))^*
$$

extends uniquely to an R-algebra homomorphism (i.e., R-linear ring homomorphism)

$$
\widehat{\rho} : RG \;\to\; \mathrm{End}_R(M) \, .
$$

Then the R-module M becomes an RG-module under the scalar multiplication $x \cdot m = \widehat{\rho}(x)(m)$; the module axioms follow from the homomorphism properties of $\widehat{\rho}$ and $\widehat{\rho}(x)$. Thus each R-linear representation ρ of G on M makes M into an RG-module under:

$$
\left(\sum_{g \in G} r_g g\right) \cdot m \;=\; \widehat{\rho}\left(\sum_{g \in G} r_g g\right)(m) \;=\; \sum_{g \in G} r_g \rho(g)(m) \, .
$$

(8.4) Notation. We denote this RG-module by the pair (M, ρ) to indicate that G acts on M through ρ.

In the other direction, from each abstract RG-module N, we can retrieve an R-linear representation of G on N; and if N is free of finite free rank as an R-module, we obtain a similarity class of concrete matrix representations of G. In detail, suppose N is an RG-module. For each $x \in RG$, the map

$$
\begin{aligned}
x \cdot (-) : \; N &\to N \\
n &\mapsto x \cdot n
\end{aligned}
$$

is R-linear:

$$
\begin{aligned}
x \cdot (n_1 + n_2) &= (x \cdot n_1) + (x \cdot n_2) \, , \quad \text{and} \\
x \cdot (r \cdot n) &= (xr) \cdot n = (rx) \cdot n = r \cdot (x \cdot n) \, .
\end{aligned}
$$

Sending x to $x \cdot (-)$ defines an R-algebra homomorphism

$$\widehat{\rho_N} : RG \;\;\to\;\; \mathrm{End}_R(N) \,,$$

since

$$
\begin{aligned}
\widehat{\rho_N}(x+y) &= (x+y) \cdot (-) = x \cdot (-) + y \cdot (-) &= \widehat{\rho_N}(x) + \widehat{\rho_N}(y) \,, \\
\widehat{\rho_N}(xy) &= x \cdot (y \cdot (-)) = x \cdot (-) \circ y \cdot (-) &= \widehat{\rho_N}(x) \circ \widehat{\rho_N}(y) \,, \\
\widehat{\rho_N}(ry) &= \widehat{\rho_N}(r) \circ \widehat{\rho_N}(y) = r \cdot \widehat{\rho_N}(y) \,,
\end{aligned}
$$

for $x, y \in RG$ and $r \in R$. Since $G \subseteq RG^*$, $\widehat{\rho_N}$ restricts to a group homomorphism

$$\rho_N : G \;\;\to\;\; \mathrm{Aut}_R(N) \,.$$

We refer to ρ_N as the R-linear representation **afforded** by the RG-module N. If N has a finite R-basis $b_1, \ldots, b_n$ with coordinate map $\alpha : N \cong R^{n \times 1}$, we call ρ_N^α a matrix representation **afforded** by N. For each j, the j-column of $\rho_N^\alpha(g)$ is the column of $b_1, \ldots, b_n$-coordinates of $g \cdot b_j$.

(8.5) Examples.

(i) The ring RG is a module over itself. The **regular representation** ρ_{RG} takes each $g \in G$ to the R-linear map $g \cdot (-) : RG \to RG$, which is left multiplication by g in the ring RG. If G is finite, using its elements $g_1, \ldots, g_n$ as an R-basis, the corresponding matrix representation ρ_{RG}^α takes each $g \in G$ to the "permutation matrix" with 1 in the i, j-entry if $g g_j = g_i$, and 0 in all other entries. This is the matrix obtained from $I_n \in GL_n(R)$ by permuting its rows in the same way that $g \cdot (-)$ permutes $g_1, \ldots, g_n$. Note that ρ_{RG}^α is faithful.

(ii) Each R-module M is an RG-module under the scalar multiplication $(\sum r_g g) \cdot m = (\sum r_g) \cdot m$ in which each $g \in G$ acts like 1. This RG-module affords the **trivial representation**

$$\rho_M : G \;\;\to\;\; \mathrm{Aut}_R(M) \,,$$

which is the constant map to i_M. Whatever R-basis $b_1, \ldots, b_n$ the R-module M may have, the associated matrix representation

$$\rho_M^\alpha : G \;\;\to\;\; GL_n(R)$$

is the constant map to I_n.

(iii) Let C_n denote the cyclic group of finite order n, generated by a. Evaluation at a induces an R-algebra isomorphism

$$\frac{R[x]}{\langle x^n - 1 \rangle} \;\;\to\;\; RC_n \quad, \quad \overline{p(x)} \mapsto p(a) \,;$$

it is bijective since its domain is spanned over R by the cosets $\bar{1}, \bar{x}, \ldots, \bar{x}^{n-1}$, which are sent to the R-basis $1, a, \ldots, a^{n-1}$ of RC_n. Now suppose $R = \mathbb{Q}$ and d is a positive integer factor of n, and $\zeta_d = e^{2\pi i/d} \in \mathbb{C}$. Then ζ_d is a root of $x^n - 1$; so its minimal polynomial $\Phi_d(x)$ over $\mathbb{Q}$ divides $x^n - 1$ in $\mathbb{Q}[x]$. Then there are $\mathbb{Q}$-algebra homomorphisms

$$\mathbb{Q}C_n \cong \frac{\mathbb{Q}[x]}{\langle x^n - 1 \rangle} \to \frac{\mathbb{Q}[x]}{\langle \Phi_d(x) \rangle} \cong \mathbb{Q}(\zeta_d) \ ,$$

whose composite takes a to ζ_d. So the field $\mathbb{Q}(\zeta_d)$ is a $\mathbb{Q}C_n$-module on which each rational number acts through multiplication, and a acts as multiplication by ζ_d. The $\mathbb{Q}$-linear representation $\rho_{\mathbb{Q}(\zeta_d)}$ afforded by $\mathbb{Q}(\zeta_d)$ is faithful if and only if $d = n$, so that $\mathbb{Q}C_n \to \mathbb{Q}(\zeta_d)$ is injective on C_n.

Over the $\mathbb{Q}$-basis $1, \zeta_d, \ldots, \zeta_d^{\phi(d)-1}$, where $\phi(d) = $ degree $\Phi_d(x)$, the associated matrix representation takes a to the companion matrix

$$\begin{bmatrix} 0 & & & & -r_0 \\ 1 & 0 & & & -r_1 \\ & 1 & \ddots & & \vdots \\ & & \ddots & 0 & -r_{\phi(d)-2} \\ & & & 1 & -r_{\phi(d)-1} \end{bmatrix}$$

of the polynomial $\Phi_d(x) = \sum r_i x^i$. As shown in Lang [93, Chapter VI, §3], $\Phi_d(x)$ is the cyclotomic polynomial

$$\prod_{\substack{0 \le m < d \\ \gcd(m,d)=1}} (x - \zeta_d^m) \ ;$$

so its degree $\phi(d)$ is Euler's totient function of d. Note that $x^n - 1$ is the product of the $\Phi_d(x)$ for $d > 0$ dividing n, and these factors are pairwise relatively prime. So, by the Chinese Remainder Theorem, there is a $\mathbb{Q}$-algebra isomorphism

$$\mathbb{Q}C_n \cong \prod_{d|n} \mathbb{Q}(\zeta_d) \ .$$

The RG-modules are in bijective correspondence with the R-linear representations of G:

(8.6) Proposition. *Each R-linear representation $\rho : G \to \mathrm{Aut}_R(M)$ is afforded by one and only one RG-module, namely (M, ρ). Each RG-module is (M, ρ) for a unique R-module M and R-linear representation ρ of G on M.*

Proof. The representation afforded by the RG-module (M, ρ) is ρ, since it takes each g to $g \cdot (-) = \rho(g)$. If ρ is afforded by any RG-module N, then $N = (M, \rho)$, since N must be M as an R-module, and on N

$$\left(\sum r_g g\right) \cdot m = \sum r_g \cdot (g \cdot m) = \sum r_g \rho(g)(m) .$$

If also $N = (M', \rho')$, then $M' = M$ as R-modules, and $\rho' = \rho_N = \rho$. ∎

The description of R-linear representations as RG-modules (M, ρ) is closely parallel to the description of single endomorphisms $f : M \to M$ as $R[x]$-modules (M, f) in Examples (3.9) and (5.19). We shall pursue this analogy.

Suppose $\mathcal{C}$ is a modest full subcategory of R-$\mathcal{Mod}$ that is closed under $\oplus$ and $\otimes_R$ and includes the modules 0 and R. (For instance, $\mathcal{C}$ could be $\mathcal{M}(R)$, $\mathcal{P}(R)$, or $\mathcal{F}(R)$.) An RG-**lattice** with respect to $\mathcal{C}$ is an RG-module N with underlying R-module $_RN$ in $\mathcal{C}$. Denote by $\mathbf{lat}_G \mathcal{C}$ the full subcategory of RG-$\mathcal{Mod}$ whose objects are the RG-lattices with respect to $\mathcal{C}$. By Proposition (8.6), there is a bijective correspondence $(N \mapsto \rho_N, \ \rho \mapsto (M, \rho))$ between the objects of $\text{lat}_G \mathcal{C}$ and the R-linear representations of G on objects of $\mathcal{C}$.

An RG-linear map $\alpha : (M, \rho) \to (M', \rho')$ in $\text{lat}_G \mathcal{C}$ is the same as an R-linear map $\alpha : M \to M'$ in $\mathcal{C}$ for which $\alpha(g \cdot m) = g \cdot \alpha(m)$ for all $g \in G$ and $m \in M$, that is, for which the square

$$
\begin{array}{ccc}
M & \xrightarrow{\rho(g)} & M \\
\alpha \downarrow & & \downarrow \alpha \\
M' & \xrightarrow{\rho'(g)} & M'
\end{array}
$$

commutes for each $g \in G$. If $\beta : M \to M'$ is an isomorphism in $\mathcal{C}$ and $\rho : G \to \text{Aut}_R(M)$ is an R-linear representation on M, then

$$
\begin{aligned}
\beta \rho \beta^{-1} : G &\to \text{Aut}_R(M') \\
g &\mapsto \beta \circ \rho(g) \circ \beta^{-1}
\end{aligned}
$$

is an R-linear representation on M' and β is an isomorphism

$$(M, \rho) \cong (M', \beta \rho \beta^{-1})$$

in $\text{lat}_G \mathcal{C}$, since β is bijective and

$$
\begin{array}{ccc}
M & \xrightarrow{\rho(g)} & M \\
\beta \downarrow & & \downarrow \beta \\
M' & \xrightarrow{\beta \circ \rho(g) \circ \beta^{-1}} & M'
\end{array}
$$

commutes for each $g \in G$. Therefore, if S is a set of objects representing the isomorphism classes in $\mathcal{C}$, then the set of all (M, ρ) with $M \in S$ represents the isomorphism classes in $\text{lat}_G \, \mathcal{C}$; so $\text{lat}_G \, \mathcal{C}$ is a modest category.

If $\rho : G \to \text{Aut}_R(M)$ and $\rho' : G \to \text{Aut}_R(M')$ are R-linear representations, there are R-linear representations

$$\rho \oplus \rho' \; : \; G \to \text{Aut}_R(M \oplus M') \;\; \text{and}$$
$$\rho \otimes \rho' \; : \; G \to \text{Aut}_R(M \otimes_R M')$$

defined by $(\rho \oplus \rho')(g) = \rho(g) \oplus \rho'(g)$ and $(\rho \otimes \rho')(g) = \rho(g) \otimes \rho'(g)$. So we can define the direct sum and tensor product of objects in $\text{lat}_G \, \mathcal{C}$ by

$$(M, \rho) \; \oplus \; (M', \rho') \; = \; (M \oplus M', \; \rho \oplus \rho') \, ,$$

$$(M, \rho) \; \overline{\otimes} \; (M', \rho') \; = \; (M \otimes_R M', \; \rho \otimes \rho') \, .$$

The direct sum defined here coincides with the direct sum in the surrounding category $RG\text{-}\mathfrak{Mod}$, since $(M \oplus M', \; \rho \oplus \rho')$ is $M \oplus M'$ as an R-module, with G-action:

$$g \cdot (m, m') \; = \; (\rho \oplus \rho')(g)(m, m') \; = \; (\rho(g)(m), \; \rho'(g)(m')) \; = \; (g \cdot m, \; g \cdot m').$$

However, the tensor product $\overline{\otimes}$ does not coincide with $\otimes_{RG}$; rather, $(M \otimes_R M', \; \rho \otimes \rho')$ is the R-module $M \otimes_R M'$ with diagonal action of G:

$$g \cdot (m \otimes m') \; = \; (\rho \otimes \rho')(g)(m \otimes m') \; = \; \rho(g)(m) \otimes \rho'(g)(m') \; = \; g \cdot m \otimes g \cdot m'.$$

The operations $\oplus$ and $\overline{\otimes}$ are commutative, associative binary operations on the category $\text{lat}_G \, \mathcal{C}$ (in the sense of (3.3)), with identity objects $(0, \text{trivial})$ for $\oplus$ and $(R, \text{trivial})$ for $\overline{\otimes}$, and with $\overline{\otimes}$ distributive over $\oplus$: The proofs that $\overline{\otimes}$ is well-defined on isomorphism classes, and that the arithmetic axioms hold, are just the isomorphisms of R-modules in the proof of (5.16), together with the verification that each isomorphism commutes with the diagonal action of G. *So the set of isomorphism classes in* $\text{lat}_G \, \mathcal{C}$ *is a commutative semiring* $\mathfrak{I}(\text{lat}_G \, \mathcal{C}, \oplus, \overline{\otimes})$.

The special case with $\mathcal{C} = \mathcal{F}(R)$ is the context for matrix representations. The category $\text{lat}_G \, \mathcal{F}(R)$ is modest because each isomorphism class includes a matrix representation $(R^{n \times 1}, \mu(-)\cdot)$:

(8.7) Proposition. *Suppose* $N \in \text{lat}_G \, \mathcal{F}(R)$. *Then a matrix representation* $\mu : G \to GL_n(R)$ *is afforded by* N *if and only if* $N \cong (R^{n \times 1}, \mu(-)\cdot)$. *And for matrix representations* μ *and* μ' *of* G *over* R,

$$(R^{m \times 1}, \; \mu(-)\cdot) \; \cong \; (R^{n \times 1}, \; \mu'(-)\cdot)$$

if and only if $m = n$ *and* μ *is similar to* μ'.

Proof. Recall from (8.6) that $N = ({}_R N, \rho_N)$. Suppose μ is a matrix representation of G afforded by N, which is to say $\alpha : N \cong R^{n \times 1}$ is R-linear and $\mu = \rho_N^\alpha$. Then

$$\alpha : ({}_R N, \rho_N) \cong (R^{n \times 1}, \mu(-)\cdot)$$

is RG-linear because, for each $g \in G$, $\mu(g) = \mathrm{Mat}_\alpha^\alpha(\rho_N(g))$; so the square

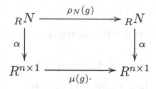

commutes. On the other hand, if α is an RG-linear isomorphism from N to some $(R^{n \times 1}, \mu(-)\cdot)$, then the preceding square commutes, and $\mu(g) = \mathrm{Mat}_\alpha^\alpha(\rho_N(g)) = \rho_N^\alpha(g)$ for each $g \in G$; so $\mu = \rho_N^\alpha$ is afforded by N.

If $(R^{m \times 1}, \mu(-)\cdot) \cong (R^{n \times 1}, \mu'(-)\cdot)$ as RG-lattices, then $R^{m \times 1} \cong R^{n \times 1}$ as R-modules. Since the commutative ring R has IBN, $m = n$. And left multiplication by $C \in GL_n(R)$ is an isomorphism between these RG-lattices if and only if

$$\begin{array}{ccc}
R^{n \times 1} & \xrightarrow{\ \mu(g)\cdot\ } & R^{n \times 1} \\
{\scriptstyle C\cdot}\downarrow & & \downarrow{\scriptstyle C\cdot} \\
R^{n \times 1} & \xrightarrow[\ \mu'(g)\cdot\ ]{} & R^{n \times 1}
\end{array}$$

commutes for each $g \in G$, which is to say, if and only if $\mu'(g) = C\mu(g)C^{-1}$ for each $g \in G$. ∎

(8.8) Corollary.

   (i) *The matrix representations ρ_N^α afforded by $N \in \mathrm{lat}_G \, \mathcal{F}(R)$ form a similarity class.*

   (ii) *In $\mathrm{lat}_G \, \mathcal{F}(R)$, $N \cong N'$ if and only if N and N' afford the same similarity class of matrix representations.*

   (iii) *There is a bijection*

$$\mathfrak{I}(\mathrm{lat}_G \, \mathcal{F}(R)) \ \to \ \mathrm{sim}_G(R) \ ,$$

taking the isomorphism class of N to the similarity class of matrix representations afforded by N, and taking

$$c(R^{n \times 1}, \mu(-)\cdot) \ \ to \ \ s(\mu)$$

for each matrix representation $\mu : G \to GL_n(R)$. ∎

The bijection (8.8) (iii) becomes a semiring isomorphism if $\text{sim}_G(R)$ is made a semiring by operations corresponding to those in $\mathfrak{I}(\text{lat}_G \ \mathcal{F}(R))$: If $M \in \text{lat}_G \ \mathcal{F}(R)$ affords the matrix representation $\mu : G \to GL_m(R)$ with respect to the R-basis $v_1, \dots, v_m$ of M, and $N \in \text{lat}_G \ \mathcal{F}(R)$ affords $\nu : G \to GL_n(R)$ over the R-basis $w_1, \dots, w_n$ of N, let $\mu \oplus \nu$ denote the matrix representation of degree $m + n$ afforded by $M \oplus N$ over the R-basis

$$(v_1, 0), \dots, (v_m, 0), (0, w_1), \dots, (0, w_n) ,$$

and let $\mu \otimes \nu$ denote the matrix representation of degree mn afforded by $M \overline{\otimes}_R N$ over the R-basis

$$\{v_i \otimes w_j \ : \ 1 \le i \le m , \ 1 \le j \le n\}$$

in lexicographic order. It is straightforward to check that if $\mu(g) = A$ and $\nu(g) = B$, then

$$(\mu \oplus \nu)(g) \ = \ A \oplus B \ = \ \begin{bmatrix} A & 0 \\ 0 & B \end{bmatrix} \in \ GL_{m+n}(R) ,$$

and by (5.21),

$$(\mu \otimes \nu)(g) \ = \ A \otimes B \ = \ \begin{bmatrix} a_{11}B & \cdots & a_{1m}B \\ \vdots & & \vdots \\ a_{m1}B & \cdots & a_{mm}B \end{bmatrix} \in \ GL_{mn}(R) .$$

(Actually, (5.21) is stated for the representation of endomorphisms as right multiplication by matrices on $R^{1\times m}$, but it works as well for left multiplication on $R^{m\times 1}$, since $(A \otimes B)^\tau = A^\tau \otimes B^\tau$, where $(-)^\tau$ denotes the transpose.)

So for each $g \in G$, $(\mu \oplus \nu)(g) = \mu(g) \oplus \nu(g)$ and $(\mu \otimes \nu)(g) = \mu(g) \otimes \nu(g)$. And the operations making $\text{sim}_G(R)$ a commutative semiring isomorphic to $\mathfrak{I}(\text{lat}_G \ \mathcal{F}(R))$ are

$$s(\mu) + s(\nu) \ = \ s(\mu \oplus \nu) ,$$
$$s(\mu) \cdot s(\nu) \ = \ s(\mu \otimes \nu) .$$

The identities are the similarity classes of the constant map $G \to GL_0(R) = \{\emptyset\}$ for addition, and the constant map to 1_R, $G \to GL_1(R) = R^*$, for multiplication.

Because the "affording map" $\mathfrak{I}(\text{lat}_G \ \mathcal{F}(R)) \cong \text{sim}_G(R)$ is such a perfect correspondence, the adjectives applying to a lattice M are also applied to the representations afforded by M. So a linear representation ρ, or matrix representation μ, is called **simple, indecomposable,** or **projective** if it is afforded by a lattice M that is a simple, indecomposable, or projective RG-module, respectively. Simple representations are more commonly known as **irreducible** representations; they play a role in representation theory over a field of zero characteristic that is analogous to the role of irreducibles in a factorial ring. Over an arbitrary field, the relationships among irreducible, indecomposable, and projective representations are spelled out in Proposition (8.14) below.

(8.9) Definition. The **representation ring** of G on $\mathcal{C}$ is the Grothendieck ring

$$K_0(\text{lat}_G \ \mathcal{C}, \oplus, \overline{\otimes}) \ ,$$

which is the group completion of the semiring $\mathfrak{J}(\text{lat}_G \ \mathcal{C})$ with respect to $\oplus$. In simpler terms, $K_0(\text{lat}_G \ \mathcal{C}, \oplus, \overline{\otimes})$ is the quotient of the free $\mathbb{Z}$-module based on the set of isomorphism classes in $\text{lat}_G \ \mathcal{C}$, modulo the $\mathbb{Z}$-submodule generated by all

$$c(M \oplus M') \ - \ c(M) \ - \ c(M')$$

for $M, M' \in \text{lat}_G \ \mathcal{C}$. Denoting the coset of an isomorphism class $c(M)$ by $[M]$, a typical element of the representation ring has the form $[M] - [N]$ for $M, N \in \text{lat}_G \ \mathcal{C}$. The ring operations are determined by

$$[M] + [N] \ = \ [M \oplus N] \ ,$$

$$[M] \cdot [N] \ = \ [M \overline{\otimes}_R N] \ .$$

And $[M] = [N]$ if and only if there exists $M' \in \text{lat}_G \ \mathcal{C}$ with $M \oplus M' \cong N \oplus M'$. In the literature, the representation ring on $\mathcal{P}(R)$

$$K_0(\text{lat}_G \ \mathcal{P}(R), \oplus, \overline{\otimes})$$

is often studied and is variously denoted as

$$a(RG), \quad a_R(G) \quad \text{or} \quad \boldsymbol{\mathcal{R}_R(G)} \ .$$

Sometimes it is called the **Green ring**, after the representation theorist J. A. Green.

When $\mathcal{C} = \mathcal{F}(R)$, we can describe the representation ring as the group completion of $\text{sim}_G(R)$ under $\oplus$; so its elements are differences $[\mu] - [\nu]$ for matrix representations μ and ν of G over R, the operations are determined by

$$[\mu] + [\nu] \ = \ [\mu \oplus \nu] \ ,$$

$$[\mu] \cdot [\nu] \ = \ [\mu \otimes \nu] \ ,$$

and $[\mu] = [\nu]$ if and only if there exists a matrix representation μ' with $\mu \oplus \mu'$ similar to $\nu \oplus \mu'$.

8A. Exercises

1. Suppose D_n is the dihedral group of order $2n$, generated by a and b, with defining relations $a^n = 1$, $b^2 = 1$, $bab^{-1} = a^{-1}$. Show, for each positive divisor d of n, there is a group homomorphism $\rho : D_n \to \text{Aut}_{\mathbb{R}}(\mathbb{C})$, with $\rho(a) =$

multiplication by $\zeta_d = e^{2\pi i/d}$, and $\rho(b) = $ complex conjugation. Assuming $d > 2$, show $1, \zeta_d$ is an $\mathbb{R}$-basis of $\mathbb{C}$, and if $\alpha : \mathbb{C} \cong \mathbb{R}^{2\times 1}$ is the associated coordinate map, determine $\rho^\alpha(a)$ and $\rho^\alpha(b)$. *Hint:* For the existence of the homomorphism ρ, just show $\rho(a)^n = i_{\mathbb{C}}$, $\rho(b)^2 = i_{\mathbb{C}}$, and $\rho(b) \circ \rho(a) \circ \rho(b)^{-1} = \rho(a)^{-1}$ in $\mathrm{Aut}_{\mathbb{R}}(\mathbb{C})$. For the matrix representation, note that $\zeta_d^2 = -1 + (\zeta_d + \zeta_d^{-1})\zeta_d$ and $\zeta_d + \zeta_d^{-1} \in \mathbb{R}$.

2. How does the preceding representation, with $n = 4$ and $d = 4$, compare to the representation in Example (8.2)?

3. Suppose $\mathbb{Q} \subseteq F$ is a Galois field extension of finite degree n and $\mathcal{O}_F$ is the ring of algebraic integers in F. By §7D, Exercise 11, $\mathcal{O}_F$ has a $\mathbb{Z}$-basis with n elements. The Galois group $G = \mathrm{Aut}(F/\mathbb{Q})$ permutes the roots of each polynomial in $\mathbb{Z}[x]$, so each $\sigma \in G$ restricts to an automorphism of $\mathcal{O}_F$. This restriction defines a $\mathbb{Z}$-linear representation $\rho : G \to \mathrm{Aut}_{\mathbb{Z}}(\mathcal{O}_F)$ of G. The $\mathbb{Z}G$-module $(\mathcal{O}_F, \rho)$ is known as a **Galois module**. (It is known to be a projective $\mathbb{Z}G$-module if and only if no prime $p \in \mathbb{Z}$ divides its own ramification indices in $\mathcal{O}_F$.) Prove $(\mathcal{O}_F, \rho)$ is a free $\mathbb{Z}G$-module if and only if there exists $\theta \in F$ whose conjugates $\{\sigma(\theta) : \sigma \in G\}$ form a $\mathbb{Z}$-basis of $\mathcal{O}_F$. (The determination of those F for which these equivalent statements are true is the **normal integral basis** problem — for further reading on this problem, see Frohlich [83] and Taylor [81].) *Hint:* Compare free ranks of $\mathbb{Z}$-modules to show $(\mathcal{O}_F, \rho)$ is a free $\mathbb{Z}G$-module if and only if it has free rank 1 over $\mathbb{Z}G$. Now examine the definition of scalar multiplication on $(\mathcal{O}_F, \rho)$.

4. **Representation of a ring:** Suppose S is a ring, R is a commutative ring, and M is an S, R-bimodule ${}_S M_R$. Prove the map $\rho : S \to \mathrm{End}_R(M_R)$, taking $s \in S$ to (left) scalar multiplication by s, is an injective ring homomorphism. If M_R has a finite basis $b_1, \ldots, b_n$ with (right) R-linear coordinate map $\alpha : M_R \cong R^{n\times 1}$, prove the composite

$$\rho^\alpha = \mathrm{Mat}_\alpha^\alpha \circ \rho : S \to M_n(R)$$

is an injective ring homomorphism, and so it restricts to a faithful matrix representation of each subgroup G of the group S^* of units in S. **Note:** If $s \in S$, the j-column of $\rho^\alpha(s)$ is the list of coefficients when $s \cdot b_j$ is written as a *right* R-linear combination of $b_1, \ldots, b_n$.

5. To illustrate Exercise 4, take S to be the ring $\mathbb{H}$ of quaternions, which has $\mathbb{R}$ as a subring; has $rh = hr$ for all $r \in \mathbb{R}$, $h \in \mathbb{H}$; has $\mathbb{R}$-basis $1, i, j, ij$; and has $i^2 = j^2 = (ij)^2 = -1$, so that $ij = -ji$. Take R to be the subring $\mathbb{C} = \mathbb{R} + \mathbb{R}i$ of complex numbers in $\mathbb{H}$, and take ${}_S M_R$ to be ${}_{\mathbb{H}}\mathbb{H}_{\mathbb{C}}$ under ring multiplication. Using the basis $1, j$ of $\mathbb{H}_{\mathbb{C}}$, calculate $\rho^\alpha(x + yi)$ for $x, y \in \mathbb{R}$, and $\rho^\alpha(j)$. For each integer $n \geq 2$, the **dicyclic group** Q_n is the subgroup of $\mathbb{H}^*$ generated by $\zeta_{2n} = e^{2\pi i/2n}$ and j. Taking $a = \zeta_{2n}$ and $b = j$, show Q_n has the $4n$ different elements $a^u b^v$ ($0 \leq u \leq 2n - 1$, $0 \leq v \leq 1$) and has group presentation with generators a, b and defining relations $a^n = b^2$, $b^4 = 1$, $ba = a^{-1}b$. For the faithful representation $\rho^\alpha : \mathbb{H} \to M_2(\mathbb{C})$ above, calculate $\rho^\alpha(a)$ and $\rho^\alpha(b)$.

6. Repeat Exercise 5 with $R = \mathbb{R}$ instead of $\mathbb{C}$, and with basis $1, i, j, ij$ of $\mathbb{H}_{\mathbb{R}}$, so that ρ^α embeds $\mathbb{H}$ into $M_4(\mathbb{R})$.

7. Suppose G is a finite group, H is a normal subgroup of order m, and R is a commutative ring with $m \in R^*$. Denote by e the average of the elements of H:

$$e \;=\; \frac{1}{m} \sum_{h \in H} h$$

in RG. Prove $ee = e$, $ex = xe$ for all $x \in RG$, and $he = e$ if $h \in H$. If H has index r and the members of G/H are $g_1 H, \ldots, g_r H$, prove $g_1 e, \ldots, g_r e$ is an R-basis of RGe and the matrix representation of G afforded by RGe with respect to this basis is the canonical group homomorphism $G \to G/H$, followed by the regular representation of G/H with respect to the R-basis $g_1 H, \ldots, g_r H$ of $R[G/H]$. Use this to find a matrix representation over $\mathbb{Q}$ of degree 6 for the dihedral group D_6 of order 12. In this case, what is the idempotent e?

8. In Exercise 7, show $f = 1 - e$ satisfies $ff = f$ and $ef = fe = 0$. Prove $RG = RGe \overset{\bullet}{\oplus} RGf$. Show that if $v_1, \ldots, v_r$ is an R-basis of RGe and $w_1, \ldots, w_s$ is an R-basis of RGf, then the regular representation of G with respect to $v_1, \ldots, v_r, w_1, \ldots, w_s$ is $\mu_1 \oplus \mu_2$, where μ_1 is the representation afforded by RGe with respect to $v_1, \ldots, v_r$ and μ_2 is the representation afforded by RGf with respect to $w_1, \ldots, w_s$. Find matrix representations μ_1, μ_2 of D_6 of degree 6 over $\mathbb{Q}$ so the regular representation of D_6 over an appropriate $\mathbb{Q}$-basis of $\mathbb{Q}D_6$ is $\mu_1 \oplus \mu_2$.

8B. Representing Finite Groups Over Fields

Suppose, in this section, that G is a finite group and F is a field. An FG-module is finitely generated over FG if and only if it is finitely generated over F. So the FG-lattices with respect to $\mathcal{F}(F) = \mathcal{P}(F) = \mathcal{M}(F)$ are just the modules in $\mathcal{M}(FG)$. If L_1 is a proper submodule of $L_2 \in \mathcal{M}(FG)$, then $\dim_F(L_1) < \dim_F(L_2)$; so strictly descending chains in $\mathcal{M}(FG)$ must have finite length, and FG is a left artinian ring. This leads to useful information about the representation ring $\mathcal{R}_F(G) = K_0(\mathcal{M}(FG), \oplus)$.

For the rest of this section, assume $\mathcal{A}$ is a left artinian ring. The inclusion $\mathcal{P}(\mathcal{A}) \to \mathcal{M}(\mathcal{A})$ and the identity $\mathcal{M}(\mathcal{A}) \to \mathcal{M}(\mathcal{A})$ induce group homomorphisms

$$\text{(8.10)} \qquad
\begin{array}{ccccc}
K_0(\mathcal{P}(\mathcal{A}), \oplus) & \overset{\theta}{\longrightarrow} & K_0(\mathcal{M}(\mathcal{A}), \oplus) & \overset{\psi}{\longrightarrow} & K_0(\mathcal{M}(\mathcal{A})) , \\[2pt]
\| & & & & \| \\[2pt]
K_0(\mathcal{A}) & & & & G_0(\mathcal{A})
\end{array}$$

whose composite is the Cartan map. By the Krull-Schmidt theory (3.23)–(3.27), each $L \in \mathcal{M}(\mathcal{A})$ has an expression

$$L \cong L_1 \oplus \cdots \oplus L_n$$

as a direct sum of indecomposable modules $L_i \in \mathcal{M}(\mathcal{A})$; and any other such expression for L differs only in the order of summands and the replacement of each L_i by an isomorphic copy L'_i. That is, the monoid $\mathcal{I}(\mathcal{M}(\mathcal{A}))$ of isomorphism classes, under $\oplus$, is free abelian based on the isomorphism classes of indecomposables in $\mathcal{M}(\mathcal{A})$.

Applying the "affording" isomorphism (8.8) of semirings

$$\mathcal{I}(\mathcal{M}(FG)) \;\cong\; \mathrm{sim}_G(F) \;,$$

we see that $\mathrm{sim}_G(F)$ is, additively, the free abelian monoid based on the similarity classes of indecomposable matrix representations of G over F. So each matrix representation $\mu : G \to M_d(F)$ is similar to a direct sum $\mu_1 \oplus \cdots \oplus \mu_n$,

$$g \;\mapsto\; \begin{bmatrix} \mu_1(g) & & & \\ & \mu_2(g) & & \\ & & \ddots & \\ & & & \mu_n(g) \end{bmatrix},$$

of indecomposables $\mu_i \in \mathrm{sim}_G(F)$. And any other such decomposition of μ differs only in the order of summands and the replacement of each μ_i by a similar μ'_i. So the determination of all matrix representations of G over F depends only on the discovery of one member from each isomorphism class of indecomposables in $\mathcal{M}(FG)$.

Since $(\mathcal{I}(\mathcal{M}(\mathcal{A})), \oplus)$ is a free abelian monoid, it is cancellative; so the group completion

$$\begin{aligned} \mathcal{I}(\mathcal{M}(\mathcal{A})) &\;\to\; K_0(\mathcal{M}(\mathcal{A}), \oplus) \\ c(L) &\;\mapsto\; [L] \end{aligned}$$

is injective. Hence $[L] = [L']$ if and only if $L \cong L'$, and we can view the cosets $[L]$ as isomorphism classes $c(L)$ in $\mathcal{M}(\mathcal{A})$. By (3.19), $K_0(\mathcal{M}(\mathcal{A}), \oplus)$ is the free $\mathbb{Z}$-module with the same basis as the monoid $\mathcal{I}(\mathcal{M}(\mathcal{A}))$ — namely, the isomorphism classes $[L]$ of indecomposables $L \in \mathcal{M}(\mathcal{A})$.

If $L \cong L_1 \oplus \cdots \oplus L_n$ in $\mathcal{M}(\mathcal{A})$, then L is projective if and only if each L_i is projective. So $(\mathcal{I}(\mathcal{P}(\mathcal{A})), \oplus)$ is also a free abelian monoid, with basis the isomorphism classes $[P]$ of indecomposables $P \in \mathcal{P}(\mathcal{A})$, and the preceding paragraph remains true if $\mathcal{P}(\mathcal{A})$ replaces $\mathcal{M}(\mathcal{A})$. So the map θ in (8.10) is just the inclusion of $K_0(\mathcal{A})$ as the $\mathbb{Z}$-submodule of $K_0(\mathcal{M}(\mathcal{A}), \oplus)$ spanned by the f.g. projective indecomposables.

A first place to look for f.g. projective indecomposable $\mathcal{A}$-modules is among the summands in a Krull-Schmidt decomposition of $\mathcal{A}$ itself. To better understand these summands in terms internal to $\mathcal{A}$, we extend the notion of internal direct sum (see §2A) to more than two summands:

(8.11) Definitions. Suppose A is a ring and $M_1, \ldots, M_n$ are submodules of an A-module M. The (**finite**) **sum**

$$\sum_{i=1}^{n} M_i = M_1 + \cdots + M_n$$

is the set of all $m_1 + \cdots + m_n$ with $m_i \in M_i$ for each i. It equals the A-linear span of the union $\bigcup_{i=1}^{n} M_i$. More generally, suppose $\mathfrak{I}$ is any set and, for each $i \in \mathfrak{I}$, M_i is a submodule of M. The **sum**

$$\sum_{i \in \mathfrak{I}} M_i$$

is the A-linear span of the union $\bigcup_{i \in \mathfrak{I}} M_i$. It equals the union of all finite sums of the M_i.

A finite sum $M_1 + \cdots + M_n$ is **direct** (or is an **internal direct sum**) and is written as

$$\overset{\bullet}{\oplus}_{i=1}^{n} M_i = M_1 \overset{\bullet}{\oplus} \cdots \overset{\bullet}{\oplus} M_n \ ,$$

if each of its elements has only one expression $m_1 + \cdots + m_n$ with $m_i \in M_i$ for each i. The sum $\sum_{i=1}^{n} M_i$ is direct if and only if

$$M_i \cap (M_1 + \cdots + M_{i-1}) = 0$$

for each $i > 1$. More generally, a sum $\sum_{i \in \mathfrak{I}} M_i$ is **direct** (or is an **internal direct sum**), written

$$\overset{\bullet}{\oplus}_{i \in \mathfrak{I}} M_i \ ,$$

if for each finite subset $\mathfrak{J}$ of $\mathfrak{I}$ the finite sum $\sum_{i \in \mathfrak{J}} M_i$ is direct. The latter is true if and only if

$$M_i \cap \left(\sum_{j \in \mathfrak{I} \setminus \{i\}} M_j \right) = 0$$

for all $i \in \mathfrak{I}$.

Note: Each internal direct sum $M_1 \overset{\bullet}{\oplus} \cdots \overset{\bullet}{\oplus} M_n$ is isomorphic to the external direct sum $M_1 \oplus \cdots \oplus M_n$ of the same modules, by taking $m_1 + \cdots + m_n$ to $(m_1, \ldots, m_n)$. On the other hand, each external direct sum $P_1 \oplus \cdots \oplus P_n$ of R-modules P_i is an internal direct sum $P_1' \oplus \cdots \oplus P_n'$, where

$$P_i' = N_1 \oplus \cdots \oplus N_n \cong P_i$$

with $N_i = P_i$, and $N_j = 0$ for $j \neq i$. Moreover, any isomorphism

$$\alpha : P_1 \oplus \cdots \oplus P_n \cong M$$

defines an internal direct sum decomposition $M = M_1 \overset{\bullet}{\oplus} \cdots \overset{\bullet}{\oplus} M_n$ with $M_i = \alpha(P_i') \cong P_i$ for each i.

In this way, any decomposition $\mathcal{A} \cong P_1 \oplus \cdots \oplus P_n$ of $\mathcal{A}$ into indecomposable $\mathcal{A}$-modules P_i imposes an internal decomposition

$$\mathcal{A} = I_1 \overset{\bullet}{\oplus} \cdots \overset{\bullet}{\oplus} I_n$$

into indecomposable left ideals $I_i \cong P_i$. For a better grasp of these ideals I_i, we consider some basic facts about idempotents:

(8.12) Definitions. Recall that an element e of a ring A is **idempotent** if $ee = e$. Idempotents $e_1, \ldots, e_n \in A$ (with $n \geq 2$) are **mutually orthogonal** if $e_i e_j = 0$ whenever $i \neq j$. An idempotent in A is **primitive** if it is nonzero, and not a sum of two mutually orthogonal nonzero idempotents from A.

(8.13) Proposition. *Suppose A is a ring.*

(i) *If $A = I_1 \overset{\bullet}{\oplus} \cdots \overset{\bullet}{\oplus} I_n$ for n (≥ 2) left ideals I_i of A, and we write $1 = e_1 + \cdots + e_n$ with each $e_i \in I_i$, then $e_1, \ldots, e_n$ are mutually orthogonal idempotents, and $I_i = Ae_i$ for each i.*

(ii) *If $e_1, \ldots, e_m$ are mutually orthogonal idempotents of A, their sum $e = e_1 + \cdots + e_m$ is idempotent, and $Ae = Ae_1 \overset{\bullet}{\oplus} \cdots \overset{\bullet}{\oplus} Ae_m$. Further, $Ae = A$ if and only if $e = 1$.*

(iii) *An idempotent $e \in A$ is primitive if and only if Ae is an indecomposable left ideal of A.*

(iv) *If N is a nilpotent ideal of A and e is a primitive idempotent of A, then $\bar{e} = e + N$ is a primitive idempotent of the ring A/N.*

Proof. For (i), if $x \in I_i$, then uniqueness of the expression of $x = xe_1 + \cdots + xe_n$ in $I_1 \overset{\bullet}{\oplus} \cdots \overset{\bullet}{\oplus} I_n$ implies $x = xe_i$ and $xe_j = 0$ for $j \neq i$. In particular, each $e_i = e_i e_i$, $e_i e_j = 0$ for $i \neq j$ and $Ae_i \subseteq I_i = I_i e_i \subseteq Ae_i$ for each i.

For (ii), e is idempotent by the distributive law. Since $e_i e = e_i$ for each i, every Ae_i is a submodule of Ae. If

$$x \in Ae_i \cap (Ae_1 + \cdots + Ae_{i-1}) ,$$

then $x = xe_i = (a_1 e_1 + \cdots + a_{i-1} e_{i-1})e_i = 0$; so the sum of the Ae_i is direct. Finally, if $Ae = A$, then $A(1 - e) = 0$; so $1 - e = (1)(1 - e) = 0$, and $e = 1$.

To prove (iii), suppose $Ae = I \overset{\bullet}{\oplus} J$ for nonzero left ideals I and J of A. Since $e, 1 - e$ are mutually orthogonal idempotents adding up to 1, $A = Ae \overset{\bullet}{\oplus} A(1 - e)$. So

$$A = I \overset{\bullet}{\oplus} J \overset{\bullet}{\oplus} A(1 - e).$$

Also, $e = e_1 + e_2$ with $e_1 \in I$ and $e_2 \in J$; so $1 = e_1 + e_2 + (1 - e)$. By part (i), e_1 and e_2 are mutually orthogonal idempotents and $I = Ae_1$, $J = Ae_2$; so e_1 and e_2 are nonzero, and e is not primitive.

Conversely, suppose $e = e_1 + e_2$ for nonzero mutually orthogonal idempotents e_1 and e_2. By (ii), $Ae = Ae_1 \overset{\bullet}{\oplus} Ae_2$; so Ae is not indecomposable.

For (iv), suppose $N^k = 0$ where $k \geq 1$. In the ring $\mathbb{Z}[x]$,

$$1 = [x + (1 - x)]^{2k} = \sum_{i=0}^{2k} \binom{2k}{i} x^{2k-i}(1 - x)^i .$$

Define

$$p(x) = \sum_{i=0}^{k} \binom{2k}{i} x^{2k-i}(1 - x)^i .$$

Then $1 = p(x) + q(x)$, where x^k divides $p(x)$ and $(1 - x)^k$ divides $q(x)$. So modulo $x^k(1 - x)^k$, $1 \equiv p(x)^2 + q(x)^2$, and

$$p(x)^2 - p(x) \equiv q(x) - q(x)^2 = p(x)q(x) \equiv 0 .$$

Suppose $\overline{e}$ is not primitive; so $\overline{e} = f_1 + f_2$ for mutually orthogonal nonzero idempotents f_1 and f_2 in A/N. Choose $a \in A$ with $\overline{a} = f_1$ and let $s = eae$. Then $\overline{s} = \overline{e}\,\overline{a}\,\overline{e} = f_1$ and $se = s = es$. Let e_1 denote $p(s)$, where $p(x)$ is the polynomial defined above. Since $\overline{s} - \overline{s}^2 = 0$, $s - s^2 \in N$. So $s^k(1 - s)^k = 0$. Therefore $e_1^2 - e_1 = 0$ and $e_1^2 = e_1$. Since $p(x) = xr(x) = r(x)x$ for $r(x) \in \mathbb{Z}[x]$, $e_1 e = r(s)se = r(s)s = e_1$ and $ee_1 = esr(s) = sr(s) = e_1$. Further, $\overline{e_1} = \overline{p(s)} = p(\overline{s}) = \overline{s}^{2k} = f_1^{2k} = f_1$ since $s(1 - s) \in N$. So $e_1 \neq 0$ and $e_1 \neq e$. Take $e_2 = e - e_1$. Then $e_2^2 = e^2 - ee_1 - e_1e + e_1^2 = e - e_1 = e_2$. And $e_2 \neq 0$. Also, $e_1 e_2 = e_2 e_1 = 0$. So e is the sum $e_1 + e_2$ of nonzero mutually orthogonal idempotents, and so it is not primitive. ∎

Applying this proposition to a left artinian ring $\mathcal{A}$, one can envision a procedure to generate every decomposition of $\mathcal{A}$ as a direct sum of indecomposable left ideals: Choose any primitive idempotent $e_1 \in \mathcal{A}$. Then, in sequence, choose primitive idempotents e_i ($i = 2, 3, \ldots$) so that $e_i e_j = 0 = e_j e_i$ for all $j < i$. Eventually $e_1 + \cdots + e_n = 1$, and no further choices are possible, since $\mathcal{A} = \mathcal{A}e_1 \overset{\bullet}{\oplus} \cdots \overset{\bullet}{\oplus} \mathcal{A}e_n$ is a Krull-Schmidt decomposition of $\mathcal{A}$. Of course, this method works only if we can get our hands on the primitive idempotents of $\mathcal{A}$.

Now suppose we fix a choice of mutually orthogonal primitive idempotents $e_1, \ldots, e_n$ of $\mathcal{A}$ with $e_1 + \cdots + e_n = 1$, numbered so that $\mathcal{A}e_1, \ldots, \mathcal{A}e_m$ are pairwise nonisomorphic, and each $\mathcal{A}e_i$ for $i > m$ is isomorphic to $\mathcal{A}e_j$ for some $j \leq m$. Being direct summands of $\mathcal{A}$, these $\mathcal{A}e_i$ are singly generated projective $\mathcal{A}$-modules. As noted in (3.27), every f.g. projective indecomposable $\mathcal{A}$-module P is isomorphic to a direct summand of some $\mathcal{A}^s$ mapping onto it, and by Krull-Schmidt uniqueness to one of the $\mathcal{A}e_i$ for $1 \leq i \leq m$; so the distinct isomorphism classes of indecomposables in $\mathcal{P}(\mathcal{A})$ are $[\mathcal{A}e_1], \ldots, [\mathcal{A}e_m]$, a $\mathbb{Z}$-basis of $K_0(\mathcal{A})$.

Because they are isomorphic to principal left ideals $\mathcal{A}e_i$, the f.g. projective indecomposable $\mathcal{A}$-modules are also known as **principal indecomposable $\mathcal{A}$-modules**. They are closely related to the simple $\mathcal{A}$-modules:

(8.14) Proposition. *Suppose $\mathcal{A}$ is a left artinian ring and I is a left ideal of $\mathcal{A}$ generated by a primitive idempotent.*

(i) *There is a simple left ideal of $\mathcal{A}$ contained in I. If I is not simple, every simple left ideal inside I is nilpotent.*

(ii) *The left ideal $\mathrm{rad}(\mathcal{A})I$ of $\mathcal{A}$ is the unique maximal proper submodule of I.*

(iii) *If $\mathcal{A} = I_1 \overset{\bullet}{\oplus} \cdots \overset{\bullet}{\oplus} I_n$ for indecomposable left ideals I_i, there is also a decomposition $\mathrm{rad}(\mathcal{A}) = \mathrm{rad}(\mathcal{A})I_1 \overset{\bullet}{\oplus} \cdots \overset{\bullet}{\oplus} \mathrm{rad}(\mathcal{A})I_n$.*

(iv) *Taking P to $P/\mathrm{rad}(\mathcal{A})P$ defines a bijection $F : S \to T$ where S is the set of isomorphism classes of principal indecomposable $\mathcal{A}$-modules and T is the set of isomorphism classes of simple $\mathcal{A}$-modules. The map F carries the f.g. projective simple modules to themselves, and the f.g. projective indecomposables that are not simple to the simple modules that are not projective.*

Proof. Since I is a f.g. $\mathcal{A}$-module and $\mathcal{A}$ is left artinian, I is artinian — so it satisfies the minimum condition on its submodules. The set of nonzero submodules of I is nonempty ($I \neq 0$); so it has a minimal element, which must be simple.

Suppose M is a nonnilpotent simple left ideal of $\mathcal{A}$ contained in I. Since M is simple, $M \neq 0$ and its only submodules are M and 0. Since $M^2 \neq 0$, M^2 must be M. Then $Mx \neq 0$ for some $x \in M$. So $Mx = M$, and $ex = x$ for some $e \in M$. Then $e^2 x = ex$, and

$$e^2 - e \in \mathrm{ann}_A(x) \cap M \, .$$

Since $Mx \neq 0$, this intersection is not M, so it must be 0. So $e^2 = e$. Since $x \neq 0$, $e \neq 0$. Then
$$e = e^2 \in Ie \subseteq M$$

implies $Ie = M$. Since I is indecomposable, $I = Ie \overset{\bullet}{\oplus} I(1-e)$ implies $I = M$, which is simple. This proves (i).

For (ii), suppose J is a maximal proper submodule of I. Then I/J is simple and $\mathrm{rad}(\mathcal{A})(I/J) = 0$; so $\mathrm{rad}(\mathcal{A})I \subseteq J$. It only remains to show $I/\mathrm{rad}(\mathcal{A})I$ is simple. Say $I = \mathcal{A}e$ for a primitive idempotent e. Let $\overline{\mathcal{A}}$ denote the ring $\mathcal{A}/\mathrm{rad}(\mathcal{A})$ and $\bar{e}$ denote $e + \mathrm{rad}(\mathcal{A})$. Since $\mathrm{rad}(\mathcal{A})$ is a two-sided ideal of $\mathcal{A}$, $\mathrm{rad}(\mathcal{A})I \subseteq \mathrm{rad}(\mathcal{A})$. The surjective $\mathcal{A}$-linear map

$$\frac{I}{\mathrm{rad}(\mathcal{A})I} \longrightarrow \frac{I}{\mathrm{rad}(\mathcal{A})} \lhd \overline{\mathcal{A}}$$

is an isomorphism, since each $x \in \mathrm{rad}(\mathcal{A}) \cap I = \mathrm{rad}(\mathcal{A}) \cap \mathcal{A}e$ has the form $x = xe \in \mathrm{rad}(\mathcal{A})I$. Now

$$\frac{I}{\mathrm{rad}(\mathcal{A})} = \frac{\mathcal{A}e}{\mathrm{rad}(\mathcal{A})} = \overline{\mathcal{A}}\bar{e}.$$

Since $\mathcal{A}$ is left artinian, $\mathrm{rad}(\mathcal{A})$ is nilpotent by (6.34); so $\bar{e}$ is a primitive idempotent of $\overline{\mathcal{A}}$. Now $\overline{\mathcal{A}}$ is left artinian (see (B.8)) and $\mathrm{rad}(\overline{\mathcal{A}}) = 0$ (see §6D, Exercise 3). By (6.34), every nilpotent left ideal of $\overline{\mathcal{A}}$ lies in $\mathrm{rad}(\overline{\mathcal{A}})$; so it is 0. So by (i) above, $\overline{\mathcal{A}}\bar{e}$ is a simple $\overline{\mathcal{A}}$-module, and hence a simple $\mathcal{A}$-module.

For (iii),

$$\mathrm{rad}(\mathcal{A}) = \mathrm{rad}(\mathcal{A})(I_1 + \cdots + I_n)$$
$$\subseteq \mathrm{rad}(\mathcal{A})I_1 + \cdots + \mathrm{rad}(\mathcal{A})I_n \subseteq \mathrm{rad}(\mathcal{A}),$$

and the latter sum is direct, since each $\mathrm{rad}(\mathcal{A})I_i \subseteq I_i$, and $I_1 + \cdots + I_n$ is direct.

Finally, consider (iv). If $P \cong I$, then $P/\mathrm{rad}(\mathcal{A})P \cong I/\mathrm{rad}(\mathcal{A})I$, which is simple by (ii). Since principal indecomposables are in $\mathcal{P}(\mathcal{A})$, the map $F : S \to T$ is injective by (6.36). For surjectivity, suppose M is a simple $\mathcal{A}$-module and $\mathcal{A} = I_1 \overset{\bullet}{\oplus} \cdots \overset{\bullet}{\oplus} I_n$ for indecomposable left ideals I_i. Since $\mathcal{A}M = M \neq 0$, there exists some $I = I_i$ with $IM \neq 0$. So $Ix \neq 0$ for some $x \in M$. Then $Ix = M$. So the $\mathcal{A}$-linear map $I \to M$, $a \mapsto ax$, is surjective, with kernel J, a maximal proper submodule of I. By (ii), $J = \mathrm{rad}(\mathcal{A})I$. So $M \cong I/\mathrm{rad}(\mathcal{A})I$, proving F is surjective.

If a principal indecomposable P is simple, $\mathrm{rad}(\mathcal{A})P = 0$; so $P/\mathrm{rad}(\mathcal{A})P = P$, which is projective. Conversely, if P is a principal indecomposable and $P/\mathrm{rad}(\mathcal{A})P$ is projective, the sequence

$$0 \longrightarrow \mathrm{rad}(\mathcal{A})P \longrightarrow P \longrightarrow P/\mathrm{rad}(\mathcal{A})P \longrightarrow 0$$

splits. Since P is indecomposable, $P = P/\mathrm{rad}(\mathcal{A})P$, which is simple. ∎

Let's summarize what we have proved about the sequence (8.10):

$$K_0(\mathcal{A}) \overset{\theta}{\longrightarrow} K_0(M(\mathcal{A}), \oplus) \overset{\psi}{\longrightarrow} G_0(\mathcal{A})$$
$$[P] \longmapsto [P] \quad , \quad [M] \longmapsto [M]$$

for a left artinian ring $\mathcal{A}$. The middle group is the free $\mathbb{Z}$-module based on the isomorphism classes of indecomposables in $\mathcal{M}(\mathcal{A})$. For $M \in \mathcal{M}(\mathcal{A})$, the $[I]$th coordinate of M is the number of summands isomorphic to I in a Krull-Schmidt decomposition of M.

The map θ embeds $K_0(\mathcal{A})$ as the $\mathbb{Z}$-submodule with basis the isomorphism classes of indecomposables in $\mathcal{P}(\mathcal{A})$. This basis is finite, represented by the indecomposables in any Krull-Schmidt decomposition of $\mathcal{A}$.

Now f.g. modules over left artinian rings have finite length composition series. Using devissage, we saw in (3.44) that $G_0(\mathcal{A})$ is the free $\mathbb{Z}$-module based on the isomorphism classes $[X]$ of simple $\mathcal{A}$-modules X, and this basis is finite, represented by the composition factors of $\mathcal{A}$. By (3.41), the $[X]$-coordinate of $[M]$ for $M \in \mathcal{M}(\mathcal{A})$ is the number of composition factors of M that are isomorphic to X.

There is an additive functor $F : \mathcal{P}(\mathcal{A}) \to \mathcal{M}(\mathcal{A})$ taking each object P to $\overline{P} = P/\mathrm{rad}(\mathcal{A})P$ and each arrow $f : P \to Q$ to the induced map $\overline{f} : \overline{P} \to \overline{Q}$. Proposition (8.14) (iv) shows F restricts to a bijection $S \to T$ from isomorphism classes of indecomposables to isomorphism classes of simples; so the $\mathbb{Z}$-linear map $K_0(\mathcal{A}) \to G_0(\mathcal{A})$, $[P] \to [\overline{P}]$, is an isomorphism, carrying each principal indecomposable $\mathcal{A}$-module to its unique top composition factor.

The composite $\psi \circ \theta : K_0(\mathcal{A}) \to G_0(\mathcal{A})$ is the Cartan map $[P] \mapsto [P]$, and by (3.41), in $G_0(\mathcal{A})$, $[P]$ is the sum $[X_1] + \cdots + [X_n]$, where $X_1, \ldots, X_n$ is the list of composition factors of P. That is, the Cartan map sends each principal indecomposable to the sum of *all* of its composition factors. If $P_1, \ldots, P_m$ represent the isomorphism classes of principal indecomposables and $X_1, \ldots, X_m$ represent the isomorphism classes of simple modules, the Cartan map is represented over the $\mathbb{Z}$-bases $[P_1], \ldots, [P_m]$ of $K_0(\mathcal{A})$ and $[X_1], \ldots, [X_m]$ of $G_0(\mathcal{A})$, as right multiplication by the "Cartan matrix" (a_{ij}) over $\mathbb{Z}$, where a_{ij} is the number of composition factors isomorphic to X_j in a composition series for P_i. We considered the Cartan matrix of a left artinian ring at the end of §3C, but had not shown, at that point, that it is square.

The map θ is injective, and the map ψ is surjective (amounting only to the imposition of extra defining relations). And the $\mathbb{Z}$-modules $K_0(\mathcal{A})$ and $G_0(\mathcal{A})$ have equal free rank. But the middle group $K_0(\mathcal{M}(\mathcal{A}), \oplus)$, which is the representation ring $\mathcal{R}_F(G)$ when $\mathcal{A} = FG$, can be larger. In fact it can have infinite free rank!

(8.15) Example. (Curtis and Reiner [62, (64.3)]) Suppose F is a field of prime characteristic p and G is a group isomorphic to $(\mathbb{Z}/p\mathbb{Z}) \oplus (\mathbb{Z}/p\mathbb{Z})$; so G is generated by a and b with defining relations $a^p = 1$, $b^p = 1$, and $ba = ab$. Then FG has an indecomposable module of every odd dimension over F:

Consider the $(n + 1, n)$ by $(n + 1, n)$ partitioned matrices

$$u(A) = \begin{bmatrix} I_{n+1} & A \\ 0 & I_n \end{bmatrix} \in GL_{2n+1}(F) \, ,$$

where $A \in F^{(n+1) \times n}$. Since $u(A)u(B) = u(A + B)$, these matrices commute; and $u(A)$ has order p if the entries of A are not all zero. So there is a faithful matrix representation $\mu : G \to GL_{2n+1}(F)$ taking

$$a \mapsto u\left(\begin{bmatrix} 0 \\ I_n \end{bmatrix}\right) \ , \quad b \mapsto u\left(\begin{bmatrix} I_n \\ 0 \end{bmatrix}\right)$$

and afforded by the FG-module $M = (F^{2n+1}, \mu(-)\cdot)$. If we rename the standard basis of F^{2n+1} so $e_1, \ldots, e_{n+1}$ become $x_0, \ldots, x_n$ and $e_{n+2}, \ldots, e_{2n+1}$ become $y_1, \ldots, y_n$, then $ax_i = bx_i = x_i$ for $0 \le i \le n$, and

$$ay_i = x_i + y_i \quad (\text{so } (a-1)y_i = x_i),$$
$$by_i = x_{i-1} + y_i \quad (\text{so } (b-1)y_i = x_{i-1}),$$

for $1 \le i \le n$. Take $X = Fx_0 + \cdots + Fx_n$ and $Y = Fy_1 + \cdots + Fy_n$. As F-vector spaces, $M = X \overset{\bullet}{\oplus} Y$. Let $\pi : M \to Y$ denote projection along X (replacing each x_i by 0). Scalar multiplication by $a - 1$ and by $b - 1$ each send Y injectively into X, and send X to 0. And $(a-1)\pi(m) = (a-1)m$ for all $m \in M$.

Suppose N is a nonzero FG-submodule of M and $r = \dim_F \pi(N)$, which equals

$$\dim_F(a-1)\pi(N) = \dim_F(a-1)N .$$

We claim $\dim_F(N) \ge 2r + 1$: If $N \subseteq X$, $\pi(N) = 0$ and the claim is immediate since $N \ne 0$. Suppose $N \not\subseteq X$. Then there is a $v \in N$ with smallest i for which the y_i-coordinate of v is nonzero. So $w = (b-1)v$ does not belong to $(a-1)N$. Choose an F-basis $w_1, \ldots, w_r$ of $(a-1)N$, so each $w_i = (a-1)v_i$ for some $v_i \in N$. If $a_1 v_1 + \cdots + a_r v_r \in X$ for some $a_i \in F$, multiplying by $a-1$ shows $a_1 w_1 + \cdots + a_r w_r = 0$; so each $a_i = 0$. Therefore

$$\{w_1, \ldots, w_r, \ w, \ v_1, \ldots, \ v_r\}$$

is an F-linearly independent subset of N, since no member of this list is in the span of its predecessors. This proves the claim.

If $M = M_1 \overset{\bullet}{\oplus} M_2$ for nonzero FG-modules M_1 and M_2, the claim implies

$$\begin{aligned} \dim_F M &= \dim_F M_1 + \dim_F M_2 \\ &\ge 2[\dim_F \pi(M_1) + \dim_F \pi(M_2)] + 2 \\ &\ge 2\dim_F \pi(M) + 2 = 2n + 2 , \end{aligned}$$

a contradiction. So M is indecomposable.

Of course FG-modules with different dimensions over F cannot be isomorphic as FG-modules, since an FG-linear isomorphism is also F-linear. So, for F and G in the example, there are infinitely many isomorphism classes of indecomposables in $\mathcal{M}(FG)$. This certainly complicates the representation theory of $(\mathbb{Z}/p\mathbb{Z}) \times (\mathbb{Z}/p\mathbb{Z})$ over a field of characteristic p.

For any field F and finite group G, the projective representations of G over F are easy to describe, since they are similar to unique direct sums of the (finitely many) indecomposable summands in a decomposition of the regular representation. Therefore, the general representation theory of G over F is more manageable when all the representations are projective — that is, when $\mathcal{M}(FG) = \mathcal{P}(FG)$. Fortunately, this happens quite often:

(8.16) Maschke's Theorem. *Suppose F is a field and G is a finite group. Then $\mathcal{M}(FG) = \mathcal{P}(FG)$ if and only if the order of G is not a multiple of the characteristic of F.*

Proof. Say G has order n and $n \cdot 1_F \neq 0_F$. Consider an exact sequence

$$0 \longrightarrow K \overset{\subseteq}{\longrightarrow} M \longrightarrow P \longrightarrow 0$$

in $\mathcal{M}(FG)$. Since every F-module is projective, the sequence splits in $M(\mathcal{F})$; so there is an F-linear map $p : M \to K$ with $p(k) = k$ for all $k \in K$. Define $\pi : M \to K$ by

$$\pi(m) = (n \cdot 1_F)^{-1} \sum_{g \in G} gp(g^{-1}m) ,$$

which is also an F-linear map, since $g^{-1} \cdot (-)$ and $g \cdot (-)$ are F-linear. In fact π is FG-linear, since, for $g_0 \in G$,

$$\pi(g_0 m) = (n \cdot 1_F)^{-1} \sum_{g \in G} g_0(g_0^{-1}g)p(g^{-1}g_0 m)$$

$$= g_0(n \cdot 1_F)^{-1} \sum_{h \in G} hp(h^{-1}m) = g_0\pi(m) .$$

And, for each $k \in K$,

$$\pi(k) = (n \cdot 1_F)^{-1} \sum_{g \in G} gp(g^{-1}k) = (n \cdot 1_F)^{-1} \sum_{g \in G} gg^{-1}k$$

$$= (n \cdot 1_F)^{-1}(n \cdot 1_F)k = k .$$

So each exact sequence in $\mathcal{M}(FG)$ splits, proving each f.g. FG-module P is projective.

Conversely, suppose $n \cdot 1_F = 0_F$. If $x = \sum_{g \in G} g \in FG$, then $xg = gx = x$ for all $g \in G$. So $x^2 = nx = 0$, and x generates a nonzero nilpotent ideal N of FG. Since $NFG = N \neq 0$, we must have $NI \neq 0$ for a principal indecomposable left ideal I of FG. Being nilpotent, $N \subseteq \text{rad}(FG)$; so I is not simple, and has a proper nonzero submodule J. Since I is indecomposable, the exact sequence

$$0 \longrightarrow J \overset{\subseteq}{\longrightarrow} I \longrightarrow I/J \longrightarrow 0$$

cannot split; so I/J is finitely generated but not projective. ∎

(8.17) Note. If $\mathcal{A}$ is a left artinian ring with $\mathcal{M}(\mathcal{A}) = \mathcal{P}(\mathcal{A})$, the groups $K_0(\mathcal{A})$, $K_0(\mathcal{M}(\mathcal{A}), \oplus)$, and $G_0(\mathcal{A})$ are identical, and θ, ψ are identity maps. So if the characteristic of F does not divide the order of G, $\mathcal{R}_F(G) = K_0(FG) = G_0(FG)$; and if $e_1, \ldots, e_n$ are mutually orthogonal primitive idempotents of FG, with $e_1 + \cdots + e_n = 1$, then each left ideal FGe_i is simple, affording an irreducible representation μ_i. Numbering these so $\mu_1, \ldots, \mu_m$ represent the distinct similarity classes among the μ_i, each matrix representation μ of G over F is similar to a direct sum of these μ_i $(1 \le i \le m)$, each appearing with multiplicity uniquely determined by μ.

8B. Exercises

1. Suppose F is a field. In the left artinian ring $\mathcal{A} = M_2(F)$, consider the idempotents

$$e = \begin{bmatrix} 1 & 0 \\ 0 & 0 \end{bmatrix}, \quad f = \begin{bmatrix} 0 & 0 \\ 0 & 1 \end{bmatrix},$$

$$g = \begin{bmatrix} 1 & 1 \\ 0 & 0 \end{bmatrix}, \quad h = \begin{bmatrix} 0 & -1 \\ 0 & 1 \end{bmatrix}.$$

Show $\mathcal{A}e$, $\mathcal{A}f$, $\mathcal{A}g$, and $\mathcal{A}h$ are simple left ideals of $\mathcal{A}$ with $\mathcal{A}f = \mathcal{A}h$ but $\mathcal{A}e \ne \mathcal{A}g$ and $\mathcal{A} = \mathcal{A}e \overset{\bullet}{\oplus} \mathcal{A}f = \mathcal{A}g \oplus \mathcal{A}h$. So in an internal Krull-Schmidt decomposition of $\mathcal{A}$, the indecomposable left ideals are not unique, nor are the idempotents generating each.

2. Suppose $\mathcal{A}$ is a left artinian ring and J is a simple left ideal of $\mathcal{A}$. Prove J is a direct summand of $\mathcal{A}$ if and only if $J^2 = J$.

3. A projective simple left ideal of a left artinian ring $\mathcal{A}$ need not be a direct summand of $\mathcal{A}$: Suppose F is a field and $\mathcal{A}$ is the F-module with basis $1, e, x$. Show $\mathcal{A}$ is a left artinian F-algebra under multiplication with

·	1	e	x
1	1	e	x
e	e	e	x
x	x	0	0

(This amounts to showing this table gives an associative operation — that is, $a(bc) = (ab)c$ whenever $a, b, c \in \{1, e, x\}$.)

Now show $\mathcal{A}e = Fe$ and $\mathcal{A}x = Fx$; so these are simple F-modules, and hence simple $\mathcal{A}$-modules. Then show right multiplication by x is an $\mathcal{A}$-linear isomorphism $\mathcal{A}e \cong \mathcal{A}x$, even though $(\mathcal{A}e)^2 = \mathcal{A}e$, while $(\mathcal{A}x)^2 = 0$. Thus $\mathcal{A}x$ is projective, but not a direct summand of $\mathcal{A}$. *This shows a principal indecomposable left ideal of a left artinian ring $\mathcal{A}$ need not appear as a summand in any internal Krull-Schmidt decomposition of $\mathcal{A}$.*

4. In a ring A, an element a is **central** if $ax = xa$ for all $x \in A$. Suppose F is a field, G is a finite group, and H is a subgroup of G of order d not divisible by the characteristic of F. Prove the average

$$(d \cdot 1_F)^{-1} \sum_{g \in H} g$$

of the elements in H is an idempotent of FG and is central if and only if $H \triangleleft G$.

5. If e is an idempotent and f is a central idempotent of a ring A, prove ef is an idempotent and ef is mutually orthogonal with an idempotent $e' \in A$ if e' is mutually orthogonal with either e or f. Use this and the idempotents e in Exercise 4 (as well as their complements $1 - e$) to produce a list of four mutually orthogonal idempotents e_1, e_2, e_3, e_4 in the group ring $A = \mathbb{Q}G$, where G is the nonabelian group $(a, b : a^3, b^2, baba)$ of order 6. *Hint:* Consider $H = \{1, a, a^2\}$ and $H = \{1, b\}$.

Then use $\mathbb{Q}$-bases of Ae_1, Ae_2, Ae_3, and Ae_4 to find the associated matrix representations of G over $\mathbb{Q}$; and prove each takes A to a full matrix ring over $\mathbb{Q}$, so each Ae_i is simple.

6. Suppose A is a ring and $x, y \in A$. Prove there is an A-linear isomorphism $f : Ax \to Ay$ if and only if $Ay = Az$ for some $z \in A$ with $ann_A(x) = ann_A(z)$. In practice it is difficult to show z does or does not exist. But for a special class of rings A, there is a simple algorithm to see if $Ax \cong Ay$ — see §8C, Exercises 5 and 6.

7. If R is a commutative ring and G is a group, the **augmentation map** $\varepsilon : RG \to R$ is the surjective R-algebra homomorphism extending the trivial group homomorphism $G \to R^*, g \mapsto 1_R$. Its kernel is the **augmentation ideal**, generated by the elements $g - 1$ for $g \in G$. When R is a field, R is a simple R-module, and hence a simple RG-module through ε; so $\ker(\varepsilon)$ is a maximal left ideal of RG. Now suppose F is a field of prime characteristic p and G is a finite abelian p-group of order p^n. Prove FG is a local ring with maximal ideal $\mathrm{rad}(FG) = \ker(\varepsilon)$. *Hint:* Show first that FG is an indecomposable FG-module by taking the p^n-power of any idempotent $e \in FG$. Then use (8.14) (ii).

8. Suppose $\mathcal{A}$ is a left artinian ring with a primitive idempotent e and $M \in \mathcal{M}(\mathcal{A})$. Prove M has a composition factor isomorphic to $\mathcal{A}e/\mathrm{rad}(\mathcal{A})e$ if and only if $eM \neq 0$. If e and f are primitive idempotents of $\mathcal{A}$, say $\mathcal{A}e$ and $\mathcal{A}f$ are **linked** if there is a finite sequence of primitive idempotents $e_1, \ldots, e_n$ of $\mathcal{A}$ with $e = e_1$, $f = e_n$ and for which $\mathcal{A}e_i$ and $\mathcal{A}e_{i+1}$ have a composition factor in common (up to isomorphism) for $i = 1, \ldots, n - 1$. Being linked is

an equivalence relation on the indecomposable left ideals in an internal Krull-Schmidt decomposition of $\mathcal{A}$. The sum of the left ideals in a given equivalence class is called a **block** of $\mathcal{A}$. Prove the blocks $B_1, \ldots, B_m$ are two-sided ideals of $\mathcal{A}$ with $B_i B_j = 0$ whenever $i \neq j$ and $\mathcal{A} = B_1 \overset{\bullet}{\oplus} \cdots \overset{\bullet}{\oplus} B_m$. *Hint:* If $\mathcal{A}e$ and $\mathcal{A}f$ are not linked, use the first part of this problem to show $\mathcal{A}e\mathcal{A}f = 0$. Then show that, if $i \neq j$, B_i and B_j have no composition factor in common.

8C. Semisimple Rings

As noted at the end of §8B, the matrix representation theory of a finite group G over a field F is simplest when $\mathcal{M}(FG) = \mathcal{P}(FG)$; and by Maschke's Theorem (8.16), this is often the case, and always true when F has characteristic zero. From Proposition (8.14) (iv), $\mathcal{M}(\mathcal{A}) = \mathcal{P}(\mathcal{A})$ for a left artinian ring $\mathcal{A}$ if and only if every indecomposable in $\mathcal{P}(\mathcal{A})$ is simple, in which case $\mathcal{A}$ is a direct sum of simple left ideals.

(8.18) Definition. Suppose A is a ring. An A-module M is **semisimple** if M is a sum of simple submodules. (The zero module is considered semisimple, being a sum of an empty set of simple submodules.) An A-module M is **completely reducible** if every submodule of M is a direct summand of M.

(8.19) Lemma. *In a completely reducible A-module M, every nonzero submodule contains a simple submodule.*

Proof. First note that every submodule P of M is completely reducible: For if P has a submodule N, then $M = N \overset{\bullet}{\oplus} N'$ for a submodule N' of M, and then $P = N \overset{\bullet}{\oplus} (P \cap N')$.

Now suppose P is a submodule of M, and $0 \neq m \in P$. Let $\mathcal{S}$ denote the set of submodules N of P with $m \notin N$. Then $0 \in \mathcal{S}$, and $\mathcal{S}$ is closed under arbitrary unions. By Zorn's Lemma, $\mathcal{S}$ has a maximal member Q (under containment). Then $P = Q \overset{\bullet}{\oplus} Q'$ for an A-module Q'. Since $m \notin Q$, $Q' \neq 0$. If Q' is not simple, it equals $D \overset{\bullet}{\oplus} E$ for nonzero submodules D and E; but then $P = Q \overset{\bullet}{\oplus} D \overset{\bullet}{\oplus} E$, and by maximality of Q,

$$ m \in (Q \overset{\bullet}{\oplus} D) \cap (Q \overset{\bullet}{\oplus} E) = Q , $$

a contradiction. ∎

(8.20) Proposition. *Suppose A is a ring and $M \in \mathcal{M}(A)$. The following are equivalent:*

(i)　*M is semisimple.*
(ii)　*M is a finite direct sum of simple submodules.*
(iii)　*M is completely reducible.*

Proof. Suppose M is a sum of simple submodules M_i $(i \in \mathfrak{I})$ and X is a finite generating set of M. Each element of X belongs to a finite sum of the M_i; so M is a sum of the M_i for i in a finite subset $\mathfrak{J}$ of $\mathfrak{I}$. Suppose N is a submodule of M. Relabel the M_i so that $M_1, \ldots, M_n$ is a shortest list of M_i with

$$M \;=\; N + M_1 + \cdots + M_n \,.$$

For each $i \geq 1$, M_i is not contained in $N + M_1 + \cdots + M_{i-1}$; and since M_i is simple,

$$M_i \;\cap\; (N + M_1 + \cdots + M_{i-1}) \;=\; 0 \,.$$

So

$$M \;=\; N \stackrel{\bullet}{\oplus} M_1 \stackrel{\bullet}{\oplus} \cdots \stackrel{\bullet}{\oplus} M_n \;=\; N \stackrel{\bullet}{\oplus} (M_1 + \cdots + M_n) \,.$$

This proves that (i) implies (iii), and taking $N = 0$, that (i) implies (ii). It is a formality that (ii) implies (i), and it only remains to prove (iii) implies (i).

Assume M is completely reducible. Every submodule of M, being a direct summand, is a homomorphic image of M; so it is finitely generated. Let N denote the sum of all simple submodules of M. Then N contains every simple submodule of M. Since M is completely reducible, $M = N \stackrel{\bullet}{\oplus} P$ for a submodule P. If $P \neq 0$, then by Lemma (8.19) P contains a simple submodule Q with $N \cap Q = 0$, a contradiction. So $M = N$, which is semisimple. ∎

(8.21) Corollary. *The collection of semisimple $M \in \mathcal{M}(A)$ is closed under the formation of finite direct sums, homomorphic images, and submodules.*

Proof. A finite external direct sum $M_1 \oplus \cdots \oplus M_n$ is an internal sum of R-modules $M_i' \cong M_i$; so it is semisimple if each M_i is. Suppose $f : M \to N$ is an R-linear map and M is a sum of simple R-modules M_i. Then $f(M)$ is the sum of the $f(M_i)$. Since $\ker(f) \cap M_i$ is either 0 or M_i, either $f(M_i) = 0$ or f restricts to an isomorphism $M_i \cong f(M_i)$ and $f(M_i)$ is simple. Finally, every submodule of M is a direct summand, and hence a homomorphic image, of M. ∎

(8.22) Definition. A ring A is a **semisimple ring** if A is the sum of its simple left ideals (that is, if A is semisimple as an A-module).

(8.23) Theorem. *For a ring A, the following are equivalent.*

(i)   *A is a semisimple ring.*

(ii) *Every f.g. A-module is semisimple.*

(iii) *Every f.g. A-module is projective.*

(iv) *The ring A is left artinian and every f.g. indecomposable A-module is projective.*

(v)   *The ring A is left artinian and every f.g. indecomposable A-module is simple.*

(vi) *The ring A is left artinian and* rad$(A) = 0$.

Proof. If A is semisimple as an A-module, so is $A^n = A \oplus \cdots \oplus A$ for each $n \geq 1$, and so is each homomorphic image of each A^n; so (i) implies (ii).

If every A^n is semisimple, it is also completely reducible; so every short exact sequence

$$0 \longrightarrow K \overset{\subseteq}{\longrightarrow} A^n \longrightarrow P \longrightarrow 0$$

splits (the inclusion splits by the projection to K). Then every f.g. A-module P is projective, and (ii) implies (iii).

Assume each f.g. A-module is projective; so each short exact sequence

$$0 \longrightarrow N \overset{\subseteq}{\longrightarrow} M \longrightarrow M/N \longrightarrow 0$$

with $M \in \mathcal{M}(A)$ splits. Then each $M \in \mathcal{M}(A)$ is completely reducible. In particular, each left ideal of A is a homomorphic image of A, so it is finitely generated; so A is left noetherian. If $I \subsetneqq J$ are left ideals of A and $A = J \overset{\bullet}{\oplus} J'$, then $J = I \overset{\bullet}{\oplus} K$, where $K \neq 0$; so $A = I \overset{\bullet}{\oplus} K \overset{\bullet}{\oplus} J'$, and $K \overset{\bullet}{\oplus} J'$ is a complement to I properly containing the given complement J' of J. Suppose S is a nonempty set of left ideals of A and $\mathcal{T}$ is the set of their complements in A. Since A is left noetherian, $\mathcal{T}$ has a maximal element J' complementary to a left ideal $J \in S$. Since J' is maximal in $\mathcal{T}$, J is minimal in S. So A is left artinian, and (iii) implies (iv).

Assuming (iv), every simple A-module is projective; so by (8.14) (iv), every projective indecomposable A-module is simple, and (iv) implies (v).

If A is left artinian and f.g. indecomposable A-modules are simple, then rad$(A) = 0$ by (8.14) (ii) and (iii). Finally, if A is left artinian and rad$(A) = 0$, then every indecomposable left ideal in a Krull-Schmidt internal decomposition of A is simple by (8.14) (ii), and A is a semisimple ring. ∎

Note: This theorem completes an elegant analogy: From the proof of (1.42), $\mathcal{M}(A) = \mathcal{F}(A)$ if and only if A is a simple A-module. Now we see $\mathcal{M}(A) = \mathcal{P}(A)$ if and only if A is a semisimple A-module.

Like semisimple modules, the class of semisimple rings is closed under finite direct sums: If $A_1, \ldots, A_n$ are rings, the direct sum of their additive groups $A_1 \oplus \cdots \oplus A_n$ is a ring under coordinatewise multiplication

$$(a_1, \ldots, a_n) \cdot (a_1', \ldots, a_n') = (a_1 a_1', \ldots, a_n a_n') .$$

We call this ring the **product**

$$\prod_{i=1}^n A_i = A_1 \times \cdots \times A_n$$

of the rings $A_1, \ldots, A_n$. If I is a simple left ideal of A_i, then the cartesian product

$$0 \times \cdots \times 0 \times I \times 0 \times \cdots \times 0$$

(with I in the i-coordinate) is a simple left ideal of $A_1 \times \cdots \times A_n$. So a finite length product of semisimple rings is a semisimple ring.

(8.24) Example. For $1 \leq i \leq r$, suppose D_i is a division ring and n_i is a positive integer. The product

$$B = M_{n_1}(D_1) \times \cdots \times M_{n_r}(D_r)$$

is a semisimple ring: For $1 \leq i \leq r$ and $1 \leq j \leq n_i$, let $\boldsymbol{L_{ij}}$ denote the left ideal of B consisting of the r-tuples of matrices $(C_1, \ldots, C_r)$ with $C_k = 0$ for $k \neq i$, and

$$C_i = \left[\, 0 \, \left| \begin{array}{c} c_1 \\ \vdots \\ c_{n_i} \end{array} \right| \, 0 \, \right]$$

having all but its j-column equal to zero. Evidently B is the direct sum of the L_{ij}; so if $1_R = \sum_{i,j} e_{ij}$ with $e_{ij} \in L_{ij}$, then the elements e_{ij} are mutually orthogonal idempotents of B and $L_{ij} = Be_{ij}$ for each i and j. Here e_{ij} has ith coordinate matrix $C_i = \epsilon_{jj}$, whose only nonzero coordinate is a 1 in the j, j-position.

Each L_{ij} is a simple left ideal because, if D is a division ring, each nonzero vector in $D^{n \times 1}$ is left unimodular; so it can be transformed, through left multiplication by a matrix in $M_n(D)$, to any other vector in $D^{n \times 1}$. So the idempotents e_{ij} are primitive and B is a semisimple ring, as claimed.

For $1 \leq i \leq r$, let α_i denote the map inserting $M_{n_i}(D_i)$ in the i-coordinate of B. Let B_i denote the image

$$0 \oplus \cdots \oplus 0 \oplus M_{n_i}(D_i) \oplus 0 \oplus \cdots \oplus 0$$

of α_i. Each B_i is a (two-sided) ideal of B. These ideals are known as the **blocks** of B.

Note that $L_{ij} \cong L_{i'j'}$ as B-modules if and only if $i = i'$: For if $i = i'$, an isomorphism is given by right multiplication with $\alpha_i(P_{jj'})$ where $P_{jj'}$ is the permutation matrix obtained from the identity by switching the j and j' columns. And if $i \neq i'$, scalar multiplication by $\alpha_i(I_{n_i})$ fixes the elements of L_{ij}, but annihilates $L_{i'j'}$; so $L_{ij} \not\cong L_{i'j'}$. By (8.23), the semisimple ring B is left artinian, and each simple B-module S is a f.g. projective indecomposable B-module; hence it is isomorphic to one of the L_{ij}. Therefore B has r isomorphism classes of simple modules, represented by $L_{11}, L_{21}, \ldots, L_{r1}$.

Each ideal I of B is a (direct) sum of some of the simple left ideals $L_{ij} = Be_{ij}$, since

$$ I = I\sum e_{ij} \subseteq \sum Ie_{ij} \subseteq I \, , $$

and Ie_{ij} is a B-submodule of L_{ij} and so is 0 or L_{ij}. If L_{ij} appears in this sum, so does each $L_{ij'} = L_{ij}\alpha_i(P_{jj'})$ for $1 \leq j' \leq n_i$. So every ideal of B is a (direct) sum of some of the blocks $B_1, \ldots, B_r$. In particular, the only ideals of B contained in a block B_i are 0 and B_i.

Each block B_i is a ring under the operations in B, and is isomorphic as a ring to $M_{n_i}(D_i)$. But B_i is not a subring of B if $r > 1$, since it does not share the same multiplicative identity. The ideals of B_i are the ideals of B that are contained in B_i; so the ring B_i has no ideals aside from 0 and B_i.

We next set about proving that Example (8.24) is typical of semisimple rings. This was the first general structure theorem for noncommutative rings. In the 19th century, matrix rings and the division ring of quaternions had been well studied, and the group ring of a finite group G over a field had been introduced. In 1893, T. Molien proved $\mathbb{C}G$ is a product of matrix rings over $\mathbb{C}$. In a sweeping generalization, in 1908, J. H. M. Wedderburn showed every finite-dimensional algebra over a field, which has no nonzero nilpotent ideal, is a product of matrix rings over division rings. And in 1927, E. Artin extended Wedderburn's Theorem to rings with the descending chain condition on left (or right) ideals — today known as left (or right) artinian rings. The Wedderburn-Artin Theorem is (8.28), below.

The first step toward the description of all semisimple rings is a basic observation about simple modules:

(8.25) Schur's Lemma. *Suppose A is a ring and M, N are simple A-modules. Each nonzero A-linear map $f : M \to N$ is an isomorphism. So $\mathrm{End}_A M$ is a division ring.*

Proof. If $f(M) \neq 0$, then $f(M)$ must be N. So $\ker(f) \neq M$, and $\ker(f)$ must be 0. The set $\mathrm{End}_A M$ is a ring under pointwise addition and composition of maps, and $f \in \mathrm{End}_A M$ is invertible under composition exactly when it is an isomorphism. ∎

Recall that for any A-modules M and N, the set $\mathrm{Hom}_A(M, N)$ is an additive abelian group under pointwise addition, and that composition distributes over this addition.

(8.26) Proposition. *If A is a ring and $M_1, \ldots, M_s$ are A-modules, then the set*

$$
H \;=\; \begin{bmatrix} \mathrm{Hom}_A(M_1, M_1) & \cdots & \mathrm{Hom}_A(M_s, M_1) \\ \vdots & & \vdots \\ \mathrm{Hom}_A(M_1, M_s) & \cdots & \mathrm{Hom}_A(M_s, M_s) \end{bmatrix}
$$

of all matrices (f_{ij}), with $f_{ij} \in \mathrm{Hom}_R(M_j, M_i)$ for $1 \le i, j \le m$, is a ring, isomorphic to

$$
\mathrm{End}_A(M_1 \oplus \cdots \oplus M_s) \ .
$$

Proof. The usual proofs for the axioms of matrix arithmetic (see (1.26)) show H is a ring under addition

$$
(f_{ij}) \;+\; (g_{ij}) \;=\; (f_{ij} + g_{ij})
$$

and multiplication

$$
(f_{ij})(g_{ij}) \;=\; \left(\sum_{k=1}^{m} f_{ik} \circ g_{kj} \right) \ .
$$

Writing $M_1 \oplus \cdots \oplus M_s$ as a set of columns

$$
\begin{bmatrix} m_1 \\ \vdots \\ m_s \end{bmatrix} \quad (m_i \in M_i) \ ,
$$

there is an evaluation action, making $M_1 \oplus \cdots \oplus M_s$ into an H-module:

$$
(f_{ij}) \begin{bmatrix} m_1 \\ \vdots \\ m_s \end{bmatrix} = \begin{bmatrix} f_{11}(m_1) + \cdots + f_{1s}(ms) \\ \vdots \\ f_{s1}(m_1) + \cdots + f_{ss}(m_s) \end{bmatrix} \ .
$$

Again, the module axioms follow from the usual proofs of properties of matrix operations.

By the discussion in Example (1.31), taking (f_{ij}) to this scalar multiplication by (f_{ij}) defines an isomorphism of rings $H \cong \mathrm{End}_A(M_1 \oplus \cdots \oplus M_s)$. ∎

Combining this with Schur's Lemma, we get:

(8.27) Proposition. *Suppose A is a ring, $S_1, \ldots, S_r$ are pairwise nonisomorphic simple A-modules, and $n_1, \ldots, n_r$ are positive integers. Let D_i denote the division ring $\mathrm{End}_A S_i$. Then there is an isomorphism of rings*

$$
\mathrm{End}_A(S_1^{m_1} \oplus \cdots \oplus S_r^{n_r}) \;\cong\; M_{n_1}(D_1) \times \cdots \times M_{n_r}(D_r) \ .
$$

Proof. By Schur's Lemma, for $i \neq j$, $\mathrm{Hom}_A(S_i, S_j) = 0$. In Proposition (8.26), if $M_1, \ldots, M_s$ are

$$S_1, \ldots, S_1, S_2, \ldots, S_2, \ldots, S_r, \ldots, S_r ,$$

with each S_i repeated n_i times, then the matrix ring H takes the block diagonal form

$$\begin{bmatrix} M_{n_1}(D_1) & & & \\ & M_{n_2}(D_2) & & \\ & & \ddots & \\ & & & M_{n_r}(D_r) \end{bmatrix} ,$$

which is isomorphic to the product of the rings $M_{n_i}(D_i)$, by

$$C_1 \oplus \cdots \oplus C_r \ \mapsto \ (C_1, \ldots, C_r) . \qquad \blacksquare$$

(8.28) Wedderburn-Artin Theorem. *Every semisimple ring A is isomorphic to a product of finitely many full matrix rings over division rings. If*

$$A \ \cong \ M_{n_1}(D_1) \times \cdots \times M_{n_r}(D_r) ,$$

for division rings D_i and positive integers n_i, then A has exactly r isomorphism classes of simple A-modules; and if these classes are represented by $S_1, \ldots, S_r$ in an appropriate order, then for each i,

$$D_i \ \cong \ (\mathrm{End}_A S_i)^{op}$$

and n_i is the number of summands isomorphic to S_i in a Krull-Schmidt decomposition of $_A A$. So the above decomposition of A is unique except for the order of factors and the replacement of each division ring D_i by an isomorphic copy.

Proof. As a left A-module, A is free with basis 1_R. So each $f \in \mathrm{End}_A A$ is right multiplication by $f(1_A)$, and the map

$$\phi : \mathrm{End}_A A \ \rightarrow \ A , \quad f \mapsto f(1_A)$$

is bijective. If $i_A : A \rightarrow A$ is the identity map and $f, g \in \mathrm{End}_A A$, then

$$\phi(i_A) \ = \ i_A(1_A) \ = \ 1_A ,$$
$$\phi(f + g) \ = \ f(1_A) + g(1_A) \ = \ \phi(f) + \phi(g) ,$$

and

$$\phi(f \circ g) \ = \ f(g(1_A)) \ = \ g(1_A)f(1_A) \ = \ \phi(g)\phi(f) .$$

So ϕ is a ring isomorphism from the opposite ring of $\mathrm{End}_A A$ to A.

Now suppose A is a semisimple ring, with

$$A \cong S_1^{n_1} \oplus \cdots \oplus S_r^{n_r}$$

for pairwise nonisomorphic simple A-modules $S_1, \ldots, S_r$ and positive integers $n_1, \ldots, n_r$. By Proposition (8.27) there are ring isomorphisms

$$A \cong (\mathrm{End}_A A)^{op} \cong (M_{n_1}(\Delta_1) \times \cdots \times M_{n_r}(\Delta_r))^{op}$$

for division rings $\Delta_i = \mathrm{End}_A S_i$. Applying the transpose to each coordinate defines a ring isomorphism

$$((M_{n_1}(\Delta_1) \times \cdots \times M_{n_r}(\Delta_r))^{op} \cong M_{n_1}(D_1) \times \cdots \times M_{n_r}(D_r) \ ,$$

where $D_i = \Delta_i^{op} = (\mathrm{End}_A S_i)^{op}$ for each i. Since Δ_i is a division ring, so is its opposite ring D_i.

For the uniqueness of n_i and D_i, assume

$$B = M_{n_1}(D_1) \times \cdots \times M_{n_r}(D_r)$$

for division rings D_i and positive integers n_i, and refer to Example (8.24). The isomorphism classes of simple B-modules are represented by $L_{11}, \ldots, L_{r1}$ and n_i is the number of summands L_{ij} isomorphic to L_{i1} in the Krull-Schmidt decomposition $B = \overset{\bullet}{\oplus} L_{ij}$.

Right multiplication of the matrix in the i-coordinate by $d \in D_i$ is a left B-linear endomorphism f_d of L_{i1}.

Since

$$f_{d+d'} = f_d + f_{d'} \ ,$$
$$f_1 = \text{identity map,}$$
$$\text{and } f_{dd'} = f_{d'} \circ f_d \ ,$$

the function

$$\psi : D_i \rightarrow (\mathrm{End}_B L_{i1})^{op}$$
$$d \mapsto f_d$$

is a ring homomorphism. Since $L_{i1} \neq 0$, f_d is not the zero map unless $d = 0$. So ψ is injective.

Recall from Example (8.24) that $L_{i1} = Be$ for a primitive idempotent $e = e_{i1} \in L_{i1}$ whose ith coordinate is the matrix unit ϵ_{11}. Suppose $f \in \mathrm{End}_B L_{i1}$. Since f is B-linear, for each $x \in L_{i1}$,

$$f(x) = f(xe) = xf(e) \ .$$

Now $f(e) \in Be$; so

$$f(e) \; = \; f(e)e \; = \; f(ee)e \; = \; ef(e)e \, .$$

Therefore the i-coordinate of $f(e)$ belongs to $\epsilon_{11}M_{n_i}(D_i)\epsilon_{11}$, and so it has $1,1$-entry $d \in D_i$ and all other entries zero. That is, f is right multiplication by $f(e)$, and so it is right multiplication by d in the i-coordinate. Then $f = f_d$, proving ψ is surjective.

Altogether, ψ is a ring isomorphism

$$D_i \; \cong \; (\mathrm{End}_B L_{i1})^{op} \, .$$

If $A \cong B$ as rings, the B-modules are the A-modules through this isomorphism, the B-submodules are the A-submodules, and the B-linear maps are the A-linear maps. So $L_{11}, \ldots, L_{r1}$ represent the isomorphism classes of simple A-modules. If $S_1, \ldots, S_r$ also represent these classes, ordered so that $S_i \cong L_{i1}$ as A-modules, then conjugation by this isomorphism defines a ring isomorphism

$$\mathrm{End}_A S_i \; \cong \; \mathrm{End}_A L_{i1} \; = \; \mathrm{End}_B L_{i1} \, ,$$

which is necessarily also an isomorphism between the opposite rings. So

$$D_i \; \cong \; (\mathrm{End}_A S_i)^{op} \, .$$

Also, $A \cong \oplus L_{ij}$ is a Krull-Schmidt decomposition and n_i is the number of summands L_{ij} isomorphic to S_i. ∎

(8.29) Note. In the proof of (8.28), we found that the opposite ring of a product of matrix rings over division rings is isomorphic to a product of matrix rings over division rings; so the opposite of a semisimple ring is semisimple. This means a ring A is a sum of simple left ideals if and only if it is a sum of simple right ideals. It also means that a ring A with $\mathrm{rad}(A) = 0$ is left artinian if and only if it is right artinian.

(8.30) Definition. If R is a commutative ring, a **division R-algebra** is an R-algebra that, as a ring, is a division ring.

(8.31) Note. Suppose R is a commutative ring and A is an R-algebra. If S is an A-module, the ring $\Delta = \mathrm{End}_A S$ is an R-algebra: For $r \in R$ and $f \in \Delta$, $rf : S \to S$ is f followed by scalar multiplication by $r 1_A$.

Now suppose, as in the proof of (8.28), there is an A-linear isomorphism

$$\lambda : A \; \cong \; S_1^{n_1} \oplus \cdots \oplus S_r^{n_r}$$

for pairwise nonisomorphic simple A-modules S_i and positive integers n_i, and let Δ_i denote $\mathrm{End}_A S_i$. The opposite ring $D_i = \Delta_i^{op}$ is an R-algebra with the same scalar multiplication as Δ_i, and

$$B \;=\; M_{n_1}(D_1) \times \cdots \times M_{n_r}(D_r)$$

is an R-algebra with scalars $r \in R$ multiplied into every entry of every coordinate of $b \in B$. *Then the Wedderburn-Artin isomorphism $B \cong A$ is an R-algebra isomorphism* since each step

$$M_{n_1}(D_1) \times \cdots \times M_{n_r}(D_r) \xrightarrow{\text{transpose}} (M_{n_1}(\Delta_1) \times \cdots \times M_{n_r}(\Delta_r))^{op}$$

$$\xrightarrow{\text{diagonal}} \begin{bmatrix} M_{n_1}(\Delta_1) & & \\ & \ddots & \\ & & M_{n_r}(\Delta_r) \end{bmatrix}^{op}$$

$$\xrightarrow{(f_{ij}) \mapsto (f_{ij})\cdot} (\mathrm{End}_A\,(S_1^{n_1} \oplus \cdots \oplus S_r^{n_r}))^{op}$$

$$\xrightarrow{\lambda^{-1} \circ (-) \circ \lambda} (\mathrm{End}_A A)^{op} \xrightarrow{f \mapsto f(1_A)} A$$

is R-linear. If R is a field F and $\dim_F A$ is finite, it follows that each $\dim_F D_i$ is finite and

$$\dim_F A \;=\; \sum_{i=1}^{r} n_i^2 \, \dim_F D_i \;.$$

If there is an R-algebra isomorphism

$$A \;\cong\; B \;=\; M_{n_1}(D_1) \times \cdots \times M_{n_r}(D_r)$$

for positive integers n_i and division R-algebras D_i, and if $S_1, \ldots, S_r$ represent the isomorphism classes of simple A-modules in the order with $S_i \cong L_{i1}$, then the ring isomorphism $D_i \cong (\mathrm{End}_A S_i)^{op}$ obtained in the Wedderburn-Artin Theorem is R-linear. So if there is another R-algebra isomorphism

$$A \;\cong\; M_{m_1}(D_1') \times \cdots \times M_{m_s}(D_s')$$

for positive integers m_i and division R-algebras D_i', then $s = r$ and, after a permutation of factors, $m_i = n_i$ and $D_i' \cong D_i$ *as R-algebras* for each i.

The simplest semisimple rings are those with only one block:

(8.32) Definition. A ring A is a **simple ring** if $A \neq 0$ and the only (two-sided) ideals of A are 0 and A.

(8.33) Theorem. *For a ring A, the following are equivalent:*

  (i) A *is simple and left artinian.*
  (ii) $A \cong M_n(D)$ *for a division ring D and positive integer n.*
  (iii) A *is semisimple, with only one isomorphism class of simple modules.*
  (iv) $\mathcal{M}(A) = \mathcal{P}(A)$ *and* $K_0(A) \cong \mathbb{Z}$.

Proof. If A is simple and left artinian, then the two-sided ideal rad(A) is either 0 or A, and is nilpotent; so rad$(A) = 0$. Therefore A is semisimple and is isomorphic to $B = B_1 \overset{\bullet}{\oplus} \cdots \overset{\bullet}{\oplus} B_r$, where each ideal B_i is isomorphic to a matrix ring over a division ring. Since A is simple, B has only one block B_1 and $A \cong B_1$. Conversely, if $A \cong M_n(D)$ for a division ring D, then A is semisimple, is a simple ring by Example (8.24), and is left artinian by (8.23). So (i) and (ii) are equivalent.

Example (8.24) shows (ii) implies (iii). If A is semisimple, the Wedderburn-Artin Theorem shows A is isomorphic to a product of r matrix rings over division rings, where r is the number of isomorphism classes of simple A-modules. So (iii) implies (ii).

By (8.23), A is semisimple if and only if $\mathcal{M}(A) = \mathcal{P}(A)$; and, in that case, A is left artinian. So $K_0(A) = G_0(A) \cong \mathbb{Z}^r$, where r is the number of isomorphism classes of simple A-modules. That is, (iii) and (iv) are equivalent. ∎

Note: A simple left artinian ring is just called a **simple artinian ring**, since it is semisimple, and hence both left and right artinian.

The Wedderburn-Artin isomorphism matches the internal structure of any semisimple ring A to that of the ring B in Example (8.24).

(8.34) Corollary. *Suppose A is a semisimple ring. The ring A has only finitely many minimal nonzero ideals $A_1, \ldots, A_r$, and $A = A_1 \overset{\bullet}{\oplus} \cdots \overset{\bullet}{\oplus} A_r$ while $A_i A_j = 0$ for $i \neq j$. Each A_i is a simple artinian ring under the operations in A and has a simple left ideal L_i. Then $L_1, \ldots, L_r$ represent the distinct isomorphism classes of simple A-modules. The ideal A_i is the sum of all simple left ideals of A that are isomorphic to L_i. Every ideal I of A is the direct sum of those A_i contained in I, and it is a ring under the operations in A. Every surjective ring homomorphism $f : A \rightarrow A'$ is the projection onto an ideal I of A followed by a ring isomorphism $I \cong A'$.*

Proof. Take A_i to be the image of B_i and L_i to be the image of L_{i1} under the Wedderburn-Artin isomorphism $B \cong A$. Through restriction of scalars

$L_{11}, \ldots, L_{r1}$ are pairwise nonisomorphic simple A-modules and $L_i \cong L_{i1}$ as A-modules for each i. Each simple A-module M is, by restriction of scalars, a simple B-module, which is isomorphic to some L_{i1}. So $M \cong L_i$ as A-modules.

If L is a simple left ideal of A,

$$L = AL \subseteq A_1 L + \cdots + A_r L \subseteq L ,$$

and this sum is direct since each $A_i L \subseteq A_i$. Since L is simple, $L = A_i L \subseteq A_i$ for exactly one i and $A_j L = 0$ for $j \neq i$. Then L is a simple left ideal of A_i; so $L \cong L_i$ by (8.33). If a left ideal $L' \cong L_i$, it follows that $L' \not\cong L_j$ and $L' \not\subseteq A_j$ for $j \neq i$, and so $L' \subseteq A_i$. By Example (8.24), A_i is the sum of its simple left ideals; so A_i is the sum of all left ideals of A in the isomorphism class of L_i.

The description of ideals in A as direct sums of A_i comes from the corresponding description of ideals in B. Since the ideals in B are rings with multiplicative identity, the same holds for ideals in A. If $f : A \to A'$ is a surjective ring homomorphism, its kernel J is a sum of the A_i. And if I is the sum of the remaining A_i, then f is the projection to I followed by the ring isomorphisms $I \cong A/J \cong A'$. ∎

The simple rings A_i above are known as the **simple components** of A. Like the Krull-Schmidt decomposition of a left artinian ring, the decomposition of a semisimple ring into its simple components can be described with idempotents.

(8.35) Definitions. In a ring A an element a is **central** if $ax = xa$ for all $x \in A$. The set of all central elements of A is called the **center** of A and it is denoted $\boldsymbol{Z(A)}$. Evidently $Z(A)$ is a commutative subring of A. A central idempotent e of A is **centrally primitive** if e is not a sum of two nonzero, mutually orthogonal, central idempotents of A.

(8.36) Corollary. *Suppose a semisimple ring A has simple component decomposition $A_1 \overset{\bullet}{\oplus} \cdots \overset{\bullet}{\oplus} A_r$, and $1_A = e_1 + \cdots + e_r$ for $e_i \in A_i$. Then e_i is the multiplicative identity of $A_i = Ae_i$ and $e_1, \ldots, e_r$ are nonzero, mutually orthogonal central idempotents of A. Further, each e_i is centrally primitive. Every central idempotent in A is a sum of some of the $e_1, \ldots, e_r$ (each taken at most once); so the only centrally primitive central idempotents of A are $e_1, \ldots, e_r$. Every ideal of A is generated by a central idempotent. Every surjective ring homomorphism $A \to A'$ is multiplication by a central idempotent e, followed by a ring isomorphism $Ae \cong A'$.*

Proof. The first assertions about $e_1, \ldots, e_r$ follow from a comparison with Example (8.24), where the element $\alpha_i(I_{n_i})$ corresponds to e_i. To see that e_i is centrally primitive, suppose $e_i = e + f$ for nonzero, mutually orthogonal, central idempotents e and f in A. Then $A_i = Ae_i = Ae \overset{\bullet}{\oplus} Af$, and Ae and Af are nonzero ideals of A, contradicting the minimality of A_i.

If e is any central idempotent of A, then so is $1 - e$, and

$$A = Ae \overset{\bullet}{\oplus} A(1 - e) \, ,$$

where Ae and $A(1-e)$ are sums of complementary sets of simple components of A. Expressing e and $1 - e$ in terms of their $A_1, \ldots, A_r$-coordinates, we thereby express their sum 1_A. So e is a sum of some of the $e_1, \ldots, e_r$, each taken at most once. And if e is centrally primitive, $e = e_i$ for some i.

Now suppose $f_1, \ldots, f_s$ are s different elements of $\{e_1, \ldots, e_r\}$. Multiplication by the central idempotent $e = f_1 + \cdots + f_s$ is the projection from A onto $Af_1 \overset{\bullet}{\oplus} \cdots \overset{\bullet}{\oplus} Af_s$. So every ideal of A is Ae for some central idempotent e, and the final assertion follows from (8.34). ∎

The Wedderburn-Artin Theorem has direct applications to the matrix representations of groups. *For the rest of this section, suppose F is a field, G is a finite group, and FG satisfies the condition of Maschke's Theorem, and so it is semisimple. In particular, suppose*

$$\theta : FG \; \cong \; B \; = \; M_{n_1}(D_1) \times \cdots \times M_{n_r}(D_r)$$

is an F-algebra isomorphism, where the n_i are positive integers and the D_i are division F-algebras.

For any nonzero F-algebra A the ring homomorphism $F \to Z(A)$, $f \mapsto f \cdot 1_A$, must be injective, since its kernel is a proper ideal of F. So it defines a ring isomorphism $F \cong F \cdot 1_A$. When it does not result in any ambiguity, we denote $f \cdot 1_A$ by f. Then F becomes a subfield of $Z(A)$ with scalar multiplication $F \times A \to A$ restricting the ring multiplication $A \times A \to A$.

Suppose μ is an irreducible matrix representation of G over F. Two f.g. FG-modules are isomorphic if and only if they afford the same similarity class of matrix representations. So the FG-modules that afford μ constitute an isomorphism class of simple modules. Since FG is semisimple, these simple modules are f.g. projective indecomposables, and so they are isomorphic to left ideals of FG. So the left ideals of FG that afford μ form an isomorphism class and sum to a simple component of FG.

(8.37) Definition. If μ is an irreducible representation of G over F, the **simple component of FG associated with** μ is that simple component S_μ whose simple left ideals afford μ.

The affording map restricts to a bijection (8.8) (iii) from the set of isomorphism classes of simple FG-modules to the set of similarity classes of irreducible matrix representations of G over F. Its inverse, followed by the map adding up the left ideals in each isomorphism class, defines a bijection $s(\mu) \mapsto S_\mu$ from the set of similarity classes of irreducible representations to the set of simple components. This proves:

(8.38) Corollary. *The number of similarity classes of irreducible matrix representations of G over F is the number of simple components in FG.* ∎

In the notation of Example (8.24), the simple components of FG are the preimages $\theta^{-1}(B_1), \ldots, \theta^{-1}(B_r)$, where

$$B_i \;=\; 0 \times \cdots \times 0 \times M_{n_i}(D_i) \times 0 \times \cdots \times 0 \,,$$

with $M_{n_i}(D_i)$ in the i-coordinate. If S_μ is the ith simple component $\theta^{-1}(B_i)$, then θ, followed by the projection π_i of B to its i-coordinate, restricts to an F-algebra isomorphism $S_\mu \cong M_{n_i}(D_i)$. We can measure D_i in terms of μ.

(8.39) Lemma. *Suppose μ is an irreducible matrix representation of G over F of degree d, and there is an F-algebra isomorphism $S_\mu \cong M_n(D)$ for a division F-algebra D. Then*

$$\dim_F D \;=\; \frac{d^2}{\dim_F S_\mu} \,.$$

Proof. An F-basis of $M_n(D)$ is given by multiplying the matrix units by an F-basis of D; so $\dim_F S_\mu = n^2 \dim_F D$. The given isomorphism restricts to an F-linear isomorphism $L \cong M_n(D)\epsilon_{11} \; (\cong D^{n \times 1})$ for a simple left ideal L of S_μ affording μ. So $d = \dim_F L = n \dim_F D$. ∎

How can we read off the irreducible matrix representations of G over F directly from the isomorphism θ? When one of the division F-algebras equals F, the corresponding irreducible representation is easy to find: Assume $D_i = F$ and π_i is the projection of B to its i-coordinate. Then $\pi_i \circ \theta : FG \to M_{n_i}(F)$ is an F-algebra homomorphism, restricting to a matrix representation $\mu_i : G \to GL_{n_i}(F)$ that is "full" in the following sense:

(8.40) Definition. A matrix representation $\mu : G \to GL_n(F)$ is **full** if its F-linear extension $\widehat{\mu} : FG \to M_n(F)$ is surjective (which is to say, if the matrices $\mu(g)$ for $g \in G$ span $M_n(F)$ as an F-vector space).

Since conjugation by any $P \in GL_n(F)$ is an F-linear automorphism of $M_n(F)$, any representation similar to a full representation is full. If μ and ν are representations of G over F of degrees ≥ 1, evidently $\mu \oplus \nu$ is not full. So full representations are irreducible. In particular, when $D_i = F$, the representation μ_i restricting $\pi_i \circ \theta$ is irreducible.

(8.41) Proposition. *If $D_i = F$, the representation μ_i restricting $\pi_i \circ \theta$ is afforded by the simple left ideal L_{i1} of B over its standard basis. The ith simple component $\theta^{-1}(B_i)$ of FG is S_{μ_i}.*

Proof. By restriction of scalars through θ, the simple left ideal L_{i1} of B becomes a simple FG-module with standard F-basis $v_1, \ldots, v_{n_i}$, where the i-coordinate of v_j is the matrix unit ϵ_{j1}. Sending these to the standard basis vectors e_j of $F^{n_i \times 1}$ defines an F-linear coordinate map

$$\alpha : L_{i1} \cong F^{n_i \times 1}$$

projecting each element to the first column of the i-coordinate. Taking $L = L_{i1}$ and $g \in G$, the j-column of $\mu_L^\alpha(g)$ is

$$\begin{aligned}
\alpha(g \cdot v_j) &= \alpha(\theta(g)v_j) \\
&= \text{the first column of } (\pi_i \circ \theta)(g)\epsilon_{j1} \\
&= \text{the } j\text{-column of } (\pi_i \circ \theta)(g) \ .
\end{aligned}$$

So $\mu_L^\alpha = \mu_i$. Now θ restricts to an F-linear isomorphism from a simple left ideal of $\theta^{-1}(B_i)$ to L_{i1}, which affords μ_i. So $S_{\mu_i} = \theta^{-1}(B_i)$. ∎

When $D_i = F$, the representation restricting $\pi_i \circ \theta$ is irreducible because it is full. But as we see in (8.45) below, being full is equivalent to a stronger kind of irreducibility:

(8.42) Definition. A matrix representation $\mu : G \to GL_n(F)$ is **absolutely irreducible** if, for each field extension $F \subseteq E$, the composite

$$G \xrightarrow{\mu} GL_n(F) \subseteq GL_n(E)$$

is irreducible as a matrix representation over E.

(8.43) Lemma. *If e is an idempotent of FG and $F \subseteq E$ is a field extension, then each F-basis of FGe is also an E-basis of EGe. So if a matrix representation $\mu : G \to GL_n(F)$ is afforded by FGe, then μ followed by $GL_n(F) \subseteq GL_n(E)$ is afforded by EGe.*

Proof. Since EG is the E-linear span of FG, EGe is the E-linear span of FGe, and hence of each F-basis of FGe. So each F-basis of FGe contains an E-basis of EGe and

$$\dim_E EGe \leq \dim_F FGe \ .$$

The same is true with $1 - e$ in place of e. If either inequality is strict, so is their sum: $\dim_E EG < \dim_F FG$. But both sides equal the order of G. ∎

As a consequence, an irreducible representation afforded by FGe for an idempotent e is absolutely irreducible if and only if e remains primitive in EG for each field extension $F \subseteq E$.

(8.44) Lemma. *Suppose F is an algebraically closed field. Then the only finite-dimensional division F-algebra D is F itself.*

Proof. Since $F \subseteq Z(D)$, if $d \in D$, the subring $F[d]$ of D, generated by F and d, is commutative. Since D is a division ring, $F[d]$ has no zero-divisors. So the kernel of evaluation at $d : F[x] \to F[d]$ is a prime ideal. Since $\dim_F D$ is finite, this kernel is nonzero. In the principal ideal domain $F[x]$, nonzero prime ideals are maximal. So $F[d]$ is a finite-dimensional field extension of F. Since F is algebraically closed, $d \in F$.   ∎

(8.45) Proposition. *For a matrix representation $\mu : G \to GL_n(F)$ the following are equivalent:*

   (i)   *μ is irreducible and $S_\mu \cong M_n(F)$ as F-algebras.*
  (ii)   *μ is full.*
 (iii)   *μ is absolutely irreducible.*

Proof. Assume μ is irreducible. For some i, $S_\mu = \theta^{-1}(B_i) \cong M_{n_i}(D_i)$. If (i) holds, $M_{n_i}(D_i) \cong M_n(F)$ as F-algebras; so $n_i = n$, $D_i \cong F$ as F-algebras, $\dim_F D_i = 1$, and hence $D_i = F$. By (8.41), $S_\mu = S_{\mu_i}$, where μ_i restricts $\pi_i \circ \theta$. So μ is similar to the full representation μ_i, and so it is full.

Next suppose μ is full; so its F-linear extension $\widehat{\mu} : FG \to M_n(F)$ is surjective. Its E-linear extension $\bar{\mu} : EG \to M_n(E)$ is also F-linear; so $\bar{\mu}(FG) = \widehat{\mu}(FG)$, which contains the matrix units. So $\bar{\mu}$ is surjective and μ followed by $GL_n(F) \subseteq GL_n(E)$ is full, and hence is irreducible.

Finally, assume μ is absolutely irreducible and take E to be an algebraic closure of F. Let $\mu' : G \to GL_n(E)$ denote μ followed by $GL_n(F) \subseteq GL_n(E)$. For some primitive idempotent e of FG, μ is afforded by FGe. By (8.43), μ' is afforded by the simple left ideal EGe. For some centrally primitive central idempotent ε of FG, $S_\mu = FG\varepsilon$. Then ε is central in EG; so $EG\varepsilon$ is an ideal of EG. By Maschke's Theorem EG is also semisimple. So $EG\varepsilon$ is a direct sum of simple components of EG. Since $FGe \subseteq S_\mu = FG\varepsilon$, we also have $EGe \subseteq EG\varepsilon$; so the simple component $S_{\mu'}$ of EG lies in $EG\varepsilon$.

For some division F-algebra D, $S_\mu \cong M_\ell(D)$ as F-algebras. Since E is algebraically closed, $S_{\mu'} \cong M_m(E)$ as E-algebras. Both μ and μ' have degree n. So, using (8.43),

$$\dim_E S_{\mu'} \;\leq\; \dim_E EG\varepsilon \;=\; \dim_F FG\varepsilon \;=\; \dim_F S_\mu \,.$$

By (8.39),

$$\dim_F D \;=\; \frac{n^2}{\dim_F S_\mu} \;\leq\; \frac{n^2}{\dim_E S_{\mu'}} \;=\; \dim_E E \;=\; 1 \,,$$

and hence $D = F$, $\dim_F S_\mu = n^2$, $\ell = n$, and $S_\mu \cong M_n(F)$ as F-algebras.   ∎

(8.46) Definition. The field F is a **splitting field** for the finite group G if there is an F-algebra isomorphism

$$\theta : FG \ \cong \ M_{n_1}(F) \times \cdots \times M_{n_r}(F)$$

for positive integers n_i. Equivalently, F is a splitting field for G if every irreducible matrix representation of G over F is absolutely irreducible.

Over a splitting field, the irreducible representations of G can be read off from θ as the projections $\pi_i \circ \theta$ restricted to G. In fact, the isomorphisms θ correspond to the lists of dissimilar irreducible representations.

(8.47) Proposition. *Suppose F is a splitting field for G. If $\nu_1, \ldots, \nu_r$ represent the distinct similarity classes of irreducible matrix representations of G over F, listed in an appropriate order, and $n_1, \ldots, n_r$ are their respective degrees, then the map*

$$[\widehat{\nu}_1 \cdots \widehat{\nu}_r] : FG \to B \ = \ M_{n_1}(F) \times \cdots \times M_{n_r}(F)$$
$$x \mapsto (\widehat{\nu}_1(x), \ldots, \widehat{\nu}_r(x))$$

is an F-algebra isomorphism. Every F-algebra isomorphism $\theta : FG \cong B$ is obtained in this way.

Proof. Since F is a splitting field, there is an F-algebra isomorphism θ as in (8.46), and by (8.41),

$$\theta \ = \ [\pi_1 \circ \theta \ \cdots \ \pi_r \circ \theta] \ = \ [\widehat{\mu}_1 \ \cdots \ \widehat{\mu}_r]$$

where μ_i is afforded by L_{i1}. If $\nu_1, \ldots, \nu_r$ are listed in the order for which ν_i is similar to μ_i, then $[\widehat{\nu}_1 \ \cdots \ \widehat{\nu}_r]$ is θ followed by conjugation by an element of B^*; so it is an F-algebra isomorphism. ∎

(8.48) Proposition. *If FG is semisimple, the algebraic closure $\overline{F}$ of F is a splitting field for G. If F is a splitting field for G and $\sigma : F \to E$ is a ring homomorphism to a field E, then for each F-algebra isomorphism $\theta : FG \cong M_{n_1}(F) \times \cdots \times M_{n_r}(F)$ there is an E-algebra isomorphism $\psi : EG \cong M_{n_1}(E) \times \cdots \times M_{n_r}(E)$ making the square of ring homomorphisms*

$$
\begin{array}{ccc}
FG & \xrightarrow{\ \theta\ } & M_{n_1}(F) \times \cdots \times M_{n_r}(F) \\
{\scriptstyle \sigma_c}\downarrow & & \downarrow{\scriptstyle \sigma_e} \\
EG & \xrightarrow{\ \psi\ } & M_{n_1}(E) \times \cdots \times M_{n_r}(E)
\end{array}
$$

commute, where σ_c applies σ to the coefficients and σ_e applies σ to the matrix entries. So E is also a splitting field for G, and the dissimilar irreducible matrix *representations $\mu_1, \ldots, \mu_r$ of G over F, followed by entrywise application of σ, become the dissimilar irreducible matrix representations $\mu'_1, \ldots, \mu'_r$ of G over E.*

Proof. Since F and $\overline{F}$ share the same characteristic, $\overline{F}G$ is semisimple. The division $\overline{F}$-algebras in the Wedderburn-Artin decomposition of $\overline{F}G$ are finite dimensional over $\overline{F}$, and so they equal $\overline{F}$ by Lemma (8.44).

The restriction of θ to G, followed by σ_e, extends to an E-algebra homomorphism ψ. Then $\psi \circ \sigma_c$ agrees with $\sigma_e \circ \theta$ on G, and both are F-linear when F acts on E-modules through σ. So the square commutes. Since the image of $\sigma_e \circ \theta$ contains the r-tuples of matrix units, ψ is surjective. Since θ is an F-linear isomorphism, the order of G is $\sum n_i^2$; so the domain and codomain of ψ have equal dimension over E, forcing ψ to be injective as well.   ∎

(8.49) Definition. Suppose $F \subseteq E$ is a field extension. A matrix representation $\mu : G \to GL_n(E)$ is **defined over** F if μ is similar to a representation ν with $\nu(G) \subseteq GL_n(F)$.

(8.50) Proposition. *Suppose $F \subseteq E$ is a field extension.*

    (i) *If F is a splitting field for G, every matrix representation of G over E is defined over F.*

    (ii) *If E is a splitting field for G and every matrix representation of G over E is defined over F, then F is a splitting field for G.*

Proof. From the extension of θ to ψ in Proposition (8.48), a full list of dissimilar irreducible representations of G over F is, when followed by inclusion in invertible matrices over E, a full list of dissimilar irreducible representations of G over E. Every matrix representation of G over E is similar to a direct sum of these, proving (i).

For (ii), choose a full list $\nu_1, \ldots, \nu_r$ of dissimilar irreducible representations of G over E whose values are matrices over F. By (8.47) they define an E-algebra isomorphism

$$[\widehat{\nu}_1 \ \cdots \ \widehat{\nu}_r] : EG \ \cong \ M_{n_1}(E) \times \cdots \times M_{n_r}(E) \ ,$$

carrying FG injectively into $M_{n_1}(F) \times \cdots \times M_{n_r}(F)$. Comparing dimensions over E, the order of G is $\sum n_i^2$. Comparing dimensions over F, the image of FG is all of $M_{n_1}(F) \times \cdots \times M_{n_r}(F)$. So F is a splitting field for G.   ∎

We conclude this section with an estimate of the number r of simple components of FG (= the number of dissimilar irreducible matrix representations of G over F). The idea is to use the Wedderburn-Artin isomorphism to compute the F-dimension of the center of FG in two different ways. Recall from Definition (8.35) that the center of a ring A is denoted by $Z(A)$.

(8.51) Proposition.

(i)   *Every ring isomorphism* $f : A \cong B$ *restricts to a ring isomorphism* $Z(A) \cong Z(B)$.

(ii) *If D is a division ring, $Z(D)$ is a field.*

(iii) *If A is a ring and $n > 0$, $Z(M_n(A)) = Z(A) \cdot I_n$.*

(iv) *If $A_1, \ldots, A_r$ are rings, $Z(A_1 \times \cdots \times A_r) = Z(A_1) \times \cdots \times Z(A_r)$.*

(v)   *If R is a commutative ring and A is an R-algebra, $Z(A)$ is an R-subalgebra of A (that is, a subring and R-submodule of A).*

(vi) *If R is a commutative ring and G is a finite group, $Z(RG)$ is free as an R-module, based on the set of the sums of all elements in each conjugacy class of G.*

Proof. For (i), each ring homomorphism carries the center of its domain into the center of its image. Apply this to f and f^{-1}.

The center $Z(D)$ of a division ring D is a commutative subring of D containing $0_D \neq 1_D$. If $d \in Z(D)$ and $d \neq 0$, then for all $x \in D$, $dx = xd$; so $xd^{-1} = d^{-1}x$ and $d^{-1} \in Z(D)$, proving (ii).

For (iii), let $\epsilon_{ij} \in M_n(A)$ denote the matrix unit that is 1 in the i,j-position and 0 elsewhere. Recall that $\epsilon_{ij}\epsilon_{kl} = 0$ if $j \neq k$ and $\epsilon_{ij}\epsilon_{jl} = \epsilon_{il}$. If $\alpha = (a_{ij}) \in M_n(A)$, then

$$\alpha = \sum_{k,l} a_{kl}\epsilon_{kl} \, .$$

Suppose $\alpha \in Z(M_n(A))$. Then it commutes with ϵ_{ij}; so

$$\sum_l a_{jl}\epsilon_{il} = \sum_k a_{ki}\epsilon_{kj} \, .$$

Comparing entries, $a_{ii} = a_{jj}$ and $a_{jl} = 0$ whenever $l \neq j$; so $\alpha = aI_n$ for some $a \in A$. Also, α commutes with bI_n for all $b \in A$; so $a \in Z(A)$. Conversely, if $a \in Z(A)$ and $\beta \in M_n(A)$, $(aI_n)\beta = a\beta = \beta a = \beta(aI_n)$; so aI_n is central.

In (iv), for $a_i, b_i \in A_i$, $(a_1, \ldots, a_r)$ commutes with $(b_1, \ldots, b_r)$ under multiplication if and only if $a_i b_i = b_i a_i$ for each i. So $(a_1, \ldots, a_r) \in Z(A_1 \times \cdots \times A_r)$ if and only if each $a_i \in Z(A_i)$.

For (v), scalar multiplication by $r \in R$ is ring multiplication by the central element $r \cdot 1_A$ of A; so $R \cdot Z(A) \subseteq Z(A)$.

Finally, $x = \sum_g r_g g$ is central in RG if and only if $h^{-1}xh = x$ for all $h \in G$. But

$$h^{-1}xh = \sum_{g \in G} r_g h^{-1}gh = \sum_{g \in G} r_{hgh^{-1}}g \, ;$$

so x is central if and only if $r_{hgh^{-1}} = r_g$ for all $g, h \in G$. So $Z(RG)$ is the R-linear span of the sums of elements in each conjugacy class of G. These sums are R-linearly independent since the elements of G are. ∎

(8.52) Proposition. *Suppose F is a field, G is a finite group, and for each i with $1 \leq i \leq r$, n_i is a positive integer, and D_i is a division F-algebra with center F_i. If*

$$\theta : FG \;\cong\; M_{n_1}(D_1) \times \cdots \times M_{n_r}(D_r)$$

is an F-algebra isomorphism, then the number of conjugacy classes in G is

$$m \;=\; \sum_{i=1}^{r} \dim_F F_i \;.$$

So $r \leq m$. And $r = m$ if and only if $F = F_i$ for each i.

Proof. The isomorphism θ restricts to an F-algebra isomorphism between centers; so

$$\dim_F Z(FG) \;=\; \sum_{i=1}^{r} \dim_F Z(M_{n_i}(D_i)) \;.$$

For each i, the injective F-algebra homomorphism $D_i \to M_{n_i}(D_i)$, $d \mapsto dI_{n_i}$, restricts to an F-algebra isomorphism between centers $F_i \cong F_i I_{n_i}$. Comparing dimensions, $\dim_F Z(M_{n_i}(D_i)) = \dim_F F_i$. For the last assertion, $\dim_F F_i = 1$ if and only if $1_{D_i} = 1_{F_i}$ spans F_i as an F-module. ∎

(8.53) Corollary. *Suppose G is a finite group of order n with exactly r dissimilar irreducible matrix representations over a splitting field F. Suppose the degrees of these representations are $d_1, \ldots, d_r$. Then*

$$FG \;\cong\; M_{d_1}(F) \times \cdots \times M_{d_r}(F)$$

as F-algebras. Comparing dimensions over F for these algebras and for their centers,

$$n \;=\; \sum_{i=1}^{r} d_i^2$$

and r is the number of conjugacy classes in G. ∎

(8.54) Example. The dihedral group D_3 of order 6, with generators a, b and defining relations $a^3 = 1$, $b^2 = 1$, $bab^{-1} = a^{-1}$, has three conjugacy classes:

$$\{1\}, \quad \{a, a^2\}, \quad \text{and} \quad \{b, \, ab, \, a^2 b\}.$$

Expressing 6 as a sum of three squares, $6 = 2^2 + 1^2 + 1^2$. So over the complex number field $\mathbb{C}$, the group D_3 has one degree 2 irreducible matrix representation $\rho : D_3 \to M_2(\mathbb{C})$ and two irreducible representations $\sigma_1, \sigma_2 : D_3 \to \mathbb{C}$, up to similarity. The standard $\mathbb{R}$-linear representation of D_3 as rotations and reflections is $\mu(-) \cdot : D_3 \to GL_2(\mathbb{R}) \subseteq GL_2(\mathbb{C})$, with

$$\mu(a) = \begin{bmatrix} -1/2 & -\sqrt{3}/2 \\ \sqrt{3}/2 & -1/2 \end{bmatrix} , \quad \mu(b) = \begin{bmatrix} 1 & 0 \\ 0 & -1 \end{bmatrix} .$$

Then

$$\mu(a^2) = \begin{bmatrix} -1/2 & \sqrt{3}/2 \\ -\sqrt{3}/2 & -1/2 \end{bmatrix} ;$$

so $\mu(a)$, $\mu(b)$, $\mu(a^2)$ are $\mathbb{C}$-linearly independent. Therefore $\rho = \mu(-) \cdot$ is full, and therefore absolutely irreducible. The degree 1 representations are group homomorphisms from D_3 into an abelian group $GL_1(\mathbb{C}) = \mathbb{C}^*$; so they take $a = bab^{-1}a^{-1}$ to 1 and the order 2 element b to 1 or -1. One of them is determined by $a \mapsto 1$, $b \mapsto 1$, and the other by $a \mapsto 1$, $b \mapsto -1$. Note that σ_1, σ_2, and μ are defined over $\mathbb{R}$, and even over $\mathbb{Q}[\sqrt{3}]$; so by (8.50) these are splitting fields of the group D_3.

8C. Exercises

1. Suppose F is a field, G is a finite group, $D_1, \ldots, D_r$ are division F-algebras, $n_1, \ldots, n_r$ are positive integers, and

$$\theta : FG \cong M_{n_1}(D_1) \times \cdots \times M_{n_r}(D_r)$$

is an F-algebra isomorphism. For each i with $1 \leq i \leq r$, suppose σ_i is a matrix representation of the ring D_i over F obtained (as in §8A, Exercise 4) from an F-basis of the D_i, F-bimodule D_i. Define μ_i to be θ, followed by the projection to the i-coordinate, followed by the application of σ_i to each entry. Prove $\mu_1, \ldots, \mu_r$ is a full set of dissimilar irreducible matrix representations of G over F. *Hint:* In the notation of (8.24), show μ_i is afforded by L_{i1} via a particular basis constructed from the basis of D_i used to obtain σ_i.

2. Prove a nontrivial ring A is a division ring if and only if A is left artinian and has no zero-divisors.

3. If A is a simple artinian ring with a simple left ideal L and $\Delta = \mathrm{End}_A L$, prove $K_0(A) \cong K_0(\Delta) \cong \mathbb{Z}$; and if A has zero-divisors, prove $\widetilde{K}_0(A) \not\cong \widetilde{K}_0(\Delta)$. In the latter case compute $\widetilde{K}_0(A)$.

4. Suppose A is a semisimple ring and M is a simple A-module. Suppose $0 \neq m \in M$. The map $A \to Am = M$, $a \mapsto am$, is A-linear. Without using Krull-Schmidt uniqueness, prove this map restricts to an A-linear isomorphism $L \cong M$ for some simple left ideal L of A. If also L' is a simple left ideal of A, prove $L' \cong M$ or $L'M = 0$, but not both.

5. Suppose e_1 and e_2 are mutually orthogonal primitive idempotents of a semisimple ring A. Prove $Ae_1 \cong Ae_2$ as A-modules if and only if there exists $\alpha \in A$ with $e_1\alpha = \alpha e_2 \neq 0$. *Hint:* If α exists, show right multiplication by α is an A-linear nonzero map from Ae_1 to Ae_2. For the converse, show e_1 and e_2 are the beginning of a list $e_1, \ldots, e_r$ of mutually orthogonal primitive central idempotents in A adding up to 1. Then there is a Wedderburn isomorphism

$$\theta : A \ \cong \ M_{n_1}(D_1) \times \cdots \times M_{n_r}(D_r)$$

with $\theta(e_1) = (\epsilon_{11}, 0, \ldots, 0)$ and $\theta(e_2) = (\epsilon_{22}, 0, \ldots, 0)$, where ϵ_{ij} is the matrix unit with 1 in the i, j-entry and 0 elsewhere. Find a matrix P with $\epsilon_{11}P = P\epsilon_{22} \neq 0$, and take α to be $\theta^{-1}(P, 0, \ldots, 0)$.

6. As Example (8.54) shows, there are only three isomorphism classes of simple $\mathbb{Q}D_3$-modules. So, in §8B, Exercise 5, two of the modules $\mathbb{Q}D_3e_i$ must be isomorphic.

   (i)   Which two must they be?
   (ii) Use Exercise 5 above to find an isomorphism between them.
   (iii) Given the matrix representations μ_i and μ_j afforded by these isomorphic modules, find an invertible matrix C over $\mathbb{Q}$ with $C\mu_i(-)C^{-1} = \mu_j(-)$.

Hint: For (ii), write the unknown element α as

$$x_1 + x_2a + x_3a^2 + x_4b + x_5ab + x_6a^2b \ ,$$

expand out $e_i\alpha$ and αe_j, and equate corresponding coefficients to get a system of linear equations over $\mathbb{Q}$. Test the vectors in a basis of the solution space to find one with $e_1\alpha \neq 0$. Then right multiplication by α is an isomorphism from $\mathbb{Q}D_3e_i$ to $\mathbb{Q}D_3e_j$. (If all basis vectors yield α with $e_i\alpha = 0$, then $\mathbb{Q}D_3e_i$ is not isomorphic to $\mathbb{Q}D_3e_j$.) For (iii), C can be the matrix representing the isomorphism $\mathbb{Q}D_3e_i \cong \mathbb{Q}D_3e_j$ over the bases chosen to produce μ_i and μ_j.

7. Prove the following generalization of Proposition (8.20):

(8.20)′ Proposition. *Suppose A is a ring and $M \in A\text{-}\mathfrak{Mod}$. The following are equivalent:*

   (i)   *M is semisimple.*
   (ii) *M is a direct sum of simple submodules.*
   (iii) *M is completely reducible.*

Hint: Suppose N is a submodule of M and $\mathcal{S}$ is the set of all sets S of simple submodules of M for which the sum

$$N \ + \ \sum_{M_i \in S} M_i$$

is direct. Use the definition of arbitrary internal direct sums (8.11) and Zorn's Lemma to show $\mathcal{S}$ has a maximal member T. Then prove

$$M = N \overset{\bullet}{\oplus} \sum_{M_i \in T} M_i .$$

8. Prove the following analog of Theorem (1.42). *Suppose A is a ring. The following are equivalent:*

    (i)   *Every A-module is projective.*
    (ii) *Every f.g. A-module is projective.*
    (iii) *Every cyclic A-module is projective.*
    (iv) *Every simple A-module is projective.*
    (v) *A is a semisimple ring.*

Hint: To prove (iv) implies (v), suppose the sum S of all simple left ideals of A is proper. Then S is contained in a maximal left ideal M, and A/M is a simple A-module.

To prove (v) implies (i), suppose F is a free A-module with M as a homomorphic image. Then F is a sum of the $Ax \cong A$, where x runs through a basis of F; so F is a sum of simple A-modules. By $(8.20)'$ in the preceding exercise, F is completely reducible.

9. Reversing arrows in the characterizations of projective modules, we obtain "injective" modules: An A-module I is **injective** if, for each exact sequence $0 \longrightarrow N \overset{g}{\longrightarrow} M$ in A-$\mathcal{M}\mathfrak{o}\mathfrak{d}$ and each A-linear map $j\colon N \to I$, there exists an A-linear map $h : M \to I$ making the diagram

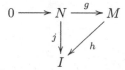

commute. Equivalently, I is injective if and only if each short exact sequence of A-linear maps

$$0 \longrightarrow I \longrightarrow M \longrightarrow N \longrightarrow 0$$

splits. Prove a ring A is semisimple if and only if every A-module is injective.

10. Suppose F is a field, G is a finite group, and the order of G is not a multiple of the characteristic of F. If μ_1 and μ_2 are dissimilar irreducible matrix representations of G over F, and $F \subseteq E$ is a field extension, prove μ_1 and μ_2 are also dissimilar as matrix representations over E. *Hint:* Show there is a central idempotent e in FG for which $\widehat{\mu}_1(e)$ is a zero matrix and $\widehat{\mu}_2(e)$ is an identity matrix.

11. If A is a ring and $S \subseteq A$, the **centralizer** of S in A is the set

$$Z_A(S) \;=\; \{a \in A : \forall s \in S, \;\; as = sa\}$$

of elements in A that commute with every element of S. For the duration of this exercise, let S' denote $Z_A(S)$. If $S \subseteq T \subseteq A$, evidently $T' \subseteq S'$. Also, $S \subseteq S''$. Putting these together, $S''' = S'$; so iterating the centralizer construction produces at most three different sets: S, S', S''. Say S has the **double centralizer property** in A if $S'' = S$, or equivalently, if $S'' \subseteq S$.

Now suppose $\mathcal{A}$ is a simple artinian ring, M is a simple $\mathcal{A}$-module, and $\rho : \mathcal{A} \to \mathrm{End}_{\mathbb{Z}}(M)$ takes $a \in \mathcal{A}$ to scalar multiplication by $a : m \mapsto a \cdot m$. Prove ρ is an injective ring homomorphism and $\rho(\mathcal{A})$ has the double centralizer property in $A = \mathrm{End}_{\mathbb{Z}}(M)$. *Hint:* First take $\mathcal{A} = M_n(D)$ for a division ring D, and $M = D^{n \times 1}$. By imitating the proof of uniqueness of the division rings in the Wedderburn-Artin Theorem, prove that, for the $\mathcal{A}, D$-bimodule M, the $\mathcal{A}$-linear endomorphisms of M are the (right) scalar multiplications by elements of D. Each (right) D-linear endomorphism of $M = D^{n \times 1}$ is, as in (1.27), left multiplication by some matrix $(a_{ij}) \in \mathcal{A} = M_n(D)$. In the ring $A = \mathrm{End}_{\mathbb{Z}}(M)$, $\rho(\mathcal{A})'$ consists of the $\mathcal{A}$-linear endomorphisms of M, and $\rho(\mathcal{A})''$ is the set of D-linear endomorphisms of M. Finally, generalize to $M \cong D^{n \times 1}$, and then to $\mathcal{A} \cong M_n(D)$.

Note: The double centralizer property in this context implies $\mathcal{A} \cong \mathrm{End}_D(M)$, where $D = \mathrm{End}_{\mathcal{A}}(M)$ acts by evaluation. An alternate proof of the double centralizer property for $\rho(\mathcal{A})$ comes from the more general Jacobson Density Theorem (see §16B, Exercise 6). This provides another proof that $\mathcal{A} \cong M_n(D)$ for a division ring D. Also notice that the proof you found above, in case D is a field F, amounts to the fact that $Z(M_n(F)) = F \cdot I_n$.

8D. Characters

Throughout this section we assume F is a field of characteristic zero, and we use the notational convention of writing m for $m \cdot 1_F$ whenever $m \in \mathbb{Z}$. So $\mathbb{Q}$ is a subfield of F. This has the advantage that the number of rows of an identity matrix over F equals its trace. Further fixing notation, G denotes a finite group of order n, and $e_1, \ldots, e_r$ are the r different centrally primitive central idempotents in the semisimple ring FG. Then FG has r simple components FGe_i, each with multiplicative identity e_i, and each with a simple left ideal L_i affording a matrix representation μ_i of G over F, of degree $d_i = \dim_F L_i$. So $\mu_1, \ldots, \mu_r$ is a complete list of pairwise dissimilar irreducible representations.

By (8.8), f.g. FG-modules M and N are isomorphic if and only if they afford similar matrix representations of G over F. How do we recognize when two matrix representations of the same degree d are similar? From experience with linear algebra, we might imagine computing a complicated set of invariants to decide if one list of matrices is obtained from another by a single inner automorphism of $GL_d(F)$. But the image of a representation $G \to GL_d(F)$ is far from an arbitrary list of matrices, and we find, in (8.62) below, that the only required invariant is the trace.

(8.55) **Definition.** Suppose R is a commutative ring. The **trace** of a matrix $A = (a_{ij}) \in M_n(R)$ is

$$tr(A) = \sum_{i=1}^{n} a_{ii} ,$$

the sum of the entries on the main diagonal of A. If $\emptyset \in M_0(R)$ is the empty matrix, we take $tr(\emptyset) = 0_R$.

(8.56) **Properties.** *If R is a commutative ring, $r \in R$, A and $B \in M_n(R)$, and $C \in M_m(R)$, then*

(i) $tr(A + B) = tr(A) + tr(B)$,

(ii) $tr(rA) = r \cdot tr(A)$,

(iii) $tr(AB) = tr(BA)$,

(iv) $tr(A \oplus C) = tr(A) + tr(C)$,

(v) $tr(A \otimes C) = tr(A) tr(C)$, *and*

(vi) $tr(I_n) = n \cdot 1_R$.

Proof. All but (iii) and (v) are routine. For (iii),

$$tr(AB) = \sum_{i=1}^{n} \left(\sum_{k=1}^{n} a_{ik} b_{ki} \right)$$

$$= \sum_{k=1}^{n} \left(\sum_{i=1}^{n} b_{ki} a_{ik} \right) = tr(BA) .$$

For (v),

$$tr\left(\begin{bmatrix} a_{11}C & \cdots & a_{1n}C \\ \vdots & & \vdots \\ a_{n1}C & \cdots & a_{nn}C \end{bmatrix} \right) = \sum_{i=1}^{n} tr(a_{ii}C)$$

$$= \sum_{i=1}^{n} a_{ii} tr(C) = tr(A) tr(C) . \qquad \blacksquare$$

(8.57) Definitions. An F-**character** of G is the composite $\chi : G \to F$ of a matrix representation $\mu : G \to GL_d(F)$, followed by the trace. Since $\chi(1_G) = tr(I_d) = d$, the degree d of the representation μ is determined by χ and is called the **degree** of the character χ. By property (iii) of the trace, similar representations yield the same character; so each $M \in \mathcal{M}(FG)$ **affords** a unique F-character

$$\chi_M \;=\; tr \circ \rho_M^\alpha \, ,$$

which is the trace of all matrix representations afforded by M. Each F-linear representation, and hence each F-character, is afforded by some $M \in \mathcal{M}(FG)$. We call χ_M **irreducible** if M is a simple FG-module. *Taking*

$$\chi_i \;=\; \chi_{L_i} \;=\; tr \circ \mu_i \, ,$$

G has r irreducible F-characters $\chi_1, \ldots, \chi_r$ of degrees $d_1, \ldots, d_r$.

Each matrix representation $\mu : G \to GL_d(F)$ extends uniquely to an F-algebra homomorphism $\widehat{\mu} : FG \to M_d(F)$. By properties (i) and (ii), the trace is F-linear; so

$$\widehat{\chi} \;=\; tr \circ \widehat{\mu} : FG \to F$$

is the unique F-linear map extending χ.

For any ring A, a **class function** $\phi : G \to A$ is any function that is constant on each conjugacy class in G. Under pointwise operations

$$
\begin{aligned}
(\phi_1 + \phi_2)(g) &= \phi_1(g) + \phi_2(g) \, , \\
(\phi_1 \phi_2)(g) &= \phi_1(g)\phi_2(g) \, , \\
(a\phi)(g) &= a(\phi(g)) \, ,
\end{aligned}
$$

the set $\mathbf{cf}(G, A)$ of class functions from G to A is a ring and an A-module; since F is commutative, $\mathrm{cf}(G, F)$ is a commutative F-algebra. By property (iii) of the trace, each F-character is a class function:

$$
\begin{aligned}
tr(\mu(ghg^{-1})) &= tr(\mu(g)\mu(hg^{-1})) \\
= tr(\mu(hg^{-1})\mu(g)) &= tr(\mu(h)) \, .
\end{aligned}
$$

So each character is specified by its value on each conjugacy class in G.

(8.58) Examples.

(i) Since the trace of a 1×1 matrix is its entry, the degree 1 F-characters of G are the degree 1 matrix representations $\mu : G \to F^*$. So if χ has degree 1, $\chi(gh) = \chi(g)\chi(h)$ for all $g, h \in G$, a property not shared by characters in general, since the trace is not multiplicative on $GL_d(F)$ for $d > 1$. Since F^* is commutative, the degree 1 characters of G are just the canonical map $G \to G^{ab} = G/[G, G]$, followed by the degree 1 characters of the abelian group

G^{ab}. These are described in Exercise 1. Of course, degree 1 characters χ_M are irreducible, since $\dim_F(M) = 1$.

(ii) The trace of the trivial representation of degree d (taking g to I_d for all $g \in G$) is the **trivial character** of degree d (taking g to d for all $g \in G$).

(iii) The dihedral group D_3 of order 6, with generators a, b and defining relations $a^3 = 1$, $b^2 = 1$, $bab^{-1} = a^{-1}$, has a matrix representation ν over $\mathbb{C}$ with

$$\nu(a) = \begin{bmatrix} 0 & -1 \\ 1 & -1 \end{bmatrix} , \quad \nu(b) = \begin{bmatrix} 1 & -1 \\ 0 & -1 \end{bmatrix} .$$

To see this is a well-defined homomorphism, note $\nu(a)^3 = I_2$, $\nu(b)^2 = I_2$ and $\nu(b)\nu(a)\nu(b)^{-1} = \nu(a)^{-1}$. The $\mathbb{C}$-character $\chi = tr \circ \nu$ has values:

conjugacy class:	$\{1\}$	$\{a, a^2\}$	$\{b, ab, a^2b\}$
$\chi :$	2	-1	0

Compare the trace of the representation μ of D_3 in Example (8.54).

(iv) In a group G, $gh = h$ if and only if $g = 1_G$. So the trace of the regular representation of G over F is the **regular character**

$$\chi_{reg} = \chi_{FG} : g \mapsto \begin{cases} n & \text{if } g = 1_G \\ 0 & \text{otherwise} , \end{cases}$$

where n is the order of G.

(8.59) Definition. The **character ring** of G over F is the subring $\mathbf{char}_F(G)$ of $cf(G, F)$ generated by the F-characters of G.

The elements of $\mathbf{char}_F(G)$ have a simple form:

(8.60) Lemma. *Suppose* $M, N \in \mathcal{M}(FG)$ *and* F *is regarded as an* FG-module *with trivial action of* G.

(i) *If* $M \cong N$, $\chi_M = \chi_N$.

(ii) *The multiplicative identity in* $cf(G, F)$ *is* χ_F .

(iii) $\chi_M + \chi_N = \chi_{M \oplus N}$.

(iv) $\chi_M \chi_N = \chi_{M \overline{\otimes}_F N}$.

(v) *Every element of* $\mathbf{char}_F(G)$ *is a pointwise difference* $\chi_M - \chi_N$ *of characters.*

(vi) *The ring* $\mathbf{char}_F(G)$ *is the* $\mathbb{Z}$-linear span of the irreducible F-characters $\chi_1, \ldots, \chi_r$.

Proof. Isomorphic modules afford similar matrix representations, which have equal trace, proving (i). As in Example (8.58) (ii), χ_F is the trivial character of degree 1; so it is the constant map to 1, proving (ii). If M and N afford matrix representations μ and ν, respectively, the discussion following (8.8) shows $M \oplus N$ affords $\mu \oplus \nu$ and $M \overline{\otimes}_F N$ affords $\mu \otimes \nu$. So (iii) and (iv) follow from properties (8.56) (iv) and (v) of the trace.

Now it follows that the set S of differences $\chi_M - \chi_N$ is closed under addition, subtraction, and multiplication and includes each F-character χ_M in the form $\chi_{M \oplus M} - \chi_M$; so it includes χ_F. That is, S is a subring of $cf(G, F)$ and is generated by the F-characters of G; so $S = \mathrm{char}_F(G)$.

Finally, each $M \in \mathcal{M}(FG)$ has a Krull-Schmidt decomposition

$$M \cong L_1^{n_1} \oplus \cdots \oplus L_r^{n_r}$$

for nonnegative integers n_i; so

$$\chi_M = n_1 \chi_1 + \cdots + n_r \chi_r ,$$

proving (vi). ∎

(8.61) Definitions. A **virtual character** of G over F is a difference $\chi_M - \chi_N$ of F-characters of G. So the character ring $\mathrm{char}_F(G)$ is sometimes called the **ring of virtual characters** of G over F.

This ring of virtual characters is a simplification of the representation ring $\mathcal{R}_F(G)$, described in (8.9) as the differences $[\mu] - [\nu]$ of similarity classes of matrix representations. To be more precise, the fact that F is a field, G is a finite group, and FG is semisimple implies

$$\mathrm{lat}_F \, \mathcal{F}(FG) = \mathcal{M}(FG) = \mathcal{P}(FG) .$$

The isomorphism classes in this first category form a commutative semiring under $\oplus$ and $\overline{\otimes}_F$, which is isomorphic by the affording map (8.8) (iii) to the semiring $(\mathrm{sim}_G(F), \oplus, \otimes)$. The group completion of this semiring is the representation ring

$$\mathcal{R}_F(G) = G_0(FG) = K_0(FG) ,$$

with typical element $[M] - [N]$ for f.g. FG-modules M and N, and with operations determined by

$$[M] + [N] = [M \oplus N] \quad \text{and} \quad [M][N] = [M \overline{\otimes}_F N] .$$

The multiplicative identity is $[F]$.

By Lemma (8.60), sending the isomorphism class $c(M)$ to χ_M defines a semiring homomorphism from $\mathcal{I}(\mathcal{P}(FG))$ to the ring $\mathrm{char}_F(G)$, inducing a surjective ring homomorphism

$$\alpha : K_0(FG) \to \mathrm{char}_F(G)$$
$$[M] - [N] \mapsto \chi_M - \chi_N .$$

(8.62) Theorem.

   (i)   *If A is a nonzero F-algebra, composing each member of $cf(G, F)$ with the inclusion $F \subseteq A$ embeds $cf(G, F)$ as an F-submodule of the A-module $cf(G, A)$. There, the irreducible F-characters $\chi_1, \ldots, \chi_r$ are A-linearly independent.*

  (ii) *The irreducible characters $\chi_1, \ldots, \chi_r$ are a $\mathbb{Z}$-basis of $\mathrm{char}_F(G)$. If $\chi = n_1\chi_1 + \cdots + n_r\chi_r$ for integers n_i, then for each i, $n_i = \widehat{\chi}(e_i)/d_i$.*

 (iii) *Matrix representations of G over F are similar if and only if their traces agree. That is, $M \cong N$ in $\mathcal{M}(FG)$ if and only if $\chi_M = \chi_N$.*

 (iv) *The map $\alpha : K_0(FG) \to \mathrm{char}_F(G)$, with $\alpha([M] - [N]) = \chi_M - \chi_N$, is a ring isomorphism.*

Proof. For each i, L_i is a left ideal of the simple component FGe_i with multiplicative identity e_i. So multiplication by e_i is the identity map on L_i and the zero map on L_j for $j \neq i$. Taking the trace,

$$\widehat{\chi}_j(e_i) \;=\; \begin{cases} d_i & \text{if } j = i \\ 0 & \text{if } j \neq i . \end{cases}$$

Now suppose $\sum a_j\chi_j = 0$ for $a_1, \ldots, a_r \in A$. Then $\sum a_j\widehat{\chi}_j = 0$, since it is F-linear. Evaluating at e_i, $a_id_i = 0$. Since $d_i \in F^*$, $a_i = 0$. This works for each i; so $\chi_1, \ldots, \chi_r$ are A-linearly independent.

Since $\mathbb{Z} \subseteq A$, they are also $\mathbb{Z}$-linearly independent. And they span $\mathrm{char}_F(G)$ over $\mathbb{Z}$ by (8.60) (vi). For the second assertion in (ii), evaluate $\widehat{\chi} = \sum n_j\widehat{\chi}_j$ at the idempotent e_i.

For (iii), matrix representations μ and ν afforded by M and N are similar if and only if $M \cong N$; and their traces agree if and only if $\chi_M = \chi_N$. So the first assertion follows from the second. Suppose $M \cong L_1^{n_1} \oplus \cdots \oplus L_r^{n_r}$ for nonnegative integers n_i. Then $\chi_M = n_1\chi_1 + \cdots + n_r\chi_r$. So by (ii), each $n_i = d_i^{-1}\widehat{\chi}_M(e_i)$. If $\chi_M = \chi_N$, then $\widehat{\chi}_M = \widehat{\chi}_N$, and the integers n_i are the same for M and N, forcing $M \cong N$.

If $\alpha([M] - [N]) = 0$, then $\chi_M = \chi_N$; so $M \cong N$ and $[M] - [N] = 0$. ∎

The table of values of the irreducible F-characters of G, also known as the F-**character table** of G, is arranged in the form

(8.63)

	c_1	c_2	$\cdots$	c_m
	g_1	g_2	$\cdots$	g_m
χ_1				
χ_2				
$\vdots$		$\chi_i(g_j)$		
χ_r				

with rows headed by the irreducible characters χ_i and columns headed by the conjugacy classes in G, each denoted by a class member g_j with the class size c_j written above it. By convention, χ_1 is the trivial character of degree 1; so the first row consists of 1's. And the first column is headed by the class of 1_G; so it lists the degrees $\chi_i(1_G) = d_i$. Apart from this, there is no standard order to the conjugacy classes or the irreducible characters; so the matrix of entries is unique except for permutations of rows and of columns.

According to (8.52), the number r of rows is less than or equal to the number m of columns, and $r = m$ if F is a splitting field of G. To some extent, the character table of G depends on the choice of field F; but if F is a splitting field of G and a subfield of a field E, the tables are the same over F and E, since irreducible matrix representations of G over F are absolutely irreducible, and the table over F is already square.

(8.64) Examples.

(i) For G a cyclic group generated by a of order 3, the $\mathbb{Q}$-character table obtained from the representations (8.5) (iii) is

	1	1	1
	1	a	a^2
χ_1	1	1	1
χ_2	2	-1	-1

while the $\mathbb{C}$-character table is

	1	1	1
	1	a	a^2
χ_1	1	1	1
χ_2	1	ζ	ζ^2
χ_3	1	ζ^2	ζ

where $\zeta = e^{2\pi i/3}$. The sum of χ_2 and χ_3 in the second table is χ_2 in the first table, because the $\mathbb{Q}$-representation

$$\mu_2 : a \; \mapsto \; \begin{bmatrix} 0 & -1 \\ 1 & -1 \end{bmatrix}$$

is similar over $\mathbb{C}$ to the direct sum of the degree 1 complex representations $\mu_2 : a \mapsto \zeta$ and $\mu_3 : a \mapsto \zeta^2$.

(ii) From Example (8.54), the $\mathbb{R}$-character table and the $\mathbb{C}$-character table of the dihedral group D_3 are both

	1	2	3
	1	a	b
χ_1	1	1	1
χ_2	1	1	-1
χ_3	2	-1	0 .

Examination of an assortment of $\mathbb{C}$-character tables reveals some recurrent patterns:

(8.65) Theorem. *Consider the matrix of entries in a $\mathbb{C}$-character table of G.*

  (i) *Suppose d is an exponent of G. All entries lie in $\mathbb{Z}[\zeta_d]$, where $\zeta_d = e^{2\pi i/d}$.*

  (ii) *The complex conjugate of each row is a row. The complex conjugate of each column is a column.*

  (iii) *Under the standard inner product over $\mathbb{C}$*

$$(v_1,\ldots,v_s) \bullet (w_1,\ldots,w_s) \;=\; v_1\overline{w}_1 + \cdots + v_s\overline{w}_s \;,$$

where $\overline{z}$ is the complex conjugate of z, the columns are pairwise orthogonal, and the inner product of the j-column with itself is the group order n divided by the conjugacy class size c_j. If we list out the rows with one column heading for each $g \in G$, the rows become pairwise orthogonal, and the inner product of each row with itself is n.

  (iv) *The degrees d_i divide n.*

Proof. Suppose $\mu : G \to GL_t(\mathbb{C})$ is a representation of G, and $g \in G$. Since $g^d = 1_G$, $\mu(g)^d = I_t$ and the minimal polynomial of $\mu(g)$ over $\mathbb{C}$ divides $x^d - 1$; hence it is separable. So $\mu(g)$ is similar to a diagonal matrix $\delta = diag(\lambda_1,\ldots,\lambda_t)$. Then $\delta^d = I_t$, so each λ_i is a power of ζ_d. If $\chi = tr \circ \mu$, it follows that

$$\chi(g) \;=\; \lambda_1 + \cdots + \lambda_t \in \mathbb{Z}[\zeta_d] \;,$$

proving (i).

If $A = (a_{ij}) \in M_t(\mathbb{C})$, let $\overline{A}$ denote $(\overline{a_{ij}})$. The entrywise complex conjugation map $\sigma : GL_t(\mathbb{C}) \to GL_t(\mathbb{C})$, $A \mapsto \overline{A}$, is a group automorphism. It takes similar matrices to similar matrices and direct sums to direct sums. If μ is an irreducible matrix representation, it follows that $\sigma \circ \mu$ is too. So the complex conjugate of each irreducible character χ_i is also an irreducible character χ_j.

From the geometry of complex multiplication, $\overline{\zeta_d} = \zeta_d^{-1}$. So

$$\delta^{-1} \;=\; diag(\lambda_1^{-1},\ldots,\lambda_t^{-1}) \;=\; \overline{\delta} \;.$$

Since $\mu(g)$ is similar to δ, $\mu(g^{-1}) = \mu(g)^{-1}$ is similar to $\delta^{-1} = \overline{\delta}$. Taking traces,

(8.66) $$\chi(g^{-1}) \;=\; \overline{\chi(g)} \;.$$

So the complex conjugate of the column headed by the class of g is the column headed by the class of g^{-1}, proving (ii).

The assertions in (iii) are known as the **orthogonality relations** for complex characters. We prove these in a more general form in (8.69) and (8.70) below, using (8.66).

(8.67) Lemma. *Suppose F is a splitting field for G, and the idempotent $e_i = \sum_{g \in G} a_g g$ for $a_g \in F$. Then*

$$a_g = \frac{1}{n} \widehat{\chi}_{reg}(e_i g^{-1}) = \frac{d_i}{n} \chi_i(g^{-1}) .$$

So the centrally primitive central idempotents $e_1, \ldots, e_r$ can be computed directly from the irreducible characters $\chi_1, \ldots, \chi_r$.

Proof. Since $FG \cong \oplus M_{d_i}(F) \cong \oplus L_i^{d_i}$, the regular representation has trace $\chi_{reg} = \chi_{FG} = \sum d_i \chi_i$. For each $g \in G$,

$$\widehat{\chi}_{reg}(e_i g^{-1}) = \widehat{\chi}_{reg}\left(\sum_{h \in G} a_h h g^{-1} \right) = \sum_{h \in G} a_h \chi_{reg}(h g^{-1}) = a_g n .$$

But also

$$\widehat{\chi}_{reg}(e_i g^{-1}) = \sum_{j=1}^{r} d_j \widehat{\chi}_j(e_i g^{-1}) = d_i \chi_i(g^{-1}) ,$$

because $e_i L_j = 0$ for $j \neq i$, and $e_i g^{-1} x = g^{-1} x$ for $x \in L_i$. Now divide by n. ∎

(8.68) Definition Let $(\ |\ )$ denote the $\mathbb{Z}$-bilinear map from the cartesian product $\mathrm{char}_F(G) \times \mathrm{char}_F(G)$ to F, defined by

$$(\phi | \psi) = \frac{1}{n} \sum_{g \in G} \phi(g) \psi(g^{-1}) = \frac{1}{n} \sum_{g \in G} \phi(g^{-1}) \psi(g) .$$

(8.69) Row Orthogonality. *Suppose F is a splitting field for G. With respect to the form $(\ |\ )$, the irreducible characters $\chi_1, \ldots, \chi_r$ are an orthonormal $\mathbb{Z}$-basis of $\mathrm{char}_F(G)$. So the values of $(\ |\ )$ lie in $\mathbb{Z}$.*

Proof. By Lemma (8.67),

$$e_i = \frac{d_i}{n} \sum_{g \in G} \chi_i(g^{-1}) g ;$$

so

$$\widehat{\chi}_j(e_i) = \frac{d_i}{n} \sum_{g \in G} \chi_i(g^{-1}) \chi_j(g) = d_i(\chi_i | \chi_j) .$$

But

$$\widehat{\chi}_j(e_i) = \begin{cases} d_i & \text{if } i = j \\ 0 & \text{if } i \neq j . \end{cases}$$

Dividing by d_i, the χ_i are orthonormal, and hence $\mathbb{Z}$-linearly independent, since

$$\left(\sum_{i=1}^{r} n_i \chi_i \Big| \chi_j \right) = \sum_{i=1}^{r} n_i \left(\chi_i | \chi_j \right) = n_j .$$

We know they span $\mathrm{char}_F(G)$ by (8.60). ∎

(8.70) Column Orthogonality. *Suppose F is a splitting field for G. If $g, h \in G$, then*

$$\sum_{i=1}^{r} \chi_i(g)\chi_i(h^{-1})$$

is 0 if g is not conjugate to h, and is n/c if g and h belong to the same conjugacy class with c elements.

Proof. Consider the character table (8.63). Since F is a splitting field, $m = r$. Denote by A the matrix in $M_r(F)$ whose i, j-entry is $\widehat{\chi}_i(c_j g_j)$. The inverses of conjugate elements are conjugate. Let $g_{i'}$ denote the representative chosen from the conjugacy class inverse to that of g_i. Suppose B is the matrix in $M_r(F)$ whose i, j-entry is $\chi_j(g_{i'})$. By row orthogonality, $AB = nI_r$, since

$$\frac{1}{n} \sum_{k=1}^{r} \widehat{\chi}_i(c_k g_k)\chi_j(g_{k'}) \;=\; (\chi_i | \chi_j) \,.$$

So $B = (\frac{1}{n}A)^{-1}$ and $BA = nI_r$. In terms of entries,

$$\sum_{k=1}^{r} \chi_k(g_{i'})\widehat{\chi}_k(c_j g_j) \;=\; c_j \sum_{k=1}^{r} \chi_k(g_i^{-1})\chi_k(g_j)$$

is n if $i = j$ and 0 otherwise. Now divide by c_j.   ∎

These last two results prove (8.65) (iii). To complete the proof of (8.65), we must show each n/d_i is an integer. By Lemma (8.67),

$$e_i \;=\; \sum_{g \in G} \frac{d_i}{n}\chi_i(g^{-1})g \,.$$

Multiply by $(n/d_i)e_i$ to get

$$\frac{n}{d_i}e_i \;=\; \sum_{g \in G} \chi_i(g^{-1})g e_i \,,$$

which lies in $\mathbb{Z}[\zeta_d]G e_i$ by (8.65) (i). Now $\mathbb{Q}e_i$ is a ring inside $\mathbb{C}G$, so it has characteristic 0; and the map $\mathbb{Q} \to \mathbb{Q}e_i$, $q \mapsto qe_i$, is a ring isomorphism. So $\mathbb{Z}e_i$ is integrally closed in its field of fractions $\mathbb{Q}e_i$. Since $\mathbb{Z}[\zeta_d]G e_i$ is a ring inside $\mathbb{C}G$ that is finitely generated as a $\mathbb{Z}e_i$-module, it follows from (7.6) that $(n/d)e_i$ is integral over $\mathbb{Z}e_i$; so it belongs to $\mathbb{Z}e_i$. Then n/d_i belongs to $\mathbb{Z}$.   ∎

The $\mathbb{Z}$-bilinear form $(\,|\,)$ on $\mathrm{char}_F(G)$ is handy for retrieving information on modules from characters they afford:

(8.71) Proposition. *Suppose F is a splitting field for G, and*

$$M \cong L_1^{n_1} \oplus \cdots \oplus L_r^{n_r}$$

in $\mathcal{M}(FG)$, *for nonnegative integers* n_i.

(i) *For each i,* $n_i = (\chi_M | \chi_i)$.
(ii) *The module M is simple if and only if* $(\chi_M | \chi_M) = 1$.

Proof. Since $\chi_M = \sum n_j \chi_j$, assertion (i) follows from orthonormality of $\chi_1, \ldots, \chi_n$. For (ii),

$$(\chi_M | \chi_M) = \sum_{i,j} n_i n_j (\chi_i | \chi_j) = \sum_{i=1}^{r} n_i^2 ,$$

which is 1 if and only if some $n_i = 1$ and $n_j = 0$ for all $j \neq i$. ∎

The property of $\mathbb{C}$-character tables, that complex conjugation permutes the rows, can be generalized. Suppose $\sigma : F \to E$ is a (necessarily injective) ring homomorphism between fields. Then E has characteristic zero too. Suppose $\chi = tr \circ \mu$ for a matrix representation μ of G over F. If σ_e is entrywise application of σ, then $\sigma_e \circ \mu$ is a matrix representation of G over E, and

$$tr \circ \sigma_e \circ \mu = \sigma \circ tr \circ \mu = \sigma \circ \chi$$

is an E-character of G. So composition with σ defines a ring homomorphism

$$\sigma \circ (-) : \mathrm{char}_F(G) \to \mathrm{char}_E(G) ,$$

which is injective because σ is injective, and which carries characters to characters.

(8.72) Proposition. *The map $\sigma \circ (-)$ carries an F-character table of G to an E-character table of G, if either F is a splitting field for G or $\sigma : F \to E$ is an isomorphism.*

Proof. Suppose σ is an isomorphism. Entrywise σ^{-1} carries direct sums of matrices to direct sums, and similar matrices to similar matrices. So $\sigma \circ (-)$ takes irreducible characters to irreducible characters. By the same argument, so does $\sigma^{-1} \circ (-)$. So $\sigma \circ (-)$ and $\sigma^{-1} \circ (-)$ are mutually inverse bijections between the irreducible F-characters and the irreducible E-characters of G.

On the other hand, if σ is not necessarily an isomorphism but F is a splitting field for G, then so is $\sigma(F)$ (according to (8.48)). And the isomorphism $F \cong \sigma(F)$ carries an F-character table to a $\sigma(F)$-character table, by the preceding paragraph. The latter is also an E-character table of G, since irreducible representations over $\sigma(F)$ are absolutely irreducible, and the maximum number of irreducible characters of G has already been reached over $\sigma(F)$. ∎

(8.73) Corollary. *If σ is an automorphism of the field F, then $\sigma \circ (-)$ permutes the rows of any F-character table of G.* ∎

Next, we see that character theory over characteristic zero splitting fields is the same as character theory over $\mathbb{C}$:

(8.74) Corollary. *Suppose d is an exponent for G, F is a splitting field for G, and its algebraic closure $\overline{F}$ contains $\mathbb{Q}(\zeta_d)$ as a subfield. Then an F-character table of G is a $\mathbb{C}$-character table of G.*

Proof. Denote by K the algebraic closure of $\mathbb{Q}(\zeta_d)$ within $\overline{F}$. Then $\mathbb{Q}(\zeta_d) \subseteq K$ is an algebraic field extension, and inclusion of $\mathbb{Q}(\zeta_d)$ into $\mathbb{C}$ extends to a ring homomorphism $\sigma : K \to \mathbb{C}$. Both F and K are splitting fields for G; so an F-character table is an $\overline{F}$-character table, which is, in turn, a K-character table. And $\sigma \circ (-)$ carries a K-character table to a $\mathbb{C}$-character table – having no effect on the entries, which lie in $\mathbb{Z}[\zeta_d]$. ∎

Note: For each characteristic zero field F, its algebraic closure $\overline{F}$ has a unique subfield isomorphic to $\mathbb{Q}(\zeta_d)$; so it can be made to contain $\mathbb{Q}(\zeta_d)$ by a notational change.

Back in (8.43) we found that if $i : F \to E$ is inclusion of a subfield F in a field E, and L is a left ideal of FG, with E-linear span EL in EG, then every F-basis of L is an E-basis of EL. So

(8.75) $$\chi_{EL} = i \circ \chi_L \ .$$

Replacing L by a f.g. FG-module M, the natural extension to an EG-module is $EG \otimes_{FG} M$. Even more generally, suppose $\sigma : F \to E$ is a ring homomorphism between characteristic zero fields. Then E is a right F-module under scalar multiplication $x \cdot y = x\sigma(y)$. Let $\sigma_c : FG \to EG$ denote the ring homomorphism applying σ to coefficients. Then EG is a right FG-module by $x \cdot y = x\sigma_c(y)$.

The map $E \times FG \to EG$, taking (a, y) to $a\sigma_c(y)$, is F-balanced, and the induced additive homomorphism on $E \otimes_F FG$ is an isomorphism of E, FG-bimodules

$$E \otimes_F FG \ \cong \ EG$$
$$a \otimes y \ \mapsto \ a\sigma_c(y)$$

with inverse taking $\sum a_g g$ to $\sum a_g \otimes g$. Therefore there is a composite E-linear isomorphism

$$E \otimes_F M \ \cong \ E \otimes_F (FG \otimes_{FG} M) \ \cong \ (E \otimes_F FG) \otimes_{FG} M$$
$$\cong \ EG \otimes_{FG} M$$

taking $a \otimes m$ to $a \otimes m$ for $a \in E$, $m \in M$. If $m_1, \ldots, m_d$ is an F-basis of M, then $1 \otimes m_1, \ldots, 1 \otimes m_d$ spans $E \otimes_F M$ over E and is an E-basis because

$$E \otimes_F M \;\cong\; E \otimes_F F^d \;\cong\; (E \otimes_F F)^d \;\cong\; E^d$$

as E-modules, so that $\dim_E(E \otimes_F M) = d$. Then $1 \otimes m_1, \ldots, 1 \otimes m_d$ is also an E-basis of $EG \otimes_{FG} M$.

Suppose $g \in G$ and $gm_j = \sum_i a_{ij} m_i$. Then

$$\begin{aligned}
g(1 \otimes m_j) \;=\; 1 \otimes gm_j \;&=\; 1 \otimes \sum_i a_{ij} m_i \\
&=\; \sum_i (1 \otimes a_{ij} m_i) \;=\; \sum_i \sigma(a_{ij}) \otimes m_i \\
&=\; \sum_i \sigma(a_{ij})(1 \otimes m_i) \,.
\end{aligned}$$

So, if a matrix representation μ of G over F is afforded by M, then $\sigma_e \circ \mu$ is afforded by $EG \otimes_{FG} M$, where σ_e is entrywise application of σ. Taking traces,

$$\chi_{EG \otimes_{FG} M} \;=\; \sigma \circ \chi_M \,.$$

When σ is inclusion and M is a left ideal L of FG, this equation and (8.75) imply

(8.76) $$EG \otimes_{FG} L \;\cong\; EL \,.$$

And for general σ, it implies the square

$$\begin{array}{ccc}
K_0(FG) & \xrightarrow{\;\cong\;} & \mathrm{char}_F(G) \\
{\scriptstyle K_0(\sigma_c)}\big\downarrow & & \big\downarrow{\scriptstyle \sigma \circ (-)} \\
K_0(EG) & \xrightarrow{\;\cong\;} & \mathrm{char}_E(G)
\end{array}$$

commutes, making $K_0(\sigma_c)$ a ring homomorphism. So by (8.72), $K_0(\sigma_c)$: $K_0(FG) \to K_0(EG)$ is a ring isomorphism if σ is an isomorphism or F is a splitting field for G.

If $\mathfrak{Field}_0$ is the category of characteristic zero fields and ring homomorphisms between them, the above square shows the affording map is a natural isomorphism between the functors $K_0((-)G)$ and $\mathrm{char}_{(-)}(G)$ from $\mathfrak{Field}_0$ to $\mathfrak{Ring}$. To see how $K_0(FG)$ is a functor in the variable G, read Bass [68, Chapter XI], Serre [77], or Swan [60], [70] and Lam [68A], [68B]. This theory is important for the computation of $K_n(\mathbb{Z}G)$ for all n.

8D. Exercises

1. Suppose G is an abelian group of finite order n and F has exactly m roots of $x^n - 1$. If $G \cong C_1 \times \cdots \times C_s$, where, for each i, C_i is a cyclic group of order n_i generated by a_i, prove the number of degree 1 F-characters of G is the product

$$\prod_{i=1}^{s} gcd(n_i, m) \, ,$$

where gcd stands for "greatest common divisor." *Hint:* Show each group homomorphism $C_i \to F^*$ is determined by the image of a_i, which can be any of the $gcd(n_i, m)$ elements of order dividing n_i in F^*. If $\psi_i : C_i \to F^*$ is a group homomorphism for each i, show

$$(\psi_1 \cdots \psi_s) : C_1 \times \cdots \times C_s \;\; \to \;\; F^*$$
$$(c_1, \ldots, c_s) \;\; \mapsto \;\; \psi_1(c_1) \cdots \psi_s(c_s)$$

is a group homomorphism and every group homomorphism from $C_1 \times \cdots \times C_s$ to F^* is $\psi_1 \cdots \psi_s$ for exactly one list $\psi_1, \ldots, \psi_s$.

2. Suppose F is a characteristic zero splitting field for the finite group G and $\chi_1, \ldots, \chi_r$ are the irreducible F-characters of G with respective degrees $d_1, \ldots, d_r$. If $1_G \neq g \in G$, prove

$$\sum_{i=1}^{r} d_i \chi_i(g) \;\; = \;\; 0 \, .$$

What version of this remains true when F is not a splitting field of G? *Hint:* Consider the trace of the regular representation.

3. From the $\mathbb{C}$-character table of D_3 given in Example (8.64) (ii), determine the centrally primitive central idempotents of $\mathbb{C}D_3$ using Lemma (8.67). Then use Theorem (8.62) (ii) to express the character χ with $\chi(1) = 11$, $\chi(a) = 5$, $\chi(b) = 1$ as a sum of the irreducible characters in the table. Achieve the same sum by using the orthogonality relations.

4. Demonstrate with an example that an F-character table can be square even when F is not a splitting field for G. *Hint:* Consider the generalized quaternion group Q_2, generated by a and b, with defining relations $a^4 = 1$, $b^2 = a^2$, and $ba = a^3 b$. It has eight elements $a^i b^j$ ($0 \leq i \leq 3$, $0 \leq j \leq 1$). Show Q_2 has four degree 1 $\mathbb{R}$-characters. Also show there is a surjective $\mathbb{R}$-algebra homomorphism $\mathbb{R}Q_2 \to \mathbb{H}$, where $\mathbb{H}$ is Hamilton's division ring of quaternions (with $\mathbb{R}$-basis $1, i, j, ij$ and $i^2 = j^2 = (ij)^2 = -1$). Conclude $\mathbb{R}Q_2 \cong \mathbb{H} \times \mathbb{R} \times \mathbb{R} \times \mathbb{R} \times \mathbb{R}$. Using $Z(\mathbb{H}) = \mathbb{R}$, apply Proposition (8.52).

5. Determine the $\mathbb{C}$-character table of Q_2. Begin with the four degree 1 characters referred to in Exercise 4, and use $n = \sum d_i^2$ to get the degree of

the fifth irreducible character (the table is square and Q_2 has five conjugacy classes). Then use column orthogonality to fill out the table. (You can confirm the values of this last character by representing $\mathbb{H}$ through right multiplication on $_\mathbb{C}\mathbb{H}_\mathbb{H}$ with $\mathbb{C}$-basis $1, j$.) Do the same thing to generate the $\mathbb{C}$-character table of the dihedral group D_4 of order 8, with generators a and b, and defining relations $a^4 = 1$, $b^2 = 1$, and $ba = a^3b$. How do these two tables compare?

6. The alternating group A_4 of even permutations of $\{1,2,3,4\}$ has generators $a = (1,2,3)$ and $b = (1,2)(3,4)$ and defining relations $a^3 = 1, b^2 = 1$, $ba^2b = aba$ and $a^2ba^2 = bab$. There are four conjugacy classes, represented by $1, b, a$, and a^2. The commutator subgroup H is the unique subgroup of order 4, consisting of the identity and the three elements of order 2. The quotient A_4/H is $\{\overline{1}, \overline{a}, \overline{a}^2\}$.

  (i)   Find the degree 1 $\mathbb{C}$-characters of A_4, and use column orthogonality to complete the $\mathbb{C}$-character table of A_4.
  (ii) Use it to find the centrally primitive central idempotents of $\mathbb{C}A_4$.
  (iii) Use averages of subgroups (as in §8B, Exercise 4) to find an idempotent e of $\mathbb{C}A_4$ for which $\mathbb{C}A_4e$ affords the irreducible character of degree > 1 generated in part (i). Why would this method not work to produce the irreducible $\mathbb{C}$-character of degree > 1 for the group Q_2? *Hint:* All subgroups of Q_2 are normal.

7. Prove a finite group G is simple if and only if, in the $\mathbb{C}$-character table of G, no nontrivial character has its degree as a value at a conjugacy class other than $\{1_G\}$. *Hint:* If $H \lhd G$ but $H \neq G$, show G/H has a nontrivial irreducible $\mathbb{C}$-character, leading to a full representation of G over $\mathbb{C}$, taking elements of H to the identity matrix. For the converse, show that a root of $x^n - 1$ in $\mathbb{C}$ other than 1 has real part less than 1; so a matrix A of finite order in $GL_d(\mathbb{C})$ has trace d if and only if $A = I_d$. Therefore the conjugacy classes that χ_i takes to d make up the kernel of μ_i. If $i \neq 1$, this kernel is a proper subgroup of G, by linear independence of characters.

8. How many conjugacy classes are in the dihedral group D_n of order $2n$? (The cases n even and n odd are different.) Complete the character table of D_{12} using the four degree 1 characters $D_{12} \to D_{12}^{ab} = \{\overline{1}, \overline{a}, \overline{b}, \overline{ab}\} \to \mathbb{C}^*$, the traces of the irreducible representations over $\mathbb{Q}(\zeta_d)$ found in §8A, Exercise 1, and the characters obtained from these by applying $\sigma \in \text{Aut}(\mathbb{Q}(\zeta_d)/\mathbb{Q})$ as in (8.73). Does this method always complete the $\mathbb{C}$-character table of D_n?

9. Suppose $\sigma : F \to E$ is a ring homomorphism between fields of characteristic zero and χ, χ' are two different irreducible F-characters of G. Prove $\sigma \circ \chi$ and $\sigma \circ \chi'$ share no summands when they are expressed as sums of irreducible E-characters of G; so

$$(\sigma \circ \chi | \sigma \circ \chi') = 0 .$$

Hint: Say χ and χ' are afforded by the simple left ideals L and L' of FG, respectively. Since $L \not\cong L'$, the simple components of FG containing L, L' are

FGe, FGe', where e and e' are mutually orthogonal central idempotents. If F is a subfield of E, and $M_1, \ldots, M_s$ represent the isomorphism classes of simple EG-modules, and if $EL \cong \oplus M_i^{a_i}$ and $EL' \cong \oplus M_i^{b_i}$, then $a_i b_i = 0$ for all i — otherwise some simple left ideal of EG contained in EGe is isomorphic to a simple left ideal of EG contained in EGe'; so $EGe \cap EGe'$ contains a simple component of EG, which is impossible since x in this intersection implies $x = xe = xee' = 0$. If σ is not an inclusion, it is an isomorphism followed by an inclusion.

10. If $\sigma : F \to E$ is a ring homomorphism between characteristic zero fields, M and N are nonisomorphic simple FG-modules, and EG is regarded as a right FG-module through σ, prove no composition factor of $EG \otimes_{FG} M$ is isomorphic to a composition factor of $EG \otimes_{FG} N$. *Hint:* Use Exercise 9 and semisimplicity of EG.

11. Suppose $F \subseteq E$ is a field extension and F has characteristic zero. Show a matrix representation afforded by $N \in \mathcal{M}(EG)$ is defined over F if and only if $N \cong EG \otimes_{FG} M$ for some $M \in \mathcal{M}(FG)$.

12. If F is a field (not necessarily of characteristic zero) and FG is semisimple, and if μ is a matrix representation of G over F afforded by $M \in \mathcal{M}(FG)$, prove μ is absolutely irreducible if and only if $EG \otimes_{FG} M$ is simple for each field extension E of F. (For fields of characteristic p dividing the order of G, simplicity of each $EG \otimes_{FG} M$ is used as a definition of absolute irreducibility of a representation over F afforded by M, and F is called a **splitting field** for G if every irreducible matrix representation of G over F is absolutely irreducible. Since FG is not semisimple, in this case, being a splitting field is *not* equivalent to FG being a product of matrix rings over F.)

PART III

Groups of Matrices: K_1

Algebraic K-theory is the study of a sequence of abelian groups $K_n(R)$ associated to each ring R. The group $K_1(R)$ is to $K_0(R)$ as invertible linear transformations are to vector spaces. The invertible matrices with entries in R form a nonabelian group $GL(R)$ under matrix multiplication. Reducing modulo a normal subgroup $E(R)$, associated to certain elementary row operations used to solve systems of linear equations, we get the initial abelian quotient $K_1(R) = GL(R)/E(R)$ of $GL(R)$. Its elements are row-equivalence classes of invertible matrices. Just as group homomorphisms $K_0(R) \to G$ correspond to generalized ranks of f.g. projective R-modules, so group homomorphisms $K_1(R) \to G$ correspond to generalized determinants on invertible matrices over the ring R.

Chapter 9 introduces $K_1(R)$, demonstrating links between commutativity and determinants. In Chapter 10 the stable rank of R leads to a bound on the dimensions of matrices needed to represent $K_1(R)$. Chapter 11 is devoted to the role of K_1 in the solution of the congruence subgroup problem over Dedekind rings of arithmetic type.

9

Definition of K_1

In §9A we review the invertible matrices associated with elementary row operations in basic linear algebra. In §9B these are considered in the context of the group $GL(R)$ of invertible matrices of all sizes over R. Elementary matrices of one type are shown to form the commutator subgroup of $GL(R)$, and the quotient is $K_1(R)$. This K_1 is shown to share some properties with K_0. Determinants are surveyed in §9C, where the canonical map $GL_n(R) \to K_1(R)$ is shown to be the initial determinant map over R, and K_1 of a commutative ring is decomposed into R^* and matrices of determinant 1. In §9D we develop the Bass K_1 of a category, which coincides with $K_1(R)$ when applied to $\mathcal{F}(R)$ or $\mathcal{P}(R)$. This construction is similar to that of the Grothendieck group and proves useful in connecting K_0 with K_1 in the exact sequences of Chapter 13. It also defines the determinant of any automorphism of a f.g. projective R-module.

9A. Elementary Matrices

Suppose R is a ring, and ϵ_{ij} is the matrix in $R^{s \times t}$ with i, j-entry 1_R and all other entries 0_R. As in §1B, Exercise 1, the "matrix units "ϵ_{ij} form a basis of $R^{s \times t}$ as both a left and a right R-module, they commute with scalars ($r\epsilon_{ij} = \epsilon_{ij}r$ for $r \in R$), and they multiply according to the rule:

$$
(\textbf{9.1}) \qquad \epsilon_{ij}\epsilon_{kl} \; = \; \begin{cases} 0 & \text{if } \; j \neq k \,, \\ \epsilon_{il} & \text{if } \; j = k \,. \end{cases}
$$

Writing each matrix (a_{ij}) as $\sum a_{ij}\epsilon_{ij}$, many facts of matrix arithmetic can be proved by using the preceding properties of matrix units.

In the ring $M_n(R)$, the multiplicative identity is $I_n = \epsilon_{11} + \epsilon_{22} + \ldots + \epsilon_{nn}$. The general linear group $GL_n(R)$ is the group $M_n(R)^*$ of invertible elements of the ring $M_n(R)$. If R is a commutative ring, a matrix $A \in M_n(R)$ is invertible if and only if its determinant d belongs to R^* — in that case, $A^{-1} = d^{-1}A^*$, where A^* is the transpose of the matrix of cofactors of A. (For details, see p. 518 of Lang [93], pp. 95–96 of Jacobson [85], or p. 303 of MacLane and Birkhoff [88].) But when R is a noncommutative ring, it is harder to determine whether a matrix over R is invertible.

Aside from I_n, the simplest invertible matrices in $M_n(R)$ are the **elementary transvections**

$$e_{ij}(r) \;=\; I_n + r\epsilon_{ij} \;,$$

where $i \neq j$, $1 \leq i \leq n$, $1 \leq j \leq n$, and $r \in R$. The matrix $e_{ij}(r)$ is obtained from I_n by replacing the 0 in the i,j-position by r. Since

$$
\begin{aligned}
e_{ij}(r)e_{ij}(s) &= (I_n + r\epsilon_{ij})(I_n + s\epsilon_{ij}) \\
&= I_n + (r+s)\epsilon_{ij} \;=\; e_{ij}(r+s) \;,
\end{aligned}
$$

each elementary transvection has an inverse $e_{ij}(r)^{-1} = e_{ij}(-r)$ that is also an elementary transvection. So the finite length products $t_1 \ldots t_m$, where each t_i is an elementary transvection from $GL_n(R)$, form a subgroup $\boldsymbol{E_n(R)}$ of $GL_n(R)$, called the group of **elementary matrices of degree** n over R.

If $A = (a_{ij}) \in M_n(R)$, the product

$$
\begin{aligned}
e_{ij}(r)A &= (I_n + r\epsilon_{ij}) \sum_{k,l} a_{kl}\epsilon_{kl} \\
&= \sum_{k,l} a_{kl}\epsilon_{kl} + \sum_{l} r a_{jl}\epsilon_{il} \\
&= \sum_{k \neq i} \sum_{l} a_{kl}\epsilon_{kl} + \sum_{l} (a_{il} + r a_{jl})\epsilon_{il}
\end{aligned}
$$

is obtained from A by adding r times the j-row to the i-row. This row operation, also called an **elementary row transvection**, is one of three types of row operations used in Gauss-Jordan elimination to simplify a system of linear equations with coefficients in R. The other two row operations are also accomplished by left multiplying A with an invertible matrix: If $i \neq j$, left multiplication by

$$P_{(i,j)} \;=\; I_n - \epsilon_{ii} - \epsilon_{jj} + \epsilon_{ij} + \epsilon_{ji}$$

switches the i-row with the j-row; so $P_{(i,j)}$ is its own inverse. If $u \in R^*$, left multiplication by

$$d_i(u) \;=\; I_n + (u - 1)\epsilon_{ii}$$

left multiplies the i-row by u; so $d_i(u)$ has inverse $d_i(u^{-1})$. The subgroup $\boldsymbol{P_n}$ of $GL_n(R)$ generated by the matrices $P_{(i,j)}$ is the group of **permutation matrices**

$$P_\sigma \;=\; \sum \epsilon_{\sigma(i)i},$$

for $\sigma \in S_n$. The subgroup of $GL_n(R)$ generated by the $d_i(u)$, with $1 \le i \le n$ and $u \in R^*$, is the group $\boldsymbol{D_n(R)}$ of invertible **diagonal matrices**

$$diag(u_1, \ldots, u_n) = \begin{bmatrix} u_1 & 0 & \cdots & 0 \\ 0 & u_2 & \cdots & 0 \\ \vdots & \vdots & \ddots & \vdots \\ 0 & 0 & \cdots & u_n \end{bmatrix},$$

where each $u_i \in R^*$.

When R is a commutative ring, the determinant

$$det : GL_n(R) \; \to \; R^*$$

is a group homomorphism, with $det(e_{ij}(r)) = 1$, $det(P_{(i,j)}) = -1$, and $det(d_i(u)) = u$. If $1 \ne -1$ in R, then among these three types of row operations, only elementary transvections preserve the determinant.

Whether R is commutative or not, say A, $B \in M_n(R)$ are t-**row equivalent** if B is obtained from A by a finite sequence of elementary row transvections. This is an equivalence relation, and the t-row equivalence class of A is the set $E_n(R)A$. When A is invertible, so is its entire t-row equivalence class, which forms a right coset of $E_n(R)$ in $GL_n(R)$.

9A. Exercises

1. Describe the column operations on $A \in M_n(R)$ that result from right multiplication by $e_{ij}(r)$, $P_{(i,j)}$, and $d_i(u)$ with $u \in R^*$. Define t-**column equivalence** by analogy with t-row equivalence, and describe the t-column equivalence class of a matrix $A \in GL_n(R)$.

2. Show every finite group G is isomorphic to a group of permutation matrices in $GL_n(R)$ with $n = |G|$, provided $1 \ne 0$ in R.

3. Prove P_n normalizes both $D_n(R)$ and $E_n(R)$, and $D_n(R)$ normalizes $E_n(R)$. So the products

$$GE_n(R) \; = \; D_n(R)E_n(R) \,,$$
$$ML_n(R) \; = \; D_n(R)P_n$$

are subgroups of $GL_n(R)$ with $E_n(R) \lhd GE_n(R)$ and $D_n(R) \lhd ML_n(R)$. The elements of $ML_n(R)$ are known as **monomial matrices** over R.

4. If $u \in R^*$, calculate the products

$$p_{ij}(u) \; = \; e_{ij}(u)e_{ji}(-u^{-1})e_{ij}(u) \,, \quad and$$
$$q_{ij}(u) \; = \; p_{ij}(u)p_{ij}(-1) \,,$$

and determine the effect of left multiplication by the matrices $p_{ij}(1)$ and $q_{ij}(u)$. Show $ML_n(R) \subseteq GE_n(R)$, as defined in Exercise 3.

5. A ring R is **generalized euclidean** if $GE_n(R) = GL_n(R)$ for all $n \geq 1$. Use Exercise 4 to prove the following are equivalent:

   (i) $GE_n(R) = GL_n(R)$.
   (ii) Each $A \in GL_n(R)$ can be reduced to I_n by row operations of the three types: (a) add r times the j-row to the i-row ($r \in R$), (b) switch two rows, and (c) left multiply a row by $u \in R^*$.
   (iii) Each $A \in GL_n(R)$ is t-row equivalent to a matrix $d_1(u) = diag(u, 1, \ldots, 1)$.

6. If R is generalized euclidean, prove t-row equivalence and t-column equivalence are the same in $GL_n(R)$ for each $n \geq 1$.

7. Prove R is a generalized euclidean ring if R is commutative euclidean, R is a division ring, or R is local.

9B. Commutators and $K_1(R)$

In this section we study noncommutativity in the abstract, and then show that noncommutativity of matrix multiplication is the fault of elementary transvections.

For elements x and y of a group G, $xy = yx$ if and only if the **commutator** $[x, y] = xyx^{-1}y^{-1}$ equals the identity. Since

$$[x, y]^{-1} = [y, x] ,$$

the finite length products of commutators $[x, y]$ with $x, y \in G$ form a subgroup of G; this subgroup is denoted by $[G, G]$ and is called the **commutator subgroup** of G. Since

$$z[x, y]z^{-1} = [zxz^{-1}, zyz^{-1}] ,$$

$[G, G]$ is a normal subgroup of G. The quotient

$$G_{ab} = G/[G, G]$$

is known as the **abelianization** of G. Since

$$[x[G, G], y[G, G]] = [x, y][G, G] = [G, G] ,$$

the group G_{ab} is abelian. It is the first abelian quotient of G:

(9.2) Proposition. *Suppose G is a group. A subgroup H contains $[G,G]$ if and only if $H \triangleleft G$ and G/H is abelian.*

Proof. If $H \triangleleft G$ and G/H is abelian, then for all $x, y \in G$, $[x,y]H = [xH, yH] = H$ in G/H; so $[G,G] \subseteq H$. Conversely, if $[G,G] \subseteq H$, then $H/[G,G]$ is a normal subgroup of the abelian group $G/[G,G]$; so $H \triangleleft G$. And G/H is a homomorphic image of $G/[G,G]$; so it is abelian. ∎

Every group homomorphism $f : G \to H$ carries $[G,G]$ into $[H,H]$; so it induces a group homomorphism from G_{ab} to H_{ab}. More generally, suppose $f : G \to H$ is a group homomorphism or **antihomomorphism** ($f(xy) = f(y)f(x)$). If $c : H \to H_{ab}$ is the canonical map, the composite $c \circ f : G \to H_{ab}$ is a group homomorphism with kernel containing $[G,G]$. The **abelianization** of f is the unique induced homomorphism f_{ab} making the square

$$
\begin{array}{ccc}
G & \xrightarrow{\ f\ } & H \\[4pt]
{\scriptstyle c}\big\downarrow & & \big\downarrow{\scriptstyle c} \\[4pt]
G_{ab} & \xrightarrow{\ f_{ab}\ } & H_{ab}
\end{array}
$$

commute, so that $f_{ab} \circ c = c \circ f$ and

$$f_{ab}(x[G,G]) = f(x)[H,H] .$$

(9.3) Proposition. *Suppose $\mathcal{C}$ is the category of groups and group homomorphisms and antihomomorphisms. There is an "abelianization" functor $(-)_{ab} : \mathcal{C} \to \mathcal{Ab}$.*

Proof. If $f : G \to H$ and $f' : H \to K$ are arrows in $\mathcal{C}$, then $c \circ f' \circ f = f'_{ab} \circ c \circ f = f'_{ab} \circ f_{ab} \circ c$; so

$$(f' \circ f)_{ab} = f'_{ab} \circ f_{ab} .$$

And $c \circ i_G = i_{G_{ab}} \circ c$; so $(i_G)_{ab} = i_{G_{ab}}$. ∎

Next consider commutators of matrices:

(9.4) Lemma. *Suppose R is a ring, $r, s \in R$, and $n \geq 1$. In the group $GL_n(R)$,*

$$[e_{ij}(r),\ e_{kl}(s)] = \begin{cases} I_n & \text{if } i \neq l, \ j \neq k, \\ e_{il}(rs) & \text{if } i \neq l, \ j = k. \end{cases}$$

Proof. For the given transvections to be defined, $i \neq j$ and $k \neq l$. If also $i \neq l$, multiply out

$$(I_n + r\epsilon_{ij})(I_n + s\epsilon_{kl})(I_n - r\epsilon_{ij})(I_n - s\epsilon_{kl})$$

using the matrix unit identities (9.1). ∎

(9.5) Corollary. *If $n \geq 3$, $[E_n(R), E_n(R)] = E_n(R)$.*

Proof. If $i, j \in \{1, \ldots, n\}$ and $r \in R$, there exists $k \in \{1, \ldots, n\}$ different from i and j. Then $e_{ij}(r) = [e_{ik}(r), e_{kj}(1)]$.   ∎

Allowing more elbow room, we can get hold of all commutators in $GL_n(R)$:

(9.6) Definitions. For each ring R, the **infinite-dimensional general linear group $GL(R)$** is the multiplicative group of $\mathbb{P} \times \mathbb{P}$ matrices (entries indexed by pairs of positive integers) that are obtained from the matrix

$$I_\infty = (\delta_{ij}) = \begin{bmatrix} 1 & 0 & \cdots \\ 0 & 1 & \cdots \\ \vdots & \vdots & \ddots \end{bmatrix}$$

by replacing an upper left corner I_n by any invertible matrix $A = (a_{ij}) \in GL_n(R)$:

$$A \oplus I_\infty = \begin{bmatrix} a_{11} & \cdots & a_{1n} & 0 & \cdots \\ \vdots & & \vdots & \vdots & \\ a_{n1} & \cdots & a_{nn} & 0 & \cdots \\ 0 & \cdots & 0 & 1 & \\ \vdots & & \vdots & & \ddots \end{bmatrix}.$$

The set $GL(R)$ is a subset of the ring $M(R)$ of "column-finite" $\mathbb{P} \times \mathbb{P}$ matrices over R (discussed in (1.37)). If $A, B \in GL_n(R)$, then $(A \oplus I_\infty)(B \oplus I_\infty) = AB \oplus I_\infty$. By identifying A with $A \oplus I_\infty$, $GL_n(R)$ can be identified with the subgroup

$$\{A \oplus I_\infty \ : \ A \in GL_n(R)\}$$

of the group of units $M(R)^*$. With these identifications,

$$R^* = GL_1(R) \subseteq GL_2(R) \subseteq GL_3(R) \subseteq \cdots ,$$

$$\text{and} \qquad GL(R) = \bigcup_{n=1}^{\infty} GL_n(R) .$$

Therefore $GL(R)$ is a subgroup of $M(R)^*$, with identity I_∞ and with $(A \oplus I_\infty)^{-1} = A^{-1} \oplus I_\infty$.

The group $E(R)$ of **elementary matrices** is the subgroup of $GL(R)$ generated by the elementary transvections

$$e_{ij}(r) = I_\infty + r\epsilon_{ij} \quad (i \neq j, \ r \in R) ,$$

obtained from I_∞ by replacing an off-diagonal 0 by r. Under the identifications of $A \in GL_n(R)$ with $A \oplus I_\infty \in GL(R)$, the $e_{ij}(r) \in GL_n(R)$ become the $e_{ij}(r) \in GL(R)$; so

$$1 = E_1(R) \subseteq E_2(R) \subseteq E_3(R) \subseteq \cdots ,$$

$$\text{and} \qquad E(R) = \bigcup_{n=1}^{\infty} E_n(R) .$$

(9.7) Whitehead Lemma. *For each ring R and integer* $n \geq 3$,

$$E_n(R) \subseteq [GL_n(R), \ GL_n(R)] \subseteq E_{2n}(R) \ ,$$

so that $[GL(R), \ GL(R)] = E(R)$.

Proof. Recall from (1.33) that block addition and multiplication of matrices agrees with ordinary matrix addition and multiplication. Also note that, if $A \in M_n(R)$, the matrices

$$\begin{bmatrix} I_n & A \\ 0 & I_n \end{bmatrix} \quad \text{and} \quad \begin{bmatrix} I_n & 0 \\ A & I_n \end{bmatrix}$$

belong to $E_{2n}(R)$, since

$$\prod_{\substack{1 \leq i \leq n \\ n < j \leq 2n}} (I_{2n} + a_{ij}\epsilon_{ij}) \quad = \quad I_{2n} + \sum_{\substack{1 \leq i \leq n \\ n < j \leq 2n}} a_{ij}\epsilon_{ij} \ ,$$

and similarly for the second matrix.

Now, for each $A \in GL_n(R)$,

$$\begin{bmatrix} 0 & A \\ -A^{-1} & 0 \end{bmatrix} = \begin{bmatrix} I_n & A \\ 0 & I_n \end{bmatrix} \begin{bmatrix} I_n & 0 \\ -A^{-1} & I_n \end{bmatrix} \begin{bmatrix} I_n & A \\ 0 & I_n \end{bmatrix},$$

$$\begin{bmatrix} A & 0 \\ 0 & A^{-1} \end{bmatrix} = \begin{bmatrix} 0 & A \\ -A^{-1} & 0 \end{bmatrix} \begin{bmatrix} 0 & -I_n \\ I_n & 0 \end{bmatrix},$$

and for $A, B \in GL_n(R)$,

$$\begin{bmatrix} ABA^{-1}B^{-1} & 0 \\ 0 & I_n \end{bmatrix} = \begin{bmatrix} A & 0 \\ 0 & A^{-1} \end{bmatrix} \begin{bmatrix} B & 0 \\ 0 & B^{-1} \end{bmatrix} \begin{bmatrix} (BA)^{-1} & 0 \\ 0 & BA \end{bmatrix}.$$

So $[GL_n(R), \ GL_n(R)] \subseteq E_{2n}(R)$. The other containments are direct consequences of (9.5) and the equations $GL(R) = \cup GL_n(R)$, $E(R) = \cup E_n(R)$. ∎

(9.8) Definition. For each ring R, the **Bass-Whitehead group** of R is the algebraic K-group

$$\boldsymbol{K_1(R)} \ = \ GL(R)_{ab} \ = \ GL(R)/E(R) \ .$$

If $\phi : R \to S$ is a ring homomorphism, entrywise application of ϕ defines a ring homomorphism $M(R) \to M(S)$, $(r_{ij}) \mapsto (\phi(r_{ij}))$, restricting to a group homomorphism

$$GL(\phi) : GL(R) \ \to \ GL(S) \ .$$

In this way, GL is a functor from $\mathfrak{Ring}$ to $\mathfrak{Group}$, or if we like, into the category $\mathcal{C} \supseteq \mathfrak{Group}$ considered in (9.3). Composing functors

$$\mathfrak{Ring} \xrightarrow{\ GL\ } \mathcal{C} \xrightarrow{\ (-)_{ab}\ } \mathcal{A}b$$

we obtain:

(9.9) Proposition. *There is a functor $K_1 : \mathfrak{Ring} \to \mathcal{Ab}$; for each ring homomorphism $\phi : R \to S$, the group homomorphism*

$$K_1(\phi) : K_1(R) \quad \to \quad K_1(S)$$

takes $(a_{ij})E(R)$ to $(\phi(a_{ij}))E(S)$.   ∎

(9.10) Proposition. *For each ring R, $K_1(R) \cong K_1(R^{op})$.*

Proof. The transpose defines mutually inverse antihomomorphisms between $GL(R)$ and $GL(R^{op})$. Applying $(-)_{ab}$, we get mutually inverse homomorphisms between $K_1(R)$ and $K_1(R^{op})$.   ∎

(9.11) Proposition. *For each ring R and each $n \geq 1, K_1(M_n(R)) \cong K_1(R)$.*

Proof. By (1.33), erasing internal matrix brackets defines a group isomorphism $GL(M_n(R)) \cong GL(R)$, inducing the desired isomorphism on K_1.   ∎

(9.12) Proposition. *If R and S are rings and $R \times S$ is the ring with coordinatewise operations, then $K_1(R \times S) \cong K_1(R) \times K_1(S)$.*

Proof. Taking $((a_{ij}, \ b_{ij}))$ to $((a_{ij}), \ (b_{ij}))$ defines a group isomorphism $GL(R \times S) \cong GL(R) \times GL(S)$, taking $E(R \times S)$ onto $E(R) \times E(S)$ since

$$e_{ij}((r, s)) \ \mapsto \ (e_{ij}(r), \ e_{ij}(s)) \ ,$$

and the latter pairs generate $E(R) \times E(S)$. So there are induced isomorphisms

$$K_1(R \times S) \ \cong \ \frac{GL(R) \times GL(S)}{E(R) \times E(S)} \ \cong \ K_1(R) \times K_1(S)$$

with composite

$$(a_{ij}, b_{ij})E(R \times S) \ \mapsto \ ((a_{ij})E(R), \ (b_{ij})E(S)) \ . \qquad ∎$$

Note that the last proposition could also be proved using abelianizations, since $(G \times H)_{ab} \cong G_{ab} \times H_{ab}$ for groups G and H. Also note the parallel between the last four propositions and the properties (6.6), (6.7), (6.10), (6.22) of K_0.

9B. Exercises

1. In $GL(R)$, show that t-row equivalence is the same as t-column equivalence.

2. Prove that a matrix $A \in M_n(R)$ is invertible if and only if its rows form an R-basis of R^n. Just as $K_0(R)$ is a natural setting in which to study the question of *existence* of bases in each f.g. projective R-module, so too $K_1(R)$ can be seen as a measure of the *uniqueness* of bases in each f.g. free R-module R^n: Two n-element bases of R^n can be regarded as equivalent if they are the rows of matrices $A, B \in GL_n(R)$ that are t-row equivalent in $GL(R)$ – that is, which become equal in $K_1(R)$.

3. Prove $[GL_n(R),\ GL_n(R)] \subseteq E_{3n}(R)$ by computing the commutator

$$[A \oplus I_n \oplus A^{-1},\ B \oplus B^{-1} \oplus I_n]$$

for $A, B \in GL_n(R)$. This provides an alternate proof that $[GL(R),\ GL(R)] = E(R)$.

4. Find a noncommutative ring R, with 128 elements, for which $K_1(R) = 1$.

5. Prove $E : \mathfrak{Ring} \to \mathfrak{Group}$ is a functor, where, for each ring homomorphism $f : R \to S$, the group homomorphism $E(f) : E(R) \to E(S)$ is entrywise application of f. If f is surjective, prove $E(f)$ is also surjective, but $GL(f)$ need not be surjective.

6. Show the inclusion and canonical maps $E(R) \to GL(R) \to K_1(R)$ define natural transformations $E \to GL \to K_1$, as defined in (0.13).

7. A group G is **solvable** if there is a finite length chain

$$1 \ = \ G_n \ \lhd \ G_{n-1} \ \lhd \ \cdots \ \lhd \ G_0 \ = \ G \, ,$$

in which each quotient G_i/G_{i+1} is abelian. For each group G, the commutator subgroup $[G, G]$ is often denoted by G' and called the **derived subgroup** of G. The chain $G \supseteq G' \supseteq G'' \supseteq \cdots$ is called the **derived series** of G. Prove G is solvable if and only if its derived series reaches 1 in finitely many steps.

8. If R is a ring other than $\{0\}$, prove that $GL_n(R)$ (for $n \geq 3$) and $GL(R)$ are not solvable.

9. Under the identifications of $GL_n(R)$ with $GL_n(R) \oplus I_\infty \subseteq GL(R)$, prove $ML_n(R) \subseteq GL_1(R)E_n(R)$, where $ML_n(R)$ is the group of monomial matrices in $GL_n(R)$. *Hint:* See §9A, Exercise 4.

9C. Determinants

Over a field F, there are three equivalent commonly used definitions of the determinant det: $M_n(F) \to F$:

(9.13) Definitions.

(i)   Regarding each matrix in $M_n(F)$ as a row of columns in $(F^{n \times 1})^n$, det is multilinear (linear in each column), alternating (zero if two columns are equal), and takes I_n to 1.

(ii)
$$\det((a_{ij})) = \sum_{\sigma \in S_n} (-1)^{p(\sigma)} a_{1\sigma(1)} \cdots a_{n\sigma(n)} ,$$

where $p(\sigma)$ is the number of transpositions that compose to give the permutation σ.

(iii) Elementary transvections on rows reduce $A \in GL_n(F)$ to the diagonal form

$$u \oplus I_{n-1} = \begin{bmatrix} u & 0 & \cdots & 0 \\ 0 & 1 & \cdots & 0 \\ \vdots & \vdots & \ddots & \vdots \\ 0 & 0 & \cdots & 1 \end{bmatrix};$$

take $\det(A) = u$. For noninvertible $A \in M_n(F)$, take $\det(A) = 0$.

Suppose $A, B \in M_n(F)$. Properties of det that follow from these definitions include:

(9.14) Properties.

(iv) $\det(AB) = \det(A) \det(B)$.
(v) $\det(e_{ij}(r)) = 1$ for each elementary transvection $e_{ij}(r)$.
(vi)   For each $m \geq 1$, $\det(A \oplus I_m) = \det(A)$.
(vii) The matrix A is invertible if and only if $\det(A)$ is a unit.

If we replace F by an arbitrary commutative ring R, definitions (i) and (ii) are still equivalent, the determinant they define can also be computed by cofactor expansion, and properties (9.14) (iv)–(vii) remain true. If this commutative ring R is euclidean (or even generalized euclidean, as defined in §9A, Exercise 5), definition (iii) produces the same determinant of an invertible matrix as do (i) and (ii).

But if F a *noncommutative* ring R, there is no function det:$M_2(R) \to R$ satisfying the conditions in (i); for if det is multilinear, alternating and takes

I_n to 1, then, for all $x, y \in R$,

$$xy = xy \det \begin{bmatrix} 1 & 0 \\ 0 & 1 \end{bmatrix} = x \det \begin{bmatrix} 1 & 0 \\ 0 & y \end{bmatrix}$$

$$= \det \begin{bmatrix} x & 0 \\ 0 & y \end{bmatrix} = y \det \begin{bmatrix} x & 0 \\ 0 & 1 \end{bmatrix}$$

$$= yx \det \begin{bmatrix} 1 & 0 \\ 0 & 1 \end{bmatrix} = yx .$$

And for a noncommutative ring R, the formula (ii) defines a function $\det : M_n(R) \to R$, but

$$\det \left(\begin{bmatrix} 1 & 0 \\ 0 & y \end{bmatrix} \begin{bmatrix} x & 0 \\ 0 & 1 \end{bmatrix} \right) = xy , \quad \text{while} \quad \det \begin{bmatrix} 1 & 0 \\ 0 & y \end{bmatrix} \det \begin{bmatrix} x & 0 \\ 0 & 1 \end{bmatrix} = yx ;$$

so the important property (iv) fails. If R is a noncommutative ring, any row-reduction determinant (iii) is not single-valued: for by Whitehead's Lemma (9.7),

$$\begin{bmatrix} xy & 0 \\ 0 & 1 \end{bmatrix} \quad \text{and} \quad \begin{bmatrix} yx & 0 \\ 0 & 1 \end{bmatrix}$$

are t-row equivalent. Apparently the definitions (9.13) (i), (ii), and (iii) are unsuitable for determinants of matrices over arbitrary rings.

In 1943, J. Dieudonné [43] showed that, over a division ring D, the row-reduction determinant can be made single-valued by taking its values to be cosets in $D^*_{ab} = D^*/[D^*, D^*]$. He also proved the resulting determinant induces a group isomorphism $GL_n(D)_{ab} \cong D^*_{ab}$. The group $GL_n(D)_{ab}$ foreshadowed the development of $K_1(R)$.

Also in the 1940s, Whitehead published a series of papers (Whitehead [41], [49], [50]) developing a classification of topological spaces X constructed from cells (CW complexes), employing a determinant over the group ring $\mathbb{Z}G$, where G is the fundamental group $\pi_1(X)$ of the space X. The group G need not be abelian, so $\mathbb{Z}G$ need not be a commutative ring. This "Whitehead determinant" is the canonical map

$$GL_n(\mathbb{Z}G) \to Wh(G) = \frac{GL(\mathbb{Z}G)}{\pm G \cdot E(\mathbb{Z}G)} ,$$

where $\pm G$ denotes those units of $\mathbb{Z}G$ having the form g or $-g$, with $g \in G$. The value group $Wh(G)$ is known as the **Whitehead group** of G and is very near to being $K_1(\mathbb{Z}G)$.

In an announcement (Bass and Schanuel [62]) and a subsequent detailed exposition (Bass [64A]) Bass introduced the algebraic K-group $K_1(R)$ for an arbitrary ring R and established its connections to $K_0(R)$. The Dieudonné and Whitehead determinants satisfy properties (9.14) (iv), (v), and (vi), but, for *all* rings R, the group $K_1(R)$ can be characterized as the value group of the *initial* determinant-like maps

$$s_n : GL_n(R) \; \rightarrow \; K_1(R)$$
$$A \; \mapsto \; (A \oplus I_\infty)E(R)$$

with those three properties:

(9.15) Definition. A **Whitehead-Bass determinant** (or **WB determinant**) over a ring R is a sequence of maps $\delta_n \; : \; GL_n(R) \to G$ into a group G satisfying properties (9.14) (iv), (v), and (vi).

(9.16) Proposition. *Suppose R is a ring and G is a group. For each group homomorphism $f : K_1(R) \to G$, the maps $f \circ s_n : GL_n(R) \to G$ form a WB determinant. Each WB determinant arises exactly once in this way: For each WB determinant*

$$\{\delta_n : GL_n(R) \to G\}$$

there is a unique group homomorphism $\overline{\delta} : K_1(R) \to G$ making each triangle

$$GL_n(R) \xrightarrow{\;s_n\;} K_1(R)$$

with δ_n and $\overline{\delta}$ mapping to G.

commute.

Proof. The maps $f \circ s_n$ have properties (iv), (v), and (vi) because f is a homomorphism and the s_n have these properties. If $\{\delta_n\}$ is a WB determinant and $\overline{\delta}$ makes the triangles commute, then $\overline{\delta}(AE(R)) = \overline{\delta}(s_n(A)) = \delta_n(A)$ for each $A \in GL_n(R)$; so $\overline{\delta}$ is uniquely determined.

If $\{\delta_n\}$ is any WB determinant over R with values in G, the maps δ_n extend to a group homomorphism $\delta : GL(R) \to G$ taking $E(R)$ to 1. So there is an induced homomorphism $\overline{\delta} : K_1(R) \to G$ with $\overline{\delta}(s_n(A)) = \overline{\delta}(AE(R)) = \delta(A) = \delta_n(A)$ whenever $A \in GL_n(R)$. ∎

One approach to the calculation of $K_1(R)$ is to begin with some WB determinant $\{\delta_n\}$ over R and attempt to calculate the kernel and cokernel of the induced group homomorphism $\overline{\delta} : K_1(R) \to G$.

(9.17) Definitions. Suppose R is a commutative ring, and det : $GL_n(R) \to R^*$ is the usual determinant (from definitions (9.13) (i) or (ii)). The group homomorphisms

$$\det : GL_n(R) \; \rightarrow \; R^*$$
$$\det : GL(R) \; \rightarrow \; R^*$$
$$\overline{\det} : K_1(R) \; \rightarrow \; R^*$$

have kernels $\boldsymbol{SL_n(R)}$, $\boldsymbol{SL(R)}$, and $\boldsymbol{SK_1(R)} = SL(R)/E(R)$, respectively.

Consider the group homomorphism $s_1 : R^* \to K_1(R)$, taking each unit $u \in R^*$ to the coset of

$$u \oplus I_\infty = \begin{bmatrix} u & 0 & 0 & \cdots \\ 0 & 1 & 0 & \cdots \\ 0 & 0 & 1 & \\ \vdots & \vdots & & \ddots \end{bmatrix}.$$

(9.18) Proposition. *If R is a commutative ring, there is a split short exact sequence*

$$1 \longrightarrow SK_1(R) \overset{\subseteq}{\longrightarrow} K_1(R) \underset{s_1}{\overset{\overline{\det}}{\underset{\longleftarrow}{\longrightarrow}}} R^* \longrightarrow 1$$

where $\det : GL(R) \to R^*$ *is the ordinary determinant. So*

$$K_1(R) \cong SK_1(R) \times R^* \,,$$

and $SK_1(R)$ is trivial if and only if s_1 is surjective – that is, if and only if every $A \in GL(R)$ is a t-row equivalent to some $u \oplus I_\infty$ with $u \in R^*$.

Proof. Since R is commutative, $\det : GL_n(R) \to R^*$ is a WB determinant, inducing $\overline{\det}$. For each $u \in R^*$, $(\overline{\det} \circ s_1)(u) = \det(u \oplus I_\infty) = u$. So $\overline{\det}$ is surjective, the indicated sequence is split exact, and s_1 is injective. Then the following are equivalent:

(i) $SK_1(R) = 1$,

(ii) $\overline{\det} : K_1(R) \to R^*$ is an isomorphism,

(iii) $s_1 \circ \overline{\det}$ is the identity on $K_1(R)$,

(iv) $s_1 : R^* \to K_1(R)$ is an isomorphism. $\blacksquare$

In light of this proposition, if R is a commutative ring, $SK_1(R)$ can be viewed as the obstruction to the existence of a row-reduction determinant (9.13) (iii) over R that allows the use of all rows in $GL(R)$.

9C. Exercises

1. If R is a commutative generalized euclidean ring (see §9A, Exercise 5), prove $SK_1(R) = 1$ and $\overline{\det} : K_1(R) \to R^*$ is an isomorphism. Conclude that $K_1(R) \cong R^*$ if the commutative ring R is a field, is euclidean, or is local. (See §9A, Exercise 7.)

2. If R is a noncommutative generalized euclidean ring, prove that the map $s_1 : R^* \to K_1(R)$ is surjective with kernel containing $[R^*, R^*]$. Assuming it is well-defined, prove the Dieudonné determinant defines an isomorphism

$$K_1(D) \cong D^*/[D^*, D^*]$$

when D is a division ring.

3. If R is a commutative ring, show the determinant (9.13) (ii) is a natural transformation from GL to GL_1, and conclude that SK_1 is a functor from $\mathcal{C}$Ring to $\mathcal{A}b$.

4. Suppose A is an R-algebra having a basis $b_1, \ldots, b_n$ as an R-module. For each $a \in A$, the map $\mu(a) : A \to A$, $x \mapsto xa$, is R-linear. Prove

$$\mu : A \quad \to \quad \mathrm{End}_R(A)$$

is a ring homomorphism. By way of the basis $b_1, \ldots, b_n$ there is a ring homomorphism

$$\rho : \mathrm{End}_R(A) \cong M_n(R)$$

representing each linear map by a matrix, as in (1.29). Show the composite group homomorphism

$$K_1(A) \xrightarrow{K_1(\rho \circ \mu)} K_1(M_n(R)) \cong K_1(R)$$

is independent of the choice of R-basis $b_1, \ldots, b_n$ of A. This composite $\tau : K_1(A) \to K_1(R)$ is called the **transfer map** on K_1. How is τ related to the norm $N_{A/R} : A^* \to R^*$ when $R \subseteq A$ are fields?

5. Suppose R is a ring, $A \in M_s(R)$, $B \in R^{s \times t}$, and $D \in M_t(R)$. Let

$$U = \begin{bmatrix} A & B \\ 0 & D \end{bmatrix} \in M_{s+t}(R) .$$

Prove that if any two of the matrices A, D, and U are invertible, then so is the third.

6. Under the hypotheses of Exercise 5, assume R is commutative as well. If U is invertible, use the determinant (9.13) (i) or (ii) to prove both A and D are invertible. *Hint:* Show $\det(U) = \det(A) \det(D)$.

7. If G is a group, the group homomorphism $G \to G_{ab}$ extends to a ring homomorphism $\mathbb{Z}G \to \mathbb{Z}G_{ab}$. Prove there are group homomorphisms

$$\pm G_{ab} \xrightarrow{f} K_1(\mathbb{Z}G) \xrightarrow{g} \pm G_{ab}$$

whose composite is the identity. (Here $\pm G_{ab}$ denotes $\{\pm 1\} \times G_{ab}$.) Conclude that

$$K_1(\mathbb{Z}G) \cong \pm G_{ab} \times Wh(G) .$$

8. Suppose R is a commutative ring and $A \in M_n(R)$. If A^τ denotes the transpose of A, prove $\det(A) = \det(A^\tau)$, where det is defined by (9.13) (ii). For which rings R can you be sure that $s_n(A) = s_n(A^\tau)$, where $s_n : GL_n(R) \to K_1(R)$ is the canonical map $A \mapsto AE(R)$? (Over these rings R, invertible matrices A, A^τ have the same value under every WB determinant.)

9. If R is a commutative ring and $A \in SL_2(R)$, prove

$$A^{-1}E(R) \;=\; A^\tau E(R) \,.$$

Conclude that the WB determinant $\{s_n : GL_n(R) \to K_1(R)\}$ distinguishes A from A^τ if $A \in SL_2(R)$ represents an element of order > 2 in $SK_1(R)$. *Hint:* The matrices

$$\begin{bmatrix} 0 & 1 \\ -1 & 0 \end{bmatrix} \quad \text{and} \quad \begin{bmatrix} 0 & -1 \\ 1 & 0 \end{bmatrix}$$

belong to $E(R)$, and $E(R) \lhd GL(R)$.

9D. The Bass K_1 of a Category

Even in his early expositions of algebraic K-theory, in which $K_1(R)$ was first defined, Bass generalized $K_1(R)$ to a modified Grothendieck group $K_1(\mathcal{C})$ of a category $\mathcal{C}$ with exact sequences and proved

$$K_1(\mathcal{P}(R)) \;\cong\; K_1(\mathcal{F}(R)) \;\cong\; K_1(R) \,.$$

The distinction between f.g. free and f.g. projective modules, so important for K_0, becomes invisible to K_1.

This Grothendieck-style definition enabled Bass to develop parallel theorems for K_0 and K_1, such as the Devissage and Resolution Theorems, and to construct exact sequences connecting $K_1(R)$ and $K_0(R)$. The presentation given here for $K_1(\mathcal{C})$ specializes the general treatment of Bass [74] to the case in which $\mathcal{C}$ is a subcategory of R-$\mathcal{Mod}$.

On the polynomial ring $R[x]$, multiplication by each x^i is injective. So, by (6.43), $R[x]$ embeds as a subring of the ring $S^{-1}(R[x])$, where

$$S \;=\; \{x^i : i \geq 1\} \,.$$

The elements $x^{-n}(r_0 + r_1 x + \cdots + r_m x^m)$ of the latter ring are known as **Laurent polynomials** over R in one variable x, and $S^{-1}(R[x])$ is more commonly denoted by $\boldsymbol{R[x, x^{-1}]}$.

Automorphisms $\alpha : M \to M$ in R-$\mathcal{Mod}$ correspond to $R[x, x^{-1}]$-modules M : For each $\alpha \in \mathrm{End}_R(M)$, the R-module M becomes an $R[x]$-module via

$$\left(\sum r_i x^i \right) \cdot m \;=\; \sum r_i \cdot \alpha^i(m) \,.$$

If α is an automorphism, the elements of S act through bijections α^i on M; so by (6.43), localization $M \to S^{-1}M$ is an $R[x]$-linear isomorphism. Through this isomorphism, M becomes an $R[x, x^{-1}]$-module, with scalar multiplication extending that by $R[x]$; so $x \cdot m = \alpha(m)$ and $x^{-1} \cdot (m) = \alpha^{-1}(m)$. Denote this $R[x, x^{-1}]$-module by (M, α). Every $R[x, x^{-1}]$-module M arises exactly once in this way, as $({}_R M, \alpha)$ where α is the R-linear automorphism $m \mapsto x \cdot m$.

An $R[x, x^{-1}]$-linear map f from (M, α) to (M', α') is the same as an R-linear map $f : M \to M'$ for which the square

$$
\begin{array}{ccc}
M & \xrightarrow{\ f\ } & M' \\
{\scriptstyle \alpha}\downarrow & & \downarrow{\scriptstyle \alpha'} \\
M & \xrightarrow{\ f\ } & M'
\end{array}
$$

commutes, so that $f(x \cdot m) = x \cdot f(m)$. So $R[x, x^{-1}]$-$\mathfrak{Mod}$ can be thought of as the category of automorphisms taken from R-$\mathfrak{Mod}$ and R-linear maps commuting with them. Note that an $R[x, x^{-1}]$-linear map is an isomorphism if and only if it is bijective; so it is an R-linear isomorphism. In particular, $(M, \alpha) \cong (R^n, \cdot A)$, for a matrix $A \in GL_n(R)$, if and only if there is an R-linear isomorphism $f : M \to R^n$ for which the square

$$
\begin{array}{ccc}
M & \xrightarrow{\ f\ } & R^n \\
{\scriptstyle \alpha}\downarrow & & \downarrow{\scriptstyle \cdot A} \\
M & \xrightarrow{\ f\ } & R^n
\end{array}
$$

commutes, which is to say, if and only if A represents α over some R-basis of the module M.

(9.19) Definitions. Suppose $\mathcal{C}$ is a subcategory of R-$\mathfrak{Mod}$. Then **aut** $\mathcal{C}$ is the subcategory of $R[x, x^{-1}]$-$\mathfrak{Mod}$ consisting of those objects and arrows that belong to $\mathcal{C}$ when scalars are restricted to R. Just as in (3.9), aut $\mathcal{C}$ is modest if $\mathcal{C}$ is modest. In that case, the **Bass K_1** of $\mathcal{C}$ is the $\mathbb{Z}$-module quotient

$$K_1(\mathcal{C}) \;=\; \mathcal{F}/\langle E \rangle \,,$$

where $\mathcal{F}$ is the free $\mathbb{Z}$-module based on the set of isomorphism classes in aut $\mathcal{C}$, and E consists of the elements

$$(M, \alpha \circ \beta) \;-\; (M, \alpha) \;-\; (M, \beta) \,,$$

and those

$$(M, \alpha) \;-\; (M', \alpha') \;-\; (M'', \alpha'')$$

for which there is an exact sequence

(9.20) $0 \longrightarrow (M', \alpha') \longrightarrow (M, \alpha) \longrightarrow (M'', \alpha'') \longrightarrow 0$

in aut $\mathcal{C}$. So $K_1(\mathcal{C})$ is the additive abelian group with generators, the cosets

$$[M, \alpha] = c((M, \alpha)) + \langle E \rangle ,$$

and defining relations

(i) $[M, \alpha \circ \beta] = [M, \alpha] + [M, \beta]$, and

(ii) $[M, \alpha] = [M', \alpha'] + [M'', \alpha'']$ for each short exact sequence (9.20) in aut $\mathcal{C}$.

From relation (i), $[M, i_M] = [M, i_M^2] = 2[M, i_M]$; so

(iii) $[M, i_M] = 0$.

If Obj $\mathcal{C}$ includes 0 and is closed under $\oplus$, then for each pair (L, α), (N, β) of objects in aut $\mathcal{C}$, the standard insertion and projection maps i and π define an exact sequence

$$0 \longrightarrow (L, \alpha) \overset{i}{\longrightarrow} (L \oplus N, \alpha \oplus \beta) \overset{\pi}{\longrightarrow} (N, \beta) \longrightarrow 0 ,$$

$$\|$$

$$(L, \alpha) \oplus (N, \beta)$$

so that

(iv) $[L, \alpha] + [N, \beta] = [L \oplus N, \alpha \oplus \beta]$.

If $\mathcal{C}$ is a full subcategory of R-$\mathcal{M}\mathfrak{od}$, then aut $\mathcal{C}$ is a full subcategory of $R[x, x^{-1}]$-$\mathcal{M}\mathfrak{od}$, and an arrow in aut $\mathcal{C}$ is an isomorphism in aut $\mathcal{C}$ if and only if it is an isomorphism in $R[x, x^{-1}]$-$\mathcal{M}\mathfrak{od}$, that is, if and only if it is bijective. As in $R[x, x^{-1}]$-$\mathcal{M}\mathfrak{od}$, if $R^n \in \mathcal{C}$, then $(M, \alpha) \cong (R^n, \cdot A)$ for $A \in GL_n(R)$ if and only if A represents α over an R-basis of M.

(9.21) Theorem. *For each ring R, the matrix representation of R-linear automorphisms defines a group isomorphism $K_1(\mathcal{F}(R)) \cong K_1(R)$.*

Proof. Suppose $\alpha : M \to M$ is an R-linear automorphism, represented by $A \in GL_m(R)$ over the R-basis $v_1, \ldots, v_m$ and by $B \in GL_n(R)$ over the R-basis $w_1, \ldots, w_n$. Then on $M \oplus M$ the R-linear map $\alpha \oplus i_M$ is represented by $A \oplus I_n$ over the R-basis

$$(v_1, 0), \ldots, (v_m, 0), (0, w_1), \ldots, (0, w_n) ,$$

and by $B \oplus I_m$ over the R-basis

$$(w_1, 0), \ldots, (w_n, 0), (0, v_1), \ldots, (0, v_m) \; .$$

Since $A \oplus I_n$ and $B \oplus I_m$ belong to $GL_{m+n}(R)$ and represent the same R-linear map, they are similar matrices. So if

$$\overline{X} \; = \; XE(R) \; ,$$

then $\overline{A} = \overline{B}$ in $K_1(R)$, and the matrices representing α determine a unique element $\rho(M, \alpha)$ of $K_1(R)$.

Suppose there is an exact sequence

$$0 \longrightarrow (M', \alpha') \overset{f}{\longrightarrow} (M, \alpha) \overset{g}{\longrightarrow} (M'', \alpha'') \longrightarrow 0$$

in aut $\mathcal{F}(R)$ and α' is represented by $A = (a_{ij})$ over a basis $v_1, \ldots, v_m$ of M', while α'' is represented by $B = (b_{ij})$ over a basis $w_1, \ldots, w_n$ of M''. For each i with $1 \le i \le n$, choose $u_i \in M$ with $g(u_i) = w_i$. Then

$$f(v_1), \ldots, f(v_m), u_1, \ldots, u_n$$

is a basis of M. Now

$$\alpha f(v_i) \; = \; f\alpha'(v_i) \; = \; f(\sum_j a_{ij} v_j) \; = \; \sum_j a_{ij} f(v_j) + \sum_j 0 u_j \; .$$

And, if

$$\alpha(u_i) \; = \; \sum_j c_{ij} f(v_j) + \sum_j c'_{ij} u_j \; ,$$

then

$$\alpha''(w_i) \; = \; \alpha'' g(u_i) \; = \; g\alpha(u_i) \; = \; \sum_j c'_{ij} w_j \; ;$$

so $c'_{ij} = b_{ij}$. Therefore α is represented by the matrix

$$D \; = \; \begin{bmatrix} A & 0 \\ C & B \end{bmatrix} \; = \; \begin{bmatrix} A & 0 \\ 0 & B \end{bmatrix} \begin{bmatrix} I_m & 0 \\ B^{-1}C & I_n \end{bmatrix}.$$

Now

$$\begin{bmatrix} A & 0 \\ 0 & B \end{bmatrix} = \begin{bmatrix} A & 0 \\ 0 & I_n \end{bmatrix} \begin{bmatrix} I_m & 0 \\ 0 & B \end{bmatrix} = \begin{bmatrix} A & 0 \\ 0 & I_n \end{bmatrix} \begin{bmatrix} 0 & I_m \\ I_n & 0 \end{bmatrix} \begin{bmatrix} B & 0 \\ 0 & I_m \end{bmatrix} \begin{bmatrix} 0 & I_n \\ I_m & 0 \end{bmatrix} ;$$

so

$$\rho(M, \alpha) \; = \; \overline{D} \; = \; \overline{A \oplus B} \; = \; \overline{A}\,\overline{B} \; = \; \rho(M', \alpha')\rho(M'', \alpha'')$$

in $K_1(R)$.

This proves ρ is a generalized rank on aut $\mathcal{F}(R)$ and so is constant on isomorphism classes in aut $\mathcal{F}(R)$. If automorphisms α and β of M are represented over the same basis by matrices A and B, then $\alpha \circ \beta$ is represented by BA; so

$$\rho(M, \alpha \circ \beta) \;=\; \overline{BA} \;=\; \overline{B}\,\overline{A} \;=\; \overline{A}\,\overline{B} \;=\; \rho(M,\alpha)\rho(M,\beta) \;,$$

and ρ induces a group homomorphism

$$\overline{\rho} : K_1(\mathcal{F}(R)) \;\to\; K_1(R) \;.$$

For each $A \in GL_n(R)$, there is an element $\delta(A) = [R^n, \cdot A]$ of $K_1(\mathcal{F}(R))$. For each $m \geq 1$,

$$\delta(A \oplus I_m) \;=\; [R^{n+m}, \cdot(A \oplus I_m)] \;=\; [R^n \oplus R^m, (\cdot A) \oplus (\cdot I_m)]$$

$$=\; [R^n, \cdot A] \;+\; [R^m, i_{R^m}] \;=\; \delta(A) \;;$$

so δ is well-defined on $GL(R)$. If $A, B \in GL_n(R)$,

$$\delta(AB) \;=\; [R^n, \cdot AB] \;=\; [R^n, (\cdot B) \circ (\cdot A)] \;=\; \delta(B) + \delta(A) \;;$$

so δ is a group homomorphism. Since $K_1(\mathcal{F}(R))$ is abelian, δ induces a homomorphism

$$\overline{\delta} : K_1(R) \;\to\; K_1(\mathcal{F}(R)) \;,$$

which is evidently inverse to $\overline{\rho}$.   ∎

Through the isomorphism (9.21), each WB determinant over R defines a determinant of automorphisms in $\mathcal{F}(R)$, which is independent of a choice of basis. Bass extended this to a determinant on automophisms of f.g. projective R-modules:

(9.22) Theorem. *For each ring R, there is a group isomorphism* $K_1(\mathcal{P}(R)) \cong K_1(\mathcal{F}(R))$.

Proof. If $(P, \alpha) \in$ aut $\mathcal{P}(R)$, there exists $Q \in \mathcal{P}(R)$ with $P \oplus Q \in \mathcal{F}(R)$. Define

$$\sigma(P, \alpha) \;:=\; [P \oplus Q, \alpha \oplus i_Q] \;\in\; K_1(\mathcal{F}(R)) \;.$$

This $\sigma(P, \alpha)$ is independent of the choice of complement Q: For if $P \oplus Q' \in \mathcal{F}(R)$, then in aut $\mathcal{F}(R)$,

$$((P \oplus Q) \oplus (P \oplus Q'), (\alpha \oplus i_Q) \oplus i_{P \oplus Q'}) \;\cong\; ((P \oplus Q') \oplus (P \oplus Q), (\alpha \oplus i_{Q'}) \oplus i_{P \oplus Q}) \;;$$

so in $K_1(\mathcal{F}(R))$,

$$[P \oplus Q, \alpha \oplus i_Q] \;=\; [P \oplus Q', \alpha \oplus i_{Q'}] \;.$$

If there is an exact sequence

$$0 \longrightarrow (P', \alpha') \xrightarrow{f} (P, \alpha) \xrightarrow{g} (P'', \alpha'') \longrightarrow 0$$

in aut $\mathcal{P}(R)$, then $P \cong P' \oplus P''$. So if $P' \oplus Q' \in \mathcal{F}(R)$ and $P'' \oplus Q'' \in \mathcal{F}(R)$, then $P \oplus Q' \oplus Q'' \in \mathcal{F}(R)$ and there is an exact sequence

$$0 \to (P' \oplus Q', \alpha' \oplus i_{Q'}) \xrightarrow{f \oplus i} (P \oplus Q' \oplus Q'', \alpha \oplus i_{Q'} \oplus Q'') \xrightarrow{g \oplus \pi} (P'' \oplus Q'', \alpha'' \oplus i_{Q''}) \to 0$$

in aut $\mathcal{F}(R)$, so that $\sigma(P, \alpha) = \sigma(P', \alpha') + \sigma(P'', \alpha'')$. If α and β are automorphisms of $P \in \mathcal{P}(R)$ and $P \oplus Q \in \mathcal{F}(R)$, then

$$
\begin{aligned}
[P \oplus Q, (\alpha \circ \beta) \oplus i_Q] &= [P \oplus Q, (\alpha \oplus i_Q) \circ (\beta \oplus i_Q)] \\
&= [P \oplus Q, \alpha \oplus i_Q] \; + \; [P \oplus Q, \beta \oplus i_Q]
\end{aligned}
$$

in $K_1(\mathcal{F}(R))$; so $\sigma(P, \alpha \circ \beta) = \sigma(P, \alpha) + \sigma(P, \beta)$. Therefore σ induces a group homomorphism

$$
\begin{aligned}
\overline{\sigma} : K_1(\mathcal{P}(R)) &\to K_1(\mathcal{F}(R)) \, . \\
[P, \alpha] &\mapsto [P \oplus Q, \alpha \oplus i_Q]
\end{aligned}
$$

Since $\mathcal{F}(R)$ is a subcategory of $\mathcal{P}(R)$, the inclusion of aut $\mathcal{F}(R)$ into aut $\mathcal{P}(R)$ induces a group homomorphism

$$
\begin{aligned}
K_1(\mathcal{F}(R)) &\to K_1(\mathcal{P}(R)) \\
[M, \alpha] &\mapsto [M, \alpha]
\end{aligned}
$$

inverse to $\overline{\sigma}$.   ∎

As we saw in (9.9), K_1 is a functor from $\mathfrak{Ring}$ to $\mathcal{Ab}$: K_1 of a ring homomorphism f applies f to each entry of a matrix.

(9.23) Theorem. *There is a functor $K_1(\mathcal{P}(-)) : \mathfrak{Ring} \to \mathcal{Ab}$; for each ring homomorphism $\phi : R \to S$ (making S an S, R-bimodule),*

$$K_1(\mathcal{P}(\phi)) : K_1(\mathcal{P}(R)) \to K_1(\mathcal{P}(S))$$

takes $[P, \alpha]$ to $[S \otimes_R P, i_S \otimes \alpha]$. The same is true with $\mathcal{P}$ replaced by $\mathcal{F}$, and the isomorphisms of (9.21) and (9.22) define natural transformations:

$$K_1(\mathcal{P}(-)) \cong K_1(\mathcal{F}(-)) \cong K_1(-) \, .$$

Proof. Each short exact sequence in aut $\mathcal{P}(R)$ amounts to a commutative diagram

$$
\begin{array}{ccccccccc}
0 & \longrightarrow & P' & \longrightarrow & P & \longrightarrow & P'' & \longrightarrow & 0 \\
 & & \downarrow{\scriptstyle\alpha'} & & \downarrow{\scriptstyle\alpha} & & \downarrow{\scriptstyle\alpha''} & & \\
0 & \longrightarrow & P' & \longrightarrow & P & \longrightarrow & P'' & \longrightarrow & 0
\end{array}
$$

in $\mathcal{P}(R)$ with split exact rows. Applying the additive functor $S \otimes_R (-)$ yields a similar diagram in $\mathcal{P}(S)$; so

$$[S \otimes_R P, i_S \otimes \alpha] = [S \otimes_R P', i_S \otimes \alpha'] + [S \otimes_R P'', i_S \otimes \alpha'']$$

in $K_1(\mathcal{P}(S))$. If α and β are R-linear automorphisms of $P \in \mathcal{P}(R)$, then

$$
\begin{aligned}
[S \otimes_R P, i_S \otimes (\alpha \circ \beta)] &= [S \otimes_R P, (i_S \otimes \alpha) \circ (i_S \otimes \beta)] \\
&= [S \otimes_R P, i_S \otimes \alpha] + [S \otimes_R P, i_S \otimes \beta] .
\end{aligned}
$$

So $(P, \alpha) \mapsto [S \otimes_R P, i_S \otimes \alpha]$ is a generalized rank on aut $\mathcal{P}(R)$ and so is constant on isomorphism classes, inducing the desired group homomorphism $K_1(\mathcal{P}(\phi))$.

If $(P, \alpha) \in$ aut $\mathcal{P}(R)$, scalar multiplication defines an isomorphism $(R \otimes_R P, i_R \otimes \alpha) \cong (P, \alpha)$; so

$$K_1(\mathcal{P}(i_R)) = i_{K_1(\mathcal{P}(R))} .$$

If $\psi : S \to T$ is another ring homomorphism, making T a T, S-bimodule via ψ and a T, R-bimodule via $\psi \circ \phi$, there is an isomorphism

$$(T \otimes_S (S \otimes_R P), i_T \otimes (i_S \otimes \alpha)) \cong (T \otimes_R P, i_T \otimes \alpha)$$

given by $t \otimes (s \otimes p) \mapsto t\psi(s) \otimes p$. So

$$K_1(\mathcal{P}(\psi \circ \phi)) = K_1(\mathcal{P}(\psi)) \circ K_1(\mathcal{P}(\phi)) ,$$

and $K_1(\mathcal{P}(-))$ is a functor. The same proof works for $K_1(\mathcal{F}(-))$.

To see that the isomorphism in (9.22) is natural, we must check that the square

$$
\begin{array}{ccc}
K_1(\mathcal{P}(R)) & \cong & K_1(\mathcal{F}(R)) \\
\downarrow & & \downarrow \\
K_1(\mathcal{P}(S)) & \cong & K_1(\mathcal{F}(S))
\end{array}
$$

commutes for each ring homomorphism $\phi : R \to S$. This follows from the isomorphism

$$(S \otimes_R (P \oplus Q), i_S \otimes (\alpha \oplus i_Q)) \cong ((S \otimes_R P) \oplus (S \otimes_R Q), (i_S \otimes \alpha) \oplus (i_S \otimes i_Q))$$

given by $s \otimes (p,q) \mapsto (s \otimes p, s \otimes q)$.

To see the isomorphism given by matrix representation in (9.21) is natural, we check that

$$
\begin{array}{ccc}
K_1(\mathcal{F}(R)) & \cong & K_1(R) \\
\downarrow & & \downarrow \\
K_1(\mathcal{F}(S)) & \cong & K_1(S)
\end{array}
$$

commutes for each ring homomorphism $\phi : R \to S$. Each automorphism $\alpha : M \to M$ in $\mathcal{F}(R)$ is represented over an R-basis of M by some matrix $A \in GL_n(R)$; so $(M, \alpha) \cong (R^n, \cdot A)$. Then it is enough to check the commutativity of the square on $[R^n, \cdot A]$. Distributivity of $\otimes$ over $\oplus$ defines an isomorphism

$$
(S \otimes_R R^n, i_S \otimes (\cdot A)) \cong (S^n, \cdot \phi(A))
$$

in aut $\mathcal{F}(S)$; so, in the above square,

$$
\begin{array}{ccc}
[R^n, \cdot A] & \longrightarrow & AE(R) \\
\downarrow & & \downarrow \\
[S^n, \cdot \phi(A)] & \longrightarrow & \phi(A)E(S)
\end{array}
$$

where $\phi(A) = \phi((a_{ij})) = (\phi(a_{ij}))$. ∎

The naturality of the isomorphisms (9.21), (9.22) is a beautiful illustration that, in a less technical sense of the word, there is a "natural" connection between $K_0(R)$ and $K_1(R)$.

9D. Exercises

1. Suppose R is a ring, $P \in \mathcal{P}(R)$, and $A = \text{End}_R(P)$. Example (1.31) establishes, for each positive integer n, a ring isomorphism

$$
\delta_n : M_n(A) \cong \text{End}_R(P^n) .
$$

Show there is a group homomorphism from $K_1(A)$ to $K_1(\mathcal{P}(R)) \cong K_1(R)$, taking the coset of each $\alpha \in GL_n(A)$ to $[P^n, \delta_n(\alpha)]$. Show by example that this homomorphism need not be injective or surjective.

2. (Bass) Suppose $\mathcal{C}$ is a modest full subcategory of R-$\mathcal{M}\text{od}$ and $M \in \mathcal{C}$. Suppose α is an R-linear automorphism of M, and M has a $\mathcal{C}$-filtration

$$
M = M_0 \supseteq M_1 \supseteq \cdots \supseteq M_n = \{0_M\}
$$

in $\mathcal{C}$ with $\alpha(M_i) = M_i$ for each i. Show

$$[M, \alpha] = \sum_{i=0}^{n-1} [M_i/M_{i+1}, \alpha_i]$$

in $K_1(\mathcal{C})$, where α_i is the automorphism of M_i/M_{i+1} induced by α. *Hint:* Apply Lemma (3.41) to aut $\mathcal{C}$, and note that $K_1(\mathcal{C})$ is a quotient of K_0 (aut $\mathcal{C}$).

3. (Bass) Suppose $\mathcal{C}$ is a modest full subcategory of R-$\mathcal{M}$od containing as objects the kernels of all its arrows, and

$$0 \longrightarrow (M_n, \alpha_n) \longrightarrow \cdots \longrightarrow (M_0, \alpha_0) \longrightarrow 0$$

is an exact sequence in aut $\mathcal{C}$. Prove

$$\sum_{i=0}^{n} (-1)^i [M_i, \alpha_i] = 0$$

in $K_1(\mathcal{C})$. *Hint:* Recall the proof of the Resolution Theorem (3.52).

10

Stability for $K_1(R)$

For any nontrivial ring R, the group $GL(R)$ is vastly complex. Even its subgroup of permutation matrices contains a copy of every finite group! For the study of its abelianization $K_1(R) = GL(R)/E(R)$ it would be reassuring and useful if each coset were represented by a matrix $A \in GL_n(R)$ of small dimension n. It would be better still if the representative matrices have uniform dimension, which is to say, if the canonical map $GL_n(R) \to K_1(R)$ were surjective for some small n. This amounts to the equation $GL(R) = GL_n(R)E(R)$. In the very papers introducing $K_1(R)$ for the first time, Bass proved this "surjective stability" when n is at least as large as the stable rank of R (see §4D) and proved that $E_{n+1}(R) \triangleleft GL_{n+1}(R)$ under the same conditions. Subsequently, Suslin proved $E_n(R) \triangleleft GL_n(R)$ for all $n \geq 3$, when R is a commutative ring. We present these proofs in §10A. An immediate application of surjective stability, using the determinant as in (9.18), is that $K_1(R) \cong R^*$ (and $SK_1(R) = 1$) for any commutative ring R of stable rank 1.

Once n is large enough so that $E_n(R) \triangleleft GL_n(R)$, one may hope that the canonical map $GL_n(R) \to K_1(R)$ induces an isomorphism

$$(\mathbf{10.1}) \qquad \frac{GL_n(R)}{E_n(R)} \cong K_1(R) \ .$$

If this were true, then t-row equivalence in $GL(R)$ would be the same as t-row equivalence in $GL_n(R)$. The injectivity of the map (10.1) amounts to the equation $GL_n(R) \cap E(R) = E_n(R)$. Conjectured for certain algebras R in Bass [64B], this injective stability was proved by Vaserstein, for all rings R, when n is greater than the stable rank of R. We present Vaserstein's proof in §10B. It constructs the inverse to the map (10.1) as a row-reduction determinant, generalizing the determinant of Dieudonné over division rings.

Although we shall not give the details here, the stability theorems in this chapter have been generalized to stability theorems for each of the higher algebraic K-groups $K_n(R)$. These are also abelian groups, functorially dependent on the ring R, and each can be described as a limit, as m goes to infinity, of groups formed from $GL_m(R)$. These groups become stable, as with K_1, when m reaches n plus the stable rank of R.

10A. Surjective Stability

Recall from §4D that the **stable rank** $sr(R)$ of a ring R is the least positive integer n for which each right unimodular row $a = (a_1, \ldots, a_{n+1})$ in $R^{1 \times (n+1)}$ can be shortened to a right unimodular row

$$(a_1 + a_{n+1}b_1 , \ldots , a_n + a_{n+1}b_n)$$

for some $b_1, \ldots, b_n \in R$. By (4.21) all longer right unimodular rows over R can be shortened in the same way.

At the outset, we cannot assume $E_n(R)$ is a normal subgroup of $GL_n(R)$. Let $GL_n(R)/E_n(R)$ denote the *set* of all cosets $AE_n(R)$ where $A \in GL_n(R)$. We refer to the functions

$$s_{n,n+1} : \frac{GL_n(R)}{E_n(R)} \longrightarrow \frac{GL_{n+1}(R)}{E_{n+1}(R)}$$

$$AE_n(R) \longmapsto AE_{n+1}(R)$$

$$\text{and} \quad s_n : \frac{GL_n(R)}{E_n(R)} \longrightarrow \frac{GL(R)}{E(R)} = K_1(R)$$

$$AE_n(R) \longmapsto AE(R)$$

as **stabilization maps,** expressing the expectation that they become bijective for sufficiently large n. In the examination of $s_{n,n+1}$, it will be convenient to work with block multiplication of partitioned matrices:

(10.2) Lemma. *Suppose $n \geq 1$, $A \in M_n(R)$, $v \in R^{n \times 1}$, and $u \in R^{1 \times n}$. Consider the set of $(n + 1) \times (n + 1)$ matrices*

$$S = \left\{ \begin{bmatrix} 1 & 0 \\ v & A \end{bmatrix}, \begin{bmatrix} 1 & u \\ 0 & A \end{bmatrix}, \begin{bmatrix} A & v \\ 0 & 1 \end{bmatrix}, \begin{bmatrix} A & 0 \\ u & 1 \end{bmatrix} \right\}.$$

If any of the matrices in S belongs to $GL_{n+1}(R)$, then $A \in GL_n(R)$. If $A \in GL_n(R)$, all the matrices in S are in $GL_{n+1}(R)$, with inverses

$$\begin{bmatrix} 1 & 0 \\ v & A \end{bmatrix}^{-1} = \begin{bmatrix} 1 & 0 \\ -A^{-1}v & A^{-1} \end{bmatrix}, \quad \begin{bmatrix} 1 & u \\ 0 & A \end{bmatrix}^{-1} = \begin{bmatrix} 1 & -uA^{-1} \\ 0 & A^{-1} \end{bmatrix}$$

$$\begin{bmatrix} A & v \\ 0 & 1 \end{bmatrix}^{-1} = \begin{bmatrix} A^{-1} & -A^{-1}v \\ 0 & 1 \end{bmatrix}, \quad \begin{bmatrix} A & 0 \\ u & 1 \end{bmatrix}^{-1} = \begin{bmatrix} A^{-1} & 0 \\ -uA^{-1} & 1 \end{bmatrix}.$$

Proof. Suppose

$$B = \begin{bmatrix} 1 & 0 \\ v & A \end{bmatrix} \quad \text{and} \quad B^{-1} = \begin{bmatrix} x_1 & x_2 \\ x_3 & x_4 \end{bmatrix}$$

as $(1, n) \times (1, n)$ partitioned matrices. From $BB^{-1} = 1 \oplus I_n$, $x_1 = 1$, and $x_2 = 0$. Also, $Ax_4 = vx_2 + Ax_4 = I_n$. From $B^{-1}B = 1 \oplus I_n$, $x_4 A = I_n$; so A has inverse x_4. A similar argument works for each of the other three matrices in S.

If $A \in GL_n(R)$, block multiplication shows that the matrices in S have the given inverses. ∎

Note: If $A = I_n$, the matrices in S belong to $E_{n+1}(R)$, since each can be reduced to I_{n+1} by row or column transvections.

The following is a special case of Theorem (4.2) in Bass [64A].

(10.3) Theorem. *Suppose $sr(R) \le n < \infty$. Then*

(i) *By right multiplication, $E_{n+1}(R)$ acts transitively on the set of right unimodular rows in $R^{1 \times (n+1)}$.*

(ii) *$GL_{n+1}(R) = GL_n(R)E_{n+1}(R)$ and the maps $s_{n,n+1}$ and s_n are surjective.*

(iii) *$E_{n+1}(R) \triangleleft GL_{n+1}(R)$ and the maps $s_{n+1,n+2}$ and s_{n+1} are group homomorphisms.*

(iv) *$[GL_{n+1}(R), GL_1(R)E_{n+1}(R)] \subseteq E_{n+1}(R)$.*

Proof. Suppose $u \in R^{1 \times n}$, $u_{n+1} \in R$, and the row $[u \ \ u_{n+1}]$ in $R^{1 \times (n+1)}$ is right unimodular. Since $n + 1 > sr(R)$, this row can be shortened; that is, $u' = u + u_{n+1}b$ is right unimodular for some $b \in R^{1 \times n}$. Then we can choose $c \in R^{n \times 1}$ with $u'c = 1 - u_{n+1}$. So

$$[u \ \ u_{n+1}] \begin{bmatrix} I_n & 0 \\ b & 1 \end{bmatrix} \begin{bmatrix} I_n & c \\ 0 & 1 \end{bmatrix} \begin{bmatrix} I_n & 0 \\ -u' & 1 \end{bmatrix} = [0 \ \ 1],$$

the last 0 being the zero row in $R^{1 \times n}$. So the only orbit under this action is the orbit of $[0 \ \ 1]$, proving (i).

Suppose $B \in GL_{n+1}(R)$. Since B has a right inverse, its last row is right unimodular. When B is partitioned as an $(n, 1) \times (n, 1)$ matrix, suppose

$$B = \begin{bmatrix} X & y \\ u & u_{n+1} \end{bmatrix}.$$

In the notation of the first paragraph,

$$B \begin{bmatrix} I_n & 0 \\ b & 1 \end{bmatrix} \begin{bmatrix} I_n & c \\ 0 & 1 \end{bmatrix} \begin{bmatrix} I_n & 0 \\ -u' & 1 \end{bmatrix} = \begin{bmatrix} A & v \\ 0 & 1 \end{bmatrix}$$

for some $A \in M_n(R)$ and $v \in R^{n \times 1}$. By (10.2), $A \in GL_n(R)$. Then

$$\begin{bmatrix} A & v \\ 0 & 1 \end{bmatrix} \begin{bmatrix} I_n & -A^{-1}v \\ 0 & 1 \end{bmatrix} = \begin{bmatrix} A & 0 \\ 0 & 1 \end{bmatrix}.$$

So

$$B = \begin{bmatrix} A & 0 \\ 0 & 1 \end{bmatrix} \begin{bmatrix} I_n & A^{-1}v \\ 0 & 1 \end{bmatrix} \begin{bmatrix} I_n & 0 \\ u' & 1 \end{bmatrix} \begin{bmatrix} I_n & -c \\ 0 & 1 \end{bmatrix} \begin{bmatrix} I_n & 0 \\ -b & 1 \end{bmatrix}$$

belongs to $GL_n(R)E_{n+1}(R)$, proving $GL_{n+1}(R) = GL_n(R)E_{n+1}(R)$. From this, $s_{n,n+1}$ is surjective. Each element of $K_1(R)$ has the form

$$AE(R) = s_m(AE_m(R))$$

with $A \in GL_m(R)$ for some finite $m > n$. Since

$$s_n = s_m \circ s_{m-1,m} \circ \cdots \circ s_{n,n+1} \,,$$

surjectivity of s_n follows from surjectivity of each $s_{n,n+1}$, proving (ii).

(10.4) Note. The above expression for B shows every matrix in $GL_{n+1}(R)$ is a product of matrices whose last row or last column is that of I_{n+1}. We use this in the proof of (10.15) below.

Assertion (iii) follows from (ii) once it is shown that $GL_n(R)$ normalizes $E_{n+1}(R)$. Suppose $A \in GL_n(R)$. For $u \in R^{1 \times n}$ and $v \in R^{n \times 1}$, the products

$$\begin{bmatrix} A & 0 \\ 0 & 1 \end{bmatrix} \begin{bmatrix} I_n & 0 \\ u & 1 \end{bmatrix} \begin{bmatrix} A^{-1} & 0 \\ 0 & 1 \end{bmatrix} = \begin{bmatrix} I_n & 0 \\ uA^{-1} & 1 \end{bmatrix} \quad \text{and}$$

$$\begin{bmatrix} A & 0 \\ 0 & 1 \end{bmatrix} \begin{bmatrix} I_n & v \\ 0 & 1 \end{bmatrix} \begin{bmatrix} A^{-1} & 0 \\ 0 & 1 \end{bmatrix} = \begin{bmatrix} I_n & Av \\ 0 & 1 \end{bmatrix}$$

belong to $E_{n+1}(R)$. It remains to prove $Ae_{ij}(r)A^{-1} \in E_{n+1}(R)$ when $1 \le i \le n$, $1 \le j \le n$, $i \ne j$, and $r \in R$. Let a denote the i-column of A and b the j-row of A^{-1}. Then $ba = 0$, the j,i-entry of I_n. And

$$\begin{bmatrix} A & 0 \\ 0 & 1 \end{bmatrix} \begin{bmatrix} e_{ij}(r) & 0 \\ 0 & 1 \end{bmatrix} \begin{bmatrix} A^{-1} & 0 \\ 0 & 1 \end{bmatrix} = \begin{bmatrix} I_n + Ar\epsilon_{ij}A^{-1} & 0 \\ 0 & 1 \end{bmatrix} = \begin{bmatrix} I_n + arb & 0 \\ 0 & 1 \end{bmatrix}$$

$$= \begin{bmatrix} I_n & 0 \\ -rb(I_n + arb)^{-1} & 1 \end{bmatrix} \left(\begin{bmatrix} I_n & a \\ 0 & 1 \end{bmatrix} \begin{bmatrix} I_n & 0 \\ rb & 1 \end{bmatrix} \begin{bmatrix} I_n & -a \\ 0 & 1 \end{bmatrix} \right)$$

belongs to $E_{n+1}(R)$.

For (iv), suppose $u \in R^* = GL_1(R)$. Then $E_1 = u \oplus I_{n-1} \oplus u^{-1} =$

$$e_{1,n+1}(u)e_{n+1,1}(-u^{-1})e_{1,n+1}(u)e_{1,n+1}(-1)e_{n+1,1}(1)e_{1,n+1}(-1)$$

belongs to $E_{n+1}(R)$, and $u \oplus I_n = (I_n \oplus u)E_1$. If $B \in GL_{n+1}(R)$, then, by (ii), $B = (A \oplus 1)E_2$, where $A \in GL_n(R)$ and $E_2 \in E_{n+1}(R)$. So, for each $E \in E_{n+1}(R)$,

$$[B, (u \oplus I_n)E] = [(A \oplus 1)E_2, (I_n \oplus u)E_1 E] \equiv [A \oplus 1, I_n \oplus u] = I_{n+1}$$

modulo $E_{n+1}(R)$.   ∎

Note: The containment (iv) is equality for $n \geq \max\{2, sr(R)\}$, since $E_{n+1}(R) = [E_{n+1}(R), E_{n+1}(R)]$ for $n \geq 2$ by (9.5).

If R is a *commutative* ring, then $E_n(R) \triangleleft GL_n(R)$ for all $n \geq 3$. This improvement of (10.3) (iii) is from Suslin [77], drawing together ideas of Bass and Vaserstein. We now present Suslin's proof. First, we need a lemma from Vaserstein [69]:

(10.5) Lemma. *Suppose R is any ring, $X \in R^{s \times t}$, $Y \in R^{t \times s}$, and $I_s + XY \in GL_s(R)$. Then $I_t + YX \in GL_t(R)$ and*

$$\begin{bmatrix} I_s + XY & 0 \\ 0 & (I_t + YX)^{-1} \end{bmatrix} \in E_{s+t}(R) .$$

Proof. As in (6.31), if $V = (I_s + XY)^{-1}$, then $I_t + YX$ has inverse $I_t - YVX$. Then

$$\begin{bmatrix} I_s + XY & 0 \\ 0 & I_t - YVX \end{bmatrix}$$

$$= \begin{bmatrix} I_s & X(I_t + YX) \\ 0 & I_t \end{bmatrix} \left(\begin{bmatrix} I_s & 0 \\ YV & 1 \end{bmatrix} \left(\begin{bmatrix} I_s & -X \\ 0 & I_t \end{bmatrix} \begin{bmatrix} I_s & 0 \\ -Y & I_t \end{bmatrix} \right) \right) ,$$

and the factors on the right are in $E_{s+t}(R)$, since they are t-row equivalent to I_{s+t}.   ∎

(10.6) Corollary. *Suppose R is any ring, $a \in R^{n \times 1}$, $b \in R^{1 \times n}$, $ba = 0$, and some coordinate of b is 0. Then $I_n + ab \in E_n(R)$.*

Proof. First assume the last coordinate of b is 0. Then

$$a = \begin{bmatrix} a' \\ a_n \end{bmatrix} \quad \text{and} \quad b = [b' \ 0]$$

with $a' \in R^{(n-1) \times 1}$ and $b' \in R^{1 \times (n-1)}$. Then $b'a' = 0$ and $1 + b'a' = 1 \in GL_1(R)$. Let I denote I_{n-1}. By (10.5), $I + a'b'$ is invertible, and

$$I_n + ab = \begin{bmatrix} I + a'b' & 0 \\ a_n b' & 1 \end{bmatrix} = \begin{bmatrix} I + a'b' & 0 \\ 0 & 1 \end{bmatrix} \begin{bmatrix} I & 0 \\ a_n b' & 1 \end{bmatrix} \in E_n(R) .$$

Now assume, instead, that the m-coordinate of b is 0, where $1 \leq m < n$. If $\sigma \in S_n$, the permutation matrix $P = \sum \epsilon_{\sigma(i)i}$ has inverse $P^{-1} = \sum \epsilon_{i\sigma(i)}$; so $P\epsilon_{ij}P^{-1} = \epsilon_{\sigma(i)\sigma(j)}$, and

$$P e_{ij}(r) P^{-1} = I_n + r P \epsilon_{ij} P^{-1} = e_{\sigma(i)\sigma(j)}(r) .$$

So P normalizes $E_n(R)$. Now suppose $\sigma(m) = n$. Then $bP^{-1}Pa = ba = 0$, and bP^{-1} has its last coordinate 0. So

$$I_n + ab = P^{-1}(I_n + PabP^{-1})P \in P^{-1}E_n(R)P = E_n(R) . \qquad \blacksquare$$

(10.7) Lemma. *Suppose R is a commutative ring, $a \in R^{n \times 1}$, $b, c \in R^{1 \times n}$, $ca = 1$, and $ba = 0$. If*

$$a = \begin{bmatrix} a_1 \\ \vdots \\ a_n \end{bmatrix},$$

then $b = \sum_{i<j} d_{ij}(a_j e_i - a_i e_j)$ for some $d_{ij} \in R$.

Proof. Say $b = [b_1 \ \cdots \ b_n]$ and $c = [c_1 \ \cdots \ c_n]$. For each i,

$$\begin{aligned}
b_i &= b_i - 0c_i = b_i - \left(\sum_j b_j a_j\right) c_i \\
&= b_i - b_i a_i c_i - \sum_{j \neq i} b_j a_j c_i \\
&= b_i\left(\sum_{j \neq i} c_j a_j\right) - \sum_{j \neq i} b_j a_j c_i \\
&= \sum_{j \neq i} a_j(b_i c_j - b_j c_i) .
\end{aligned}$$

Let d_{ij} denote $b_i c_j - b_j c_i$. Note that $d_{ji} = -d_{ij}$. So

$$\begin{aligned}
b &= \sum_i b_i e_i = \sum_i \sum_{j \neq i} a_j d_{ij} e_i \\
&= \sum_{i<j} d_{ij} a_j e_i + \sum_{j<i} d_{ij} a_j e_i \\
&= \sum_{i<j} d_{ij}(a_j e_i - a_i e_j) . \qquad \blacksquare
\end{aligned}$$

(10.8) Theorem. *If R is a commutative ring and $n \geq 3$, then $E_n(R) \triangleleft GL_n(R)$.*

Proof. It is enough to show $Ae_{ij}(r)A^{-1} \in E_n(R)$ for each generator $e_{ij}(r)$ of $E_n(R)$ and each matrix $A \in GL_n(R)$. Take

$$a = i\text{-column of } A ,$$
$$b = j\text{-row of } A^{-1},$$
$$c = i\text{-row of } A^{-1}.$$

Then $ca = 1$ and $ba = 0$. Just as in the proof of (10.3) (iii), $A(e_{ij}(r))A^{-1} = I_n + arb$. By (10.7),

$$b = \sum_{i<j} \beta_{ij} , \quad \text{where} \quad \beta_{ij} = d_{ij}(a_j e_i - a_i e_j)$$

and $d_{ij} \in R$. Since $n \geq 3$, some coordinate of each β_{ij} is 0. And

$$\beta_{ij} a = d_{ij}(a_j a_i - a_i a_j) = 0 .$$

So $(r\beta_{ij})a = 0$ while $ca = 1$. By (10.6), each $I_n + ar\beta_{ij}$ belongs to $E_n(R)$, as does their product:

$$\prod_{i<j}(I_n + ar\beta_{ij}) = I_n + \sum_{i<j} ar\beta_{ij} = I_n + ar\sum_{i<j}\beta_{ij} = I_n + arb . \quad \blacksquare$$

Note: Theorem (10.8) has been generalized to rings R that are almost commutative — that is, finitely generated as modules over their center (see Tulenbaev [81]).

10B. Injective Stability

Theorem (10.3) of Bass shows the maps

$$\frac{GL_n(R)}{E_n(R)} \rightarrow \frac{GL_{n+1}(R)}{E_{n+1}(R)} , \quad \frac{GL_n(R)}{E_n(R)} \rightarrow \frac{GL(R)}{E(R)} ,$$

induced by inclusions, are surjective for $n \geq sr(R)$ and are group homomorphisms between groups for $n > sr(R)$. The theorem (10.15) of Vaserstein, presented below, shows these are isomorphisms for $n > sr(R)$. As with the surjective stability theorem of Bass, we only present a special case of Vaserstein's theorem here. The general form of both theorems involves the K_1-theory of a ring R relative to an ideal J and will be discussed in Chapter 11.

Our presentation follows Vaserstein [69] closely. The first step is reminiscent of Lemma (6.31), related to the Jacobson radical of a ring.

(10.9) Lemma. *Assume* $sr(R) \le n < \infty$. *Suppose there are* $(1, n)$ *by* $(1, n)$
partitioned matrices

$$Y = \begin{bmatrix} y & 0 \\ 0 & I_n \end{bmatrix}, \quad X = \begin{bmatrix} x_1 & x_2 \\ x_3 & x_4 \end{bmatrix}$$

over R *with* $I_{n+1} + XY \in GL_{n+1}(R)$. *Then*

$$(I_{n+1} + XY)^{-1}(I_{n+1} + YX) \in E_{n+1}(R) .$$

Proof. The first row of the right invertible matrix

$$I_{n+1} + XY = \begin{bmatrix} 1 + x_1 y & x_2 \\ x_3 y & I_n + x_4 \end{bmatrix}$$

is right unimodular with $n + 1 > sr(R)$ entries. So there exists $u \in R^{1 \times n}$ with

$$x_2' = x_2 + (1 + x_1 y)u$$

right unimodular. Then there exists $v \in R^{n \times 1}$ with $x_2' v = -x_1$. So

(10.10) $\quad (I_{n+1} + XY) \begin{bmatrix} 1 & u \\ 0 & I_n \end{bmatrix} \begin{bmatrix} 1 & 0 \\ vy & I_n \end{bmatrix} \begin{bmatrix} 1 & -x_2' \\ 0 & I_n \end{bmatrix} = \begin{bmatrix} 1 & 0 \\ w & A \end{bmatrix} ,$

with the last three factors on the left side belonging to $E_{n+1}(R)$. Now

$$I_{n+1} + YX = \begin{bmatrix} 1 + yx_1 & yx_2 \\ x_3 & I_n + x_4 \end{bmatrix}.$$

Since $yx_2' = yx_2 + y(1 + x_1 y)u = yx_2 + (1 + yx_1)yu$, we also have

(10.11) $\quad (I_n + YX) \begin{bmatrix} 1 & yu \\ 0 & I_n \end{bmatrix} \begin{bmatrix} 1 & 0 \\ v & I_n \end{bmatrix} \begin{bmatrix} 1 & -yx_2' \\ 0 & I_n \end{bmatrix} = \begin{bmatrix} 1 & 0 \\ w' & A' \end{bmatrix} ,$

with the last three factors on the left side in $E_{n+1}(R)$. By direct calculation in
(10.10) and (10.11), $A = A'$. So, for $E, E' \in E_{n+1}(R)$,

$$(I_{n+1} + XY)^{-1}(I_{n+1} + YX) = E \begin{bmatrix} 1 & 0 \\ w & A \end{bmatrix}^{-1} \begin{bmatrix} 1 & 0 \\ w' & A \end{bmatrix} E'$$

$$= E \begin{bmatrix} 1 & 0 \\ -A^{-1}w & A^{-1} \end{bmatrix} \begin{bmatrix} 1 & 0 \\ w' & A \end{bmatrix} E'$$

$$= E \begin{bmatrix} 1 & 0 \\ A^{-1}(w' - w) & I_n \end{bmatrix} E' \in E_{n+1}(R) . \quad \blacksquare$$

(10.12) Proposition. *If $sr(R) < n < \infty$, every matrix in $GL_{n+1}(R)$ is a product*

$$\begin{bmatrix} 1 & 0 \\ u & A \end{bmatrix} e_{1,n+1}(x) \begin{bmatrix} B & 0 \\ v & 1 \end{bmatrix} \ ,$$

where $A, B \in GL_n(R)$, $x \in R$, and v has last coordinate 0.

Proof. Suppose $C \in GL_{n+1}(R)$ has first row $[c \quad x]$ where $c \in R^{1 \times n}$. Since $n > sr(R)$, by the Skipping Lemma (4.24) there is some $v \in R^{1 \times n}$, with last coordinate 0, for which $c' = c - xv$ is right unimodular in $R^{1 \times n}$. Since $n > sr(R)$, by (10.3) (i) there exists $B \in E_n(R)$ with $e_1 B = c'$, where $e_1 = [1 \ 0 \ \cdots \ 0]$. Then

$$[c \quad x] \begin{bmatrix} I_n & 0 \\ -v & 1 \end{bmatrix} \begin{bmatrix} B^{-1} & 0 \\ 0 & 1 \end{bmatrix} e_{1,n+1}(-x) \ = \ [e_1 \ x] e_{1,n+1}(-x) \ = \ [e_1 \ 0] \ .$$

So

$$C \begin{bmatrix} I_n & 0 \\ -v & 1 \end{bmatrix} \begin{bmatrix} B^{-1} & 0 \\ 0 & 1 \end{bmatrix} e_{1,n+1}(-x) \ = \ \begin{bmatrix} 1 & 0 \\ u & A \end{bmatrix},$$

where $A \in M_n(R)$ and $u \in R^{n \times 1}$. Then

$$C \ = \ \begin{bmatrix} 1 & 0 \\ u & A \end{bmatrix} e_{1,n+1}(x) \begin{bmatrix} B & 0 \\ 0 & 1 \end{bmatrix} \begin{bmatrix} I_n & 0 \\ v & 1 \end{bmatrix}$$

$$= \ \begin{bmatrix} 1 & 0 \\ u & A \end{bmatrix} e_{1,n+1}(x) \begin{bmatrix} B & 0 \\ v & 1 \end{bmatrix}.$$

Since C is invertible, so are A and B, by (10.2). ∎

The factorization in (10.12) is not unique. But, in any two factorizations of the same matrix from $GL_{n+1}(R)$, the products AB are congruent modulo $E_n(R)$:

(10.13) Proposition. *Suppose $sr(R) < n < \infty$. Let $\overline{X}$ denote the coset $XE_n(R)$ in the group $GL_n(R)/E_n(R)$. For $i = 0$ or 1, let A_i, B_i denote matrices in $GL_n(R)$. In $GL_{n+1}(R)$, if*

$$\begin{bmatrix} 1 & 0 \\ u_1 & A_1 \end{bmatrix} e_{1,n+1}(x) \begin{bmatrix} B_1 & 0 \\ v_1 & 1 \end{bmatrix} = \begin{bmatrix} 1 & 0 \\ u_2 & A_2 \end{bmatrix} e_{1,n+1}(y) \begin{bmatrix} B_2 & 0 \\ v_2 & 1 \end{bmatrix},$$

then $\overline{A}_1 \overline{B}_1 = \overline{A}_2 \overline{B}_2$.

Proof. By (10.2),

$$\begin{bmatrix} 1 & 0 \\ u_2 & A_2 \end{bmatrix}^{-1} \begin{bmatrix} 1 & 0 \\ u_1 & A_1 \end{bmatrix} = \begin{bmatrix} 1 & 0 \\ A_2^{-1}(u_1 - u_2) & A_2^{-1}A_1 \end{bmatrix},$$

and

$$\begin{bmatrix} B_2 & 0 \\ v_2 & 1 \end{bmatrix} \begin{bmatrix} B_1 & 0 \\ v_1 & 1 \end{bmatrix}^{-1} = \begin{bmatrix} B_2 B_1^{-1} & 0 \\ (v_2 - v_1) B_1^{-1} & 1 \end{bmatrix}.$$

So

(10.14)
$$\begin{bmatrix} 1 & 0 \\ u & A_2^{-1} A_1 \end{bmatrix} e_{1,n+1}(x) = e_{1,n+1}(y) \begin{bmatrix} B_2 B_1^{-1} & 0 \\ v & 1 \end{bmatrix},$$

where $u \in R^{n \times 1}$, $v \in R^{1 \times n}$. Let $A = A_2^{-1} A_1$ and $B = B_2 B_1^{-1}$. Comparing $(1, n+1)$-entries in (10.14), $x = y$. Comparing $(n+1, 1)$-entries, the same $w \in R$ is the last entry of u and the first entry of v. Say

$$u = \begin{bmatrix} u' \\ w \end{bmatrix} \quad \text{and} \quad v = [w \ v']$$

with $u' \in R^{(n-1) \times 1}$, $v' \in R^{1 \times (n-1)}$. Partition A and B so

$$A = \begin{bmatrix} A' & b \\ c & d \end{bmatrix} \quad \text{and} \quad B = \begin{bmatrix} e & f \\ g & B' \end{bmatrix}$$

with $A', B' \in M_{n-1}(R)$. Then

$$\begin{bmatrix} 1 & 0 & 0 \\ u' & A' & b \\ w & c & d \end{bmatrix} e_{1,n+1}(x) = e_{1,n+1}(x) \begin{bmatrix} e & f & 0 \\ g & B' & 0 \\ w & v' & 1 \end{bmatrix}$$

implies, on consideration of the row and column transvections corresponding to $e_{1,n+1}(x)$, that

$$A = \begin{bmatrix} A' & -u'x \\ v' & 1 - wx \end{bmatrix} \quad \text{and} \quad B = \begin{bmatrix} 1 - xw & -xv' \\ u' & A' \end{bmatrix}.$$

Define

$$X = \begin{bmatrix} -x & 0 \\ 0 & I \end{bmatrix} \quad \text{and} \quad Y = \begin{bmatrix} w & v' \\ u' & A' - I \end{bmatrix},$$

where $I = I_{n-1}$. Then

$$I_n + XY = I_n + \begin{bmatrix} -xw & -xv' \\ u' & A' - I \end{bmatrix} = B, \quad \text{and}$$

$$I_n + YX = I_n + \begin{bmatrix} -wx & v' \\ -u'x & A' - I \end{bmatrix}$$

$$= \begin{bmatrix} 1 - wx & v' \\ -u'x & A' \end{bmatrix} = \sigma A \sigma^{-1},$$

$$\text{where} \quad \sigma = \begin{bmatrix} 0 & 1 \\ I & 0 \end{bmatrix} \quad \text{and} \quad \sigma^{-1} = \begin{bmatrix} 0 & I \\ 1 & 0 \end{bmatrix}.$$

Since σ is a permutation matrix, $\sigma \in GL_1(R)E_n(R)$. By (10.3) (iv), $\overline{\sigma}$ lies in the center of the group $GL_n(R)/E_n(R)$. By (10.9), $(I_n+XY)^{-1}(I_n+YX) \in E_n(R)$. So, in $GL_n(R)/E_n(R)$,

$$\overline{A} = \overline{\sigma A \sigma^{-1}} = \overline{I_n + YX} = \overline{I_n + XY} = \overline{B}.$$

That is, $\overline{A}_2^{-1}\overline{A}_1 = \overline{B}_2\overline{B}_1^{-1}$; so $\overline{A}_1\overline{B}_1 = \overline{A}_2\overline{B}_2$. ∎

(10.15) Theorem. *If $sr(R) < n < \infty$, then $GL_n(R) \cap E(R) = E_n(R)$ and the inclusions of $GL_n(R)$ into $GL_m(R)$ (for $m > n$) and $GL(R)$ induce isomorphisms*

$$s_{n,m} : \frac{GL_n(R)}{E_n(R)} \quad \to \quad \frac{GL_m(R)}{E_m(R)}, \quad and$$

$$s_n : \frac{GL_n(R)}{E_n(R)} \quad \to \quad \frac{GL(R)}{E(R)} = K_1(R).$$

Proof. By (10.3), s_n and $s_{n,m} = s_{m-1,m} \circ \cdots \circ s_{n,n+1}$ are surjective. The kernel of s_n is

$$\frac{GL_n(R) \cap E(R)}{E_n(R)}.$$

Since

$$GL_n(R) \cap E(R) = \bigcup_{m>n} (GL_n(R) \cap E_m(R)),$$

the kernel of s_n is the union of the kernels of all $s_{n,m}$ for $n < m < \infty$. So it will suffice to prove each $s_{n,n+1}$ is injective. For this, we only need a map

$$\delta : \frac{GL_{n+1}(R)}{E_{n+1}(R)} \quad \to \quad \frac{GL_n(R)}{E_n(R)}$$

with $\delta \circ s_{n,n+1}$ the identity map on $GL_n(R)/E_n(R)$. To this end we prove the map

$$D : GL_{n+1}(R) \quad \to \quad \frac{GL_n(R)}{E_n(R)},$$

defined by

$$D(\begin{bmatrix} 1 & 0 \\ u & A \end{bmatrix} e_{1,n+1}(x) \begin{bmatrix} B & 0 \\ v & 1 \end{bmatrix}) = \overline{AB},$$

has the determinant-like properties:

(i) D is a group homomorphism,

(ii) $D(e_{ij}(r)) = 1$ for $i \neq j$ and $r \in R$,

(iii) $D(A \oplus 1) = \overline{A}$ for $A \in GL_n(R)$.

From these it follows that D induces a homomorphism δ left inverse to $s_{n,n+1}$, as required. Assertions (ii) and (iii) are immediate consequences of:

(10.16) Lemma. *Suppose* $sr(R) < n < \infty$, $A \in GL_n(R)$, $u \in R^{n \times 1}$, *and* $v \in R^{1 \times n}$. *Then*

$$D\left(\begin{bmatrix} A & u \\ 0 & 1 \end{bmatrix}\right) = D\left(\begin{bmatrix} A & 0 \\ v & 1 \end{bmatrix}\right) = D\left(\begin{bmatrix} 1 & v \\ 0 & A \end{bmatrix}\right) = D\left(\begin{bmatrix} 1 & 0 \\ u & A \end{bmatrix}\right)$$

$= \overline{A}$. *And if* $B \in GL_n(R)$ *and* $C \in GL_{n+1}(R)$, *then*

$$D\left(\begin{bmatrix} 1 & 0 \\ u & A \end{bmatrix} C \begin{bmatrix} B & 0 \\ v & 1 \end{bmatrix}\right) = \overline{A} D(C) \overline{B} .$$

Proof. Evidently,

$$\begin{bmatrix} A & 0 \\ v & 1 \end{bmatrix} = \begin{bmatrix} 1 & 0 \\ 0 & I \end{bmatrix} e_{1,n+1}(0) \begin{bmatrix} A & 0 \\ v & 1 \end{bmatrix}, \quad \text{and}$$

$$\begin{bmatrix} 1 & 0 \\ u & A \end{bmatrix} = \begin{bmatrix} 1 & 0 \\ u & A \end{bmatrix} e_{1,n+1}(0) \begin{bmatrix} I & 0 \\ 0 & 1 \end{bmatrix} .$$

Say $u = \begin{bmatrix} x \\ x' \end{bmatrix}$ and $v = [y' \ y]$ with $x, y \in R$. Then the matrices

$$E = \begin{bmatrix} I & x' \\ 0 & 1 \end{bmatrix} \quad \text{and} \quad E' = \begin{bmatrix} 1 & y' \\ 0 & I \end{bmatrix}$$

belong to $E_n(R)$, and with $(1, n-1, 1)$ by $(1, n-1, 1)$ partitioned matrices, one can directly verify:

$$\begin{bmatrix} A & u \\ 0 & 1 \end{bmatrix} = \begin{bmatrix} 1 & 0 \\ 0 & E \end{bmatrix} e_{1,n+1}(x) \begin{bmatrix} A & 0 \\ 0 & 1 \end{bmatrix}, \quad \text{and}$$

$$\begin{bmatrix} 1 & v \\ 0 & A \end{bmatrix} = \begin{bmatrix} 1 & 0 \\ 0 & A \end{bmatrix} e_{1,n+1}(y) \begin{bmatrix} E' & 0 \\ 0 & 1 \end{bmatrix} .$$

The final assertion of the lemma follows from the identities:

$$\begin{bmatrix} 1 & 0 \\ u & A \end{bmatrix} \begin{bmatrix} 1 & 0 \\ u' & A' \end{bmatrix} = \begin{bmatrix} 1 & 0 \\ u'' & AA' \end{bmatrix},$$

$$\begin{bmatrix} B' & 0 \\ v' & 1 \end{bmatrix} \begin{bmatrix} B & 0 \\ v & 1 \end{bmatrix} = \begin{bmatrix} B'B & 0 \\ v'' & 1 \end{bmatrix} . \qquad \blacksquare$$

On the way to proving (i), we need a technical property of the map D:

(10.17) Lemma. *Suppose $sr(R) < n < \infty$, $B \in GL_n(R)$, $v \in R^{1 \times n}$ with last coordinate 0, and $x, y \in R$. Then*

$$D(e_{n,n+1}(y)e_{1,n+1}(x) \begin{bmatrix} B & 0 \\ v & 1 \end{bmatrix} e_{n,n+1}(y)^{-1}) = \overline{B} .$$

Proof. Consider partitioned matrices

$$B = \begin{bmatrix} b_{11} & b_1 & b_{1n} \\ b_2 & B' & b_3 \\ b_{n1} & b_4 & b_{nn} \end{bmatrix} \in GL_n(R) , \quad \text{and}$$

$$v = [d_1 \; d \; 0] \in R^{1 \times n},$$

where $B' \in M_{n-2}(R)$ and $d \in R^{1 \times (n-2)}$. Block multiplying $(1, n-2, 1, 1)$ by $(1, n-2, 1, 1)$ partitioned matrices,

$$e_{n,n+1}(y)e_{1,n+1}(x) \begin{bmatrix} B & 0 \\ v & 1 \end{bmatrix} e_{n,n+1}(y)^{-1}$$

$$= \begin{bmatrix} b_{11} + xd_1 & b_1 + xd & b_{1n} & x - b_{1n}y \\ b_2 & B' & b_3 & -b_3 y \\ b_{n1} + yd_1 & b_4 + yd & b_{nn} & y - b_{nn}y \\ d_1 & d & 0 & 1 \end{bmatrix} .$$

Clearing the last column with row transvections yields a matrix with upper left $n \times n$ corner

$$\begin{bmatrix} b_{11} + b_{1n}yd_1 & b_1 + b_{1n}yd & b_{1n} \\ b_2 + b_3 yd_1 & B' + b_3 yd & b_3 \\ b_{n1} + b_{nn}yd_1 & b_4 + b_{nn}yd & b_{nn} \end{bmatrix} = BE ,$$

where

$$E = \begin{bmatrix} 1 & 0 & 0 \\ 0 & I_{n-2} & 0 \\ yd_1 & yd & 1 \end{bmatrix} \in E_n(R) .$$

So

$$e_{n,n+1}(y)e_{1,n+1}(x) \begin{bmatrix} B & 0 \\ v & 1 \end{bmatrix} e_{n,n+1}(y)^{-1} = \begin{bmatrix} 1 & 0 \\ 0 & E' \end{bmatrix} e_{1,n+1}(x - b_{1n}y) \begin{bmatrix} BE & 0 \\ v & 1 \end{bmatrix} ,$$

where

$$E' = \begin{bmatrix} I_{n-2} & 0 & -b_3 y \\ 0 & 1 & y - b_{nn}y \\ 0 & 0 & 1 \end{bmatrix} \in E_n(R) .$$

Apply D to get $\overline{E'}\,\overline{BE} = \overline{B}$. ∎

(10.18) Corollary. *Suppose* $sr(R) < n < \infty$. *If* $C \in GL_{n+1}(R)$ *and* $y \in R$, *then*

$$D(e_{n,n+1}(y)Ce_{n,n+1}(y)^{-1}) = D(C) .$$

Proof. By (10.12),

$$C = \begin{bmatrix} 1 & 0 \\ u & A \end{bmatrix} e_{1,n+1}(x) \begin{bmatrix} B & 0 \\ v & 1 \end{bmatrix} ,$$

where $A, B \in GL_n(R)$ and the last coordinate of v is 0. Now

$$e_{n,n+1}(y)Ce_{n,n+1}(y)^{-1} = XY,$$

where

$$X = e_{n,n+1}(y) \begin{bmatrix} 1 & 0 \\ u & A \end{bmatrix} e_{n,n+1}(y)^{-1} = \begin{bmatrix} 1 & 0 \\ u' & A' \end{bmatrix}$$

with $A' = e_{n-1,n}(y)Ae_{n-1,n}(y)^{-1}$, and

$$Y = e_{n,n+1}(y)e_{1,n+1}(x) \begin{bmatrix} B & 0 \\ v & 1 \end{bmatrix} e_{n,n+1}(y)^{-1}.$$

By Lemmas (10.16) and (10.17), $D(XY) = D(X)D(Y) = \overline{A'}\overline{B} = \overline{AB} = D(C)$. ∎

We complete the proof of Theorem (10.15) by proving that D is a homomorphism: Let H denote the set of all matrices $M \in GL_{n+1}(R)$ for which $D(CM) = D(C)D(M)$ for all $C \in GL_{n+1}(R)$. If $M_1, M_2 \in H$, then

$$
\begin{aligned}
D(CM_1M_2) &= D(CM_1)D(M_2) \\
&= D(C)D(M_1)D(M_2) \\
&= D(C)D(M_1M_2) ;
\end{aligned}
$$

so $M_1M_2 \in H$. And $I_{n+1} \in H$, since $D(CI) = D(C) = D(C)D(I)$. If $M \in H$, then

$$\overline{I_n} = D(I_{n+1}) = D(M^{-1}M) = D(M^{-1})D(M) ;$$

so $D(M^{-1}) = D(M)^{-1}$. Then

$$D(CM^{-1})D(M) = D(CM^{-1}M) = D(C) ; \quad \text{so}$$

$$D(CM^{-1}) = D(C)D(M)^{-1} = D(C)D(M^{-1}) ,$$

and $M^{-1} \in H$. Therefore H is a subgroup of $GL_{n+1}(R)$.

By Lemma (10.16), H includes each matrix

$$\begin{bmatrix} B & 0 \\ v & 1 \end{bmatrix} \in GL_{n+1}(R) .$$

Let e denote $e_{n,n+1}(y)$. By (10.18), if $M \in H$, then for each $C \in GL_{n+1}(R)$,

$$D(CeMe^{-1}) \;=\; D(ee^{-1}CeMe^{-1}) \;=\; D(e^{-1}CeM)$$

$$=\; D(eCe^{-1})D(M) \;=\; D(C)D(eMe^{-1}) ;$$

so $eMe^{-1} \in H$. Thus, for each $y \in R$,

$$e_{n,n+1}(y)He_{n,n+1}(y)^{-1} \;\subseteq\; H .$$

Since $GL_n(R) \oplus 1 \subseteq H$, for each $x \in R$ and $1 \le j \le n-1$, H includes

$$[e_{jn}(x), e_{n,n+1}(1)] \;=\; e_{j,n+1}(x) ,$$

and then H includes

$$[e_{n-1,n}(1)e_{n,n-1}(-x)e_{n-1,n}(-1), e_{n,n+1}(1)]e_{n-1,n+1}(-x) \;=\; e_{n,n+1}(x) .$$

Therefore H includes all matrices of the form

$$\begin{bmatrix} I_n & u \\ 0 & 1 \end{bmatrix}\begin{bmatrix} B & 0 \\ 0 & 1 \end{bmatrix} = \begin{bmatrix} B & u \\ 0 & 1 \end{bmatrix}$$

in $GL_{n+1}(R)$. By (10.4), the matrices

$$\begin{bmatrix} B & 0 \\ v & 1 \end{bmatrix} \quad \text{and} \quad \begin{bmatrix} B & u \\ 0 & 1 \end{bmatrix}$$

generate $GL_{n+1}(R)$. So $H = GL_{n+1}(R)$ and D is a homomorphism. ∎

11

Relative K_1

As Dedekind's work shows so well, an understanding of ideals is a powerful tool for the solution of problems formulated in a ring. Although proper ideals do not contain units, the structure of unit groups can be revealed by considering units belonging to $1+J$ for an ideal J. In this chapter we discuss the characterization of normal subgroups in $GL_n(R)$ in terms of "congruence subgroups" consisting of the units in $M_n(R)$ belonging to $I_n + M_n(J)$ for an ideal J of R. The work of Kubota, and of Bass, Milnor, and Serre on this characterization, in the mid-1960s, laid the foundation for a good deal of algebraic K-theory, culminating in a substantial portion of the encyclopedic text *Algebraic K-Theory* by Bass in 1968.

In §11A we extend the stability theory from Chapter 10 to the K_1 group relative to an ideal. Then §11B focuses on groups of determinant 1 matrices relative to an ideal. In §11C we present the complete proof of the presentation of the relative SK_1 of a ring of "arithmetic type" in terms of "Mennicke symbols" and discuss applications. Later, in Chapter 13, we present the exact sequence for relative K-theory, in which a relative K_1 acts as a bridge between $K_1(R)$ and $K_2(R)$.

11A. Congruence Subgroups of $GL_n(\mathbf{R})$

The choice of a ring R greatly affects the assortment of subgroups in each general linear group $GL_n(R)$. As we see in this section, there is a lovely interplay between subgroups of $GL_n(R)$ and (two-sided) ideals of the ring R. We shall make free use of the notation and properties of matrix units, commutators, and elementary transvections, as discussed in §§9A and B. For finite n, we continue to identify $GL_n(R)$ with a subgroup of $GL(R)$, by setting each matrix A equal to $A \oplus I_\infty$. In the definitions below, it is convenient to treat $GL_n(R)$ for all finite n, and $GL(R)$, at the same time. So we use the convention that the n in $GL_n(R)$ is either a positive integer or the number ∞, exceeding all integers, and that $GL_\infty(R) = GL(R)$. We use the same convention for the n in groups

$GL_n(R, J)$, $E_n(R, J)$, and $SL_n(R, J)$, defined below.

Problem: *What are the normal subgroups of $GL_n(R)$?*

For each ideal J of R, applying the canonical map $R \to R/J$ to each entry defines a group homomorphism

$$c_J : GL_n(R) \quad \to \quad GL_n(R/J) .$$

One of the normal subgroups of $GL_n(R)$ is the kernel of c_J, which we denote by $\boldsymbol{GL_n(R, J)}$. Since it consists of all matrices in $GL_n(R)$ whose entries are congruent mod J to 1 on the diagonal and to 0 off the diagonal, $GL_n(R, J)$ is known as a **congruence subgroup** of $GL_n(R)$.

For each ideal J of R, denote by $\boldsymbol{E_n(J)}$ the group of finite length products of transvections $e_{ij}(a) \in GL_n(R, J)$ — that is, $n \times n$ transvections $e_{ij}(a)$ with $i \neq j$ and $a \in J$. Denote by $\boldsymbol{E_n(R, J)}$ the normal subgroup of $E_n(R)$ generated by $E_n(J)$; so $E_n(R, J)$ consists of the finite length products of conjugates $Ae_{ij}(a)A^{-1}$, where $A \in E_n(R)$ and $e_{ij}(a)$ is a transvection in $GL_n(R, J)$. Each such conjugate lies in the kernel of c_J; so

$$E_n(J) \quad \subseteq \quad E_n(R, J) \quad \subseteq \quad GL_n(R, J) .$$

Note in particular that $GL_n(R, R) = GL_n(R)$ and $E_n(R, R) = E_n(R)$. Note also that $\boldsymbol{GL(R, J)} = GL_\infty(R, J)$ is the union of the nested sequence

$$GL_1(R, J) \quad \subseteq \quad GL_2(R, J) \quad \subseteq \quad GL_3(R, J) \quad \subseteq \quad \cdots \ ,$$

and $\boldsymbol{E(R, J)} = E_\infty(R, J)$ is the union of

$$1 = E_1(R, J) \quad \subseteq \quad E_2(R, J) \quad \subseteq \quad E_3(R, J) \quad \subseteq \quad \cdots \ .$$

The advantage of the stable case ($n = \infty$) is evident in the proofs of the following two results of Bass, which first appeared in one of the founding papers of algebraic K-theory, Bass [64A]. To state them, we need a little terminology:

If G is a group with subgroups H and K, say K **normalizes** H if $xHx^{-1} \subseteq H$ for all $x \in K$. The **mixed commutator subgroup** $[H, K]$ is the subgroup of G consisting of all finite length products of commutators $[x, y] = xyx^{-1}y^{-1}$ with $x \in H$, $y \in K$ and their inverses $[x, y]^{-1} = [y, x]$. So $[K, H] = [H, K]$. A brief computation shows K normalizes H if and only if H contains $[H, K]$.

(11.1) Relative Whitehead Lemma. *For each ideal J of a ring R,*

$$E(R, J) \quad = \quad [E(R), E(R, J)] \quad = \quad [GL(R), GL(R, J)] .$$

Proof. Suppose $3 \leq n < \infty$. If $a \in J$, $i \neq j$ and $i, j \in \{1, \ldots, n\}$, there exists $k \in \{1, \ldots, n\}$ different from i and j, and

$$e_{ij}(a) \quad = \quad [e_{ik}(1), e_{kj}(a)] \quad \in \quad [E_n(R), E_n(R, J)] .$$

The latter group is normal in $E_n(R)$, since $z[x,y]z^{-1} = [zxz^{-1}, zyz^{-1}]$. So

$$E_n(R,J) \subseteq [E_n(R), \ E_n(R,J)] \ .$$

The reverse containment is true because $E_n(R,J)$ is normalized by $E_n(R)$.

If $z \in M_n(R)$ has entries in J, then in $M_2(M_n(R)) = M_{2n}(R)$, $e_{12}(z)$ and $e_{21}(z)$ belong to $E_{2n}(J)$. Suppose $x \in GL_n(R)$ and $y \in GL_n(R,J)$. The product

$$\varepsilon(x,y) \ = \ \begin{bmatrix} xy & 0 \\ 0 & 1 \end{bmatrix} \begin{bmatrix} x & 0 \\ 0 & y \end{bmatrix}^{-1} \ = \ \begin{bmatrix} 1 & 0 \\ (1-y) & 1 \end{bmatrix} \begin{bmatrix} 1 & x \\ 0 & 1 \end{bmatrix}$$

$$\begin{bmatrix} 1 & 0 \\ (y-1)x^{-1} & 1 \end{bmatrix} \begin{bmatrix} 1 & -x \\ 0 & 1 \end{bmatrix} \begin{bmatrix} 1 & x(y-1)y^{-1} \\ 0 & 1 \end{bmatrix} \begin{bmatrix} 1 & 0 \\ y(y-1)(xy-1)x^{-1} & 1 \end{bmatrix}$$

belongs to $E_{2n}(R,J)$. In particular, $E_{2n}(R,J)$ contains

$$\varepsilon(y,y^{-1}) \ = \ \begin{bmatrix} y^{-1} & 0 \\ 0 & y \end{bmatrix}.$$

So it also includes

$$\varepsilon(x,y) \begin{bmatrix} x & 0 \\ 0 & y \end{bmatrix} \begin{bmatrix} x^{-1} & 0 \\ 0 & y^{-1} \end{bmatrix} \varepsilon(y,y^{-1})$$

$$= \ \begin{bmatrix} xy & 0 \\ 0 & 1 \end{bmatrix} \begin{bmatrix} x^{-1}y^{-1} & 0 \\ 0 & 1 \end{bmatrix} = \begin{bmatrix} [x,y] & 0 \\ 0 & 1 \end{bmatrix}.$$

To summarize,

$$E_n(R,J) \ = \ [E_n(R), E_n(R,J)] \ \subseteq \ [GL_n(R), GL_n(R,J)] \ \subseteq \ E_{2n}(R,J) \ ,$$

for all finite $n \geq 3$. The equations in the stable case now follow from $E(R,J) = \bigcup_{n<\infty} E_n(R,J)$ and $GL(R,J) = \bigcup_{n<\infty} GL_n(R,J)$. $\blacksquare$

(11.2) Corollary. *For each ring R and ideal J of R, $E(R,J)$ is a normal subgroup of $GL(R)$.*

Proof. By the Whitehead Lemma,

$$E(R,J) \ = \ [GL(R), GL(R,J)] \ \supseteq \ [GL(R), E(R,J)] \ . \qquad \blacksquare$$

(11.3) Stable Normal Structure Theorem. *If H is a subgroup of $GL(R)$, the following three statements are equivalent:*

    (i)   *H is normal in $GL(R)$,*
    (ii) *H is normalized by $E(R)$,*
    (iii) *For some ideal J of R, $E(R, J) \subseteq H \subseteq GL(R, J)$.*

Proof (adapted from Hahn and O'Meara [89, p. 43]). That (i) implies (ii) is purely formal. Assume (ii) and let J denote the ideal of R generated by the entries of every matrix $A - I_\infty$ with $A \in H$. Thus J is minimal with $H \subseteq GL(R, J)$.

Claim. Suppose k and ℓ are two different positive integers, and a is an entry of some $A - I_\infty$ with $A \in H$. Then $e_{k\ell}(b) \in H$ for each b in the ideal RaR generated by a.

Indeed, choose $n < \infty$ large enough so that $A \in GL_n(R)$ and $k, \ell < n + 1$. Say a appears in the i, j-entry of $A - I_\infty$ and $u, v \in R$. Consider the scalar multiples of matrix units $B = v\epsilon_{j1}$ and $C = u\epsilon_{1i}$. Since H is normalized by $E(R)$, $H \cap GL_{2n}(R)$ includes

$$\begin{bmatrix} I & -B \\ 0 & I \end{bmatrix} \begin{bmatrix} A & 0 \\ 0 & I \end{bmatrix} \begin{bmatrix} I & B \\ 0 & I \end{bmatrix} \begin{bmatrix} A^{-1} & 0 \\ 0 & I \end{bmatrix} = \begin{bmatrix} I & (A-I)B \\ 0 & I \end{bmatrix},$$

and $H \cap GL_{3n}(R)$ includes

$$\begin{bmatrix} I & (A-I)B & 0 \\ 0 & I & 0 \\ 0 & 0 & I \end{bmatrix}^{-1} \begin{bmatrix} I & 0 & 0 \\ 0 & I & 0 \\ C & 0 & I \end{bmatrix} \begin{bmatrix} I & (A-I)B & 0 \\ 0 & I & 0 \\ 0 & 0 & I \end{bmatrix} \begin{bmatrix} I & 0 & 0 \\ 0 & I & 0 \\ C & 0 & I \end{bmatrix}^{-1}$$

$$= \begin{bmatrix} I & 0 & 0 \\ 0 & I & 0 \\ 0 & C(A-I)B & I \end{bmatrix} = e_{2n+1, n+1}(uav) .$$

Multiplying such transvections, for various $u, v \in A$, shows $e_{2n+1, n+1}(b)$ is in H for each $b \in RaR$. Then H includes

$$[e_{k, 2n+1}(1), \ e_{2n+1, n+1}(b)] = e_{k, n+1}(b) ,$$

and

$$[e_{k, n+1}(b), \ e_{n+1, \ell}(1)] = e_{k\ell}(b) ,$$

proving the claim.

Now every element c of J belongs to a finite sum $\sum Ra_i R$, where the elements a_i are entries of matrices $A - I_\infty$ with $A \in H$. Multiplying the $e_{k\ell}(b_i)$ with

$b_i \in Ra_iR$ shows H includes $e_{k\ell}(c)$ for each $c \in J$. This works for all pairs k, ℓ with $k \neq \ell$; so $E(J) \subseteq H$. Since $E(R)$ normalizes H, $E(R, J) \subseteq H$, completing the proof of (iii).

Finally, suppose there is an ideal J of R with

$$E(R, J) \subseteq H \subseteq GL(R, J).$$

By (11.1), $E(R, J) = [GL(R), H]$. Since this is contained in H, H is normal in $GL(R)$, proving (i). ∎

Since $E(R, J)$ is normal in $GL(R)$, it is also normal in $GL(R, J)$. It also contains $[GL(R, J), GL(R, J)]$; so the quotient group $GL(R, J)/E(R, J)$ is a homomorphic image of the abelianization of $GL(R, J)$.

(11.4) Definition. If R is a ring and J is an ideal of R, the K_1 **group of** R **relative to** J is the abelian group

$$K_1(R, J) = \frac{GL(R, J)}{E(R, J)}.$$

Under the canonical map from $GL(R, J)$ to $K_1(R, J)$, the poset of subgroups H with

$$(11.5) \qquad\qquad E(R, J) \subseteq H \subseteq GL(R, J)$$

is isomorphic to the poset of subgroups of $K_1(R, J)$. Further, if

$$E(R, J_i) \subseteq H \subseteq GL(R, J_i)$$

for two ideals J_1, J_2 of R, then each $e_{k\ell}(a)$ with $a \in J_1$ belongs to $GL(R, J_2)$; so $J_1 \subseteq J_2$. Similarly, $J_2 \subseteq J_1$; so the ideal J in (11.5) is uniquely determined by H. Following the language used by Bass, J is called the **level** of H. Thus *each normal subgroup H of $GL(R)$ determines, and is determined by, its level J and a subgroup of $K_1(R, J)$.*

A useful connection between the three relative groups $K_1(R, J)$, $GL_n(R, J)$, $E_n(R, J)$ and their "absolute" counterparts $K_1(R)$, $GL_n(R)$, $E_n(R)$ is the construction of the "double" of a ring:

(11.6) Definition. If J is an ideal of a ring R, the **double of** R **along** J is the subset D of $R \times R$ consisting of those $(a, b) \in R \times R$ with $a - b \in J$. Evidently, D is a subring of $R \times R$. If $X = (x_{ij})$ and $Y = (y_{ij})$ belong to $R^{m \times n}$, denote by (X, Y) the matrix in $(R \times R)^{m \times n}$ with i, j-entry (x_{ij}, y_{ij}) for all i, j. In this notation, matrix multiplication is coordinatewise: If also $X', Y' \in R^{n \times p}$, then $(X, Y)(X', Y') = (XX', YY')$. Of course $(X, Y) \in D^{m \times n}$ if and only if each $x_{ij} - y_{ij} \in J$.

(11.7) Proposition. *For each $n \geq 1$, there is an exact sequence*

$$1 \longrightarrow GL_n(R, J) \xrightarrow{\ g\ } GL_n(D) \xrightarrow{\ f\ } GL_n(R) \longrightarrow 1 \ ,$$

with $g(X) = (X, I_n)$ and $f((X,Y)) = Y$. It restricts to an exact sequence

$$1 \longrightarrow E_n(R, J) \xrightarrow{\ g\ } E_n(D) \xrightarrow{\ f\ } E_n(R) \longrightarrow 1 \ .$$

Proof. If $a \in J$, then $(a, 0) \in D$ and $(1+a, 1) \in D$; so for $X \in GL_n(R, J)$, $g(X) = (X, I_n) \in M_n(D)$. Since $X^{-1} \in GL_n(R, J)$, $(X, I_n)^{-1} = (X^{-1}, I_n) \in M_n(D)$ as well; so $g(X) \in GL_n(D)$. Plainly, f and g are group homomorphisms.

Each $Y \in GL_n(R)$ is $f((Y, Y))$; so f is surjective. If $(X, I_n) \in GL_n(D)$, then $X - I_n$ has entries in J; so $X \in GL_n(R, J)$, proving the first sequence is exact.

The map f carries $E_n(D)$ onto $E_n(R)$ since $f(e_{ij}(a, a)) = e_{ij}(a)$. If $e_{ij}(a)$ is any transvection in $GL_n(R)$, then $(e_{ij}(a), e_{ij}(a)) = e_{ij}(a, a) \in E_n(D)$. So if $T \in E_n(R)$ and $e_{ij}(b)$ is a transvection in $GL_n(R, J)$, then

$$g(Te_{ij}(b)T^{-1}) \ = \ (T, T)(e_{ij}(b), I_n)(T, T)^{-1}$$

belongs to $E_n(D)$. Therefore g carries $E_n(R, J)$ into $E_n(D)$, and the second sequence is defined. To prove its exactness, we need only check that, for each (X, I_n) in $E_n(D)$, the matrix X belongs to $E_n(R, J)$.

Each transvection $e_{ij}(a, b) \in GL_n(D)$ equals

$$(e_{ij}(b)e_{ij}(a - b), e_{ij}(b))$$

with $a - b \in J$. If $(X, I_n) \in E_n(D)$, it is a product

$$\prod_{i=1}^{m}(S_iT_i, \ S_i) \ ,$$

where, for each i, S_i is a transvection in $GL_n(R)$ and T_i is a transvection in $GL_n(R, J)$. Then $\prod S_i = I_n$; so

$$X \ = \ \prod_{i=1}^{m} S_iT_i \ = \ (S_1T_1S_1^{-1})(S_1S_2T_2S_2^{-1}S_1^{-1})\cdots(S_1\cdots S_mT_mS_m^{-1}\cdots S_1^{-1}),$$

an element of $E_n(R, J)$. ∎

We can combine the two sequences in (11.7) when $n = \infty$ to get an alternate description of $K_1(R, J)$. The device we will use is a common algebraic tool that we will also need in later chapters. Its proof is a typical example of the technique known as "a diagram chase":

(11.8) Snake Lemma for Groups. *Suppose*

$$
\begin{array}{ccccccc}
H_1 & \xrightarrow{g_1} & H_2 & \xrightarrow{f_1} & H_3 & \longrightarrow & 1 \\
\downarrow{\alpha_1} & & \downarrow{\alpha_2} & & \downarrow{\alpha_3} & & \\
1 & \longrightarrow & G_1 & \xrightarrow{g_2} & G_2 & \xrightarrow{f_2} & G_3
\end{array}
$$

is a commutative diagram of group homomorphisms with exact rows and assume $\alpha_i(H) \lhd G_i$ *for each* i. *There is an induced exact sequence of group homomorphisms:*

$$
\ker(\alpha_1) \xrightarrow{g_1} \ker(\alpha_2) \xrightarrow{f_1} \ker(\alpha_3)
$$

$$
\xrightarrow{\partial} \operatorname{coker}(\alpha_1) \xrightarrow{\overline{g_2}} \operatorname{coker}(\alpha_2) \xrightarrow{\overline{f_2}} \operatorname{coker}(\alpha_3) \ .
$$

If g_1 *in the diagram is injective, so is* g_1 *in the sequence; if* f_2 *in the diagram is surjective, so is* $\overline{f_2}$ *in the sequence.*

Proof. That g_1 carries $\ker(\alpha_1)$ into $\ker(\alpha_2)$ and f_1 carries $\ker(\alpha_2)$ into $\ker(\alpha_3)$ follows from commutativity of the squares. Similarly, g_2 carries $\alpha_1(H_1)$ into $\alpha_2(H_2)$ and f_2 carries $\alpha_2(H_2)$ into $\alpha_3(H_3)$. So g_2 and h_2 induce homomorphisms $\overline{g_2}$ and $\overline{h_2}$ between the cokernels.

The map ∂ "snakes" through the diagram: If $x \in \ker(\alpha_3)$, choose $y \in H_2$ with $f_1(y) = x$. Then $f_2\alpha_2(y) = \alpha_3(x) = 1$; so $\alpha_2(y) = g_2(z)$ for some $z \in G_1$. Define $\partial(x)$ to be $\overline{z} \in G_1/\alpha_1(H_1)$. This does not depend on the choice of y with $f_1(y) = x$, since any other choice is $g_1(w)y$ for some $w \in H_1$, and $\alpha_2(g_1(w)y) = g_2(\alpha_1(w)z)$ with $\overline{\alpha_1(w)z} = \overline{z}$. This map ∂ is a homomorphism: If $f_1(y_1) = x_1$, $f_2(y_2) = x_2$, $g_2(z_1) = \alpha_2(y_1)$, and $g_2(z_2) = \alpha_2(y_2)$, then $f_1(y_1y_2) = x_1x_2$ and $g_2(z_1z_2) = \alpha_2(y_1y_2)$; so $\partial(x_1x_2) = \overline{z_1z_2} = \overline{z_1}\,\overline{z_2} = \partial(x_1)\partial(x_2)$.

The composites of consecutive maps are easily seen to be the trivial map. If g_1 is injective on H_1, its restriction is injective on $\ker(\alpha_1)$. For exactness at $\ker(\alpha_2)$, suppose $f_1(y) = 1$, where $\alpha_2(y) = 1$. Then $y = g_1(w)$ for some $w \in H_1$ and $g_2\alpha_1(w) = \alpha_2(y) = 1$; so $\alpha_1(w) = 1$ and $w \in \ker(\alpha_1)$. For exactness at $\ker(\alpha_3)$, suppose $\partial(x) = \overline{1}$, where $\alpha_3(x) = 1$. Then there exist $w \in H_1$ and $y \in H_2$ with $g_2\alpha_1(w) = \alpha_2(y)$ and $f_1(y) = x$. So $x = f_1(g_1(w)^{-1}y)$ and $g_1(w)^{-1}y \in \ker(\alpha_2)$.

For exactness at $\operatorname{coker}(\alpha_1)$, suppose $z \in G_1$ and $\overline{g}_2(\overline{z}) = \overline{1}$. Then $g_2(z) = \alpha_2(y)$ for some $y \in H_2$. Then $f_1(y) \in \ker(\alpha_3)$ and $\partial f_1(y) = \overline{z}$. For exactness at $\operatorname{coker}(\alpha_2)$, suppose $u \in G_2$ and $\overline{f_2}(\overline{u}) = \overline{1}$. Then $f_2(u) = \alpha_3(x)$ for some $x \in H_3$. Choose $y \in H_2$ with $f_1(y) = x$. Then $f_2\alpha_2(y) = \alpha_3(x) = f_2(u)$; so $\alpha_2(y)^{-1}u = g_2(z)$ for some $z \in G_1$, and $\overline{u} = \overline{\alpha_2(y)^{-1}u} = \overline{g_2}(\overline{z})$. Finally, if f_2 is surjective, so is the map $\overline{f_2}$ induced by the composite of f_2 and the canonical map $G_3 \to G_3/\alpha_3(H_3)$. ∎

(11.9 Corollary). *If D is the double of R along J, there is an exact sequence*

$$1 \longrightarrow K_1(R,J) \xrightarrow{\bar{g}} K_1(D) \xrightarrow{\bar{f}} K_1(R) \longrightarrow 1 .$$

That is, $K_1(R,J)$ is isomorphic to the kernel of K_1 of the projection $D \to R$ to the second coordinate.

Proof. Apply the Snake Lemma to the diagram with vertical maps being inclusions:

$$
\begin{array}{ccccccccc}
1 & \longrightarrow & E(R,J) & \xrightarrow{g} & E(D) & \xrightarrow{f} & E(R) & \longrightarrow & 1 \\
& & \downarrow & & \downarrow & & \downarrow & & \\
1 & \longrightarrow & GL(R,J) & \xrightarrow{g} & GL(D) & \xrightarrow{f} & GL(R) & \longrightarrow & 1
\end{array}
\quad \blacksquare
$$

For finite n, Proposition (11.7) enables us to extend to relative groups the stability theorems of Chapter 10, following the exposition in Curtis and Reiner [87, §44].

(11.10) Stability Theorem for Relative K_1. *Suppose $sr(R) \leq n < \infty$ and J is an ideal of R.*

   (i) $GL_{n+1}(R,J) \;=\; GL_n(R,J)E_{n+1}(R,J)$.
  (ii) $GL_{n+1}(R) \cap E(R,J) \;=\; E_{n+1}(R,J)$.
 (iii) $E_{n+1}(R,J) \;\supseteq\; [GL_{n+1}(R), GL_{n+1}(R,J)]$.
 (iv) $E_{n+1}(R,J) \;\triangleleft\; GL_{n+1}(R)$.
  (v) *Inclusions induce a surjective group homomorphism*

$$GL_n(R,J) \;\to\; \frac{GL_{n+1}(R,J)}{E_{n+1}(R,J)}$$

and group isomorphisms

$$\frac{GL_{n+1}(R,J)}{E_{n+1}(R,J)} \;\simeq\; \frac{GL_{n+2}(R,J)}{E_{n+2}(R,J)} \;\simeq\; K_1(R,J) .$$

Proof. Suppose $B \in GL_{n+1}(R,J)$ with last row $[u_1 \; u]$ where $u \in R^{1 \times n}$. Then $B^{-1} \in GL_{n+1}(R,J)$ with last column $\begin{bmatrix} v_1 \\ v \end{bmatrix}$ where $v \in R^{n \times 1}$, and

$$[u_1 v_1 \; u] \begin{bmatrix} 1 \\ v \end{bmatrix} \;=\; [u_1 \; u] \begin{bmatrix} v_1 \\ v \end{bmatrix} \;=\; 1 .$$

Since $sr(R) \leq n$, there exists $b \in R^{1 \times n}$ for which $u' = u + u_1 v_1 b$ is right unimodular. So we can choose $c \in R^{n \times 1}$ with $u'c = 1$. Note that $u_1, v_1 \in J$. For each $x \in J$, the last row of

$$B_1 = B \begin{bmatrix} 1 & v_1 b \\ 0 & I_n \end{bmatrix} \begin{bmatrix} 1 & 0 \\ c(x - u_1) & I_n \end{bmatrix}$$

is $[x \ u']$. The last entry of u' has the form $1 + \alpha$ with $\alpha \in J$. Choose x to be α. Then $[x \ u'] = [\alpha \ a \ 1 + \alpha]$ with $a \in J^{1 \times (n-1)}$. Since right multiplication by $e_{ij}(r)$ adds the i-column times r to the j-column, the last row of

$$B_2 = B_1 e_{1,n+1}(-1) e_{n+1,1}(-\alpha) e_{1,n+1}(1)$$

is $[0 \ a \ 1]$. Then

$$B_3 = B_2 \begin{bmatrix} 1 & 0 & 0 \\ 0 & I_{n-1} & 0 \\ 0 & -a & 1 \end{bmatrix} = \begin{bmatrix} A & w \\ 0 & 1 \end{bmatrix}$$

with $A \in M_n(R)$ and $w \in J^{n \times 1}$. By Lemma (10.2), $A \in GL_n(R)$. Then

$$B_4 = B_3 \begin{bmatrix} I_n & -A^{-1}w \\ 0 & 1 \end{bmatrix} = \begin{bmatrix} A & 0 \\ 0 & 1 \end{bmatrix} \in GL_n(R, J).$$

So we found $E \in E_{n+1}(R, J)$ with $BE \in GL_n(R, J)$. So

$$B \in GL_n(R, J)E^{-1} \subseteq GL_n(R, J)E_{n+1}(R, J),$$

proving (i).

By the Injective Stability Theorem (10.15), $GL_{n+1}(R) \cap E(R) = E_{n+1}(R)$. We extend this to the relative case by using the double D of R along J. First, we need:

(11.11) Lemma. *If a matrix over D has a right (resp. left) inverse over $R \times R$, it has a right (resp. left) inverse over D. The stable rank of D equals that of R.*

Proof. We prove the first statement for right inverses, the left being similar. Suppose $X, Y \in R^{m \times n}$ with $(X, Y) \in D^{m \times n}$, and $S, T \in R^{n \times m}$ with $(X, Y)(S, T) = (I_m, I_m)$. Then $M_m(D)$ includes

$$(I_m, I_m) - (X, Y)(S, S) = (0, I_m - YS),$$

and

$$(X, Y)[(S, S) + (T, T)(0, I_m - YS)] = (I_m, I_m).$$

Now suppose $n > sr(R)$ and $(d_1, \ldots, d_n) \in D^n$ is right unimodular over D. If each $d_i = (a_i, b_i)$ in $R \times R$, then $(a_1, \ldots, a_n)$ is right unimodular over R. So for some $c_2, \ldots, c_n \in R$,

$$(u_2, \ldots, u_n) \; = \; (a_2 + a_1 c_2, \ldots, a_n + a_1 c_n)$$

is right unimodular over R. These are the first coordinates in elements

$$d_i' \; = \; d_i + d_1(c_i, c_i) \; = \; (u_i, v_i)$$

for $2 \le i \le n$. So there exist $r_2, \ldots, r_n \in R$ with

(11.12) $$\sum_{i=2}^{n} d_i'(r_i, r_i) \; = \; (1, s) \; \in \; D \; .$$

But $(d_1, d_2', \ldots, d_n')$ is also right unimodular over D; so there are $e_1, \ldots, e_n \in D$ with

$$d_1 e_1 + \sum_{i=2}^{n} d_i' e_i \; = \; (1, 1) \; .$$

Subtracting this equation times $(0, s - 1)$ from equation (11.12) shows

$$(d_1 e_1(0, s - 1), \; d_2', \ldots, d_n')$$

is right unimodular over D. Its vector of first coordinates is $(0, u_2, \ldots, u_n)$, and its vector of second coordinates is right unimodular over R. So for some $f_2, \ldots, f_n \in R$, $(d_2'', \ldots, d_n'')$ is right unimodular over $R \times R$, where

$$\begin{aligned} d_i'' \; &= \; d_i' + d_1 e_1(0, s - 1)(f_i, f_i) \\ &= \; d_i + d_1[(c_i, c_i) + e_1(0, s - 1)(f_i, f_i)] \end{aligned}$$

for $2 \le i \le n$. By the first assertion, $(d_2'', \ldots, d_n'')$ is also right unimodular over the double D. ∎

Now we can prove (11.10) (ii). Suppose X belongs to

$$GL_{n+1}(R) \cap E(R, J) \; = \; GL_{n+1}(R, J) \cap E(R, J) \; .$$

By (11.7), $g(X) = (X, I_{n+1})$ belongs to $GL_{n+1}(D) \cap E(D)$. By the Injective Stability Theorem (10.15), this intersection is $E_{n+1}(D)$. Again by (11.7), X belongs to $E_{n+1}(R, J)$. So $GL_{n+1}(R) \cap E(R, J) \subseteq E_{n+1}(R, J)$. The reverse containment is immediate, proving part (ii).

For (iii), by (11.1),

$$\begin{aligned} {[GL_{n+1}(R), GL_{n+1}(R, J)]} \; &\subseteq \; GL_{n+1}(R) \cap [GL(R), GL(R, J)] \\ &= \; GL_{n+1}(R) \cap E(R, J) \; = \; E_{n+1}(R, J) \; . \end{aligned}$$

Part (iv) is also quick: By (11.2), $E(R, J) \triangleleft GL(R)$; intersecting both groups with $GL_{n+1}(R)$ yields $E_{n+1}(R, J) \triangleleft GL_{n+1}(R)$. The surjectivity in part (v) is just a restatement of part (i); the injectivity follows from

$$GL_{n+1}(R, J) \cap E_{n+2}(R, J) \subseteq GL_{n+1}(R) \cap E(R, J) = E_{n+1}(R, J) . \quad \blacksquare$$

Now that we know $E_n(R, J) \triangleleft GL_n(R, J)$ for finite n beyond $sr(R)$, we can consider other normal subgroups H of $GL_n(R)$ to see if they lie between $E_n(R, J)$ and $GL_n(R, J)$ for some ideal J, as in the stable case. But even for the simplest rings R we encounter a problem:

(11.13) Example. Suppose n is a positive integer. In $GL_n(\mathbb{Z})$, let H denote the subgroup consisting of those invertible matrices in $I_n + 3M_n(\mathbb{Z})$ or $-I_n + 3M_n(\mathbb{Z})$. This H is the kernel of the composite of entrywise reduction mod 3 followed by the canonical map

$$GL_n(\mathbb{Z}/3\mathbb{Z}) \quad \rightarrow \quad GL_n(\mathbb{Z}/3\mathbb{Z})/\{\pm I_n\} .$$

So $H \triangleleft GL_n(\mathbb{Z})$. If $E_n(J) \subseteq H$, then $J = m\mathbb{Z}$, where m is a multiple of 3. But $-I_n \in H$, and entrywise reduction mod m takes $-I_n$ to $-I_n \neq I_n$ in $GL_n(\mathbb{Z}/m\mathbb{Z})$; so $H \not\subseteq GL_n(\mathbb{Z}, m\mathbb{Z})$.

We must adjust our expectations. In general the group $GL_n(R/J)$ has center C consisting of the scalar matrices rI_n with r a unit in the center of the ring R/J, since the elements of C must commute with the matrix units. Denote the kernel of the group homomorphism

$$GL_n(R) \quad \rightarrow \quad GL_n(R/J)/C$$

by $\widetilde{GL}_n(R, J)$. (In the group theory literature, if C is the center of $GL_n(A)$, then $GL_n(A)/C$ is denoted by $PGL_n(A)$.)

Many rings R satisfy the following modified condition for large enough n:

(11.14) Sandwich Condition. *A subgroup H of $GL_n(R)$ is normalized by $E_n(R)$ if and only if*

$$E_n(R, J) \subseteq H \subseteq \widetilde{GL}_n(R, J)$$

for some ideal J of R.

Note: Exactly as in the stable case, the ideal J is uniquely determined by H and is called the **level** of H.

The most general result we know along these lines is due to L. Vaserstein:

(11.15) Unstable Normal Structure Theorem. *Suppose R is a ring with center R_1, $n \geq 3$, and for each maximal ideal P of R_1, the stable rank of $R_P = (R_1 - P)^{-1}R$ is at most $n - 1$. Then the Sandwich Condition is true for R and n.*

Proof. For the intricate matrix computations involved, we send the reader to the original article of Vaserstein [81]. ∎

Vaserstein's theorem (11.15) generalized earlier results of Wilson [72] and of Borevich and Vavilov [85], which we can now present as a special case:

(11.16) Corollary. *Every commutative ring R has the Sandwich Condition for all $n \geq 3$.*

Proof. Since $R = R_1$ is commutative, R_P is local. As in §4D, Exercise 5, the stable rank of a local ring is 1. ∎

11A. Exercises

1. Suppose R is a ring and $\mathcal{A}$ is a set of matrices in $GL(R)$. If H is the smallest normal subgroup of $GL(R)$ containing $\mathcal{A}$, prove the level of H is the ideal J generated by the entries of all $A - I_\infty$ with $A \in \mathcal{A}$. *Hint:* Show J also contains the entries of all $A - I_\infty$ with $A \in H$; then use the proof of (11.3).

2. Although $K_1(R, J)$ is a homomorphic image of the abelianization of $GL(R, J)$, prove it is not always equal to that abelianization. *Hint:* Prove

$$[GL(\mathbb{Z}, 2\mathbb{Z}), GL(\mathbb{Z}, 2\mathbb{Z})] \subsetneq GL(\mathbb{Z}, 4\mathbb{Z}) ,$$

which does not contain $E(\mathbb{Z}, 2\mathbb{Z})$. So $E(\mathbb{Z}, 2\mathbb{Z})$ does not become trivial in the abelianization of $GL(\mathbb{Z}, 2\mathbb{Z})$.

3. If J is an ideal of a ring R, and D is the double of R along J, there is a commutative square in $\mathfrak{Ring}$:

$$
\begin{array}{ccc}
D & \xrightarrow{\pi_2} & R \\
{\scriptstyle \pi_1}\downarrow & & \downarrow{\scriptstyle c} \\
R & \xrightarrow{c} & R/J
\end{array}
$$

where π_i is projection to the i-coordinate and c is the canonical map $r \mapsto r + J$. Of course $\ker(\pi_2) = J \times \{0\}$ and $\ker(c) = J$; so π_1 is an isomorphism (of rings without unit) from ker(π_2) to ker(c). If $F : \mathfrak{Ring} \to \mathfrak{Group}$ is a functor, then applying F to this square creates another commutative square. Show ker $F(\pi_2) \cong$ ker $F(c)$ when F is GL_n, E_n, and K_1. *Hint:* Use (11.7) and (11.9).

4. If A is a ring and n a positive integer, prove the center C of $GL_n(A)$ consists of the scalar matrices rI_n with r a unit in the center of A. If $\mathbb{F}_q$ is a finite field with q elements, what is the order of the group $PGL_2(\mathbb{F}_q)$?

5. Suppose A and B are rings and $\pi_A : A \times B \to A$, $\pi_B : A \times B \to B$ are the coordinate projections. A subring D of $A \times B$ is a **subdirect sum** of $A \times B$ if $\pi_A(D) = A$ and $\pi_B(D) = B$. Identifying $(A \times B)^{m \times n}$ with $A^{m \times n} \times B^{m \times n}$, any matrix in the former can be written as (X, Y) with $X \in A^{m \times n}$ and $Y \in B^{m \times n}$. Then matrix multiplication is coordinatewise:

$$(X, Y)(X', Y') = (XX', YY') .$$

And $(X, Y) \in D^{m \times n}$ if and only if $(x_{ij}, y_{ij}) \in D$ for each pair (i, j).

(i) Prove a matrix over D has a right (resp. left) inverse over $A \times B$ if and only if it has a right (resp. left) inverse over D.
(ii) Prove $sr(D) \le \max(sr(A), sr(B))$.

6. If C_p is a cyclic group of prime order p, prove the group ring $\mathbb{Z}C_p$ is isomorphic to a subdirect sum of $\mathbb{Z} \times \mathbb{Z}[\zeta_p]$ where $\zeta_p = e^{2\pi i/p}$.

11B. Congruence Subgroups of $SL_n(R)$

Throughout this section R is a commutative ring; so the determinant det : $GL_n(R) \to R^*$ is a group homomorphism with kernel $SL_n(R)$, the matrices of determinant 1. For each ideal J of R, entrywise reduction mod J restricts to a group homomorphism

$$c_J : SL_n(R) \to SL_n(R/J)$$

whose kernel

$$\boldsymbol{SL_n(R, J)} = SL_n(R) \cap (I_n + M_n(J))$$

is known as a **congruence subgroup** of $SL_n(R)$, since it consists of those matrices in $SL_n(R) = SL_n(R, R)$ whose entries are congruent mod J to 1 on the diagonal and 0 off the diagonal. The map c_J induces an embedding

$$\frac{SL_n(R)}{SL_n(R, J)} \to SL_n(R/J) .$$

(11.17) Definition. A commutative ring R is **just infinite** if R is infinite but R/J is finite for each nonzero ideal J of R. Note that a just infinite commutative ring R is necessarily noetherian of Krull dimension at most 1: For, if $J_1 \subsetneq J_2 \subsetneq J_3 \subsetneq \ldots$ were a chain of ideals in R, the finite ring R/J_2 would have infinitely many ideals; and if $J_1 \subsetneq J_2 \subsetneq J_3$ were prime ideals of R, then R/J_2 would have a nonzero prime ideal J_3/J_2 — but R/J_2 is a finite commutative domain, and hence a field. By (4.34), R has stable rank ≤ 2.

For just infinite commutative rings R and nonzero ideals J of R, $SL_n(R/J)$ is finite for each positive integer n. So every subgroup H of $SL_n(R)$ containing a congruence subgroup $SL_n(R, J)$ with $J \neq 0$ is necessarily of finite index in $SL_n(R)$. The converse condition is not always true:

(11.18) Congruence Subgroup Property (CSP). *Every subgroup H of finite index in $SL_n(R)$ contains $SL_n(R, J)$ for some nonzero ideal J of R.*

In fact the CSP fails rather spectacularly even in $SL_2(\mathbb{Z})$. The term "congruence subgroup" was coined by Klein and Fricke [92] in 1890–92. The subgroups of finite index in $SL_2(\mathbb{Z})$ have long been of special interest because of their actions on the upper half of the complex plane as linear fractional transformations

$$\begin{bmatrix} a & b \\ c & d \end{bmatrix} \bullet z \;=\; \frac{az+b}{cz+d}$$

and the resulting automorphisms of hyperbolic 2-space. In this form, congruence subgroups appear in the theory of modular functions in number theory. Even Klein and Fricke knew of subgroups of finite index in $SL_2(\mathbb{Z})$ that contain no $SL_2(\mathbb{Z}, m\mathbb{Z})$ with $m \neq 0$. The enumeration of such subgroups is still a work in progress; see Jones [86] for a survey.

So it was a striking discovery by Mennicke [65], and independently by Bass, Lazard, and Serre [64], that $\mathbb{Z}$ has the CSP for all $n \geq 3$. This is a testament to the advantage of "elbow room" in matrix manipulations. Brenner [60] discovered the surjective stability of the quotients

$$\frac{SL_n(\mathbb{Z}, m\mathbb{Z})}{E_n(\mathbb{Z}, m\mathbb{Z})} \;,$$

conjectured these quotients to be trivial for $n \geq 3$, and verified this conjecture for $m < 6$. Following the 1965 papers of Mennicke and Bass-Lazard-Serre, referred to above, Bass, Milnor, and Serre [67] showed R has the CSP for $n \geq 3$ if R is the ring of algebraic integers in a number field F with a real embedding. This was the first major application of the stability theorems for relative K_1, and many of the concepts of algebraic K-theory sprang from this collaboration of Bass, Milnor, and Serre. In this and the next section, we trace the ideas in their proof.

In the context of $GL(R)$,

$$1 = SL_1(R, J) \ \subseteq \ SL_2(R, J) \ \subseteq \ SL_3(R, J) \ \subseteq \ \cdots \ ,$$

and their union is $SL(R, J) = SL_\infty(R, J)$. Since elementary transvections $e_{ij}(r)$ have determinant 1, $E_n(R, J)$ is a subgroup of $SL_n(R, J)$ for all n. In particular, $E(R, J) \lhd SL(R, J)$ and the quotient

$$\boldsymbol{SK_1(R, J)} \ = \ \frac{SL(R, J)}{E(R, J)}$$

is a subgroup of the abelian group $K_1(R, J) = GL(R, J)/E(R, J)$.

Restricting to matrices of determinant 1, the Relative Stability Theorem (11.10) implies

(11.19) Corollary. *Suppose* $sr(R) \le n < \infty$ *and* J *is an ideal of* R.

  (i) $SL_{n+1}(R, J) \ = \ SL_n(R, J)E_{n+1}(R, J)$.
 (ii) $SL_{n+1}(R) \cap E(R, J) \ = \ E_{n+1}(R, J)$.
(iii) $E_{n+1}(R, J) \ \supseteq \ [SL_{n+1}(R), SL_{n+1}(R, J)]$.
 (iv) $E_{n+1}(R, J) \ \lhd \ SL_{n+1}(R)$.
  (v) *Inclusions induce a surjective group homomorphism*

$$SL_n(R, J) \ \longrightarrow \ \frac{SL_{n+1}(R, J)}{E_{n+1}(R, J)}$$

  and group isomorphisms
$$\frac{SL_{n+1}(R, J)}{E_{n+1}(R, J)} \ \cong \ \frac{SL_{n+2}(R, J)}{E_{n+2}(R, J)} \ \cong \ SK_1(R, J) \ . \qquad \blacksquare$$

In the spirit of the Sandwich Condition of the preceding section, we have:

(11.20) Proposition. *If* R *is an infinite commutative ring and* $n \ge 3$, *each subgroup* H *of finite index in* $SL_n(R)$ *contains* $E_n(R, J)$ *for some nonzero ideal* J *of* R.

To reduce the proof to the case of normal subgroups, we recall a basic fact from group theory:

(11.21) Lemma. *If* H *is a subgroup of finite index in a group* G, *then* H *has only finitely many conjugates in* G, *and their intersection is a normal subgroup of finite index in* G.

Proof. This is an exercise in group actions (see §5B). Using the conjugation action of G on its subgroups, the number of conjugates of H is the index in G of the normalizer K of H. Since $H \subseteq K$, that index is finite.

It only remains to prove that if H_1 and H_2 are subgroups of finite index in G, so is $H_1 \cap H_2$. But G acts on the finite set $X = (G/H_1) \times (G/H_2)$ by

$$g \bullet (xH_1, \, yH_2) \;=\; (gxH_1, \, gyH_2) \, ,$$

and $H_1 \cap H_2$ contains the kernel of the permutation representation $G \to \text{Aut}(X)$. Since $\text{Aut}(X)$ is finite, $H_1 \cap H_2$ has finite index in G. ∎

Proof of the Proposition. Define

$$N \;=\; \bigcap_{g \in SL_n(R)} gHg^{-1}.$$

For distinct $i, j \in \{1, , \ldots, n\}$, define

$$J_{ij} \;=\; \{r \in R : e_{ij}(r) \in N\} \, .$$

Since R is infinite and N has finite index in $SL_n(R)$, there exist $s, t \in R$ with $s \neq t$ and

$$e_{ij}(s - t) \;=\; e_{ij}(s) e_{ij}(t)^{-1} \in N \, .$$

So $s - t \in J_{ij}$ and $J_{ij} \neq \{0\}$.

Since N is a subgroup of $SL_n(R)$, J_{ij} is closed under addition. Suppose $1 \leq k \leq n$ and $k \notin \{i, j\}$. For $r, s \in R$,

$$[e_{ij}(r), \, e_{jk}(s)] \;=\; e_{ik}(rs) \, .$$

Therefore, since N is normal in $SL_n(R)$, it contains $e_{ik}(rs)$ whenever it contains either $e_{ij}(r)$ or $e_{jk}(s)$. So

$$J_{ij}R \;\subseteq\; J_{ik} \quad \text{and} \quad RJ_{jk} \;\subseteq\; J_{ik}$$

for all triples $\{i, j, k\}$. So all J_{ij} equal the same nonzero ideal J of R, and $E_n(J) \subseteq N$. Since N is normalized by $E_n(R)$,

$$E_n(R, J) \;\subseteq\; N \;\subseteq\; H \, . \qquad\qquad ∎$$

In light of (11.19) and (11.20), we see a connection to $SK_1(R, J)$:

(11.22) Proposition. *Suppose R is a just infinite commutative ring. If the group $SK_1(R, J) = 1$ for all nonzero ideals J of R, then R has the CSP for all $n \geq 3$. If, on the other hand, $SK_1(R, J)$ is finite but not 1 for some nonzero ideal J of R, the CSP fails for R for all $n \geq 3$.*

Proof. By (11.19),

$$\frac{SL_n(R, J)}{E_n(R, J)} \;\cong\; SK_1(R, J)$$

for all $n \geq 3$. So the first assertion follows from (11.20). Now assume $SK_1(R, J)$ is finite and nontrivial, and $n \geq 3$. Then $E_n(R, J)$ has finite index in $SL_n(R)$ but does not contain $SL_n(R, J)$. Suppose J' is another nonzero ideal of R. Denote R/J' by $\widetilde{R}$ and $(J + J')/J'$ by $\widetilde{J}$. Since finite integral domains are fields, every prime ideal in the finite ring $\widetilde{R}$ is maximal; so $\widetilde{R}$ has Krull dimension 0. By (4.34) and (11.19) (i),

$$
\begin{aligned}
SL_n(\widetilde{R}, \widetilde{J}) &= SL_{n-1}(\widetilde{R}, \widetilde{J})E_n(\widetilde{R}, \widetilde{J}) \\
&= SL_{n-2}(\widetilde{R}, \widetilde{J})E_n(\widetilde{R}, \widetilde{J}) \\
&= \cdots = SL_1(\widetilde{R}, \widetilde{J})E_n(\widetilde{R}, \widetilde{J}) \\
&= E_n(\widetilde{R}, \widetilde{J}) \ .
\end{aligned}
$$

Entrywise reduction mod J' defines a group homomorphism

$$
SL_n(R, J + J') \quad \rightarrow \quad SL_n(\widetilde{R}, \widetilde{J})
$$

with kernel $SL_n(R, J')$ and carrying $E_n(R, J)$ onto $E_n(\widetilde{R}, \widetilde{J}) = SL_n(\widetilde{R}, \widetilde{J})$. Therefore

$$
SL_n(R, J + J') \quad = \quad SL_n(R, J')E_n(R, J) \ .
$$

Since $E_n(R, J)$ does not contain $SL_n(R, J)$, this equation shows it cannot contain $SL_n(R, J')$ either. ∎

11B. Exercises

1. Prove every just infinite commutative ring R is a domain. *Hint:* If $a \in R$, multiplication by a is an R-linear map $R \to aR$; so its kernel is an ideal J of R.

2. Give an example of a commutative noetherian domain R of Krull dimension at most 1 that is infinite but not just infinite.

3. If J is an ideal of a commutative ring R, prove

$$
K_1(R, J) \quad \cong \quad SK_1(R, J) \times GL_1(R, J) \ ,
$$

generalizing (9.18).

4. If R is a commutative ring, the linear groups $GL_n(R)$ and $SL_n(R)$ share the same center $Z(GL_n(R)) = Z(SL_n(R))$ consisting of the scalar matrices rI_n with $r \in R^*$. For a group G, define $PG = G/Z(G)$. Following (11.13), we defined $\widetilde{GL}_n(R, J)$ for each ideal J of R to be the kernel of the composite

$$
GL_n(R) \quad \rightarrow \quad GL_n(R/J) \quad \rightarrow \quad PGL_n(R/J) \ .
$$

Now define $\widetilde{SL}_n(R, J)$ to be the kernel of the composite

$$SL_n(R) \quad \rightarrow \quad SL_n(R/J) \quad \rightarrow \quad PSL_n(R, J) .$$

So $\widetilde{SL}_n(R, J)$ consists of the $n \times n$ matrices of determinant 1 that reduce mod J to a scalar matrix over R/J.

(i) Using (11.16), prove that if $n \geq 3$, each normal subgroup H of $SL_n(R)$ satisfies

$$E_n(R, J) \quad \subseteq \quad H \quad \subseteq \quad \widetilde{SL}_n(R, J)$$

for an ideal J of R.

(ii) For each $i \neq j$ with $i, j \in \{1, \ldots, n\}$, prove

$$J = \{r \in R : e_{ij}(r) \in H\} .$$

(iii) If $A = (a_{ij}) \in H$, prove J contains a_{ij} and $a_{ii} - a_{jj}$ whenever $i \neq j$. Prove J is generated by these elements as A varies through H.

11C. Mennicke Symbols

Suppose throughout this section that R is a commutative ring of stable rank $sr(R) \leq 2$, and J is an ideal of R. Our goal here is a presentation of the abelian group $SK_1(R, J)$ by generators and defining relations. Inspired by ideas of Mennicke [65] and Kubota [66], such a presentation was established by Bass, Milnor, and Serre [67] for rings R of "arithmetic type," including the ring of all algebraic integers in a number field. The generators come from the first row of matrices in $SL_2(R, J)$, and the relations among them resemble those among Legendre symbols and their generalizations in number theory. Bass, Milnor, and Serre went on to compute $SK_1(R, J)$, and thereby determine which rings R of arithmetic type have the Congruence Subgroup Property (see (11.22)). To conclude this section, we state their theorem as (11.33) and provide a sketch of the proof. A sketch is all that we can provide at this point, since the proof depends on some deep number theory beyond the scope of the current discussion.

(11.23) **Definition.** Suppose J is an ideal of R. Then $\mathbf{W}_J$ denotes the set of all pairs (a, b) with $a \in 1 + J$, $b \in J$ and $aR + bR = R$.

(11.24) Lemma. *If e_1 denotes the matrix $[1 \ 0]$, then $W_J = e_1 GL_2(R, J) = e_1 SL_2(R, J)$. That is, W_J consists of the first rows of the matrices in $GL_2(R, J)$, and these coincide with the first rows of matrices in $SL_2(R, J)$.*

Proof. If $\begin{bmatrix} x & * \\ y & * \end{bmatrix}^{-1} = \begin{bmatrix} a & b \\ * & * \end{bmatrix}$ in $GL_2(R, J)$, then $a \in 1 + J$, $b \in J$, and $ax + by = 1$; so $(a, b) \in W_J$.

Conversely, if $a \in 1 + J$, $b \in J$, and $ax + by = 1$ for some $x, y \in R$, then

$$A = \begin{bmatrix} a & b \\ -by^2 & x + bxy \end{bmatrix}$$

has determinant $ax + by(ax + by) = 1$, and $-by^2 \in J$ while $x + bxy \in 1 + J$, the latter because

$$x + bxy + J = (a + J)^{-1} = (1 + J)^{-1} = 1 + J$$

in R/J. So $A \in SL_2(R, J)$. ∎

(11.25) Proposition. *If two matrices in $SL_2(R, J)$ have the same first row, they represent the same element of $SK_1(R, J)$. If $(a, b) \in W_J$ and*

$$\begin{bmatrix} a & b \\ c & d \end{bmatrix} \in SL_2(R, J) ,$$

let

$$[a, b]_J = \overline{\begin{bmatrix} a & b \\ c & d \end{bmatrix}} = \begin{bmatrix} a & b \\ c & d \end{bmatrix} E(R, J)$$

in $SK_1(R, J)$. If (a, b), (a_i, b) and (a, b_i) belong to W_J, then

(i) $[a, b]_J = [a, b + ra]_J$ *if* $r \in J$,

(ii) $[a, b]_J = [a + rb, b]_J$ *if* $r \in R$,

(iii) $[a_1, b]_J [a_2, b]_J = [a_1 a_2, b]_J$, *and*

(iv) $[a, b_1]_J [a, b_2]_J = [a, b_1 b_2]_J$.

Proof. Suppose

$$\begin{bmatrix} a & b \\ c & d \end{bmatrix} \text{ and } \begin{bmatrix} a & b \\ c' & d' \end{bmatrix}$$

belong to $SL_2(R, J)$. Then

$$e_1 = e_1 \begin{bmatrix} a & b \\ c & d \end{bmatrix} \begin{bmatrix} a & b \\ c & d \end{bmatrix}^{-1} = e_1 \begin{bmatrix} a & b \\ c' & d' \end{bmatrix} \begin{bmatrix} a & b \\ c & d \end{bmatrix}^{-1} ;$$

so

$$\begin{bmatrix} a & b \\ c' & d' \end{bmatrix} \begin{bmatrix} a & b \\ c & d \end{bmatrix}^{-1} = \begin{bmatrix} 1 & 0 \\ * & 1 \end{bmatrix} \in E(R, J) ,$$

proving the first assertion.

Note: If $x \in SL(R, J)$ and $y \in SL(R)$, then

$$x^{-1}yxy^{-1} \in [SL(R, J), SL(R)] = E(R, J) ;$$

so x and yxy^{-1} represent the same elements of $SK_1(R, J)$. We use this fact more than once below.

For (i), in $SK_1(R, J)$,

$$\overline{\begin{bmatrix} a & b \\ c & d \end{bmatrix}} = \overline{\begin{bmatrix} a & b \\ c & d \end{bmatrix} \begin{bmatrix} 1 & r \\ 0 & 1 \end{bmatrix}} = \overline{\begin{bmatrix} a & b+ra \\ c & * \end{bmatrix}} .$$

For (ii), by the "Note" above.

$$\overline{\begin{bmatrix} a & b \\ c & d \end{bmatrix}} = \overline{\begin{bmatrix} 1 & 0 \\ -r & 1 \end{bmatrix} \begin{bmatrix} a & b \\ c & d \end{bmatrix} \begin{bmatrix} 1 & 0 \\ r & 1 \end{bmatrix}}$$

$$= \overline{\begin{bmatrix} 1 & 0 \\ -r & 1 \end{bmatrix} \begin{bmatrix} a+rb & b \\ c+rd & d \end{bmatrix}} = \overline{\begin{bmatrix} a+rb & b \\ * & * \end{bmatrix}} .$$

To prove (iii), suppose

$$A_1 = \begin{bmatrix} a_1 & b \\ c_1 & d_1 \end{bmatrix} , \quad A_2 = \begin{bmatrix} a_2 & b \\ c_2 & d_2 \end{bmatrix}$$

in $SL_2(R, J)$, and suppose E is the matrix

$$\begin{bmatrix} -1 & 0 & 0 \\ 0 & 0 & 1 \\ 0 & 1 & 0 \end{bmatrix} \in SL_3(R) .$$

Then modulo $E(R, J)$, $A_1 A_2$ is congruent to the matrix

$$A = e_{23}(-a_1)A_1(EA_2E^{-1})e_{23}(a_1) = \begin{bmatrix} a_1a_2 & b & 0 \\ * & * & 1-a_1d_2 \\ * & * & d_2 \end{bmatrix} ,$$

where we used $\det(A_1) = 1$ to simplify the $2, 3$-entry. So it is also congruent mod $E(R, J)$ to

$$B = e_{32}(1)(e_{23}(a_1 - 1)A)e_{32}(-1) = \begin{bmatrix} a_1a_2 & b & 0 \\ * & * & 1-d_2 \\ s & t & 1 \end{bmatrix} ,$$

and to

$$C \;=\; (e_{23}(d_2-1)B)e_{32}(-t)e_{31}(-s) \;=\; \begin{bmatrix} a_1a_2 & b & 0 \\ u & v & 0 \\ 0 & 0 & 1 \end{bmatrix}.$$

So $[a_1,b]_J[a_2,b]_J = [a_1a_2,b_J]$.

Finally, suppose

$$B_1 \;=\; \begin{bmatrix} a & b_1 \\ c_1 & d_1 \end{bmatrix}, \quad B_2 \;=\; \begin{bmatrix} a & b_2 \\ c_2 & d_2 \end{bmatrix}$$

belong to $SL_2(R,J)$, and consider the matrices

$$U \;=\; \begin{bmatrix} 0 & -1 & 0 \\ 0 & 0 & -1 \\ 1 & 0 & 0 \end{bmatrix}, \quad V \;=\; \begin{bmatrix} 0 & 0 & 1 \\ 0 & 1 & 0 \\ -1 & 0 & 0 \end{bmatrix}$$

in $SL_3(R)$. Then B_1B_2 is congruent modulo $E(R,J)$ to

$$A \;=\; e_{31}(-c_2)V(B_1(UB_2U^{-1}))V^{-1} \;=\; \begin{bmatrix} a & 0 & b_2 \\ * & * & -c_1d_2 \\ 0 & -b_1 & 1 \end{bmatrix},$$

where the $3,3$-entry was simplified by using $\det(B_2)=1$. So it is also congruent mod $E(R,J)$ to

$$B \;=\; (e_{23}(c_1d_2)e_{13}(-b_2)A)e_{32}(b_1) \;=\; \begin{bmatrix} a & b_1b_2 & 0 \\ * & * & 0 \\ 0 & 0 & 1 \end{bmatrix},$$

and $[a,b_1]_J[a,b_2]_J = [a,b_1b_2]_J$. ∎

(11.26) Definition. A **Mennicke symbol** on W_J is any function f from W_J to an abelian group $(A,\cdot)$ satisfying

MS1 : $f(a,b) \;=\; f(a,b+ra) \quad$ if $r \in J$,

$\qquad\quad f(a,b) \;=\; f(a+rb,b) \quad$ if $r \in R$,

MS2 : $f(a_1,b)\cdot f(a_2,b) \;=\; f(a_1a_2,b)$,

$\qquad\quad f(a,b_1)\cdot f(a,b_2) \;=\; f(a,b_1b_2)$,

whenever (a,b), (a_i,b), (a,b_i) belong to W_J.

Proposition (11.25) shows that the map

$$\mu : W_J \;\to\; SK_1(R,J)$$
$$(a,b) \;\mapsto\; [a,b]_J$$

is a Mennicke symbol. Since the stable rank of R is assumed to be ≤ 2, every element of $SK_1(R, J)$ is represented by a matrix in $SL_2(R, J)$ by (11.19). So μ is surjective.

Let $F(W_J)$ denote the multiplicative free abelian group with basis W_J, and suppose M is the subgroup generated by the relators:

$$(a, b)(a, b + ra)^{-1} \text{ for } r \in J \quad , \quad (a, b)(a + rb, b)^{-1} \text{ for } r \in R,$$

$$(a_1, b)(a_2, b)(a_1 a_2, b)^{-1} \text{ and } (a, b_1)(a, b_2)(a, b_1 b_2)^{-1}.$$

Define C_J to be the quotient group $F(W_J)/M$. If $(a, b) \in W_J$, denote its coset by $[a, b] = (a, b)M$. By the construction of C_J, the map $(a, b) \mapsto [a, b]$ is a Mennicke symbol $\phi : W_J \to C_J$.

This Mennicke symbol ϕ is initial among all Mennicke symbols on W_J: Each Mennicke symbol $f : W_J \to (\mathcal{A}, \cdot)$ extends to a homomorphism $F(W_J) \to \mathcal{A}$, inducing a homomorphism $\overline{f} : C_J \to \mathcal{A}$, and f is $\overline{f} \circ \phi$. This defines a bijection $f \leftrightarrow \overline{f}$ between the set of Mennicke symbols $W_J \to \mathcal{A}$ and the set of group homomorphisms $C_J \to \mathcal{A}$.

Since the Mennicke symbol $\mu : W_J \to SK_1(R, J)$ is surjective, so is the group homomorphism $\overline{\mu} : C_J \to SK_1(R, J)$, taking $[a, b]$ to $[a, b]_J$. The main theorem (11.27) in this section says that $\overline{\mu}$ is an isomorphism when R is a commutative noetherian domain of Krull dimension ≤ 1. Following Kubota [66] and Bass [68], we construct an inverse to $\overline{\mu}$ in a sequence of steps:

1. Define $k : GL_2(R, J) \to C_J$, taking $\begin{bmatrix} a & b \\ c & d \end{bmatrix}$ to $[a, b]$.

2. If H is the set of $B \in GL_2(R, J)$ with $k(AB) = k(A)k(B)$ for all $A \in GL_2(R, J)$, and N is the set of $T \in GL_2(R)$ with $k(TAT^{-1}) = k(A)$ for all $A \in GL_2(R, J)$, show H and N are subgroups of $GL_2(R)$ and N normalizes H.

3. Show N contains $E_2(R)$ and each matrix $\begin{bmatrix} 1 & 0 \\ 0 & u \end{bmatrix} \in GL_2(R)$. Show H contains $E_2(R, J)$ and every matrix $\begin{bmatrix} 1 & 0 \\ 0 & u \end{bmatrix} \in GL_2(R, J)$. Show $k(E_2(J)) = 1$.

4. Show k is a homomorphism.

5. Show each element of $GL_3(R, J)$ has an expression as a product

$$\begin{bmatrix} 1 & 0 & 0 \\ * & a & b \\ * & c & d \end{bmatrix} \begin{bmatrix} 1 & 0 & x \\ 0 & 1 & 0 \\ 0 & 0 & 1 \end{bmatrix} \begin{bmatrix} a' & b' & 0 \\ c' & d' & 0 \\ * & * & 1 \end{bmatrix}$$

with all three factors in $GL_3(R, J)$.

6. Show there is a single-valued function $\widehat{k} : GL_3(R, J) \to C_J$, taking each product in step 5 to

$$k\left(\begin{bmatrix} a & b \\ c & d \end{bmatrix}\right) \; k\left(\begin{bmatrix} a' & b' \\ c' & d' \end{bmatrix}\right) \; .$$

7. If $\widehat{H}$ is the set of $B \in GL_3(R, J)$ with $\widehat{k}(AB) = \widehat{k}(A)\widehat{k}(B)$ for all $A \in GL_3(R, J)$, and $\widehat{N}$ is the set of $T \in GL_3(R)$ with $\widehat{k}(TAT^{-1}) = \widehat{k}(A)$ for all $A \in GL_3(R, J)$, show that $\widehat{H}$ and $\widehat{N}$ are subgroups of $GL_3(R)$ and that $\widehat{N}$ normalizes $\widehat{H}$.

8. If

$$\tau = \begin{bmatrix} 0 & 1 & 0 \\ 1 & 0 & 0 \\ 0 & 0 & 1 \end{bmatrix}$$

and S is the set of matrices A in $GL_3(R, J)$ with $\widehat{k}(\tau A \tau^{-1}) = \widehat{k}(A)$, show S includes all matrices

$$\begin{bmatrix} 1 & 0 & 0 \\ e & a & b \\ 0 & c & d \end{bmatrix} e_{13}(x)$$

in $GL_3(R, J)$, where $x \in J$ and $d \neq 0$.

9. Show S includes all of $GL_3(R, J)$; so $\tau \in N$.

10. Show $E_3(R) \subseteq N$.

11. Prove $H = GL_3(R, J)$; so $\widehat{k}$ is a homomorphism.

12. Show $\widehat{k}(E_3(R, J)) = 1$.

We will then have proved:

(11.27) Theorem. *If R is a commutative noetherian domain of Krull dimension ≤ 1, and J is an ideal of R, there is a group homomorphism $K_1(R, J) \to C_J$ whose restriction to $SK_1(R, J)$ is an isomorphism inverse to $\overline{\mu}$. Therefore $SK_1(R, J)$ has an abelian group presentation with one generator $[a, b]$ for each pair $(a, b) \in W_J$, and defining relations*

MS1 : $[a, b] = [a, b + ra]$ *if* $r \in J$,

$\qquad\quad\; [a, b] = [a + rb, b]$ *if* $r \in R$,

MS2 : $[a_1, b][a_2, b] = [a_1 a_2, b]$, *and*

$\qquad\quad\; [a, b_1][a, b_2] = [a, b_1 b_2]$.

Proof. By (4.34), $sr(R) \leq 2$. If J is a nonzero ideal of R, then the (prime) ideals of R/J correspond to the (prime) ideals of R that contain J; so R/J is noetherian of Krull dimension 0. By (4.34) again, $sr(R/J) = 1$.

Step 1. By (11.24), the first rows of the matrices in $GL_2(R, J)$ constitute W_J. So there is a function $k : GL_2(R, J) \to C_J$ with

$$k\left(\begin{bmatrix} a & b \\ c & d \end{bmatrix}\right) = [a, b] .$$

The relations MS1 and MS2 do hold for these elements $[a, b]$ in C_J.

Step 2. Define

$$H = \{B \in GL_2(R, J) : k(A)k(B) = k(AB) \quad \text{for all} \quad A \in GL_2(R, J)\} ,$$
$$N = \{T \in GL_2(R) : k(TAT^{-1}) = k(A) \quad \text{for all} \quad A \in GL_2(R, J)\} .$$

By MS2, $[1, 0]$ is its own square, so it is the identity in C_J. If $B \in H$, then

$$[1, 0] = k(I_2) = k(B^{-1}B) = k(B^{-1})k(B) .$$

So $k(B^{-1}) = k(B)^{-1}$. If $A \in GL_2(R, J)$, $k(A) = k(AB^{-1}B) = k(AB^{-1})k(B)$; so $k(AB^{-1}) = k(A)k(B^{-1})$ and $B^{-1} \in H$. If $B_1, B_2 \in H$, $k(AB_1B_2) = k(AB_1)k(B_2) = k(A)k(B_1)k(B_2) = k(A)k(B_1B_2)$; so $B_1B_2 \in H$ and H is a subgroup of $GL_2(R)$.

If $A \in GL_2(R, J)$ and $T \in N$, $k(T^{-1}AT) = k(TT^{-1}ATT^{-1}) = k(A)$; so $T^{-1} \in N$. If $T_1, T_2 \in N$, then $k(T_1T_2AT_2^{-1}T_1^{-1}) = k(T_2AT_2^{-1}) = k(A)$; so $T_1T_2 \in N$ and N is a subgroup of $GL_2(R)$.

If $T \in N$, $B \in H$, and $A \in GL_2(R, J)$, the definitions of H and N imply $k(ATBT^{-1}) = k(T(T^{-1}ATB)T^{-1}) = k(T^{-1}ATB) = k(T^{-1}AT)k(B) = k(A)k(TBT^{-1})$. So $TBT^{-1} \in H$, proving N normalizes H.

(11.28) Lemma.

    (i)   If $(a, b) \in W_J$ and $a \in R^*$, then $[a, b] = 1$.

    (ii) If $(a, b) \in W_J$ and $b + ra \in R^*$ for some $r \in R$, then $[a, b] = 1$.

    (iii) If $\begin{bmatrix} a & b \\ c & d \end{bmatrix} \in GL_2(R, J)$, then $[a, b] = [d, c] = [d, b]^{-1} = [a, c]^{-1}$.

Proof. If $a \in R^*$, every element $c \in J$ has the form $ca^{-1}a$; so $Ja = J$ and $b + Ja = J$. By MS1, therefore, $[a, b] = [a, 1 - a] = [1, 1 - a]$. By MS2, the latter is its own square; so it equals 1.

If $b + ra \in R^*$, $[a, b] = [a, b - ab] = [a, (1 - a)b] = [a, (1 - a)b + (1 - a)ra] = [a, (1 - a)(b + ra)] = [1, (1 - a)(b + ra)] = 1$.

For part (iii), $ad - bc = 1$. So by (ii), $[a, bc] = 1$ and $[d, bc] = 1$. And $[ad, b] = [ad - bc, b] = [1, b] = 1$ by (i). So, by MS2, $[a, b] = [ad, b][d, b]^{-1} = [d, b]^{-1} = [d, bc][d, b]^{-1} = [d, c]$. And $[a, b] = [a, bc][a, c]^{-1} = [a, c]^{-1}$. ∎

Step 3. Suppose $A = \begin{bmatrix} a & b \\ c & d \end{bmatrix} \in GL_2(R, J)$. We **claim**

(i) $E_2(R) \subseteq N$.

(ii) If $u \in R^*$, $\begin{bmatrix} 1 & 0 \\ 0 & u \end{bmatrix} \in N$.

(iii) $k(AE) = k(A) = k(A)k(E)$ for all $E \in E_2(J)$. So $E_2(J) \subseteq H$.

(iv) If $D = \begin{bmatrix} 1 & 0 \\ 0 & u \end{bmatrix} \in GL_2(R, J)$, then $k(AD) = k(A) = k(A)k(D)$. So $D \in H$.

(v) $E_2(R, J) \subseteq H$.

Proof of Claim. Suppose $r \in R$. Then

$$e_{12}(r)Ae_{12}(-r) = \begin{bmatrix} a + rc & * \\ c & * \end{bmatrix},$$

$$e_{21}(r)Ae_{21}(-r) = \begin{bmatrix} a - rb & b \\ * & * \end{bmatrix}.$$

So

$$k(e_{12}(r)Ae_{12}(-r)) = [a + rc, c]^{-1} = [a, c]^{-1} = k(A),$$
$$k(e_{21}(r)Ae_{21}(-r)) = [a - rb, b] = [a, b] = k(A),$$

and hence $e_{12}(r)$, $e_{21}(r) \in N$. Since N is a subgroup of $GL_2(R)$, $E_2(R) \subseteq N$.

Now suppose $D = \begin{bmatrix} 1 & 0 \\ 0 & u \end{bmatrix} \in GL_2(R)$. Then $D^{-1}AD$ has first row (a, ub). So $k(D^{-1}AD) = [a, ub] = [a, u(1-a)b] = [a, u(1-a)][a, b] = [1, u(1-a)][a, b] = [a, b] = k(A)$. So $D^{-1} \in N$. Since N is a subgroup of $GL_n(R)$, $D \in N$, proving assertion (ii).

Now suppose $r \in J$. The matrix $Ae_{12}(r)$ has first row $(a, b + ra)$; so $k(Ae_{12}(r)) = [a, b + ra] = [a, b] = k(A) = k(A)[1, r] = k(A)k(e_{12}(r))$. And $Ae_{21}(r)$ has first row $(a + rb, b)$; so $k(Ae_{21}(r)) = [a + rb, b] = [a, b] = k(A) = k(A)[1, 0] = k(A)k(e_{21}(r))$. This proves $E_2(J) \subseteq H$ and $k(E_2(J)) = 1$, as desired for (iii).

If $D = \begin{bmatrix} 1 & 0 \\ 0 & u \end{bmatrix} \in GL_2(R, J)$, then AD has first row (a, ub). As in the proof of (ii), $k(AD) = [a, ub] = k(A) = k(A)[1, 0] = k(A)k(D)$. So $D \in H$, proving assertion (iv).

Finally, N normalizes H, $E_2(R) \subseteq N$ and $E_2(J) \subseteq H$; so $E_2(R, J) \subseteq H$. ∎

(11.29) Lemma. *Suppose* $t \in J$,

$$A = \begin{bmatrix} a & b \\ c & d \end{bmatrix} \quad and \quad A' = \begin{bmatrix} a' & b' \\ c' & d' \end{bmatrix}$$

belong to $GL_2(R, J)$, *d and a' belong to* $1 + tR$, *and* $a'R + dR = R$. *Then* $k(A'A) = k(A')k(A)$.

Proof. Note that (d, t) and (a', t) belong to W_J, and $[d, -t] = [d, t] = [a', t] = [1, t] = 1$. Multiplying the expression for 1 in $a'R + dR$ by its expression in $bR + dR$, we see that $(d, a'b) \in W_J$. Similarly, $(a', b'd) \in W_J$. And $ad - bc = u \in R^* \cap (1 + J)$.

Now, considering the first row of $A'A$, $k(A'A) = [a'a + b'c, \ a'b + b'd]$

$$
\begin{aligned}
&= [(a'a + b'c)d, \ a'b + b'd][d, \ a'b + b'd]^{-1} \\
&= [a'(ad - bc) + c(a'b + b'd), \ a'b + b'd][d, a'b]^{-1} \\
&= [a'u, a'b + b'd][d, a'b]^{-1}[d, t]^{-1} \\
&= [a'u, b'd][d, a't]^{-1}[d, b]^{-1} \\
&= [u, b'd][a', b'd][d, a't]^{-1}k(A) \\
&= [a', b'd][a', t][d, a't]^{-1}k(A) \\
&= [a', b'][a', dt][d, a't]^{-1}k(A) \\
&= k(A')[a', dt][d, a't]^{-1}k(A) \ .
\end{aligned}
$$

For some $x \in R$, $d - a' = tx$. Then

$$
\begin{aligned}
[a', dt] &= [a', dt - a't] = [a', t^2 x] \\
&= [a', tx][a', t] = [a', tx] \\
&= [d, tx] = [d, tx][d, -t] \\
&= [d, -t^2 x] = [d, t(a' - d)] \\
&= [d, a't] \ .
\end{aligned}
$$
∎

Step 4. We **claim** *k is a group homomorphism (i.e., $H = GL_2(R, J)$).*

Proof. Say $A = \begin{bmatrix} a & b \\ c & d \end{bmatrix}$ and $A' = \begin{bmatrix} a' & b' \\ c' & d' \end{bmatrix}$ belong to $GL_2(R, J)$. If $a' = 1$, take $t = d - 1 \in J$; then d and a' belong to $1 + tR$ and $a'R + dR = R$. By Lemma (11.29), $k(A'A) = k(A')k(A)$.

Now suppose instead that $a' \neq 1$, and take $t = a' - 1$. Then $t \neq 0$ and $t \in J$. If

$$D = \begin{bmatrix} 1 & 0 \\ 0 & (ad - bc)^{-1} \end{bmatrix},$$

then $AD \in SL_2(R, J)$.

By the opening remarks of this proof, $sr(R/tR) = 1$. By (11.19) (i),

$$SL_2(R/tR, \ J/tR) \quad = \quad E_2(R/tR, \ J/tR) \ .$$

Using the same argument as appeared at the end of the proof of (11.22), but with $J' = tR$,

$$SL_2(R, J) \quad = \quad SL_2(R, tR)E_2(R, J) \ .$$

So there exists $B \in E_2(R, J)$ with

$$\begin{bmatrix} a_1 & b_1 \\ c_1 & d_1 \end{bmatrix} \quad = \quad ADB \quad \in \quad SL_2(R, tR) \ .$$

Squaring the expression for 1 in $c_1 R + d_1 R$, we find $c_1^2 R + d_1 R = R$. If $a' = 0$, $J = R$ and $A' \in H$ by Step 3. Supposing $a' \neq 0$, $sr(R/a'R) = 1$. So, for some $r \in R$, $d_2 = d_1 + c_1^2 r$ represents a unit of $R/a'R$.

If

$$C \quad = \quad \begin{bmatrix} 1 & c_1 r \\ 0 & 1 \end{bmatrix} ,$$

then $ADBC$ has 2, 2-entry d_2. Now $a'R + d_2 R = R$ and $a', d_2 \in 1 + tR$. So by Lemma (11.29),

$$k(A'ADBC) \quad = \quad k(A')k(ADBC) \ .$$

But by Step 3 (iii) and (iv),

$$k(A'ADBC) \quad = \quad k(A'A) \quad \text{and} \quad k(ADBC) \quad = \quad k(A) \ ,$$

proving the claim. ∎

(11.30) Lemma. *If $a \in 1 + J$, $b \in J$, $c \in J$, and (a, b, c) is unimodular over R, there exist $r_1, r_2 \in J$ with $(a + cr_1, \ b + cr_2)$ unimodular over R.*

Proof. Say $ax + by + cz = 1$ with $x, y, z \in R$. Multiplying by $1 - a$ and adding a, $ax' + by' + cz' = 1$ with $x' = x(1 - a) + 1 \in 1 + J$ and $y' = y(1 - a)$, $z' = z(1 - a) \in J$. Then (a, b, cz') is unimodular. Since $sr(R) \leq 2$, there exist $d_1, d_2 \in R$ with $(a + cz'd_1, b + cz'd_2)$ unimodular, and $z'd_1, z'd_2 \in J$. ∎

Step 5. We **claim** *every element of $GL_3(R, J)$ has the "standard form"*

$$\begin{bmatrix} 1 & 0 \\ u & A \end{bmatrix} e_{13}(x) \begin{bmatrix} B & 0 \\ v & 1 \end{bmatrix}$$

with $A, B \in GL_2(R, J)$, $u \in J^{2 \times 1}$, and $v \in J^{1 \times 2}$.

Proof. Suppose $C \in GL_3(R, J)$ has first row $[c \; x]$ where $c \in R^{1 \times 2}$. By the preceding lemma, there exists $v \in J^{1 \times 2}$ with $c' = c - xv$ unimodular. By (11.24), c' is the first row of a matrix $B \in SL_2(R, J)$. So $e_1 B = c'$, where $e_1 = [1 \; 0] \in R^{1 \times 2}$. Then

$$[c \; x] \begin{bmatrix} I_2 & 0 \\ -v & 1 \end{bmatrix} \begin{bmatrix} B^{-1} & 0 \\ 0 & 1 \end{bmatrix} e_{13}(-x) = [1 \; 0 \; 0].$$

So

$$C \begin{bmatrix} I_2 & 0 \\ -v & 1 \end{bmatrix} \begin{bmatrix} B^{-1} & 0 \\ 0 & 1 \end{bmatrix} e_{13}(-x) = \begin{bmatrix} 1 & 0 \\ u & A \end{bmatrix} \in GL_3(R, J),$$

where $A \in I_2 + M_2(J)$ and $u \in J^{2 \times 1}$. By (10.2), $A \in GL_2(R, J)$. Now

$$C = \begin{bmatrix} 1 & 0 \\ u & A \end{bmatrix} e_{13}(x) \begin{bmatrix} B & 0 \\ 0 & 1 \end{bmatrix} \begin{bmatrix} I_2 & 0 \\ v & 1 \end{bmatrix}$$

$$= \begin{bmatrix} 1 & 0 \\ u & A \end{bmatrix} e_{13}(x) \begin{bmatrix} B & 0 \\ v & 1 \end{bmatrix}. \qquad \blacksquare$$

Step 6. We claim *if*

$$\begin{bmatrix} 1 & 0 \\ u_1 & A_1 \end{bmatrix} e_{13}(x) \begin{bmatrix} B_1 & 0 \\ v_1 & 1 \end{bmatrix} = \begin{bmatrix} 1 & 0 \\ u_2 & A_2 \end{bmatrix} e_{13}(y) \begin{bmatrix} B_2 & 0 \\ v_2 & 1 \end{bmatrix},$$

with all six matrices in $GL_3(R, J)$, then $k(A_1)k(B_1) = k(A_2)k(B_2)$.

Proof. By (10.2), these matrices are in $GL_3(R, J)$ if and only if $x, y \in J$, $u_i \in J^{2 \times 1}$, $v_i \in J^{1 \times 2}$, and $A_i, B_i \in GL_2(R, J)$ for $i = 1, 2$. Exactly as in the proof of (10.13), $x = y$, and if $A = A_2^{-1} A_1$ and $B = B_2 B_1^{-1}$, then

$$A = \begin{bmatrix} a & -u'x \\ v' & 1 - wx \end{bmatrix} \text{ and } B = \begin{bmatrix} 1 - xw & -xv' \\ u' & a \end{bmatrix},$$

with $x, u', v', w \in J$. Then

$$\begin{aligned} k(B) &= [1 - xw, -xv'] \\ &= [1 - xw, -x][1 - xw, v'] \\ &= [1, -x][1 - xw, v'] \\ &= [1 - xw, v'] = k(A), \end{aligned}$$

the last step by (11.28). Since k is a homomorphism on $GL_2(R, J)$, it follows that $k(A_2)^{-1} k(A_1) = k(B_2)k(B_1)^{-1}$, as required. $\qquad \blacksquare$

Now define $\widehat{k} : GL_3(R, J) \to C_J$ by

$$\widehat{k}\left(\begin{bmatrix} 1 & 0 \\ u & A \end{bmatrix} e_{13}(x) \begin{bmatrix} B & 0 \\ v & 1 \end{bmatrix}\right) = k(A)k(B)$$

if $x \in J$, $u \in J^{2\times1}$, $v \in J^{1\times2}$, and $A, B \in GL_2(R, J)$. Taking $x = 0$, $u = 0, v = 0$, and $A = I_2$, we find that $\widehat{k}(B) = k(B)$ for all $B \in GL_2(R, J)$. So $\widehat{k}$ extends k.

Step 7. Define

$$\widehat{H} = \{B \in GL_3(R, J) : \widehat{k}(A)\widehat{k}(B) = \widehat{k}(AB) \text{ for all } A \in GL_3(R, J)\},$$

$$\widehat{N} = \{T \in GL_3(R) : \widehat{k}(TAT^{-1}) = \widehat{k}(A) \text{ for all } A \in GL_3(R, J)\}.$$

Then $\widehat{H}$ and $\widehat{N}$ are subgroups of $GL_3(R)$ and $\widehat{N}$ normalizes $\widehat{H}$. The proof is the same as in Step 2, but with hats on.

Step 8. Suppose $\tau = \begin{bmatrix} 0 & 1 & 0 \\ 1 & 0 & 0 \\ 0 & 0 & 1 \end{bmatrix}$ and S is the set of matrices A in $GL_3(R, J)$

with $\widehat{k}(\tau A \tau^{-1}) = \widehat{k}(A)$. We **claim** *that S contains all matrices*

$$M = \begin{bmatrix} 1 & 0 & 0 \\ e & a & b \\ 0 & c & d \end{bmatrix} e_{13}(x) = \begin{bmatrix} 1 & 0 & x \\ e & a & b+ex \\ 0 & c & d \end{bmatrix}$$

in $GL_3(R, J)$ with $x \in J$ and $d \neq 0$.

Proof. This matrix M is in our standard form, with

$$\begin{bmatrix} a & b \\ c & d \end{bmatrix} \in GL_2(R, J),$$

with $ad - bc = \Delta \in GL_1(R, J)$. So $\widehat{k}(M) = k\left(\begin{bmatrix} a & b \\ c & d \end{bmatrix}\right)$.

If $J = 0$, $M = I_3$ and there is nothing to prove. Suppose $J \neq 0$. Since $d \neq 0$, $sr(R/dR) = 1$. The last column of M, $(x, b + ex, d)$, is unimodular; so $(\overline{x}, \overline{b + ex})$ is unimodular over R/dR. So, for some $r \in R$,

$$\overline{x} + \overline{r}(\overline{b + ex}) \in (R/dR)^*.$$

Say $s = 1 - d \in J$. Since $\overline{s} = \overline{1}$ in R/dR,

$$\overline{x} + \overline{sr}(\overline{b + ex}) \in (R/dR)^*.$$

Now R is an integral domain, $d \neq 0$, and $J \neq 0$. So for some $t \in J$, $dt \neq 0$. Then

$$\overline{x} \;+\; \overline{(sr + dt)}\overline{(b + ex)} \;\in\; (R/dR)^*.$$

If $e \neq 0$, $1 + sre \neq 1 + (sr + dt)e$. So whether or not $e = 0$, we can choose $m \in J$ with $1 + me \neq 0$ and

$$\overline{x} \;+\; \overline{m}\overline{(b + ex)} \;\in\; (R/dR)^*.$$

Then $(d, x + m(b + ex))$ is unimodular over R; so by (11.24) it is the first row of a matrix

$$\begin{bmatrix} \alpha & \beta \\ \gamma & \delta \end{bmatrix} \;\in\; SL_2(R, J) .$$

Hence the column $\begin{bmatrix} x + m(b + ex) \\ d \end{bmatrix}$ is the second column of the matrix

$$A \;=\; \begin{bmatrix} \delta & \beta \\ \gamma & \alpha \end{bmatrix} \;\in\; SL_2(R, J) .$$

We shall use A in a standard form for $\tau M \tau^{-1}$.
Say

$$A^{-1} \;=\; \begin{bmatrix} a_{11} & a_{12} \\ a_{21} & a_{22} \end{bmatrix}.$$

Then

$$e_{13}(-b - ex)\begin{bmatrix} 1 & 0 \\ 0 & A^{-1} \end{bmatrix} e_{21}(m)\tau M \tau^{-1}$$

$$= \; e_{13}(-b - ex)\begin{bmatrix} 1 & 0 & 0 \\ 0 & a_{11} & a_{12} \\ 0 & a_{21} & a_{22} \end{bmatrix}\begin{bmatrix} a & e & b + ex \\ ma & 1 + me & x + m(b + ex) \\ c & 0 & d \end{bmatrix}$$

$$= \; e_{13}(-b - ex)\begin{bmatrix} a & e & b + ex \\ f & g & 0 \\ h & i & 1 \end{bmatrix} = \begin{bmatrix} * & * & 0 \\ f & g & 0 \\ h & i & 1 \end{bmatrix}$$

$$= \; \begin{bmatrix} B & 0 \\ v & 1 \end{bmatrix},$$

where $B \in GL_2(R, J)$. So

$$\tau M \tau^{-1} = \; e_{21}(-m)\begin{bmatrix} 1 & 0 \\ 0 & A \end{bmatrix} e_{13}(b + ex)\begin{bmatrix} B & 0 \\ v & 1 \end{bmatrix}$$

$$= \; \begin{bmatrix} 1 & 0 \\ u & A \end{bmatrix} e_{13}(b + ex)\begin{bmatrix} B & 0 \\ v & 1 \end{bmatrix},$$

and $\widehat{k}(\tau M \tau^{-1}) = k(A)k(B) = k(A^{-1})^{-1}k(B).$

Now, in the multiplication carried out above,

$$\begin{bmatrix} a_{11} & a_{12} \\ a_{21} & a_{22} \end{bmatrix} \begin{bmatrix} 1+me & x+m(b+ex) \\ 0 & d \end{bmatrix} = \begin{bmatrix} g & 0 \\ i & 1 \end{bmatrix}.$$

Taking determinants, $g = d(1 + me)$. Comparing $1, 1$-entries, $g = a_{11}(1 + me)$. By choice of m, $1 + me \neq 0$. So $a_{11} = d$. Comparing $1, 2$-entries,

$$d(x + m(b + ex)) + a_{12}d = 0.$$

Since $d \neq 0$, $a_{12} = -x - m(b + ex) = -mb - x(1 + me)$. So

$$\begin{aligned} f &= a_{11}ma + a_{12}c \\ &= dma + (-mb - x(1 + me))c \\ &= m(ad - bc) - x(1 + me)c \\ &= m\Delta - x(1 + me)c . \end{aligned}$$

So

$$A^{-1} = \begin{bmatrix} d & -mb - x(1 + me) \\ * & * \end{bmatrix}$$

and

$$B = \begin{bmatrix} * & * \\ f & g \end{bmatrix} = \begin{bmatrix} * & * \\ m\Delta - x(1 + me)c & d(1 + me) \end{bmatrix}.$$

Then

$$\begin{aligned} k(A^{-1}) &= [d, -mb - x(1 + me)] \\ &= [d, -mbc - x(1 + me)c][d, c]^{-1} \\ &= [d, m(\Delta - ad) - x(1 + me)c][d, c]^{-1} \\ &= [d, m\Delta - x(1 + me)c][d, c]^{-1}, \end{aligned}$$

and

$$\begin{aligned} k(B) &= [d(1 + me), m\Delta - x(1 + me)c] \\ &= [d, m\Delta - x(1 + me)c][1 + me, m\Delta - x(1 + me)c] \\ &= [d, m\Delta - x(1 + me)c][1 + me, m\Delta] \\ &= [d, m\Delta - x(1 + me)c][1, m\Delta] \\ &= [d, m\Delta - x(1 + me)c] . \end{aligned}$$

So $\widehat{k}(\tau M \tau^{-1}) = [d, c] = k\left(\begin{bmatrix} a & b \\ c & d \end{bmatrix} \right) = \widehat{k}(M).$ ∎

(11.31) Lemma. *If* $M \in GL_3(R, J)$ *and*

$$\alpha = \begin{bmatrix} 1 & 0 \\ u' & A' \end{bmatrix} , \quad \beta = \begin{bmatrix} B' & 0 \\ v' & 1 \end{bmatrix}$$

belong to $GL_3(R, J)$, *then* $\widehat{k}(\alpha M \beta) = \widehat{k}(\alpha)\widehat{k}(M)\widehat{k}(\beta)$.

Proof. Say M has the standard form

$$\begin{bmatrix} 1 & 0 \\ u & A \end{bmatrix} e_{13}(x) \begin{bmatrix} B & 0 \\ v & 1 \end{bmatrix} .$$

Then

$$
\begin{aligned}
\widehat{k}(\alpha M \beta) &= \widehat{k}\left(\begin{bmatrix} 1 & 0 \\ u'' & A'A \end{bmatrix} e_{13}(x) \begin{bmatrix} BB' & 0 \\ v'' & 1 \end{bmatrix} \right) \\
&= k(A'A)k(BB') \\
&= k(A')(k(A)k(B))k(B') \\
&= \widehat{k}(\alpha)\widehat{k}(M)\widehat{k}(\beta) .
\end{aligned}
$$
∎

Step 9. We **claim** $S = GL_3(R, J)$; *so* $\tau \in \widehat{N}$.

Proof. Suppose $M \in GL_3(R, J)$ has the standard form

$$\begin{bmatrix} 1 & 0 \\ u & A \end{bmatrix} e_{13}(x) \begin{bmatrix} B & 0 \\ v & 1 \end{bmatrix} .$$

Since A is invertible, its last column has a nonzero entry; so for some $y \in J$,

$$e_{32}(y) \begin{bmatrix} 1 & 0 \\ u & A \end{bmatrix} = \begin{bmatrix} 1 & 0 & 0 \\ e & a & b \\ f & c & d \end{bmatrix}$$

with $d \neq 0$. Then

$$
\begin{aligned}
M' &= e_{31}(-f)e_{32}(y) \begin{bmatrix} 1 & 0 \\ u & A \end{bmatrix} e_{13}(x) \\
&= \begin{bmatrix} 1 & 0 & 0 \\ e & a & b \\ 0 & c & d \end{bmatrix} e_{13}(x)
\end{aligned}
$$

belongs to S, by Step 8. And

$$
\begin{aligned}
M &= e_{32}(-y)e_{31}(f)M' \begin{bmatrix} B & 0 \\ v & 1 \end{bmatrix} \\
&= \begin{bmatrix} 1 & 0 & 0 \\ 0 & 1 & 0 \\ f & -y & 1 \end{bmatrix} M' \begin{bmatrix} b_{11} & b_{12} & 0 \\ b_{21} & b_{22} & 0 \\ * & * & 1 \end{bmatrix} .
\end{aligned}
$$

By the preceding lemma,

$$\widehat{k}(M) = k\left(\begin{bmatrix} 1 & 0 \\ -y & 1 \end{bmatrix}\right) \widehat{k}(M')k\left(\begin{bmatrix} b_{11} & b_{12} \\ b_{21} & b_{22} \end{bmatrix}\right)$$

$$= [1,0]\widehat{k}(M')[b_{22}, b_{21}] ,$$

and conjugating each factor,

$$\tau M \tau^{-1} = \begin{bmatrix} 1 & 0 & 0 \\ 0 & 1 & 0 \\ -y & f & 1 \end{bmatrix} \tau M' \tau^{-1} \begin{bmatrix} b_{22} & b_{21} & 0 \\ b_{12} & b_{11} & 0 \\ * & * & 1 \end{bmatrix} .$$

Again, by the lemma,

$$\widehat{k}(\tau M \tau^{-1}) = k\left(\begin{bmatrix} 1 & 0 \\ f & 1 \end{bmatrix}\right) \widehat{k}(\tau M' \tau^{-1})k\left(\begin{bmatrix} b_{22} & b_{21} \\ b_{12} & b_{11} \end{bmatrix}\right)$$

$$= [1,0]\widehat{k}(\tau M' \tau^{-1})[b_{22}, b_{21}] .$$

Since $M' \in S$, we also have $M \in S$. ∎

Step 10. Claim $E_3(R) \subseteq \widehat{N}$.

Proof. If $M \in GL_3(R, J)$ has the standard form

$$\begin{bmatrix} 1 & 0 \\ u & A \end{bmatrix} e_{13}(x) \begin{bmatrix} B & 0 \\ v & 1 \end{bmatrix} ,$$

and $r \in R$, then

$$\widehat{k}(e_{21}(r)Me_{21}(-r)) = \widehat{k}\left(\begin{bmatrix} 1 & 0 \\ u' & A \end{bmatrix} e_{13}(x) \begin{bmatrix} B e_{21}(-r) & 0 \\ v' & 1 \end{bmatrix}\right)$$

$$= k(A)k(Be_{21}(-r)) = k(A)k(B) = \widehat{k}(M).$$

So $e_{21}(r) \in \widehat{N}$. Since $\widehat{N}$ is closed under multiplication, $e_{12}(r) = \tau e_{21}(r)\tau^{-1} \in \widehat{N}$, as well.

If $C \in GL_n(R)$, denote its transpose by C^t. If

$$\sigma = \begin{bmatrix} & & & 1 \\ & & \cdot{\cdot}^{\cdot} & \\ & 1 & & \\ 1 & & & \end{bmatrix} \in GL_n(R) ,$$

the **skew transpose** of C is

$$C^s = \sigma C^t \sigma \ .$$

Note that

$$\begin{bmatrix} c_{11} & c_{12} \\ c_{21} & c_{22} \end{bmatrix}^s = \begin{bmatrix} c_{22} & c_{12} \\ c_{21} & c_{11} \end{bmatrix} \ ,$$

$$\begin{bmatrix} c_{11} & c_{12} & c_{13} \\ c_{21} & c_{22} & c_{23} \\ c_{31} & c_{32} & c_{33} \end{bmatrix}^s = \begin{bmatrix} c_{33} & c_{23} & c_{13} \\ c_{32} & c_{22} & c_{12} \\ c_{31} & c_{21} & c_{11} \end{bmatrix} \ ,$$

and the skew transpose is an antiautomorphism of $GL_n(R, J)$:

$$(CD)^s \ = \ D^s C^s \ ; \ (C^s)^s \ = \ C \ .$$

If $C \in GL_2(R, J)$, then by (11.28), $k(C^s) = k(C)^{-1}$. So if $M \in GL_3(R, J)$, with standard form

$$M \ = \ \begin{bmatrix} 1 & 0 \\ u & A \end{bmatrix} e_{13}(x) \begin{bmatrix} B & 0 \\ v & 1 \end{bmatrix} \ ,$$

then

$$\widehat{k}(M^s) \ = \ \widehat{k}\left(\begin{bmatrix} 1 & 0 \\ v^t & B^s \end{bmatrix} e_{13}(x) \begin{bmatrix} A^s & 0 \\ u^t & 1 \end{bmatrix} \right)$$

$$= \ k(B^s)k(A^s) \ = \ k(B)^{-1}k(A)^{-1} \ = \ \widehat{k}(M)^{-1} \ .$$

If $C \in \widehat{N}$, then $C^s \in \widehat{N}$, for

$$\widehat{k}(C^s M (C^s)^{-1}) \ = \ \widehat{k}(C^s M (C^{-1})^s)$$
$$= \ \widehat{k}((C^{-1} M^s C)^s) \ = \ \widehat{k}(C^{-1} M^s C)^{-1}$$
$$= \ \widehat{k}(M^s)^{-1} \ = \ \widehat{k}(M) \ .$$

Since $e_{21}(r)$, $e_{12}(r) \in \widehat{N}$, it follows that $e_{32}(r)$, $e_{23}(r) \in \widehat{N}$, and then $\widehat{N}$ also includes

$$e_{31}(r) \ = \ [e_{32}(r), e_{21}(1)] \ , \ \text{and}$$
$$e_{13}(r) \ = \ [e_{12}(r), e_{23}(1)] \ .$$

So $E_3(R) \subseteq \widehat{N}$. ∎

Step 11. By Lemma (11.31), $\widehat{H}$ contains all matrices of the form

$$\begin{bmatrix} B & 0 \\ v & 1 \end{bmatrix}$$

with $B \in GL_2(R, J)$ and $v \in J^{1 \times 2}$. In particular, $\widehat{H}$ includes $e_{12}(r), e_{21}(r),$ $e_{31}(r)$, and $e_{32}(r)$ for each $r \in J$. Since $E_3(R) \subseteq \widehat{N}$, and $\widehat{N}$ normalizes $\widehat{H}$, $\widehat{H}$ also includes the mixed commutators

$$e_{13}(r) = [e_{12}(r), e_{23}(1)] \quad \text{and} \quad e_{23}(r) = [e_{21}(r), e_{13}(1)]$$

from $[\widehat{H}, \widehat{N}]$. So $E_3(J) \subseteq \widehat{H}$. Again, since $\widehat{N}$ normalizes $\widehat{H}$ and contains $E_3(R)$, $E_3(R, J) \subseteq \widehat{H}$. Also, $GL_2(R, J) \subseteq \widehat{H}$. So

$$GL_3(R, J) \;=\; GL_2(R, J)E_3(R, J) \;\subseteq\; \widehat{H}$$

by Theorem (11.10).

We have proved $\widehat{H} = GL_3(R, J)$; so $\widehat{k}$ is a homomorphism.

Step 12. By Step 3 (iii), $k(E_2(J)) = 1$. So for $r \in J$,

$$\widehat{k}(e_{31}(r)) \;=\; \widehat{k}\left(\begin{bmatrix} 1 & 0 \\ re_2 & I_2 \end{bmatrix}\right) \;=\; k(I_2) \;=\; 1 \,,$$

$$\widehat{k}(e_{21}(r)) \;=\; \widehat{k}\left(\begin{bmatrix} 1 & 0 \\ re_1 & I_2 \end{bmatrix}\right) \;=\; k(I_2) \;=\; 1 \,,$$

$$\widehat{k}(e_{32}(r)) \;=\; \widehat{k}\left(\begin{bmatrix} 1 & 0 \\ 0 & e_{21}(r) \end{bmatrix}\right) \;=\; k(e_{21}(r)) \;=\; 1 \,,$$

$$\widehat{k}(e_{12}(r)) \;=\; \widehat{k}\left(\begin{bmatrix} e_{12}(r) & 0 \\ 0 & 1 \end{bmatrix}\right) \;=\; k(e_{12}(r)) \;=\; 1 \,,$$

$$\widehat{k}(e_{23}(r)) \;=\; \widehat{k}\left(\begin{bmatrix} 1 & 0 \\ 0 & e_{12}(r) \end{bmatrix}\right) \;=\; k(e_{12}(r)) \;=\; 1 \,;$$

and because $e_{23}(1) \in \widehat{N}$,

$$\begin{aligned} \widehat{k}(e_{13}(r)) \;&=\; \widehat{k}(e_{12}(r)e_{23}(1)e_{12}(-r)e_{23}(-1)) \\ &=\; \widehat{k}(e_{12}(r))\widehat{k}(e_{23}(1)e_{12}(-r)e_{23}(-1)) \\ &=\; \widehat{k}(e_{12}(r))\widehat{k}(e_{12}(-r)) \;=\; 1 \,. \end{aligned}$$

So $\widehat{k}(E_3(J)) = 1$. If $x \in E_3(R) \subseteq \widehat{N}$ and $y \in E_3(J)$, then

$$\widehat{k}(xyx^{-1}) \;=\; \widehat{k}(y) \;=\; 1 \,.$$

So $\widehat{k}(E_3(R, J)) = 1$, completing Step 12.

Now we have an induced group homomorphism

$$\overline{k} : K_1(R, J) \;\cong\; \frac{GL_3(R, J)}{E_3(R, J)} \;\rightarrow\; C_J$$

taking $\begin{bmatrix} a & b \\ c & d \end{bmatrix} E(R, J)$ to $[a, b]$. On $SK_1(R, J)$, $\overline{k}$ restricts to the inverse map
to the map $\overline{\mu}$ induced by the Mennicke symbol $\mu = [\ ,\ ]_J$. This completes the
proof of Theorem (11.27). ∎

We are now in a position to compute with Mennicke symbols.

(11.32) Corollary. *If R is a just infinite commutative domain, $SK_1(R, J)$ is
a torsion group for each ideal J of R.*

Proof. If $(a, b) \in W_J$, either $a \in R^*$ or $b \neq 0$. If $a \in R^*$,

$$
\begin{aligned}
[a, b] \;&=\; [a, b + aa^{-1}(1 - a - b)] \\
&=\; [a, 1 - a] \\
&=\; [1, 1 - a] \;=\; 1 \ .
\end{aligned}
$$

If, instead, $b \neq 0$, then R/bR is a finite ring, and $aR + bR = R$ implies $\overline{a} \in (R/bR)^*$. If n is the order of $(R/bR)^*$, then $\overline{a}^n = \overline{1}$, and for some $c \in R$,
$a^n + bc = 1$. Then

$$[a, b]^n \;=\; [a^n, b] \;=\; [1, b] \;=\; 1 \ . \qquad \blacksquare$$

The most famous computation of $SK_1(R, J)$ is that of Bass, Milnor, and
Serre, for R a ring of "arithmetic type." We now present the theorem and a
sketch of its proof. For the complete details of proof, see the original paper of
Bass, Milnor, and Serre [67].

First, we define some number-theoretic terms. Suppose R is a Dedekind
domain. If P is a maximal ideal of R and Q is a nonzero fractional ideal (see
§7C), let $v_P(Q)$ denote the exponent of P in the maximal ideal factorization
of Q. If $a \in F^*$, $v_P(aR)$ will be shortened to $v_P(a)$. If p is a positive prime in
$\mathbb{Z}$ and q is a nonzero rational number, we use the special notation $\mathrm{ord}_p(q)$ for
the exponent of p in the prime factorization of q. Now suppose F is a number
field, meaning a finite degree field extension of $\mathbb{Q}$. Suppose R is the ring of
algebraic integers in F and $\mathcal{M}$ is a set of all, or all but finitely many, of the
maximal ideals of R. The set

$$A \;=\; \{a \in F : v_P(a) \geq 0 \ \forall P \in \mathcal{M}\} \ \cup \ \{0\}$$

is a subring of F, called a ring of "arithmetic type." There is a similar definition
of rings of "arithmetic type" in the field $k(x)$ of rational functions in one variable

over a finite field k. For clarity in the present context, we shall use an equivalent definition: A **ring of arithmetic type** is either the ring of algebraic integers in a number field, or the ring $k[x]$ of polynomials in one variable over a finite field k, or is the localization of one of these two types of rings by inversion of a single nonzero element (and its powers). By (7.27), a ring of arithmetic type is a just infinite Dedekind domain.

If F is a number field, then from Galois theory we know there are exactly $[F : \mathbb{Q}]$ embeddings (i.e., ring homomorphisms) from F into the field $\mathbb{C}$ of complex numbers. If none of these embeddings take F into the field $\mathbb{R}$ of real numbers, F is **totally imaginary**. This is the case, for instance, if F^* has an element of finite order $n > 2$. Note that the torsion subgroup of F^* is a finite cyclic group for each number field F.

(11.33) Bass-Milnor-Serre Theorem. *Suppose F is a totally imaginary number field and m is the number of elements of finite order in F^*. For each nonzero ideal J of the ring R of algebraic integers in F, $SK_1(R, J)$ is a cyclic group of finite order r. For each prime p, $ord_p(r)$ is the nearest integer in the interval $[0, ord_p(m)]$ to*

$$\min_{P | pR} \left[\!\!\left[\frac{v_P(J)}{v_P(pR)} - \frac{1}{p-1} \right]\!\!\right],$$

where $[\![x]\!]$ denotes the greatest integer $\leq x$. In particular, $SK_1(R, 4R)$ is finite and nontrivial; so R does not have the Congruence Subgroup Property for any $n \geq 3$.

If R is any other ring of arithmetic type, $SK_1(R, J) = 1$ for all ideals J of R; so R has the Congruence Subgroup Property for all $n \geq 3$.

Before indicating the proof, we define "power residue symbols." Suppose F is a number field and $\zeta_d \in F^*$ has finite order $d > 1$. Since $\zeta_d^d - 1 = 0$, ζ_d is integral over $\mathbb{Z}$; so it belongs to $R = $ alg. int.(F). If P is a maximal ideal of R not containing d, we show in (15.40) below that $\overline{\zeta_d}$ has order d in (R/P^*); so the canonical map $R \to R/P$ restricts to an isomorphism of cyclic groups $\langle \zeta_d \rangle \to \langle \overline{\zeta_d} \rangle$. Say the finite field R/P has q elements. Then $(R/P)^*$ is cyclic of order $q - 1$ and has only one subgroup of order d. So every root of $x^d - 1$ in R/P lies in $\langle \overline{\zeta_d} \rangle$. In particular, if $b \in R$ and $b \notin P$, then $\overline{b}^{(q-1)/d} \in \langle \overline{\zeta_d} \rangle$, and there is a unique element

$$\left(\frac{b}{P} \right)_d \in \langle \zeta_d \rangle$$

congruent mod P to $b^{(q-1)/d}$. Note that $\left(\frac{b}{P} \right)_d = 1$ if and only if $\overline{b}$ is a dth power in $(R/P)^*$.

If $a \in R$ with $aR + dbR = R$, then each maximal ideal P of R containing a contains neither d nor b. In that case, define the d**th power residue symbol**

$$\binom{b}{a}_d = \prod_{P|aR} \binom{b}{P}_d^{v_P(aR)}.$$

If aR is square-free (meaning each $v_P(aR)$ is 1 or 0), then each factor

$$\binom{b}{P}_d^{v_P(aR)} = 1$$

if and only if $\bar{b}$ is a dth power in $(R/aR)^*$.

Proof of the first paragraph of (11.33). If J is a nonzero ideal of R and every maximal ideal of R containing d also contains J, we have a map

$$\kappa : W_J \ \rightarrow \ \langle \zeta_d \rangle \ \subseteq \ F^* \ ,$$

$$(a,b) \ \mapsto \ \binom{b}{a}_d$$

provided we introduce the convention that

$$\binom{0}{a}_d = 1 \ .$$

From the definition of $\binom{b}{a}_d$, the map κ is multiplicative in both a and b, and $\kappa(a,b) = \kappa(a, b+ra)$ for all $r \in J$.

If we make the stronger assumption about J that, for each prime p dividing d and each maximal ideal P dividing pR,

$$\frac{v_P(J)}{v_P(pR)} - \frac{1}{p-1} \geq \operatorname{ord}_p(d) \ ,$$

and assume F is totally imaginary, then $\kappa(a,b) = \kappa(a+rb,b)$ for all $r \in R$, and κ is a Mennicke symbol.

Given any nonzero ideal J of R, this inequality holds when d is the number r in assertion (i). The resulting Mennicke symbol κ is shown to be initial among all Mennicke symbols on W_J; so it induces an isomorphism $SK_1(R,J) \cong \langle \zeta_r \rangle$, proving part (i). ∎

Note: We have used the letter κ for this Mennicke symbol, since it was discovered by Kubota [66].

In preparation for the proof of the second part, we review two theorems from number theory. In these, we use the convention that $a \in \mathbb{Z}$ is "prime" if $a\mathbb{Z}$ is a maximal ideal of $\mathbb{Z}$. So primes can be either positive or negative. Of course $a\mathbb{Z} = b\mathbb{Z}$ if and only if $a = \pm b$.

Dirichlet's Theorem. *If a and b are relatively prime integers, the sequence $a + b\mathbb{Z}$ contains infinitely many positive primes and infinitely many negative primes.*

Proof. See Serre [73, Chapter VI]. ∎

Define the **real symbol**

$$(\ , \ )_\infty : \mathbb{R}^* \times \mathbb{R}^* \ \to \ \{\pm 1\}$$

by $(a, b)_\infty = -1$ if both a and b are negative, and $(a, b)_\infty = +1$ if one or both are positive.

Quadratic Reciprocity. *If a and b are odd primes with $a\mathbb{Z} \neq b\mathbb{Z}$, then*

$$\left(\frac{a}{b}\right)_2 \cdot \left(\frac{b}{a}\right)_2 \cdot (-1)^{\left(\frac{a-1}{2}\right)\left(\frac{b-1}{2}\right)} \cdot (a, b)_\infty \ = \ 1 \ .$$

Proof. See (15.32) below. ∎

We also need:

(11.34) Lemma. *If R is a noetherian commutative domain of Krull dimension at most 1, then for each nonzero element c of an ideal J of R, and each $(a, b) \in W_J$, there exists $(a', b') \in W_{cR}$ with $[a, b]_J = [a', b']_J$.*

Proof. Since $c \neq 0$, R/cR is noetherian of Krull dimension 0; so $sr(R/cR) = 1$. Over R/cR, $(\overline{a}, \overline{b})$ is unimodular; so for some $t \in R$, $\overline{a} + \overline{tb} \in (R/cR)^*$. Say $a_1 = a + tb$ and $u \in R$ with $\overline{a}_1\overline{u} = \overline{1}$. By MS1, $[a, b]_J = [a_1, b]_J$.

Now $u(1 - a_1 - b) \in J$. Define $b_1 = b + a_1u(1 - a_1 - b)$. Then $\overline{b}_1 = \overline{1} - \overline{a}_1$, and by MS1, $[a_1, b]_J = [a_1, b_1]_J$. If $a_2 = a_1 + b_1$, then $\overline{a}_2 = \overline{1}$ and $a_2 \in 1 + cR$. By MS1, $[a_1, b_1]_J = [a_2, b_1]_J$. Finally, if $b_2 = b_1 - b_1a_2$, then $\overline{b}_2 = \overline{0}$ and $b_2 \in cR$. By MS1, $[a_2, b_1]_J = [a_2, b_2]_J$. ∎

Proof of the last assertion of (11.33). We only present the argument for $R =$ alg. int.(F) for a number field F with an embedding $\varepsilon : F \to \mathbb{R}$. The arguments for other rings of arithmetic type differ only in a few technicalities.

(11.35) Lemma. *With this R and any nonzero ideal J of R, each element of $SK_1(R, J)$ is a square.*

Proof. Suppose $(a, b) \in W_J$ and $0 \neq c \in 2J$. By the preceding lemma, $[a, b]_J = [a', b']_J$ with $a' \in 1 + cR$ and $b' = rc$ for some $r \in R$. Then $a'R + rR = R$ and $a'R + P = R$ for each maximal ideal P of R containing 2. So by an extension of the Dirichlet Theorem to rings of arithmetic type (see Bass, Milnor, and Serre

[67, Theorem A10]) there exists $q \in r + a'R$ with qR a maximal ideal of R, $\varepsilon(q) < 0$, $\delta(q) > 0$ for all other real embeddings $\delta : F \to \mathbb{R}$, $2 \notin qR$, and q a square in each "local field" F_P where P is a maximal ideal of R containing 2. (For an exposition of local fields, see §15D below.) By MS1, $[a', b']_J = [a', rc]_J = [a', qc]_J$.

Again, $a'R + qcR = R$; so by the extended Dirichlet Theorem, there exists $p \in a' + qcR$ with pR a maximal ideal, $2 \notin pR$, and $\varepsilon(p)$ having the same sign as the 2-power residue symbol $\left(\frac{a'}{q}\right)_2$. By MS1, $[a', qc]_J = [p, qc]_J$.

With these choices of p and q, Hilbert's product formula (see (15.34) below), extending quadratic reciprocity to $R = \text{alg. int.}(F)$, shows that the reciprocal 2-power residue symbol $\left(\frac{q}{p}\right)_2 = 1$. Since pR is square-free, this implies $\bar{q}$ is a square in $(R/pR)^*$. Say $\bar{q} = \bar{d}^2$ for $d \in R$. Then $q + px = d^2$ for some $x \in R$, and $qc + pxc = d^2c$ with $xc \in J$. So by MS1, $[p, qc]_J = [p, d^2c]_J$.

Now $p \in a' + cR = 1 + cR$. So $[p, c]_J = 1$, and

$$
\begin{aligned}
[p, d^2 c]_J &= [p, d^2 c]_J [p, c]_J \\
&= [p, (dc)^2]_J = [p, dc]_J^2
\end{aligned}
$$

by MS2. Altogether, $[a, b]_J = [p, dc]^2$.   ∎

(11.36) Lemma. *For each $a \in 1 + J$, there is a group homomorphism*

$$
\phi : (R/aR)^* \to SK_1(R, J)
$$

taking $\bar{b}$ to $[a, b(1 - a)]_J$.

Proof. If $b \in R$ and $\bar{b} \in (R/aR)^*$, no maximal ideal of R contains both a and $b(1 - a)$; so $(a, b(1 - a)) \in W_J$. Let $c = 1 - a$; so $c \in J$.

If $\bar{b} = \bar{b}'$ in $(R/aR)^*$, for $b, b' \in R$, then $b' = b + at$ for some $t \in R$. So by MS1, $[a, bc]_J = [a, b'c]_J$. Thus ϕ is well-defined. Since $[a, c]_J = [a, 1 - a]_J = [1, 1 - a]_J = 1$, for $\bar{b}_1, \bar{b}_2 \in (R/aR)^*$ we have

$$
\begin{aligned}
[a, b_1 b_2 c]_J &= [a, b_1 b_2 c]_J [a, c]_J \\
&= [a, b_1 c]_J [a, b_2 c]_J ,
\end{aligned}
$$

by MS2. So ϕ is a homomorphism.   ∎

Now if $(a, b) \in W_J$, then $[a, b]_J = [a, b - ab]_J = [a, b(1 - a)]_J$ is in a homomorphic image of $(R/aR)^*$. So the order of $[a, b]_J$ divides the order of $(R/aR)^*$. If ℓ is 4 or an odd positive prime, Lemma (11.34) and the extended Dirichlet Theorem can be used to find $(a', b') \in W_J$ with $[a, b]_J = [a', b']_J$ and $(R/a'R)^*$ having no element of order ℓ. So the torsion group $SK_1(R, J)$ has no element of order ℓ. That is, each element has order 1 or 2. By Lemma (11.35), $SK_1(R, J) = 1$.   ∎

(11.37) Corollary. *For each ring R of arithmetic type, the group $SK_1(R) = SK_1(R, R) = 1$.*

Proof. We need only consider the case $R = \text{alg. int.}(F)$, where F is a number field with no real embedding. When $J = R$, each $v_P(J) = 0$. So $SK_1(R, R)$ is cyclic of order $r = 1$. ∎

11C. Exercises

1. (Lam) Over a commutative ring R with an ideal J, suppose $[\,,\,] : W_J \to \mathcal{A}$ is a function into a multiplicative abelian group $\mathcal{A}$, and $[\,,\,]$ satisfies MS1 and is multiplicative in the first coordinate. Prove $[\,,\,]$ must also be multiplicative in the second coordinate. *Hints:*

   (i)   If $(a, b) \in W_J$ and $t = 1 - a$, prove $[a, b] = [a, bt^n]$ for all positive integers n.
   (ii) Show $[1 + bt, bt^2] = [1 + bt, -t] = 1$.
   (iii) Use (ii) to show $[a, bt^2] = [a + bt, -at]$.
   (iv) Use (i), with $n = 2$, and (iii) to prove $[a, b_1][a, b_2] = [a, b_1 b_2]$.

2. (Kervaire) Over a commutative ring R with an ideal J, suppose $[\,,\,] : W_J \to \mathcal{A}$ is a Mennicke symbol. If $t \in J$, $a, b \in 1 + tR$, and $aR + bR = R$, prove $[a, bt] = [b, at]$. *Hint:* Say $b - a = st$. Using $[a, t] = [b, -t] = 1$, show $[a, bt] = [a, st^2] = [b, -st^2] = [b, at]$.

3. Suppose J is an ideal of a commutative ring R and $sr(R) \leq 2$.

   (i)   If $SL_2(R, J) = E_2(J)$, prove $SL_n(R, J) = E_n(J)$ for all $n \geq 2$. *Hint:* Extend Lemma (11.30) to rows of length $n \geq 3$. Then right multiply any $A \in SL_n(R, J)$ by a matrix in $E_n(J)$ to transform its last row to $[0 \cdots 0\ 1]$. Using (10.2), right multiply by another matrix in $E_n(J)$ to reach a matrix in $SL_{n-1}(R, J)$. Repeat these steps, to finally reach the matrix I_n.
   (ii) Prove $SL_2(\mathbb{Z}, 2\mathbb{Z}) = E_2(2\mathbb{Z})$. *Hint:* First establish a "weighted" division algorithm. That is, show, for $a, b \in \mathbb{Z}$ not both odd with $gcd(a, b) = 1$, that there exists $q \in 2\mathbb{Z}$ for which $|a - bq| < |b|$. Now right multiply any matrix Y in $SL_2(\mathbb{Z}, 2\mathbb{Z})$ by a sequence of transvections from $E_2(2\mathbb{Z})$ to reach I_2, thereby performing a weighted Euclid's algorithm on the first row.
   (iii) If J is an ideal of a commutative ring R, $SL_n(R) = E_n(R)$, and $SL_n(R, J) = E_n(J)$, prove $SL_n(D) = E_n(D)$, where D is the double of R along J. *Hint:* For $(X, Y) \in SL_n(D)$, multiply by $(Z, Z) \in E_n(D)$ to reach (I_n, Y'); then multiply by (I_n, W) to reach (I_n, I_n).

(iv)   Show $SL_n(D) = E_n(D)$ if $n \geq 2$ and D is the double of $\mathbb{Z}$ along $2\mathbb{Z}$. (This proves D is a "generalized euclidean ring," as defined in §9A, Exercise 5.)

4.   Prove the Dirichlet Theorem, as we have stated it, follows from the Dirichlet Theorem for positive primes alone. Prove Quadratic Reciprocity, as we have stated it, is equivalent to Quadratic Reciprocity for positive odd primes $a = p, b = q$, together with the equation

$$\left(\frac{-1}{q}\right)_2 = (-1)^{\frac{(q-1)}{2}} .$$

5.   If $p \neq q$ are positive odd primes, prove Quadratic Reciprocity for $a = p$ and $b = q$ is equivalent to the assertion:

$$\left(\frac{p}{q}\right)_2 = \begin{cases} -\left(\dfrac{q}{p}\right)_2 & \text{if } p \equiv q \equiv 3 \ (\text{mod } 4), \\[2ex] \left(\dfrac{q}{p}\right)_2 & \text{otherwise.} \end{cases}$$

6.   Using Dirichlet's Theorem and Quadratic Reciprocity, as we have stated them, give the complete proof of Lemma (11.35) for the case $R = \mathbb{Z}$. (In place of the condition that q be "a square in each local field F_P where P is a maximal ideal of R containing 2," use the condition "$q \equiv 1 \ (\text{mod } 8)$.")

7.   Finish the proof that $SK_1(\mathbb{Z}, m\mathbb{Z}) = 1$ by using the Dirichlet Theorem as follows: By (11.34), $[a, b]_{m\mathbb{Z}} = [a', b']_{m\mathbb{Z}}$, where $a' \in 1 + 4m\mathbb{Z}$ and $b' \in 4m\mathbb{Z}$. Choose a positive prime p with $-p \equiv a' \ (\text{mod } b')$. Then $[a', b']_{m\mathbb{Z}} = [-p, b']_{m\mathbb{Z}}$. Show $(\mathbb{Z}/(-p)\mathbb{Z})^*$ has no element of order 4. Next, choose a positive prime $q \equiv a(\text{mod } b)$; so $[a, b]_{m\mathbb{Z}} = [q, b]_{m\mathbb{Z}}$. Show $(\mathbb{Z}/q\mathbb{Z})^*$ has no element of odd prime order ℓ not dividing $q - 1$. Finally, choose two different odd primes s, t with $s \equiv -1 \ (\text{mod } b(q - 1))$ and $t \equiv -q(\text{mod } b(q - 1))$. So $st \equiv q \ (\text{mod } b)$ and $[q, b]_{m\mathbb{Z}} = [st, b]_{m\mathbb{Z}}$. Show $(\mathbb{Z}/st\mathbb{Z})^*$ has no element of odd prime order ℓ dividing $q - 1$, by computing $(s - 1)(t - 1)$ modulo ℓ.

8.   Suppose F is a totally imaginary number field with exactly m roots of unity, $R =$ alg. int.(F), and J is an ideal of R. Prove $SK_1(R, J) = 1$, provided J is not contained in $4R$ and is not properly contained in pR for each positive odd prime p dividing m. *Hint:* Use (11.33).

9.   Suppose R is a ring of arithmetic type, but is not alg. int.(F) for a totally imaginary number field F. For $n \geq 3$, prove every normal subgroup of $SL_n(R)$ either is of finite index or lies in the center $\{rI_n : r \in R, r^n = 1\}$. *Hint:* Use §11B, Exercise 4.

PART IV

Relations Among Matrices: K_2

We have associated to each ring R the abelian groups $K_0(R)$ and $K_1(R)$. The elements of $K_0(R)$ are classes of modules, and the elements of $K_1(R)$ are classes of invertible matrices. Now we add a third abelian group $K_2(R)$, whose elements are the "nonstandard" relations among the generators $e_{ij}(r)$ of the group $E(R)$ – which is to say, among the different sequences of row operations on a matrix. In spite of its arcane sounding definition, $K_2(R)$ shares many properties with $K_0(R)$ and $K_1(R)$, and all three groups are directly connected in exact sequences (see Chapter 13). But the real brilliance of $K_2(R)$ becomes evident when R is a field F. Just as group homomorphisms $K_0(R) \to G$ and $K_1(R) \to G$ correspond to generalized ranks and generalized determinants, so also do the group homomorphisms $K_2(F) \to G$ correspond to "symbol maps" $F^* \times F^* \to G$, which are multiplicative in each coordinate, and take (a, b) to 1 if $a + b = 1_F$. Examples of symbol maps include the Hilbert symbol and norm residue symbol, historically used to formulate generalizations of quadratic reciprocity in number theory, and the quaternion symbol, of importance in the classification of division rings.

Chapter 12 introduces $K_2(R)$, establishes its basic properties, and develops the method of computation by Steinberg symbols. Chapter 13 unites K_0, K_1, and K_2 in exact sequences. Using one of these sequences, Chapter 14 provides Keune's proof of Matsumoto's Theorem relating $K_2(F)$ to symbol maps. Then Part V will explore the rich mathematics of symbol maps in number theory and noncommutative algebra.

12

$K_2(R)$ and Steinberg Symbols

12A. Definition and Properties of $K_2(R)$

We begin with a couple of comments on group presentations. If a group G is generated by a set X, a set of equations $u_i = v_i$ between "words" $\ell_1 \cdots \ell_n$, where each "letter" ℓ_i is x or x^{-1} for $x \in X$, is called a set of **defining relations** among these generators if every equation between words that is true in G is a consequence of the equations $u_i = v_i$ and the basic fact $xx^{-1} = 1_G = x^{-1}x$. Under these circumstances we shall say G is generated by X *subject only to the relations $u_i = v_i$*.

Given any set X, there is a **free group** $\mathcal{F}(X)$ based on X, with generating set X and an empty set of defining relations. Given any set of equations $u_i = v_i$ between words over X, there is a normal subgroup $\mathcal{D}$ of $\mathcal{F}(X)$ generated by the conjugates of the **relators** $u_i v_i^{-1}$; and $\mathcal{F}(X)/\mathcal{D}$ is a group with generators $\overline{x}$ for $x \in X$ and defining relations $\overline{u}_i = \overline{v}_i$ obtained from $u_i = v_i$ by substituting $\overline{x}$ for x in each letter x or x^{-1} occurring in u_i or v_i. So *there exist groups with any chosen generators and defining relations.*

If G is a group with generating set X and defining relations $u_i = v_i$, there is an isomorphism $\mathcal{F}(X)/\mathcal{D} \cong G$ taking $\overline{x}$ to x for each $x \in X$. Now $\mathcal{F}(X)$ has the following universal property: If H is any group, each function $f : X \to H$ extends uniquely to a group homomorphism $\mathcal{F}(X) \to H$. So, *to define a group homomorphism $f : G \to H$, it suffices to choose a function $f : X \to H$ for which, when x is replaced by $f(x)$ and x^{-1} by $f(x)^{-1}$, the defining relations $u_i = v_i$ for G become true equations in H:* For then the homomorphism $\mathcal{F}(X) \to H$ extending f takes $\mathcal{D}$ to 1_H; so it induces a homomorphism from $\mathcal{F}(X)/\mathcal{D}$ to H taking $\overline{x}$ to $f(x)$ for $x \in X$. The composite

$$ G \ \cong \ \mathcal{F}(X)/\mathcal{D} \ \to \ H $$

is a homomorphism taking each $x \in X$ to $f(x) \in H$.

For R a ring, we seek a presentation of the group $E(R)$ by generators and defining relations. Recall some notation and facts from Chapter 9. A "matrix

401

unit" is a matrix $\epsilon_{ij} \in GL(R)$ with i,j-entry 1_R and all other entries 0_R. These multiply according to:

$$\epsilon_{ij}\epsilon_{k\ell} = \begin{cases} 0 & \text{if } j \neq k, \\ \epsilon_{i\ell} & \text{if } j = k. \end{cases}$$

If $i \neq j$ and $r \in R$, the "elementary transvection" $e_{ij}(r) = I + r\epsilon_{ij}$ is obtained from the identity matrix I by replacing the 0 in the i,j-entry by r. Every matrix $B = (b_{ij}) \in GL(R)$ is I plus a finite sum

$$\sum_i (b_{ii} - 1)\epsilon_{ii} + \sum_{i \neq j} b_{ij}\epsilon_{ij} .$$

Using the matrix unit identity above, one can show that left multiplication of a matrix B by $e_{ij}(r)$ adds r times the j-row of B to the i-row of B, and right multiplication by $e_{ij}(r)$ adds the i-column of B times r to the j-column of B.

Each elementary transvection $e_{ij}(r)$ is invertible, with inverse $e_{ij}(-r)$. So for each positive integer n, the group $E_n(R)$, generated by those elementary transvections in $GL_n(R)$, consists of the finite length products of $n \times n$ elementary transvections over R. Likewise $E(R)$ is the finite length products of elementary transvections over R.

Using matrix units, one can derive two handy relations among transvections:

EL1 : $\qquad e_{ij}(r)e_{ij}(s) = e_{ij}(r+s),$ and

EL2 : $\qquad$ the commutator

$$[e_{ij}(r), e_{k\ell}(s)] = \begin{cases} 1 & \text{if } i \neq \ell, \ j \neq k, \\ e_{i\ell}(rs) & \text{if } i \neq \ell, \ j = k. \end{cases}$$

We consider these the **standard relations** among elementary transvections. Another relation is

EL3 : $\qquad (e_{12}(1)e_{21}(-1)e_{12}(1))^4 = 1,$ since

$$e_{12}(1)e_{21}(-1)e_{12}(1) = \begin{bmatrix} 0 & 1 \\ -1 & 0 \end{bmatrix}.$$

From the recent historical perspective of K-theory, the following theorem is "classical:"

(12.1) Theorem. *For $n \geq 3$, $E_n(\mathbb{Z})$ is generated by the $e_{ij}(r)$ with $r \in \mathbb{Z}$, $i \neq j$, and $1 \leq i, j \leq n$, subject only to the defining relations EL1, EL2, and EL3.*

Proof. This was proved for $n = 3$ by Nielsen [24] in 1924, and for $n > 3$ by Magnus [35] in 1935. Actually Nielsen and Magnus did not need EL1 and used only $r = s = 1$ in EL2. For a modern version of their proof due to Silvester see Milnor [71, §10]. ∎

(12.2) Corollary. *The group $E(\mathbb{Z})$ is generated by the $e_{ij}(r)$ with $r \in \mathbb{Z}$, $i \neq j$, and $1 \leq i, j$, subject only to the relations EL1, EL2, and EL3.*

Proof. Any relation among the transvections in $E(\mathbb{Z})$ involves only finitely many transvections, all belonging to $E_n(\mathbb{Z})$ for some positive integer n. So it is a consequence of EL1, EL2, and EL3 in $E_n(\mathbb{Z}) \subseteq E(\mathbb{Z})$. ∎

As we shall see, relation EL3 may or may not be a consequence of relations EL1 and EL2 in $E(R)$, depending on the ring R. And for many rings R, other relations must supplement EL1 and EL2 to present the group $E_n(R)$.

(12.3) Definitions. To assess the power of the standard relations EL1 and EL2, Steinberg [62] defined a group, now known as the **Steinberg group** $St_n(R)$, generated by the expressions $x_{ij}(r)$, with $i \neq j$, $1 \leq i \leq n$, $1 \leq j \leq n$, and $r \in R$, subject only to the defining relations:

ST1 : $\quad x_{ij}(r)x_{ij}(s) = x_{ij}(r+s)$, and

ST2 : the commutator

$$[x_{ij}(r), x_{k\ell}(s)] = \begin{cases} 1 & \text{if } i \neq \ell, \ j \neq k \\ x_{i\ell}(rs) & \text{if } i \neq \ell, \ j = k. \end{cases}$$

Removing the restrictions $i \leq n$ and $j \leq n$ on the generators $x_{ij}(r)$, the same presentation defines the **Steinberg group** $St(R) = St_\infty(R)$. Because the $e_{ij}(r)$ generate $E_n(R)$ and $E(R)$, and obey the standard relations, substituting e for x throughout each product of generators defines surjective group homomorphisms

$$\phi_n : St_n(R) \to E_n(R) \quad \text{and} \quad \phi : St(R) \to E(R) .$$

The kernels $K_{2,n}(R)$ and $K_2(R)$ consist of 1 and the nonstandard relators in a presentation of the groups $E_n(R)$ and $E(R)$, respectively.

Note: Steinberg's paper [62] assumed R is a field F but discussed these concepts for several algebraic groups besides $SL_n(F) = E_n(F)$. Many of the notations and computations below come from this paper. But it was Milnor who extended these ideas and computations to arbitrary rings R and chose the notation $St_n(R)$, $K_{2,n}(R)$, $St(R)$, and $K_2(R)$. In 1971, in his beautifully written book *Introduction to Algebraic K-Theory* [71], Milnor amply justified the choice of $K_2(R)$ to join $K_0(R)$ and $K_1(R)$ within a general algebraic K-theory. Much of our presentation of this section closely follows his treatment.

Increasing the codomain of ϕ from $E(R)$ to $GL(R)$, we obtain an exact sequence

$$1 \longrightarrow K_2(R) \overset{i}{\longrightarrow} St(R) \overset{\phi}{\longrightarrow} GL(R) \overset{c}{\longrightarrow} K_1(R) \longrightarrow 1 ,$$

where i is inclusion and c is the abelianization map. This sequence has a certain symmetry – both ends being related to the middle groups in terms of commutativity:

(12.4) Theorem. *For each ring R, $K_2(R)$ is the center of $St(R)$.*

Proof. Suppose $c \in Z(St(R))$. Since ϕ is surjective, $\phi(c) \in Z(E(R))$. So $\phi(c)$ commutes with each $e_{ij}(1)$, and hence $\phi(c) = rI_\infty$ for some $r \in R$. But in $GL(R)$, diagonal entries are eventually 1; so $r = 1$, $\phi(c) = I_\infty$, and $c \in K_2(R)$.

Conversely, suppose $d \in K_2(R)$ and m is an integer large enough so that d is a product of generators $x_{ij}(r)$ with $i, j < m$. Denote by P_m the subgroup of $St(R)$ generated by the elements $x_{im}(r)$ with $1 \le i \le m-1$ and $r \in R$. By ST1 and ST2, P_m is an abelian group, and $x_{im}(r)^{-1} = x_{im}(-r)$. So every element of P_m can be written in the form

$$x_{1m}(r_1)x_{2m}(r_2)\cdots x_{m-1,m}(r_{m-1}).$$

Since ϕ of this element is

$$e_{1m}(r_1)e_{2m}(r_2)\cdots e_{m-1,m}(r_{m-1}) \;=\; \begin{bmatrix} 1 & & & & r_1 \\ & 1 & & & r_2 \\ & & \ddots & & \vdots \\ & & & 1 & r_{m-1} \\ & & & & 1 \end{bmatrix}$$

in $E_m(R)$, ϕ is injective on P_m.

If $i \ne j$ and $\{i, j, k\} \subseteq \{1, \ldots, m-1\}$, and $r, s \in R$, then

$$\begin{aligned} x_{ij}(r)x_{km}(s)x_{ij}(r)^{-1} &= [x_{ij}(r), x_{km}(s)]x_{km}(s) \\ &= x_{km}(s) \text{ or } x_{im}(rs)x_{km}(s) \,, \end{aligned}$$

both in P_m. So $x_{ij}(r)P_m x_{ij}(r)^{-1} \subseteq P_m$. Since d is a product of such $x_{ij}(r)$, $dP_m d^{-1} \subseteq P_m$. If $p \in P_m$, so that $dpd^{-1} \in P_m$ too, then

$$\phi(dpd^{-1}) \;=\; \phi(d)\phi(p)\phi(d)^{-1} \;=\; \phi(p) \,.$$

Injectivity of ϕ on P_m then implies $dpd^{-1} = p$. So d commutes with every element of P_m.

By similar argument, d commutes with every element of the subgroup P_{-m} of $St(R)$ generated by those $x_{mi}(r)$ with $1 \le i \le m-1$ and $r \in R$. So if $i \ne j$ and $i, j \in \{1, \ldots, m-1\}$, and $r \in R$, then d commutes with

$$x_{ij}(r) \;=\; [x_{im}(r), x_{mj}(1)] \,.$$

Increasing m by any amount, the same argument works. So d commutes with every generator, and hence every element, of $St(R)$.   ∎

If $f : R \to R'$ is a ring homomorphism, then replacing r by $f(r)$ and s by $f(s)$ in the defining relations ST1 and ST2 for $St(R)$ yields equations that are true in $St(R')$. So $x_{ij}(r) \mapsto x_{ij}(f(r))$ determines a group homomorphism

$$St(f) : St(R) \to St(R') \, .$$

The square

(12.5)

$$
\begin{array}{ccc}
St(R) & \xrightarrow{\ \phi\ } & E(R) \\
{\scriptstyle St(f)}\downarrow & & \downarrow{\scriptstyle E(f)} \\
St(R') & \xrightarrow[\ \phi\ ]{} & E(R')
\end{array}
$$

commutes; so $St(f)$ restricts to a homomorphism between kernels of ϕ,

$$\mathbf{K_2}(f) : K_2(R) \to K_2(R') \, .$$

If $f : R \to R$ is the identity map, $St(f)$ is the identity on $St(R)$, and its restriction $K_2(f)$ is the identity on $K_2(R)$. If $f : R \to R'$ and $g : R' \to R''$ are ring homomorphisms, then

$$St(g \circ f)(x_{ij}(r)) \ = \ x_{ij}(g(f(r))) \ = \ St(g) \circ St(f)(x_{ij}(r)) \, .$$

So $St(g \circ f) = St(g) \circ St(f)$. Restricting to $K_2(R)$, we also have $K_2(g \circ f) = K_2(g) \circ K_2(f)$. This and the commutativity of (12.5) prove:

(12.6) Proposition. *Both St and K_2 are functors from $\mathfrak{Ring}$ to $\mathfrak{Group}$, and ϕ defines a natural transformation from St to K_2. In fact, if $f : R \to R'$ is a ring homomorphism, the diagram*

$$
\begin{array}{ccccccccc}
1 & \longrightarrow & K_2(R) & \xrightarrow{\ i\ } & St(R) & \xrightarrow{\ \phi\ } & GL(R) & \xrightarrow{\ c\ } & K_1(R) & \longrightarrow & 1 \\
& & {\scriptstyle K_2(f)}\downarrow & & {\scriptstyle St(f)}\downarrow & & {\scriptstyle GL(f)}\downarrow & & {\scriptstyle K_1(f)}\downarrow & & \\
1 & \longrightarrow & K_2(R') & \xrightarrow{\ i\ } & St(R') & \xrightarrow{\ \phi\ } & GL(R') & \xrightarrow{\ c\ } & K_1(R') & \longrightarrow & 1
\end{array}
$$

commutes. ∎

Of course, since $K_2(R)$ is abelian, we can also think of K_2 as a functor from $\mathfrak{Ring}$ to the category $\mathcal{Ab}$ of abelian groups (or to the category $\mathbb{Z}\text{-}\mathfrak{Mod}$ of $\mathbb{Z}$ modules).

Like K_0 and K_1, K_2 is symmetric with respect to R:

(12.7) Proposition. *If R is a ring with opposite ring R^{op}, then $K_2(R) \cong K_2(R^{op})$.*

Proof. If G is a group, let G^{op} denote the opposite group, with operation $x \cdot y = yx$. If we replace each $x_{ij}(r)$ by $x_{ji}(r)$, relations ST1 and ST2 become true equations in the group $(St(R^{op}))^{op}$:

$$
\begin{aligned}
x_{ji}(r) \cdot x_{ji}(s) \;&=\; x_{ji}(s)x_{ji}(r) \\
&=\; x_{ji}(s+r) \;=\; x_{ji}(r+s) \;; \\
x_{ji}(r) \cdot x_{\ell k}(s) \cdot x_{ji}(r)^{-1} \cdot x_{\ell k}(s)^{-1} & \\
&=\; [x_{\ell k}(s)^{-1}, x_{ji}(r)^{-1}] \;=\; [x_{\ell k}(-s), x_{ji}(-r)] \\
&=\; \begin{cases} 1 \ \text{ if } \ i \neq \ell, j \neq k \\ x_{\ell i}((-s)*(-r)) \;=\; x_{\ell i}(rs) \ \text{ if } \ i \neq \ell, \ j = k \;. \end{cases}
\end{aligned}
$$

So there is a group antihomomorphism f from $St(R)$ to $St(R^{op})$, taking $x_{ij}(r)$ to $x_{ji}(r)$. Since $(R^{op})^{op} = R$, the map f is evidently its own inverse and so is an antiisomorphism. If $a, b \in St(R)$ with $ab = ba$, then $f(a)f(b) = f(ba) = f(ab) = f(b)f(a)$. So f restricts to an isomorphism between centers: $K_2(R) \cong K_2(R^{op})$. $\blacksquare$

Like K_0 and K_1, K_2 commutes with cartesian products:

(12.8) Proposition. *For rings R and S, there is an isomorphism*

$$ St(R \times S) \;\cong\; St(R) \times St(S) $$

taking each $x_{ij}((r,s))$ to $(x_{ij}(r), x_{ij}(s))$ and restricting to an isomorphism

$$ K_2(R \times S) \;\cong\; K_2(R) \times K_2(S) \;. $$

Proof. The projections of $R \times S$ to R and to S are ring homomorphisms, inducing group homomorphisms

$$
\begin{array}{cc}
\pi_1 : St(R \times S) \to St(R) \;, & \pi_2 : St(R \times S) \to St(S) \;. \\
x_{ij}((r,s)) \mapsto x_{ij}(r) & x_{ij}((r,s)) \mapsto x_{ij}(s)
\end{array}
$$

So $y \mapsto (\pi_1(y), \pi_2(y))$ defines a homomorphism

$$ (\pi_1, \pi_2) : St(R \times S) \to St(R) \times St(S) \;. $$

Replacing $x_{ij}(r) \in St(R)$ by $x_{ij}(r,0)$, or $x_{ij}(s) \in St(S)$ by $x_{ij}(0,s)$, takes ST1 and ST2 to true equations in $St(R \times S)$; so they determine homomorphisms

$$
\begin{array}{cc}
i_1 : St(R) \to St(R \times S) \;, & i_2 : St(S) \to St(R \times S) \;. \\
x_{ij}(r) \mapsto x_{ij}((r,0)) & x_{ij}(s) \mapsto x_{ij}((0,s))
\end{array}
$$

The generators of image(i_1) commute with the generators of image(i_2):

$$x_{ij}((r,0))x_{ij}((0,s)) \;=\; x_{ij}((r,s)) \;=\; x_{ij}((0,s))x_{ij}((r,0)) \;;$$

$$[x_{ij}((r,0)), x_{k\ell}((0,s))] \;=\; 1 \text{ if } i \neq \ell, j \neq k \;;$$
$$[x_{ij}((r,0)), x_{j\ell}((0,s))] \;=\; x_{i\ell}((0,0)) \;=\; 1 \text{ if } i \neq \ell \;;$$
$$[x_{j\ell}((r,0)), x_{ij}((0,s))] \;=\; [x_{ij}((0,s)), x_{j\ell}((r,0))]^{-1}$$
$$=\; x_{i\ell}((0,0))^{-1} \;=\; 1 \text{ if } i \neq \ell \;;$$

and so $x_{ij}((r,0))$ also commutes with

$$x_{ji}((0,s)) = [x_{jk}((0,s)), x_{ki}((0,1))] \;,$$

where $k \notin \{i,j\}$. So $z \mapsto i_1(z)i_2(z)$ defines a group homomorphism

$$(i_1 \cdot i_2) : St(R) \times St(S) \to St(R \times S) \;.$$

Now $(i_1 \cdot i_2) \circ (\pi_1, \pi_2)$ is the identity on $St(R \times S)$, since it takes each $x_{ij}((r,s))$ to itself. And $(\pi_1, \pi_2) \circ (i_1 \cdot i_2)$ is the identity on $St(R) \times St(S)$, since it takes $(x_{ij}(r), x_{ij}(0))$ and $(x_{ij}(0), x_{ij}(s))$ to themselves, and these generate $St(R) \times St(S)$. So (π_1, π_2) is an isomorphism, as required.

The natural transformation ϕ provides the vertical maps in a commutative square

$$
\begin{array}{ccc}
St(R \times S) & \xrightarrow{(\pi_1,\pi_2)} & St(R) \times St(S) \\
\downarrow & & \downarrow \\
E(R \times S) & \longrightarrow & E(R) \times E(S) \;,
\end{array}
$$

where the bottom is an isomorphism $((a_{ij}, b_{ij})) \mapsto ((a_{ij}), (b_{ij}))$. So (π_1, π_2) restricts to an isomorphism of kernels $K_2(R \times S) \to K_2(R) \times K_2(S)$. ∎

The group $K_2(R)$ is determined by the map $\phi : St(R) \to E(R)$. But there is a way to retrieve $K_2(R)$ from $E(R)$ without reference to transvections, as the kernel of a "universal central extension" of $E(R)$.

(12.9) Definition. A **central extension** of a group H is a surjective group homomorphism $\theta : G \to H$ with $\ker(\theta) \subseteq Z(G)$.

There is a category $\mathfrak{Cen}(H)$ whose objects are the central extensions θ of H, and in which an arrow from $\theta_1 : G_1 \to H$ to $\theta_2 : G_2 \to H$ is just a group homomorphism $f : G_1 \to G_2$ making the triangle

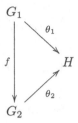

commute. Composite and identity arrows are defined as usual for functions f.

(12.10) Definition. A **universal central extension** of a group H is an initial object in $\mathfrak{Cen}(H)$. That is, it is a central extension $u : U \to H$ for which, for each central extension $\theta : G \to H$, there is one and only one group homomorphism $f : U \to G$ with $\theta \circ f = u$.

(12.11) Theorem. *For each ring R, the map $\phi : St(R) \to E(R)$ is a universal central extension.*

Proof. Given a central extension $\theta : G \to E(R)$, we must show there is a unique homomorphism $f : St(R) \to G$ for which

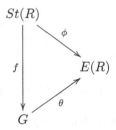

commutes. Since f will be determined by its effect on the generators $x_{ij}(r)$, we just need to prove there is exactly one way to choose $y_{ij}(r)$ from each set $\theta^{-1}(e_{ij}(r))$ satisfying the Steinberg relations

$$y_{ij}(r)y_{ij}(s) \;=\; y_{ij}(r+s) \, ,$$

$$[y_{ij}(r), y_{k\ell}(s)] \;=\; \begin{cases} 1 & \text{if } i \neq \ell \, , \ j \neq k \, , \\ y_{i\ell}(rs) & \text{if } i \neq \ell \, , \ j = k \, . \end{cases}$$

First we prove uniqueness. If there are two such choices $y_{ij}(r)$, $y'_{ij}(r)$, then θ sends both to $e_{ij}(r)$.

Note: If $a, a', b, b' \in G$ and $\theta(a) = \theta(a')$ while $\theta(b) = \theta(b')$, then $u = a^{-1}a'$, $v = b^{-1}b'$ belong to $\ker(\theta)$, and hence to $Z(G)$. So $[a', b'] = [au, bv] = uu^{-1}vv^{-1}[a, b] = [a, b]$. This device is used frequently in the rest of the proof.

Now if $k \neq i$ and $k \neq j$,

$$y_{ij}(r) = [y_{ik}(r), y_{kj}(1)] = [y'_{ik}(r), y'_{kj}(1)] = y'_{ij}(r).$$

So if f exists, it is unique. To prove f exists, we employ a few commutator identities. The last one is the **Jacobi identity**:

(12.12) Lemma. *Suppose G is a group and $a, b, c \in G$.*

    (i) $[ab, c] = [a, [b, c]] [b, c] [a, c]$.
    (ii) *If $[a, b]$ and $[a, c]$ are in $Z(G)$, then a commutes with $[b, c]$.*
    (iii) *If $[G, G]$ is abelian, $[a, [b, c]] [b, [c, a]] [c, [a, b]] = 1$.*

Proof. For (i), expand and simplify. For (ii), $a[b, c]a^{-1} = [aba^{-1}, aca^{-1}] = [[a, b]b, [a, c]c] = [b, c]$. Using (i) the expression in (iii) simplifies to the product $[ab, c] [ca, b] [bc, a] = 1$. ∎

Now, to show f exists, we first make a blind choice of one element $z_{ij}(r)$ from each set $\theta^{-1}(e_{ij}(r))$. One of the Steinberg relations does hold: If $i \neq \ell$, $j \neq k$ and $m \notin \{i, j, k, \ell\}$, then by (12.12) (ii),

$$[z_{ij}(r), z_{k\ell}(s)] = [z_{ij}(r), [z_{km}(s), z_{m\ell}(1)]] = 1.$$

But the other Steinberg relations need not hold for the $z_{ij}(r)$. If there is a choice of $y_{ij}(r) \in \theta^{-1}(e_{ij}(r))$ satisfying the Steinberg relations, we would have $y_{ij}(r) = [y_{im}(r), y_{mj}(1)] = [z_{im}(r), z_{mj}(1)]$ for each m not equal to i or j. So we define

$$y_{ij}^m(r) = [z_{im}(r), z_{mj}(1)] \in \theta^{-1}(e_{ij}(r)).$$

It only remains to prove $y_{ij}^m(r)$ is independent of m and obeys the Steinberg relations. Suppose i, j, m, and n are four different positive integers, and $r \in R$. Taking $u = z_{in}(r)$, $v = z_{nm}(s)$, and $w = z_{mj}(t)$, and G the group generated by u, v, and w, we see by the Steinberg relation verified above that $[u, w] = 1$. So [G,G] is generated by the elements $gxg^{-1} = [g, x]x$ where $g \in G$ and x is $[u, v]$ or $[v, w]$. Now $\theta([u, v]) = e_{im}(rs)$ and $\theta([v, w]) = e_{nj}(st)$; so $[u, v]$ commutes with $[v, w]$. Similarly $[u, v]$ commutes with u, u^{-1}, v, and v^{-1}, and for $i = 1$ or -1, $[w^i, [u, v]]$ commutes with u, v, and w, so lies in $Z(G)$. Taking $c = [u, v], a \in G$, and $b = u^i, v^i$, or w^i in (12.12) (i),

$$[ab, [u, v]] = [b, [u, v]] [a, [u, v]].$$

Repeating this, $[g, [u, v]] \in Z(G)$ for all $g \in G$. Similarly each $[g, [v, w]] \in Z(G)$. So $[G, G]$ is abelian. Replacing a, b, c by u, v, w in the Jacobi identity, and recalling that $[u, w] = 1$, we get

(12.13) $$[u, [v, w]] \;=\; [[u, v], w] \,.$$

With $s = t = 1$, this becomes

$$
\begin{aligned}
y_{ij}^m(r) &= [[z_{in}(r), z_{nm}(1)], z_{mj}(1)] \\
&= [z_{in}(r), [z_{nm}(1), z_{mj}(1)]] \;=\; y_{ij}^n(r) \,.
\end{aligned}
$$

So $y_{ij}^m(r)$ is independent of m, and we rename it $y_{ij}(r)$. Now suppose i, j, k, ℓ are positive integers with $i \neq j, k \neq \ell, i \neq \ell$, and $j \neq k$. Replacing our original choice of $z_{ij}(r)$ by $y_{ij}(r)$, we have seen that

$$[y_{ij}(r), y_{k\ell}(s)] \;=\; 1 \,.$$

By definition of $y_{ij}^m(r)$, we also know

$$[y_{im}(r), y_{mj}(1)] \;=\; y_{ij}(r)$$

whenever $m \neq i$ and $m \neq j$. Suppose i, j, m, and n are four different positive integers. Take u, v, w, as above with $t = 1$. Then by (12.13),

$$
\begin{aligned}
[y_{in}(r), y_{nj}(s)] &= [y_{in}(r), [y_{nm}(s), y_{mj}(1)]] \\
&= [[y_{in}(r), y_{nm}(s)], y_{mj}(1)] \\
&= [y_{im}(rs), y_{mj}(1)] \\
&= y_{ij}(rs) \,.
\end{aligned}
$$

For the remaining relation, if i, j, m are three different positive integers,

$$
\begin{aligned}
y_{ij}(r + s) &= y_{ij}(s + r) \\
&= [y_{im}(s + r), y_{mj}(1)] \\
&= [y_{im}(s) y_{im}(r), y_{mj}(1)] \\
&= [y_{im}(r), y_{mj}(1)] \, [y_{im}(s), y_{mj}(1)] \\
&= y_{ij}(r) y_{ij}(s) \,,
\end{aligned}
$$

where the penultimate equality is (12.12)(i) with $a = y_{im}(s), b = y_{im}(r)$, and $c = y_{mj}(1)$, so that $[a, [b, c]] = [y_{im}(s), y_{ij}(r)] = 1$. This completes the proof that ϕ is a universal central extension. ∎

(12.14) Lemma.

(i) *If $\sigma : \widehat{H} \to H$ is a universal central extension, and $\psi : H \to K$ is an isomorphism, then $\psi \circ \sigma : \widehat{H} \to K$ is also a universal central extension.*

(ii) *If also $\tau : \widehat{K} \to K$ is a universal central extension, there is a unique isomorphism $\widehat{\psi} : \widehat{H} \to \widehat{K}$ making the square*

$$
\begin{array}{ccc}
\widehat{H} & \xrightarrow{\ \sigma\ } & H \\
{\scriptstyle \widehat{\psi}}\downarrow & & \downarrow{\scriptstyle \psi} \\
\widehat{K} & \xrightarrow{\ \tau\ } & K
\end{array}
$$

commute. And $\widehat{\psi}$ restricts to an isomorphism from $\ker(\sigma)$ to $\ker(\tau)$.

Proof. Since ψ is injective, $\psi \circ \sigma$ is a central extension with the same kernel as σ. Suppose $\theta : G \to K$ is also a central extension. Then $\psi^{-1} \circ \theta : G \to H$ is a central extension, and since σ is universal, there is a unique homomorphism $f : \widehat{H} \to G$ with $\psi^{-1} \circ \theta \circ f = \sigma$. That is, there is a unique f with $\theta \circ f = \psi \circ \sigma$; so $\psi \circ \sigma$ is universal.

For part (ii), $\psi \circ \sigma$ and τ are both initial objects in $\mathfrak{Cen}(K)$; so there is a unique isomorphism $\widehat{\psi} : \widehat{H} \to \widehat{K}$ with $\tau \circ \widehat{\psi} = \psi \circ \sigma$. So we have a commutative diagram with exact rows

$$
\begin{array}{ccccccccc}
1 & \longrightarrow & \ker(\sigma) & \longrightarrow & \widehat{H} & \xrightarrow{\ \sigma\ } & H & \longrightarrow & 1 \\
& & & & {\scriptstyle \widehat{\psi}}\downarrow & & \downarrow{\scriptstyle \psi} & & \\
1 & \longrightarrow & \ker(\tau) & \longrightarrow & \widehat{K} & \xrightarrow{\ \tau\ } & K & \longrightarrow & 1
\end{array}
$$

and vertical isomorphisms. Then $\widehat{\psi}$ restricts to an injective homomorphism from $\ker(\sigma)$ to $\ker(\tau)$, which is surjective by a diagram chase. ∎

(12.15) Proposition. *For each ring R and positive integer n, there is an isomorphism $f : K_2(M_n(R)) \to K_2(R)$.*

Proof. By (1.33), erasing internal brackets is an isomorphism of linear groups $GL(M_n(R)) \cong GL(R)$, restricting to an isomorphism between commutator subgroups $E(M_n(R)) \cong E(R)$. By the preceding lemma, there is an isomorphism $St(M_n(R)) \to St(R)$ having the same effect on x_{ij} as bracket erasure has on e_{ij}, and restricting to an isomorphism f from $K_2(M_n(R))$ to $K_2(R)$. ∎

12A. Exercises

1. Suppose $\phi : F \to H$ and $\psi : G \to H$ are group homomorphisms and ψ is surjective. If F is a free group based on a set X, prove there is a group homomorphism $f : F \to G$ with $\psi \circ f = \phi$.

2. A group G is **perfect** if $[G, G] = G$. If there is a universal central extension $\phi : G \to H$, prove H is perfect. *Hint:* If H is not perfect, show G cannot be perfect. If $A = G/[G, G]$, show projection to the first coordinate is a central extension $\pi : H \times A \to H$, but there are two different homomorphisms, f_1 and f_2, from G to $H \times A$ with $\pi \circ f_i = \phi$.

3. If H is a perfect group and $\phi : G \to H$ is a central extension, prove $[G, G]$ is perfect and ϕ restricts to a central extension $[G, G] \to H$. *Hint:* To show $[G, G]$ is contained in $[[G, G], [G, G]]$, choose $x_1, x_2 \in G$; since $\phi(x_i)$ is a product of commutators in H, it equals $\phi(x'_i)$ for some x'_i in $[G, G]$. So $x_i = c_i x'_i$ for an element c_i in the center of G. Prove $[x_1, x_2] = [x'_1, x'_2]$.

4. If $\phi : P \to H$, $\psi : G \to H$ are central extensions and P is perfect, prove there is at most one homomorphism $f : P \to G$ with $\psi \circ f = \phi$. *Hint:* Show any two such homomorphisms must agree on commutators.

5. Prove every perfect group H has a universal central extension. *Hint:* Show there is a free group F with a surjective homomorphism $\phi : F \to H$. Let R denote its kernel. Show $[R, F]$ is a normal subgroup of F contained in R; so ϕ induces a homomorphism

$$\overline{\phi} : F/[R, F] \;\; \to \;\; H \; .$$

Prove $\overline{\phi}$ is a central extension. Since H is perfect, use Exercise 2 to show $[F, F]/[R, F]$ is perfect, and $\overline{\phi}$ restricts to a central extension

$$\overline{\phi} : [F, F]/[R, F] \;\; \to \;\; H \; .$$

This one is universal: For each central extension $\psi : G \to H$ there is at most one homomorphism f from $[F, F]/[R, F]$ to G with $\psi \circ f = \overline{\phi}$ by Exercise 4. To construct such an f, use Exercise 1 to get a group homomorphism $h : F \to G$ with $\psi \circ h = \phi$, prove $h([R, F]) = 1_G$, and restrict the induced homomorphism on $F/[R, F]$ to

$$f \;\; = \;\; \overline{h} : [F, F]/[R, F] \;\; \to \;\; G \; .$$

Note: The kernel of this universal central extension $\overline{\phi}$ is

$$\frac{R \cap [F, F]}{[R, F]} \; ,$$

commonly called the **Schur multiplier** of H, in honor of J. Schur, who introduced universal central extensions in 1904 (see Schur [04]). So $K_2(R)$ is the Schur multiplier of the perfect group $E(R)$.

12B. Elements of $St(R)$ and $K_2(R)$

Again, suppose R is a ring. If $a, b, c, d \in R$ and $i \neq j$ are positive integers, denote by

$$\begin{bmatrix} a & b \\ c & d \end{bmatrix}^{(i,j)}$$

the matrix obtained from the identity $I \in GL(R)$ by replacing the i, i-coordinate with a, the i, j-coordinate with b, the j, i-coordinate with c, and the jj-coordinate with d. There is an invertible matrix P, depending only on i and j, with

$$\begin{bmatrix} a & b \\ c & d \end{bmatrix}^{(i,j)} = P \begin{bmatrix} a & b \\ c & d \end{bmatrix} P^{-1}.$$

(For the exact matrix P and verification of this equation, see Exercise 1 below.) So these matrices multiply like ordinary 2×2 matrices: $A^{(i,j)} B^{(i,j)} = (AB)^{(i,j)}$.

Following Steinberg [62] and Milnor [71], we define some elements of $St(R)$ that ϕ sends to matrices in $E(R)$ of this type.

For positive integers $i \neq j$ and units $a, b \in R^*$, define

$$w_{ij}(a) = x_{ij}(a)x_{ji}(-a^{-1})x_{ij}(a) , \quad \text{and}$$
$$h_{ij}(a) = w_{ij}(a)w_{ij}(-1) .$$

Then

$$\phi(x_{ij}(a)) = \begin{bmatrix} 1 & a \\ 0 & 1 \end{bmatrix}^{(i,j)} ; \quad \text{so}$$

$$\phi(w_{ij}(a)) = \begin{bmatrix} 1 & a \\ 0 & 1 \end{bmatrix}^{(i,j)} \begin{bmatrix} 1 & 0 \\ -a^{-1} & 1 \end{bmatrix}^{(i,j)} \begin{bmatrix} 1 & a \\ 0 & 1 \end{bmatrix}^{(i,j)}$$

$$= \begin{bmatrix} 0 & a \\ -a^{-1} & 0 \end{bmatrix}^{(i,j)} , \quad \text{and}$$

$$\phi(h_{ij}(a)) = \begin{bmatrix} 0 & a \\ -a^{-1} & 0 \end{bmatrix}^{(i,j)} \begin{bmatrix} 0 & -1 \\ 1 & 0 \end{bmatrix}^{(i,j)}$$

$$= \begin{bmatrix} a & 0 \\ 0 & a^{-1} \end{bmatrix}^{(i,j)} .$$

These matrices are the nearest $E(R)$ has to the elementary matrices used in Gauss-Jordan elimination to solve a system of equations: Left multiplication by $\phi(x_{ij}(a)) = e_{ij}(a)$ adds a times the j-row to the i-row. Left multiplication by $\phi(w_{ij}(1))$ multiplies the i-row by -1 and then switches the i-row with the j-row. Left multiplication by $\phi(h_{ij}(a))$ multiplies the i-row by a and the j-row by a^{-1}. When R is commutative, these are the natural adjustments to the Gauss-Jordan operations, to make them determinant-preserving.

(12.16) Lemma. $w_{ij}(a)^{-1} = w_{ij}(-a)$, and hence $h_{ij}(1) = 1$.

Proof.

$$
\begin{aligned}
(x_{ij}(a)x_{ji}(-a^{-1})x_{ij}(a))^{-1} &= x_{ij}(a)^{-1}x_{ji}(-a^{-1})^{-1}x_{ij}(a)^{-1} \\
&= x_{ij}(-a)x_{ji}(-(-a)^{-1})x_{ij}(-a) . \quad \blacksquare
\end{aligned}
$$

(12.17) Definitions. Let W denote the subgroup of $St(R)$ generated by the $w_{ij}(a)$ for positive integers $i \neq j$ and units $a \in R^*$. By the lemma, each element of W is a finite length product of these generators. Let H denote the subgroup of $St(R)$ generated by the $h_{ij}(a)$. By construction, H is a subgroup of W.

The subgroups W and H of $St(R)$ are related by ϕ to some standard subgroups of $GL(R)$. For each positive integer n, there is a homomorphism $\boldsymbol{P}$ from the symmetric group S_n to $GL(R)$, taking σ to the **permutation matrix**

$$
P(\sigma) = P_\sigma = \sum_{i=1}^{n} \epsilon_{\sigma(i)i} ,
$$

since

$$
P(\sigma)P(\tau) = \sum \epsilon_{\sigma(i)i} \sum \epsilon_{\tau(j)j} = \sum \epsilon_{\sigma\tau(j)j} = P(\sigma \circ \tau) .
$$

The image $\boldsymbol{P_n}(R)$ is a subgroup of $GL_n(R)$.

Also, there is a homomorphism $\boldsymbol{\Delta}$ from the n-fold cartesian product of groups $R^* \times \cdots \times R^*$ to $GL_n(R)$, taking $(d_1, \ldots, d_n)$ to the **diagonal matrix**

$$
\Delta(d_1, \ldots, d_n) = \sum_{i=1}^{n} d_i \epsilon_{ii} .
$$

The image $\boldsymbol{D_n}(R)$ is also a subgroup of $GL_n(R)$ and is abelian if R^* is abelian.

As in §9A, Exercise 3, $P_n(R)$ normalizes $D_n(R)$:

$$
\Delta(d_1, \ldots, d_n)P(\sigma) = \sum d_i \epsilon_{ii} \sum \epsilon_{\sigma(j)j}
$$

$$
= \sum d_{\sigma(j)} \epsilon_{\sigma(j)j} = \sum \epsilon_{\sigma(i)i} \sum d_{\sigma(j)} \epsilon_{jj}
$$

$$
= P(\sigma)\Delta(d_{\sigma(1)}, \ldots, d_{\sigma(n)}) .
$$

So $P_n(R)D_n(R)$ is a subgroup of $GL_n(R)$. Its members are the $n \times n$ **monomial matrices**, namely, the $n \times n$ matrices in which each row and each column has exactly one nonzero entry, and these nonzero entries come from R^*.

In $GL(R)$, these subgroups are nested: $P_n(R) \subseteq P_{n+1}(R)$, $D_n(R) \subseteq D_{n+1}(R)$, and their unions

$$P(R) \ = \ \bigcup_{n=1}^{\infty} P_n(R) \ , \quad D(R) \ = \ \bigcup_{n=1}^{\infty} D_n(R)$$

are subgroups of $GL(R)$. The members of $P(R)$ are the matrices obtained from I_∞ by permuting finitely many rows; the matrices in $D(R)$ are the diagonal matrices with diagonal entries from R^*, all but finitely many of which are 1. Then $P(R)$ normalizes $D(R)$, and

$$P(R)D(R) \ = \ \bigcup_{n=1}^{\infty} (P_n(R)D_n(R)) \ ,$$

which is the group of all monomial matrices over R.

Regard each $\sigma \in S_n$ as a permutation of $\mathbb{N} = \{1, 2, 3, \cdots\}$; so $S_1 \subseteq S_2 \subseteq \cdots$. The union S_∞ of these symmetric groups is the group of permutations of $\mathbb{N}$ moving only finitely many elements of $\mathbb{N}$. Since $P(R) \cap D(R) = \{I_\infty\}$, as we see by comparing entries, each monomial matrix has exactly one expression

$$P(\sigma)\Delta(d_1, d_2, \cdots)$$

with $\sigma \in S_\infty$ and $d_i \in R^*$, with $d_i = 1$ for all but finitely many i. Refer to σ as the **underlying permutation** of this monomial matrix. If $\sigma_1, \ldots, \sigma_m \in S_\infty$ and $\Delta_1, \ldots, \Delta_m \in D(R)$, then

$$(P(\sigma_1)\Delta_1)\,(P(\sigma_2)\Delta_2)\,\cdots\,(P(\sigma_m)\Delta_m)$$
$$= P(\sigma_1)\,\cdots\,P(\sigma_m)\Delta' \ = \ P(\sigma_1 \circ \cdots \circ \sigma_m)\,\Delta'$$

for some $\Delta' \in D(R)$. So there is a group homomorphism

$$u : P(R)D(R) \ \rightarrow \ S_\infty$$

taking each monomial matrix to its underlying permutation.

For any positive integers $i \neq j$ and any $a \in A^*$,

$$\phi(w_{ij}(a)) \ = \ \begin{bmatrix} 0 & 1 \\ 1 & 0 \end{bmatrix}^{(i,j)} \begin{bmatrix} -a^{-1} & 0 \\ 0 & a \end{bmatrix}^{(i,j)} ,$$

a monomial matrix. So $\phi(W)$ is a group of monomial matrices:

$$\phi(W) \ \subseteq \ (P(R)D(R)) \cap E(R) \ .$$

(12.18) Definition. Let $\psi : W \to S_\infty$ denote the homomorphism taking each w to the underlying permutation of $\phi(w)$.

Since $\psi(w_{ij}(a))$ is the transposition (i,j), and transpositions generate S_∞, ψ is surjective. Since

$$\phi(h_{ij}(a)) = \begin{bmatrix} a & 0 \\ 0 & a^{-1} \end{bmatrix}^{(i,j)} ,$$

$\phi(H)$ is a group of diagonal matrices:

$$\phi(H) \subseteq D(R) \cap E(R) .$$

So H lies in the kernel of ψ. We say more about this in (12.24) below.

The relations ST1 and ST2 tell us how to move the x_{ij} past each other. For, if we know $[x,y] = z$, then

$$xy = zyx \quad \text{and} \quad yx = z^{-1}xy .$$

Using this, we can learn how to move $w_{k\ell}$ past x_{ij}:

(12.19) Lemma. *Suppose $a \in R^*$ and $b \in R$. Assume $i, j, k,$ and ℓ are four different positive integers. Then*

(i) $w_{k\ell}(a)x_{ij}(b)w_{k\ell}(a)^{-1} = x_{ij}(b)$,

(ii) $w_{ki}(a)x_{ij}(b)w_{ki}(a)^{-1} = x_{kj}(ab)$,

(iii) $w_{kj}(a)x_{ij}(b)w_{kj}(a)^{-1} = x_{ik}(ba^{-1})$,

(iv) $w_{i\ell}(a)x_{ij}(b)w_{i\ell}(a)^{-1} = x_{\ell j}(-a^{-1}b)$,

(v) $w_{j\ell}(a)x_{ij}(b)w_{j\ell}(a)^{-1} = x_{i\ell}(-ba)$,

(vi) $w_{ij}(a)x_{ij}(b)w_{ij}(a)^{-1} = x_{ji}(-a^{-1}ba^{-1})$,

(vii) $w_{ji}(a)x_{ij}(b)w_{ji}(a)^{-1} = x_{ji}(-aba)$.

Proof. For (i) just notice that $x_{ij}(b)$ commutes with $x_{k\ell}(a)$ and $x_{\ell k}(-a^{-1})$. For (ii) we use $xy = [x,y]yx$:

$$
\begin{aligned}
w_{ki}(a)x_{ij}(b) &= x_{ki}(a)\, x_{ik}(-a^{-1})\, x_{ki}(a)\, x_{ij}(b) \\
&= x_{ki}(a)\, x_{ik}(-a^{-1})\, x_{kj}(ab)\, x_{ij}(b)\, x_{ki}(a) \\
&= x_{ki}(a)\, x_{ij}(\!\!\!\diagup b)\, x_{kj}(ab)\, x_{ik}(-a^{-1})\, x_{ij}(\!\!\!\diagup b)\, x_{ki}(a) \\
&= x_{ki}(a)\, x_{kj}(ab)\, x_{ik}(-a^{-1})\, x_{ki}(a) \\
&= x_{kj}(ab)\, w_{ki}(a) ,
\end{aligned}
$$

where, in the fourth equation, we canceled $x_{ij}(-b)$ and $x_{ij}(b)$ because they commute with the factors between them

For (iii) use $yx = [x,y]^{-1}xy$:

$$
\begin{aligned}
w_{kj}(a)\, x_{ij}(b) &= x_{kj}(a)\, x_{jk}(-a^{-1})\, x_{kj}(a)\, x_{ij}(b) \\
&= x_{kj}(a)\, x_{jk}(-a^{-1})\, x_{ij}(b)\, x_{kj}(a) \\
&= x_{kj}(a)\, x_{ik}(ba^{-1})\, x_{ij}(b)\, x_{jk}(-a^{-1})\, x_{kj}(a) \\
&= x_{ij}(-b)\, x_{ik}(ba^{-1})\, x_{kj}(a)\, x_{ij}(b)\, x_{jk}(-a^{-1})\, x_{kj}(a) \\
&= x_{ik}(ba^{-1})\, w_{kj}(a)\ .
\end{aligned}
$$

From (ii) we get (iv):

$$
\begin{aligned}
x_{\ell j}(-a^{-1}b)w_{i\ell}(a) &= (w_{i\ell}(-a)x_{\ell j}(a^{-1}b))^{-1} \\
= (x_{ij}(-b)w_{i\ell}(-a))^{-1} &= w_{i\ell}(a)x_{ij}(b)\ .
\end{aligned}
$$

And using (iii), we get (v):

$$
\begin{aligned}
x_{i\ell}(-ba)w_{j\ell}(a) &= (w_{j\ell}(-a)x_{i\ell}(ba))^{-1} \\
= (x_{ij}(-b)w_{j\ell}(-a))^{-1} &= w_{j\ell}(a)x_{ij}(b)\ .
\end{aligned}
$$

To prove (vi) we employ (iii), (iv), and the identity $z[x,y]z^{-1} = [zxz^{-1}, zyz^{-1}]$:

$$
\begin{aligned}
w_{ij}(a)x_{ij}(b)w_{ij}(a)^{-1} &= w_{ij}(a)[x_{ik}(b), x_{kj}(1)]w_{ij}(a)^{-1} \\
&= [x_{jk}(-a^{-1}b), x_{ki}(a^{-1})] = x_{ji}(-a^{-1}ba^{-1})\ .
\end{aligned}
$$

Similarly for (vii), using (ii) and (v):

$$
\begin{aligned}
w_{ji}(a)x_{ij}(b)w_{ji}(a)^{-1} &= w_{ji}(a)[x_{ik}(b), x_{kj}(1)]w_{ji}(a)^{-1} \\
&= [x_{jk}(ab), x_{ki}(-a)] = x_{ji}(-aba)\ ,
\end{aligned}
$$

completing the proof. ∎

Having done the work to prove these equations, we will not need to remember them — they can be recovered from a simple pattern. If $w \in W$, the lemma does tell us that

$$
wx_{ij}(b)w^{-1} = x_{st}(c)
$$

for some positive integers $s \neq t$ and some $c \in R$. So

$$
\phi(w)e_{ij}(b)\phi(w)^{-1} = e_{st}(c)\ ,
$$

and we can recover s, t, and c by calculating the matrix product on the left side. We know $\phi(w)$ is a monomial matrix $P(\sigma)\Delta(d_1, \ldots, d_n)$ in $GL_n(R)$ for some n. And

$$
\begin{aligned}
\Delta(d_1, \ldots, d_n)e_{ij}(b) &= \sum d_k \epsilon_{kk}(I + b\epsilon_{ij}) \\
&= \left(\sum d_k \epsilon_{kk}\right) + d_i b \epsilon_{ij} \\
&= (I + d_i b d_j^{-1}\epsilon_{ij}) \sum d_k \epsilon_{kk} \\
&= e_{ij}(d_i b d_j^{-1})\Delta(d_1, \ldots, d_n) .
\end{aligned}
$$

(This, by the way, shows $E_n(R)$ normalizes $D_n(R)$.) So

$$
\begin{aligned}
\phi(w)e_{ij}(b)\phi(w)^{-1} &= P_\sigma e_{ij}(d_i b d_j^{-1})P_\sigma^{-1} \\
&= I + d_i b d_j^{-1} P_\sigma \epsilon_{ij} P_{\sigma^{-1}} \\
&= I + d_i b d_j^{-1} \epsilon_{\sigma(i)\sigma(j)} \\
&= e_{\sigma(i)\sigma(j)}(d_i b d_j^{-1}) .
\end{aligned}
$$

This and two short calculations prove:

(12.20) **Proposition.** *Suppose R is a ring, $b \in R$, $a \in R^*$, and $i \neq j$ are positive integers. For $w \in W$, write $\phi(w)$ as $P(\sigma)\Delta(d_1, \ldots, d_n)$ in $P_n(R)D_n(R)$ for some integer $n \geq \max\{i, j\}$. Then*

(i) $wx_{ij}(b)w^{-1} = x_{\sigma(i)\sigma(j)}(d_i b d_j^{-1})$,

(ii) $ww_{ij}(a)w^{-1} = w_{\sigma(i)\sigma(j)}(d_i a d_j^{-1})$,

(iii) $wh_{ij}(a)w^{-1} = h_{\sigma(i)\sigma(j)}(d_i a d_j^{-1})h_{\sigma(i)\sigma(j)}(d_i d_j^{-1})^{-1}$. ∎

The reader is invited to recover the equations in (12.19) by applying part (i) of this proposition.

To identify elements of $K_2(R)$, we consider how the h_{ij} move past each other. For $h, h' \in H$, the diagonal entries of $\phi(h)$ commute with the corresponding diagonal entries of $\phi(h')$ if and only if

$$
\phi([h, h']) = [\phi(h), \phi(h')] = 1 ,
$$

that is, if and only if $[h, h']$ is an element of $K_2(R)$. With the aid of Proposition (12.20), we can compute such commutators:

(12.21) Proposition. *Suppose $a, b \in R^*$ and $ab = ba$. For any three different positive integers $i, j,$ and k,*

$$[h_{ik}(a), h_{ij}(b)] \quad = \quad h_{ij}(ab)h_{ij}(a)^{-1}h_{ij}(b)^{-1}.$$

Proof. In (12.20) (iii), take w to be $h_{ik}(a)$; so the underlying permutation of $\phi(w)$ is $\sigma = 1$, and the diagonal entries $d_1, d_2, \ldots$ have $d_i = a$ and $d_j = 1$. So

$$
\begin{aligned}
h_{ik}(a)h_{ij}(b)h_{ik}(a)^{-1}h_{ij}(b)^{-1} &= (wh_{ij}(b)w^{-1})h_{ij}(b)^{-1} \\
&= h_{ij}(ab)h_{ij}(a)^{-1}h_{ij}(b)^{-1}. \quad \blacksquare
\end{aligned}
$$

Using this computation and similar ones, it is short work to show that, for any positive integers $p \neq q$ and $i \neq j$ with $[h_{pq}(a), h_{ij}(b)] \in K_2(R)$, this commutator can be rewritten in the form 1 or $[h_{ik}(\alpha), h_{ij}(\beta)]$ for commuting units α and β. For example,

$$
\begin{aligned}
[h_{ij}(a), h_{ij}(b)] &= h_{ij}(a)h_{ij}(b)h_{ij}(a)^{-1}h_{ij}(b)^{-1} \\
&= h_{ij}(aba)h_{ij}(aa)^{-1}h_{ij}(b)^{-1} \\
&= h_{ij}(a^2b)h_{ij}(a^2)^{-1}h_{ij}(b)^{-1} \\
&= [h_{ik}(a^2), h_{ij}(b)] .
\end{aligned}
$$

Proposition (12.21) shows the commutator $[h_{ik}(a), h_{ij}(b)]$ is independent of k. Denote it by $\{a, b\}_{ij}$. It is also independent of i and j:

(12.22) Proposition. *If $a, b \in R^*$ with $ab = ba$, and $i \neq j$ are any positive integers, $\{a, b\}_{ij} = \{a, b\}_{12}$.*

Proof. Conjugating a diagonal matrix by

$$
\phi(w_{st}(1)) \quad = \quad \begin{bmatrix} 0 & 1 \\ 1 & 0 \end{bmatrix}^{(s,t)} \begin{bmatrix} -1 & 0 \\ 0 & 1 \end{bmatrix}^{(s,t)}
$$

has the same effect as conjugating by the first factor; so it switches the s, s-entry with the t, t-entry. So, conjugating by an appropriate $w \in W$ permutes the diagonal by any desired σ in S_∞. Choose such a w so that $\sigma(1) = i, \sigma(2) = j$ and $\sigma(3) = k$. Then

$$
\begin{aligned}
\phi(wh_{13}(a)w^{-1}) &= \phi(h_{ik}(a)) , \quad \text{and} \\
\phi(wh_{12}(a)w^{-1}) &= \phi(h_{ij}(a)) .
\end{aligned}
$$

Elements of $St(R)$ with the same image under ϕ differ only by a central factor; so they have the same effect in commutators. So

$$
\begin{aligned}
\{a,b\}_{ij} &= [h_{ik}(a), h_{ij}(b)] \\
&= [wh_{13}(a)w^{-1}, wh_{12}(b)w^{-1}] \\
&= w[h_{13}(a), h_{12}(b)]w^{-1} \\
&= w\{a,b\}_{12}w^{-1} \;=\; \{a,b\}_{12} \,,
\end{aligned}
$$

since $\{a,b\}_{12} \in K_2(R) = Z(St(R))$. ∎

(12.23) Definition. Suppose R is a ring. A **Steinberg symbol** over R is an element

$$
\begin{aligned}
\{a,b\}_R &= \{a,b\}_{12} \\
&= h_{12}(ab)h_{12}(a)^{-1}h_{12}(b)^{-1}
\end{aligned}
$$

of $K_2(R)$, where a and b are commuting units of R. When the choice of ring R is evident from the context, we write $\{a,b\}$ for $\{a,b\}_R$. By the preceding propositions,

$$
\begin{aligned}
\{a,b\} &= [h_{ik}(a), h_{ij}(b)] \\
&= h_{ij}(ab)h_{ij}(a)^{-1}h_{ij}(b)^{-1}
\end{aligned}
$$

whenever i, j, and k are three different positive integers.

Let **SYM** denote the subgroup of $St(R)$ generated by Steinberg symbols. So with arrows indicating containments, we have

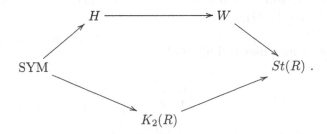

Our next goal is to prove that, when R^* is abelian,

$$
W \cap K_2(R) \;=\; H \cap K_2(R) \;=\; \text{SYM} \,.
$$

Then $K_2(R)$ will be generated by Steinberg symbols exactly when $K_2(R) \subseteq W$. We complete this section by proving this containment for all fields R.

(12.24) Proposition. *The kernel of $\psi : W \to S_\infty$ is H. That is, H is the set of $w \in W$ for which $\phi(w)$ is diagonal.*

Proof. If $w \in W$, then for positive integers $i \neq j$ and a unit $a \in R^*$, $w h_{ij}(a) w^{-1} \in H$ by (12.20) (iii). So $wHw^{-1} \subseteq H$ and H is normal in W.

On W, let "$\equiv$" denote congruence modulo H. Since $h_{ij}(a) = w_{ij}(a) w_{ij}(-1) = w_{ij}(a) w_{ij}(1)^{-1}$, it follows that

$$w_{ij}(a) \quad \equiv \quad w_{ij}(1) \, .$$

Since $w_{ji}(1) w_{ij}(1) w_{ji}(1)^{-1} = w_{ji}(-1) = w_{ji}(1)^{-1}$, we see that

$$w_{ji}(1) \quad \equiv \quad w_{ij}(1)^{-1} \quad = \quad w_{ij}(-1) \quad \equiv \quad w_{ij}(1) \, .$$

So each $w \in W$ is congruent modulo H to a product of factors $w_{ij}(1)$ with $i < j$, and the square of each factor is $\equiv 1$.

Now suppose w is congruent to such a product and $\psi(w) = 1$.

Step 1. Suppose ℓ is the largest subscript involved in this product. If σ is the transposition (i, ℓ), then

$$w_{i\ell}(1) w_{jk}(1) w_{i\ell}(1)^{-1} \quad \equiv \quad w_{\sigma(j)\sigma(k)}(1) \, ; \quad \text{so}$$
$$w_{i\ell}(1) w_{jk}(1) \quad \equiv \quad w_{\sigma(j)\sigma(k)}(1) w_{i\ell}(1) \, .$$

Consider a factor involving ℓ as a subscript to be "out of order" if it is to the left of a factor not involving ℓ. Using the preceding congruence and staying within the same coset modulo H, we can move the right-most factor that is out of order to the right, just until it is no longer out of order. Repeat this until the product consists of a first segment not involving ℓ, followed by a second segment in which every factor involves ℓ.

Step 2. If $i \neq j$ and $i, j < \ell$,

$$w_{i\ell}(1)^2 \quad \equiv \quad w_{i\ell}(1) w_{i\ell}(-1) \quad = \quad 1 \, , \quad \text{and}$$

$$w_{i\ell}(1) w_{j\ell}(1) \quad \equiv \quad w_{ji}(1) w_{i\ell}(1) \, .$$

Staying within the same coset modulo H, move the first factor $w_{i\ell}(1)$ involving ℓ to the right past each $w_{j\ell}(1)$ with $j \neq \ell$, leaving in its wake only factors $w_{ji}(1)$ not involving ℓ, until the factor being moved arrives next to a copy of itself, and these two equal factors drop out. (This must occur; otherwise, we reach a product in which only the last factor $w_{i\ell}(1)$ involves ℓ. Since $\psi(H) = 1$, ψ of the product $= \psi(w) = 1$, but ψ of this product would take i to ℓ.) Repeat this until no remaining factors involve ℓ. The modified product is now shorter than the original one.

Now repeat Steps 1 and 2 (with a new largest subscript ℓ each time) until the entire product vanishes. So $w \equiv 1$ modulo H and $w \in H$. ∎

(12.25) Corollary. *The map ψ, taking $w \in W$ to the underlying permutation of $\phi(w)$, induces an isomorphism of groups $W/H \cong S_\infty$.* ∎

Note: If we restrict indices to be at most n, we obtain subgroups $H_n \subseteq W_n \subseteq St_n(R)$. For $n \geq 3$, the same arguments prove $W_n/H_n \cong S_n$. If R is a field, then, in the language of algebraic groups, W_n/H_n is the **Weyl group** associated to $SL_n(R)$, named for Hermann Weyl. This accounts for the traditional notation W, W_n, and $w_{ij}(a)$.

(12.26) Corollary. *For each ring R, $K_2(R) \cap W = K_2(R) \cap H$.*

Proof. Of course $H \subseteq W$. For the reverse containment, any $w \in W$ with $\phi(w) = I_\infty$ has $\psi(w) = 1$; so $w \in H$. ∎

(12.27) Proposition. *Suppose $i, j,$ and k are three different positive integers and $a \in R^*$. Then*

> (i) $h_{ij}(a) = h_{kj}(a)h_{ki}(a)^{-1}$,
> (ii) $h_{ij}(a) = h_{ji}(a)^{-1}$, and
> (iii) H is generated by the elements $h_{1i}(a)$ for $1 < i$ and $a \in R^*$.

Proof. From (12.20) we have the two identities:

$$h_{kj}(a)w_{ij}(1)h_{kj}(a)^{-1} = w_{ij}(a), \quad \text{and}$$
$$w_{ij}(1)h_{kj}(a)w_{ij}(1)^{-1} = h_{ki}(a)h_{ki}(1)^{-1} = h_{ki}(a) .$$

With these substitutions,

$$\begin{aligned}
h_{ij}(a) &= w_{ij}(a)w_{ij}(-1) \\
&= h_{kj}(a)w_{ij}(1)h_{kj}(a)^{-1}w_{ij}(1)^{-1} \\
&= h_{kj}(a)h_{ki}(a)^{-1},
\end{aligned}$$

proving (i). From this (ii) is immediate. For (iii), if $i \neq 1$,

$$h_{ij}(a) = \begin{cases} h_{1j}(a)h_{1i}(a)^{-1} & \text{if } j \neq 1 , \\ h_{1i}(a)^{-1} & \text{if } j = 1 , \end{cases}$$

by parts (i) and (ii). ∎

(12.28) Theorem. *If R^* is abelian, then $W \cap K_2(R) = SYM$. So $K_2(R)$ is generated by Steinberg symbols if and only if $K_2(R) \subseteq W$.*

Proof. Since SYM lies in the center of $St(R)$, it is normal in H. By the preceding proposition, H/SYM is generated by the cosets $\overline{h}_{1j}(a)$ with $1 < j$ and $a \in R^*$. If $a, b \in R^*$, then $ab = ba$ and

$$\begin{aligned} \{a, b\} &= [h_{1k}(a),\ h_{1j}(b)] \\ &= h_{1j}(ab)h_{1j}(a)^{-1}h_{1j}(b)^{-1}. \end{aligned}$$

So $\overline{h}_{1k}(a)$ commutes with $\overline{h}_{1j}(b)$, and for each j, $h_{1j}(-)$ is a homomorphism from R^* to H/SYM.

Therefore, every element of H/SYM can be written in the form

$$\overline{h}_{12}(a_2)\overline{h}_{13}(a_3)\cdots\overline{h}_{1n}(a_n)$$

for a positive integer n and units $a_2, \ldots, a_n \in R^*$. Because SYM lies in the kernel of ϕ, there is an induced homomorphism

$$\overline{\phi}: \frac{H}{SYM} \to E(R)$$

with $\overline{\phi}(\overline{h}) = \phi(h)$. Since

$$\begin{aligned} \overline{\phi}(\overline{h}_{12}(a_2)\cdots\overline{h}_{1n}(a_n)) &= \phi(h_{12}(a_2)\cdots h_{1n}(a_n)) \\ &= \Delta((a_2\cdots a_n), a_2^{-1}, \ldots, a_n^{-1}, 1, 1, \ldots), \end{aligned}$$

the map $\overline{\phi}$ is injective. If $w \in W \cap K_2(R) = H \cap K_2(R)$, then $\overline{\phi}(\overline{w}) = \phi(w) = I_\infty$; so $\overline{w} = \overline{1}$ and $w \in SYM$. ∎

Let T denote the subgroup of $St(R)$ generated by all $x_{ij}(a)$ with $1 \le i < j$ and $a \in R$.

(12.29) Lemma. *Each element of T can be written as a product in lexicographic order*

$$\begin{aligned} x_{12}(a_{12})x_{13}(a_{13}) &\quad\cdots\quad x_{1n}(a_{1n}) \\ \cdot\ x_{23}(a_{23}) &\quad\cdots\quad x_{2n}(a_{2n}) \\ &\quad\ \ \vdots \\ &\quad\quad \cdot\ x_{n-1,n}(a_{n-1,n}) \end{aligned}$$

with $a_{ij} \in R$, and also as a product in the reverse of this order. So ϕ takes T isomorphically onto the group of upper triangular matrices in $GL(R)$ with 1's on the diagonal.

Note: We need the reverse lex order for the second part of the lemma, but the lex order is so much nicer to write down!

Proof. For each i, let T_i denote the subgroup of $St(R)$ generated by those $x_{ij}(a)$ with $j > i$ and $a \in R$. These generators commute; so T_i is abelian. By ST1, each element of T_i can be written in the form

$$x_{i,i+1}(a_{i,i+1}) \cdots x_{in}(a_{in}) \, ,$$

and in the reverse of this order.

Suppose $i < k < \ell$ and $i < j$. For $a, b \in R$, ST2 implies

$$x_{k\ell}(a)x_{ij}(b) \;=\; \begin{cases} x_{i\ell}(-ba)x_{ij}(b)x_{k\ell}(a), & \text{if } j = k \\ x_{ij}(b)x_{k\ell}(a), & \text{if } j \neq k \, , \end{cases}$$

and

$$x_{ij}(b)x_{k\ell}(a) \;=\; \begin{cases} x_{k\ell}(a)x_{i\ell}(ba)x_{ij}(b), & \text{if } j = k \\ x_{k\ell}(a)x_{ij}(b), & \text{if } j \neq k \, . \end{cases}$$

So if $i < k < \ell$, then $x_{k\ell}(a)T_i = T_i x_{k\ell}(a)$. Therefore $T_k T_i = T_i T_k$ if $i < k$, and hence for all pairs i, k.

Suppose $t \in T$. Then t belongs to some product

$$\begin{aligned} T_{i(1)}T_{i(2)} \cdots T_{i(r)} \;&\subseteq\; T_1 T_2 \cdots T_{n-1} \\ &=\; T_{n-1} \cdots T_2 T_1 \end{aligned}$$

for sufficiently large n. So t has the desired expressions.

The final assertion follows from the application of ϕ to the reverse lex expression of t to get

$$\begin{bmatrix} 1 & a_{12} & a_{13} & \cdots & a_{1n} \\ & 1 & a_{23} & \cdots & a_{2n} \\ & & \ddots & \ddots & \vdots \\ & & & \ddots & a_{n-1,n} \\ & & & & 1 \end{bmatrix} \oplus I_\infty \, . \qquad \blacksquare$$

(12.30) Theorem. *If F is a field, $K_2(F) \subseteq W$; so $K_2(F)$ is generated by Steinberg symbols.*

Proof. By ST2, the group $St(F)$ is generated by those $x_{ij}(a)$ with $j = i \pm 1$. Since

$$x_{ji}(a) \;=\; w_{ij}(1)x_{ij}(-a)w_{ij}(1)^{-1},$$

it is also generated by the elements $x_{i,i+1}(a)$ and $w_{i,i+1}(1)$.

Claim. $St(F) = TWT$.

To prove the claim, it is enough to show that TWT contains the elements $w_{i,i+1}(1)$, for it certainly contains the elements $x_{i,i+1}(a) \in T$. Since $1 \in TWT$, it is enough to prove

$$TWTw_{i,i+1}(1) \subseteq TWT .$$

Say $j = i + 1$, and suppose $t_1, t_2 \in T$ and $w \in W$, so that $t_1 w t_2 \in TWT$. As in the proof of the preceding lemma, the element t_2 of T can be written $x_{ij}(a)t_3$, where t_3 is a product of factors $x_{k\ell}(b)$ with $k < \ell$ and $(k, \ell) \neq (i, j)$. So if σ is the transposition (i, j), then in every case, $\sigma(k) < \sigma(\ell)$; so

$$t_4 \;=\; w_{ij}(-1)t_3 w_{ij}(-1)^{-1} \in T .$$

Now

$$
\begin{aligned}
t_1 w t_2 w_{ij}(1) \;&=\; t_1 w x_{ij}(a) t_3 w_{ij}(1) \\
&=\; t_1 w x_{ij}(a) w_{ij}(1) t_4 .
\end{aligned}
$$

So it is enough to prove the middle $w x_{ij}(a) w_{ij}(1)$ belongs to TWT. Say $\sigma = \psi(w)$.

Case 1. If $\sigma(i) < \sigma(j)$, then $w x_{ij}(a) w_{ij}(1) = x_{\sigma(i)\sigma(j)}(a') w w_{ij}(1)$ belongs to $TW \subseteq TWT$.

Case 2. If $\sigma(i) > \sigma(j)$ and $a \in F^*$, then

$$x_{ij}(a) \;=\; w_{ij}(a)x_{ij}(-a)x_{ji}(a^{-1}) ,$$

and substituting this,

$$
\begin{aligned}
w x_{ij}(a) w_{ij}(1) \;&=\; (w w_{ij}(a)x_{ij}(-a))(x_{ji}(a^{-1})w_{ij}(1)) \\
&=\; x_{\sigma(j)\sigma(i)}(a') w w_{ij}(a) w_{ij}(1) x_{ij}(-a^{-1}) ,
\end{aligned}
$$

which belongs to TWT.

Case 3. If $a \notin F^*$, then $a = 0$ because F is a field. So

$$w x_{ij}(a) w_{ij}(1) \;=\; w w_{ij}(1) \;\in\; W \;\subseteq\; TWT ,$$

completing the proof of the claim.

Now suppose $x \in K_2(F)$. By the claim, $x = t_1 w t_2$ with $t_1, t_2 \in T$ and $w \in W$. So

$$\phi(w) \;=\; \phi(t_1^{-1}xt_2^{-1}) \;=\; \phi(t_1^{-1})\phi(t_2^{-1}) \;=\; \phi(t_1^{-1}t_2^{-1})$$

is both monomial and upper triangular with a diagonal of $1's$; so it equals I_∞. Then w lies in the center of $St(F)$, and since ϕ is injective on T, $t_2 = t_1^{-1}$. Thus

$$x \;=\; t_1 w t_1^{-1} \;=\; w \;\in\; W . \qquad \blacksquare$$

For each ring R, a generating set for $K_2(R)$ is a set of relators supplementing ST1 and ST2 in a presentation of the group $E(R)$. But there are a lot of Steinberg symbols, one for each ordered pair (a, b) of commuting units in R. Are they all really necessary in an efficient presentation of $E(R)$?

The next theorem shows they are not. And it begins a list of defining relations among Steinberg symbols in a presentation of the abelian group $\mathrm{SYM} \subseteq K_2(R)$.

(12.31) Theorem. *Suppose $a, b,$ and c are commuting units of a ring R. Then in $K_2(R)$,*

  (i) $\{a, b\}^{-1}$ = $\{b, a\}$,
  (ii) $\{ab, c\}$ = $\{a, c\}\{b, c\}$,
  (iii) $\{a, bc\}$ = $\{a, b\}\{a, c\}$, *and*
  (iv) $\{a, b\}$ = 1 *if* $a + b = 1_R$.

Proof.   Part (i) is a commutator identity:

$$\{a, b\}^{-1} = [h_{13}(a), h_{12}(b)]^{-1}$$
$$= [h_{12}(b), h_{13}(a)] = \{b, a\}.$$

Part (ii) also relies on a commutator identity:

$$[xy, z] = [x, [y, z]]\,[y, z]\,[x, z].$$

Since $ab = ba$,

$$\phi(h_{13}(ab)) = \begin{bmatrix} ab & 0 \\ 0 & (ab)^{-1} \end{bmatrix}^{(1,3)} = \begin{bmatrix} ab & 0 \\ 0 & a^{-1}b^{-1} \end{bmatrix}^{(1,3)}$$

$$= \phi(h_{13}(a))\phi(h_{13}(b)) = \phi(h_{13}(a)h_{13}(b)).$$

Because elements with the same image under ϕ differ by a central factor,

$$\{ab, c\} = [h_{13}(ab), h_{12}(c)]$$
$$= [h_{13}(a)h_{13}(b), h_{12}(c)]$$
$$= [h_{13}(a), \{b, c\}]\{b, c\}\{a, c\}$$
$$= \{a, c\}\{b, c\}\,,$$

proving (ii). For multiplicativity in the second coordinate (iii), combine relations (i) and (ii).

Now suppose $a + b = 1_R$. Other equivalent versions of this equation include $ab(a^{-1} + b^{-1}) = 1$ and $ab = a - a^2 = b - b^2$. If $u \in R^*$, computing $w_{ij}(u)w_{ij}(u)w_{ij}(u)^{-1}$ in two ways yields

(12.32) $$w_{ij}(u) = w_{ji}(-u^{-1})\,.$$

So

$$h_{12}(b)h_{12}(a)w_{12}(1) = w_{12}(b)w_{12}(-1)w_{12}(a)$$
$$= w_{12}(b)w_{21}(1)w_{12}(a)$$
$$= w_{12}(b)x_{21}(1)x_{12}(-1)x_{21}(1)w_{12}(a)\,;$$

and by (12.20) applied to the first two and last two factors, this equals

$$x_{12}(-b^2)w_{12}(b)x_{12}(-1)w_{12}(a)x_{12}(-a^2)$$
$$= x_{12}(b-b^2)x_{21}(-b^{-1})x_{12}(b-1+a)x_{21}(-a^{-1})x_{12}(a-a^2)$$
$$= x_{12}(ab)x_{21}(-b^{-1})x_{12}(0)x_{21}(-a^{-1})x_{12}(ab)$$
$$= x_{12}(ab)x_{21}(-(b^{-1}+a^{-1}))x_{12}(ab)$$
$$= x_{12}(ab)x_{21}(-(ab)^{-1})x_{12}(ab)$$
$$= w_{12}(ab) = h_{12}(ab)w_{12}(1) \ .$$

Canceling $w_{12}(1)$,

$$\{a,b\} = h_{12}(ab)h_{12}(a)^{-1}h_{12}(b)^{-1} = 1 \ . \qquad \blacksquare$$

(12.33) Corollary. *If R is a finite field, $K_2(R) = 1$.*

Proof. Say $R = \mathbb{F}_q$ is a field with q elements. Then $\mathbb{F}_q^*$ is a cyclic group of order $q - 1$, generated by an element v. Then $\{v^n, v^m\} = \{v, v^m\}^n = \{v, v\}^{mn}$; so $K_2(\mathbb{F}_q)$ is cyclic, generated by $\{v, v\}$. Since $\{v, v\}^{-1} = \{v, v\}$ by antisymmetry, $\{v, v\}^2 = 1$. It only remains to prove $\{v, v\}^n = 1$ for an odd integer n.

If q is even, $\{v, v\}^{q-1} = \{v^{q-1}, v\} = \{1, v\} = 1$, the last equation because $\{1, v\}^2 = \{1^2, v\} = \{1, v\}$. Suppose instead that q is odd. Exactly half of the cyclic group $\mathbb{F}_q^*$ consists of squares: $1, v^2, v^4, \ldots, v^{q-3}$. Deleting 1, the set $\mathbb{F}_q^* - \{1\}$ has more nonsquares than squares. The function

$$f : \mathbb{F}_q^* - \{1\} \to \mathbb{F}_q^* - \{1\} \ , \quad x \longmapsto 1 - x$$

is its own inverse; so it is bijective. So for some nonsquare u, $f(u) = 1 - u$ is also a nonsquare. Then $u = v^r$ and $1 - u = v^s$ for odd integers r and s; so

$$1 = \{u, 1-u\} = \{v^r, v^s\} = \{v, v\}^{rs},$$

and rs is odd. $\qquad \blacksquare$

12B. Exercises

1. For positive integers i and j, let $P(i, j)$ denote the matrix

$$I - \epsilon_{ii} - \epsilon_{jj} + \epsilon_{ij} + \epsilon_{ji}$$

obtained from I by switching the i-row with the j-row. If $j \neq 1$, let Q_{ij} denote $P(2, j)P(1, i)$; but if $j = 1$, let $Q_{ij} = Q_{i1}$ denote $P(2, i)P(1, i)$. If $i \neq j$, show

$$Q_{ij} \, \epsilon_{ii} \, Q_{ij}^{-1} = \epsilon_{11} \ , \quad Q_{ij} \, \epsilon_{ij} \, Q_{ij}^{-1} = \epsilon_{12} \ ,$$
$$Q_{ij} \, \epsilon_{ji} \, Q_{ij}^{-1} = \epsilon_{21} \ , \quad Q_{ij} \, \epsilon_{jj} \, Q_{ij}^{-1} = \epsilon_{22} \ ,$$

and hence

$$Q_{ij} \begin{bmatrix} a & b \\ c & d \end{bmatrix}^{(i,j)} Q_{ij}^{-1} = \begin{bmatrix} a & b \\ c & d \end{bmatrix}^{(1,2)} .$$

2. As in the note following (12.25), let W_n denote the subgroup of $St_n(R)$ generated by the $w_{ij}(a)$ with $i \neq j$, $1 \leq i, j \leq n$, and $a \in R^*$, and H_n denote the subgroup generated by the $h_{ij}(a)$ under the same conditions on i, j, and a. Suppose R is a commutative ring. Prove

    (i) $\phi_n(H_n)$ is a normal subgroup of $P_n(R)D_n(R)$.
    (ii) Modulo $\phi_n(H_n)$, elements of $P_n(R)$ commute with elements of $D_n(R)$.
    (iii) If $\mathrm{sgn}(\sigma) = \det P(\sigma)$, prove the map $S_n \to \phi_n(W_n)/\phi_n(H_n)$, taking σ to the coset of $P(\sigma)\Delta$ $(\mathrm{sgn}(\sigma), 1, \ldots, 1)$, is a group isomorphism.

3. Suppose R is a ring and the kernel of the stabilization map $R^* \to K_1(R)$ is the commutator subgroup $[R^*, R^*]$. Prove $\phi(W) = (P(R)D(R)) \cap E(R)$ and $\phi(H) = D(R) \cap E(R)$.

4. If $i \neq j$ and $p \neq q$ are positive integers, and a, b are commuting units of a ring R, prove there is a positive integer k distinct from i and j, and there are commuting units α, β in R with

$$[h_{pq}(a), h_{ij}(b)] = [h_{ik}(\alpha), h_{ij}(\beta)] .$$

5. Show the relator in $St(R)$ corresponding to the relation

$$\mathrm{EL3}: \qquad (e_{12}(1)e_{21}(-1)e_{12}(1))^4 = 1 ,$$

mentioned at the beginning of §12A, is the Steinberg symbol $\{-1, -1\}$. Since every ring R has -1 as a unit, this symbol is defined over every ring. And $\{-1, -1\}_R = 1$ if and only if EL3 is a consequence of EL1 and EL2 over R. Also prove $\{-1, -1\}_R^2 = 1$; and if -1 is a square in R, prove $\{-1, -1\}_R = 1$.

6. Use the theorem (12.2) of Nielsen and Magnus to prove $K_2(\mathbb{Z})$ is generated by the Steinberg symbol $\{-1, -1\}$ and so is cyclic of order 1 or 2.

7. If $f : R \to R'$ is a ring homomorphism and a, b are commuting units of R, show $K_2(f)$ takes $\{a, b\}_R$ to $\{f(a), f(b)\}_{R'}$.

8. Suppose R is a ring and $a \in R^*$. In $K_2(R)$, prove $\{a, -a\} = 1$.

9. This exercise develops Herstein's simplified proof of Wedderburn's Theorem, that every finite division ring is a field (see Herstein [75]).

    (i) Suppose D is a finite division ring with center F. Show F is a field. If F has q elements, show D has q^n elements for some integer $n \geq 1$. *Hint*: D is an F-vector space.
    (ii) Suppose $a \in D^*$ but $a \notin F^*$. Let $C(a)$ denote the set of elements in D that commute with a under multiplication. Prove $C(a)$ is a

division ring containing F in its center and so has $q^{m(a)}$ elements for an integer $m(a)$ with $n > m(a) \geq 1$.

(iii) By Lagrange's Theorem applied to the unit groups, $q^{m(a)} - 1$ divides $q^n - 1$. Prove $m(a)$ divides n. *Hint:* Use the division algorithm in $Q[x]$ to divide $x^{m(a)} - 1$ into $x^n - 1$, and show the quotient and remainder lie in $\mathbb{Z}[x]$. Now evaluate at $x = q$.

(iv) Show the Class Equation for D^* has the form

$$q^n - 1 \;=\; (q - 1) \;+\; \sum \frac{q^n - 1}{q^{m(a)} - 1} \,,$$

where the sum is over one a from each nontrivial conjugacy class, and each $m(a)$ is a proper divisor of n.

(v) The roots of $x^n - 1$ in the field $\mathbb{C}$ of complex numbers form a group of order n under multiplication. The **cyclotomic polynomial** $\Phi_n(x)$ is the product of all $x - \zeta$, where $\zeta \in \mathbb{C}^*$ has order n. Therefore

$$x^n - 1 \;=\; \prod_{d \mid n} \Phi_d(x) \,.$$

Prove, by induction on n, that $\Phi_n(x)$ is a monic polynomial in $\mathbb{Z}[x]$.

(vi) Using (iv), show $\Phi_n(q)$ divides $q - 1$, but that $\Phi_n(q) > q - 1$ unless $n = 1$. *Hint:* Each element $\zeta \in \mathbb{C}^*$ of order n lies on the unit circle and q is an integer > 1.

(vii) Conclude that $D = F$, a field.

10. In view of Wedderburn's Theorem (in the preceding problem), prove $K_2(R) = 1$ if R is a finite semisimple ring. Then prove $K_2((\mathbb{Z}/n\mathbb{Z})G) = 1$ if G is a finite group and n is a square-free integer relatively prime to the order of G. *Hint:* For the first part, use properties of K_2 given in §12A and the Wedderburn-Artin Structure Theorem for semisimple rings (8.28). For the second part, extend the Chinese Remainder Theorem, and use Maschke's Theorem (see (8.16) and (8.23)).

13

Exact Sequences

The functors K_0, K_1, and K_2 from rings to abelian groups are similar in properties and related by their algebraic constructions, but they are really woven together in a single tapestry of algebraic K-theory by three exact sequences: the relative sequence, the Mayer-Vietoris sequence, and the localization sequence. Each of these sequences offers considerable computational power — a group can be determined in many cases by the groups and maps before and after it in the sequence. This general principle was first formulated in homological algebra, but it yields striking results in K-theory.

In §13A we develop the relative sequence (13.20), connecting the maps $K_n(c)$ where $c : R \to R/J$ is reduction modulo an ideal J. The bridges between these maps, for different n, are the relative K-groups $K_n(R, J)$. We discussed $K_1(R, J)$ in (11.4), and define $K_0(R, J)$ in (13.15) and $K_2(R, J)$ in (13.17). On the way to the relative sequence we consider fiber squares of rings, which are used to construct f.g. projective modules over certain subrings of direct products of rings.

Section 13B is devoted to the Mayer-Vietoris sequence (13.33), which is derived from relative sequences using their naturality, and an "excision," altering a relative K-group by changing R without changing J. Applications to the K-theory of the integral group ring of a finite cyclic group are given in (13.34) and (13.35).

Finally, the localization sequence (13.38) in §13C connects the maps $K_n(f)$ where $f : A \to S^{-1}A$ is a localization map of a ring A, inverting a submonoid S of central elements that are not zero-divisors in A. The classical $K_1 - K_0$ sequence is given in (13.38), and its extension to higher K-groups is described in (13.39). An application in (13.40) is known as the "Fundamental Theorem of Algebraic K-Theory." For left regular rings R, it says that K_n of the Laurent polynomial ring $R[t, t^{-1}]$ is isomorphic to $K_n(R) \oplus K_{n-1}(R)$. In the discussion prior to (13.39) we also describe the negative algebraic K-groups $K_{-n}(R)$ defined by Bass.

13A. The Relative Sequence

The most pervasive influence on the development of algebraic K-theory has been the algebraic topology developed in the 1940s and 1950s. In 1952, the text by Eilenberg and Steenrod [52] united several homology/cohomology theories of topological spaces within one axiomatic framework. Such theories assign to each pair (X, A) of spaces $A \subseteq X$ a sequence of abelian groups $H_n(X, A)$ (or $H^n(X, A)$ for cohomology), so that each H_n is a functor on the category of pairs and continuous maps between them. If $A = \emptyset$, $H_n(X, A)$ is just written $H_n(X)$. Among the axioms is the existence of a natural long exact sequence

$$\cdots \longrightarrow H_n(A) \longrightarrow H_n(X) \longrightarrow H_n(X, A) \longrightarrow H_{n-1}(A) \longrightarrow \cdots$$

of group homomorphisms.

In the 1960s Adams, Atiyah, and Hirzebruch extended Grothendieck's $K_0(X)$ for X a scheme to a topological K-theory $K_n(X, A)$ satisfying all but one of the Eilenberg-Steenrod axioms; and Serre, Bass, Swan, and others translated it into a parallel algebraic K-theory of rings, replacing X by the ring $C(X)$ of continuous functions $X \to \mathbb{R}$ or $\mathbb{C}$, and later by an arbitrary ring R. Essential to this parallel is the existence of long exact sequences like those in homology theory. The algebraic K-theory version of the long exact sequence above is the **relative sequence**

$$\cdots \longrightarrow K_n(R) \xrightarrow{K_n(c)} K_n(R/J) \longrightarrow K_{n-1}(R, J) \longrightarrow K_{n-1}(R) \xrightarrow{K_{n-1}(c)} \cdots ,$$

where J is an ideal of the ring R and c is the canonical map $R \to R/J$.

We have already seen the relative group $K_1(R, J)$ in connection with the congruence subgroup problem (see (11.4)). We now develop the relative groups $K_0(R, J)$ and $K_2(R, J)$ following the work of Swan, Stein, and Keune, and construct the relative exact sequence for $n = 2, 1, 0$.

Our first relative K-group $K_1(R, J)$ was characterized in (11.9) in terms of the subring

$$D = R \times_J R = \{(x, y) \in R \times R : \overline{x} = \overline{y} \text{ in } R/J\}$$

of $R \times R$, as the kernel of the map $K_1(D) \to K_1(R)$ induced by projection to the second coordinate. A similar construction yields other relative groups.

The ring D is called the "double of R along J." If π_j $(j = 1, 2)$ denotes projection to the j-coordinate $D \to R$, and $c : R \to R/J$ is the canonical map, the square

$$
\begin{array}{ccc}
D & \xrightarrow{\pi_2} & R \\
{\scriptstyle \pi_1} \downarrow & & \downarrow {\scriptstyle c} \\
R & \xrightarrow{c} & R/J
\end{array}
$$

commutes; so π_1 carries $\ker(\pi_2)$ into $\ker(c)$. Since $\ker(\pi_1) \cap \ker(\pi_2) = \{0_D\}$, π_1 is injective on $\ker(\pi_2)$. And since $\pi_1((r, 0)) = r$ for $r \in J = \ker(c)$, $\pi_1(\ker(\pi_2)) = \ker(c)$. So π_1 restricts to an additive, multiplicative bijection from $\ker(\pi_2)$ to J, and we can regard J as an ideal of two rings, D and R.

Now suppose F is a functor from $\mathfrak{Ring}$ to $\mathfrak{Group}$. Define $\boldsymbol{F^S(R, J)}$, $\boldsymbol{F'(R, J)}$, and π_1^* by exactness of the rows and commutativity of the diagram:

$$
\begin{array}{ccccccc}
1 & \longrightarrow & F^S(R, J) & \overset{\subseteq}{\longrightarrow} & F(D) & \overset{F(\pi_2)}{\longrightarrow} & F(R) \\
 & & \Big\downarrow{\pi_1^*} & & \Big\downarrow{F(\pi_1)} & & \Big\downarrow{F(c)} \\
1 & \longrightarrow & F'(R, J) & \underset{i}{\overset{\subseteq}{\longrightarrow}} & F(R) & \underset{F(c)}{\longrightarrow} & F(R/J) \ .
\end{array}
$$

Here π_1^* is just the restriction of $F(\pi_1)$ to the kernels of $F(\pi_2)$ and $F(c)$, and carries the first kernel into the second because the right square commutes.

Following π_1^* by the inclusion i into $F(R)$ defines the first map in a sequence

$$
F^S(R, J) \overset{i\pi_1^*}{\longrightarrow} F(R) \overset{F(c)}{\longrightarrow} F(R/J)
$$

with composite zero. This sequence is exact if and only if π_1^* is surjective. For some functors F, π_1^* will always be surjective; we will demonstrate this property when $F = K_0, K_1$, and K_2, and then define connecting homomorphisms

$$
\partial \ : \ K_{n+1}(R/J) \longrightarrow K_n^S(R, J) \quad (n = 0, 1)
$$

to create a nine-term exact sequence involving all three functors.

The maps $i\pi_1^*$ and $F(c)$ define natural transformations: There is a category $\mathfrak{Ideal}$ with objects (R, J), where R is a ring and J is an ideal of R, and arrows $(R, J) \to (R', J')$ that are ring homomorphisms $R \to R'$ carrying J into J'. Such an arrow f induces a ring homomorphism $\widehat{f}$ from $D = R \times_J R$ to $D' = R' \times_{J'} R'$ with $\widehat{f}((r_1, r_2)) = (f(r_1), f(r_2))$. Then $f \circ \pi_2 = \pi_2 \circ \widehat{f}$; so applying F yields a commutative square

$$
\begin{array}{ccc}
F(D) & \overset{F(\pi_2)}{\longrightarrow} & F(R) \\
\Big\downarrow{F(\widehat{f})} & & \Big\downarrow{F(f)} \\
F(D') & \overset{F(\pi_2)}{\longrightarrow} & F(R') \ ,
\end{array}
$$

and we define $F^S(f)$ to be the restriction of $F(\widehat{f})$ to kernels:

$$
F^S(f) \ : \ F^S(R, J) \longrightarrow F^S(R', J') \ .
$$

Since F preserves composites and identities, so does F^S. The functor $F^S : \mathfrak{Ideal} \to \mathfrak{Group}$ is known as the **Stein relativization** of F, after M. Stein, who developed it in Stein [71].

The arrow $f : (R, J) \to (R', J')$ also induces a homomorphism of quotient rings $\bar{f} : R/J \to R'/J'$. Taking f to $F(f)$ and $F(\bar{f})$ also defines functors from $\mathfrak{Ideal}$ to $\mathfrak{Group}$, and the diagram

$$
\textbf{(13.1)} \qquad
\begin{array}{ccccc}
F^S(R, J) & \xrightarrow{\ i\pi_1^*\ } & F(R) & \xrightarrow{\ F(c)\ } & F(R/J) \\
\ {\scriptstyle F^S(f)}\downarrow & & \ {\scriptstyle F(f)}\downarrow & & \ {\scriptstyle F(\bar{f})}\downarrow \\
F^S(R', J') & \xrightarrow{\ i\pi_1^*\ } & F(R') & \xrightarrow{\ F(c)\ } & F(R'/J')
\end{array}
$$

commutes, since the left square restricts F of a commutative square

$$
\begin{array}{ccc}
D & \xrightarrow{\ \pi_1\ } & R \\
{\scriptstyle \hat{f}}\downarrow & & \downarrow{\scriptstyle f} \\
D' & \xrightarrow{\ \pi_1\ } & R'
\end{array}
$$

and the right square is F of the commutative square defining $\bar{f}$. So $i\pi_1^*$ and $F(c)$ are natural, as claimed.

In Chapter 11 we already considered the cases $F = GL, E$, and K_1. The exact sequences (11.7) and (11.9) prove the inclusions and induced map

$$
\begin{aligned}
GL(R, J) &\subseteq GL(R) \\
E(R, J) &\subseteq E(R) \\
K_1(R, J) &\to K_1(R)
\end{aligned}
$$

factor as isomorphisms

$$
\begin{aligned}
GL(R, J) &\cong GL^S(R, J), && X \mapsto (X, I), \\
E(R, J) &\cong E^S(R, J), && X \mapsto (X, I), \\
K_1(R, J) &\cong K_1^S(R, J), && XE(R, J) \mapsto (X, I)E(D),
\end{aligned}
$$

followed by $i\pi_1^*$ taking (X, I) to X in the first two cases and $(X, I)E(D)$ to $XE(R)$ in the third.

(13.2) Lemma. *There is a natural exact sequence*

$$
K_1(R, J) \to K_1(R) \to K_1(R/J)
$$

with the first map induced by inclusion of $GL(R, J)$ in $GL(R)$ and the second by entrywise reduction mod J.

Proof. The composite is zero even on the GL level. If $A \in GL(R)$ and $AE(R)$ goes to 1 in $K_1(R/J)$, then $\overline{A} \in E(R/J)$. Since $E(R) \to E(R/J)$ is surjective,

$\overline{A} = \overline{E}$ for some $E \in E(R)$. Then $\overline{AE^{-1}} = \overline{A}\,\overline{E}^{-1} = I$ in $GL(R/J)$. So $AE^{-1} \in GL(R, J)$, and the first map takes its coset to $AE^{-1}E(R) = AE(R)$, proving exactness.

If $f : (R, J) \to (R', J')$ is an arrow in $\mathfrak{Ideal}$, entrywise application of f induces the left vertical map in a commutative diagram

$$
\begin{array}{ccccc}
K_1(R, J) & \longrightarrow & K_1(R) & \longrightarrow & K_1(R/J) \\
\downarrow & & \downarrow{\scriptstyle K_1(f)} & & \downarrow{\scriptstyle K_1(\bar f)} \\
K_1(R', J') & \longrightarrow & K_1(R') & \longrightarrow & K_1(R'/J') \,,
\end{array}
$$

proving naturality. ∎

In this way, the relative K_1 becomes a functor on $\mathfrak{Ideal}$, and the isomorphism to K_1^S defined above is natural. So commutativity of this diagram also follows from that of (13.1).

Note: If we restrict ourselves to commutative rings, the preceding discussion carries over verbatim when GL is replaced by SL and K_1 by SK_1, to prove there is a natural exact sequence

$$SK_1(R, J) \to SK_1(R) \to SK_1(R/J)$$

restricting that in (13.2). The reader who is only interested in the $K_1 - K_2$ parts of the K-theory exact sequences will still need to read (13.3) and (13.4) (i)–(vii) below, but may then skip (13.6) to (13.16).

To prove π_1^* carries $K_0^S(R, J)$ onto $K_0'(R, J)$, we will use a description due to Milnor [71] of the f.g. projective modules over rings like D, following a refinement of Milnor's results in §42 of Curtis and Reiner [87]. This somewhat more general approach paves the way for "Mayer-Vietoris sequences" in §13B and provides a class of examples of nonfree projective modules (see Corollary (13.14)).

(13.3) Definition. Suppose C is a subcategory of R-$\mathfrak{Mod}$ or $\mathfrak{Ring}$. A square of arrows in $\mathcal{C}$

$$
\begin{array}{ccc}
A & \overset{f_2}{\longrightarrow} & R_2 \\
{\scriptstyle f_1}\downarrow & & \downarrow{\scriptstyle g_2} \\
R_1 & \underset{g_1}{\longrightarrow} & R'
\end{array}
$$

is a **fiber square** if it commutes and, for each pair $(r_1, r_2) \in R_1 \times R_2$ with $g_1(r_1) = g_2(r_2)$, there is exactly one $a \in A$ with both $f_1(a) = r_1$ and $f_2(a) = r_2$.

Equivalently, the above square is a fiber square if the sequence of additive groups

$$0 \longrightarrow A \xrightarrow{\begin{bmatrix} f_1 \\ f_2 \end{bmatrix}} R_1 \oplus R_2 \xrightarrow{[g_1 \ -g_2]} R'$$

is exact.

Since such a fiber square commutes, f_1 carries $\ker(f_2)$ into $\ker(g_1)$. Since only $0 \in A$ goes to $(0,0) \in R_1 \oplus R_2$, $\ker(f_1) \cap \ker(f_2) = \{0\}$ and f_1 is injective on $\ker(f_2)$. If $r_1 \in R_1$ goes to $0 \in R'$, then it has the same image in R' as $0 \in R_2$. So there exists $a \in \ker(f_2)$ with $f_1(a) = r_1$. Altogether, f_1 restricts to an additive, multiplicative bijection from $\ker(f_2)$ to $\ker(g_1)$; so $\ker(g_1)$ can be regarded as an ideal of both R_1 and A.

(13.4) Examples.

(i) If R is a ring with an ideal J, then

$$\begin{array}{ccc} D & \xrightarrow{\pi_2} & R \\ {\scriptstyle \pi_1}\downarrow & & \downarrow{\scriptstyle c} \\ R & \xrightarrow{c} & R/J \end{array}$$

is a fiber square of rings.

(ii) If I and J are ideals of a ring R, the square of canonical maps

$$\begin{array}{ccc} R/(I \cap J) & \longrightarrow & R/J \\ \downarrow & & \downarrow \\ R/I & \longrightarrow & R/(I+J) \end{array}$$

is a fiber square in $\mathfrak{Ring}$: It certainly commutes. If $a + I$ and $b + J$ have the same image $a+I+J = b+I+J$, then $a-b = x+y$, where $x \in I$ and $y \in J$. Then $c = a - x = b + y$ has a coset $c + I \cap J$ with images $c + I = a + I$ and $c+J = b+J$. If $d+I$ also has these images, $d-c \in I \cap J$; so $c+I \cap J = d+I \cap J$.

(iii) If each corner of a fiber square is replaced by an isomorphic copy and the maps are replaced by the induced composites, we obtain another fiber square. Every fiber square of surjective ring homomorphisms is obtained in this way from a square from Example (ii), as shown in Exercise 2.

(iv) Suppose B_1 and B_2 are rings and A is a subring of $B_1 \times B_2$. Projections to the first and second coordinates restrict to the top and left surjective maps in a fiber square,

$$
\begin{array}{ccc}
A & \longrightarrow & \pi_2(A) \\
\downarrow & & \downarrow \\
\pi_1(A) & \longrightarrow & R'
\end{array}
$$

since the kernels I, J of these two maps intersect in $\{0_A\}$ and $A/(I \cap J) = A$, $\pi_1(A) \cong A/I$, $\pi_2(A) \cong A/J$, and we can take $R' = A/(I + J)$.

(v) Suppose G is a finite group with a normal subgroup H of order m. The canonical map $G \to G/H$ extends to a $\mathbb{Q}$-algebra homomorphism from $\mathbb{Q}G$ onto $\mathbb{Q}[G/H]$, which is multiplication by the central idempotent

$$
e_H = \frac{1}{m} \sum_{h \,\epsilon\, H} h \;,
$$

followed by an isomorphism. So $\mathbb{Q}G \cong \mathbb{Q}[G/H] \times (1 - e_H)\mathbb{Q}G$ and there is a fiber square of surjective ring homomorphisms

$$
\begin{array}{ccc}
g & \mathbb{Z}G \longrightarrow (1 - e_H)\mathbb{Z}G \\
\downarrow & \quad \downarrow \qquad\qquad \downarrow \\
gH & \mathbb{Z}[G/H] \longrightarrow A' \;.
\end{array}
$$

The kernel of the top map is $\mathbb{Z}G \cap e_H \mathbb{Q}G = m e_H \mathbb{Z}G$, and the left map takes this to the kernel $m\mathbb{Z}[G/H]$ of the bottom map. So we can replace the bottom map by reduction mod m and the top map by reduction mod $m e_H \mathbb{Z}G$ to get the fiber square

(13.5)

$$
\begin{array}{ccc}
\mathbb{Z}G & \longrightarrow & \mathbb{Z}G/(\Sigma_{h\in H}\, h)\mathbb{Z}G \\
\downarrow & & \downarrow \\
\mathbb{Z}[G/H] & \longrightarrow & (\mathbb{Z}/m\mathbb{Z})[G/H] \;.
\end{array}
$$

For instance, when $G = H$ is a cyclic group generated by α of prime order p, $\mathbb{Q}G \cong \mathbb{Q} \times \mathbb{Q}(\zeta_p)$, where $\zeta_p = e^{2\pi i/p}$ (see Example (8.5) (iii)). In this case $\mathbb{Q}G \to \mathbb{Q}(\zeta_p)$, $\alpha \mapsto \zeta_p$, is reduction mod $(1 + \alpha + \cdots + \alpha^{p-1})$ followed by an isomorphism. Thus we obtain the **Rim square**

$$
\begin{array}{ccccc}
\mathbb{Z}G \longrightarrow \mathbb{Z}[\zeta_p] & & \alpha \longmapsto \zeta_p \\
\downarrow \qquad\quad \downarrow & , & \downarrow \qquad\quad \downarrow & , \\
\mathbb{Z} \longrightarrow \mathbb{Z}/_p\mathbb{Z} & & 1 \longmapsto \bar{1}
\end{array}
$$

named for Rim, who used it in one of the first papers (Rim [59]) on algebraic
K-theory. We recover Rim's calculation with this square in (13.34), in the next
section.

(vi) Suppose B is a ring, A is a subring of B, and J is an ideal of B contained
in A. Inclusions and canonical maps constitute a fiber square

$$
\begin{array}{ccc}
A & \overset{\subseteq}{\longrightarrow} & B \\
\downarrow & & \downarrow \\
A/J & \overset{\subseteq}{\longrightarrow} & B/J
\end{array}
$$

in $\mathfrak{Ring}$, since $b + J = a + J$ for $a \in A$ implies $b \in A$. This is known as a
conductor square, since the most useful example for computations uses the
largest B-ideal in A, and that ideal

$$ J = \{a \in A : aB \subseteq A\} $$

is called the **conductor** from B to A. This is our first example involving some
nonsurjective maps.

(vii) Suppose $\mathcal{C}$ is R-$\mathfrak{Mod}$ or $\mathfrak{Ring}$, and $R_1 \overset{g_1}{\longrightarrow} R' \overset{g_2}{\longleftarrow} R_2$ are arrows
in $\mathcal{C}$. The set

$$ F = \{x, y \in R_1 \times R_2 : g_1(x) = g_2(y)\} $$

is a submodule (if $\mathcal{C} = R$-$\mathfrak{Mod}$) or subring (if $\mathcal{C} = \mathfrak{Ring}$) of $R_1 \times R_2$. So F is an
object of $\mathcal{C}$. If $p_i : F \to R_i$ is projection to the i-coordinate, then

$$
\begin{array}{ccc}
F & \overset{p_2}{\longrightarrow} & R_2 \\
{\scriptstyle p_1}\downarrow & & \downarrow{\scriptstyle g_2} \\
R_1 & \overset{}{\underset{g_1}{\longrightarrow}} & R'
\end{array}
$$

is a fiber square in $\mathcal{C}$. We call F the **fiber product** of g_1 and g_2.

Given any commutative square

$$
\begin{array}{ccc}
B & \overset{f_2}{\longrightarrow} & R_2 \\
{\scriptstyle f_1}\downarrow & & \downarrow{\scriptstyle g_2} \\
R_1 & \overset{}{\underset{g_1}{\longrightarrow}} & R'
\end{array}
$$

in $\mathcal{C}$ sharing the same maps g_1 and g_2, there is a unique arrow $h : B \to$
F in $\mathcal{C}$ with $p_1 \circ h = f_1$ and $p_2 \circ h = f_2$. This h is necessarily given by
$h(b) = (f_1(b), f_2(b))$. The second square is a fiber product if and only if h is
an isomorphism in $\mathcal{C}$. So the first square is a terminal object in the category,

with objects the commutative squares in $\mathcal{C}$ sharing these g_1 and g_2, and with an arrow between such squares being an arrow in $\mathcal{C}$ between their upper left corners, commuting with the maps to R_1 and R_2. An object in this category is terminal if and only if it is a fiber square. So the fiber squares over g_1 and g_2 are called **pullbacks** or **colimits** of g_1 and g_2.

Consider a fiber square in $\mathfrak{Ring}$:

$$
\begin{array}{ccc}
A & \xrightarrow{f_2} & R_2 \\
\downarrow{\scriptstyle f_1} & & \downarrow{\scriptstyle g_2} \\
R_1 & \xrightarrow[g_1]{} & R' \ .
\end{array}
$$

Milnor constructed the modules in $\mathcal{P}(A)$ from pairs of modules $P_1 \in \mathcal{P}(R_1)$ and $P_2 \in \mathcal{P}(R_2)$, which become isomorphic under extension of scalars to R', thereby showing the induced square

$$
\begin{array}{ccc}
K_0(A) & \longrightarrow & K_0(R_2) \\
\downarrow & & \downarrow \\
K_0(R_1) & \longrightarrow & K_0(R')
\end{array}
$$

is very nearly a fiber square of additive abelian groups.

(13.6) Definition. Via maps f_i, g_i of the fiber square, regard R'-modules as R_1- and R_2-modules, and R_1-, R_2-, or R'-modules as A-modules. Suppose $P_1 \in \mathcal{P}(R_1)$, $P_2 \in \mathcal{P}(R_2)$, and there is an R'-linear isomorphism α from

$$\overline{P_1} \;=\; R' \otimes_{R_1} P_1 \quad \text{to} \quad \overline{P_2} \;=\; R' \otimes_{R_2} P_2.$$

For $x \in P_i$, let $\overline{x}$ denote $1 \otimes x \in \overline{P}_i$. Then $c_i : P_i \to \overline{P}_i$, $x \mapsto \overline{x}$, are A-linear maps, as is α; and we define

$$(P_1, P_2, \alpha) \;=\; \{(x, y) \in P_1 \times P_2 : \alpha(\overline{x}) = \overline{y}\}.$$

This is the fiber product F of $\alpha \circ c_1$ and c_2, in the fiber square of A-modules

$$
\begin{array}{ccc}
F & \xrightarrow{\ p_2\ } & P_2 \\
\downarrow{\scriptstyle p_1} & & \downarrow{\scriptstyle c_2} \\
P_1 & \xrightarrow{c_1} \overline{P}_1 \xrightarrow{\alpha} & \overline{P}_2 \ ,
\end{array}
$$

as described in Example (vii) above. So $F = (P_1, P_2, \alpha)$ is an A-module via $a \cdot (x, y) = (ax, ay)$.

(13.7) Lemma. *Every $P \in \mathcal{P}(A)$ is isomorphic to (P_1, P_2, α) where $P_1 = R_1 \otimes_A P$, $P_2 = R_2 \otimes_A P$, and $\alpha : \overline{P}_1 \cong \overline{P}_2$ is an R'-linear isomorphism.*

Proof. Apply the exact functor $(-) \otimes_A P$ to the square

$$
\begin{array}{ccc}
A & \xrightarrow{\ f_2\ } & R_2 \\
{\scriptstyle f_1}\downarrow & & \downarrow{\scriptstyle g_2} \\
R_1 & \xrightarrow[\ g_1\ ]{} & R'
\end{array}
\qquad \text{to get} \qquad
\begin{array}{ccc}
A \otimes_A P & \xrightarrow{\ f_2 \otimes 1\ } & P_2 \\
{\scriptstyle f_1 \otimes 1}\downarrow & & \downarrow{\scriptstyle g_2 \otimes 1} \\
P_1 & \xrightarrow[\ g_1 \otimes 1\ ]{} & P' .
\end{array}
$$

Then $(-) \otimes_A P$ of the exact sequence showing the first square is a fiber square induces (via $(R_1 \oplus R_2) \otimes_A P \cong P_1 \oplus P_2$) an exact sequence showing the second square is a fiber square. Each map $g_i \otimes 1$ factors as $c_i : P_i \to \overline{P}_i = R' \otimes_{R_i} P_i$ followed by an isomorphism

$$
\alpha_i : R' \otimes_{R_i} R_i \otimes_A P \ \cong \ R' \otimes_A P = P'.
$$

So

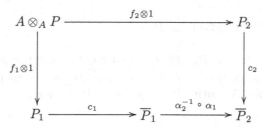

is a fiber square and therefore $P \cong A \otimes_A P$ is isomorphic to the fiber product $(P_1, P_2, \alpha_2^{-1} \circ \alpha_1)$. ∎

By this lemma, every isomorphism class in $\mathcal{P}(A)$ includes a fiber product (P_1, P_2, α). But it does not say every fiber product (P_1, P_2, α) belongs to $\mathcal{P}(A)$. To approach this question, we need to see how such fiber products behave with regard to isomorphisms, free modules, and direct sums. We begin with isomorphisms.

Suppose we have fiber products (P_1, P_2, α) , (Q_1, Q_2, β) and R_i-linear maps $u_i : P_i \to Q_i$. Let $\overline{u}_i : \overline{P}_i \to \overline{Q}_i$ denote the R'-linear map $1 \otimes u_i$. Now $u_1 \oplus u_2$ is an $R_1 \oplus R_2$-linear, hence A-linear, map from $P_1 \oplus P_2$ to $Q_1 \oplus Q_2$. The fiber products are A-submodules of these, so $u_1 \oplus u_2$ restricts to an A-linear map

$$
(u_1, u_2) \ : \ (P_1, P_2, \alpha) \ \to \ (Q_1, Q_2, \beta)
$$

if and only if it carries (P_1, P_2, α) into (Q_1, Q_2, β), or equivalently, if and only if the square

(13.8)

$$
\begin{array}{ccc}
\overline{P}_1 & \xrightarrow{\ \overline{u}_1\ } & \overline{Q}_1 \\
{\scriptstyle \alpha}\downarrow & & \downarrow{\scriptstyle \beta} \\
\overline{P}_2 & \xrightarrow[\ \overline{u}_2\ ]{} & \overline{Q}_2
\end{array}
$$

commutes. If the u_i are also R_i-linear isomorphisms, we call (u_1, u_2) a **diagonal isomorphism** from (P_1, P_2, α) to (Q_1, Q_2, β). A diagonal isomorphism (u_1, u_2) is also an A-linear isomorphism with inverse (u_1^{-1}, u_2^{-1}), since the square with $\overline{u}_i$ replaced by the reverse arrows $\overline{u}_i^{-1} = \overline{u_i^{-1}}$ also commutes.

Now suppose (P_1, P_2, α) and (Q_1, Q_2, β) are fiber products as defined in (13.6). Distributivity of $\otimes$ over $\oplus$ defines an R'-linear isomorphism $\delta_i : \overline{P_i \oplus Q_i} \cong \overline{P}_i \oplus \overline{Q}_i$, taking $\overline{(x, y)} = 1 \otimes (x, y)$ to $(1 \otimes x, 1 \otimes y) = (\overline{x}, \overline{y})$. For brevity we write

$$(P_1 \oplus Q_1, P_2 \oplus Q_2, \alpha \oplus \beta) \text{ for } (P_1 \oplus Q_1, P_2 \oplus Q_2, \delta_2^{-1} \circ (\alpha \oplus \beta) \circ \delta_1) \ .$$

(13.9) Lemma. *There is an A-linear isomorphism*

$$(P_1, P_2, \alpha) \oplus (Q_1, Q_2, \beta) \ \cong \ (P_1 \oplus Q_1, P_2 \oplus Q_2, \alpha \oplus \beta)$$

taking $((x, y), (s, t))$ to $((x, s), (y, t))$.

Proof. With $x \in P_1$, $y \in P_2$, $s \in Q_1$, and $t \in Q_2$, $\alpha(\overline{x}) = \overline{y}$ and $\beta(\overline{s}) = \overline{t}$ if and only if $\delta_2^{-1} \circ (\alpha \oplus \beta) \circ \delta_1$ takes $\overline{(x, s)}$ to $\overline{(y, t)}$. So the map is defined, and evidently invertible. A routine check proves it is A-linear. ∎

If we denote the module $(P_1 \oplus Q_1, P_2 \oplus Q_2, \alpha \oplus \beta)$ by $(\boldsymbol{P_1}, \boldsymbol{P_2}, \boldsymbol{\alpha}) * (\boldsymbol{Q_1}, \boldsymbol{Q_2}, \boldsymbol{\beta})$, each associativity isomorphism for $\oplus$ on A-modules induces, through the isomorphism of the lemma, a diagonal isomorphism giving the corresponding associativity for $*$.

The distributivity and multiplication isomorphisms for $\otimes$ compose to give an R'-linear isomorphism

$$\boldsymbol{\Delta_i} : \overline{R_i^n} \ = \ R' \otimes_{R_i} R_i^n \ \cong \ (R' \otimes_{R_i} R_i)^n \ \cong \ R'^n$$

taking $\overline{x} = \overline{(x_1, \ldots, x_n)} = 1 \otimes (x_1, \ldots, x_n)$ to $(g_i(x_1), \ldots, g_i(x_n))$, which we denote by $g_i(x)$. There is a bijection between the set of R'-linear isomorphisms $\alpha : \overline{R_1^n} \cong \overline{R_2^n}$ and the set of matrices $C \in GL_n(R')$ given by commutativity of the square

$$
\begin{array}{ccc}
\overline{R_1^n} & \xrightarrow{\ \alpha\ } & \overline{R_2^n} \\
{\scriptstyle \Delta_1} \downarrow & & \downarrow {\scriptstyle \Delta_2} \\
R'^n & \xrightarrow[\ \cdot C\ ]{} & R'^n,
\end{array}
$$

and we write $(\boldsymbol{R_1^n}, \boldsymbol{R_2^n}, \boldsymbol{C})$ for (R_1^n, R_2^n, α).

(13.10) Lemma.

(i) *There is an A-linear isomorphism $A^n \cong (R_1^n, R_2^n, I_n)$ taking v to the pair $(f_1(v), f_2(v))$.*

(ii) *There is a diagonal isomorphism $(P_1, P_2, \alpha) \cong (R_1^n, R_2^n, C)$ if and only if there is an R_1-basis $v_1, \ldots, v_n$ of P_1 and an R_2-basis $w_1, \ldots, w_n$ of P_2 so that C represents α over the R'-bases $\overline{v}_1, \ldots, \overline{v}_n$ of $\overline{P}_1$ and $\overline{w}_1, \ldots, \overline{w}_n$ of $\overline{P}_2$.*

(iii) *There is a diagonal isomorphism*

$$(\varepsilon, \varepsilon) : (R_1^m, R_2^m, C) * (R_1^n, R_2^n, C') \;\cong\; (R_1^{m+n}, R_2^{m+n}, C \oplus C') \,,$$

where $\varepsilon : R_i^m \oplus R_i^n \cong R_i^{m+n}$ is the erasure of parentheses.

Proof. By definition of a fiber product (13.6), a pair (x, y) belongs to

$$(R_1^n, R_2^n, I_n) \;=\; (R_1^n, R_2^n, \Delta_2^{-1} \circ \Delta_1)$$

if and only if $\Delta_2^{-1} \circ \Delta_1(\overline{x}) = \overline{y}$. The latter condition $\Delta_1(\overline{x}) = \Delta_2(\overline{y})$ amounts to $g_1(x) = g_2(y)$. So $(R_1, R_2, 1)$ is the fiber product F of rings in Example (13.4) (vii), and there is an A-linear ring isomorphism

$$A \;\cong\; (R_1, R_2, 1) \,, \quad a \mapsto (f_1(a), f_2(a)) \,.$$

The desired isomorphism (i) is the composite

$$A^n \;\cong\; (R_1, R_2, 1)^n \;\cong\; (R_1^n, R_2^n, I_n) \,,$$

the last map taking

$$((x_1, y_1), \ldots, (x_n, y_n)) \;\text{to}\; ((x_1, \ldots, x_n), (y_1, \ldots, y_n)).$$

For (ii), a diagonal isomorphism from (P_1, P_2, α) to (R_1^n, R_2^n, C) is a pair of coordinate maps $u_i : P_i \cong R_i^n$ associated with bases $u_i^{-1}(e_1), \ldots, u_i^{-1}(e_n)$ of P_i, for which the diagram

$$
\begin{array}{ccccc}
\overline{P}_1 & \xrightarrow{\;\overline{u}_1\;} & \overline{R_1^n} & \xrightarrow{\;\Delta_1\;} & R'^n \\
{\scriptstyle \alpha}\downarrow & & & & \downarrow{\scriptstyle \cdot C} \\
\overline{P}_2 & \xrightarrow[\;\overline{u}_2\;]{} & \overline{R_2^n} & \xrightarrow[\;\Delta_2\;]{} & R'^n
\end{array}
$$

commutes. Now $\Delta_i(\overline{e}_j) = e_j$; so $\Delta_i \circ \overline{u}_i$ is a coordinate isomorphism associated with the R'-basis $u_i^{-1}(e_1), \ldots, u_i^{-1}(e_n)$ of $\overline{P}_i$. So the existence of this commutative diagram amounts to the existence of bases $\{v_j\}$ of P_1 and $\{w_j\}$ of P_2 (for $1 \le j \le n$), so that C represents α over the bases $\{\overline{v}_j\}$ of $\overline{P}_1$ and $\{\overline{w}_j\}$ of $\overline{P}_2$.

For (iii), let $\varepsilon : R^m \oplus R^n \cong R^{m+n}$ denote erasure of parentheses. There is a commutative diagram of R'-linear isomorphisms

$$
\begin{array}{ccccccccc}
\overline{R_1^m \oplus R_1^n} & \xrightarrow{\delta} & \overline{R_1^m} \oplus \overline{R_1^n} & \xrightarrow{\Delta \oplus \Delta} & R'^m \oplus R'^n & \xrightarrow{\varepsilon} & R'^{m+n} & \xleftarrow{\Delta} & R_1^{m+n} \\
\downarrow{\mu} & & \downarrow{(\cdot C)\oplus(\cdot C')} & & \downarrow{\cdot (C \oplus C')} & & & & \downarrow{\nu} \\
\overline{R_2^m \oplus R_2^n} & \xrightarrow{\delta} & \overline{R_2^m} \oplus \overline{R_2^n} & \xrightarrow{\Delta \oplus \Delta} & R'^m \oplus R'^n & \xrightarrow{\varepsilon} & R'^{m+n} & \xleftarrow{\Delta} & R_2^{m+n}
\end{array}
$$

defining μ and ν, in which the rows, from left to right, compose to $\overline{\varepsilon}$. So $(\varepsilon, \varepsilon)$ is a diagonal isomorphism from

$$(R_1^m \oplus R_1^n, \ R_2^m \oplus R_2^n, \ \mu) \ = \ (R_1^m, \ R_2^m, C) \ * \ (R_1^n, \ R_2^n, C')$$

to

$$(R_1^{m+n}, \ R_2^{m+n}, \ \nu) \ = \ (R_1^{m+n}, \ R_2^{m+n}, \ C \oplus C') \ . \qquad \blacksquare$$

(13.11) Lemma. *For $C, C' \in GL_n(R')$, $M_1 \in GL_n(R_1)$, and $M_2 \in GL_n(R_2)$, there is a diagonal isomorphism*

$$(\cdot M_1, \cdot M_2) \ : \ (R_1^n, \ R_2^n, \ C) \ \cong \ (R_1^n, \ R_2^n, C')$$

if and only if $C' = g_1(M_1^{-1}) C g_2(M_2)$, where g_i is applied entrywise.

Proof. The isomorphism $\Delta_i : \overline{R_i^n} \to R'^n$ takes $\overline{x}$ to $g_i(x)$. So

$$\Delta_i \circ (\overline{\cdot M_i}) \circ \Delta_i^{-1}(e_i) \ = \ \Delta_i \circ (\overline{\cdot M_i})(\overline{e_i})$$

$$= \ \Delta_i(\overline{e_i M_i}) \ = \ g_i(e_i M_i) \ = \ e_i \, g_i(M_i),$$

and $\Delta_i \circ (\overline{\cdot M_i}) \circ \Delta_i^{-1}$ is right multiplication by $g_i(M_i)$. Then

$$
\begin{array}{ccccccc}
R'^n & \xleftarrow{\Delta_1} & \overline{R_1^n} & \xrightarrow{\overline{\cdot M_1}} & \overline{R_1^n} & \xrightarrow{\Delta_1} & R'^n \\
\downarrow{\cdot C} & & & & & & \downarrow{\cdot C'} \\
R'^n & \xleftarrow{\Delta_2} & \overline{R_2^n} & \xrightarrow{\overline{\cdot M_2}} & \overline{R_2^n} & \xrightarrow{\Delta_2} & R'^n
\end{array}
$$

commutes if and only if $C' = g_1(M_1^{-1}) C g_2(M_2)$. $\qquad \blacksquare$

(13.12) Proposition. *Suppose g_1 or g_2 is surjective. Then each fiber product (P_1, P_2, α) belongs to $\mathcal{P}(A)$, and for $i = 1$ and 2 there are R_i-linear isomorphisms*

$$\theta_i : R_i \otimes_A (P_1, P_2, \alpha) \;\cong\; P_i \,,$$

taking $1 \otimes (x_1, x_2)$ to x_i.

Proof. We prove the first assertion in the process of proving the second. The function from $R_i \times (P_1, P_2, \alpha)$ to P_i, taking $(r_i, (x_1, x_2))$ to $r_i x_i$, is balanced over A, inducing an R_i-linear map

$$\theta_i : R_i \otimes_A (P_1, P_2, \alpha) \to P_i$$

taking $1 \otimes (x_1, x_2)$ to x_i. We need only prove θ_i is bijective.

If there is a diagonal isomorphism (u_1, u_2) from (P_1, P_2, α) to (Q_1, Q_2, β), the square

$$
\begin{array}{ccc}
R_i \otimes_A (P_1, P_2, \alpha) & \xrightarrow{\;1 \otimes (u_1, u_2)\;} & R_i \otimes_A (Q_1, Q_2, \beta) \\
{\scriptstyle \theta_i}\big\downarrow & & \big\downarrow {\scriptstyle \theta_i} \\
P_i & \xrightarrow{\qquad u_i \qquad} & Q_i
\end{array}
$$

commutes. So θ_i is bijective for (P_1, P_2, α) if and only if it is bijective for (Q_1, Q_2, β).

For fiber products $X = (P_1, P_2, \alpha)$ and $Y = (Q_1, Q_2, \beta)$, the square

$$
\begin{array}{ccc}
R_i \otimes_A (X \oplus Y) & \xrightarrow{\;\cong\;} & R_i \otimes_A (X * Y) \\
\big\downarrow & & \big\downarrow {\scriptstyle \theta_i} \\
(R_i \otimes_A X) \oplus (R_i \otimes_A Y) & \xrightarrow{\;\theta_i \oplus \theta_i\;} & P_i \oplus Q_i
\end{array}
$$

commutes, where the left map is the distributivity isomorphism. So θ_i bijects for $X * Y$ if and only if it bijects for both X and Y.

The square

$$
\begin{array}{ccc}
R_i \otimes_A A^n & \xrightarrow{\;\cong\;} & R_i \otimes_A (R_1^n, R_2^n, I_n) \\
\big\downarrow & & \big\downarrow {\scriptstyle \theta_i} \\
R_i^n & \xrightarrow{\qquad = \qquad} & R_i^n
\end{array}
$$

also commutes, where the left map is the distributivity and multiplication isomorphism, taking $1 \otimes (a_1, \ldots, a_n)$ to $(f_i(a_1), \ldots, f_i(a_n))$, and the top map is $1 \otimes f$ where f is the isomorphism $v \mapsto (f_1(v), f_2(v))$ of Lemma (13.10) (i). So θ_i is bijective for (R_1^n, R_2^n, I_n).

Now suppose we have a fiber product (P_1, P_2, α), and we choose $Q_i \in \mathcal{P}(R_i)$ with $P_i \oplus Q_i \cong R_i^n$ (same n for $i = 1$ and 2). There are R'-linear isomorphisms $\overline{R}_1 \cong \overline{R}_2$ and $\overline{P}_1 \cong \overline{P}_2$; so there is a composite isomorphism

$$
\begin{aligned}
\overline{R_1^n \oplus Q_1} \;\cong\; & \overline{R}_1^n \oplus \overline{Q}_1 \;\cong\; \overline{R}_2^n \oplus \overline{Q}_1 \\
\cong\; & \overline{P}_2 \oplus (\overline{Q}_2 \oplus \overline{Q}_1) \;\cong\; \overline{P}_1 \oplus (\overline{Q}_1 \oplus \overline{Q}_2) \\
\cong\; & \overline{R}_1^n \oplus \overline{Q}_2 \;\cong\; \overline{R}_2^n \oplus \overline{Q}_2 \\
\cong\; & \overline{R_2^n \oplus Q_2} \;,
\end{aligned}
$$

which we call β. For $i = 1$ and 2 there is an associativity R_i-linear isomorphism $u_i : P_i \oplus (R_i^n \oplus Q_i) \cong R_i^{2n}$. And for some $C \in GL_{2n}(R')$ there is a diagonal isomorphism

$$
(u_1, u_2) : (P_1, P_2, \alpha) * (R_1^n \oplus Q_1, R_2^n \oplus Q_2, \beta) \;\cong\; (R_1^{2n}, R_2^{2n}, C) \;.
$$

There is, by (13.10) (iii), a diagonal isomorphism

$$
(R_1^{2n}, R_2^{2n}, C) * (R_1^{2n}, R_2^{2n}, C^{-1}) \;\cong\; (R_1^{4n}, R_2^{4n}, C \oplus C^{-1}) \;.
$$

By the proof of Whitehead's Lemma (9.7), $C \oplus C^{-1} \in E_{4n}(R')$. Since some $g_i : R_i \to R'$ is surjective, elementary transvections over R' lift to R_i; so $C \oplus C^{-1}$ belongs to the image of

$$
GL_{4n}(R_i) \;\to\; GL_{4n}(R') \;.
$$

By Corollary (13.11), there is a diagonal isomorphism

$$
(R_1^{4n}, R_2^{4n}, C \oplus C^{-1}) \;\cong\; (R_1^{4n}, R_2^{4n}, I) \;.
$$

Now for both $i = 1$ and $i = 2$, θ_i is bijective for $X = (R_1^{4n}, R_2^{4n}, I)$, and hence for $Y = (R_1^{2n}, R_2^{2n}, C)$, and finally for (P_1, P_2, α). And (P_1, P_2, α) is isomorphic to a direct summand of Y, which is isomorphic to a direct summand of X, a f.g. free A-module. So $(P_1, P_2, \alpha) \in \mathcal{P}(A)$. ∎

(13.13) Corollary. *Suppose g_1 or g_2 is surjective. If fiber products (P_1, P_2, α) and (Q_1, Q_2, β) are A-linearly isomorphic, there is a diagonal isomorphism between them.*

Proof. Say $P = (P_1, P_2, \alpha)$ and $Q = (Q_1, Q_2, \beta)$ and $h : P \cong Q$ is an A-linear isomorphism. Let $\theta_i : R_i \otimes_A P \cong P_i$, $\theta_i : R_i \otimes_A Q \cong Q_i$ denote the isomorphisms from Proposition (13.12). There is a composite R'-linear isomorphism

$$
\begin{aligned}
R' \otimes_{R_1} (R_1 \otimes_A P) \;\cong\; & (R' \otimes_{R_1} R_1) \otimes_A P \;\cong\; R' \otimes_A P \\
\cong\; & (R' \otimes_{R_2} R_2) \otimes_A P \;\cong\; R' \otimes_{R_2} (R_2 \otimes_A P)
\end{aligned}
$$

taking R-linear generators $1 \otimes (1 \otimes p)$ to $1 \otimes (1 \otimes p)$. Call this composite u, and let v denote the same composite with Q in place of P, taking $1 \otimes (1 \otimes q)$ to $1 \otimes (1 \otimes q)$. We claim the diagram

$$
\begin{array}{ccccccc}
\overline{P}_1 & \xleftarrow{\overline{\theta}_1} & R_1 \otimes_A P & \xrightarrow{1 \otimes h} & R_1 \otimes_A Q & \xrightarrow{\overline{\theta}_1} & \overline{Q}_1 \\
{\scriptstyle \alpha}\downarrow & & {\scriptstyle u}\downarrow & & {\scriptstyle v}\downarrow & & {\scriptstyle \beta}\downarrow \\
\overline{P}_2 & \xleftarrow{\overline{\theta}_2} & R_2 \otimes_A P & \xrightarrow{1 \otimes h} & R_2 \otimes_A Q & \xrightarrow{\overline{\theta}_2} & \overline{Q}_2
\end{array}
$$

commutes. For, if $p = (p_1, p_2) \in P$ with $h(p) = (q_1, q_2) \in Q$, then $\alpha(\overline{p}_1) = \overline{p}_2$ and $\beta(\overline{q}_1) = \overline{q}_2$; and

$$
\begin{array}{ccccccc}
\overline{p}_1 & \longleftarrow & 1 \otimes (p_1, p_2) & \longrightarrow & 1 \otimes (q_1, q_2) & \longrightarrow & \overline{q}_1 \\
\downarrow & & \downarrow & & \downarrow & & \downarrow \\
\overline{p}_2 & \longleftarrow & 1 \otimes (p_1, p_2) & \longrightarrow & 1 \otimes (q_1, q_2) & \longrightarrow & \overline{q}_2.
\end{array}
$$

So $(\theta_1 \circ (1 \otimes h) \circ \theta_1^{-1}, \theta_2 \circ (1 \otimes h) \circ \theta_2^{-1})$ is the desired diagonal isomorphism from P to Q. ∎

(13.14) Corollary. *Suppose g_1 or g_2 is surjective, and R_1 or R_2 has IBN. Then (R_1^n, R_2^n, C) is a free A-module if and only if $C = g_1(M_1) g_2(M_2)$ for some matrices $M_i \in GL_n(R_i)$.*

Proof. We know $(R_1^n, R_2^n, C) \in \mathcal{P}(A)$, so it is finitely generated. If it is isomorphic to $A^m \cong (R_1^m, R_2^m, I)$, there is a diagonal isomorphism. Since R_1 or R_2 has IBN, this forces $m = n$. So (R_1^n, R_2^n, C) is free if and only if it has a diagonal isomorphism to (R_1^n, R_2^n, I). Now apply (13.11). ∎

Now we are ready to produce the K_0 piece of the relative sequence.

(13.15) Definition. Let $K_0(R, J)$ denote the Stein relativization $K_0^S(R, J)$ of K_0.

(13.16) Proposition. *The sequence*

$$
K_0(R, J) \xrightarrow{i\pi_1^*} K_0(R) \xrightarrow{K_0(c)} K_0(R/J)
$$

is exact and natural.

Proof. As shown for all functors from $\mathfrak{Ring}$ to $\mathfrak{Group}$ at the beginning of this section, K_0 of the fiber square of rings

$$
\begin{array}{ccc}
D & \xrightarrow{\pi_2} & R \\
{\scriptstyle \pi_1}\downarrow & & \downarrow{\scriptstyle c} \\
R & \xrightarrow{c} & R/J
\end{array}
$$

commutes; and so $K_0(\pi_1)$ restricts to a homomorphism $\pi_1^* : K_0^S(R, J) \to K_0'(R, J)$ between kernels of the horizontal maps. Hence $K_0(c) \circ i\pi_1^*$ is zero. Also, naturality of this sequence is a special case of commutativity of the diagram (13.1). It remains to prove π_1^* is surjective.

Suppose $P, Q \in \mathcal{P}(R)$ and $[P] - [Q] \in K_0'(R, J)$; that is,

$$K_0(c)([P] - [Q]) = [\overline{P}] - [\overline{Q}] = 0 .$$

Then $\overline{P}$ and $\overline{Q}$ are stably isomorphic: There is a positive integer n and an $\overline{R}$-linear isomorphism from $\overline{P} \oplus \overline{R}^n$ to $\overline{Q} \oplus \overline{R}^n$. Then $P_1 = P \oplus R^n$ and $Q_1 = Q \oplus R^n$ belong to $\mathcal{P}(R)$, and there is a composite isomorphism

$$\alpha : \overline{P}_1 \cong \overline{P} \oplus \overline{R}^n \cong \overline{Q} \oplus \overline{R}^n \cong \overline{Q}_1 .$$

By (13.12), $(P_1, Q_1, \alpha) \in \mathcal{P}(D)$. Say

$$x = [(P_1, Q_1, \alpha)] - [(Q_1, Q_1, 1)] .$$

By the isomorphisms in (13.12), $K_0(\pi_2)(x) = [Q_1] - [Q_1] = 0$ while $K_0(\pi_1)(x) = [P_1] - [Q_1] = [P] - [Q]$. That is, $x \in K_0^S(R, S)$ and $i\pi_1^*(x) = [P] - [Q]$. ∎

Finally we turn to relative K_2. Here it is useful to know that Quillen's higher algebraic K-theory provides functors

$$K_n^Q : \mathfrak{Ring} \to \mathbb{Z}\text{-}\mathfrak{Mod}$$

$$K_n^Q : \mathfrak{Ideal} \to \mathbb{Z}\text{-}\mathfrak{Mod}$$

for all positive integers n, with $K_n^Q(R) = K_n(R)$ for $n = 1$ and 2, and provides an exact sequence

$$\cdots \longrightarrow K_3^Q(R) \xrightarrow{K_3^Q(c)} K_3^Q(R/J)$$

$$\longrightarrow K_2^Q(R, J) \longrightarrow K_2(R) \xrightarrow{K_2(c)} K_2(R/J) \longrightarrow \cdots$$

in $\mathbb{Z}\text{-}\mathfrak{Mod}$. But the Quillen construction of the algebraic K-groups is complicated, passing from rings to categories, to topological spaces, and finally to their homotopy groups. For the purposes of computation, one seeks an algebraic definition of $K_2^Q(R, J)$. As Swan [71] proved, the Stein relativization $K_2^S(R, J)$ is not quite it. We briefly sketch Swan's argument, because it is the inspiration for the correct algebraically defined $K_2(R, J)$.

Recall the square

$$
\begin{array}{ccc}
D & \xrightarrow{\pi_2} & R \\
{\scriptstyle \pi_1}\downarrow & & \downarrow{\scriptstyle c} \\
R & \xrightarrow{c} & R/J
\end{array}
$$

and the definition of $K_2^S(R, J)$ as the kernel of $K_2(\pi_2)$. Suppose, for $i = 1, 2$, that x_i is in the kernel of $St(\pi_i)$. By commutativity of

$$St(D) \xrightarrow{St(\pi_i)} St(R)$$
$$\phi \downarrow \qquad \qquad \downarrow \phi$$
$$GL(D) \xrightarrow{GL(\pi_i)} GL(R) \ ,$$

each $\phi(x_i)$ is in the kernel of $GL(\pi_i)$; so $\phi(x_1) = (I, Y)$ and $\phi(x_2) = (X, I)$ for some $X, Y \in GL(R, J)$. These two matrices over D commute; so $\phi([x_1, x_2]) = [\phi(x_1), \phi(x_2)] = 1$ in $GL(D)$, putting $[x_1, x_2]$ in $K_2(D)$. By the choice of x_i and the fact that $[1, y] = 1 = [z, 1]$ in $St(R)$, this commutator $[x_1, x_2]$ belongs to the kernels of both maps $K_2(\pi_i) : K_2(D) \to K_2(R)$. So it lies in the kernel of $K_2^S(R, J) \to K_2(R)$.

If there is to be an exact sequence

$$K_3(R) \xrightarrow{K_3(c)} K_3(R/J) \longrightarrow K_2^S(R, J) \longrightarrow K_2(R)$$

for a functor $K_3 : \mathfrak{Ring} \to \mathbb{Z}\text{-}\mathfrak{Mod}$ and for each pair $(R, J) \in \mathfrak{Ideal}$, these elements $[x_1, x_2] \in K_2(D)$ must vanish whenever $K_3(c)$ is surjective. Swan pointed out that if F is a field, R is the polynomial ring $F[t]$, and $J = tF[t]$, then the identity on F factors as

$$F \xrightarrow{\subseteq} R \xrightarrow{c} R/J \cong F \ ;$$

so the identity on $K_3(F)$ factors as

$$K_3(F) \longrightarrow K_3(R) \xrightarrow{K_3(c)} K_3(R/J) \cong K_3(F)$$

and $K_3(c)$ is surjective. However, he proved that $[x_{12}(0, t), x_{21}(t, 0)]$ is nontrivial in $K_2(D)$, a contradiction. (We commend the proof of nontriviality in Swan [71, §6] to the reader, as a very pretty application of Mennicke and norm residue symbols.)

So we adopt the definition due to Keune [78]:

(13.17) Definition. If $K(R, J)$ is the group of mixed commutators

$$[\ker St(\pi_1), \ker St(\pi_2)] \ ,$$

then $K(R, J)$ is a normal subgroup of $St(D)$ contained in $K_2^S(R, J)$. Define

$$\mathbf{K_2(R, J)} \quad = \quad K_2^S(R, J)/K(R, J).$$

Note: If $f : (R, J) \to (R', J')$ is an arrow in $\mathfrak{Ideal}$, $D = R \times_J R$ and $D' = R' \times_{J'} R'$, the ring homomorphism $\widehat{f} : D \to D', (a, b) \mapsto (f(a), f(b))$, induces a homomorphism $St(\widehat{f})$ carrying $K_2^S(R, J)$ into $K_2^S(R', J')$ as $K_2^S(f)$, and carrying $K(R, J)$ into $K(R', J')$ because the squares

$$
\begin{array}{ccc}
St(D) & \xrightarrow{\;St(\pi_i)\;} & St(R) \\
{\scriptstyle St(\widehat{f})}\downarrow & & \downarrow{\scriptstyle St(f)} \\
St(D') & \xrightarrow{\;St(\pi_i)\;} & St(R')
\end{array}
$$

commute. So $St(\widehat{f})$ induces a homomorphism $K_2(R, J) \to K_2(R', J')$, making $K_2(\ ,\ )$ a functor from $\mathfrak{Ideal}$ to $\mathbb{Z}\text{-}\mathfrak{Mod}$.

The following lemma shows $St'(R, J)$ is the analog in $St(R)$ to the group $E(R, J)$ in $E(R)$.

(13.18) Lemma. *If J is an ideal of a ring R, the kernel $St'(R, J)$ of*

$$St(c) : St(R) \quad \to \quad St(R/J)$$

is the normal subgroup N of $St(R)$ generated by all $x_{ij}(a)$ with $a \in J$.

Proof. Certainly $N \subseteq St'(R, J)$, since $x_{ij}(\overline{a}) = x_{ij}(0) = 1$ in $St(R/J)$. In the quotient $St(R)/N$, if $a \in J$, we have $\overline{x_{ij}(r + a)} = \overline{x_{ij}(r)\, x_{ij}(a)} = \overline{x_{ij}(r)}$. So it is not ambiguous to denote the coset of $x_{ij}(r)$ by $y_{ij}(\overline{r})$, where $\overline{r} = r + J$. These $y_{ij}(\overline{r})$ obey the Steinberg relations ST1 and ST2; so there is an induced homomorphism $f : St(R/J) \to St(R)/N$, taking $x_{ij}(\overline{r})$ to $y_{ij}(\overline{r})$, that is left inverse to the homomorphism $\overline{St(c)}$ from $St(R)/N$ to $St(R/J)$. So $\overline{St(c)}$ is injective, and $N = St'(R, J)$. ∎

Applying the Snake Lemma to the second and third rows, we obtain the first row in each of the commutative diagrams with exact rows and columns:

$$
\begin{array}{ccccccccc}
& & 1 & & 1 & & 1 & & \\
& & \downarrow & & \downarrow & & \downarrow & & \\
1 & \longrightarrow & K_2^S(R, J) & \longrightarrow & St^S(R, J) & \longrightarrow & E^S(R, J) & & \\
& & {\scriptstyle\cap}\downarrow & & {\scriptstyle\cap}\downarrow & & {\scriptstyle\cap}\downarrow & & \\
1 & \longrightarrow & K_2(D) & \xrightarrow{\;\subseteq\;} & St(D) & \xrightarrow{\;\phi\;} & E(D) & \longrightarrow & 1 \\
& & {\scriptstyle\pi_2}\downarrow & & {\scriptstyle\pi_2}\downarrow & & {\scriptstyle\pi_2}\downarrow & & \\
1 & \longrightarrow & K_2(R) & \xrightarrow{\;\subseteq\;} & St(R) & \xrightarrow{\;\phi\;} & E(R) & \longrightarrow & 1,
\end{array}
$$

$$
\begin{array}{ccccc}
1 & & 1 & & 1 \\
\downarrow & & \downarrow & & \downarrow \\
1 \longrightarrow K_2'(R,J) \longrightarrow & St'(R,J) \longrightarrow & E'(R,J) \\
\cap\big| & \cap\big| & \cap\big| \\
1 \longrightarrow K_2(R) \overset{\subseteq}{\longrightarrow} & St(R) \overset{\phi}{\longrightarrow} & E(R) \longrightarrow 1 \\
c\big\downarrow & c\big\downarrow & c\big\downarrow \\
1 \longrightarrow K_2(R/J) \overset{\subseteq}{\longrightarrow} & St(R/J) \overset{\phi}{\longrightarrow} & E(R/J) \longrightarrow 1,
\end{array}
$$

where the maps labeled π_2 and c are induced by $\pi_2 : D \to R$ and $c : R \to R/J$. Regard the two diagrams as layers in a commutative diagram where the maps induced by π_1 connect the second rows and the maps induced by c connect the third rows. The entire diagram commutes because $K_2, St,$ and E are functors, and the $K_2 - St - E$ exact sequence is natural. So we have a commutative diagram of kernels connecting the top rows:

$$
\begin{array}{ccccc}
1 \longrightarrow K_2^S(R,J) \overset{\subseteq}{\longrightarrow} & St^S(R,J) \longrightarrow & E^S(R,J) \\
\pi_1^*\big\downarrow & \pi_1^*\big\downarrow & \pi_1^*\big\downarrow \\
1 \longrightarrow K_2'(R,J) \overset{\subseteq}{\longrightarrow} & St'(R,J) \longrightarrow & E'(R,J) \ .
\end{array}
$$

Now $St^S(R,J)$ is a normal subgroup of $St(D)$ containing all $x_{ij}(a,0)$ with $a \in J$; so the surjective $St(\pi_1)$ carries it onto $St'(R,J)$ by Lemma (13.18), and the middle π_1^* is surjective. By (11.7), the right π_1^* is an isomorphism to $E(R,J)$ followed by inclusion; so it is injective. A diagram chase now shows the left π_1^* is surjective. With the naturality (13.1) inherent in the Stein relativization, we now have a natural exact sequence

$$
K_2^S(R,J) \overset{i\pi_1^*}{\longrightarrow} K_2(R) \overset{K_2(c)}{\longrightarrow} K_2(R/J) \ .
$$

The first map kills $K(R,J)$, so it induces a natural map

$$
\theta : K_2(R,J) \longrightarrow K_2(R)
$$

with the same image $K_2'(R,J)$. We have proved:

(13.19) Proposition. *The sequence*

$$
K_2(R,J) \overset{\theta}{\longrightarrow} K_2(R) \overset{K_2(c)}{\longrightarrow} K_2(R/J)
$$

is exact and natural. ∎

Stitching the $K_0, K_1,$ and K_2 pieces together with connecting homomorphisms ∂_i, we obtain the **relative sequence** of algebraic K-theory.

(13.20) Theorem. *If J is an ideal of a ring R, there is an exact sequence of abelian groups*

$$K_2(R,J) \longrightarrow K_2(R) \longrightarrow K_2(R/J)$$
$$\xrightarrow{\partial_1} K_1(R,J) \longrightarrow K_1(R) \longrightarrow K_1(R/J)$$
$$\xrightarrow{\partial_0} K_0(R,J) \longrightarrow K_0(R) \longrightarrow K_0(R/J) \ .$$

Proof. For $C \in GL_n(R/J)$, define

$$d(C) \;=\; [(R^n, R^n, C)] - [(R^n, R^n, I_n)]$$

in $K_0(D)$. By the isomorphisms in (13.12), both $K_0(\pi_1)$ and $K_0(\pi_2)$ take this to $[R^n] - [R^n] = 0$ in $K_0(R)$; so $d(C)$ lies in the kernel of the map $i\pi_1^* : K_0(R,J) \to K_0(R)$ in the relative sequence. By (13.10) (iii), $d(C \oplus I_m) = d(C)$; so d is a well-defined function on $GL(R/J)$. For $C, D \in GL_n(R/J)$, $CD \oplus I_n = (C \oplus D)(D \oplus D^{-1})$ and $D \oplus D^{-1} \in E(R/J)$, which lies in the image of $GL(R) \to GL(R/J)$. So by (13.11) and (13.10) (iii), $d(CD) = d(CD \oplus I_n) = d(C \oplus D) = d(C) + d(D)$ and d is a homomorphism. Since $K_0(R,J)$ is abelian, d induces a homomorphism

$$\partial_0 : K_1(R/J) \;\to\; K_0(R,J)$$

whose image is that of d, so it lies in the kernel of the next map $i\pi^*$ in the relative sequence.

To prove exactness at $K_0(R,J)$ suppose $P, Q \in \mathcal{P}(D)$ and $K_0(\pi_i)$ takes $[P] - [Q]$ to zero for both $i = 1$ and 2. By (13.7) we may assume $P = (P_1, P_2, \alpha)$ and $Q = (Q_1, Q_2, \beta)$; then, by (13.12), $[P_1] - [Q_1] = [P_2] - [Q_2] = 0$ in $K_0(R)$. Choose n large enough so $P_i \oplus R^n \cong Q_i \oplus R^n$ for $i = 1, 2$. Replacing P and Q by $P \oplus D^n$ and $Q \oplus D^n$, we may assume $P_i \cong Q_i$ for $i = 1, 2$. Then $Q \cong (P_1, P_2, \gamma)$ for some $\gamma : \overline{P}_1 \cong \overline{P}_2$, so we may assume

$$P \;=\; (P_1, P_2, \alpha) \ , \quad Q \;=\; (P_1, P_2, \gamma) \ .$$

As in the proof (13.12), there is a fiber product $M \in \mathcal{P}(D)$ and a diagonal isomorphism $Q * M \cong (R^n, R^n, I_n)$. Then $P * M \cong (R^n, R^n, C)$ for some $C \in GL_n(R/J)$, and

$$[P] - [Q] \;=\; [P * M] - [Q * M] \;=\; d(C) \;=\; \partial_0(CE(R/J)) \ .$$

The composite map at $K_1(R/J)$ is zero by (13.11). For exactness here, suppose $C \in GL_n(R/J)$ and $d(C) = 0$ in $K_0(D)$. Adding $D^m - D^m$ for sufficiently large m,

$$(R^{n+m}, R^{n+m}, C \oplus I_m) \;\cong\; (R^{n+m}, R^{n+m}, I_{m+n}) \ ;$$

so by (13.11), $C \oplus I_m = c(B)$ for a matrix $B \in GL_{m+n}(R)$. Then $CE(R/J)$ comes from $BE(R)$ in $K_1(R)$.

We get ∂_1 and exactness of the rest of the sequence from the Snake Lemma (11.8) applied to the commutative diagram with exact rows:

$$
\begin{array}{ccccccccc}
1 & \longrightarrow & St'(R,J) & \overset{\subseteq}{\longrightarrow} & St(R) & \longrightarrow & St(R/J) & \longrightarrow & 1 \\
& & \downarrow & & \phi\downarrow & & \phi\downarrow & & \\
1 & \longrightarrow & GL(R,J) & \overset{\subseteq}{\longrightarrow} & GL(R) & \longrightarrow & GL(R/J) & &
\end{array}
$$

Here the image of $St'(R,J)$ in $GL(R,J)$ is $E(R,J)$ by the description of $St'(R,J)$ in Lemma (13.18). The kernel of this left vertical map is also the kernel $K_2'(R,J)$ of the map $K_2(R) \to K_2(R/J)$, by a diagram chase in:

$$
\begin{array}{ccccc}
& & K_2(R) & \longrightarrow & K_2(R/J) \\
& & \cap\,\downarrow & & \cap\,\downarrow \\
St'(R,J) & \overset{\subseteq}{\longrightarrow} & St(R) & \longrightarrow & St(R/J) \\
\downarrow & & \downarrow & & \downarrow \\
GL(R,J) & \overset{\subseteq}{\longrightarrow} & GL(R) & \longrightarrow & GL(R/J)\, .
\end{array}
$$

So the Snake Lemma yields an exact sequence

$$
K_2'(R,J) \overset{\subseteq}{\longrightarrow} K_2(R) \longrightarrow K_2(R/J)
$$
$$
\overset{\partial_1}{\longrightarrow} K_1(R,J) \longrightarrow K_1(R) \longrightarrow K_1(R/J)\, .
$$

By (13.19), $K_2(R,J) \to K_2'(R,J)$ is surjective; so we obtain the relative sequence all the way back to $K_2(R,J)$. ∎

(13.21) Proposition. *In the relative sequence, the connecting homomorphisms* $\delta_1 : K_2(\ /\ ) \to K_1(\ ,\ )$ *and* $\delta_0 : K_1(\ /\ ) \to K_0(\ ,\ )$ *are natural transformations between functors from* ℐ𝔡𝔢𝔞𝔩 *to* 𝔊𝔯𝔬𝔲𝔭*. So the entire relative sequence is natural.*

Proof. Suppose $f : (R,J) \to (R',J')$ is an arrow in ℐ𝔡𝔢𝔞𝔩. We first show the induced map $\varepsilon_0 : K_0(R,J) \to K_0(R',J')$ takes fiber products $[(P_0,P_1,\alpha)]$ to fiber products $[(P_0',P_1',\alpha')]$ in a simple way. For $i = 1,2$, suppose $P_i \in \mathcal{P}(R)$; so $P_i' = R' \otimes_R P_i \in \mathcal{P}(R')$. There exist R'/J'-linear isomorphisms

$$
h_i : R'/J' \otimes_{R'} (R' \otimes_R P_i) \;\cong\; R'/J' \otimes_{R/J} (R/J \otimes_R P_i)\, ,
$$

taking $\overline{1} \otimes 1 \otimes p$ to $\overline{1} \otimes \overline{1} \otimes p$, obtained from the associativity and multiplication tensor product identities. If

$$
\alpha : (R/J) \otimes_R P_1 \;\cong\; (R/J) \otimes_R P_2
$$

is an R/J-linear isomorphism, so that (P_1, P_2, α) is a fiber product over the square

$$D = R \times_J R \longrightarrow R$$
$$\downarrow \qquad\qquad \downarrow$$
$$R \longrightarrow R/J \ ,$$

there is an R'/J'-linear isomorphism

$$\alpha' : (R'/J') \otimes_{R'} P_1' \cong (R'/J') \otimes_{R'} P_2'$$

given by $\alpha' = h_2^{-1} \circ (1 \otimes \alpha) \circ h_1$, and an associated fiber product (P_1', P_2', α') over the square

$$D' = R' \times_{J'} R' \longrightarrow R'$$
$$\downarrow \qquad\qquad\qquad \downarrow$$
$$R' \longrightarrow R'/J' \ .$$

Now f induces a ring homomorphism $D \to D'$ taking (r_1, r_2) to $(f(r_1), f(r_2))$; so D' is a D, D-bimodule. The map from $D' \times (P_1, P_2, \alpha)$ to (P_1', P_2', α'), taking $((r', s'), (x, y))$ to $(r' \otimes x, \ s' \otimes y)$, is defined and balanced over D, inducing a D'-linear map

$$\psi : D' \otimes_D (P_1, P_2, \alpha) \longrightarrow (P_1', P_2', \alpha')$$

taking $1 \otimes (x, y)$ to $(1 \otimes x, \ 1 \otimes y)$. As in the proof of (13.12), one can show ψ is an isomorphism for $(P_1, P_2, \alpha) = (R^n, R^n, I)$ for any (Q_1, Q_2, β) diagonally isomorphic to (R^n, R^n, I), and for direct summands (using $*$) of (Q_1, Q_2, β). So ψ is always a D'-linear isomorphism, and the map $K_0(D) \to K_0(D')$ induced by f takes $[(P_1, P_2, \alpha)]$ to $[(P_1', P_2', \alpha')]$.

Recall that the connecting homomorphism $\partial_0 : K_1(R/J) \to K_0(R, J)$ takes the coset of a matrix $C \in GL_n(R/J)$ to

$$[(R^n, R^n, C)] - [(R^n, R^n, I)] \ .$$

For $(P_1, P_2, \alpha) = (R^n, R^n, C)$, the isomorphism $u : R' \otimes_R R^n \cong R'^n$, taking $1 \otimes (r_1, \cdots, r_n)$ to $(f(r_1), \cdots, f(r_n))$, determines a diagonal isomorphism

$$(u, u) : (P', Q', \alpha') \cong (R'^n, R'^n, \overline{f}(C)) \ ,$$

where $\overline{f}$ is applied entrywise. So the map $K_0(R, J) \to K_0(R', J')$, restricting $K_0(D) \to K_0(D')$, takes

$$[(R^n, R^n, C)] - [(R^n, R^n, I)] \quad \text{to} \quad [(R'^n, R'^n, \overline{f}(C)] - [(R'^n, R'^n, I)] \ ,$$

and hence the square

$$
\begin{array}{ccc}
K_1(R/J) & \xrightarrow{\ \partial_0\ } & K_0(R, J) \\
\downarrow & & \downarrow \\
K_1(R'/J') & \xrightarrow{\ \partial_0\ } & K_0(R', J')
\end{array}
$$

commutes, proving ∂_0 is natural.

The map f is taken, by the functors involved, to group homomorphisms from the groups in

$$
K_2(R/J)
$$
$$
\downarrow \cap
$$
$$
St(R) \longrightarrow St(R/J)
$$
$$
\downarrow
$$
$$
GL(R, J) \xrightarrow{\ \subseteq\ } GL(R)
$$
$$
\downarrow
$$
$$
K_1(R, J)
$$

to the corresponding groups with R and J replaced by R' and J', respectively. The squares created by these maps commute. If there exists $z \in St(R)$ going to $x \in K_2(R/J)$ and to $y \in GL(R, J)$, then the image of z in $St(R')$ goes to the image of x in $K_2(R'/J')$ and to the image of y in $GL(R', J')$. Thus the square

$$
\begin{array}{ccc}
K_2(R/J) & \xrightarrow{\ \partial_1\ } & K_1(R, J) \\
\downarrow & & \downarrow \\
K_2(R', J') & \xrightarrow{\ \partial_1\ } & K_1(R', J')
\end{array}
$$

commutes, proving ∂_1 is natural. ∎

(13.22) Example. As argued in Exercise 8, the relative sequence restricts to an exact sequence

$$
K_2(R, J) \longrightarrow K_2(R) \longrightarrow K_2(R/J)
$$
$$
\longrightarrow SK_1(R, J) \longrightarrow SK_1(R) \longrightarrow SK_1(R/J)
$$

when R is a commutative ring. If R is the ring of algebraic integers in a totally imaginary number field (e.g., $R = \mathbb{Z}[i]$), then $SK_1(R) = 1$ but $SK_1(R, 4R) \neq 1$

by (11.33). Exactness of the above sequence shows $K_2(R/4R) \neq 1$. This is our first example where K_2 is nontrivial.

13A. Exercises

1. For J an ideal of a ring R, prove $E'(R, J)$ need not equal $E(R, J)$. *Hint:* If $E'(R, J) = E(R, J)$, prove the standard map $K_1(R, J) \to K_1(R)$ is injective. Find a pair (R, J) for which this map is not injective.

2. Suppose

$$
\begin{array}{ccc}
A & \xrightarrow{\ f_2\ } & R_2 \\
{\scriptstyle f_1}\big\downarrow & & \big\downarrow{\scriptstyle g_2} \\
R_1 & \xrightarrow[\ g_1\ ]{} & R'
\end{array}
$$

is a fiber square of surjective ring homomorphisms. Find ideals I and J of A with $I \cap J = 0$, and isomorphisms from the corners of the square of canonical maps

$$
\begin{array}{ccc}
A & \longrightarrow & R/J \\
\big\downarrow & & \big\downarrow \\
R/I & \longrightarrow & R/(I + J)
\end{array}
$$

to the corresponding corners of the first square, making a cube with commuting faces.

3. In Example (13.4) (iv), prove $I + J$ is the conductor from $\pi_1(A) \times \pi_2(A)$ into A — that is,

$$
I + J = \{a \in A : a(\pi_1(A) \times \pi_2(A)) \subseteq A\}.
$$

In particular, $I + J$ is an ideal of both A and $\pi_1(A) \times \pi_2(A)$.

4. For fiber squares (13.3), prove g_1 injective implies f_2 injective, and g_1 surjective implies f_2 surjective.

5. If R_1, R_2 are subrings of R' and g_1, g_2 are inclusions, prove there is a fiber square in $\mathfrak{Ring}$ (13.3) if and only if $A \cong R_1 \cap R_2$. *Hint:* Show the square of inclusions with $A = R_1 \cap R_2$ is a fiber square, and apply the universal property in (13.4) (vii).

6. Suppose $D = \mathbb{Z} \times_{n\mathbb{Z}} \mathbb{Z}$, where n is a positive integer. If $n \notin \{1, 2, 3, 4, 6\}$, prove D has a nonfree f.g. projective module $(\mathbb{Z}, \mathbb{Z}, \alpha)$, defined as a fiber product

over the fiber square

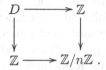

7. In a fiber square of ring homomorphisms (13.3), suppose $g_1(R_1) = R'$, $g_i(R_i^*) = R'^*$ for either $i = 1$ or 2, and R' is a finite ring. Prove each fiber product (R_1^n, R_2^n, α) over this square is a free A-module of free rank n. *Hint:* Use Theorem (10.3) (ii) and the fact that finite rings have stable rank 1.

8. Suppose R is a commutative ring with an ideal J. Prove the relative sequence (13.20) restricts to an exact sequence

$$K_2(R, J) \to K_2(R) \to K_2(R/J)$$

$$\to SK_1(R, J) \to SK_1(R) \to SK_1(R/J) .$$

Hint: The image of the connecting homomorphism ∂_1 is the kernel of $K_1(R, J) \to K_1(R)$, which is induced by an inclusion $GL(R, J) \to GL(R)$.

9. For each integer $n > 1$, prove $K_2(\mathbb{Z}/n\mathbb{Z})$ is cyclic of order 1 or 2, generated by $\{-\bar{1}, -\bar{1}\}$. *Hint:* Use the sequence in Exercise 8, the Bass-Milnor-Serre Theorem (11.33), and the generator of $K_2(\mathbb{Z})$ from §12B, Exercise 6.

10. If p is an odd prime and $n \geq 1$, the group $(\mathbb{Z}/p^n\mathbb{Z})^*$ is cyclic (see Stewart and Tall [87, Proposition A.8]). Prove $K_2(\mathbb{Z}/p^n\mathbb{Z}) = 1$. Then prove $K_2(\mathbb{Z}/m\mathbb{Z}) = 1$ if $m \in \mathbb{Z}$ and $m \notin 4\mathbb{Z}$. *Hint:* Use the argument proving (12.33), and for the second assertion use (12.8).

11. The group $K_2(\mathbb{Z}/4\mathbb{Z})$ is cyclic of order 2 (see Dennis and Stein [75] or Silvester [81]). Prove $K_2(\mathbb{Z}/m\mathbb{Z})$ is cyclic of order 2 for each positive integer $m \in 4\mathbb{Z}$.

12. Suppose J is an ideal of finite index in a ring R and $R^* \to (R/J)^*$ is surjective. Suppose $D = R \times_J R$ and $P, Q \in \mathcal{P}(D)$. Now R is a D-module R_1 via projection $D \to R$ to the first coordinate, and is a D-module R_2 under projection $D \to R$ to the second coordinate. If $R_1 \otimes_D P$ is stably isomorphic to $R_1 \otimes_D Q$ and $R_2 \otimes_D P$ is stably isomorphic to $R_2 \otimes_D Q$, prove P is stably isomorphic to Q. *Hint:* Use the relative sequence and the hint in Exercise 7.

13. Show the conclusion of Exercise 12 can fail if we delete the hypothesis $R^* \to (R/J)^*$ is surjective. *Hint:* Take $R = \mathbb{Z}$ and $J = 5\mathbb{Z}$, and show there exist $(\mathbb{Z}^m, \mathbb{Z}^m, C)$ and $(\mathbb{Z}^m, \mathbb{Z}^m, C')$ in $\mathcal{P}(D)$ that are not stably isomorphic; for this, begin with exactness of the relative sequence.

13B. Excision and the Mayer-Vietoris Sequence

Use of the relative sequence requires the often difficult computation of the relative K-groups $K_n(R, J)$. Following a parallel to algebraic topology, we can use a simplification called "excision," replacing one relative K-group $K_n(R, J)$ by another $K_n(R', J')$, resulting in a "Mayer-Vietoris" exact sequence (13.33) skipping over the relative groups altogether.

(13.23) Definition. An arrow $f : (R, J) \to (R', J')$ in $\mathfrak{Ideal}$ is an **excision** if it restricts to a bijection $J \to J'$.

In the discussion following (13.3) we saw that any map of f_1 in a fiber square of rings

$$
\begin{array}{ccc}
A & \xrightarrow{\ f_2\ } & S \\
{\scriptstyle f_1}\downarrow & & \downarrow{\scriptstyle g_2} \\
R & \xrightarrow{\ g_1\ } & T
\end{array}
$$

is an excision $(A, J) \to (R, \mathfrak{J})$, where $J = \ker(f_2)$ and $\mathfrak{J} = \ker(g_1)$. Conversely, each excision f from (R, J) to (R', J') occurs in the fiber square of rings

$$
\begin{array}{ccc}
R & \longrightarrow & R/J \\
{\scriptstyle f}\downarrow & & \downarrow{\scriptstyle \overline{f}} \\
R' & \longrightarrow & R'/J'.
\end{array}
$$

The word "excision" comes from an earlier use in algebraic topology. To excise is to remove by cutting away. If X is a topological space with a subset Y and an open subset Z whose closure is contained in the interior of Y, the inclusion of pairs

$$(X - Z,\ Y - Z) \longrightarrow (X, Y)$$

is called an excision, the subset Z having been cut away from the second pair to form the first. One of the axioms for a homology theory (Eilenberg and Steenrod [52]) says this inclusion induces an isomorphism of nth homology groups for each n. Now there is a contravariant functor C from the poset of subsets of X to the category of commutative rings, taking each Y to the ring $C(Y)$ of complex-valued continuous functions on Y (with pointwise operations) and taking an inclusion $i : Z \to Y$ to the restriction homomorphism $C(Y) \to C(Z)$, $f \mapsto f \circ i$. Applying C to an excision between pairs of spaces yields an excision in $\mathfrak{Ideal}$

$$(C(X), J) \longrightarrow (C(X - Z), J'),$$

where J and J' are kernels of the horizontal maps in the fiber square of rings

$$
\begin{array}{ccc}
C(X) & \longrightarrow & C(Y) \\
\downarrow & & \downarrow \\
C(X - Z) & \longrightarrow & C(Y - Z),
\end{array}
$$

explaining our use of the word.

Notation: To study an excision $f_1 : (A, J) \to (R, \mathcal{J})$ we put it in a fiber square of rings

$$
\begin{array}{ccc}
A & \xrightarrow{\ f_2\ } & \overline{A} \\
\downarrow{\scriptstyle f_1} & & \downarrow{\scriptstyle g_2} \\
R & \xrightarrow[\ g_1\ ]{} & \overline{R} ,
\end{array}
$$

(13.24)

where $\overline{A} = A/J$, $\overline{R} = R/\mathcal{J}$, f_2 and g_1 are canonical maps, and $g_2 = \overline{f_1}$. Let $D = A \times_J A$ and $D' = R \times_{\mathcal{J}} R$. The excision f_1 induces **excision maps**

$$\varepsilon_0 : K_0(A, J) \ \to \ K_0(R, \mathcal{J}) , \quad [P] \ \mapsto \ [D' \otimes_D P] ,$$

$$\varepsilon_1 : K_1(A, J) \ \to \ K_1(R, \mathcal{J}) , \quad \overline{(a_{ij})} \ \mapsto \ \overline{(f_1(a_{ij}))} ,$$

$$\varepsilon_2 : K_2^S(A, J) \ \to \ K_2^S(R, \mathcal{J}) , \quad \text{restricting}$$

$$St(D) \ \to \ St(D') , \quad x_{ij}((a,b)) \ \mapsto \ x_{ij}(f_1(a), f_1(b)) .$$

By naturality of the relative sequence, f_1 induces the vertical maps in a commutative diagram with exact rows:

$$
\begin{array}{ccccc}
K_2(A, J) & \longrightarrow & K_2(A) & \longrightarrow & K_2(\overline{A}) \\
\downarrow{\scriptstyle \varepsilon_2} & & \downarrow & & \downarrow \\
K_2(R, \mathcal{J}) & \longrightarrow & K_2(R) & \longrightarrow & K_2(\overline{R})
\end{array}
$$

$$
\begin{array}{ccccc}
\longrightarrow K_1(A, J) & \longrightarrow & K_1(A) & \longrightarrow & K_1(\overline{A}) \\
\downarrow{\scriptstyle \varepsilon_1} & & \downarrow & & \downarrow \\
\longrightarrow K_1(R, \mathcal{J}) & \longrightarrow & K_1(R) & \longrightarrow & K_1(\overline{R})
\end{array}
$$

$$
\begin{array}{ccccc}
\longrightarrow K_0(A, J) & \longrightarrow & K_0(A) & \longrightarrow & K_0(\overline{A}) \\
\downarrow{\scriptstyle \varepsilon_0} & & \downarrow & & \downarrow \\
\longrightarrow K_0(R, \mathcal{J}) & \longrightarrow & K_0(R) & \longrightarrow & K_0(\overline{R}) .
\end{array}
$$

The squares in this diagram not involving relative terms are K_n ($n = 0, 1, 2$) of the fiber square (13.24).

From the analogy with homology theories, we might expect the excision maps ε_n to be isomorphisms, showing the relative K-groups to be dependent on the ideal, but somewhat independent of its ambient ring. When the excision maps are well-behaved, we obtain an exact sequence skipping the relative terms:

(13.25) Lemma. *Suppose there is a commutative diagram in* $\mathbb{Z}$-$\mathfrak{Mod}$ *with exact rows*

$$
\begin{array}{ccccccccccc}
E_{n+1} & \xrightarrow{\theta} & A_{n+1} & \xrightarrow{\alpha} & B_{n+1} & \xrightarrow{\sigma} & E_n & \xrightarrow{\theta} & A_n & \xrightarrow{\alpha} & B_n \\
\downarrow{\varepsilon_{n+1}} & & \downarrow{\beta} & & \downarrow{\gamma} & & \downarrow{\varepsilon_n} & & \downarrow{\beta} & & \downarrow{\gamma} \\
F_{n+1} & \xrightarrow{\tau} & C_{n+1} & \xrightarrow{\delta} & D_{n+1} & \xrightarrow{\phi} & F_n & \xrightarrow{\tau} & C_n & \xrightarrow{\delta} & D_n
\end{array}
$$

and ε_n *is an isomorphism. Defining* $\lambda = \theta \circ \varepsilon_n^{-1} \circ \phi$, *there is a sequence*

$$
A_{n+1} \xrightarrow{\begin{bmatrix} \alpha \\ \beta \end{bmatrix}} B_{n+1} \oplus C_{n+1} \xrightarrow{[\gamma \; -\delta]} D_{n+1} \xrightarrow{\lambda} A_n \xrightarrow{\begin{bmatrix} \alpha \\ \beta \end{bmatrix}} B_n \oplus C_n
$$

that is exact at D_{n+1} *and* A_n, *and has zero composite at* $B_{n+1} \oplus C_{n+1}$. *If* ε_{n+1} *is surjective, the sequence is also exact at* $B_{n+1} \oplus C_{n+1}$.

Proof. This is just a diagram chase. Since $\gamma \circ \alpha = \delta \circ \beta$, the composite is zero at $B_{n+1} \oplus C_{n+1}$. Since ε_n is an isomorphism, $\sigma = \varepsilon_n^{-1} \circ \phi \circ \gamma$. So $\lambda \circ \gamma = \theta \circ \sigma = 0$ $= \theta \circ \varepsilon_n^{-1} \circ (\phi \circ \delta) = \lambda \circ \delta$, and the composite is zero at D_{n+1}. And $\alpha \circ \lambda$ $= (\alpha \circ \theta) \circ \varepsilon_n^{-1} \circ \phi = 0$, while $\beta \circ \lambda = \tau \circ \phi = 0$; so the composite is zero at A_n.

If $(b, c) \in B_{n+1} \oplus C_{n+1}$ and $\gamma(b) - \delta(c) = 0$, then $\varepsilon_n \circ \sigma(b) = \phi \circ \gamma(b) = \phi \circ \delta(c) = 0$; so $\sigma(b) = 0$ and $b = \alpha(a)$ for some $a \in A_{n+1}$. Now $\delta(c - \beta(a)) = \delta(c) - \gamma(b) = 0$; so $c - \beta(a) = \tau(f)$ for some $f \in F_{n+1}$. If ε_{n+1} is surjective, $f = \varepsilon_{n+1}(e)$ with $e \in E_{n+1}$. Then $c - \beta(a) = \tau \circ \varepsilon_{n+1}(e) = \beta \circ \theta(e)$, and $\alpha \circ \theta(e) = 0$. Therefore $\begin{bmatrix} \alpha \\ \beta \end{bmatrix}$ takes $a + \theta(e)$ to (b, c), and the sequence is exact at $B_{n+1} \oplus C_{n+1}$.

Suppose $d \in D_{n+1}$ and $\lambda(d) = 0$. Then $\varepsilon_n^{-1} \circ \phi(d) = \sigma(b)$ for some $b \in B_{n+1}$. Now $\phi(d - \gamma(b)) = \phi(d) - \varepsilon_n \circ \sigma(b) = 0$; so $d - \gamma(b) = \delta(c)$ for some $c \in C_{n+1}$. Then $d = d - \gamma(b) + \gamma(b) = \delta(c) + \gamma(b) = [\gamma \; - \delta]((b, -c))$, proving exactness at D_{n+1}.

Finally, suppose $a \in A_n$ and $a \in \ker(\alpha) \cap \ker(\beta)$. Then $a = \theta(e)$ for some $e \in E_n$. Now $\tau \circ \varepsilon_n(e) = \beta \circ \theta(e) = \beta(a) = 0$; so $\varepsilon_n(e) = \phi(d)$ for some $d \in D_{n+1}$. Therefore $\lambda(d) = \theta \circ \varepsilon_n^{-1} \circ \phi(d) = a$, and the sequence is exact at the module A_n. ∎

From here to (13.31) we develop the proof that the excision map ε_0 is an isomorphism. Excision for K_1 and K_2' is discussed in (13.31) and (13.32), leading to the Mayer-Vietoris sequence (13.33).

Consider the excision map ε_0, which restricts $K_0(D) \rightarrow K_0(D')$. Each f.g. projective D-module P is isomorphic to a fiber product (P_1, P_2, α) over the square

$$
\begin{array}{ccc}
D & \longrightarrow & A \\
\downarrow & & \downarrow \\
A & \longrightarrow & \overline{A} \ ,
\end{array}
$$

and both f.g. projective A-modules P_i are isomorphic to fiber products over the square

$$
\begin{array}{ccc}
A & \longrightarrow & \overline{A} \\
\downarrow & & \downarrow \\
R & \longrightarrow & \overline{R} \ .
\end{array}
$$

So P is isomorphic to a fiber product of fiber products.

Notation: For $C \in GL_n(\overline{R})$, there is a fiber product $(R^n, \overline{A}^n, C) \in \mathcal{P}(A)$, and, by (13.12), there are isomorphisms

$$\theta_1 : R \otimes_A (R^n, \overline{A}^n, C) \cong R^n,$$
$$\theta_2 : \overline{A} \otimes_A (R^n, \overline{A}^n, C) \cong \overline{A}^n,$$

with $\theta_i(1 \otimes (x_1, x_2)) = x_i$. If $B, C \in GL_n(\overline{R})$ and $H \in GL_n(\overline{A})$, there is a fiber product

$$\pi(\mathbf{B}, \mathbf{C}, \mathbf{H}) = ((R^n, \overline{A}^n, B), \ (R^n, \overline{A}^n, C), \ \theta_2^{-1} \circ (\cdot H) \circ \theta_2)$$

in $\mathcal{P}(D)$, defined over the square

$$
\begin{array}{ccc}
D & \longrightarrow & A \\
\downarrow & & \downarrow \\
A & \longrightarrow & \overline{A} \ .
\end{array}
$$

Here $\alpha = \theta_2^{-1} \circ (\cdot H) \circ \theta_2$ is an $\overline{A}$-linear isomorphism making the square

$$
\begin{array}{ccc}
\overline{A} \otimes_A (R^n, \overline{A}^n, B) & \xrightarrow{\ \theta_2\ } & \overline{A}^n \\
{\scriptstyle \alpha}\downarrow & & \downarrow{\scriptstyle \cdot H} \\
\overline{A} \otimes_A (R^n, \overline{A}^n, C) & \xrightarrow{\ \theta_2\ } & \overline{A}^n
\end{array}
$$

commute; so H represents α over the coordinates provided by the maps θ_2. Thus every fiber product

$$((R^n, \overline{A}^n, B),\ (R^n, \overline{A}^n, C),\ \beta)$$

is $\pi(B, C, H)$ for a unique matrix H representing β.

These iterated fiber products $\pi(B, C, H)$ need not represent every isomorphism class in $\mathcal{P}(D)$, but they will be useful in the proof that ε_0 is bijective. Here are three lemmas listing their basic properties:

(13.26) Lemma. *If $B, B', C, C' \in GL_n(\overline{R})$ and $H, H' \in GL_n(\overline{A})$, then matrices $M_1, N_1 \in GL_n(R)$ and $M_2, N_2 \in GL_n(\overline{A})$ define a diagonal isomorphism*

$$((\cdot M_1, \cdot M_2),\ (\cdot N_1, \cdot N_2))\ :\ \pi(B, C, H) \longrightarrow \pi(B', C', H')$$

if and only if

$$(\cdot M_1, \cdot M_2)\ :\ (R^n, \overline{A}^n, B) \longrightarrow (R^n, \overline{A}^n, B'),$$

$$(\cdot N_1, \cdot N_2)\ :\ (R^n, \overline{A}^n, C) \longrightarrow (R^n, \overline{A}^n, C')$$

are diagonal isomorphisms of A-modules and $H N_2 = M_2 H'$.

Note: The condition on $(\cdot M_1, \cdot M_2)$ and $(\cdot N_1, \cdot N_2)$ amounts to a relation between B' and B, and between C' and C, by (13.11).

Proof. The given map carries $\pi(B, C, H)$ into $\pi(B', C', H')$ if and only if the diagram

$$\overline{A}^n \xleftarrow{\theta_2} \overline{A} \otimes_A (R^n, \overline{A}^n, B) \xrightarrow{1 \otimes (\cdot M_1, \cdot M_2)} \overline{A} \otimes_A (R^n, \overline{A}^n, B') \xrightarrow{\theta_2} \overline{A}^n$$

$$\downarrow{\cdot H} \qquad\qquad\qquad\qquad\qquad\qquad\qquad\qquad\qquad\qquad\qquad \cdot H' \downarrow$$

$$\overline{A}^n \xleftarrow{\theta_2} \overline{A} \otimes_A (R^n, \overline{A}^n, C) \xrightarrow{1 \otimes (\cdot N_1, \cdot N_2)} \overline{A} \otimes_A (R^n, \overline{A}^n, C') \xrightarrow{\theta_2} \overline{A}^n$$

commutes. Tracing e_i, we find the top composite is $\cdot M_2$ and the bottom composite is $\cdot N_2$. ∎

(13.27) Lemma. *If $B, C \in GL_n(\overline{R}), H \in GL_n(\overline{A}), B', C' \in GL_m(\overline{R})$ and $H' \in GL_m(\overline{A})$, there is a D-linear isomorphism*

$$\pi(B, C, H)\ \oplus\ \pi(B', C', H')\ \cong\ \pi(B \oplus B', C \oplus C', H \oplus H')\ .$$

Proof. For a matrix $M \in GL_r(\overline{R})$ denote $(R^r, \overline{A}^r, M)$ by $F(M)$. If we also have $M' \in GL_s(\overline{R})$, recall the isomorphisms

$$\tau : F(M) \oplus F(M') \cong F(M) * F(M')$$
$$((x_1, y_1), (x_2, y_2)) \longmapsto ((x_1, x_2), (y_1, y_2))$$

from (13.9) and

$$\varepsilon : F(M) * F(M') \cong F(M \oplus M')$$
$$((x_1, x_2), (y_1, y_2)) \longmapsto (x, y)$$

from (13.10) (iii), where x concatenates the coordinates of x_1 with those of x_2, and y does the same for y_1 and y_2. We have a commutative diagram of $\overline{A}$-linear isomorphisms

$$
\begin{array}{ccc}
\overline{A} \otimes_A F(B \oplus B') & \xrightarrow{\;\;\beta\;\;} & \overline{A} \otimes_A F(C \oplus C') \\
\uparrow{\scriptstyle 1 \otimes \varepsilon} & & \uparrow{\scriptstyle 1 \otimes \varepsilon} \\
\overline{A} \otimes_A (F(B) * F(B')) & & \overline{A} \otimes_A (F(C) * F(C')) \\
\uparrow{\scriptstyle 1 \otimes \tau} & & \uparrow{\scriptstyle 1 \otimes \tau} \\
\overline{A} \otimes_A (F(B) \oplus F(B')) & \xrightarrow{\;\;\beta'\;\;} & \overline{A} \otimes_A (F(C) \oplus F(C')) \\
\downarrow{\scriptstyle \delta} & & \downarrow{\scriptstyle \delta} \\
(\overline{A} \otimes_A F(B)) \oplus (\overline{A} \otimes_A F(B')) & \xrightarrow{\;\alpha \oplus \alpha'\;} & (\overline{A} \otimes_A F(C)) \oplus (\overline{A} \otimes_A F(C')) \\
\downarrow{\scriptstyle \theta_2 \oplus \theta_2} & & \downarrow{\scriptstyle \theta_2 \oplus \theta_2} \\
\overline{A}^n \oplus \overline{A}^m & \xrightarrow{(\cdot H) \oplus (\cdot H')} & \overline{A}^n \oplus \overline{A}^m \\
\downarrow & & \downarrow \\
\overline{A}^{n+m} & \xrightarrow{\;\cdot(H \oplus H')\;} & \overline{A}^{n+m} \,,
\end{array}
$$

where $\alpha = \theta_2^{-1} \circ (\cdot H) \circ \theta_2$, $\alpha' = \theta_2^{-1} \circ (\cdot H') \circ \theta_2$ are the isomorphisms in the fiber products

$$\pi(B, C, H) = (F(B), F(C), \alpha) \,,$$
$$\pi(B', C', H') = (F(B'), F(C'), \alpha') \,,$$

and β' is the isomorphism in the fiber product

$$\pi(B, C, H) * \pi(B', C', H') = (F(B) \oplus F(B'), F(C) \oplus F(C'), \beta') \,.$$

The map β at the top is defined to make the top square commute. This top square shows $(\varepsilon \circ \tau, \varepsilon \circ \tau)$ is a diagonal isomorphism from $\pi(B, C, H) *$ $\pi(B', C', H')$ to a fiber product

$$(F(B \oplus B'), F(C \oplus C'), \beta) \ ,$$

which will be $\pi(B \oplus B', C \oplus C', H \oplus H')$ if $\beta = \theta_2^{-1} \circ (\cdot(H \oplus H')) \circ \theta_2$.

Tracing an element $1 \otimes (x, y)$ of the upper left corner $\overline{A} \otimes_A F(B \oplus B')$ down the left side of the diagram,

$$\begin{aligned}
1 \otimes (x, y) &\longmapsto 1 \otimes ((x_1, x_2), (y_1, y_2)) \\
&\longmapsto 1 \otimes ((x_1, y_1), (x_2, y_2)) \\
&\longmapsto (1 \otimes (x_1, y_1), 1 \otimes (x_2, y_2)) \\
&\longmapsto (y_1, y_2) \longmapsto y \ .
\end{aligned}$$

The same thing happens down the right side, so both downward composites are θ_2. Commutativity of the diagram shows $\beta = \theta_2^{-1} \circ (\cdot(H \oplus H')) \circ \theta_2$. ∎

(13.28) Lemma. *If* $B, C \in GL_n(\overline{R})$ *and* $H \in GL_n(\overline{A})$, *there is a* $(R \times_{\mathfrak{g}} R =)$ D'*-linear isomorphism*

$$D' \otimes_D \pi(B, C, H) \ \cong \ (R^n, R^n, Bg_2(H)C^{-1}) \ ,$$

where $g_2 : \overline{A} \to \overline{R}$ *is applied entrywise.*

Proof. Say $F(B) = (R^n, \overline{A}^n, B)$ and $F(C) = (R^n, \overline{A}^n, C)$. There is a commutative diagram of $\overline{R}$-linear isomorphisms:

$$
\begin{array}{ccc}
\overline{R}^n & \xrightarrow{\ \cdot g_2(H)\ } & \overline{R}^n \\
{\scriptstyle \Delta_2}\big\uparrow & & \big\uparrow{\scriptstyle \Delta_2} \\
\overline{R} \otimes_{\overline{A}} \overline{A}^n & \xrightarrow{\ 1\otimes(\cdot H)\ } & \overline{R} \otimes_{\overline{A}} \overline{A}^n \\
{\scriptstyle 1\otimes\theta_2}\big\uparrow & & \big\uparrow{\scriptstyle 1\otimes\theta_2} \\
\overline{R} \otimes_{\overline{A}} (\overline{A} \otimes_A F(B)) & \xrightarrow{\ 1\otimes\alpha\ } & \overline{R} \otimes_{\overline{A}} (\overline{A} \otimes_A F(C)) \\
{\scriptstyle h_1}\big\uparrow & & \big\uparrow{\scriptstyle h_2} \\
\overline{R} \otimes_R (R \otimes_A F(B)) & \xrightarrow{\ \beta\ } & \overline{R} \otimes_R (R \otimes_A F(C)) \\
{\scriptstyle 1\otimes\theta_1}\big\downarrow & & \big\downarrow{\scriptstyle 1\otimes\theta_1} \\
\overline{R} \otimes_R R^n & \xrightarrow{\ \gamma\ } & \overline{R} \otimes_R R^n \\
{\scriptstyle \Delta_1}\big\downarrow & & \big\downarrow{\scriptstyle \Delta_1} \\
\overline{R}^n & \xrightarrow{\ \cdot X\ } & \overline{R}^n
\end{array}
$$

defining α, β, γ, and $X \in GL_n(\overline{R})$ by its commutativity. As in the proof of (13.21),

$$D' \otimes_D \pi(B, C, H) = D' \otimes_D (F(B), F(C), \alpha)$$

$$\cong (R \otimes_A F(B), R \otimes_A F(C), \beta) .$$

Now $(x, y) \in F(B) = (R^n, \overline{A}^n, B)$ if and only if $g_1(x)B = g_2(y)$, where the g_i are applied coordinatewise. Since $g_1 : R \to \overline{R}$ is surjective and B is invertible, for each $y \in \overline{A}^n$ there is an $x \in R^n$ with $(x, y) \in F(B)$. Tracing $g_2(y) \in \overline{R}^n$ along the left side of the diagram from top to bottom,

$$g_2(y) \;\mapsto\; \overline{1} \otimes y \;\mapsto\; \overline{1} \otimes (\overline{1} \otimes (x, y))$$
$$\mapsto\; \overline{1} \otimes (1 \otimes (x, y)) \;\mapsto\; \overline{1} \otimes x \;\mapsto\; g_1(x) .$$

So this composite is $\cdot B^{-1}$ on each $g_2(y)$. In particular, it takes $e_i = g_2(e_i)$ to $e_i B^{-1}$ for each i; so it is $\cdot B^{-1}$ on all vectors.

Similarly, the right side from top to bottom composes to $\cdot C^{-1}$. Since the perimeter of the diagram commutes, $X = Bg_2 (H)C^{-1}$, and the bottom two squares show (θ_1, θ_1) is an isomorphism

$$(R \otimes_A F(B), R \otimes_A F(C), \beta) \cong (R^n, R^n, X) . \qquad \blacksquare$$

One more lemma paves the way for the proof that ε_0 is an isomorphism:

(13.29) Lemma. *Suppose J is an ideal of a ring A and $D = A \times_J A$. Every element of $K_0(A, J)$ has the form*

$$[(P, A^n, \alpha)] - [(A^n, A^n, I)]$$

for some $P \in \mathcal{P}(A)$.

Proof. The group $K_0(A, J)$ is the kernel of the map

$$K_0(\pi_2) : K_0(D) \longrightarrow K_0(A)$$

induced by projection $\pi_2 : D \to A$ to the second coordinate. By (13.7), every object of $\mathcal{P}(D)$ is isomorphic to a fiber product (P, Q, β) where $P, Q \in \mathcal{P}(A)$. Suppose $x \in K_0(A, J)$. Like all other elements of $K_0(D)$,

$$x = [(P, Q, \beta)] - [D^n]$$

for some $P, Q \in \mathcal{P}(A)$ and $n \geq 0$. Regarding A as a D-module via projection to the second coordinate,

$$0 \;=\; K_0(\pi_2)(x) \;=\; [A \otimes_D (P,Q,\beta)] - [A \otimes_D D^n] \;=\; [Q] - [A^n] \,.$$

So Q and A^n are stably isomorphic A-modules. Adding

$$0 \;=\; [(A^m, A^m, 1)] \;-\; [D^m]$$

to x for sufficiently large m permits us to assume $Q \cong A^n$. Replacing (P,Q,β) and D^n by isomorphic copies (P, A^n, α) and (A^n, A^n, I) yields the desired expression for x. ∎

(13.30) Theorem. *Every excision $f : (A, J) \to (R, \mathcal{J})$ in* Ideal *induces an isomorphism*

$$\varepsilon_0 : K_0(A, J) \;\cong\; K_0(R, \mathcal{J}) \,.$$

Proof. Suppose (P, R^n, γ) is a fiber product over the square

$$
\begin{array}{ccc}
D' & \xrightarrow{\ p_2\ } & R \\
{\scriptstyle p_1}\downarrow & & \downarrow \\
R & \longrightarrow & \overline{R} \,.
\end{array}
$$

Then there are $\overline{R}$-linear isomorphisms

$$\overline{R} \otimes_R P \xrightarrow{\ \gamma\ } \overline{R} \otimes_R R^n \xrightarrow{\ \beta\ } \overline{R} \otimes_{\overline{A}} \overline{A}^n$$

$$\alpha$$

and fiber products $M = (P, \overline{A}^n, \alpha)$ and $N = (R^n, \overline{A}^n, \beta)$ over the square

$$
\begin{array}{ccc}
A & \xrightarrow{\ f_2\ } & \overline{A} \\
{\scriptstyle f_1}\downarrow & & \downarrow {\scriptstyle g_2} \\
R & \xrightarrow[\ g_1\ ]{} & \overline{R} \,.
\end{array}
$$

The isomorphisms θ_2 of (13.12) combine to give an $\overline{A}$-linear isomorphism

$$\tau : \overline{A} \otimes_A M \;\cong\; \overline{A}^n \;\cong\; \overline{A} \otimes_A N \,;$$

so there is a fiber product (M, N, τ) defined over the square

$$
\begin{array}{ccc}
D & \longrightarrow & A \\
\downarrow & & \downarrow \\
A & \longrightarrow & \overline{A} \,.
\end{array}
$$

Claim: $D' \otimes_D (M, N, \tau) \cong (P, R^n, \gamma)$.

Proof of Claim. Recall from the proof of (13.21) that $D' \otimes_D (M, N, \tau)$ is isomorphic to

$$(R \otimes_A M, \ R \otimes_A N, \sigma) \, ,$$

where σ is the composite down the left side of the diagram

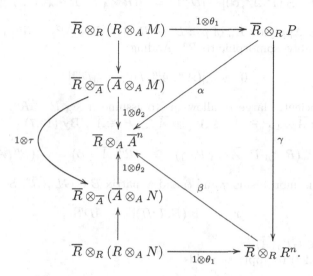

If $(x, y) \in M$, then $\alpha(1 \otimes x) = 1 \otimes y$. The upper left triangle commutes because the upper left corner is generated by elements $\overline{1} \otimes (1 \otimes (x, y))$, and these trace around the triangle as

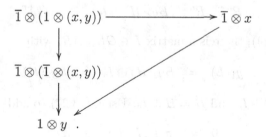

The lower left triangle commutes for the same reason. So the perimeter commutes, showing (θ_1, θ_1) is a diagonal isomorphism from $(R \otimes_A M, \ R \otimes_A N, \ \sigma)$ to (P, R^n, γ), as claimed. ∎

Returning to the proof of the theorem, by Lemma (13.29) each element of $K_0(R, \mathcal{J})$ has the form

$$[(P, R^n, \gamma)] \ - \ [(R^n, R^n, I)] \, ,$$

and both terms are in the image of ε_0 according to the claim. So ε_0 is surjective.

For the prove of injectivity, suppose $x \in \ker(\varepsilon_0)$. By (13.29),

$$x \; = \; [(P, A^n, \alpha)] \; - \; [D^n] \, ,$$

where $P \in \mathcal{P}(A)$. Applying ε_0,

$$0 \; = \; [D' \otimes_D (P, A^n, \alpha)] - [D'^n] \; = \; [(R \otimes_A P, \, R \otimes_A A^n, \; \beta)] - [D'^n] \, ,$$

as in (13.21). Applying K_0 of $p_1 : D' \to R$, $0 = [R \otimes_A P] - [R^n]$ in $K_0(R)$. So $R \otimes_A P$ is stably isomorphic to R^n. Adding

$$0 \; = \; [(A^m, A^m, I)] \; - \; [D^m]$$

to x for sufficiently large m allows us to assume $R \otimes_A P \cong R^n$. And α is an isomorphism $\overline{A} \otimes_A P \cong \overline{A} \otimes_A A^n$; so $\overline{A} \otimes_A P \cong \overline{A}^n$. By (13.7),

$$P \; \cong \; (R \otimes_A P, \overline{A} \otimes_A P, \gamma) \; \cong \; (R^n, \overline{A}^n, \phi) \; = \; (R^n, \overline{A}^n, B)$$

for $\overline{R}$-linear isomorphisms γ and ϕ and a matrix $B \in GL_n(\overline{R})$. So

$$x \; = \; [\pi(B, I, H)] \; - \; [D^n]$$

for some $H \in GL_n(\overline{A})$.

Using (13.28) to apply ε_0 yields

$$0 \; = \; [(R^n, R^n, Bg_2(H))] \; - \; [D'^n]$$

in $K_0(D')$. So, for sufficiently large r,

$$(R^{n+r}, R^{n+r}, Bg_2(H) \oplus I_r) \; \cong \; D'^{n+r}.$$

By Lemma (13.14), there is a matrix $L \in GL_{n+r}(R)$ with

$$g_1(L) \; = \; Bg_2(H) \oplus I_r \; = \; \widehat{B}g_2(\widehat{H})$$

and with $\widehat{B} = B \oplus I_r$ and $\widehat{H} = H \oplus I_r$. Using (13.27) to add

$$0 \; = \; [\pi(I_r, I_r, I_r)] \; - \; [D^r]$$

to x results in the expression

$$x \; = \; [\pi(\widehat{B}, I, \widehat{H})] \; - \; [D^{n+r}] \, ,$$

where $I = I_{n+r}$. Since $I = g_1(L^{-1})\widehat{B}g_2(\widehat{H})$, Lemma (13.11) says $(\cdot L, \cdot \widehat{H})$ is a diagonal isomorphism

$$(R^{n+r}, \overline{A}^{n+r}, \widehat{B}) \; \cong \; (R^{n+r}, \overline{A}^{n+r}, I) \, .$$

So, by Lemma (13.26), there is a diagonal isomorphism

$$((\cdot L, \cdot \widehat{H}), (\cdot I, \cdot I)) : \pi(\widehat{B}, I, \widehat{H}) \;\cong\; \pi(I, I, I) .$$

Since $\pi(I, I, I) \cong D^{n+r}$, $x = 0$, proving ε_0 is injective. ∎

Note: If $\mathcal{J}$ is an ideal of a ring R, there is a simplest ring with ideal $\mathcal{J}$. Define $\mathcal{J}^+$ to be the additive group $\mathcal{J} \oplus \mathbb{Z}$, endowed with multiplication

$$(x, m) \cdot (y, n) \;=\; (xy + my + nx, mn) ,$$

where my, nx are defined as usual for x, y in an additive group. Then $\mathcal{J}^+$ is a ring, and $J = \mathcal{J} \oplus 0$ is an ideal of $\mathcal{J}^+$. Further, the map $(x, m) \mapsto x + m1_R$ is an excision

$$(\mathcal{J}^+, J) \;\rightarrow\; (R, \mathcal{J}) .$$

By Theorem (13.30), the associated excision map is an isomorphism

$$\varepsilon_0 : K_0(\mathcal{J}^+, J) \;\cong\; K_0(R, \mathcal{J}) ,$$

showing $K_0(R, \mathcal{J})$ *depends only on the ideal* $\mathcal{J}$, *and not on the choice of ambient ring* R *with* $\mathcal{J}$ *as an ideal.*

As the following example of Swan [71] shows, K_1 is not as well-behaved as K_0 in regard to excision.

(13.31) Example. Suppose J is an ideal of a ring R, multiplication is commutative on J, but there exist $r \in R$, $x \in J$ with $rx \neq xr$. (For instance, take

$$R = \left\{ \begin{bmatrix} a & b \\ 0 & c \end{bmatrix} \in M_2(\mathbb{Z}) \right\}, \; J = \left\{ \begin{bmatrix} 0 & b \\ 0 & 0 \end{bmatrix} \in M_2(\mathbb{Z}) \right\}$$

$$r = \begin{bmatrix} 1 & 0 \\ 0 & 0 \end{bmatrix}, \; x = \begin{bmatrix} 0 & 1 \\ 0 & 0 \end{bmatrix} .)$$

Let S denote the smallest subring $J + \mathbb{Z}1_R$ of R containing J. Consider the excision $(S, J) \rightarrow (R, J)$ given by inclusion. The element

$$e_{21}(r)e_{12}(x)e_{21}(r)^{-1} \;=\; \begin{bmatrix} 1 - xr & x \\ -rxr & 1 + rx \end{bmatrix}$$

belongs to $E(R, J) \cap GL(S, J)$, so it represents a coset in the kernel of the excision map

$$\varepsilon_1 : K_1(S, J) \;\rightarrow\; K_1(R, J) .$$

But S is a commutative ring, and the determinant of this matrix is

$$
\begin{aligned}
(1 - xr)&(1 + rx) - (-rxr)x \\
&= 1 + rx - xr - (xr)(rx) + r(xr)x \\
&= 1 + rx - xr - rxxr + rxxr \neq 1 ;
\end{aligned}
$$

so this matrix does not belong to $E(S, J)$ and represents a nontrivial element of $K_1(S, J)$.

However, a couple of large classes of excisions do induce bijections on $K_1(\ ,\ )$ and surjections on $St'(\ ,\ )$ and $K_2'(\ ,\ )$, where

$$
\begin{aligned}
St'(R, J) &= \ker\ St(R) \to St(R/J) \\
K_2'(R, J) &= \ker\ K_2(R) \to K_2(R/J) .
\end{aligned}
$$

Suppose a ring B is a direct (i.e., cartesian) product $B_1 \times \cdots \times B_n$ of finitely many rings B_i. A **subdirect product** of B is a subring $A \subseteq B$ for which each projection $B \to B_i$ carries A onto B_i.

(13.32) Theorem. *An excision* $f : (A, J) \to (B, \mathcal{J})$ *in* $\mathfrak{Ideal}$ *induces an isomorphism*

$$
\varepsilon_1 : K_1(A, J) \cong K_1(B, \mathcal{J})
$$

and surjective maps

$$
\begin{aligned}
St'(A, J) &\to St'(B, \mathcal{J}) \\
K_2'(A, J) &\to K_2'(B, \mathcal{J})
\end{aligned}
$$

if $f : A \to B$ *is either surjective or the inclusion of a subdirect product.*

Proof. Since $GL(A, J)$ consists of the matrices with diagonal entries in $1 + J$ and off-diagonal entries in J, with an inverse matrix of the same form, and $GL(B, \mathcal{J})$ has the corresponding description in terms of $\mathcal{J}$, and since f restricts to a bijection $J \to \mathcal{J}$, the map $GL(f)$ restricts to an isomorphism $GL(A, J) \cong GL(B, \mathcal{J})$.

We will be done once we show f induces surjective maps on $E(\ ,\ )$ and $St'(\ ,\ )$, since f induces a commutative diagram with exact rows

$$
\begin{array}{ccccccccc}
1 & \longrightarrow & K_2'(A, J) & \longrightarrow & St'(A, J) & \longrightarrow & E(A, J) & \longrightarrow & 1 \\
& & \downarrow & & \downarrow & & \downarrow & & \\
1 & \longrightarrow & K_2'(B, \mathcal{J}) & \longrightarrow & St'(B, \mathcal{J}) & \longrightarrow & E(B, \mathcal{J}) & \longrightarrow & 1
\end{array}
$$

by (13.18) and the paragraph following it. If the middle vertical is surjective and the right vertical is an isomorphism, the left vertical must be surjective by the Snake Lemma.

Suppose first that f is surjective. Then f induces surjective maps $E(A) \to E(B)$ and $St(A) \to St(B)$, since the generators $e_{ij}(b)$, $x_{ij}(b)$ lift. Then the generators $ee_{ij}(c)e^{-1}$ ($c \in \mathfrak{J}, e \in E(B)$) of $E(B, \mathfrak{J})$ also lift to $E(A, J)$, and the generators $xx_{ij}(c)x^{-1}$ ($c \in \mathfrak{J}, x \in St(B)$) of $St'(B, \mathfrak{J})$, described in (13.18), also lift to $St'(A, J)$ under $St(f)$.

Suppose instead that $B = B_1 \times \cdots \times B_n$ with subdirect product A, and the excision f is inclusion of A in B. Then $\mathfrak{J} = f(J) = J = J_1 \times \cdots \times J_n$ for ideals $J_i \lhd B_i$. Entrywise projections to coordinates combine to form isomorphisms

$$GL(B) \quad \cong \quad GL(B_1) \times \cdots \times GL(B_n) \ ,$$
$$E(B) \quad \cong \quad E(B_1) \times \cdots \times E(B_n) \ ,$$

taking $e_{ij}((b_1, \ldots, b_n))$ to $(e_{ij}(b_1), \ldots, e_{ij}(b_n))$. By (12.8) we also have an isomorphism

$$St(B) \quad \cong \quad St(B_1) \times \cdots \times St(B_n)$$

carrying $x_{ij}((b_1, \ldots, b_n))$ to $(x_{ij}(b_1), , \ldots, x_{ij}(b_n))$.

The group $E(J)$ is generated by the transvections $e_{ij}(c_k)$ with c_k of the form $(0, \ldots, c, \ldots, 0)$, $c \in J_k$, $1 \le k \le n$. So $E(B, J)$ is generated by elements $ee_{ij}(c_k)e^{-1}$ with $e \in E(B)$. Each transvection over B has the same image in $E(B_k)$ as a transvection over A; so e has the same image in $E(B_k)$ as some $d \in E(A)$. Then $ee_{ij}(c_k)e^{-1} = de_{ij}(c_k)d^{-1} \in E(A, J)$. So $E(A, J) = E(B, J)$.

The same argument with x's replacing e's shows the generators of $St'(B, J)$ come from $St'(A, J)$. ∎

The excision theorems (13.30) and (13.32) are enough to produce the following exact **Mayer-Vietoris sequence**, paralleling an exact sequence by the same name among homology groups of spaces.

(13.33) Theorem. *Suppose we have a fiber square in* $\mathfrak{Ring}$

$$
\begin{array}{ccc}
A & \xrightarrow{f_2} & R_2 \\
{\scriptstyle f_1}\downarrow & & \downarrow{\scriptstyle g_2} \\
R_1 & \xrightarrow[g_1]{} & R'
\end{array}
$$

with g_1 surjective. If f_* denotes $K_n(f)$ with n determined by the context, there is an exact sequence

$$K_1(A) \xrightarrow{\begin{bmatrix} f_{1*} \\ f_{2*} \end{bmatrix}} K_1(R_1) \oplus K_1(R_2) \xrightarrow{[g_{2*} \ \ -g_{1*}]} K_1(R')$$

$$\xrightarrow{\lambda_0} K_0(A) \xrightarrow{\begin{bmatrix} f_{1*} \\ f_{2*} \end{bmatrix}} K_0(R_1) \oplus K_0(R_2) \xrightarrow{[g_{2*} \quad -g_{1*}]} K_0(R') \ .$$

If also g_2 is surjective or f_1 is inclusion of a subdirect product, this exact sequence extends to the left to include

$$K_2(A) \xrightarrow{\begin{bmatrix} f_{1*} \\ f_{2*} \end{bmatrix}} K_2(R_1) \oplus K_2(R_2) \xrightarrow{[g_{2*} \quad -g_{1*}]} K_2(R') \xrightarrow{\lambda_1} \ .$$

Proof. Since the square is a fiber square and g_1 is surjective, f_2 is also surjective. Say $J = \ker(f_2)$ and $\mathcal{J} = \ker(g_1)$. Then the square is the perimeter of a commutative diagram

$$
\begin{array}{ccc}
A & \longrightarrow & A/J \cong R_2 \\
f_1 \downarrow & \quad \bar{f}_1 \downarrow & \quad \downarrow g_2 \\
R_1 & \longrightarrow & R_1/\mathcal{J} \cong R' ,
\end{array}
$$

in which the left square is also a fiber square. So we can derive our sequence from the special case in which $A/J = R_2$, $R_1/\mathcal{J} = R'$, $\bar{f}_1 = g_2$, and f_2, g_1 are canonical maps. We make these assumptions.

Then the excision $f_1 : (A, J) \to (R_1, \mathcal{J})$ induces the vertical maps in a commutative diagram connecting the relative sequences for (A, J) and $(R_1, \mathcal{J})$. By (13.30) and the proof of (13.32), f_1 induces isomorphisms on $K_0(\ ,\ )$ and $GL(\ ,\ )$, and hence a surjective map on $K_1(\ ,\ )$. By the diagram chase (13.25), λ_0 is defined and our sequence is exact at $K_1(R_1) \oplus K_1(R_2)$, $K_1(R')$, and $K_0(A)$. Since K_0 of the square commutes, the composite is zero at $K_0(R_1) \oplus K_0(R_2)$.

If $x \in K_0(R_1)$ and $y \in K_0(R_2)$, we can write

$$x = [P] - [R_1^n] \ , \quad y = [Q] - [R_2^n]$$

for $P \in \mathcal{P}(R_1)$, $Q \in \mathcal{P}(R_2)$, and some $n \geq 0$. Suppose (x, y) goes to 0 in $K_0(R')$. Then

$$[R' \otimes_{R_1} P] = [R' \otimes_{R_2} Q]$$

in $K_0(R')$. Adding $0 = [R_i^m] - [R_i^m]$ to x and y, we can assume there is an R'-linear isomorphism

$$\alpha : R' \otimes_{R_1} P \cong R' \otimes_{R_2} Q \ .$$

Then $z = [(P, Q, \alpha)] - [A^n] \in K_0(A)$ maps to (x, y), proving exactness at $K_0(R_1) \oplus K_0(R_2)$.

If g_2 (hence also f_1) is surjective, or f_1 is the inclusion of a subdirect product, then f_1 induces an isomorphism on $K_1(\ ,\ )$ and a surjective map on $K_2'(\ ,\ ) = $ image of $K_2^S(\ ,\ )$. Applying the diagram chase (13.25) to

$$
\begin{array}{ccccccccc}
K_2'(A, J) & \xrightarrow{\subseteq} & K_2(A) & \longrightarrow & K_2(R_2) & \longrightarrow & K_1(A, J) & \longrightarrow & K_1(A) \to \cdots \\
\downarrow & & \downarrow & & \downarrow & & \downarrow & & \downarrow \\
K_2'(R_1, \mathfrak{J}) & \xrightarrow{\subseteq} & K_2(R_1) & \longrightarrow & K_2(R') & \longrightarrow & K_1(R_1, \mathfrak{J}) & \longrightarrow & K_1(R_1) \to \cdots,
\end{array}
$$

we get the map λ_1 and exactness at the remaining terms $K_2(R_1) \oplus K_2(R_2)$, $K_2(R')$, and $K_1(A)$. ∎

Note: When using the Mayer-Vietoris sequence, remember that K_1 is generally written as a multiplicative group; so $[g_{2*} - g_{1*}](x, y)$ is really

$$
g_{2*}(x)[g_{1*}(y)]^{-1}.
$$

The same comment applies to K_2.

To illustrate the use of the Mayer-Vietoris sequence, we recover a theorem of Rim dating back to the origins of algebraic K-theory. Projective modules made their debut in Cartan and Eilenberg's foundational text on homological algebra in 1956. In their treatment of the cohomology of a finite group G, these authors raised the question of whether or not f.g. projective $\mathbb{Z}G$-modules are free. The paper by Rim [59] contains the first appearance of the projective class group of an arbitrary ring R ($\widetilde{K}_0(R)$ in our notation) and answers this question. Rim's notation, definition, and theorem suggest that the projective class group was originally seen as a generalization of the ideal class group of Dedekind, and not as a modification of Grothendieck's K_0, which appeared in 1958 in a geometric context.

Our proof of Rim's Theorem differs from his, which predates the appearance of $K_1(R)$ in 1962. Suppose p is a prime and $\zeta_p = e^{2\pi i/p}$, an element of order p in $\mathbb{C}^*$.

Recall, from Example (13.4) (v), the Rim square

$$
\begin{array}{ccc}
\mathbb{Z}C_p & \xrightarrow{f_2} & \mathbb{Z}[\zeta_p] \\
f_1 \downarrow & & \downarrow g_2 \\
\mathbb{Z} & \xrightarrow{g_1} & \mathbb{Z}/p\mathbb{Z}
\end{array}
\qquad,\qquad
\begin{array}{ccc}
\alpha & \longrightarrow & \zeta_p \\
\downarrow & & \downarrow \\
1 & \longrightarrow & \overline{1}
\end{array}
\qquad,
$$

a fiber square of surjective ring homomorphisms.

Consider the Mayer-Vietoris sequence:

$$
\cdots \longrightarrow K_1(\mathbb{Z}) \oplus K_1(\mathbb{Z}[\zeta_p]) \xrightarrow{[g_2^* \ -g_1^*]} K_1(\mathbb{Z}/p\mathbb{Z})
$$

$$\xrightarrow{\lambda_0} K_0(\mathbb{Z}C_p) \xrightarrow{\begin{bmatrix} f_1^* \\ f_2^* \end{bmatrix}} K_0(\mathbb{Z}) \oplus K_0(\mathbb{Z}[\zeta_p]) \xrightarrow{[g_2^* \ -g_1^*]} K_0(\mathbb{Z}/p\mathbb{Z}) \ .$$

Since $\mathbb{Z}[\zeta_p]$ is the full ring of algebraic integers in the number field $\mathbb{Q}(\zeta_p)$ (see Washington [97, Proposition 1.2]), the computation of Bass-Milnor-Serre (11.33) says $SK_1(\mathbb{Z}[\zeta_p]) = 1$. So $K_1(\mathbb{Z}[\zeta_p])$ is represented by the 1×1 matrices in $\mathbb{Z}[\zeta_p]^*$. For each integer a with $0 < a < p$ these units include

$$u(a) \;=\; \frac{1 - \zeta_p^a}{1 - \zeta_p} \;=\; 1 + \zeta_p + \cdots + \zeta_p^{a-1}$$

since $ab \equiv 1 \pmod{p}$ for some integer $b > 0$ and

$$\frac{1 - \zeta_p}{1 - \zeta_p^a} \;=\; \frac{1 - \zeta_p^{ab}}{1 - \zeta_p^a} \;=\; 1 + \zeta_p^a + \cdots + \zeta_p^{a(b-1)} \ .$$

Since $\mathbb{Z}/p\mathbb{Z}$ is a field, $K_1(\mathbb{Z}/p\mathbb{Z})$ is also represented by units, and the induced map $g_2^* : K_1(\mathbb{Z}[\zeta_p]) \to K_1(\mathbb{Z}/p\mathbb{Z})$ takes $[u(a)]$ to $[\bar{a}]$ for each $\bar{a} \in \mathbb{Z}/p\mathbb{Z}^*$. So g_2^* is surjective on K_1, and λ_0 must be the zero map.

Also, $g_1^* : K_0(\mathbb{Z}) \to K_0(\mathbb{Z}/p\mathbb{Z})$ is an isomorphism, since these are infinite cyclic groups and g_1^* takes the generator $[\mathbb{Z}]$ to the generator $[\mathbb{Z}/p\mathbb{Z}]$. So the last two maps in the Mayer-Vietoris sequence form a short exact sequence and

$$
\begin{array}{ccc}
K_0(\mathbb{Z}C_p) & \xrightarrow{\ f_2^*\ } & K_0(\mathbb{Z}[\zeta_p]) \\
{\scriptstyle f_1^*}\big\downarrow & & \big\downarrow{\scriptstyle g_2^*} \\
K_0(\mathbb{Z}) & \xrightarrow[\ g_1^*\ ]{} & K_0(\mathbb{Z}/p\mathbb{Z})
\end{array}
$$

is a fiber square in $\mathbb{Z}\text{-}\mathcal{M}\mathfrak{o}\mathfrak{d}$. Since g_1^* is an isomorphism, it follows that f_2^* is an isomorphism. Of course f_2^* carries $n[\mathbb{Z}C_p]$ to $n[\mathbb{Z}[\zeta_p]]$ for each integer n. So it induces an isomorphism

$$\widetilde{K}_0(\mathbb{Z}C_p) \;\cong\; \widetilde{K}_0(\mathbb{Z}[\zeta_p]) \ .$$

As in (7.52), the latter group is isomorphic to the ideal class group $Cl(\mathbb{Z}[\zeta_p])$, which has been known to be nontrivial for some primes p since the 19th century. (For a proof when $p = 23$, see Washington [97, p. 7].)

(13.34) Rim's Theorem. *For each prime p, there is an isomorphism of abelian groups*

$$\widetilde{K}_0(\mathbb{Z}C_p) \;\cong\; Cl(\mathbb{Z}[\zeta_p]) \ .$$

So for some primes p, there exist nonfree f.g. projective $\mathbb{Z}C_p$-modules. ∎

As with the relative sequence, the Mayer-Vietoris sequence restricts to

$$K_2(A) \to K_2(R_1) \oplus K_2(R_2) \to K_2(R')$$

$$\to SK_1(A) \to SK_1(R_1) \oplus SK_1(R_2) \to SK_1(R')$$

when the rings A, R_1, R_2, R' in a suitable fiber square are commutative rings – see Exercise 2. Applying this to the Rim square, we see that $K_2(\mathbb{Z}/p\mathbb{Z}) \to SK_1(\mathbb{Z}C_p)$ is surjective. But $K_2(\mathbb{Z}/p\mathbb{Z})$ vanishes by (12.33)! So $SK_1(\mathbb{Z}C_p) = 1$ for all primes p. We do not highlight this result, since a better computation, (13.35), follows shortly.

The following application of excision for K_1 further illustrates a general principle that K-theory computations for rings of algebraic integers lead to similar computations for $\mathbb{Z}G$ with G a finite group. Further examples of this appear in Exercise 3.

(13.35) Theorem. *If C_n is a cyclic group of finite order n, then $SK_1(\mathbb{Z}C_n) = 1$. So every matrix in $GL(\mathbb{Z}C_n)$ can be reduced by elementary row transvections to a diagonal matrix $diag(u, 1, 1, \dots)$.*

Proof. Suppose α generates C_n. Recall from (8.5) (iii) that the Chinese Remainder Theorem leads to a $\mathbb{Q}$-algebra isomorphism

$$\psi : \mathbb{Q}C_n \;\cong\; \prod_{\substack{d \mid n \\ d > 0}} \mathbb{Q}(\zeta_d)$$

taking α to $\zeta_d = e^{2\pi i/d}$ in the d-coordinate for each d. Let $\Phi_\mathbf{d}(\mathbf{x})$ denote the cyclotomic polynomial $\prod(x - \zeta_d^r)$ where r runs through the positive integers less than d relatively prime to d. We assume here some elementary properties of cyclotomic polynomials (proved, for instance, in §36 of van der Waerden [91]): For each positive integer d, $\Phi_d(x)$ is the minimal polynomial of ζ_d over $\mathbb{Q}$; so $[\mathbb{Q}(\zeta_d) : \mathbb{Q}] = \phi(d)$, where ϕ is Euler's totient function. The coefficients of $\Phi_d(x)$ are integers. For different positive integers d and e, $\Phi_d(x)$ and $\Phi_e(x)$ are nonassociate irreducibles in the factorial ring $\mathbb{Z}[x]$. The cyclotomic polynomials can be computed by applying the following identities (in which p is a positive prime):

C1 : $\qquad\qquad \Phi_p(x) \;=\; x^{p-1} + x^{p-2} + \cdots + x + 1 \;;$

C2 : $\qquad\qquad \Phi_{pn}(x) \;=\; \begin{cases} \Phi_n(x^p) & \text{if } p \mid n \\[2mm] \dfrac{\Phi_n(x^p)}{\Phi_n(x)} & \text{if } p \nmid n \;; \end{cases}$

C3 : $\qquad\qquad \Phi_n(x) \;=\; \prod_{d \mid n}(x^d - 1)^{\mu(n/d)},$

where the Möbius function μ is defined so that $\mu(m) = 0$ if m is divisible by the square of a prime, $\mu(p_1 \cdots p_k) = (-1)^k$ for distinct positive primes $p_1, \ldots, p_k$, and $\mu(1) = 1$.

The elements of $\mathbb{Z}C_n$ consist of all $p(\alpha)$ with $p(x) \in \mathbb{Z}[x]$. If D is a set of positive divisors of n, following ψ by projection to the D-coordinates, we define $\mathcal{O}(D)$ to be the image of $\mathbb{Z}C_n$ in $\prod_{d \in D} \mathbb{Q}(\zeta_d)$. Then $\mathcal{O}(d) = \mathcal{O}(\{d\})$ is $\mathbb{Z}[\zeta_d]$ and $\mathcal{O}(D)$ is a subdirect product of $\prod_{d \in D} \mathbb{Z}[\zeta_d]$.

The kernel of the surjective map $\mathbb{Z}C_n \to \mathcal{O}(D)$ is

$$\{p(\alpha) : p(x) \in \mathbb{Z}[x] \cap \prod_{d \in D} \Phi_d(x)\mathbb{Q}[x]\}$$

$$= \{p(\alpha) : p(x) \in \prod_{d \in D} \Phi_d(x)\mathbb{Z}[x]\}$$

$$= \prod_{d \in D} \Phi_d(\alpha)\mathbb{Z}C_n ,$$

a principal ideal of $\mathbb{Z}C_n$. So we define

$$\Phi_D(x) = \prod_{d \in D} \Phi_d(x) ,$$

and this kernel is generated by $\Phi_D(\alpha)$. If $E \subseteq D$, the kernel of the projection $\mathcal{O}(D) \to \mathcal{O}(E)$ is the image in $\mathcal{O}(D)$ of the kernel of $\mathbb{Z}C_n \to \mathcal{O}(E)$; so it is the principal ideal generated by $\Phi_E(\alpha_D)$, where α_D is the image of α in $\mathcal{O}(D)$; α_D has d-coordinate ζ_d for each $d \in D$.

If E, F are nonempty and nonoverlapping subsets of D with $D = E \cup F$, then $\mathcal{O}(D)$ is a subdirect product of $\mathcal{O}(E) \times \mathcal{O}(F)$. As in (13.4) (iv), there is a fiber square of rings

$$\begin{array}{ccc} \mathcal{O}(D) & \xrightarrow{\;p_F\;} & \mathcal{O}(F) \\ {\scriptstyle p_E}\downarrow & & \downarrow{\scriptstyle g_2} \\ \mathcal{O}(E) & \xrightarrow[\;g_1\;]{} & R' \end{array}$$

where p_E, p_F are projections. So $\ker(g_1) = p_E(\ker p_F)$ is generated by $\Phi_F(\alpha_E)$, with e-coordinate

$$\prod_{d \in F} \Phi_d(\zeta_e)$$

for each $e \in E$.

The following computation shows the effect of evaluating a cyclotomic polynomial at the "wrong" root of unity.

(13.36) Lemma. *Suppose* a, b, p, *and* v *are positive integers, with* $a \neq b$ *and* p *a prime.*

(i) *If* $a/b = p^v$, *then* $\Phi_a(\zeta_b)$ *is associate to* p *in* $\mathbb{Z}[\zeta_b]$.

(ii) *If* $b/a = p^v$, *then* $\Phi_a(\zeta_b)^r$ *is associate to* p *in* $\mathbb{Z}[\zeta_b]$, *where the exponent* $r = \phi(p^{u+v})/\phi(p^u)$ *and* p^u *is the largest power of* p *dividing* a.

(iii) *Otherwise* $\Phi_a(\zeta_b)$ *is a unit in* $\mathbb{Z}[\zeta_b]$.

Note: As shown, for instance, in Washington [97, Theorem 2.6], the ring $\mathbb{Z}[\zeta_d]$ is the full ring of algebraic integers in the number field $\mathbb{Q}(\zeta_d)$; so it is a Dedekind domain. So condition (ii) uniquely determines the ideal $\Phi_a(\zeta_b)\mathbb{Z}[\zeta_b]$. Also, the exponent r is p^v unless $u = 0$.

Proof of Lemma. From cyclotomic identities $C1$ and $C2$,

$$\Phi_n(1) \;=\; \begin{cases} p & \textit{if } \; n = p^v \\ 1 & \textit{if } \; n \; \textit{is composite.} \end{cases}$$

Let a' denote the product of distinct positive primes $p_1, \ldots, p_k$ dividing a. By $C2$, $\Phi_a(\zeta_b) = \Phi_{a'}(\zeta_{b'})$, where $\zeta_{b'} = \zeta_b^{a/a'}$ has order

$$b' \;=\; \frac{b}{gcd(b, a/a')} \, .$$

Suppose q is one of the primes $p_1, \ldots, p_k$. The positive divisors of a' fall into two sets, $D =$ those not divisible by q, and qD. These sets are in bijective correspondence $d \leftrightarrow qd$, with $\mu(a'/d) = -\mu(a'/qd)$. So, by $C3$, $\Phi_{a'}(x)$ is a product of the factors

$$\left(\frac{x^{dq} - 1}{x^d - 1} \right)^{\pm 1}$$

for $d \in D$.

Suppose first that $a \nmid b$ and $b \nmid a$. Then $a' \nmid b'$ and $b' \nmid a'$. Order the prime factors of a' so that $q = p_k \nmid b'$. The factors of $\Phi_{a'}(\zeta_{b'})$ are

$$\left(\frac{(\zeta_{b'}^d)^q - 1}{\zeta_{b'}^d - 1} \right)^{\pm 1} \qquad (d \mid (a'/q)) \, ,$$

where $\zeta_{b'}^d$ has order $b'/gcd(d, b')$, which is not 1 (since $b' \nmid a'$) and is relatively prime to q. So each of these factors is a unit in $\mathbb{Z}[\zeta_{b'}^d] \subseteq \mathbb{Z}[\zeta_b]$, and $\Phi_a(\zeta_b) = \Phi_{a'}(\zeta_{b'})$ is a unit.

Next assume $b|a$; so $b'|a'$. Since $a \neq b$, some prime divides a more than b; so that prime divides a' but not b'. Say $b' = p_1 \cdots p_j$ with $j < k$ and $a' = p_1 \cdots p_k$. Let a'' denote $a'/p_1 = p_2 \cdots p_k$. By $C2$, taking $q = p_1$,

$$\Phi_{a'}(\zeta_{b'}) = \frac{\Phi_{a''}(\zeta_{b'}^q)}{\Phi_{a''}(\zeta_{b'})},$$

where the denominator is a unit in $\mathbb{Z}[\zeta_{b'}] \subseteq \mathbb{Z}[\zeta_b]$ by the preceding paragraph. Then $\Phi_a(\zeta_b)$ is associate to $\Phi_{a''}(\zeta_{b''})$ where $b'' = b'/p_1 = p_2 \cdots p_j$. In effect, p_1 was canceled. Proceeding inductively, $\Phi_a(\zeta_b)$ is associate to $\Phi_d(1)$ where $d = p_{j+1} \cdots p_k$. Any prime dividing d divides a' but not b', and so it divides a/b; so if $a/b = p^v$, then $d = p$ and $\Phi_a(\zeta_b)$ is associate to $\Phi_p(1) = p$. Any prime dividing a/b divides a' but not b', and so it divides d; so if a/b is composite, d is composite and $\Phi_a(\zeta_b)$ is associate to $\Phi_d(1) = 1$.

For the remaining cases, suppose $a|b$. Consider the fiber square

$$
\begin{array}{ccc}
\mathcal{O}(\{a,b\}) & \xrightarrow{\;p_b\;} & \mathbb{Z}[\zeta_b] \\
{\scriptstyle p_a}\downarrow & & \downarrow{\scriptstyle g_2} \\
\mathbb{Z}[\zeta_a] & \xrightarrow[\;g_1\;]{} & R' \; .
\end{array}
$$

All maps here are surjective, by its construction in (13.4) (iv). Then there is a composite isomorphism of rings

$$
\theta : \frac{\mathbb{Z}[\zeta_a]}{\Phi_b(\zeta_a)\mathbb{Z}[\zeta_a]} \;\cong\; R' \;\cong\; \frac{\mathbb{Z}[\zeta_b]}{\Phi_a(\zeta_b)\mathbb{Z}[\zeta_b]} \; .
$$

This offers a way to reverse the roles of a and b. If b/a is composite, we have just shown $\Phi_b(\zeta_a)$ is a unit. So R' is trivial and $\Phi_a(\zeta_b)$ is a unit.

It only remains to deal with the case $b/a = p^v$. If R is any Dedekind domain and $M_1, \ldots, M_r$ are maximal ideals of R, and $e(1), \ldots, e(r)$ are positive integers, the Chinese Remainder Theorem isomorphism

$$
\frac{R}{M_1^{e(1)} \cdots M_r^{e(r)}} \;\cong\; \frac{R}{M_1^{e(1)}} \times \cdots \times \frac{R}{M_r^{e(r)}}
$$

decomposes the quotient as a product of local rings. There are exactly r maximal ideals $\mathcal{M}_i$ of this quotient, since they correspond to the maximal ideals $\overline{R} \times \cdots \times \overline{M_i} \times \cdots \times \overline{R}$ in the product. And $e(i)$ is the least positive integer e with $\mathcal{M}_i^e = \mathcal{M}_i^{e+1}$. So r and the exponents $e(i)$ are determined by the ring structure of the quotient.

Reducing coefficients mod p,

$$
\Phi_{p^m}(x) \;=\; \frac{x^{p^m} - 1}{x^{p^{m-1}} - 1} \;=\; (x-1)^{\phi(p^m)}
$$

in $\mathbb{F}_p[x]$ for each $m > 0$. So, by Kummer's Theorem (7.47), p is totally ramified in $\mathbb{Z}[\zeta_{p^m}]$; that is, its ramification is the degree $\phi(p^m)$ of $\mathbb{Q}(\zeta_{p^m})$ over $\mathbb{Q}$. It is also well-known that p does not ramify in $\mathbb{Z}[\zeta_t]$ if p does not divide t (see Washington [97], Proposition 2.3). If $F_1 \subseteq F_2$ is a Galois extension of number fields with rings of algebraic integers $R_1 \subseteq R_2$ and P is a maximal ideal of R_1, then, for each maximal ideal Q of R_2 dividing PR_2, the ramification index $e(Q/P) = e$

and the residue degree $f(Q/P) = f$ are the same (see §7D, Exercise 15); so by (7.37), $efg = [F_2 : F_1]$, where g is the number of Q in the factorization of PR_2. And by §7D, Exercise 1, e, f, and g are multiplicative in towers of Galois extensions.

So in $\mathbb{Z}[\zeta_a]$ the ideal generated by p has maximal ideal factorization

$$(P_1 \cdots P_g)^{\phi(p^u)} ,$$

where p^u is the largest power of p dividing a. From the isomorphism θ, we know the ideal of $\mathbb{Z}[\zeta_b]$ generated by $\Phi_a(\zeta_b)$ has factorization

$$(Q_1 \cdots Q_g)^{\phi(p^u)}$$

for distinct maximal ideals Q_i. Since $p = 0$ in R', $\Phi_a(\zeta_b)\mathbb{Z}[\zeta_b]$ divides $p\mathbb{Z}[\zeta_b]$. Since $b/a = p^v$, the ideals $P_i\mathbb{Z}[\zeta_b]$ are not split in $\mathbb{Z}[\zeta_b]$; so they are the Q_i (after a permutation) raised to some powers. Since p^{u+v} is the largest power of p dividing b,

$$p\mathbb{Z}[\zeta_b] \;=\; (Q_1 \cdots Q_g)^{\phi(p^{u+v})} \;=\; \Phi_a(\zeta_b)^r\mathbb{Z}[\zeta_b] ,$$

where $r = \phi(p^{u+v})/\phi(p^u)$. $\blacksquare$

To prove the theorem, we first show $SK_1(\mathbb{Z}C_n)$ is a finite (abelian) group whose order is a product of the prime factors of n. Then we show $SK_1(\mathbb{Z}C_n) = 1$ by induction on n.

Let D denote the set of all positive divisors of n. If $d \in D$, $\Phi_{D-\{d\}}(\alpha)$ is an element of $\mathbb{Z}C_n$ projecting to $\Phi_{D-\{d\}}(\zeta_d) \neq 0$ in the d-coordinate and to 0 in all other coordinates. So $\Phi_{D-\{d\}}(\alpha_D)$ generates the same ideal in

$$A \;=\; \prod_{d \in D} \mathbb{Z}[\zeta_d]$$

and its subdirect product $\mathcal{O}(D)$, namely, the elements whose d-coordinate belongs to

$$\mathfrak{c}_d \;=\; \Phi_{D-\{d\}}(\zeta_d)\mathbb{Z}[\zeta_d] ,$$

and whose other coordinates are 0. The ideal of $\mathcal{O}(D)$ generated by the elements $\Phi_{D-\{d\}}(\alpha_D)$ for all $d \in D$ is the ideal

$$\mathfrak{c} \;=\; \prod_{d \in D} \mathfrak{c}_d$$

of both A and $\mathcal{O}(D)$.

As in (13.32), there is an excision isomorphism

$$SK_1(\mathcal{O}(D), \mathfrak{c}) \;\cong\; SK_1(A, \mathfrak{c}) .$$

Entrywise projection to the d-coordinate defines the d-coordinate map in the decomposition

$$SK_1(A, \mathfrak{c}) \cong \prod_{d \in D} SK_1(\mathbb{Z}[\zeta_d], \mathfrak{c}_d) .$$

The latter group is finite by (11.33).

Moreover, if a prime p does not divide n, no maximal ideal of $\mathbb{Z}[\zeta_d]$ (for $d|n$) contains both p and the generator

$$\Phi_{D-\{d\}}(\zeta_d) = \prod_{\substack{d' \in D \\ d' \neq d}} \Phi_{d'}(\zeta_d)$$

of $\mathfrak{c}_d$, since each of these factors is a unit, or has a power associate to a prime factor of n. So, by the Bass-Milnor-Serre formula (11.33), primes not dividing n also do not divide the order of any $SK_1(\mathbb{Z}[\zeta_d], \mathfrak{c}_d)$ for $d \in D$; so they do not divide the order of $SK_1(\mathcal{O}(D), \mathfrak{c})$.

The ring $\mathcal{O}(D)/\mathfrak{c}$ is a finite commutative ring, since it is contained in

$$A/\mathfrak{c} \cong \prod_{d \in D} \frac{\mathbb{Z}[\zeta_d]}{\mathfrak{c}_d}$$

and each $\mathfrak{c}_d \neq 0$. So it is noetherian of Krull dimension 0 and has stable rank 1 by (4.34). Thus $K_1(\mathcal{O}(D)/\mathfrak{c})$ is represented by units, and $SK_1(\mathcal{O}(D)/\mathfrak{c}) = 1$. From the relative sequence, it follows that

$$SK_1(\mathcal{O}(D), \mathfrak{c}) \to SK_1(\mathcal{O}(D))$$

is surjective. So $SK_1(\mathbb{Z}C_n) \cong SK_1(\mathcal{O}(D))$ is finite, with no p-torsion for primes p not dividing n.

Now we proceed by induction on n. If $n = 1$, $\mathbb{Z}C_n = \mathbb{Z}$. Using Euclid's Algorithm, elementary row and column transvections reduce each matrix in $SL(\mathbb{Z})$ to the identity. So $SL(\mathbb{Z}) = E(\mathbb{Z})$ and $SK_1(\mathbb{Z}) = 1$. (For an alternate proof, note that $\mathbb{Z}$ is a ring of arithmetic type, and apply (11.33).)

Suppose now that $SK_1(\mathbb{Z}C_d) = 1$ for each $d < n$, and suppose p is a prime factor of n. It will suffice to prove $SK_1(\mathbb{Z}C_n)$ has no element of order p. For E the set of positive divisors of n/p, consider the fiber square of surjective ring homomorphisms

$$\begin{array}{ccccc}
\mathbb{Z}C_n & \cong & \mathcal{O}(D) & \xrightarrow{\ p_{D-E}\ } & \mathcal{O}(D - E) \\
& & \Big\downarrow{\scriptstyle p_E} & & \Big\downarrow{\scriptstyle g_2} \\
\mathbb{Z}C_{n/p} & \cong & \mathcal{O}(E) & \xrightarrow[\ \ g_1\ \ ]{} & R' .
\end{array}$$

By hypothesis, $SK_1(\mathcal{O}(E)) = 1$. So, by the relative sequence and K_1 excision, there is a surjective homomorphism

$$SK_1(\mathcal{O}(D - E), J) \to SK_1(\mathcal{O}(D)) \ ,$$

where $J = p_{D-E}(\ker p_E)$. The ideal

$$\mathfrak{c}' \;=\; p_{D-E}(\mathfrak{c} \cap \ker p_E) \;=\; \prod_{d \in D-E} \mathfrak{c}_d$$

is contained in J and has finite index in the ring

$$A' \;=\; \prod_{d \in D-E} \mathbb{Z}[\zeta_d] \ .$$

(13.37) Lemma. *If R is a commutative ring with ideals $J' \subseteq J$ of finite index, then inclusion on SL induces a surjective homomorphism*

$$SK_1(R, J') \to SK_1(R, J) \ .$$

Proof. Inclusion and entrywise reduction mod J' induce homomorphisms

$$SK_1(R, J') \to SK_1(R, J) \to SK_1(R/J', J/J')$$

whose composite is zero, since any $M \in SL(R, J')$ reduces mod J' to the identity. If $N \in SL(R, J)$ goes to $\overline{N} \in E(R/J', J/J')$, there exists $N_0 \in E(R, J)$ with $\overline{N}_0 = \overline{N}$. Then $\overline{NN_0^{-1}} = \overline{N} \ \overline{N}_0^{-1} = I$ in $SL(R/J', J/J')$, and NN_0^{-1} belongs to $SL(R, J')$. This proves the sequence is exact.

Now R/J' is a finite commutative ring, so it has stable rank 1; and by (11.19) (i), $SK_1(R/J', J/J') = 1$. ∎

Applying this lemma, there is a surjective homomorphism

$$SK_1(\mathcal{O}(D - E), \mathfrak{c}') \to SK_1(\mathcal{O}(D - E), J) \ .$$

And since $\mathcal{O}(D - E)$ is a subdirect product of A' and $\mathfrak{c}'$ is an ideal of A' as well, there is an excision isomorphism

$$SK_1(A', \mathfrak{c}') \;\cong\; SK_1(\mathcal{O}(D - E), \mathfrak{c}') \ .$$

The left side decomposes as

$$\prod_{d \in D-E} SK_1(\mathbb{Z}[\zeta_d], \mathfrak{c}_d) \ ,$$

where $\mathfrak{c}_d$ is generated by $\Phi_{D-\{d\}}(\zeta_d)$. Each factor $SK_1(\mathbb{Z}[\zeta_d], \mathfrak{c}_d)$ is a finite cyclic group by (11.33), and the largest power of p dividing its order is p^r, where

$$r \;\leq\; \left[\!\!\left[\frac{v_P(\mathfrak{c}_d)}{v_P(p)} - \frac{1}{p-1} \right]\!\!\right]$$

for each maximal ideal P of $\mathbb{Z}[\zeta_d]$ containing p.

Now

$$v_P(\mathfrak{c}_d) \;=\; v_P\left(\prod_{d' \in D-\{d\}} \Phi_{d'}(\zeta_d)\right) \;=\; \sum_{d' \in D-\{d\}} v_P(\Phi_{d'}(\zeta_d)) \;.$$

Say p^t is the largest power of p dividing n. Since $d \in D - E$, d does not divide n/p. So $d = p^t b$, where $p \nmid b$. If $v_P(\Phi_{d'}(\zeta_d))$ is not zero, then $p \in P$ implies the ratio d/d' must be p^v for some $v \geq 1$. So $d' = p^s b$ with $0 \leq s < t$. Then

$$v_P(\Phi_{d'}(\zeta_d)^{\phi(p^t)/\phi(p^s)}) = v_P(p)$$

and

$$\frac{v_P(\Phi_{d'}(\zeta_d))}{v_P(p)} \;=\; \frac{\phi(p^s)}{\phi(p^t)} \;.$$

The sum of these terms for $0 \leq s < t$ is $1/(p-1)$; so $r \leq 0$ and p does not divide the order of $SK_1(\mathbb{Z}[\zeta_d], \mathfrak{c}_d)$ nor, therefore, the order of $SK_1(\mathbb{Z}C_n)$. ∎

13B. Exercises

1. Prove exactness of the Mayer-Vietoris sequence (13.33) at the term $K_1(R_1) \oplus K_1(R_2)$ without using the excision map ε_0. *Hint:* Prove directly that GL of the fiber square of rings is a "fiber square of groups."

2. For a fiber square of rings (13.3) with g_1 surjective and f_1 either surjective or the inclusion of a subdirect product, show the Mayer-Vietoris sequence (13.33) restricts to

$$K_2(A) \to K_2(R_1) \oplus K_2(R_2) \to K_2(R')$$

$$\to SK_1(A) \to SK_1(R_1) \oplus SK_1(R_2) \to SK_1(R')$$

when the rings A, R_1, R_2, and R' are commutative.

3. Suppose $G \cong C_2^n = C_2 \times \cdots \times C_2$, where C_2 is the cyclic group of order 2. Prove $SK_1(\mathbb{Z}G) = 1$. *Hint:* Since $\mathbb{Q}G$ is a commutative semisimple ring, its simple components are fields. Projection to a simple component takes elements of G to 1 or -1; so each simple component is $\mathbb{Q}$. Thus $\mathbb{Z}G$ is isomorphic to a subdirect product of $\mathbb{Z} \times \cdots \times \mathbb{Z}$ (2^n copies). Some positive integer multiple of each central idempotent in $\mathbb{Q}G$ lies in $\mathbb{Z}G$. So the i-coordinates of those elements in $\mathbb{Z}G$ having all other coordinates 0 form a nonzero ideal $\mathfrak{c}_i$ of $\mathbb{Z}$, and $\mathfrak{c} = \Pi \mathfrak{c}_i$ is a $\mathbb{Z} \times \cdots \times \mathbb{Z}$-ideal in $\mathbb{Z}G$. Show $SK_1(\mathbb{Z} \times \cdots \times \mathbb{Z}, \mathfrak{c})$ is zero

and maps onto $SK_1(\mathbb{Z}G)$, by using excision and the fact that $\mathbb{Z}G/\mathfrak{c}$ is a finite commutative ring.

4. If R and S are rings, show there is a fiber square in $\mathfrak{Ring}$

$$
\begin{array}{ccc}
R \times S & \xrightarrow{\ p_2\ } & S \\
{\scriptstyle p_1}\downarrow & & \downarrow \\
R & \longrightarrow & 0
\end{array}
$$

where p_i is projection to the i-coordinate. Use the Mayer-Vietoris sequence to conclude

$$
\begin{bmatrix} p_1^* \\ p_2^* \end{bmatrix} \ : \ K_n(R \times S) \to K_n(R) \oplus K_n(S)
$$

is an isomorphism for $n = 0$ and 1. In the context of the Mayer-Vietoris sequence (13.33) for a fiber square of surjective ring homomorphisms

$$
\begin{array}{ccc}
A & \longrightarrow & R_2 \\
\downarrow & & \downarrow \\
R_1 & \longrightarrow & R' \ ,
\end{array}
$$

conclude that the image of $\lambda_n : K_{n+1}(R') \to K_n(A)$ equals the kernel of the map $K_n(A) \to K_n(R_1 \times R_2)$ induced by the embedding of A in $R_1 \times R_2$.

5. In the Mayer-Vietoris sequence for a fiber square of rings

$$
\begin{array}{ccc}
A & \longrightarrow & \overline{A} \\
\downarrow & & \downarrow \\
R & \longrightarrow & \overline{R}
\end{array}
$$

with surjective horizontal maps, prove $\lambda_0 : K_1(\overline{R}) \to K_0(A)$ takes the coset of a unit $u \in \overline{R}^*$ to $[(R, \overline{A}, u)] - [(R, \overline{A}, 1)]$. *Hint:* Combine the beginning of the proof of (13.20) with (13.30) and (13.12) to evaluate the three steps involved in the map λ_0.

13C. The Localization Sequence

Our third exact sequence connecting algebraic K-groups is based on the localization (inversion of central elements) in a ring A. Like the relative sequence, it is a long exact sequence involving the Quillen K-groups $K_n(A)$ for all $n \geq 0$. Unlike the relative sequence, Quillen K-theory is needed to describe the terms and maps involving K_2; so we describe the maps in detail only for the $K_1 - K_0$ part. Because we shall not use the localization sequence in further proofs, we refer the reader to perfectly good expositions of the proof of exactness elsewhere. But in §15D, we do give Milnor's descriptions of the maps in the $K_2 - K_1 - K_0$ part of the sequence when A is a Dedekind domain.

We employ the notation and definitions from §6E. In particular, A is a ring, R is a subring of the center of A, and S is a submonoid of the multiplicative monoid $(R, \cdot)$. We further assume S contains neither 0 nor any zero-divisors in A; so S acts through injections on A, and the localization map $f : A \to S^{-1}A$, $a \mapsto a/1$, is injective.

In (6.58) we already have the final piece of the sequence: If $\mathcal{T}_1$ is the full subcategory of A-$\mathcal{M}\mathfrak{o}\mathfrak{d}$ whose objects are the f.g. S-torsion A-modules M having a short $\mathcal{P}(A)$-resolution

$$0 \longrightarrow P_1 \longrightarrow P_0 \longrightarrow M \longrightarrow 0 ,$$

then there is an exact sequence of group homomorphisms

$$K_0(\mathcal{T}_1) \xrightarrow{\ \delta\ } K_0(A) \xrightarrow{\ K_0(f)\ } K_0(S^{-1}A)$$

where $\delta([M]) = [P_0] - [P_1]$.

Building the sequence to the left, we next define a map ∂ from $K_1(S^{-1}A)$ to $K_0(\mathcal{T}_1)$. Suppose $\alpha \in GL_n(S^{-1}A)$ and $s \in S$ is a common denominator of the entries of α; so $s\alpha \in M_n(A)$. Define

$$d_n(\alpha) \ = \ [A^n/A^n s\alpha] - [A^n/A^n s] .$$

This does not depend on the choice of s : If $t \in S$ and $t\alpha \in M_n(A)$, there are exact sequences

$$0 \longrightarrow \frac{A^n}{A^n t} \xrightarrow{\ \cdot s\alpha\ } \frac{A^n}{A^n t s\alpha} \longrightarrow \frac{A^n}{A^n s\alpha} \longrightarrow 0 ,$$

$$0 \longrightarrow \frac{A^n}{A^n t} \xrightarrow{\ \cdot s\ } \frac{A^n}{A^n t s} \longrightarrow \frac{A^n}{A^n s} \longrightarrow 0$$

in $\mathcal{T}_1$. Injectivity of the second maps in each sequence uses the hypothesis that S acts through injections on A. So, in $K_0(\mathcal{T}_1)$,

$$[A^n/A^n s\alpha] \ - \ [A^n/A^n s] \ = \ [A^n/A^n t s\alpha] \ - \ [A^n/A^n t s] .$$

Switching s and t, we get the same right side, and hence equal left sides, showing d_n is a well-defined function from $GL_n(S^{-1}A)$ to $K_0(\mathcal{T}_1)$.

Actually, d_n is a homomorphism: If $\alpha, \beta \in GL_n(S^{-1}A)$ with common denominator $s \in S$, the exact sequences

$$0 \longrightarrow \frac{A^n}{A^n s\alpha} \xrightarrow{\cdot s\beta} \frac{A^n}{A^n ss\alpha\beta} \longrightarrow \frac{A^n}{A^n s\beta} \longrightarrow 0 \ ,$$

$$0 \longrightarrow \frac{A^n}{A^n s} \xrightarrow{\cdot s} \frac{A^n}{A^n ss} \longrightarrow \frac{A^n}{A^n s} \longrightarrow 0$$

show

$$
\begin{aligned}
d_n(\alpha\beta) &= [A^n/A^n ss\alpha\beta] - [A^n/A^n ss] \\
&= [A^n/A^n s\alpha] - [A^n/A^n s] + [A^n/A^n s\beta] - [A^n/A^n s] \\
&= d_n(\alpha) + d_n(\beta) \ .
\end{aligned}
$$

If $m > 0$, projections to the first n and last m coordinates determine isomorphisms

$$\frac{A^{n+m}}{A^{n+m} s(\alpha \oplus I_m)} \cong \frac{A^n}{A^n s\alpha} \oplus \frac{A^m}{A^m s} \ ,$$

$$\frac{A^{n+m}}{A^{n+m} s} \cong \frac{A^n}{A^n s} \oplus \frac{A^m}{A^m s}$$

in $\mathfrak{T}_1$. So $d_{n+m}(\alpha \oplus I_m) = d_n(\alpha)$, and these maps extend to a homomorphism $d : GL(S^{-1}A) \to K_0(\mathfrak{T}_1)$. Since the latter group is abelian, d induces a homomorphism

$$\partial : K_1(S^{-1}A) \ \to \ K_0(\mathfrak{T}_1) \ .$$

(13.38) Theorem. *The sequence*

$$K_1(A) \xrightarrow{K_1(f)} K_1(S^{-1}A) \xrightarrow{\partial} K_0(\mathfrak{T}_1) \xrightarrow{\delta} K_0(A) \xrightarrow{K_0(f)} K_0(S^{-1}A)$$

is exact.

Proof. See Bass [74, §10] or Curtis and Reiner [87, §40B]. ∎

Note: Suppose $\mathfrak{T}_{<\infty}$ is the full subcategory of $\mathfrak{M}(A)$ consisting of those f.g. A-modules that have finite length $\mathcal{P}(A)$-resolutions. By Bass [74, Lemma (10.3)], every $M \in \mathfrak{T}_{<\infty}$ has a finite length $\mathfrak{T}_1$-resolution. So, by the Resolution Theorem (3.52), inclusion of categories induces a group isomorphism $K_0(\mathfrak{T}_1) \cong K_0(\mathfrak{T}_{<\infty})$, defined on generators by $[M] \mapsto [M]$. So we can replace $K_0(\mathfrak{T}_1)$ by $K_0(\mathfrak{T}_{<\infty})$ in the localization sequence; the map to $K_0(A)$ is still given by the Euler characteristic.

For an important application of the localization sequence, suppose R is a ring and $A = R[t]$ is the polynomial ring in the indeterminate t. For a submonoid

S of $(Z(A), \cdot)$ take $\{t^n : n \geq 0\}$. Then $S^{-1}A$ is the Laurent polynomial ring $R[t, t^{-1}]$. The sequence (13.38) becomes

$$K_1(R[t]) \xrightarrow{K_1(f)} K_1(R[t, t^{-1}]) \xrightarrow{\partial} K_0(\mathcal{T}_{<\infty})$$
$$\xrightarrow{\delta} K_0(R[t]) \xrightarrow{K_0(f)} K_0(R[t, t^{-1}]) \,,$$

where $\partial(\overline{\alpha \oplus I_\infty}) = [P] - [Q]$ with

$$P = \frac{R[t]^n}{R[t]^n t^r \alpha} \,, \quad Q = \frac{R[t]^n}{R[t]^n t^r}$$

whenever $t^r \alpha \in M_n(R[t])$. By Exercise 2, P and Q are f.g. projective R-modules. So ∂ factors as a homomorphism ∂' to $K_0(R)$ defined by the same formula as ∂, followed by the homomorphism to $K_0(\mathcal{T}_{<\infty})$ induced by the inclusion of categories $\mathcal{P}(R) \subseteq \mathcal{T}_{<\infty}$. By Exercise 3, there is a homomorphism $\psi : K_0(R) \to K_1(R[t, t^{-1}])$ with $\partial' \circ \psi = 1$. So $K_0(R)$ is isomorphic to a direct summand of $K_1(R[t, t^{-1}])$.

Inclusion $R \to R[t, t^{-1}]$ induces a group homomorphism i from $K_1(R)$ into $K_1(R[t, t^{-1}])$, with a left inverse given by sending t to 1 in each entry. So $K_1(R)$ is also isomorphic to a direct summand of $K_1(R[t, t^{-1}])$. By exactness of the localization sequence, the images of $K_0(R)$ and $K_1(R)$ intersect only in 0. So $K_1(R) \oplus K_0(R)$ is isomorphic to a direct summand of $K_1(R[t, t^{-1}])$.

For the moment, suppose R is left regular, meaning R is left noetherian, and every $M \in \mathcal{M}(R)$ has a finite length $\mathcal{P}(R)$-resolution. By the Hilbert Syzygy Theorem (see Bass [68, XII, §1]), the rings $R[t]$ and $R[t, t^{-1}]$ are also left regular. So $\mathcal{T}_{<\infty}$ is the category $\mathcal{T}$ of all f.g. S-torsion $R[t]$-modules. By a theorem of Grothendieck (see Bass [68, XII, §3]), inclusions $R \subseteq R[t] \subseteq R[t, t^{-1}]$ induce isomorphisms

$$K_0(R) \;\cong\; K_0(R[t]) \;\cong\; K_0(R[t, t^{-1}]) \,.$$

The second map is $K_0(f)$ for the localization map f; so, by exactness of the localization sequence, ∂ is surjective. Bass, Heller, and Swan [64] proved that (for R left regular) inclusion $R \subseteq R[t]$ induces an isomorphism

$$K_1(R) \;\cong\; K_1(R[t])$$

as well, and the localization sequence restricts to a split exact sequence

$$0 \longrightarrow K_1(R) \xrightarrow{i} K_1(R[t, t^{-1}]) \xrightarrow{\partial'} K_0(R) \longrightarrow 0 \,;$$

so $K_1(R[t, t^{-1}]) \cong K_1(R) \oplus K_0(R)$.

Even if R is not left regular, Bass [68, XII, §7] proved there is an exact sequence

$$K_1(R[t]) \oplus K_1(R[t^{-1}]) \xrightarrow{\tau} K_1(R[t, t^{-1}]) \longrightarrow K_0(R) \longrightarrow 0 \,,$$

where τ is the sum of maps induced by inclusions of the subrings $R[t]$ and $R[t^{-1}]$ into $R[t, t^{-1}]$. The sequence is natural in R, so it could be used to define $K_0(-)$ as a functor. More generally, Bass showed that if $F : \mathfrak{Ring} \to \mathcal{A}\mathfrak{b}$ is any covariant functor, and $LF(R)$ is defined as the cokernel of the map

$$F(R[t]) \oplus F(R[t^{-1}]) \to F(R[t, t^{-1}]) \ ,$$

then LF is also a covariant functor from $\mathfrak{Ring}$ to $\mathcal{A}\mathfrak{b}$. Iterating this construction produces a sequence of functors $L^n F$. Since LK_1 is naturally isomorphic to K_0, Bass defined the **negative algebraic K-groups** of a ring R as

$$\boldsymbol{K_{-n}(R)} \ = \ L^n K_0(R) \qquad (n \geq 1) \ .$$

Note that these vanish when R is left regular, since then $K_{-1}(R) = 0$ by Grothendieck's Theorem (above). Bass [68, XII, §8] also showed the negative K-groups are connected by a long exact extension to the right of the Mayer-Vietoris sequence (13.33) associated to a fiber square of rings.

Quillen's higher algebraic K-theory of categories includes an extension of the localization sequence (13.38) to the left:

(13.39) Theorem. *There is a long exact sequence of group homomorphisms*

$$\cdots \longrightarrow K_{n+1}(\mathfrak{I}_1) \longrightarrow K_{n+1}(A) \longrightarrow K_{n+1}(S^{-1}A)$$
$$\longrightarrow K_n(\mathfrak{I}_1) \longrightarrow K_n(A) \longrightarrow K_n(S^{-1}A) \longrightarrow \cdots$$

ending in the sequence (13.38)

Proof. See Grayson [76, p. 233]. $\blacksquare$

Note that Quillen's $K_2(R)$ agrees with Milnor's $K_2(R)$ as we have presented it in Chapter 12.

The Quillen higher algebraic K-groups are tied together in the same way as $K_n(R)$ for $n \leq 1$; namely, $LK_n(R) \cong K_{n-1}(R)$ for all integers n (see Grayson [76, p.236]). In more detail, for any functor F from $\mathfrak{Ring}$ to $\mathcal{A}\mathfrak{b}$, Bass also defined the functor NF with

$$\boldsymbol{NF(R)} \ = \ \ker(F(R[t]) \xrightarrow{\varepsilon} F(R)) \ ,$$

where ε is induced by the map fixing R and taking t to 1. If $j : F(R) \to F(R[t])$ is induced by the inclusion $R \subseteq R[t]$, then $\varepsilon \circ j = 1$. So $F(R[t]) \cong F(R) \oplus NF(R)$. When R is left regular, $NK_n(R) = 0$ for all n. For all rings R and all integers n, the image of $K_n(R[t]) \oplus K_n(R[t^{-1}])$ in $K_n(R[t, t^{-1}])$ is the image of $K_n(R)$ and two copies of $NK_n(R)$. So we have what has been called the **Fundamental Theorem** of algebraic K-theory:

(13.40) Theorem. *For all rings R and all integers $n > 1$,*

$$K_n(R[t, t^{-1}]) \;\cong\; K_n(R) \oplus NK_n(R) \oplus NK_n(R) \oplus K_{n-1}(R) \;.$$

Proof. For the case $n = 1$, see Bass [68, XII, Theorem (7.4)]. For $n > 1$, see Grayson [76, pp. 236–239]. ∎

The groups $NK_n(R)$, for all integers n, are related to the K-theory of pairs (P, ν) where $P \in \mathcal{P}(R)$ and ν is a nilpotent endomorphism of P (see Bass [68, XII, §§6–7] and Grayson [76]). Because of this connection, these groups for $n \leq 1$ serve as the targets for some useful topological invariants.

13C. Exercises

1. Verify the exactness of the localization sequence (13.38) when $A = \mathbb{Z}$ and $S = \mathbb{Z} - \{0\}$.

2. If $\beta \in GL_n(R[t, t^{-1}]) \cap M_n(R[t])$, prove $R[t]^n / R[t]^n \beta$ is a f.g. projective R-module. *Hint:* Show the sequence

$$0 \longrightarrow R[t]^n \xrightarrow{\;\cdot\beta\;} R[t]^n \longrightarrow \frac{R[t]^n}{R[t]^n \beta} \longrightarrow 0$$

is exact, and $\cdot\beta$ has an R-linear left inverse given by

$$\cdot\beta^{-1} : R[t]^n \;\rightarrow\; R[t, t^{-1}]^n$$

followed by sending each negative power of t to 0. Then show the quotient is f.g. as an R-module.

3. If $P \in \mathcal{P}(R)$, let $P[t, t^{-1}]$ denote $R[t, t^{-1}] \otimes_R P$. Writing $t^n \otimes v$ as vt^n for $v \in P$, we can think of elements in $P[t, t^{-1}]$ as Laurent polynomials with coefficients in P. Show there is a homomorphism

$$\psi : K_0(R) \;\rightarrow\; K_1(R[t, t^{-1}]) \;, \quad [P] \mapsto [P[t, t^{-1}], \cdot t] \;,$$

with $\partial' \circ \psi = 1$ on $K_0(R)$. So ψ embeds $K_0(R)$ as a direct summand of $K_1(R[t, t^{-1}])$. *Hint:* If $P \oplus Q \cong R^n$ via a basis $\{(p_i, q_i)\}$, show the matrix α representing

$$[P[t, t^{-1}], \cdot t] \;=\; [P[t, t^{-1}] \oplus Q[t, t^{-1}], \cdot t \oplus I]$$

over $\{1 \otimes p_i, 1 \otimes q_i\}$ has entries in $R[t]$.

4. Show $NK_1(R)$ is the subgroup of $K_1(R)$ consisting of all of the elements represented by matrices $I + tN$, where N is a nilpotent matrix over R. *Hint:*

If $\alpha(t) \in GL(R[t])$, show there exists $e \in E(R[t])$ and a nilpotent matrix N over R with $\alpha(t) = e(I + tN)\alpha(0)$.

5. Use a theorem from the section to show that, if F is a field, all $P \in \mathcal{P}(F[t_1, \ldots, t_n])$ are stably free. (In fact, all such P are free – see §4C, Exercise 5 and Example (4.8) (i).)

6. If R is a commutative ring and S is a submonoid of $(R, \cdot)$ acting through injections on R and meeting all but finitely many maximal ideals of R, prove the kernel of the homomorphism $K_1(R) \to K_1(S^{-1}R)$, induced by the localization map $R \to S^{-1}R$, is $SK_1(R)$. *Hint:* It is enough to show $SK_1(S^{-1}R) = 1$, and this is true if the stabilization map $GL_1(S^{-1}R) \to K_1(S^{-1}R)$ is surjective (see (9.18)). This, in turn, follows if the stable rank of $S^{-1}R$ is 1, by (10.3) (ii). Now use §6D, Exercise 11, and §4D, Exercises 4 and 5.

7. If G is a finite abelian group, prove the kernel of the homomorphism $K_1(\mathbb{Z}G) \to K_1(\mathbb{Q}G)$, induced by inclusion of $\mathbb{Z}G$ in $\mathbb{Q}G$, is $SK_1(\mathbb{Z}G)$. *Hint:* As above, show $SK_1(\mathbb{Q}G) = 1$, this time using the Wedderburn-Artin Theorem on the structure of $\mathbb{Q}G$.

8. Let $s_1 : R^* \to K_1(R)$ denote the stabilization map $u \mapsto \overline{u \oplus I_\infty}$. If R is a left regular commutative integral domain, show $R[t, t^{-1}]^* = R^* \stackrel{\bullet}{\oplus} \langle t \rangle$, where $\langle t \rangle$ is the cyclic group under multiplication generated by t and $\stackrel{\bullet}{\oplus}$ is the internal direct sum when the abelian group $R[t, t^{-1}]^*$ is written additively. Then show the isomorphism from $K_0(R) \oplus K_1(R)$ to $K_1(R[t, t^{-1}])$ carries $\langle [R] \rangle \stackrel{\bullet}{\oplus} s_1(R^*)$ onto $s_1(R[t, t^{-1}]^*)$, so there is an isomorphism of quotients:

$$\widetilde{K}_0(R) \oplus SK_1(R) \cong SK_1(R[t, t^{-1}]) .$$

So, for a Dedekind domain R of arithmetic type, $SK_1(R[t, t^{-1}]) \cong Cl(R)$.

14

Universal Algebras

In Chapter 12 we saw that K_2 of a field is generated by Steinberg symbols $\{a, b\}$ with $a, b \in F^*$, which are bimultiplicative and vanish if $a + b = 1$. The ultimate goal of this chapter is to provide a proof of Matsumoto's Theorem, that every relation among Steinberg symbols on F arises from these properties. This finally allows us to detect nontrivial elements of $K_2(F)$.

But, along the way to this proof, we encounter some very interesting scenery. In §14A we review algebras and their presentations and in §14B introduce graded algebras and their graded quotients. This prepares the way for an excursion through tensor, symmetric, and exterior algebras. In §14C the tensor algebra of an R-module M is a universal way to make M an additive part of an R-algebra $T(M)$. That is, it provides the simplest way to multiply vectors. Reducing modulo relations $xy = yx$ or $x^2 = 0$, we produce the symmetric and exterior algebras in §14D. The exterior algebra is connected with determinants over a commutative ring, and we revisit the ideal class group of a Dedekind domain in the more general form of the Picard group of a commutative ring, linked to K_0 by an "exterior power" map $\det_0$ for projective modules.

Reducing the tensor algebra of the $\mathbb{Z}$-module F^* (F a field) modulo a relation making $x \otimes y = 1$ if $x + y = 1$ in F, we come upon the Milnor ring of F in §14E. The pieces of this graded $\mathbb{Z}$-algebra constitute a higher K-theory $K_n^M(F)$ for all $n \geq 0$. We explore its relation with orderings in the field F, culminating with a glimpse of Milnor's conjecture relating $K_*^M(F)$ to the Witt ring $W(F)$. In §14F we discuss tame symbols, which are symbol maps arising from discrete valuations on F. We find all the tame symbols on $\mathbb{Q}$ coming together in Tate's computation of $K_2(\mathbb{Q})$, which leads to a proof of quadratic reciprocity in §15C. A parallel conjunction of extended tame symbols over a field $F(x)$ of rational functions prepares the way for §14G, in which a norm is defined from $K_n^M(A_p)$ to $K_n^M(F)$, where A_p is a simple field extension of F generated by an element with minimal polynomial p. In §14H we reach our destination, Keune's proof of Matsumoto's Theorem (14.69), using norms on K_2^M to complete a triangle of connections among $K_2(F)$, $K_2^M(F)$ and a relative SK_1 group.

14A. Presentation of Algebras

Suppose R is a commutative ring. Recall from Chapter 0 that an R-**algebra** is an R-module A that is also a ring and satisfies $r(ab) = (ra)b = a(rb)$ whenever $r \in R$ and $a, b \in A$. If A and B are R-algebras, a function $f : A \to B$ is an R-**algebra homomorphism** if f is an R-linear ring homomorphism, which is to say,

$$
\begin{aligned}
f(a_1 + a_2) &= f(a_1) + f(a_2) \\
f(a_1 a_2) &= f(a_1) f(a_2) \\
f(1_A) &= 1_B \\
f(ra) &= r f(a)
\end{aligned}
$$

for all $r \in R$ and $a, a_1, a_2 \in A$. Since $ra = r(1_A a) = (r1_A)a$, the fourth equation can be replaced by $f(r1_A) = r1_B$. R-algebras and R-algebra homomorphisms form a category R-$\mathcal{Alg}$.

A subset A' of an R-algebra A is an R-**subalgebra** of A if A' is both an R-submodule and a subring of A. So $A' \subseteq A$ is an R-subalgebra of A if and only if it is closed under addition and multiplication and contains $r1_A$ for all $r \in R$. The image of an R-algebra homomorphism $f : A \to B$ is an R-subalgebra of B.

If J is an ideal of an R-algebra A, then J is also an R-submodule of A, since $r \in R$ and $a \in J$ implies $ra = (r1_A)a \in J$. So the quotient A/J is an R-algebra, being an R-module and a ring with $r((a + J)(b + J)) = r(ab) + J$, $(r(a + J))(b + J) = (ra)b + J$, and $(a + J)(r(b + J)) = a(rb) + J$ all equal. If an ideal J of the domain A of an R-algebra homomorphism $f : A \to B$ is contained in the kernel of f, the induced map $\overline{f} : A/J \to B$, $a + J \mapsto f(a)$, is an R-algebra homomorphism.

(14.1) Examples.

(i) Every ring A is a $\mathbb{Z}$-algebra in one and only one way, since the unique $\mathbb{Z}$-module action

$$
na = \begin{cases}
a + \cdots + a \ (n \text{ terms}) & \text{if } n > 0 \\
0 & \text{if } n = 0 \\
(-a) + \cdots + (-a) \ (-n \text{ terms}) & \text{if } n < 0
\end{cases}
$$

satisfies the algebra property $n(ab) = (na)b = a(nb)$ for all $n \in \mathbb{Z}$ and $a, b \in A$. Every ring homomorphism is a $\mathbb{Z}$-algebra homomorphism; so $\mathbb{Z}$-$\mathcal{Alg}$ coincides with $\mathcal{Ring}$.

(ii) If R is a subring of the center $Z(A)$ of a ring A, then A is an R-algebra with ra being both scalar multiplication and ring multiplication, for $r \in R$, $a \in A$. If R is a subring of A, but does not lie in $Z(A)$, this will not work; the condition $ra = ar$ is required so that $r(ab) = (ra)b = a(rb)$. If R is also a subring of the center $Z(B)$ of a ring B, then an R-algebra homomorphism $f : A \to B$ is the same as a ring homomorphism with $f(r) = r$ for all $r \in R$.

(iii) If $\phi : R \to A$ is a ring homomorphism and $\phi(R) \subseteq Z(A)$, then A is an R-algebra with $ra = \phi(r)a$ for all $r \in R$, $a \in A$. (This generalizes (ii), in which ϕ is inclusion.) Every R-algebra A is obtained in this way, since we can define $\phi : R \to A$ by $\phi(r) = r1_A$ to obtain the desired ring homomorphism, and then $ra = (r1_A)a = \phi(r)a$. If A and B are R-algebras with associated ring homomorphisms $\phi : R \to A$, $\psi : R \to B$, then an R-algebra homomorphism $f : A \to B$ is the same as a ring homomorphism making the triangle

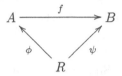

commute.

(iv) Recall from (1.16) that, if N is a monoid (written in multiplicative notation), the **monoid ring** $R[N]$ is the free R-module based on the set N, with multiplication

$$\left(\sum_{\nu \in N} r_\nu \nu \right) \left(\sum_{\nu \in N} s_\nu \nu \right) = \sum_{\nu \in N} \left(\sum_{\sigma\tau = \nu} r_\sigma s_\tau \right) \nu \ .$$

Then $R[N]$ is an R-algebra with $1_{R[N]} = 1_N$. The multiplication restricts on N to the monoid operation; that is, the inclusion $N \to R[N]$ is a monoid homomorphism into $(R[N], \cdot)$. Recall from (1.17) that each monoid homomorphism $\mu : N \to A$, from N into the multiplicative monoid of an R-algebra A, extends uniquely to an R-algebra homomorphism $\widehat{\mu} : R[N] \to A$.

Now we generalize the construction (1.20) of the free ring. Recall from (1.18) that the **free monoid** based on a set S is the set $\mathbf{Mon}(S)$ of all strings $s_1 s_2 \cdots s_m$, where $m \geq 0$ and each $s_i \in S$. Two strings are multiplied by concatenation:

$$(s_1 \cdots s_m)(s_1' \cdots s_n') = s_1 \cdots s_m s_1' \cdots s_n' \ .$$

The empty string (with $m = 0$) is the identity 1 of $\mathrm{Mon}(S)$. And by (1.19), each function from a set S into a monoid N has a unique extension to a monoid homomorphism $\mathrm{Mon}(S) \to M$.

(14.2) Definition. If S is any set and R is a commutative ring, the **free R-algebra based on** S is the monoid ring $R[\mathrm{Mon(S)}]$ of the free monoid based on S.

(14.3) Proposition. *Each function* $\theta : S \to A$ *from a set* S *into an* R-*algebra* A *has a unique extension to an* R-*algebra homomorphism* $\widehat{\theta} : R[\mathrm{Mon}(S)] \to A$.

Proof. By the universal mapping properties of $\mathrm{Mon}(S)$ and monoid rings $R[N]$, it suffices to note that each R-algebra homomorphism from $R[\mathrm{Mon}(S)]$ to A extending θ extends a monoid homomorphism on $\mathrm{Mon}(S)$ that extends θ. ∎

(14.4) Definition. Suppose T is a subset of an R-algebra A. Then T **generates** A (as an R-algebra) if the only R-subalgebra of A containing T is A itself.

We can describe a typical element of an R-algebra A in terms of a generating set $T \subseteq A$: If $S \to T$ is any surjection, the extension to an R-algebra homomorphism $R[\mathrm{Mon}(S)] \to A$ is surjective, since its image is an R-subalgebra of A containing T. So each element of A is an R-linear combination of products $t_1 \cdots t_n$ $(n \geq 0, \ t_i \in T)$, with the empty product understood to be 1_A.

(14.5) Definitions. An R-algebra A has R-**algebra presentation** $(S : D)$, with **generators** S and **defining relators** D, if there is an R-algebra isomorphism $F/J \cong A$, where F is the free R-algebra based on S and J is an ideal of F generated by D.

The set S need not be in this algebra A; it is a kind of external generating set. But the composite $f : F \to F/J \cong A$ carries S to a generating set $f(S) = T$ of A in the sense of (14.4). An equation

$$\sum r_{i,s_1 \cdots s_n} f(s_1) \cdots f(s_n) \ = \ 0$$

is true in A if and only if the expression

$$\sum r_{i,s_1 \cdots s_n} (s_1 \cdots s_n)$$

lies in the kernel J of f; it is called a **defining relation** for $(S : D)$ if this expression lies in D. Since D generates J as an ideal, every true equation in A is a consequence of the defining relations.

(14.6) Examples.

(i) The free R-algebra based on S has presentation $(S : \emptyset)$; so it is *free* of relations.

(ii) The free R-algebra based on a single element set $S = \{x\}$ is the free R-module based on $\{1, x, x^2, \dots\}$, with a multiplication that makes it the polynomial ring $R[x]$. This is a commutative ring. But the free R-algebra based on a set S including at least two elements x and y is noncommutative (as long as $R \neq \{0\}$) since the two strings xy and yx are R-linearly independent.

(iii) If $R = \mathbb{R}$ is the field of real numbers, the division ring $\mathbb{H}$ of Hamilton's quaternions has an $\mathbb{R}$-algebra presentation

$$(i, j : i^2 + 1, \ j^2 + 1, \ (ij)^2 + 1) \ .$$

Here $S = \{i, j\}$ is inside (and generates) $\mathbb{H}$, and the isomorphism $F/J \cong \mathbb{H}$ is induced by $f : F = \mathbb{R}[\text{Mon}(S)] \to \mathbb{H}$ extending the inclusion of S in $\mathbb{H}$. Note that there is a composite $\mathbb{R}$-algebra homomorphism

$$\mathbb{C} \ \cong \ \frac{\mathbb{R}[x]}{(x^2 + 1)\mathbb{R}[x]} \ \to \ \mathbb{H}$$

that takes $i \mapsto \bar{x} \mapsto i$. Since $\mathbb{C}$ is a field, this map is injective and we can identify $\mathbb{C}$ with a subfield $\mathbb{R} + \mathbb{R}i$ of $\mathbb{H}$. Since $ijij = -1$ in $\mathbb{H}$, $ij \neq 0$ and $ji = i^2 j i j^2 = i(-1)j = -ij$; so j does not commute with i, and $j \notin \mathbb{C}$. So 1 and j are $\mathbb{C}$-linearly independent, and $\{1, i, j, ij\}$ is an $\mathbb{R}$-basis of $\mathbb{H}$.

(iv) Suppose $f : R[\text{Mon}(S)] \to A$ is a surjective R-algebra homomorphism with kernel J generated by D. Suppose $E \subseteq R[\text{Mon}(S)]$ and $f(E)$ generates an ideal L of A. The composite

$$R[\text{Mon}(S)] \ \xrightarrow{\ f\ } \ A \ \xrightarrow{\ c\ } \ A/L$$

is a surjective R-algebra homomorphism, with kernel $f^{-1}(L)$. Since the ideal K of $R[\text{Mon}(S)]$ generated by E maps onto L, we have $f^{-1}(L) = J + K$, which is generated by $D \cup E$. So A/L has R-algebra presentation $(S : D \cup E)$. Therefore, *reducing modulo an ideal (generated by $f(E)$) is the same as adding more relators (E).*

14A. Exercises

1. Suppose A is an R-algebra for a commutative ring R. If $a \in A$, prove left multiplication by a is an R-linear map $A \to A$. Then show the map $\rho : A \to \text{End}_R(A)$, taking a to left multiplication by a, is an R-algebra homomorphism and is injective. If A has a finite R-basis $a_1, \ldots, a_n$, prove there is an injective R-algebra homomorphism $\mu : A \to M_n(R)$. So if F is a field, all finite-dimensional F-algebras can be expressed as algebras of matrices over F.

2. Suppose A is an F-vector space (F a field), with basis $b_1, \ldots, b_n$. Choose scalars $c_{ij}^k \in F$ and define

$$\{b_1, \ldots, b_n\} \times \{b_1, \ldots, b_n\} \ \to \ A$$

by

$$(b_i, b_j) \ \mapsto \ b_i b_j \ = \ c_{ij}^1 b_1 + \cdots + c_{ij}^n b_n \ .$$

Suppose b_1 acts as an identity under this multiplication and $(b_i b_j) b_k = b_i (b_j b_k)$ for all i, j, k if we take

$$\left(\sum c_\ell b_\ell \right) b_k = \sum c_\ell (b_\ell b_k) ,$$
$$b_i \left(\sum c_\ell b_\ell \right) = \sum c_\ell (b_i b_\ell) .$$

Show A is an F-algebra under a multiplication extending this product on $\{b_1, \ldots, b_n\}$. The scalars c_{ij}^k are called **structure constants** for this algebra and basis.

14B. Graded Rings

The free R-algebra $A(S) = R[\text{Mon}(S)]$ based on a nonempty set S consists of R-linear combinations of strings $s_1 \cdots s_m$ of elements $s_i \in S$. For each integer $n \geq 0$, let $A(S)_n$ denote the R-linear span of the strings $s_1 \cdots s_n$ of length n in $\text{Mon}(S)$. Since the strings are R-linearly independent,

$$A(S) = \overset{\bullet}{\underset{n \geq 0}{\bigoplus}} A(S)_n ,$$

meaning each element of $A(S)$ has a unique expression as a sum $a_0 + a_1 + a_2 + \cdots$ with $a_n \in A(S)_n$ for all $n \geq 0$, and $a_n = 0$ for all but finitely many n. For each pair of nonnegative integers i and j,

$$A(S)_i A(S)_j \subseteq A(S)_{i+j} .$$

The elements of $A(S)_n$ are called "homogeneous of degree n" because each term of such an element has the same length n.

Similarly, the polynomial ring $R[x_1, \ldots, x_m]$ in m indeterminates over R can be written as

$$R[x_1, \ldots, x_m] = \overset{\bullet}{\underset{n \geq 0}{\bigoplus}} P_n ,$$

where P_n is the R-linear span of the monomials

$$x_1^{e_1} \cdots x_m^{e_m} , \quad e_1 + \cdots + e_m = n .$$

For nonnegative integers i and j, $P_i P_j \subseteq P_{i+j}$. The elements of P_n are called "homogeneous of degree n," since they consist of R-linear combinations of monomials of the same total degree $e_1 + \cdots + e_m = n$.

Let $\mathbb{N}$ denote the nonnegative integers. A **graded ring** of type $\mathbb{N}$ is a ring A whose additive group is an internal direct sum

$$A \;=\; A_0 \overset{\bullet}{\oplus} A_1 \overset{\bullet}{\oplus} A_2 \overset{\bullet}{\oplus} \cdots \;=\; \overset{\bullet}{\underset{n \geq 0}{\bigoplus}} A_n$$

of additive subgroups A_n, with $A_i A_j \subseteq A_{i+j}$ for each pair $i, j \in \mathbb{N}$. If A is an R-algebra and $RA_n \subseteq A_n$ for each $n \in \mathbb{N}$, then A is a **graded R-algebra** of type $\mathbb{N}$. The additive subgroups A_n of A are the **graded components** of A. An element of A_n is **homogeneous** of **degree** n.

The set $\cup_n A_n$ of all homogeneous elements in a graded ring A is closed under products, but it is usually not closed under sums. On the other hand, each A_n is closed under sums, but it is usually not closed under products. However, A_0 is closed under both operations. In fact, we have:

(14.7) Proposition. *In a graded ring A of type $\mathbb{N}$, A_0 is a subring of A.*

Proof. Since A_0 is an additive subgroup of A, closed under multiplication, it suffices to prove $1_A \in A_0$. Say

$$1_A \;=\; e_0 + e_1 + e_2 + \cdots$$

with $e_i \in A_i$ for each i. Then for each $n \in \mathbb{N}$, $e_0 e_n = e_n$, since

$$e_n - e_0 e_n \;=\; \sum_{i \geq 1} e_i e_n$$

belongs to $A_n \cap (\overset{\bullet}{\underset{i > n}{\oplus}} A_i) \;=\; \{0\}$. Then

$$e_0 \;=\; e_0 1 \;=\; \sum_{n \geq 0} e_0 e_n \;=\; \sum_{n \geq 0} e_n \;=\; 1 \,. \qquad \blacksquare$$

An ideal J of a graded ring A is **homogeneous** if J is generated by homogeneous elements of A.

(14.8) Theorem. *Suppose A is a graded ring of type $\mathbb{N}$, with a homogeneous ideal J, generated by a set T of homogeneous elements.*

(i)   *For each $n \in \mathbb{N}$, $J \cap A_n$ is the set of finite length sums of elements $xty \in A_n$, where x and y are homogeneous and $t \in T$.*

(ii) $J = \overset{\bullet}{\oplus}_{n \geq 0} (J \cap A_n)$ *as an additive group.*

(iii) A/J *is a graded ring of type $\mathbb{N}$, with graded components*

$$(A/J)_n \;=\; A_n/J \;\cong\; A_n/(J \cap A_n)$$

for all $n \in \mathbb{N}$, where the latter is an isomorphism of additive groups.

Proof. The ideal J generated by T consists of all finite length sums of terms atb with $a, b \in A$ and $t \in T$. By the distributive laws, J is all finite length sums of terms xty, where x and y are homogeneous and $t \in T$. Each term xty lies in $J \cap A_n$ for some n, proving (ii). Subtracting from any $z \in J \cap A_n$ its terms $xty \in J \cap A_n$ leaves 0, since $A_n \cap \sum_{i \neq n} A_i = \{0\}$, proving (i).

The external direct sum $\oplus_{n \geq 0} A_n$ is the additive group of all eventually zero sequences $(a_0, a_1, a_2, \dots)$ with $a_n \in A_n$ for all n, under coordinatewise addition. There is an isomorphism of additive groups from $\overset{\bullet}{\oplus}_{n \geq 0} A_n$ to $\oplus_{n \geq 0} A_n$ given by

$$a_0 + a_1 + a_2 + \cdots \;\mapsto\; (a_0, a_1, a_2, \dots) \,.$$

Applying the canonical map

$$A_n \longrightarrow A_n/(J \cap A_n)$$

in each coordinate defines a surjective composite homomorphism of additive groups

$$A \;=\; \overset{\bullet}{\bigoplus_{n \geq 0}} A_n \;\to\; \bigoplus_{n \geq 0} A_n \;\to\; \bigoplus_{n \geq 0} (J \cap A_n)$$

with kernel J. The induced isomorphism

$$f : A/J \;\cong\; \bigoplus_{n \geq 0} A_n/(J \cap A_n)$$
$$\overline{a_0 + a_1 + \cdots} \;\mapsto\; (\bar{a}_0, \bar{a}_1, \dots)$$

restricts to isomorphisms

$$f_n : A_n/J \;\cong\; A_n/(J \cap A_n) \,,$$

where the right side is identified with its insertion into the nth coordinate. So A/J is the internal direct sum $\overset{\bullet}{\oplus}_{n \geq 0} (A_n/J)$, and for each pair $i, j \in \mathbb{N}$, multiplication in the ring A/J restricts to

$$(A_i/J) \times (A_j/J) \;\to\; A_{i+j}/J \,. \qquad \blacksquare$$

Note: If A is a graded R-*algebra*, the action of $r \in R$ coincides with multiplication by $r1_R \in A$. So an ideal J is an R-submodule of A, each $J \cap A_n$ is an R-submodule of A_n, and the isomorphism

$$f : A/J \;\cong\; \bigoplus_{n \geq 0} A_n/(J \cap A_n)$$

is R-linear. So A/J is a graded R-*algebra*.

14B. Exercises

1. If the set $H = \cup_{n \geq 0} A_n$ of homogeneous elements in the graded ring A of type $\mathbb{N}$ is closed under sums, prove $A_n = 0$ for all $n > 0$.

2. If a homogeneous component A_n of a graded ring A of type $\mathbb{N}$ is closed under products and $n > 0$, prove $A_n = \{0\}$.

3. Suppose F is a field and $m > 0$. Describe the graded components of the **truncated polynomial ring** $F[x]/x^m F[x]$, induced by the grading on $F[x]$ given by the polynomial degree. If F has characteristic p (a prime), show

$$\frac{F[x]}{x^p F[x]} \cong \frac{F[x]}{(x^p - 1)F[x]} ;$$

so reduction modulo an ideal J that is not homogeneous can also yield a graded quotient ring.

4. Suppose A is a graded ring $\overset{\bullet}{\oplus} A_n$ of type $\mathbb{N}$. A **graded A-module** is an A-module M with a direct sum decomposition $\overset{\bullet}{\oplus}_{n \geq 0} M_n$ into additive subgroups M_n for which $A_i M_j \subseteq M_{i+j}$ for all $i, j \in \mathbb{N}$. Elements of M_n are **homogeneous** of **degree** n. If N is an A-submodule of M, prove the following are equivalent:

  (i) N can be generated by homogeneous elements of M.
  (ii) If $x \in N$, each homogeneous coordinate of x belongs to N.
  (iii) $N = \sum_{n \geq 0} (N \cap M_n)$.

If these equivalent conditions are true, N is called a **homogeneous submodule** of M, and M/N is a graded A-module with homogeneous components $M_n/N \cong M_n/N_n$.

5. If R is a commutative ring with an ideal J, consider the graded ring A consisting of those polynomials $r_0 + r_1 x + r_2 x^2 + \cdots$ with $r_n \in J^n$ for each n. (Here $J^0 = R$.) The homogeneous component of degree n is $J^n x^n$. So $A \cong R \oplus J \oplus J^2 \oplus \cdots$ via $r_0 + r_1 x + r_2 x^2 + \cdots \mapsto (r_0, r_1, r_2, \dots)$, and this makes the external direct sum a graded ring under the multiplication

$$(r_0, r_1, \dots)(s_0, s_1, \dots) = (t_0, t_1, \dots),$$

where $t_n = \sum_{i+j=n} r_i s_j$ for $n \geq 0$. Then A is an R-module (by multiplication in A), and JA is both an R-submodule and a homogeneous ideal of A (generated by the homogeneous elements $J \subseteq A_0$). Describe A/JA as a graded ring by giving a simple external direct sum isomorphic to it, and describe the multiplication in that external direct sum. A/JA is known as the **associated graded ring** to the ideal J of R.

14C. The Tensor Algebra

In the presence of the associative law, we can drop the use of parentheses in a product of several terms. But the associativity of tensor products is an isomorphism rather than an equality; so dropping parentheses introduces some ambiguity. We can resolve this difficulty by introducing a tensor product constructed from several modules at once.

Throughout this section R is a commutative ring, and each R-module M is regarded as an R, R-bimodule with $rm = mr$ for $m \in M$, $r \in R$.

(14.9) Definition. If $M_1, \ldots, M_n$ are R-modules, define $\boldsymbol{M_1 \otimes_R \cdots \otimes_R M_n}$ to be the quotient $\mathcal{F}/\mathcal{R}$, where $\mathcal{F}$ is the free $\mathbb{Z}$-module based on $M_1 \times \cdots \times M_n$ and $\mathcal{R}$ is the $\mathbb{Z}$-submodule generated by the set D of relators

$$(m_1, \ldots, m_i + m_i', \ldots, m_n) \; - \; (m_1, \ldots, m_i, \ldots, m_n) \; - \; (m_1, \ldots, m_i', \ldots, m_n)$$

and

$$(m_1, \ldots, rm_i, m_{i+1}, \ldots, m_n) \; - \; (m_1, \ldots, m_i, rm_{i+1}, \ldots, m_n)$$

for $m_i, m_i' \in M_i$ and $1 \leq i \leq n$ in the first relator, and $m_i \in M_i$, $r \in R$, and $1 \leq i < n$ in the second. Denote the coset of a single n-tuple $(m_1, \ldots, m_n)$ in $M_1 \times \cdots \times M_n$ by

$$\boldsymbol{m_1 \otimes \cdots \otimes m_n} \; = \; (m_1, \ldots, m_n) \; + \; \langle D \rangle.$$

Then $M_1 \otimes_R \cdots \otimes_R M_n$ is generated as a $\mathbb{Z}$-module by the **little tensors** $m_1 \otimes \cdots \otimes m_n$, subject to the defining relations:

$$m_1 \otimes \cdots \otimes (m_i + m_i') \otimes \cdots \otimes m_n =$$
$$(m_1 \otimes \cdots \otimes m_i \otimes \cdots \otimes m_n) + (m_1 \otimes \cdots \otimes m_i' \otimes \cdots \otimes m_n)$$

and

$$m_1 \otimes \cdots \otimes rm_i \otimes m_{i+1} \otimes \cdots \otimes m_n = m_1 \otimes \cdots \otimes m_i \otimes rm_{i+1} \otimes \cdots \otimes m_n .$$

This generalizes the tensor product $M_1 \otimes_R M_2$ defined in (5.9).

With essentially the same proof as in the case $n = 2$, the n-fold tensor product $M_1 \otimes_R \cdots \otimes_R M_n$ is also an R-module, with

$$r(m_1 \otimes \cdots m_n) \; = \; (rm_1 \otimes \cdots \otimes m_n)$$

for each little tensor $m_1 \otimes \cdots \otimes m_n$ and $r \in R$. There is an alternate construction that takes the R-module structure into account from the beginning:

(14.10) Definition. If $M_1, \ldots, M_n$ are R-modules, define $\boldsymbol{M_1 \otimes'_R \cdots \otimes'_R M_n}$ to be the quotient $\mathcal{F}'/\mathcal{R}'$, where $\mathcal{F}'$ is the free R-module based on $M_1 \times \cdots \times M_n$ and $\mathcal{R}'$ is the R-submodule generated by the set D' of relators

$$(m_1, \ldots, m_i + m'_i, \ldots, m_n) \ - \ (m_1, \ldots, m_i, \ldots, m_n) \ - \ (m_1, \ldots, m'_i, \ldots, m_n)$$

and

$$r(m_1, \ldots, m_n) \ - \ (m_1, \ldots, rm_i, \ldots, m_n)$$

for $m_i, m'_i \in M$, $r \in R$ and $1 \leq i \leq n$.

Note: The first relators are the same in D and D'. The second relator in D' does not exist in the free $\mathbb{Z}$-module based on $M_1 \times \cdots \times M_n$, since it has a coefficient $r \in R$. This second construction has the usual action of R on a quotient module.

Denote the coset of $(m_1, \ldots, m_n)$ by

$$m_1 \otimes' \cdots \otimes' m_n \ = \ (m_1, \ldots, m_n) + \langle D' \rangle \ .$$

(14.11) Proposition. *There is an R-linear isomorphism*

$$M_1 \otimes_R \cdots \otimes_R M_n \ \cong \ M_1 \otimes'_R \cdots \otimes'_R M_n$$

taking $m_1 \otimes \cdots \otimes m_n$ to $m_1 \otimes' \cdots \otimes' m_n$ for each $(m_1, \ldots, m_n) \in M_1 \times \cdots \times M_n$.

Proof. On the right side of the equation,

$$\begin{aligned}
m_1 \otimes' \cdots \otimes' rm_i \otimes' m_{i+1} \otimes' \cdots \otimes' m_n \ \\
= \ r(m_1 \otimes' \cdots \otimes' m_n) \ \\
= \ m_1 \otimes' \cdots \otimes' m_i \otimes' rm_{i+1} \otimes' \cdots \otimes' m_n \ ;
\end{aligned}$$

so the $\mathbb{Z}$-linear extension to $F_{\mathbb{Z}}(M_1 \times \cdots \times M_n)$ of the map $(m_1, \ldots, m_n) \mapsto m_1 \otimes' \cdots \otimes' m_n$ induces an additive homomorphism θ from the left side to the right side with the desired effect on little tensors. By the second relation in D', θ is R-linear.

On the left side of the equation, iteration of the second type of relation from D yields

$$r(m_1 \otimes \cdots \otimes m_n) \ = \ (rm_1 \otimes \cdots \otimes m_n) \ = \ (m_1 \otimes \cdots \otimes rm_i \otimes \cdots \otimes m_n)$$

for each i. So the R-linear extension to $F_R(M_1 \times \cdots \times M_n)$ of the map taking $(m_1, \ldots, m_n)$ to $m_1 \otimes \cdots \otimes m_n$ induces an R-linear homomorphism ψ inverse to θ. ∎

In view of this isomorphism, we drop the special notation $\otimes'$ and just write $\otimes$; so both constructions yield the same multiple tensor product. Just as $M_1 \otimes_R M_2$ is an initial target for a multiplication of vectors, so too is $M_1 \otimes_R \cdots \otimes_R M_n$ a first stop for a parenthesis-free product of n vectors:

(14.12) Definition. Suppose $M_1, \ldots, M_n$ and N are R-modules. A function

$$f : M_1 \times \cdots \times M_n \to N$$

is R-**multilinear** if, for each i and $m_1, \ldots, m_{i-1}, m_{i+1}, \ldots, m_n$ with $m_j \in M_j$, the function

$$M_i \to N, \quad m \mapsto f(m_1, \ldots, m, \ldots, m_n) \,,$$

is R-linear. Put another way, f is R-multilinear if

$$f(m_1, \ldots, m_i + m_i', \ldots, m_n)$$
$$= \; f(m_1, \ldots, m_i, \ldots, m_n) \; + \; f(m_1, \ldots, m_i', \ldots, m_n)$$

and

$$f(m_1, \ldots, rm_i, \ldots, m_n) \; = \; rf(m_1, \ldots, m_n)$$

whenever $m_i, m_i' \in M$, $r \in R$ and $1 \le i \le n$.

(14.13) Proposition. *Suppose $M_1, \ldots, M_n$, N are R-modules. The map*

$$c : M_1 \times \cdots \times M_n \quad \to \quad M_1 \otimes_R \cdots \otimes_R M_n$$
$$(m_1, \ldots, m_n) \quad \mapsto \quad m_1 \otimes \cdots \otimes m_n$$

is R-multilinear. If $f : M_1 \times \cdots \times M_n \to N$ is R-multilinear, there is one and only one R-linear map h from $M_1 \otimes_R \cdots \otimes_R M_n$ to N with $f = h \circ c$.

Proof. The first assertion is immediate from the relators D'. For the second assertion, h is induced by the R-linear extension of f, which kills D' because f is R-multilinear. The map h is uniquely determined by its effect on the little tensors:

$$h(m_1 \otimes \cdots \otimes m_n) \; = \; h \circ c((m_1, \ldots, m_n))$$
$$= \; f((m_1, \ldots, m_n)) \,. \qquad \blacksquare$$

Just as in the case $n = 2$, we use this proposition to construct R-linear maps with domain $M_1 \otimes_R \cdots \otimes_R M_n$.

Every R-algebra is an R-module if we forget multiplication. In the other direction, every R-module M is part of an R-algebra A in which the vector addition from M restricts the ring addition of A, and each of the little tensors $m_1 \otimes \cdots \otimes m_n \in M \otimes_R \cdots \otimes_R M$ is a product $m_1 \cdots m_n$ in the ring A. We now proceed to construct the simplest such A.

Suppose M is an R-module. Duplicate M as a set $M' = \{(m) : m \in M\}$, and let $A(M)$ denote the free R-algebra $R[\mathrm{Mon}(M')]$ based on this copy M'. The elements of $A(M)$ are the R-linear combinations of strings $(m_1)\cdots(m_n)$ of finite lengths $n \geq 0$, where $m_1,\ldots,m_n \in M$. The empty string $(n = 0)$ is denoted by 1 and serves as a multiplicative identity for $A(M)$. Multiplication in $A(M)$ is determined by the R-algebra axioms and concatenation of strings.

The module M is represented in $A(M)$ as the set M' of strings of length 1. But these representatives are R-linearly independent; so M is not in $A(M)$ as a submodule.

(14.14) Definition. The **tensor algebra** of a module M over a commutative ring R is the quotient ring $\boldsymbol{T}(M) = A(M)/J$, where J is the ideal of $A(M)$ generated by the elements

$$(m_1 + m_2) \;-\; (m_1) \;-\; (m_2) \quad \text{and} \quad (rm) \;-\; r(m)$$

for $m, m_1, m_2 \in M$ and $r \in R$, where the operations within parentheses are those of the R-module M.

Since $A(M)$ is the free R-algebra based on M', $T(M)$ has the R-algebra presentation with generators

$$[m] \;=\; (m) \;+\; J \quad (m \in M)$$

and defining relations

$$[m_1 + m_2] \;=\; [m_1] \;+\; [m_2]$$
$$[rm] \;=\; r[m]$$

for $m, m_1, m_2 \in M$ and $r \in R$. A typical element of $T(M)$ is an R-linear combination of products $[m_1]\cdots[m_n]$ of finite lengths n, where each $m_i \in M$. The empty product $(n = 0)$ is 1. So these products span $T(M)$ as an R-module. In fact, by the second relation, these products together with the elements $r1$ (for $r \in R$) span $T(M)$ as a $\mathbb{Z}$-module.

(14.15) Definition. For $n \geq 0$, the nth **tensor power** of M is the R-submodule $\boldsymbol{T}^n(M)$ spanned by the products $[m_1]\cdots[m_n]$ of length n.

(14.16) Proposition. *For each module M over a commutative ring R, the tensor algebra $T(M)$ is a graded R-algebra of type $\mathbb{N}$:*

$$T(M) \;=\; \bigoplus_{n \geq 0}^{\bullet} T^n(M) .$$

The map $R \to T^0(M)$, $r \mapsto r1$, *is a ring isomorphism. The map* $M \to$
$T^1(R)$, $m \mapsto [m]$, *is an* R-*linear isomorphism. For* $n > 1$, *there is an* R-*linear
isomorphism*

$$M \otimes_R \cdots \otimes_R M \ \to \ T^n(M)$$

(n factors of M in the domain) taking $m_1 \otimes \cdots \otimes m_n$ to $[m_1] \cdots [m_n]$.

Proof. The R-algebra $A(M)$ is graded as

$$A(M) \ = \ \overset{\bullet}{\bigoplus_{n \geq 0}} A^n(M) \, ,$$

where $A^n(M)$ is the free R-module with basis given by the length n products $(m_1) \cdots (m_n)$ with each $m_i \in M$. In terms of this grading, the ideal
J (in the definition of $T(M)$) is homogeneous, generated by elements of degree 1. By (14.8) (iii), $T(M)$ is a graded R-algebra with nth graded component $A^n(M)/J = T^n(M)$. By (14.8) (i), $J \cap A^0(M) = \{0\}$; so $A^0(M) =$
$A^0(M)/(J \cap A^0(M))$. Then we have R-linear isomorphisms

$$R \ \cong \ A^0(M) \ \cong \ \frac{A^0(M)}{J} \ = \ T^0(M)$$
$$r \ \mapsto \ r1 \ \mapsto \ r1 + J \ = \ r(1 + J) \, .$$

The composite is actually a ring isomorphism.

The map $f : M \to T^1(M)$, taking m to $[m]$, is R-linear by the defining
relations for $T(M)$. Erasing parentheses defines a bijection $M' \to M$ extending
to an R-linear map $A^1(M) \to M$, which sends the generators of J to 0. So it
induces an R-linear map $T^1(M) \to M$ inverse to f.

Now suppose $n > 1$. The R-module $J \cap A^n(M)$ is additively generated by
the elements $xty \in A^n(M)$, where x and y are homogeneous elements of $A(M)$
and t is a relator:

$$(m + m') \ - \ (m) \ - \ (m') \quad \text{or} \quad (rm) \ - \ r(m) \, .$$

By the distributive law, $J \cap A^n(M)$ is additively generated by such xty where
$x = (m_1) \cdots (m_{i-1})$ and $y = (m_{i+1}) \cdots (m_n)$ for some i. So $T^n(M)$ is the free
R-module $A^n(M)$, based on the strings $(m_1) \cdots (m_n)$ of length n, modulo the
submodule generated by the elements

$$(m_1) \cdots (m_{i-1})(m + m')(m_{i+1}) \cdots (m_n)$$
$$- (m_1) \cdots (m_{i-1})(m)(m_{i+1}) \cdots (m_n)$$
$$- (m_1) \cdots (m_{i-1})(m')(m_{i+1}) \cdots (m_n) \, ,$$

and

$$(m_1) \cdots (m_{i-1})(rm)(m_{i+1}) \cdots (m_n)$$
$$- r(m_1) \cdots (m_{i-1})(m)(m_{i+1}) \cdots (m_n)$$

for $m_j, m, m' \in M$, $r \in R$, and $1 \le i \le n$.

The map $M^n \to A^n(M)$ taking $(m_1, \ldots, m_n)$ to $(m_1) \cdots (m_n)$ extends to an R-linear map on the free R-module $F_R(M^n)$, carrying the relators for the group $M \otimes_R \cdots \otimes_R M$ to those for $T^n(M)$. So it induces an isomorphism

$$M \otimes_R \cdots \otimes_R M \quad \to \quad T^n(M)$$

taking $m_1 \otimes \cdots \otimes m_n$ to $[m_1] \cdots [m_n]$.   ∎

In view of these isomorphisms, we can use an alternate notation in $T(M)$, replacing $r1$ by r, $[m]$ by m, and using $\otimes$ for the multiplication sign between elements of degree ≥ 1 and scalar multiplication for the multiplication of any element by an $r \in R$. In this notation,

$$T(M) \quad = \quad R \overset{\bullet}{\oplus} M \overset{\bullet}{\oplus} (M \otimes_R M) \overset{\bullet}{\oplus} \cdots \ .$$

We shall use either notation, as fits the purpose at hand.

The tensor algebra is the simplest R-algebra constructed around an R-module M:

(14.17) Theorem. *Let $e : M \to T(M)$ denote the R-linear embedding $m \mapsto [m]$. For each R-linear map $f : M \to A$, from an R-module M to an R-algebra A, there is one and only one R-algebra homomorphism $\widehat{f} : T(M) \to A$ with $\widehat{f} \circ e = f$. If we use e to identify M with $e(M)$, so $[m]$ is replaced by m, this says each R-linear map $f : M \to A$ into an R-algebra A extends uniquely to an R-algebra homomorphism $\widehat{f} : T(M) \to A$.*

Proof. Since $e(M)$ generates $T(M)$ as an R-algebra, there can be at most one R-algebra homomorphism $\widehat{f} : T(M) \to A$ with $\widehat{f} \circ e = f$. The function $M' \to A$, $(m) \to f(m)$, extends to an R-algebra homomorphism $\phi : A(M) \to A$, taking the generators

$$(m + m') - (m) - (m') , \quad (rm) - r(m) ,$$

of J to 0 because f is R-linear. So ϕ induces an R-algebra homomorphism $\widehat{f} : T(M) \to A$ with

$$\begin{aligned}
\widehat{f} \circ e \, (m) \quad &= \quad \widehat{f}([m]) \quad = \quad \widehat{f}((m) + J) \\
&= \quad \phi((m)) \quad = \quad f(m) \ .
\end{aligned}$$

Identifying $[m]$ with m, we see that $\widehat{f}$ extends f.   ∎

Note: The $\widehat{f}$ in this theorem takes $m_1 \otimes \cdots \otimes m_n$ to $f(m_1) \cdots f(m_n)$ in A.

(14.18) Corollary. *There are functors $T : R\text{-}\mathfrak{Mod} \to R\text{-}\mathfrak{Alg}$ and, for each $n \geq 1$, $T^n : R\text{-}\mathfrak{Mod} \to R\text{-}\mathfrak{Mod}$, defined on objects as the tensor algebra and nth tensor power, and for each arrow $f : M \to N$ in $R\text{-}\mathfrak{Mod}$, by*

$$T(f)(m_1 \otimes \cdots \otimes m_n) \;=\; T^n(f)(m_1 \otimes \cdots \otimes m_n) \;=\; f(m_1) \cdots f(m_n)$$

whenever $n \geq 1$ and each $m_i \in M$.

Proof. If $f : M \to N$ is R-linear, there is a unique R-algebra homomorphism $T(f) : T(M) \to T(N)$ making the square

$$
\begin{array}{ccc}
M & \xrightarrow{\;f\;} & N \\
{\scriptstyle e}\downarrow & & \downarrow{\scriptstyle e} \\
T(M) & \xrightarrow{\;T(f)\;} & T(N)
\end{array}
$$

commute. Then $T(f)(m_1 \otimes \cdots \otimes m_n) = T(f)[m_1] \otimes \cdots \otimes T(f)[m_n]$, and this is $f(m_1) \otimes \cdots \otimes f(m_n)$. Evidently $T(f)$ restricts to an R-linear map $T^n(f)$ from $T^n(M)$ to $T^n(N)$. Checking the following equations on generators, we see that $T(i_M) = i_{T(M)}$ and $T(g \circ f) \;=\; T(g) \circ T(f)$, and the same is true for T^n. ∎

Another consequence of the universal property (14.17) of $T(M)$ is a connection between R-module and R-algebra presentations:

(14.19) Corollary. *If an R-module M has presentation $(S : D)$, then $T(M)$ has R-algebra presentation $(S : D)$.*

Proof. There is an R-linear surjection $f : F_R(S) \to M$ with kernel $\langle D \rangle$. The elements of D belong to $R[\mathrm{Mon}(S)]$ as R-linear combinations of length 1 strings, and to $T(F_R(S))$ as elements of $F_R(S)$. There are unique R-algebra homomorphisms

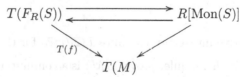

with those on top fixing each $s \in S$ and those below extending $f : S \to T(M)$. Since S generates each of the two top R-algebras, the triangles commute, and the top maps are mutually inverse isomorphisms fixing $R1$ and S, so fixing D. So it will suffice to prove D generates the kernel K of $T(f)$ as an ideal.

Suppose $\mathcal{J}$ is the ideal of $T(F_R(S))$ generated by D. Each element of $\mathcal{J}$ is a sum of terms xdy with $x, y \in T(F_R(S))$ and $d \in D$. Since $T(f)$ is an R-algebra homomorphism extending f, and $f(d) = 0$, $\mathcal{J} \subseteq K$. Then $T(f)$ induces an R-algebra homomorphism

$$g : \frac{T(F_R(S))}{\mathcal{J}} \quad \to \quad T(M)$$

taking $s + \mathcal{J}$ to $f(s)$ for each $s \in S$.

Since $D \subseteq \mathcal{J}$, its R-linear span $\langle D \rangle$ is contained in $\mathcal{J}$. So the inclusion of $F_R(S)$ into $T(F_R(S))$ induces an R-linear map

$$\phi : \frac{F_R(S)}{\langle D \rangle} \quad \to \quad \frac{T(F_R(S))}{\mathcal{J}} \quad .$$

Let

$$\overline{f} : \frac{F_R(S)}{\langle D \rangle} \quad \to \quad M$$

denote the R-linear isomorphism induced by f. The composite $\phi \circ \overline{f}^{-1}$ extends to an R-algebra homomorphism

$$h : T(M) \quad \to \quad \frac{T(F_R(S))}{\mathcal{J}}$$

with

$$
\begin{aligned}
h(f(s)) &= (\phi \circ \overline{f}^{-1})(f(s)) \\
&= \phi(s + \langle D \rangle) \quad = \quad s + \mathcal{J} \; .
\end{aligned}
$$

So $h \circ g$ and $g \circ h$ are R-algebra homomorphisms fixing the generators $s + \mathcal{J}$, $f(s)$, respectively, and so are identity maps. In particular, g is injective and the ideal $\mathcal{J} = K$.   ∎

14C. Exercises

1. For each positive integer m, compute $T(\mathbb{Z}/m\mathbb{Z})$ for the $\mathbb{Z}$-module $\mathbb{Z}/m\mathbb{Z}$.

2. If M is a cyclic R-module, prove $T(M)$ is a commutative ring.

3. If $f : M \to N$ is a surjective R-linear map, prove $T(f)$ is surjective. If $g : 2\mathbb{Z}/4\mathbb{Z} \to \mathbb{Z}/4\mathbb{Z}$ is the inclusion of $\mathbb{Z}$-modules, prove $T(g)$ is not injective. *Hint:* Look in degree 2.

4. Regarding $T(M)$ just as an R-module, T is a functor from $R\text{-}\mathcal{Mod}$ to $R\text{-}\mathcal{Mod}$. If M is a free R-module, show $T(M)$ is also a free R-module. If M is projective, prove $T(M)$ is projective. If $M \in \mathcal{P}(R)$, prove each $T^n(M) \in \mathcal{P}(R)$.

Note that T is generally not an additive functor from R-$\mathfrak{Mod}$ to R-$\mathfrak{Mod}$.

5. Using the notation $T^n(M) = M \otimes_R \cdots \otimes_R M$ for $n \geq 1$, and using the multiplication in $T(M)$, prove $T^m(M) \otimes_R T^n(M) \cong T^{m+n}(M)$ as R-modules. Iterating tensor products of these isomorphisms proves erasure of parentheses is a general "associativity" isomorphism

$$M \otimes_R (M \otimes_R (\cdots (M \otimes_R M) \cdots)) \;\cong\; M \otimes_R \cdots \otimes_R M \,.$$

14D. Symmetric and Exterior Algebras

Suppose M is a module over a commutative ring R, and $T(M)$ is the tensor algebra of M, defined in the last section.

(14.20) Definition. The **symmetric algebra** $S(M)$ of the R-module M is the quotient $T(M)/J$, where J is the ideal of $T(M)$ generated by the elements $x \otimes y - y \otimes x$ for all $x, y \in M$.

In terms of the type $\mathbb{N}$ grading,

$$T(M) \;=\; \bigoplus_{n \geq 0}^{\bullet} T^n(M) \,,$$

the ideal J is homogeneous, with all its generators in $T^2(M)$. By our theorem (14.8) on graded rings, $S(M)$ is a graded R-algebra:

$$S(M) \;=\; \bigoplus_{n \geq 0}^{\bullet} S^n(M) \,,$$

where, for each $n \geq 0$,

$$S^n(M) \;=\; T^n(M)/J \;\cong\; T^n(M)/(J \cap T^n(M))$$

as R-modules. For each n, $S^n(M)$ is called the nth **symmetric power** of M. By the graded ring theorem (14.8) (i), $J \cap T^0(M) = J \cap T^1(M) = 0$; so $S^0(M) = T^0(M) = R$ as a ring, and $S^1(M) = T^1(M) = M$ as an R-module. The scalar multiplication $R \times M \to M$ restricts the ring multiplication in $S(M)$.

Since M generates $T(M)$ as an R-algebra, it also generates $S(M)$ as an R-algebra. So the elements of $S(M)$ are sums of products $m_1 \cdots m_n$ $(m_i \in M)$. But $S(M)$ is commutative since $m_i m_j = m_j m_i$ by the relators defining J. If we choose a linear order "$\prec$" on M, a typical element of $S(M)$ can be written as a sum of some $r \in R$ and the products

$$m_1^{e_1} m_2^{e_2} \cdots m_t^{e_t}$$

with $t \geq 1$, each $m_i \in M$, each $e_i > 0$, and $m_1 \prec \cdots \prec m_t$. For each $n > 0$, the nth symmetric power $S^n(M)$ is the set of sums of such products having $e_1 + \cdots + e_t = n$.

(14.21) Theorem. *Let $e : M \to S(M)$ denote the R-linear inclusion of M as $S^1(M)$. For each R-linear map $f : M \to A$ to a commutative R-algebra A, there is one and only one R-algebra homomorphism $\widehat{f} : S(M) \to A$ with $\widehat{f} \circ e = f$ (that is, extending f).*

Proof. Since M generates $S(M)$ as an R-algebra, there is at most one such $\widehat{f}$. The extension of f to an R-algebra homomorphism $T(M) \to A$ kills each difference $x \otimes y - y \otimes x$ because A is commutative; so it induces an R-algebra homomorphism $\widehat{f} : S(M) \to A$ extending f. ∎

(14.22) Example. Suppose M is a free R-module with basis S. By the distributive laws in $S(M)$, each element in $S(M)$ is an R-linear combination of products

$$s_1^{e_1} s_2^{e_2} \cdots s_t^{e_t}$$

with $t \geq 0$, $e_i > 0$, $s_i \in S$, and $s_1 \prec \cdots \prec s_t$ under the linear order chosen for M. This much follows if S is any R-linear spanning set of M. As on p. 74, the above expressions $s_1^{e_1} \cdots s_t^{e_t}$ constitute the free abelian monoid $\mathcal{A}(S)$ based on S, and the monoid ring $A = R[\mathcal{A}(S)]$ is a commutative R-algebra. Since S is an R-basis of M, the inclusion $S \to R[\mathcal{A}(S)]$ extends to an R-linear map $M \to R[\mathcal{A}(S)]$ and then to an R-algebra map $S(M) \to R[\mathcal{A}(S)]$. Since the distinct products $s_1^{e_1} \cdots s_t^{e_t}$ are sent to themselves as R-linearly independent elements of $\mathcal{A}(S)$, they are also R-linearly independent in $S(M)$, and $S(M) \cong R[\mathcal{A}(S)]$. In particular, if M has a finite R-basis $\{s_1, \ldots, s_n\}$, then $S(M)$ is the polynomial ring $R[s_1, \ldots, s_n]$ in n indeterminates.

Just as with the tensor algebra, an R-linear map $f : M \to N$ determines an R-algebra homomorphism

$$S(f) : S(M) \quad \to \quad S(N)$$

and R-linear maps

$$S^n(f) : S^n(M) \quad \to \quad S^n(N)$$

for each $n \geq 0$ taking $m_1^{e_1} \cdots m_t^{e_t}$ to $f(m_1)^{e_1} \cdots f(m_t)^{e_t}$ or, in the case $n = 0$, taking each $r = r1_{S(M)}$ to $r = r1_{S(N)}$ since each ring homomorphism $S(M) \to S(N)$ takes $1_{S(M)}$ to $1_{S(N)}$. Since $T(-)$ and $T^n(-)$ are functors,

$$S(-) : R\text{-}\mathcal{M}\mathfrak{od} \quad \to \quad R\text{-}\mathcal{A}\mathfrak{lg} \, ,$$

$$S^n(-) : R\text{-}\mathcal{M}\mathfrak{od} \quad \to \quad R\text{-}\mathcal{M}\mathfrak{od}$$

are also functors, although S can be taken as a functor to the full subcategory of commutative R-algebras.

By (14.21), if a commutative multiplication of vectors restricts the multiplication in an R-algebra A with M as a submodule, then the subalgebra generated by M is a quotient of $S(M)$. As an example, the algebra of complex numbers is the quotient of $S(\mathbb{R}^2) \cong \mathbb{R}[e_1, e_2]$ by the ideal generated by $e_1 - 1$ and $e_2^2 + 1$. This is the idea of the complex plane.

But, as Hamilton discovered, the cross product in $\mathbb{R}^3$ does not occur in any commutative ring, for there is an anticommutative property $v \times w = -(w \times v)$, related to the identity $v \times v = 0$. Unfortunately, the cross product is also nonassociative; so it does not restrict the multiplication in any quotient of $T(\mathbb{R}^3)$. The determinant provides an example of such anticommutativity that can be accommodated within an (associative) algebra. Recall that if the determinant of a square matrix is regarded as a kind of product of its rows

$$\det(A) = f(e_1 A, \ldots, e_n A) ,$$

then $f : (R^n)^n \to R$ is R-multilinear and "alternating," so that a switch of two rows multiplies the determinant by -1; and if two rows are equal, the determinant is 0.

(14.23) Definition. The **exterior algebra** $\wedge(M)$ of the R-module M is the quotient $T(M)/J$, where J is the ideal of $T(M)$ generated by the elements $x \otimes x$ for all $x \in M$.

Like the symmetric algebra construction, the ideal J is homogeneous, generated by elements of $T^2(M)$. So $\wedge(M)$ is a graded R-algebra:

$$\wedge(M) = \overset{\bullet}{\underset{n \geq 0}{\bigoplus}} \wedge^n(M) ,$$

where, for each $n \geq 0$,

$$\wedge^n(M) = T^n(M)/J \cong T^n(M)/(J \cap T^n(M))$$

as R-modules, and $\wedge^0(M) = T^0(M) = R$ as a ring while $\wedge^1(M) = T^1(M) = M$ as an R-module. For each n, $\wedge^n(M)$ is the nth **exterior power** of M. Scalar multiplication $R \times M \to M$ agrees with the ring multiplication of $\wedge^0(M)$ times $\wedge^1(M)$ in $\wedge(M)$.

As an R-module, M has a presentation $(M' : D)$, where D consists of all $(m_1 + m_2) - (m_1) - (m_2)$ and all $(rm) - r(m)$ for $m, m_i \in M$ and $r \in R$. By (14.19), $T(M)$ has presentation $(M' : D)$ as an R-algebra. Also, by Example (14.6) (iv), $\wedge(M)$ has presentation $(M' : D \cup E)$, where E is the set of relators $(m)(m)$ with $m \in M$. So $\wedge(M)$ is the R-linear span of strings $(m_1) \cdots (m_n)$ with $m_i \in M$, $n \geq 0$, that are not linearly independent but satisfy $(m_1) + (m_2) = (m_1 + m_2)$, $r(m) = (rm)$ and $(m)(m) = 0$. A more conventional notation erases the parentheses $(m) \mapsto m$ and uses the multiplication sign "$\wedge$". So a typical element of $\wedge(M)$ is a sum of an $r \in R$ and terms $m_1 \wedge m_2 \wedge \cdots \wedge m_n$ with all $m_i \in M$ and $n > 0$. And $m \wedge m = 0$ while $r \wedge m = rm$.

An antisymmetry applies here, since

$$(m_1 \wedge m_2) \; + \; (m_2 \wedge m_1) \; = \; (m_1 + m_2) \wedge (m_1 + m_2)$$
$$- \; (m_1 \wedge m_1) \; - \; (m_2 \wedge m_2) \; = \; 0 \, .$$

So if $\sigma \in S_n =$ the permutation group of $\{1, \ldots, n\}$, then

$$m_{\sigma(1)} \wedge \cdots \wedge m_{\sigma(n)} \; = \; \mathrm{sgn}(\sigma) m_1 \wedge \cdots \wedge m_n \, ,$$

where $\mathrm{sgn}(\sigma)$ is 1 if σ is even and -1 if σ is odd. Suppose $X = (x(1), \ldots, x(a))$ and $Y = (y(1), \ldots, y(b))$ are disjoint subsequences of $(1, \ldots, n)$. Say

$$\mathbf{sgn(X, Y)} \; = \; (-1)^\alpha \, ,$$

where α is the number of ordered pairs in $X \times Y$ with first coordinate larger than the second coordinate. If $m_1, \ldots, m_n \in M$, then α is the number of switches of adjacent numbers in the sequence $(x(1), \ldots, x(a), y(1), \ldots, y(b))$ needed to reorder this concatenation of subsequences as a subsequence

$$XY \; = \; (xy(1), \ldots, xy(a+b))$$

of $(1, \ldots, n)$; for α is the number of switches needed to move $x(a)$, then $x(a-1)$, etc., to the right into their correct positions. It follows that if $m_1, \ldots, m_n \in M$, then

$$(m_{x(1)} \wedge \cdots \wedge m_{x(a)}) \wedge (m_{y(1)} \wedge \cdots \wedge m_{y(b)})$$
$$= \; \mathrm{sgn}(X, Y) m_{xy(1)} \wedge \cdots \wedge m_{xy(a+b)} \, .$$

Like the tensor and symmetric algebras, the exterior algebra has a universal property:

(14.24) Theorem. *Let $e : M \to \wedge(M)$ denote the inclusion of M as $\wedge^1(M)$. For each R-linear map $f : M \to A$ to an R-algebra A with $f(m)^2 = 0$ whenever $m \in M$, there is one and only one R-algebra homomorphism $\widehat{f} : \wedge(M) \to A$ with $\widehat{f} \circ e = f$ (that is, extending f).*

Proof. Since $\wedge(M)$ is generated as an R-algebra by $e(M) = \wedge^1(M)$, there is at most one such $\widehat{f}$. The extension of f to an R-algebra homomorphism $T(M) \to A$ kills each $m \otimes m$; so it induces an R-algebra homomorphism $\widehat{f} : \wedge(M) \to A$ extending f. ∎

Just as with the tensor and symmetric algebras, this theorem provides, for each R-linear map $f : M \to N$, an extension to an R-algebra homomorphism

$$\wedge(f) : \wedge(M) \to \wedge(N) \, ,$$

restricting to R-linear maps

$$\wedge^n(f) : \wedge^n(M) \;\rightarrow\; \wedge^n(N)$$

for each $n \geq 0$, and taking $m_1 \wedge \cdots \wedge m_n$ to $f(m_1) \wedge \cdots \wedge f(m_n)$, or in the case $n = 0$, fixing the elements of R. Evidently

$$\wedge(-) : R\text{-}\mathcal{M}o\mathfrak{d} \;\rightarrow\; R\text{-}\mathcal{A}\mathfrak{lg} \;,$$
$$\wedge^n(-) : R\text{-}\mathcal{M}o\mathfrak{d} \;\rightarrow\; R\text{-}\mathcal{M}o\mathfrak{d}$$

are functors.

If J is the ideal of $T(M)$ generated by all $m \otimes m$ with $m \in M$, then by our graded rings theorem (14.8), $J \cap T^n(M)$ is (for $n \geq 2$) the set of finite length sums of elements $m_1 \otimes \cdots \otimes m_n$, where $m_i = m_{i+1}$ for some i with $1 \leq i < n$.

(14.25) Definition. Suppose M and N are R-modules and $n \geq 2$. An R-multilinear function $f : M^n \to N$ is **alternating** if

$$f((m_1, \ldots, m_n)) \;=\; 0$$

whenever $m_i = m_{i+1}$ for some i with $1 \leq i < n$.

(14.26) Proposition. *Suppose M and N are R-modules, and $n \geq 2$. Let $e_n : M^n \to \wedge^n(M)$ denote the alternating function taking $(m_1, \ldots, m_n)$ to $m_1 \wedge \cdots \wedge m_n$. For each alternating function $f : M^n \to N$, there is one and only one R-linear map $\widehat{f} : \wedge^n(M) \to N$ with $\widehat{f} \circ e_n = f$.*

Proof. Since $\wedge^n(M)$ is generated as an R-module by $e_n(M^n)$, there is at most one $\widehat{f}$. Since f is alternating, the induced R-linear map $T^n(M) \to N$ kills $J \cap T^n(M)$, and so induces an R-linear map $\widehat{f} : \wedge^n(M) \to N$ taking $m_1 \wedge \cdots \wedge m_n$ to $f(m_1, \ldots, m_n)$. ∎

(14.27) Exterior Algebra of R^n. Suppose S is a finite set $\{s_1, \ldots, s_m\}$ of m elements and G is the set of finite length strings $s_{i(1)} \cdots s_{i(n)}$ with $i(1) < \cdots < i(n)$ and $0 \leq n \leq m$. Denote the empty string by 1_A. Let A denote the free R-module based on G. We create a multiplication on A as follows. If two strings share a letter s_i, say their product is 0_A. If $X = (x(1), \ldots, x(a))$ and $Y = (y(1), \ldots, y(b))$ are disjoint subsequences of $(1, \ldots, m)$, say

$$(s_{x(1)} \cdots s_{x(a)})(s_{y(1)} \cdots s_{y(b)}) \;=\; \text{sgn}(X, Y) s_{xy(1)} \cdots s_{xy(a+b)} \;,$$

using the notation preceding (14.24). This defines a product $G \times G \to A$. For mutually disjoint subsequences X, Y, and Z of $(1, \ldots, m)$,

$$\text{sgn}(X, Y)\text{sgn}(XY, Z) \;=\; \text{sgn}(X, YZ)\text{sgn}(Y, Z) \;,$$

since both equal $(-1)^\alpha$ where α is the number of ordered pairs in

$$(X \times Y) \cup (X \times Z) \cup (Y \times Z)$$

with the first coordinate exceeding the second. So this product extends bilinearly to an associative multiplication with identity 1_A, making A an R-algebra.

Now suppose M is a free R-module with basis $\{s_1, \ldots, s_m\}$. Inclusion of this basis into A extends to an R-linear map $f : M \to A$ with

$$f\left(\sum_{i=1}^{m} r_i s_i\right)^2 = \left(\sum_{i=1}^{m} r_i s_i\right)^2 = \sum_{i=1}^{m} r_i^2 s_i^2 + \sum_{i<j} r_i r_j (s_i s_j + s_j s_i) = 0$$

since $s_i s_i = 0$ and $s_j s_i = -s_i s_j$ in A. So f extends to an R-algebra homomorphism $\widehat{f} : \wedge(M) \to A$, taking the strings $s_{x(1)} \wedge \cdots \wedge s_{x(a)}$ with $x(1) < \cdots < x(a)$ to the R-linearly independent strings $s_{x(1)} \cdots s_{x(a)}$ in A. The former strings span $\wedge(M)$ as an R-module. So they must be an R-basis, and $\widehat{f}$ is an isomorphism of R-algebras $\wedge(M) \cong A$.

The exterior power $\wedge^n(M)$ for $n \leq m$ therefore has an R-basis consisting of the strings $s_{x(1)} \wedge \cdots \wedge s_{x(n)}$, where $x(1) < \cdots < x(n)$. There is one such string for each subset of n elements in $\{1, \ldots, m\}$. So $\wedge^n(M)$ is a free R-module of free rank given by the binomial coefficient

$$\binom{m}{n} = \frac{m!}{n!(m-n)!} .$$

For $n > m$, every product of n basis vectors s_i has a repeated factor and so is 0. So

$$\wedge(M) = \wedge^0(M) \overset{\bullet}{\oplus} \cdots \overset{\bullet}{\oplus} \wedge^m(M)$$

is a free R-module of free rank

$$\binom{m}{0} + \cdots + \binom{m}{m} = (1+1)^m = 2^m.$$

In particular, $\wedge^m(M)$ has free rank 1. If $f : M \to M$ is an R-linear map, then $\wedge^m(f) : \wedge^m(M) \to \wedge^m(M)$ must be multiplication by a scalar $r \in R$, depending on f. In terms of the basis $s_1, \ldots, s_m$ of M, f is represented by a matrix $B = (b_{ij}) \in M_n(R)$ and

$$\wedge^m(f)(s_1 \wedge \cdots \wedge s_m) = f(s_1) \wedge \cdots \wedge f(s_m)$$

$$= \left(\sum_j b_{1j} s_j\right) \wedge \cdots \wedge \left(\sum_j b_{mj} s_j\right)$$

$$= \sum_{\sigma \in S_m} b_{1\sigma(1)} \cdots b_{m\sigma(m)} (s_{\sigma(1)} \wedge \cdots \wedge s_{\sigma(m)})$$

$$= \left(\sum_{\sigma \in S_m} \mathrm{sgn}(\sigma) b_{1\sigma(1)} \cdots b_{m\sigma(m)}\right) (s_1 \wedge \cdots \wedge s_m) .$$

So $\wedge^m(f)$ is multiplication by the determinant of B. Taking $\mathbf{det}(f)$ to be the scalar $r \in R$ for which $\wedge^m(f)$ is multiplication by r, we get a coordinate-free definition of the determinant of a linear map $f : M \to M$.

Now we turn to exterior powers of f.g. projective modules. Although these modules need not have bases, there is a rank defined at each prime ideal p of R: Over the local ring $R_p = (R - p)^{-1}R$, each f.g. projective is free with a unique finite free rank. Since the free rank is additive over direct sums, multiplicative over tensor products, and takes R_p to 1, it defines a ring homomorphism $K_0(R_p) \to \mathbb{Z}$. The standard localization map $R \to R_p$, $r \mapsto r/1$, induces a ring homomorphism $K_0(R) \to K_0(R_p)$. The composite

$$r_p : K_0(R) \quad \to \quad \mathbb{Z}$$

is known as the **local rank at** p. Taking $r_p(P) = r_p([P]) =$ free rank of P_p over R_p, we obtain a generalized rank that can vary with the choice of prime p. However, if $p \subseteq q$ are prime ideals of R, there is a localization map $R_q \to R_p$ inducing an isomorphism $K_0(R_q) \cong K_0(R_p)$ ($\cong \mathbb{Z}$), and the local ranks at p and q agree. In particular, if R is an integral domain, so $\{0\}$ is a prime ideal, then all the local ranks agree. If $P \in \mathcal{P}(R)$ and $r_p(P) = n$ for all prime ideals p of R, we say P is a **rank n projective** over R.

Localization is respected by exterior powers:

(14.28) Lemma. *Suppose S is a submonoid of $(R, \cdot)$ and M is an R-module. There is an $S^{-1}R$-linear isomorphism*

$$\wedge^n(S^{-1}M) \quad \cong \quad S^{-1} \wedge^n (M)$$

for each $n \geq 0$.

Proof. For $n = 0$ or 1 this is evident. Suppose $n \geq 2$. The map $(S^{-1}M)^n \to S^{-1} \wedge^n (M)$, taking $(m_1/s_1, \ldots, m_n/s_n)$ to $(m_1 \wedge \cdots \wedge m_n)/s_1 \cdots s_n$, is well defined, since if $m_i/s_i = n_i/t_i$, with $n_i \in M$, $t_i \in S$, there exists $u \in S$ with $ut_im_i = us_in_i$; then

$$\frac{m_1 \wedge \cdots \wedge m_i \wedge \cdots \wedge m_n}{s_1 \cdots s_i \ldots s_n} \quad - \quad \frac{m_1 \wedge \cdots \wedge n_i \wedge \cdots \wedge m_n}{s_1 \cdots t_i \cdots s_n}$$

$$= \quad \frac{(m_1 \wedge \cdots \wedge t_im_i \wedge \cdots \wedge m_n) \quad - \quad (m_1 \wedge \cdots \wedge s_in_i \wedge \cdots \wedge m_n)}{s_1 \cdots (s_it_i) \cdots s_n}$$

$$= \quad \frac{m_1 \wedge \cdots \wedge u(t_im_i - s_in_i) \wedge \cdots \wedge m_n}{us_1 \cdots (s_it_i) \cdots s_n} \quad = \quad 0 \, .$$

This map is evidently $S^{-1}R$-multilinear. If $m_i/s_i = m_{i+1}/s_{i+1}$, we may put both fractions over a common denominator and assume $s_i = s_{i+1} = s$. Then

$tsm_i = tsm_{i+1}$ for some $t \in S$; multiplying by $ts/ts = 1$, we may assume $m_i = m_{i+1}$ too. Then $(m_1 \wedge \cdots \wedge m_n)/s_1 \cdots s_n = 0$. So there is an induced $S^{-1}R$-linear map

$$f : \wedge^n(S^{-1}M) \rightarrow S^{-1} \wedge^n (M) ,$$

$$\frac{m_1}{s_1} \wedge \cdots \wedge \frac{m_n}{s_n} \mapsto \frac{m_1 \wedge \cdots \wedge m_n}{s_1 \cdots s_n} .$$

The R-linear map $M \rightarrow S^{-1}M$ induces an R-linear map

$$\gamma : \wedge^n(M) \rightarrow \wedge^n(S^{-1}M)$$

taking $m_1 \wedge \cdots \wedge m_n$ to $(m_1/1) \wedge \cdots \wedge (m_n/1)$. Now $\wedge^n(S^{-1}M)$ is an $S^{-1}R$-module; so S acts through bijections on it, and the embedding

$$\wedge^n(S^{-1}M) \rightarrow S^{-1} \wedge^n (S^{-1}M)$$
$$x \mapsto x/1$$

is an $S^{-1}R$-linear isomorphism. The composite

$$S^{-1} \wedge^n (M) \xrightarrow{S^{-1}(\gamma)} S^{-1} \wedge^n (S^{-1}M) \cong \wedge^n(S^{-1}M)$$

is an $S^{-1}R$-linear map g taking $(m_1 \wedge \cdots \wedge m_n)/s$ to $(1/s)((m_1/1) \wedge \cdots \wedge (m_n/1))$. And g is a two-sided inverse to f. ∎

(14.29) Proposition. *Suppose R is an integral domain and $r(P)$ denotes the local rank of a f.g. projective R-module P.*

   (i)   *If $P \in \mathcal{P}(R)$, each $\wedge^n(P) \in \mathcal{P}(R)$.*
   (ii) *If $P \in \mathcal{P}(R)$ and $n > r(P)$, then $\wedge^n(P) = 0$.*
   (iii) *If $P, Q \in \mathcal{P}(R)$, then $\wedge^{r(P \oplus Q)}(P \oplus Q) \cong \wedge^{r(P)}(P) \otimes_R \wedge^{r(Q)}(Q)$.*

Proof. For (i), there is a f.g. free R-module F with P as a direct summand; so there are R-linear maps $i : P \rightarrow F$ and $\pi : F \rightarrow P$ with composite $\pi \circ i$ the identity on P. Since $\wedge^n$ is a functor, there are R-linear maps $\wedge^n(i) : \wedge^n(P) \rightarrow \wedge^n(F)$ and $\wedge^n(\pi) : \wedge^n(F) \rightarrow \wedge^n(P)$ with composite $\wedge^n(\pi) \circ \wedge^n (i)$ the identity on $\wedge^n(P)$. So $\wedge^n(P)$ is isomorphic to a direct summand of the f.g. free R-module $\wedge^n(F)$.

For (ii), suppose $S = R - \{0\}$ and use the lemma: Since $\wedge^n(P)$ is projective and R has no zero-divisors, S acts through injections on $\wedge^n(P)$. So $\wedge^n(P)$ embeds in $S^{-1} \wedge^n (P) \cong \wedge^n(S^{-1}P)$, which is 0 for $n > r(P)$.

To prove (iii), we construct the isomorphism and its inverse. Denote by $i_1 : P \rightarrow P \oplus Q$ and $i_2 : Q \rightarrow P \oplus Q$ the insertions in the first and second coordinates. Suppose $r, s \geq 0$. Following $\wedge^r(i_1) \oplus \wedge^s(i_2)$ by multiplication in $\wedge(P \oplus Q)$ defines a map

$$\wedge^r(P) \times \wedge^s(Q) \quad \to \quad \wedge^{r+s}(P \oplus Q)$$

that is balanced over R; so it induces an R-linear map

$$f : \wedge^r(P) \otimes_R \wedge^s(Q) \quad \to \quad \wedge^{r+s}(P \oplus Q) ,$$

taking $(p_1 \wedge \cdots \wedge p_r) \otimes (q_1 \wedge \cdots \wedge q_s)$ to

$$(p_1, 0) \wedge \cdots \wedge (p_r, 0) \wedge (0, q_1) \wedge \cdots \wedge (0, q_s) .$$

Now suppose $r = r(P)$, $s = r(Q)$, and therefore $r + s = r(P \oplus Q)$. Define a map

$$(P \oplus Q)^{r+s} \quad \to \quad \wedge^r(P) \otimes_R \wedge^s(Q)$$

by sending $((p_1, q_1), \ldots, (p_{r+s}, q_{r+s}))$ to

$$\sum_X \mathrm{sgn}(X, Y)(p_{x(1)} \wedge \cdots \wedge p_{x(r)}) \otimes (q_{y(1)} \wedge \cdots \wedge q_{y(s)}) ,$$

where the sum is over all length r subsequences $X = (x(1), \ldots, x(r))$ of the sequence $(1, \ldots, r+s)$, with complementary subsequences $Y = (y(1), \ldots, y(s))$. This map is alternating, so it induces an R-linear map

$$g : \wedge^{r+s}(P \oplus Q) \quad \to \quad \wedge^r(P) \otimes_R \wedge^s(Q) .$$

Now $\wedge^{r+s}(P \oplus Q)$ is additively generated by the elements

$$(p_1, 0) \wedge \cdots \wedge (p_m, 0) \wedge (0, q_1) \wedge \cdots \wedge (0, q_n) ,$$

where $m + n = r + s$. But this element is 0 if $m > r$ or $n > s$, since its first part is in the image of $\wedge^m(P)$ and the second part in the image of $\wedge^n(Q)$. So the additive generators can be restricted to elements with $m = r$ and $n = s$, proving f is surjective. And g takes such elements to $(p_1 \wedge \cdots \wedge p_r) \otimes (q_1 \wedge \cdots \wedge q_s)$; so $g \circ f$ is the identity, and f is injective. ∎

(14.30) Definition. Suppose R is an integral domain, $P \in \mathcal{P}(R)$, and $r(P)$ is the local rank of P. Then

$$\mathbf{det_0}(\boldsymbol{P}) \quad = \quad \wedge^{r(P)}(P) .$$

For $S = R - \{0\}$, $S^{-1} \wedge^{r(P)}(P) \cong \wedge^{r(P)}(S^{-1}P) \cong S^{-1}P$, since the latter is free with free rank $r(P)$. So $\det_0(P)$ is a rank 1 projective R-module. In particular, $\det_0(P)$ is the last nonzero homogeneous component of $\wedge(P)$, which accounts for its name. Isomorphic modules in $\mathcal{P}(R)$ have equal local rank, so $\det_0$ takes isomorphic modules to isomorphic modules. Now property (14.29) (iii) suggests that $\det_0$ is a generalized rank on $\mathcal{P}(R)$, defining a group homomorphism from $K_0(R)$ to some group of isomorphism classes of rank 1 projectives under tensor product. But is there such a group?

(14.31) Definition. For R a commutative ring, $\mathrm{Pic}(R)$ is the set of isomorphism classes of rank 1 projectives in $\mathcal{P}(R)$. This set is known as the **Picard group** of R because of the following:

(14.32) Theorem. *For each commutative ring R, $\mathrm{Pic}(R)$ is an abelian group under the operation $c(P) \cdot c(Q) = c(P \otimes_R Q)$.*

Proof. The argument proving Lemma (14.28), in the case $n = 2$, also proves there is an $S^{-1}R$-linear isomorphism

$$\psi : S^{-1}(M \otimes_R N) \;\cong\; S^{-1}M \otimes_{S^{-1}R} S^{-1}N$$
$$\frac{m \otimes n}{s} \;\longmapsto\; \frac{1}{s}\left(\frac{m}{1} \otimes \frac{n}{1}\right)$$

for each pair M, N of R-modules and submonoid S of $(R, \cdot)$. Since the free rank is multiplicative over tensor products, so is each local rank at a prime ideal p. So the operation on $\mathrm{Pic}(R)$ is defined. By standard tensor product isomorphisms, this product is associative and commutative, with identity $c(R)$.

For inverses, we show that if P is a rank 1 projective, so is its dual $P^* = \mathrm{Hom}_R(P, R)$, and $P^* \otimes_R P \cong R$. Recall that the dual functor $(-)^*$, defined as $\mathrm{Hom}_R(-, R)$, is an additive functor from $R\text{-}\mathcal{M}\mathrm{od}$ to itself, taking R to $R^* \cong R$. So $P^* \in \mathcal{P}(R)$. As in §6E, Exercise 8, there is an $S^{-1}R$-linear isomorphism

$$\theta : S^{-1}\mathrm{Hom}_R(P, R) \;\cong\; \mathrm{Hom}_{S^{-1}R}(S^{-1}P, S^{-1}R)$$

for each submonoid S of $(R, \cdot)$, where θ extends the localization functor from $\mathrm{Hom}_R(P, R)$ to $\mathrm{Hom}_{S^{-1}R}(S^{-1}P, S^{-1}R)$, so that $\theta(f/s)(x/t) = f(x)/st$. Taking $S = R - p$ for a prime ideal p of R, $S^{-1}P \cong S^{-1}R$; so the right side of θ becomes $S^{-1}R$, and P^* is a rank 1 projective.

Evaluation $P^* \times P \to R$, $(a, b) \mapsto a(b)$, is balanced over R, inducing an R-linear map $f : P^* \otimes_R P \to R$. The image is an R-submodule of R — that is, an ideal of R.

If R happens to be a local ring, P has a basis $\{v\}$ and P^* contains $v^* : P \to R$ taking each x to its v-coefficient. Then $f(v^*, v) = v^*(v) = 1$; so f is surjective. Since R is a free R-module, the exact sequence

$$0 \longrightarrow \ker(f) \longrightarrow P^* \otimes_R P \longrightarrow R \longrightarrow 0$$

splits, and $\ker(f) \in \mathcal{P}(R)$. So $\ker(f)$ is free of free rank equal to rank $(P^* \otimes_R P)$ − rank $(R) = 1 - 1 = 0$, proving f is also injective.

If R is not local, for each prime ideal p the map f_p factors as the isomorphism

$$(\theta \otimes 1) \circ \psi : (P^* \otimes_R P)_p \quad \to \quad (P_p)^* \otimes_{R_p} P_p$$

followed by the evaluation f' from $(P_p)^* \otimes_{R_p} P_p$ to R_p. Since R_p is local, the evaluation f' is an isomorphism. So f_p is an isomorphism for each prime ideal p. By (6.45) (i), f is an isomorphism. ∎

Suppose now that R is an integral domain. Then $\det_0$ actually does define a homomorphism of abelian groups

$$\det_0 : K_0(R) \quad \to \quad \operatorname{Pic}(R) .$$

The map

$$i : \operatorname{Pic}(R) \quad \to \quad K_0(R) , \; c(P) \; \mapsto \; [P] ,$$

is a homomorphism into the group of units $K_0(R)^*$ of the ring $K_0(R)$. Since $\wedge^1(P) \cong P$, $\det_0 \circ i$ is the identity on $\operatorname{Pic}(R)$, showing i is injective. This proves

(14.33) Corollary. *If R is an integral domain, $\operatorname{Pic}(R)$ embeds as a subgroup of $K_0(R)^*$. If P and Q are rank 1 projectives in $\mathcal{P}(R)$, then for each $n \geq 0$, $P \oplus R^n \cong Q \oplus R^n$ implies $P \cong Q$.* ∎

For a generalization to arbitrary commutative rings, see §8 of Swan [68] or Chapter IX, §3 of Bass [68].

Note that $\det_0([R]) = c(R)$ is the identity in $\operatorname{Pic}(R)$; so $\det_0$ induces a surjective homomorphism

$$\overline{\det}_0 : \widetilde{K}_0(R) \quad \to \quad \operatorname{Pic}(R)$$

taking $\overline{[P]}$ to $c(P)$ for each rank 1 projective $P \in \mathcal{P}(R)$. There are examples (see Anderson [78, Example 7.1]) where $\overline{\det}_0$ is not injective. Its kernel is denoted by $SK_0(R)$, by analogy with the kernel $SK_1(R)$ of the determinant map $K_1(R) \to R^*$.

The reverse map

$$\operatorname{Pic}(R) \quad \to \quad \widetilde{K}_0(R) ,$$

taking $c(P)$ to $\overline{[P]}$, is a homomorphism if and only if, for rank 1 projectives P and Q, $P \oplus Q$ is stably equivalent to $P \otimes_R Q$. Considering the local ranks, this amounts to $P \oplus Q$ being stably isomorphic to $R \oplus (P \otimes_R Q)$. When this holds, there is even an embedding

$$i' : \operatorname{Pic}(R) \quad \to \quad K_0(R)$$
$$c(P) \quad \mapsto \quad [P] - [R]$$

of $\operatorname{Pic}(R)$ as an additive subgroup of $K_0(R)$, with $\det_0 \circ i'$ the identity on $\operatorname{Pic}(R)$.

After a reading of §7E, these conditions should look familiar. When R is a Dedekind domain, Steinitz' Theorem (7.48) says the elements of $\widetilde{K}_0(R)$ and of $\operatorname{Pic}(R)$ are represented by the nonzero ideals of R; so $SK_0(R) = 0$ and $\overline{\det}_0$ is an isomorphism

$$\widetilde{K}_0(R) \quad \cong \quad \operatorname{Pic}(R) \ .$$

And, in the Dedekind case, nonzero ideals I and J satisfy $I \otimes_R J \cong IJ$ by Exercise 3 below. So $\operatorname{Pic}(R)$ coincides with the ideal class group $\operatorname{Cl}(R)$.

14D. Exercises

1. Suppose R is a nonzero commutative ring and $n \geq 0$. If $M, N \in \mathcal{M}(R)$, adapt the proof of (14.29) (iii) to prove

$$\wedge^n(M \oplus N) \quad \cong \quad \bigoplus_{r=0}^{n} \left(\wedge^r(M) \otimes_R \wedge^{n-r}(N) \right),$$

and consequently that

$$\wedge(M \oplus N) \quad \cong \quad \wedge(M) \otimes_R \wedge(N)$$

as R-modules. Describe the multiplication of $x \otimes y \in \wedge^p(M) \otimes_R \wedge^q(N)$ times $x' \otimes y' \in \wedge^r(M) \otimes_R \wedge^s(N)$ that makes this an isomorphism of R-algebras.

2. Suppose R is an integral domain, $S = R - \{0\}$, and F is the field of fractions $S^{-1}R$. If M is an R-module, the **local rank** of M is the F-dimension of $S^{-1}M$. Prove every nonzero R-submodule of F has local rank 1.

3. Suppose R is an integral domain with field of fractions F, and I, J are projective R-submodules of F. Prove the map $I \otimes_R J \to IJ$, taking $\sum x_i \otimes y_i$ to $\sum x_i y_i$, is an R-linear isomorphism. Then show every rank 1 projective $P \in \mathcal{P}(R)$ is isomorphic to an invertible R-submodule of F. Conclude that $\operatorname{Pic}(R)$ is isomorphic to the quotient $I(R)/PI(R)$, where $I(R)$ is the group of invertible R-submodules of F and $PI(R)$ is the subgroup consisting of the modules Rx with $x \in F - \{0\}$. *Hint:* Show $I \otimes_R J \cong IJ$ locally by showing I and J are locally in $PI(R)$; then deduce the global case from (6.45) (i). For the rest, show $P \subseteq P_{\{0\}} \subseteq R_{\{0\}}$ and use (7.11) and the paragraph following (7.34).

4. Suppose R is an integral domain and P_1, P_2, Q are rank 1 projectives in $\mathcal{P}(R)$. If $P_1 \oplus P_2 \cong R \oplus Q$, prove $Q \cong P_1 \otimes_R P_2$.

5. For R an integral domain, prove the following are equivalent:

   (i) $\overline{\det}_0 : \widetilde{K}_0(R) \to \operatorname{Pic}(R)$ is a group isomorphism.
   (ii) For each $n \geq 1$, every rank n projective in $\mathcal{P}(R)$ is stably isomorphic to $R^{n-1} \oplus Q$ for a rank 1 projective $Q \in \mathcal{P}(R)$.

It is a well-known theorem of Serre [58] that condition (ii) is true when R is a noetherian commutative ring of Krull dimension ≤ 1, and more generally, that every rank n projective $P \in \mathcal{P}(R)$ is isomorphic to $R^{n-m} \oplus Q$ for a rank m projective $Q \in \mathcal{P}(R)$ with $m \leq$ the maximal spectrum dimension of R (defined in (4.32)).

6. Suppose R is a commutative ring. Prove every $P \in \mathcal{P}(R)$ with $P \oplus R^{n-1} \cong R^n$ for some $n \geq 1$ is a free R-module. Then show every stably free ideal of R is free. *Hint:* For each prime ideal p of R, $P_p \cong R_p$; so $\wedge^r(P)_p = \wedge^r(P_p) = 0$ for $r > 1$. Apply $\wedge^n$ and use Exercise 1. If J is a nonzero stably free ideal of R, $J \oplus R^m \cong R^n$ implies $m = n - 1$, since $J_p \neq 0$ for some prime ideal p.

7. If $f : R \to S$ is a homomorphism of commutative rings and P is a rank n projective in $\mathcal{P}(R)$, prove $S \otimes_R P$ is a rank n projective in $\mathcal{P}(S)$. Then show there is a group homomorphism

$$\begin{aligned} \operatorname{Pic}(f) : \operatorname{Pic}(R) &\longrightarrow \operatorname{Pic}(S) \\ c(P) &\longmapsto c(S \otimes_R P) \,, \end{aligned}$$

making Pic into a functor from $\mathcal{CRing}$ to $\mathcal{Ab}$. *Hint.* If p is a prime ideal of S, then $f^{-1}(p)$ is a prime ideal of R, and there is a commutative square of ring homomorphisms:

$$\begin{array}{ccc} R & \xrightarrow{\ f\ } & S \\ \downarrow & & \downarrow \\ R_{f^{-1}(p)} & \longrightarrow & S_p \end{array} \,.$$

8. If there is a fiber square in $\mathcal{CRing}$

$$\begin{array}{ccc} A & \xrightarrow{\ f_2\ } & R_2 \\ {\scriptstyle f_1}\downarrow & & \downarrow{\scriptstyle g_2} \\ R_1 & \xrightarrow{\ g_1\ } & R' \end{array}$$

with g_2 surjective, prove there is an exact Mayer-Vietoris sequence in $\mathcal{Ab}$:

$$A^* \longrightarrow R_1^* \times R_2^* \longrightarrow R'^* \xrightarrow{\ h\ } \operatorname{Pic}(A) \longrightarrow \operatorname{Pic}(R_1) \oplus \operatorname{Pic}(R_2) \,,$$

where all the maps but h arise by combining maps obtained from the square by the functors GL_1 and Pic.

14E. The Milnor Ring

In a ring R, addition is to multiplication as multiplication is to ... what? Can multiplication in R be addition in another ring S? Addition in S makes S an abelian group; so R would have to be a commutative ring, and zero-divisors in R could not be included in S. If R is a field, the effect of these restrictions is minimal. Even if R is just commutative, we may hope for a ring S with R^* as an additive subgroup.

Assume R is a commutative ring. Its multiplicative group of units R^* is abelian, so it is a $\mathbb{Z}$-module, although in module notation R^* should be written as an additive group. The universal way to make this $\mathbb{Z}$-module R^* into the addition in a ring is to form the tensor algebra

$$T(R^*) \;=\; \overset{\bullet}{\underset{n\geq 0}{\bigoplus}} T^n(R^*) \,,$$

where $T^0(R^*) = \mathbb{Z}$, $T^1(R^*) = R^*$, and, for $n > 1$,

$$T^n(R^*) \;=\; R^* \otimes_{\mathbb{Z}} \cdots \otimes_{\mathbb{Z}} R^* \quad (n \text{ factors}).$$

To write R^* additively, we can revert to our original notation for the tensor algebra as a quotient $A(R^*)/I$, where $A(R^*)$ is the free $\mathbb{Z}$-algebra ($=$ free ring) based on the (r) with $r \in R^*$, and I is the ideal generated by the elements

$$(r_1 r_2) \;-\; (r_1) \;-\; (r_2) \,,$$
$$(r^n) \;-\; n(r)$$

for $r, r_1, r_2, \in R^*$ and $n \in \mathbb{Z}$. Then

$$[r] \;=\; (r) \;+\; I \quad (r \in R^*)$$

generate $T(R^*)$ as ring, and these generators are subject only to the relations

(i) $[r_1 r_2] \;=\; [r_1] \;+\; [r_2] \,,$
(ii) $[r^n] \;=\; n[r]$

for $r, r_1, r_2 \in R^*$ and $n \in \mathbb{Z}$. Since the relations (ii) are consequences of the relations (i), we need only use the relations (i) in a presentation of $T(R^*)$ as a ring.

Recall from (14.16) that, for each module M, the map

$$M \to T(M) \,, \quad m \mapsto [m] \,,$$

is an injective linear map onto $T^1(M)$. In the present case, the map

$$R^* \to T(R^*) , \quad r \mapsto [r] ,$$

is an injective homomorphism from the multiplicative abelian group R^* onto the additive subgroup $T^1(R^*)$ of $T(R^*)$, rather like a logarithm. So the multiplication in $T(R^*)$ provides an answer to the question posed at the beginning of this section: $+$ is to $\cdot$ as $\cdot$ is to $\otimes$.

This is a simplest answer to the question, since any logarithm-like homomorphism $f : R^* \to (A, +)$ for a ring A factors uniquely by (14.17) as our embedding $R^* \to T(R^*)$, followed by a ring homomorphism $T(R^*) \to A$. But there is a more interesting answer:

(14.34) Definitions. Suppose R is a commutative ring and J is the ideal of $T(R^*)$ generated by all $[r][s] = r \otimes s$ with r, $s \in R^*$ and $r + s = 1$ in R. The quotient

$$K_*^M(R) \;=\; \frac{T(R^*)}{J}$$

is the **Milnor ring** of R. In terms of the grading $T(R^*) = \overset{\bullet}{\oplus} T^n(R^*)$, the ideal J is homogeneous with its generators in $T^2(R^*)$. So the Milnor ring of R is a graded ring

$$K_*^M(R) \;=\; \bigoplus_{n \geq 0}^{\bullet} K_n^M(R) ,$$

where

$$K_n^M(R) \;=\; \frac{T^n(R^*)}{J} \;\cong\; \frac{T^n(R^*)}{J \cap T^n(R^*)}$$

is the nth **Milnor K-group** of R.

The $\mathbb{Z}$-module $J \cap T^0(R^*) = 0$; so $\mathbb{Z} = T^0(R^*) \cong K_0^M(R)$. The degree 0 component is a subring of the graded ring $K_*^M(R)$, so $K_*^M(R)$ is a ring of characteristic zero. If $r \in R^*$, denote the coset of $[r]$ in $K_*^M(R)$ by

$$\ell(r) \;=\; [r] \;+\; J .$$

The composite

$$R^* \longrightarrow T(R^*) \longrightarrow K_*^M(R)$$
$$r \longmapsto [r] \longmapsto \ell(r)$$

is a logarithm-like map ℓ, which is injective because $J \cap T^1(R^*) = 0$. So ℓ embeds R^* as the additive subgroup $K_1^M(R)$ of the Milnor ring.

Since $T(R^*)$ is presented as a ring by generators $[r]$ with $r \in R^*$ and defining relations $[r_1 r_2] = [r_1] + [r_2]$, the quotient $K_*^M(R)$ is the ring with generators $\ell(r)$ with $r \in R^*$ and defining relations

Mi1 : $\ell(rs) = \ell(r) + \ell(s)$, and

Mi2 : $\ell(r)\ell(s) = 0$ if $r + s = 1$ in R .

By our theorem (14.8) on quotients of graded rings, $J \cap T^n(R^*)$ is generated as a $\mathbb{Z}$-module by the elements

$$[r_1] \cdots [r_n] = r_1 \otimes \cdots \otimes r_n$$

with $r_1, \ldots, r_n \in R^*$ and $r_i + r_{i+1} = 1$ for some $i < n$. Adding this to the relations presenting the $\mathbb{Z}$-module $T^n(R^*)$, we see that for $n \geq 2$, $K_n^M(R)$ is presented as a $\mathbb{Z}$-module by generators $\ell(r_1) \cdots \ell(r_n)$ with $r_1, \ldots, r_n \in R^*$ and defining relations

Mi3 : $\ell(r_1) \cdots \ell(r_i s_i) \cdots \ell(r_n) = \ell(r_1) \cdots \ell(r_i) \cdots \ell(r_n)$

$\qquad\qquad + \ell(r_1) \cdots \ell(s_i) \cdots \ell(r_n)$, *for* $1 \leq i \leq n$,

Mi4 : $\ell(r_1) \cdots \ell(r_n) = 0$ *if* $r_i + r_{i+1} = 1$ *in R for some* $i < n$.

Each homomorphism between commutative rings $f : R \to S$ restricts to a $\mathbb{Z}$-linear map $f : R^* \to S^*$. Then

$$T(f) : T(R^*) \to T(S^*) , \quad [r] \mapsto [f(r)] ,$$

is a ring homomorphism. Since $r + s = 1$ in R implies $f(r) + f(s) = 1$ in S, $T(f)$ induces a ring homomorphism

$$\boldsymbol{K_*^M}(f) : K_*^M(R) \to K_*^M(S) .$$
$$\ell(r) \mapsto \ell(f(r))$$

For each $n > 0$, this restricts to $\mathbb{Z}$-linear maps

$$\boldsymbol{K_n^M}(f) : K_n^M(R) \to K_n^M(S) .$$
$$\ell(r_1) \cdots \ell(r_n) \mapsto \ell(f(r_1)) \cdots \ell(f(r_n))$$

In this way we get functors

$$K_*^M(-) : \mathcal{C}\mathfrak{Ring} \to \mathfrak{Ring}, \quad \text{and}$$
$$K_n^M(-) : \mathcal{C}\mathfrak{Ring} \to \mathbb{Z}\text{-}\mathfrak{Mod}, \quad \text{for}\ n > 0 .$$

Note: We could define a functor $K_0^M(-)$ by restricting $K_*^M(-)$ to its 0th homogeneous component $\mathbb{Z}$. But $K_0^M(f) : \mathbb{Z} \to \mathbb{Z}$ would always be the identity map, since that is the only ring homomorphism from $\mathbb{Z}$ to $\mathbb{Z}$. And if we identify $K_1^M(R)$ with R^* via the map ℓ, the functor $K_1^M(-)$ becomes $GL_1(-)$.

For a field F, the Milnor K-groups $K_n^M(F)$ look like the algebraic K-groups $K_n(F)$ for $n = 0, 1$, and 2: Every f.g. projective F-module V is free of invariant free rank $\dim(V)$; so $\dim(-)$ induces a ring isomorphism $K_0(F) \cong \mathbb{Z}$. If f is a ring homomorphism between fields, $K_0(f)$ thereby induces the identity map on $\mathbb{Z}$. Every matrix of determinant 1 in $GL(F)$ can be reduced by elementary row transvections to the identity matrix; so $\det(-)$ induces an isomorphism $K_1(F) \cong F^*$, which is a natural isomorphism $K_1(-) \cong GL_1(-)$ between functors on fields. Finally, considering the defining relations Mi1, Mi2 for $K_2^M(F)$, every function

$$F^* \times F^* \to G, \quad (a, b) \mapsto \sigma(a, b) ,$$

into an abelian group $(G, \cdot)$, satisfying

$$
\begin{aligned}
\sigma(a_1 a_2, b) &= \sigma(a_1, b)\sigma(a_2, b) , \\
\sigma(a, b_1 b_2) &= \sigma(a, b_1)\sigma(a, b_2) ,
\end{aligned}
$$

and

$$\sigma(a, b) = 1_G \quad \text{if} \quad a + b = 1_F$$

defines a homomorphism from the additive group $K_2^M(F)$ to the group G, taking $\ell(a)\ell(b)$ to $\sigma(a, b)$ for each pair $a, b \in F^*$. By Theorem (12.31), Steinberg symbols $\sigma(a, b) = \{a, b\}$ have these properties in $K_2(F)$, and by (12.30) they generate $K_2(F)$. So there is a surjective homomorphism $s : K_2^M(F) \to K_2(F)$ taking the additive generators $\ell(a)\ell(b)$ of the first group to the multiplicative generators $\{a, b\}$ of the second. If f is a ring homomorphism between fields, $K_2^M(f)$ takes $\ell(a)\ell(b)$ to $\ell(f(a))\ell(f(b))$ and $K_2(f)$ takes $\{a, b\}$ to $\{f(a), f(b)\}$; so s is natural. In §14H below, we give Keune's proof of Matsumoto's Theorem, stating that s is an isomorphism for each field F. So, on fields F, the functors $K_n^M(-)$ and $K_n(-)$ coincide for $n = 0, 1$, and 2.

Since the sequence continues as $K_3^M(-)$, $K_4^M(-), \ldots$, we have our first example of a higher algebraic K-theory. To put this into context, the reader should be aware that there is a standard sequence of functors $K_n : \mathfrak{Ring} \to \mathcal{A}b$ for $n \geq 0$, coinciding with the K_0, K_1, and K_2 we have introduced in Chapters 3, 9, and 12. In the period 1968–72, definitions for higher algebraic K-theories of rings were given by Swan, Karoubi and Villamayor, Volodin, Wagoner, and Quillen. All but the Karoubi-Villamayor definition are equivalent, and are most widely known as Quillen K-theory. For left-regular rings, the Karoubi-Villamayor theory also coincides with Quillen K-theory. Milnor's K-groups of fields appeared in Milnor [70] in connection with quadratic forms. (We consider the connection to forms in (14.48) below.) Suslin [82] proved there are homomorphisms

$$K_n^M(F) \to K_n(F) \to K_n^M(F)$$

of additive abelian groups, whose composite is multiplication by $(n-1)!$; so Milnor K-groups are closely connected to the Quillen K-groups.

Let's consider some elementary consequences of the defining relations in $K_*^M(F)$ for F a field. Of course, those relations we obtain in $K_2^M(F)$ will also hold for Steinberg symbols $\{a,b\}$ over F.

(14.35) Proposition. *Suppose* $a,b \in F^*$. *In* $K_*^M(F)$,

  (i) $\ell(a)\,\ell(-a) = 0$,
  (ii) $\ell(b)\,\ell(a) = -\ell(a)\,\ell(b)$,
  (iii) $\ell(a)\,\ell(a) = \ell(a)\,\ell(-1) = \ell(-1)\,\ell(a)$.

Proof. Since $\ell(1) = \ell(1 \cdot 1) = \ell(1) + \ell(1)$, we have $\ell(1) = 0$. So, if $a = 1$, all three assertions are true. Assume $a \neq 1$. Then $a^{-1} \neq 1$ and $1-a, 1-a^{-1} \in F^*$. Since $(-a)(1-a^{-1}) = 1-a$,

$$
\begin{aligned}
\ell(a)\,\ell(-a) &= \ell(a)\,[\ell(1-a) - \ell(1-a^{-1})] \\
&= \ell(a)\,\ell(1-a) - \ell(a)\,\ell(1-a^{-1}) \\
&= -\,\ell(a)\,\ell(1-a^{-1}) \\
&= \ell(a^{-1})\,\ell(1-a^{-1}) = 0 ,
\end{aligned}
$$

proving (i). Then

$$
\begin{aligned}
\ell(a)\ell(b) + \ell(b)\ell(a) &= \ell(a)\ell(-a) + \ell(a)\ell(b) + \ell(b)\ell(a) + \ell(b)\ell(-b) \\
&= \ell(a)\ell(-ab) + \ell(b)\ell(-ab) \\
&= \ell(ab)\ell(-ab) = 0 ,
\end{aligned}
$$

proving (ii). Finally,

$$
\begin{aligned}
\ell(a)\ell(a) &= \ell(a)[\ell(-1) + \ell(-a)] = \ell(a)\ell(-1) \\
&= -\,\ell(-1)\ell(a) = \ell(-1^{-1})\ell(a) \\
&= \ell(-1)\ell(a) .
\end{aligned}
$$ ∎

(14.36) Corollary. *Suppose* $a_1, \dots, a_n \in F^*$.

  (i) *If* σ *is a composite of* r *transpositions in* S_n,

$$
\ell(a_{\sigma(1)}) \cdots \ell(a_{\sigma(n)}) = (-1)^r \ell(a_1) \cdots \ell(a_n) .
$$

(ii)   *If $x \in K_i^M(F)$ and $y \in K_j^M(F)$, then*

$$yx = (-1)^{ij}xy .$$

(iii)   *If $a_i + a_j = 0$ or 1 for some i and j with $1 \le i < j \le n$, then*

$$\ell(a_1) \cdots \ell(a_n) = 0 .$$

Proof. Apply the antisymmetry $\ell(b)\ell(a) = -\ell(a)\ell(b)$.   ∎

(14.37) Proposition. *If a, b and $a + b$ belong to F^*,*

$$\ell(a)\ell(b) = \ell(a+b)\ell(-\frac{b}{a}) = \ell(-\frac{a}{b})\ell(a+b) .$$

Proof. The second equation is a consequence of antisymmetry. For the first, use

$$\frac{a}{a+b} + \frac{b}{a+b} = 1$$

to get

$$0 = \ell(\frac{a}{a+b}) \, \ell(\frac{b}{a+b})$$

$$= [\ell(a) - \ell(a+b)] \, [\ell(b) - \ell(a+b)]$$

$$= \ell(a) \, \ell(b) - \ell(a+b) \, \ell(b) - \ell(a) \, \ell(a+b) + \ell(a+b)^2$$

$$= \ell(a) \, \ell(b) - [\ell(a+b) \, \ell(b) - \ell(a+b) \, \ell(a) + \ell(a+b) \, \ell(-1)]$$

$$= \ell(a) \, \ell(b) - \ell(a+b) \, \ell(-\frac{b}{a}) .$$ ∎

(14.38) Proposition. *Suppose $a_1, \ldots, a_n \in F^*$.*

(i)   *If $a_1 + \cdots + a_n \in F^*$, then $\ell(a_1) \cdots \ell(a_n) \in \ell(a_1 + \cdots + a_n)K_*^M(F)$.*

(ii)   *If $a_1 + \cdots + a_n = 0$ or 1, then $\ell(a_1) \cdots \ell(a_n) = 0$.*

Proof. If $n = 1$, both assertions are immediate. If $n > 1$, they are also immediate when $a_1 + a_2 = 0$, since that forces $\ell(a_1) \, \ell(a_2) = 0$. If $a_1 + a_2 \ne 0$, the preceding proposition implies

$$\ell(a_1) \cdots \ell(a_n) = \ell(-\frac{a_1}{a_2})\ell(a_1 + a_2)\ell(a_3) \cdots \ell(a_n)$$

$$= \pm \, \ell(a_1 + a_2) \, \ell(a_3) \cdots \ell(a_n)\ell(-\frac{a_1}{a_2}) ;$$

so both assertions are true by induction on n.   ∎

Combining the defining relations Mi1 and Mi2 with the consequent relations in (14.35)–(14.38), we can begin to compute Milnor rings of fields.

(14.39) Theorem. *If $\mathbb{F}_q$ is a finite field with q elements, $K_n^M(\mathbb{F}_q) = 0$ for all $n \geq 2$. So there is a ring isomorphism:*

$$\frac{\mathbb{Z}[x]}{(q-1)x\mathbb{Z}[x] + x^2\mathbb{Z}[x]} \;\cong\; K_*^M(\mathbb{F}_q) \ .$$

Proof. The proof of (12.33), that $K_2(\mathbb{F}_q) = 1$, used only the Steinberg symbol identities that $\{a, b\}$ is antisymmetric, multiplicative in each coordinate, and trivial when $a + b = 1$. The corresponding facts

$$
\begin{aligned}
\ell(b)\, \ell(a) &= -\,\ell(a)\, \ell(b) \ , \\
\ell(ab)\, \ell(c) &= \ell(a)\, \ell(c) + \ell(b)\, \ell(c) \ , \\
\ell(a)\, \ell(bc) &= \ell(a)\, \ell(b) + \ell(a)\, \ell(c) \ , \\
\ell(a)\, \ell(b) &= 0 \ \ if \ a + b = 1 \ ,
\end{aligned}
$$

are true in K_*^M of a field by (14.35) (ii), Mi1, and Mi2. So that same proof shows $\ell(a)\, \ell(b) = 0$ for all $a, b \in \mathbb{F}_q^*$, and hence that $K_n^M(\mathbb{F}_q) = 0$ when $n \geq 2$.

The group $\mathbb{F}_q^*$ is cyclic, generated by a unit u of order $q - 1$. So $T(\mathbb{F}_q^*)$ has a $\mathbb{Z}$-algebra presentation with one generator $[u]$ and one relation $(q - 1)[u] = 0$. The canonical map to the Milnor ring therefore induces a surjective ring homomorphism

$$\frac{\mathbb{Z}[x]}{(q-1)x\mathbb{Z}[x]} \rightarrow K_*^M(\mathbb{F}_q)$$

taking x to $\ell(u)$. Since $\ell(u)\, \ell(u) = 0$, this induces the surjective map

$$\frac{\mathbb{Z}[x]}{(q-1)x\mathbb{Z}[x] + x^2\mathbb{Z}[x]} \rightarrow K_*^M(\mathbb{F}_q)$$

taking a typical element $\overline{a + bx}$ with $a, b \in \mathbb{Z}$, $0 \leq b < q - 1$, to $a + b\ell(u) = a + \ell(u^b)$. Since $K_0^M(\mathbb{F}_q) = \mathbb{Z}$ and $\ell : \mathbb{F}_q^* \cong K_1^M(\mathbb{F}_q)$, this map is also injective.   ∎

For comparison, one of the first computations of the higher algebraic K-groups of a ring was the determination of $K_n(\mathbb{F}_q)$ by Quillen [72]:

(14.40) Theorem. (Quillen) *For $n > 0$,*

$$K_n(\mathbb{F}_q) \;\cong\; \begin{cases} \mathbb{Z}/(q^{(n+1)/2} - 1)\mathbb{Z} & \textit{if } n \textit{ is odd,} \\ 0 & \textit{if } n \textit{ is even.} \end{cases}$$
                                                                                      ∎

Next, consider algebraically closed fields F. In such a field, every unit is an mth power, for each positive integer m. Thinking of F^* as a $\mathbb{Z}$-module, every vector is the scalar m times a vector.

(14.41) Definition. Suppose A is an abelian group, written as a $\mathbb{Z}$-module. For each $m \in \mathbb{Z}$, let $A \xrightarrow{m} A$ denote the $\mathbb{Z}$-linear map that is multiplication by the scalar m. If $m > 0$, say A is **divisible by** m if $A \xrightarrow{m} A$ is surjective, and is **uniquely divisible by** m if $A \xrightarrow{m} A$ is bijective. Say A is **divisible** (resp. **uniquely divisible**) if A is divisible (resp. uniquely divisible) by m for all positive integers m.

If S is a submonoid of $(\mathbb{Z}, \cdot)$, note that A is uniquely divisible by all $m \in S$ if and only if S acts through bijections on A, which is to say, if and only if the standard localization map $A \to S^{-1}A$ is a $\mathbb{Z}$-linear isomorphism. *So A is uniquely divisible by all $m \in S$ if and only if A is an $S^{-1}\mathbb{Z}$-module.* In particular, A is uniquely divisible if and only if A is a $\mathbb{Q}$-vector space. Of course, in a uniquely divisible abelian group, the identity is a power of only itself, so there are no elements of finite order except the identity.

The following results and proofs (14.42)–(14.44) appeared in Bass and Tate [73]:

(14.42) Lemma. *If abelian groups A and B are divisible by m (> 0), then $A \otimes_{\mathbb{Z}} B$ is uniquely divisible by m.*

Proof. The "m-primary part" of B is the union B_m over all $r \geq 1$ of the kernels of

$$B \xrightarrow{m^r} B \ .$$

Since A is divisible by m, $A \otimes_{\mathbb{Z}} B_m = 0$. Since $A \otimes_{\mathbb{Z}} (-)$ is right exact, it follows that $A \otimes_{\mathbb{Z}} B \cong A \otimes_{\mathbb{Z}} (B/B_m)$. Multiplication by m is an isomorphism on B/B_m, and hence on $A \otimes_{\mathbb{Z}} (B/B_m)$. ∎

(14.43) Theorem. *Suppose F is a field, m is a positive integer, and each polynomial $x^m - a \in F[x]$ splits into linear factors in $F[x]$ — so F^* is divisible by m. Then $K_n^M(F)$ is uniquely divisible by m for $n \geq 2$.*

Proof. Consider the commutative diagram with exact rows:

$$
\begin{array}{ccccccccc}
0 & \longrightarrow & J & \longrightarrow & T^n(F^*) & \longrightarrow & K_n^M(F) & \longrightarrow & 0 \\
 & & \downarrow{\scriptstyle m} & & \downarrow{\scriptstyle m} & & \downarrow{\scriptstyle m} & & \\
0 & \longrightarrow & J & \longrightarrow & T^n(F^*) & \longrightarrow & K_n^M(F) & \longrightarrow & 0 \ .
\end{array}
$$

Since F^* is divisible by m and $n \geq 2$, the middle vertical map is an isomorphism. By the Snake Lemma, it will suffice to prove the left vertical map is surjective.

For the latter it is enough to prove $[a][1 - a] \in mJ$ for each $a \in F$ other than 0 and 1. But

$$x^m - a = \prod_{i=1}^{m} (x - b_i) \,,$$

where each $b_i \in F$ and $b_i^m = a$. So

$$[a]\,[1 - a] = [a]\,[\textstyle\prod (1 - b_i)] = \sum [a]\,[1 - b_i]$$
$$= \sum [b_i^m]\,[1 - b_i] = m \sum [b_i]\,[1 - b_i] \in mJ \,. \quad \blacksquare$$

(14.44) Corollary. *If F is an algebraically closed field, then, for $n \geq 2$, $K_n^M(F)$ is uniquely divisible; so it is a $\mathbb{Q}$-vector space.* $\quad \blacksquare$

In Exercise 1 below we discover that $K_n^M(F) = 0$ for all $n \geq 2$ when F is an algebraic extension of a finite field; even Corollary (14.44) allows the possibility that $K_n^M(F) = 0$ in degree $n \geq 2$. We need some tools to detect nontrivial elements of degree ≥ 2 in the Milnor ring. The simplest such tool occurs over the field $\mathbb{R}$ of real numbers. Recall from our statement of quadratic reciprocity, just before (11.34), that the **real symbol** is the function

$$(-, -)_\infty : \mathbb{R}^* \times \mathbb{R}^* \to \{\pm 1\}$$

with $(a, b)_\infty = -1$ if $a < 0$ and $b < 0$, and $(a, b)_\infty = 1$ otherwise. That is, the real symbol assigns a sign to each point in the cartesian plane (off the axes) according to the rule:

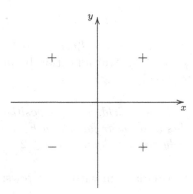

Checking cases according to the signs of a and b, it is evident that $(a, -)_\infty$ and $(-, b)_\infty$ are multiplicative homomorphisms $\mathbb{R}^* \to \{\pm 1\}$. The graph of $x + y = 1$ does not meet the third quadrant; so $(a, b)_\infty = 1$ if $a + b = 1$. Therefore the real symbol induces a homomorphism

$$K_2^M(\mathbb{R}) \longrightarrow \{\pm 1\} \,,$$
$$\ell(a)\ell(b) \longmapsto (a, b)_\infty$$

which is evidently surjective. In particular, $\ell(-1)\ell(-1) \neq 0$ in $K_2^M(\mathbb{R})$. What makes this possible is the ordering on $\mathbb{R}$.

Milnor found a generalization of the real symbol that works in all degrees, over fields with an ordering like that in $\mathbb{R}$. A **totally ordered abelian group** is an additive abelian group $(A, +)$ with a subset P including 0, for which $P + P \subseteq P$, $P \cap (-P) = \{0\}$ and $P \cup (-P) = A$. Here $-P$ denotes $\{-a : a \in P\}$. We say $a \leq b$ in A if $b - a \in P$, and $a < b$ if $a \leq b$ but $a \neq b$. Then $\leq$ is a linear ordering on A, and $P = \{a \in A : 0 \leq a\}$. Also, $a \leq b$ and $c \leq d$ in A implies $a + c \leq b + d$.

A field F is **ordered** if $(F, +)$ is a totally ordered abelian group with respect to a set P with $PP \subseteq P$ (i.e., with elements ≥ 0 closed under multiplication). Of course $\mathbb{R}$ is ordered with the usual "$\leq$." Every subfield F of an ordered field E is ordered, with $P_F = F \cap P_E$.

Note that in an ordered field, $1 \neq 0$ and $(-1)^2 = 1^2 = 1$; so $1 \in P$. Then $-1 \notin P$. For each nonzero $a \in F$, $(-a)^2 = a^2 \in P$; so -1 cannot be a sum of squares in an ordered field.

In an arbitrary field F, denote the set of finite length nonzero sums of squares by

$$S(F) = \left\{ \sum_{i=1}^{n} a_i^2 : a_1, \ldots, a_n \in F^*, n \geq 1 \right\}.$$

The set $S(F)$ is closed under addition, multiplication, and division ($a \in S(F)$ implies $a^{-1} = a(a^{-1})^2 \in S(F)$). A field F is **formally real** if $-1 \notin S(F)$.

(14.45) Proposition. *A field F is ordered if and only if it is formally real.*

Proof. We have already shown that ordered fields are formally real. Suppose F is a formally real field, with algebraic closure $\overline{F}$. Denote by $\mathcal{S}$ the set of formally real subfields of $\overline{F}$ containing F. If a nonempty subset of $\mathcal{S}$ is linearly ordered by containment, its union is also in $\mathcal{S}$, since an expression $-1 = a_1^2 + \cdots + a_n^2$ in the union also occurs in some field from the subset. By Zorn's Lemma, $\mathcal{S}$ has a maximal element E.

Take $P_E = S(E) \cup \{0\}$. Since E is formally real and $S(E)$ is closed under division, $P_E \cap -P_E = \{0\}$. And $P_E + P_E, P_E P_E$ are contained in P_E. If we can show $P_E \cup -P_E = E$, then E will be an ordered field, and hence its subfield F will be ordered.

Suppose $a \in E^*$ and $a \notin S(E)$. Then $x^2 - a$ has a root $\sqrt{a} \in \overline{F}$, but $\sqrt{a} \notin E$. So $E \subsetneq E(\sqrt{a})$ and $E(\sqrt{a})$ is not formally real. For some $b_i, c_i \in E$ with some $c_i \neq 0$,

$$-1 = \sum_{i=1}^{n} (b_i + c_i\sqrt{a})^2.$$

Since 1 and $\sqrt{a}$ are E-linearly independent,

$$-1 = \sum_{i=1}^{n} (b_i^2 + ac_i^2) \, ;$$

so

$$a \;=\; -\left(\frac{1 + \sum\limits_{i=1}^{n} b_i^2}{\sum\limits_{i=1}^{n} c_i^2} \right) \;\in\; -S(E) \,. \qquad \blacksquare$$

(14.46) Theorem. (Milnor) *Suppose F is a field. The following are equivalent:*

    (i) $-1 \in S(F)$.

   (ii) *$\ell(-1)$ is nilpotent in $K_*^M(F)$.*

  (iii) *Every element of $\overset{\bullet}{\oplus}_{n>0} K_n^M(F)$ is nilpotent.*

Proof. Suppose $-1 \notin S(F)$; so F is an ordered field under some ordering " $\leq$." As a generalization of the real symbol, define

$$\sigma : (F^*)^n \;=\; F^* \times \cdots \times F^* \to \{\pm 1\}$$

by $\sigma(a_1, \ldots, a_n) = -1$ if all $a_i < 0$, and $= 1$ if some $a_i > 0$. Then σ is multiplicative in each coordinate. If $a_i + a_{i+1} = 1$, where $i < n$, either $a_i > 0$ or $a_{i+1} > 0$; in either case, $\sigma(a_1, \ldots, a_n) = 1$. Therefore σ induces a group homomorphism

$$K_n^M(F) \to \{\pm 1\}$$
$$\ell(a_1) \cdots \ell(a_n) \mapsto \sigma(a_1, \ldots, a_n)$$

taking $\ell(-1)^n$ to -1. So $\ell(-1)^n \neq 0$ for all $n > 0$ and $\ell(-1)$ is not nilpotent, proving (ii) implies (i).

Suppose $-1 = a_1^2 + \cdots + a_r^2$ for $a_1, \ldots, a_r \in F^*$. Then $1 = (-a_1^2) + \cdots + (-a_r^2)$, and by (14.38) (ii),

$$\begin{aligned}
0 \;&=\; \ell(-a_1^2) \cdots \ell(-a_r^2) \\
&=\; (\ell(-1) + 2\ell(a_1)) \cdots (\ell(-1) + 2\ell(a_r)) \\
&=\; \ell(-1)^r + 2b \,,
\end{aligned}$$

where $b \in K_*^M(F)$. Since $2\ell(-1) = \ell((-1)^2) = 0$, multiplying by $\ell(-1)$ leaves $0 = \ell(-1)^{r+1}$.

By anticommutativity and the relation $\ell(-1)\ell(a) = \ell(a)\ell(a)$, for each $n > 0$, $s > 0$, and additive generator $\gamma = \ell(a_1) \cdots \ell(a_n)$ of $K_n^M(F)$,

$$\begin{aligned}
\gamma^s \;&=\; (\ell(a_1) \cdots \ell(a_n))^s \;=\; \pm \ell(a_1)^s \cdots \ell(a_n)^s \\
&=\; \pm (\ell(-1)^{s-1} \ell(a_1)) \cdots (\ell(-1)^{s-1} \ell(a_n)) \\
&=\; \pm \ell(-1)^{n(s-1)} \ell(a_1) \cdots \ell(a_n) \,.
\end{aligned}$$

So $\gamma^s = 0$ if $s > (r/n) + 1$.

If $\gamma_1 \in K_m^M(F)$ and $\gamma_2 \in K_n^M(F)$ are this sort of additive generator, then $\gamma_2\gamma_1 = \pm\gamma_1\gamma_2$. So if $\gamma_1, \ldots, \gamma_k$ are k of these additive generators of $\overset{\bullet}{\oplus}_{n>0} K_n^M(F)$, then $(\gamma_1 + \cdots + \gamma_k)^s$ is a $\mathbb{Z}$-linear combination of monomials

$$\gamma_1^{i_1}\gamma_2^{i_2}\cdots\gamma_k^{i_k}$$

with $i_1 + i_2 + \cdots + i_k = s$. So, for sufficiently large s, $(\gamma_1 + \cdots + \gamma_k)^s = 0$, proving (i) implies (iii). That (iii) implies (ii) is purely formal. ∎

Note: If $-1 \in S(F)$ for a field F, the smallest number of nonzero squares in F whose sum is -1 is the **stufe** of F, denoted $s(F)$. ("Stufe" is the German word for **level**.) Let $i(F)$ denote the least positive integer n with $\ell(-1)^n = 0$. The proof above shows $i(F) \le s(F) + 1$.

(14.47) Corollary. *For $\mathbb{R}$ the field of real numbers and n any positive integer, $K_n^M(\mathbb{R}) = A \overset{\bullet}{\oplus} B$, where A is the cyclic group of order 2 generated by $\ell(-1)^n$, and B is the divisible group generated by those $\ell(a_1)\cdots\ell(a_n)$ with every $a_i > 0$.*

Proof. Since $-1 \notin S(\mathbb{R})$, $\ell(-1)^n \ne 0$, but $2\ell(-1)^n = \ell(1)\ell(-1)^{n-1} = 0$. So $\ell(-1)^n$ has additive order 2. If σ is the extended real symbol used to prove (14.46), there is a split short exact sequence of abelian groups

$$0 \longrightarrow B \overset{\subseteq}{\longrightarrow} K_n^M(\mathbb{R}) \underset{\tau}{\overset{\sigma}{\rightleftarrows}} \{\pm 1\} \longrightarrow 1 \ ,$$

where $\tau(-1) = \ell(-1)^n$. Each $\ell(-a) = \ell(-1) + \ell(a)$; so $K_n^M(\mathbb{R})$ is generated by elements $\ell(-1)^s\ell(b_1)\cdots\ell(b_{n-s})$ with $0 \le s \le n$ and all b_i positive. If $s < n$, this term equals $\ell(b_1)^{s+1}\cdots\ell(b_{n-s})$. Terms with $s = n$ are $\ell(-1)^n$, and these cancel in pairs. So B is generated by the $\ell(a_1)\cdots\ell(a_n)$ with each $a_i > 0$. Finally, B is divisible, since $a_1 > 0$ has a positive mth root $c \in \mathbb{R}^*$, for each positive integer m, and

$$\ell(c^m)\ell(a_2)\cdots\ell(a_n) \ = \ m\ell(c)\ell(a_2)\cdots\ell(a_n) \ . \qquad \blacksquare$$

To close this section, we glimpse the connection of Milnor K-theory to quadratic forms, conjectured and partly verified in Milnor [70], the paper where Milnor introduced $K_*^M(F)$. Because the complete verification has turned out to be difficult, and the verified cases have proved useful in quadratic form theory (see Elman and Lam [72]), Milnor's Conjecture has become somewhat famous.

Suppose F is a field of characteristic $\ne 2$, and recall from (3.11), (3.28), and (5.24)–(5.28) that the Witt-Grothendieck ring $\widehat{W}(F)$ can be described as formal differences of congruence classes of invertible matrices over F, and each congruence class is represented by a diagonal matrix. The class of the matrix diag $(a_1, \ldots, a_n)$ is denoted by $\langle a_1, \ldots, a_n \rangle$ and represents the form $\Sigma_i \, a_i x_i y_i$. The addition and multiplication are induced by $\oplus$ and $\otimes$ of matrices:

$$\langle a_i, \ldots, a_n \rangle \perp \langle b_1, \ldots, b_m \rangle = \langle a_1, \ldots, a_n, b_1, \ldots, b_m \rangle \, ,$$

$$\langle a_1, \ldots, a_n \rangle \langle b_1, \ldots, b_m \rangle = \langle a_1 b_1, \ldots, a_1 b_m,$$

$$a_2 b_1, \ldots, a_2 b_m, \ldots, a_n b_1, \ldots, a_n b_m \rangle \, .$$

The Witt ring $W(F)$ is the quotient $\widehat{W}(F)/\widehat{W}(F) \cdot \langle 1, -1 \rangle = \widehat{W}(F)/\mathbb{Z} \cdot \langle 1, -1 \rangle$. In the Witt ring, $\langle 1 \rangle + \langle -1 \rangle \equiv 0$; so $\langle -1 \rangle = -\langle 1 \rangle = -1$, and each element

$$\langle a_1, \ldots, a_n \rangle - \langle b_1, \ldots, b_m \rangle = \langle a_1, \ldots, a_n, -b_1, \ldots, -b_m \rangle$$

is represented by a single form.

The dimension of a form is the number of rows in a representative matrix; so dim $\langle a_1, \ldots, a_n \rangle = n$. Since dimension is additive over $\oplus$ and multiplicative over $\otimes$ of matrices, it induces a ring homomorphism $\widehat{W}(F) \to \mathbb{Z}$. Following this by reduction mod 2, we get a ring homomorphism $\widehat{W}(F) \to \mathbb{Z}/2\mathbb{Z}$ taking $\langle 1, -1 \rangle$ to 0, so inducing a ring homomorphism

$$\overline{\dim} : W(F) \to \mathbb{Z}/2\mathbb{Z} \, .$$

The kernel of this mod 2 dimension on $W(F)$ is the ideal $I = I(F)$ of even-dimensional forms. In $W(F)$,

$$\begin{aligned} \langle a, b \rangle &\equiv \langle a, b \rangle \perp \langle 1, -1 \rangle \\ &\equiv \langle 1, a \rangle \perp \langle -1 \rangle \langle 1, -b \rangle \\ &\equiv \langle 1, a \rangle - \langle 1, -b \rangle \, . \end{aligned}$$

So I is the $\mathbb{Z}$-linear span of forms $\langle 1, a \rangle$ with $a \in F^*$. Then, for each $n > 0$, I^n is the $\mathbb{Z}$-linear span of forms

$$\ll a_1, \ldots, a_n \gg = \prod_{i=1}^{n} \langle 1, a_i \rangle \, ,$$

which are known as n-**fold Pfister forms**.

Now consider the map

$$\sigma : (F^*)^n = F^* \times \cdots \times F^* \to I^n/I^{n+1}$$

$$(a_1, \ldots, a_n) \mapsto \overline{\ll -a_1, \ldots, -a_n \gg} \, .$$

If $a + b = 1$ in F^*,

$$\begin{bmatrix} 1 & -1 \\ b & a \end{bmatrix} \begin{bmatrix} a & 0 \\ 0 & b \end{bmatrix} \begin{bmatrix} 1 & b \\ -1 & a \end{bmatrix} = \begin{bmatrix} 1 & 0 \\ 0 & ab \end{bmatrix} ;$$

so $\langle a, b \rangle = \langle 1, ab \rangle$ and

$$\langle 1, -a \rangle \langle 1, -b \rangle \quad = \quad \langle 1, -b, -a, ab \rangle$$

$$\equiv \quad \langle 1, ab \rangle \quad - \quad \langle a, b \rangle \quad = \quad 0 \ .$$

So $\overline{\ll -a_1, \dots, -a_n \gg} = 0$ if $a_i + a_{i+1} = 1$ for some $i < n$. Even if a and b don't sum to 1,

$$\langle 1, ab \rangle \ \perp \ \langle a, b \rangle \quad = \quad \langle 1, a \rangle \langle 1, b \rangle \ \in \ I^2 \ ;$$

so in $W(F)/I^2$,

$$
\begin{aligned}
\langle 1, -ab \rangle \quad &\equiv \quad \langle 1, -1, 1, -ab \rangle \\
&\equiv \quad \langle 1, 1 \rangle \ - \ \langle 1, ab \rangle \ \equiv \ \langle 1, 1 \rangle \ \perp \ \langle a, b \rangle \\
&\equiv \quad \langle 1, 1 \rangle \ - \ \langle -a, -b \rangle \\
&= \quad \langle 1, 1 \rangle \ \perp \ \langle -a, -b \rangle \ - \ \langle 1, 1 \rangle \langle -a, -b \rangle \\
&\equiv \quad \langle 1, -a \rangle \ \perp \ \langle 1, -b \rangle \ .
\end{aligned}
$$

Therefore σ is $\mathbb{Z}$-multilinear, inducing a homomorphism of groups

$$K_n^M(F) \quad \to \quad I^n/I^{n+1},$$

$$\ell(a_1) \cdots \ell(a_n) \quad \mapsto \quad \overline{\ll -a_1, \dots, -a_n \gg}$$

which is evidently surjective, since Pfister forms generate I^n as a $\mathbb{Z}$-module. Since multiplication by 2 in $W(F)$ is multiplication by $\langle 1, 1 \rangle$, it is the zero map in I^n/I^{n+1}. So there is an induced surjective homomorphism

$$s_n : k_n(F) \quad = \quad K_n^M(F)/2K_n^M(F) \quad \to \quad I^n/I^{n+1}$$

for each $n \geq 1$. Milnor discovered the maps s_n and posed:

(14.48) Milnor's Conjecture.

(i) s_n *is an isomorphism for all n and all fields F of characteristic $\neq 2$.*

(ii) $\bigcap_{n=1}^{\infty} I^n = 0$.

He verified (i) for $n = 1$ and 2, and proved both parts in full when F is a global field (number field or finite extension of $\mathbb{F}_q(x)$ with $\mathbb{F}_q$ a finite field of odd order q), as well as when F is a union of global fields. In 1996, Voevodsky circulated a preprint with a proof of Milnor's Conjecture using the homotopy theory of algebraic varieties. For an overview, see Voevodsky [99].

In §14B, Exercise 5, there is a discussion of the "associated graded ring" of an ideal I in a commutative ring A, having the form

$$(A/I) \oplus (I/I^2) \oplus (I^2/I^3) \oplus \cdots .$$

Milnor's conjecture would make $K_*^M(F)$ isomorphic to the associated graded ring of the ideal I of even dimensional forms in $W(F)$.

14E. Exercises

1. Suppose F is a field with a finite subfield K, and F is an algebraic extension of K. Prove $K_n^M(F) = 0$ for all $n \geq 2$. *Hint:* Show every list of units $a_1, \ldots, a_n$ of F lies in a finite subfield of F.

2. For all $n \geq 2$, prove $K_n^M(\mathbb{Z})$ is cyclic of order 2, generated by $\ell(-1)^n$. *Hint:* Use what we know about $K_n^M(\mathbb{R})$.

3. Suppose R is a commutative ring and R^* is cyclic, generated by an element ζ of finite order m. Prove the following:

  (i) For $n \geq 1$, $T^n(R^*)$ is the additive cyclic group of order m generated by $\zeta \otimes \cdots \otimes \zeta = [\zeta]^m$.
  (ii) The ring $T(R^*) \cong \mathbb{Z}[x]/(mx)\mathbb{Z}[x]$. *Hint:* Use (14.19).

4. Suppose F is a degree 2 field extension of $\mathbb{Q}$, within $\mathbb{C}$ but not contained in $\mathbb{R}$. Let R denote the ring of algebraic integers in F. Use Exercise 3 to compute $K_*^M(R)$. *Hints:*

  (i) There is a square-free positive integer d with $F = \mathbb{Q}(\sqrt{-d})$. Then

$$R = \{a + b\sqrt{-d} : a, b \in \mathbb{Z}\}$$

if $-d \not\equiv 1 \pmod 4$, and

$$R = \left\{a + b\left(\frac{1 + \sqrt{-d}}{2}\right) : a, b \in \mathbb{Z}\right\}$$

if $-d \equiv 1 \pmod 4$. So, in the complex plane, the points in R form an array in which they are regularly spaced along horizontal lines, which are regularly spaced vertically. Then $R^* = \{x \in R : |x| = 1\}$ is the intersection of R with the unit circle. So R^* is a finite subgroup of F^*. Since F is a field, this means R^* is cyclic, generated by an element ζ_m of finite order m. Since $-1 \in R^*$, m is even. Since $\mathbb{Q} \subseteq \mathbb{Q}(\zeta_m) \subseteq F$ and $[F : \mathbb{Q}] = 2$, $\phi(m) \leq 2$. So $m = 2, 4$, or 6. If $m = 6$, $F = \mathbb{Q}(\zeta_6) = \mathbb{Q}(\sqrt{-3})$ and $R = \mathbb{Z}[\zeta_6]$. If $m = 4$, $F = \mathbb{Q}(i)$ and $R = \mathbb{Z}[i]$. So if $F \notin \{\mathbb{Q}(\sqrt{-3}), \mathbb{Q}(i)\}$, then $m = 2$ and $R^* = \{\pm 1\}$.

  (ii) If there exist $a, b \in R^*$ with $a + b = 1$, then, since a and b lie on the unit circle, $a = \zeta_6^{\pm 1}$ and $R = \mathbb{Z}[\zeta_6]$. Otherwise, $K_*^M(R) = T(R^*)$, computed in Exercise 3.

5. Use Exercise 4 to prove that, in $K_*^M(\mathbb{Z}[i])$,

    (i) $\ell(i)\ell(i) \;\neq\; -\ell(i)\ell(i)$,
    (ii) $\ell(i)\ell(i) \;\neq\; \ell(-1)\ell(i) \;=\; \ell(i)\ell(-1)$,
    (iii) $\ell(i)\ell(-i) \;=\; \ell(-i)\ell(i) \;\neq\; 0$.

Use this to prove the map $K_2^M(\mathbb{Z}[i]) \to K_2(\mathbb{Z}[i])$, $\ell(a)\ell(b) \mapsto \{a, b\}$ is not injective.

    6. (Nesterenko and Suslin [90]) Prove Lemma (14.35) with the field F replaced by a commutative local ring R with a residue field k having more than three elements. *Hint:* First prove parts (i) and (ii) of (14.35) for $a, b \in R^*$ with $\bar{a} \neq \bar{1}$ and $\bar{b} \neq \bar{1}$ in k. Then show every $a \in R^*$ can be expressed as b_1/b_2 with $b_1, b_2 \in R^*$ and $\bar{b}_1 \neq \bar{1} \neq \bar{b}_2$ in k, and use this to prove the general case. Note that, in Suslin and Yarosh [91], our Propositions (14.37) and (14.38) are also proved over a commutative local ring with more than five elements in its residue field.

    7. A field F is **real closed** if each formally real algebraic extension field of F equals F. If F is real closed, prove each sum of squares in F is actually a square in F.

    8. If F is a formally real field and $a \in F$ is not a sum of squares in F, prove F is an ordered field with an ordering in which $a < 0$. *Hint:* Show $F(\sqrt{-a})$ is formally real; then use the ordering of (14.45).

    9. Suppose A is a totally ordered abelian group. A **valuation** on a field F is a surjective group homomorphism $v : F^* \to A$ for which

$$v(a + b) \;\geq\; \min\;\{v(a), v(b)\}$$

whenever $a, b \in F^*$. (We considered discrete valuations in (7.25).) Prove

    (i)   If $v(a) < v(b)$, $v(a + b) = v(a)$.
    (ii) If $a + b = 1$, then $v(a) = 0$, $v(b) = 0$, or $v(a) = v(b)$.
    (iii) If $\wedge(A)$ is the exterior algebra of the $\mathbb{Z}$-module A, there is a surjective ring homomorphism $K_*^M(F) \to \wedge(A)$ taking $\ell(a)$ to $v(a)$ for each $a \in F^*$.

This was used by Springer [72] to get lower bounds on the size of $K_n^M(F)$, by constructing valuations with $\mathbb{Q} \otimes_\mathbb{Z} A$ of large $\mathbb{Q}$-dimension. The **Kronecker dimension** $\delta(F)$ of F is the transcendence degree of F over its prime field if F has positive characteristic, and $1 +$ this transcendence degree if F has characteristic 0. Springer deduced that the local rank of the $\mathbb{Z}$-module $K_n^M(F)$ is the cardinality of F if $1 \leq n \leq \delta(F)$. A different proof of this appeared in the paper of Bass and Tate [73]. So $K_n^M(F)$ is not a torsion group if $n \leq \delta(F)$. It remains an important open question whether $K_n^M(F)$ must be torsion for $n > \delta(F)$. Bass and Tate verified this for algebraic extensions of $\mathbb{Q}$ or $\mathbb{F}_q(x)$.

14F. Tame Symbols

By using the ordering on a formally real field F, the real symbol detects nonzero elements of $K_*^M(F)$. Fields F without an ordering can still be mapped to an ordered abelian group by a valuation, detecting nonzero elements of $K_*^M(F)$ if the group of values has nonzero exterior powers (see §14E, Exercise 9). In this section we see how discrete valuations $v : F^* \to \mathbb{Z}$ can be used to construct a "tame symbol" $\tau : F^* \times F^* \to k_v^*$, detecting nonzero elements of $K_2^M(F)$, and we present Milnor's extended tame symbol on $(F^*)^n$, inducing a map on $K_n^M(F)$. In particular, we focus on the field $\mathbb{Q}$ and the field $F(x)$ of rational functions over a field F.

Recall some facts about discrete valuations from (7.25)–(7.26): A **discrete valuation** on a field F is a surjection $v : F^* \to \mathbb{Z}$ satisfying

(i) $v(ab) = v(a) + v(b)$, and
(ii) $v(a+b) \geq \min\{v(a), v(b)\}$,

for all $a, b \in F^*$. The set

$$\mathcal{O}_v = \{a \in F^* : v(a) \geq 0\} \cup \{0_F\}$$

is a subring of F known as the **discrete valuation ring** associated to v. If $v(b) < 0$, $v(b^{-1}) = -v(b) \geq 0$. So F is the field of fractions of $\mathcal{O}_v$. The units of $\mathcal{O}_v$ are the $a \in F^*$ with $v(a) = 0$. In particular, $v(-1) = v(1) = 0$ and $v(-a) = v(a)$. The set $\mathcal{O}_v - \mathcal{O}_v^*$ of nonunits is an ideal, which is necessarily the unique maximal ideal

$$\mathcal{P}_v = \{a \in F^* : v(a) > 0\} \cup \{0_F\}$$

of $\mathcal{O}_v$. The quotient $k_v = \mathcal{O}_v/\mathcal{P}_v$ is the **residue field** associated with v.

The ideal $\mathcal{P}_v$ is principal, generated by each element $\pi \in F^*$ with value $v(\pi) = 1$. Each element of F^* has a unique expression $u\pi^i$ with $u \in \mathcal{O}_v^*$ and $i \in \mathbb{Z}$; so $\mathcal{O}_v$ is a factorial ring with prime element π. Then $v(u\pi^i) = i$. So $v(a)$ is the largest integer i with $a \in \mathcal{P}_v^i$, and v *is uniquely determined by the discrete valuation ring* $\mathcal{O}_v$.

Computations with a discrete valuation frequently use the fact:

(iii) $v(a+b) = \min\{v(a), v(b)\}$ if $v(a) \neq v(b)$.

This follows from axioms (i) and (ii): If $v(a) < v(b)$, then $v(a) = v(a+b-b) \geq \min\{v(a+b), v(b)\}$, which cannot be $v(b)$; and $v(a+b) \geq \min\{v(a), v(b)\} = v(a)$.

(14.49) Definition. Suppose R is a commutative principal ideal domain with field of fractions F, and T is a set of irreducible elements of R, one generating each maximal ideal of R. Each $a \in F^*$ has a unique expression

$$a = u \prod_{p \in T} p^{v_p(a)},$$

where $u \in R^*$, each $v_p(a) \in \mathbb{Z}$, and $v_p(a) = 0$ for all but finitely many $p \in T$. For each $p \in T$,

$$v_p : F^* \to \mathbb{Z}$$

is the p-**adic valuation** on F: Axiom (i) for v_p is a law of exponents, and axiom (ii) comes from the distributive law.

Every irreducible $p \in R$ generates a maximal ideal pR, so it belongs to some such set T. For $a, b \in R$, $aR = bR$ if and only if $aR^* = bR^*$; so the p-adic valuation v_p is independent of the choice of T containing p, and $v_p = v_q$ if $pR = qR$.

(14.50) Proposition. *Suppose R is a commutative principal ideal domain with field of fractions F. The discrete valuations $v : F^* \to \mathbb{Z}$ with $R \subseteq \mathcal{O}_v$ are the p-adic valuations for irreducibles p in R. If $v = v_p$, the inclusion $R \to \mathcal{O}_v$ induces an isomorphism $R/pR \cong \mathcal{O}_v/\mathcal{P}_v = k_v$.*

Proof. If v is a p-adic valuation of F, the exponent of p in a factorization of $a \in R$ is nonnegative; so $R \subseteq \mathcal{O}_v$. Conversely, suppose v is a discrete valuation on F with $R \subseteq \mathcal{O}_v$. For some $a, b \in R - \{0\}, v(a/b) > 0$; so $v(a) > 0$ and $a \in R \cap \mathcal{P}_v$. So $R \cap \mathcal{P}_v$ is a nonzero prime ideal of R, generated by some irreducible $p \in R$. Then $\mathcal{O}_v$ consists of all lowest terms fractions a/b with $a, b \in R$ and $b \notin pR$. Thus $\mathcal{O}_v$ is the discrete valuation ring associated to the p-adic valuation v_p, forcing $v = v_p$.

If $v = v_p$, and $a, b \in R$ with $b \notin pR$, then $bR + pR = R$. So $a = br + ps$ for some $r, s \in R$, and

$$\frac{a}{b} - \frac{r}{1} \in \mathcal{P}_v .$$

Thus $R \to \mathcal{O}_v \to \mathcal{O}_v/\mathcal{P}_v$ is surjective. The kernel contains pR, and the induced homomorphism from R/pR to $\mathcal{O}_v/\mathcal{P}_v$ is also injective, since it is a ring homomorphism between fields. ∎

(14.51) Corollary. *Each discrete valuation on $\mathbb{Q}$ is a p-adic valuation $v = v_p$, where p is a positive prime in $\mathbb{Z}$. The inclusion $\mathbb{Z} \to \mathcal{O}_v$ induces an isomorphism $\mathbb{Z}/p\mathbb{Z} \cong k_v$.*

Proof. Every subring $\mathcal{O}_v$ of $\mathbb{Q}$ contains $\mathbb{Z}$. ∎

(14.52) Corollary. *If F is a field, each discrete valuation on the field $F(x)$ of rational functions with $v(F) = \{0\}$ is either a p-adic valuation, where p is a monic irreducible in $F[x]$, or is the valuation v_∞ defined by*

$$v_\infty\left(\frac{a}{b}\right) = degree(b) - degree(a)$$

for nonzero $a, b \in F[x]$. The inclusion $F[x] \to \mathcal{O}_v$ induces an isomorphism $F[x]/pF[x] \cong k_v$ if v is p-adic, and the inclusion $F \to \mathcal{O}_v$ induces an isomorphism $F \cong k_v$ if $v = v_\infty$.

Proof. If $x \in \mathcal{O}_v$, then $v(F) = \{0\}$ implies $F[x] \subseteq \mathcal{O}_v$. So v is p-adic for a monic irreducible $p \in F[x]$.

If $x \notin \mathcal{O}_v, v(y) > 0$, where $y = 1/x$. Then each polynomial in $F[y]$ with nonzero constant term a_0 has value $v(a_0) = 0$, so it belongs to $\mathcal{O}_v^*$. If $a \in F[x]$ has degree $n \geq 0, ay^n = u \in \mathcal{O}_v^*$ and $a = uy^{-n}$. Then $\mathbb{Z} = v(F(x)^*) \subseteq v(y)\mathbb{Z}$. Since $v(y) > 0$, this forces $v(y) = 1$, and $v(a) = -n = -\text{degree}(a)$. If $a, b \in F[x] - \{0\}$, it follows that $v(a/b) = \text{degree}(b) - \text{degree}(a)$.

Suppose $v = v_\infty$ is this last valuation. If $a, b \in F[x] - \{0\}$ have equal degree, then

$$\frac{a}{b} = c + \frac{d}{b} ,$$

where $c \in F^*$ and $d/b \in \mathcal{P}_v$. If a has smaller degree than b, the same equation holds with $c = 0$ and $d = a$. So the composite $F \subseteq \mathcal{O}_v \to \mathcal{O}_v/\mathcal{P}_v$ is surjective. It is injective since it is a ring homomorphism between fields. ■

(14.53) Definition. Suppose v is a discrete valuation on a field F. The **tame symbol** associated to v is the function $\tau_v : F^* \times F^* \to k_v^*$ taking each pair (a, b) to

$$\tau_v(a, b) = (-1)^{v(a)v(b)} \frac{a^{v(b)}}{b^{v(a)}} + \mathcal{P}_v .$$

Note that this formula defines an element of k_v^* because

$$v(\pm \frac{a^{v(b)}}{b^{v(a)}}) = v(b)v(a) - v(a)v(b) = 0 .$$

By axiom (i) for a discrete valuation, τ_v is $\mathbb{Z}$-bilinear. Suppose $a, b \in F^*$ and $a + b = 1$. If $v(a) > 0, v(b) = v(1 - a) = v(1) = 0$, and $\tau_v(a, b) = \overline{(1 - a)}^{-v(a)} = \bar{1}$. The same thing happens if $v(b) > 0$. If $v(a) < 0$, $v(b) = v(1 - a) = v(a)$. Then, modulo $\mathcal{P}_v$,

$$\frac{a^{v(b)}}{b^{v(a)}} = (\frac{b}{a})^{-v(a)} = (\frac{1 - a}{a})^{-v(a)}$$
$$= (a^{-1} - 1)^{-v(a)} \equiv (-1)^{-v(a)}$$
$$= (-1)^{v(a)} = (-1)^{v(a)v(a)} ;$$

so $\tau_v(a, b) = \bar{1}$. The same happens if $v(b) < 0$. Finally, if $v(a) = v(b) = 0$, $\tau_v(a, b) = \overline{a^0/b^0} = \bar{1}$. So there is an induced **tame symbol**

$$\tau_v : K_2^M(F) \to k_v^* , \quad \ell(a)\ell(b) \mapsto \tau_v(a, b) .$$

Note: The name "tame symbol" comes from the description of τ_v as a tamely ramified case of the norm residue symbol (see §15D).

(14.54) Example. Suppose p is a positive prime number and $v = v_p$ is the p-adic valuation on $\mathbb{Q}$. The map $\tau_v : K_2^M(\mathbb{Q}) \to k_v^*$ takes each $\ell(x)\ell(p)$, with x a nonzero integer of absolute value $|x| < p$, to

$$\tau_v(x,p) = (-1)^{(0)(1)} \frac{x^1}{p^0} + \mathcal{P}_v = x + \mathcal{P}_v,$$

the image of $x + p\mathbb{Z}$ under the isomorphism $\mathbb{Z}/p\mathbb{Z} \cong k_p$. So $\ell(1)\ell(p)$, $\ell(2)\ell(p)$, $\ldots, \ell(p-1)\ell(p)$ are $p-1$ different elements of $K_2^M(\mathbb{Q})$, the first one equal to 0 because $\ell(1) = 0$. If q is a positive prime and $q < p$, the distinct elements $\ell(1)\ell(q), \ell(2)\ell(q), \ldots, \ell(q-1)\ell(q)$ are sent to $\bar{1}$ by τ_v, because $v(n) = 0$ for $1 \le n < p$. So aside from $\ell(1)\ell(q) = 0 = \ell(1)\ell(p)$, these latter elements are distinct from those in the former list. Note that $\ell(-1)\ell(-1) \ne 0$ since it maps to the same element in $K_2^M(\mathbb{R})$, where the real symbol sends it to -1. But $\ell(-1)\ell(-1)$ remains undetected by all tame symbols on $\mathbb{Q}$, since each $v(-1) = 0$.

As an application of tame symbols, Tate completed the calculation of $K_2^M(\mathbb{Q})$ using steps inspired by the first proof of quadratic reciprocity by Gauss [86]. We follow the argument of Tate given in Milnor's book [71].

If p is a positive prime number, and $v = v_p$ is the p-adic valuation on $\mathbb{Q}$, (14.51) says there is an isomorphism of groups $(\mathbb{Z}/p\mathbb{Z})^* \cong k_v^*, n + p\mathbb{Z} \mapsto n + \mathcal{P}_v$. Following the tame symbol τ_v by the inverse of this isomorphism, we get a group homomorphism

$$\tau_p : K_2^M(\mathbb{Q}) \to (\mathbb{Z}/p\mathbb{Z})^*,$$

carrying $\ell(n)\ell(p)$ to $\bar{n} = n + p\mathbb{Z}$ for each integer n with $1 \le n \le p-1$. Evidently τ_p is surjective.

Define a homomorphism

$$\tau : K_2^M(\mathbb{Q}) \to \bigoplus_p (\mathbb{Z}/p\mathbb{Z})^*,$$

where p runs through the positive primes in order, by

$$\tau(\alpha) = (\tau_2(\alpha), \tau_3(\alpha), \tau_5(\alpha), \ldots).$$

For each prime p, let M_p denote the subgroup of the direct sum consisting of all elements whose q-coordinate is $\bar{1}$ for all primes $q > p$. For each positive integer m, let L_m denote the subgroup of $K_2^M(\mathbb{Q})$ generated by all $\ell(a)\ell(b)$, where a and b are nonzero integers with $|a|, |b| \le m$.

If q is a prime exceeding p, then for nonzero integers a, b with $|a|, |b| \le p$, $v_q(a) = v_q(b) = 0$; so $\tau_q(L_p) = \{\bar{1}\}$. Therefore $\tau(L_p) \subseteq M_p$. Since $(\mathbb{Z}/2\mathbb{Z})^*$ is trivial, $\tau(L_2) = M_2$. Assume $\tau(L_r) = M_r$ for the prime r preceding p. If $\beta \in M_p$ has p-coordinate $\bar{n}$ for $1 \le n \le p-1$, then

$$\frac{\beta}{\tau(\ell(n)\ell(p))} \in M_r = \tau(L_r) \subseteq \tau(L_p).$$

So $\beta \in \tau(L_p)$, and $\tau(L_p) = M_p$ for all positive primes p. Every element of the direct sum belongs to some M_p; so τ is surjective.

(14.55) Lemma. *For each positive prime* p, τ_p *induces a group isomorphism* $L_p/L_{p-1} \cong (\mathbb{Z}/p\mathbb{Z})^*$.

Proof. Note that if $a, b \in \mathbb{Q}^*$, then $\ell(a)\ell(b) \in L_p$, where p is the largest prime factor of ab; this comes from bilinearity of $\ell(-)\ell(-)$. So $K_2^M(\mathbb{Q})$ is the union of the nested groups $L_1 \subseteq L_2 \subseteq L_3 \subseteq \cdots$. And if q is the prime preceding p, then $L_q = L_{q+1} = \cdots = L_{p-1}$.

Suppose $x, y, z \in \{1, \ldots, p-1\}$ and $\bar{x}\, \bar{y} = \bar{z}$ in $(\mathbb{Z}/p\mathbb{Z})^*$. Since $xy < p^2$, $xy = z + rp$ with $0 \le r < p$. If $r = 0$, $\ell(z)\ell(p) = \ell(xy)\ell(p)$. If $r > 0$,

$$\ell(z)\ell(p) - \ell(xy)\ell(p) = \ell(\frac{z}{xy})\ell(p)$$

$$= \ell(\frac{z}{xy})\ell(\frac{rp}{xy}) - \ell(\frac{z}{xy})\ell(\frac{r}{xy})$$

$$= 0 - \ell(\frac{z}{xy})\ell(\frac{r}{xy}) \in L_{p-1}.$$

So there is a group homomorphism

$$\phi_p : (\mathbb{Z}/p\mathbb{Z})^* \to \frac{L_p}{L_{p-1}}$$

taking $\bar{x}$ to $\ell(x)\ell(p) + L_{p-1}$ if $1 \le x \le p-1$. And $\tau_p \circ \phi_p$ is the identity on $(\mathbb{Z}/p\mathbb{Z})^*$. It only remains to prove ϕ_p is surjective.

Now L_p is generated by L_{p-1} and the elements $\ell(\pm p)\ell(\pm p)$, $\ell(\pm n)\ell(\pm p)$ and $\ell(\pm p)\ell(\pm n)$ for integers n with $1 \le n \le p-1$. By antisymmetry, the last type of generator is not needed. Of course, $\ell(p)\ell(-p) = \ell(-p)\ell(p) = 0$. Modulo L_{p-1},

$$\ell(-p)\ell(-p) = (\ell(-1) + \ell(p))^2 = \ell(-1)^2 + \ell(p)^2$$

$$\equiv \ell(p)\ell(p) = \ell(-1)\ell(p).$$

So the first type of generator is not needed. Also, mod L_{p-1},

$$\ell(\pm n)\ell(-p) \equiv \ell(\pm n)\ell(p).$$

Since

$$\frac{-n}{p-n} + \frac{p}{p-n} = 1,$$

we also have

$$0 = \ell(\frac{-n}{p-n})\ell(\frac{p}{p-n})$$

$$= [\ell(-n) - \ell(p-n)]\,[\ell(p) - \ell(p-n)]$$

$$\equiv \ell(-n)\ell(p) - \ell(p-n)\ell(p)$$

mod L_{p-1}, and $1 \le p-n \le p-1$. So L_p/L_{p-1} is generated by the cosets

$$\ell(n)\ell(p) + L_{p-1} = \phi_p(\bar{n})$$

for $1 \le n \le p-1$, proving ϕ_p is surjective. ∎

(14.56) Theorem. (Tate) *Each element of $K_2^M(\mathbb{Q})$ has a unique expression*

$$\alpha = \varepsilon\ell(-1)\ell(-1) + \sum_p \ell(n_p)\ell(p) ,$$

where $\varepsilon = 0$ or 1, p runs through the positive primes, $n_p \in \{1, \ldots, p-1\}$, and $n_p = 1$ for all but finitely many primes p. There is a split short exact sequence

$$1 \longrightarrow \{\pm 1\} \underset{\longleftarrow}{\overset{\longrightarrow}{}} K_2^M(\mathbb{Q}) \overset{\tau}{\longrightarrow} \bigoplus_p (\mathbb{Z}/p\mathbb{Z})^* \longrightarrow 1 \ ;$$

so

$$K_2^M(\mathbb{Q}) \cong \{\pm 1\} \oplus \bigoplus_p (\mathbb{Z}/p\mathbb{Z})^* .$$

Proof. That each element has such an expression comes from the fact that $L_2 = L_1$ is cyclic of order 2 generated by $\ell(-1)\ell(-1)$, and from the surjectivity of each ϕ_p in the proof of the lemma. Now $\varepsilon = 0$ or 1, according to the value of the real symbol of α. And if α has a different such expression

$$\varepsilon\ell(-1)\ell(-1) + \sum_p \ell(n_p')\ell(p) ,$$

there will be a largest prime p with $n_p' \neq n_p$. But then

$$\begin{aligned}
\tau_p(0) &= \tau_p(\alpha - \alpha) = \tau_p(\ell(n_p)\ell(p) - \ell(n_p')\ell(p)) \\
&= \overline{n}_p/\overline{n}_p' \neq \overline{1}
\end{aligned}$$

in $(\mathbb{Z}/p\mathbb{Z})^*$, a contradiction.

If α is expressed as such a sum and p is the largest prime with $n_p \neq 1$, then $\tau_p(\alpha) = \overline{n}_p \neq \overline{1}$ in $(\mathbb{Z}/p\mathbb{Z})^*$. So α is in the kernel of τ if and only if every $n_p = 1$, in which case $\alpha = \varepsilon\,\ell(-1)\ell(-1)$. Thus the kernel of τ is the image of the homomorphism $\{\pm 1\} \to K_2^M(\mathbb{Q})$ taking -1 to $\ell(-1)\ell(-1)$. This map has a left inverse given by the real symbol $(\ ,\ )_\infty$, so the sequence is split exact. The splitting maps combine in an isomorphism

$$K_2^M(\mathbb{Q}) \cong \{\pm 1\} \oplus \bigoplus_p (\mathbb{Z}/p\mathbb{Z})^* . \qquad \blacksquare$$

Note: There is another splitting map $K_2^M(\mathbb{Q}) \to \{\pm 1\}$ called the "2-adic symbol," which is used in Tate's original description of $K_2^M(\mathbb{Q})$ in Milnor [71]. Although the real symbol is simpler to define, the 2-adic symbol is more like the tame symbols τ_p for p odd. We consider the 2-adic symbol, Tate's version of the above isomorphism, and Tate's use of it to prove quadratic reciprocity in Chapter 15.

For each field E with discrete valuation $v : E^* \to \mathbb{Z}$, Milnor has defined an extension of the tame symbol $\tau_v : K_2^M(E) \to k_v^* \cong K_1^M(k_v)$ to a map

$$\tau_v : K_n^M(E) \to K_{n-1}^M(k_v)$$

for each $n \geq 1$. We now present Milnor's τ_v, by means of a ring extension of $K_*^M(k_v)$ suggested by Serre (see Milnor [70, p. 323]).

For a field k, there is a $\mathbb{Z}$-linear map $k^* \to K_*^M(k)$ taking a to $-\ell(a)$. So it extends to a ring homomorphism on the tensor algebra

$$T(k^*) \to K_*^M(k) \ , \ [a] \mapsto -\ell(a) \ .$$

If $a, b \in k^*$ and $a+b = 1$, this map takes $[a][b]$ to $(-\ell(a))(-\ell(b)) = \ell(a)\ell(b) = 0$; so it induces a ring homomorphism

$$\mu : K_*^M(k) \to K_*^M(k) \ , \ \ell(a) \mapsto -\ell(a) \ ,$$

which is multiplication by $(-1)^n$ on $K_n^M(k)$. Notice that, for all $\alpha \in K_*^M(k)$,

$$\mu(\alpha\ell(-1)) \ = \ \mu(\alpha)\ell(-1) \ = \ \alpha\ell(-1) \ ,$$

since $-\ell(-1) = \ell((-1)^{-1}) = \ell(-1)$. Define

$$\boldsymbol{K_*^M(k)\langle \xi \rangle} \ = \ K_*^M(k) \cdot 1 \ \oplus \ K_*^M(k) \cdot \xi$$

to be the free $K_*^M(k)$-module with basis $\{1, \xi\}$. On this module, define a multiplication by

$$(\alpha \cdot 1 + \beta \cdot \xi) \ (\alpha' \cdot 1 + \beta' \cdot \xi)$$
$$= (\alpha\alpha') \cdot 1 + (\alpha\beta' + \beta\mu(\alpha') + \beta\mu(\beta')\ell(-1)) \cdot \xi \ .$$

Using the properties of μ from the preceding paragraph, the reader can verify directly that $K_*^M(k)\langle \xi \rangle$ is a ring under this multiplication and the module addition, and that the map $\alpha \mapsto \alpha \cdot 1 + 0 \cdot \xi$ embeds $K_*^M(k)$ as a subring. For simpler notation, we write α for $\alpha \cdot 1 + 0 \cdot \xi$; then the basis vector 1 is identified with the 1 in both rings, and scalar multiplication coincides with ring multiplication. With this identification, the multiplication in $K_*^M(k)\langle \xi \rangle$ can be carried out using only the ring axioms and the identities

$$\xi^2 \ = \ \ell(-1)\xi \ ,$$
$$\xi\ell(a) \ = \ -\ell(a)\xi \ ,$$

which hold in $K_*^M(k)\langle \xi \rangle$ for each $a \in k^*$.

For E a field and $v : E^* \to \mathbb{Z}$ a discrete valuation with residue field k_v and prime element $\pi \in E$ (with $v(\pi) = 1$), define a map

$$\theta : E^* \to K_*^M(k_v)\langle \xi \rangle \ ,$$
$$u\pi^i \mapsto \ell(\overline{u}) + i\xi$$

for $u \in \mathcal{O}_v^*$ and $i \in \mathbb{Z}$.

(14.57) Proposition. *If* $a, b \in E^*$,

(i) $\theta(ab) = \theta(a) + \theta(b)$,

(ii) $\theta(b)\theta(a) = -\theta(a)\theta(b)$,

(iii) $\theta(a)\theta(-a) = 0$,

(iv) $\theta(a^{-1})\theta(1 - a^{-1}) = -\theta(a)\theta(1 - a)$ *if* $a \neq 1$,

(v) *If* $a + b = 1$, $\theta(a)\theta(b) = 0$.

Proof. If $a = u\pi^i$ and $b = u'\pi^{i'}$ for $u, u' \in \mathcal{O}_v^*$, and $i, i' \in \mathbb{Z}$, then

$$\theta(ab) = \ell(\overline{uu'}) + (i + i')\xi$$
$$= \ell(\overline{u}) + \ell(\overline{u'}) + i\xi + i'\xi$$
$$= \theta(a) + \theta(b),$$

proving (i). And

$$\theta(b)\theta(a) = (\ell(\overline{u'}) + i'\xi)(\ell(\overline{u}) + i\xi)$$
$$= \ell(\overline{u'})\ell(\overline{u}) + \ell(\overline{u'})i\xi - i'\ell(\overline{u})\xi + i'i\ell(-1)\xi$$
$$= -(\ell(\overline{u}) + i\xi)(\ell(\overline{u'}) + i'\xi) = -\theta(a)\theta(b),$$

proving (ii).

Now $\theta(u)\theta(-u) = \ell(\overline{u})\ell(-\overline{u}) = 0$, and $\theta(\pi^i)\theta(-\pi^i) = i\xi(\ell(-1) + i\xi) = (i^2 - i)\ell(-1)\xi = 0$, since $i^2 - i = i(i - 1)$ is even, and $2\ell(-1) = \ell(1) = 0$. So, by antisymmetry,

$$\theta(u\pi^i)\theta(-u\pi^i) = (\theta(u) + \theta(\pi^i))(\theta(-1) + \theta(u) + \theta(\pi^i))$$
$$= \theta(u)\theta(-u) + \theta(\pi^i)\theta(-\pi^i) = 0,$$

proving (iii).

If $a \neq 1$, $-a = (1 - a)(1 - a^{-1})^{-1}$; so

$$0 = \theta(a)\theta(-a) = \theta(a)(\theta(1 - a) - \theta(1 - a^{-1}))$$
$$= \theta(a)\theta(1 - a) + \theta(a^{-1})\theta(1 - a^{-1}),$$

the last step using (i). So (iv) is true.

For (v), suppose first that $v(a) > 0$. Since $a + b = 1$, $v(b) = 0$ and $\theta(b) = \ell(\overline{b}) = \ell(\overline{1} - \overline{a}) = \ell(\overline{1}) = 0$. Similarly, if $v(b) > 0$, then $\theta(a) = 0$. Now suppose $v(a) < 0$; so $v(a^{-1}) > 0$. By the previous case, $0 = \theta(a^{-1})\theta(1 - a^{-1}) = -\theta(a)\theta(1 - a) = -\theta(a)\theta(b)$. The same happens if $v(b) < 0$. Finally, if $v(a) = v(b) = 0$, then $\theta(a)\theta(b) = \ell(\overline{a})\ell(\overline{b}) = 0$, since $\overline{a} + \overline{b} = \overline{a + b} = \overline{1}$. ∎

By part (i) of the proposition, θ is $\mathbb{Z}$-linear. So it extends to a ring homomorphism

$$T(E^*) \to K_*^M(k_v)\langle\xi\rangle$$

taking $[a][b]$ to $\theta(a)\theta(b) = 0$ whenever $a, b \in E^*$ and $a + b = 1$. So this map induces a ring homomorphism

$$\widehat{\theta} : K_*^M(E) \to K_*^M(k_v)\langle\xi\rangle \ .$$

On the additive generators of $K_*^M(E)$,

$$\widehat{\theta}(\ell(u_1\pi^{i_1})\cdots\ell(u_n\pi^{i_n})) = (\ell(\overline{u_1}) + i_1\xi)\cdots(\ell(\overline{u_n}) + i_n\xi)$$
$$= \ell(\overline{u_1})\cdots\ell(\overline{u_n}) + \beta\xi \ ,$$

where β is a sum of terms

$$m\ell(\overline{u_{i(1)}})\cdots\ell(\overline{u_{i(r)}})\ell(-1)^{n-r-1}$$

with $m \in \mathbb{Z}$ and $0 \le r < n$.

(14.58) Definition. Following $\widehat{\theta}$ by $\alpha + \beta\xi \mapsto \alpha$ defines a $\mathbb{Z}$-linear map

$$\sigma_\pi : K_*^M(E) \to K_*^M(k_v)$$

restricting to the identity map $\mathbb{Z} \to \mathbb{Z}$ in degree 0, and to

$$\sigma_\pi : K_n^M(E) \to K_n^M(k_v)$$

$$\ell(u_1\pi^{i_1})\cdots\ell(u_n\pi^{i_n}) \mapsto \ell(\overline{u_1})\cdots\ell(\overline{u_n})$$

for each $n > 0$. Following $\widehat{\theta}$ by $\alpha+\beta\xi \mapsto \beta$ defines the **extended tame symbol**

$$\tau_v : K_*^M(E) \to K_*^M(k_v) \ ,$$

which is $\mathbb{Z}$-linear, and restricts to the zero map on $K_0^M(E) = \mathbb{Z}$, and to

$$\tau_v : K_n^M(E) \to K_{n-1}^M(k_v)$$

$$\ell(u_1)\ldots\ell(u_{n-1})\ell(a) \mapsto \ell(\overline{u_1})\ldots\ell(\overline{u_{n-1}})v(a)$$

for each $n > 0$, where $u_1, \ldots, u_{n-1} \in \mathcal{O}_v^*$ and $a \in E^*$.

Since $\mathcal{O}_v^* \to k_v^*$ is surjective, τ_v is also surjective. By multilinearity, anticommutativity, and the relation $\ell(\pi)\ell(\pi) = \ell(-1)\ell(\pi)$, $K_n^M(E)$ is generated as a $\mathbb{Z}$-module by the elements $\ell(u_1)\cdots\ell(u_{n-1})\ell(a)$; so, unlike σ_π, τ_v is independent of the choice of prime element $\pi \in \mathcal{O}_v$.

The extended tame symbol really does coincide with the tame symbol on $K_2^M(E)$, since it takes $\ell(u\pi^i)\ell(v\pi^j)$ to the ξ-coefficient of

$$(\ell(\overline{u}) + i\xi)(\ell(\overline{v}) + j\xi) = \ell(\overline{u})\ell(\overline{v}) + (ij\ell(-\overline{1}) + j\ell(\overline{u}) - i\ell(\overline{v}))\xi$$
$$= \ell(\overline{u})\ell(\overline{v}) + \ell((-1)^{ij}\overline{u}^j\pi^{ij}/\overline{v}^i\pi^{ij})\xi \ .$$

And the extended tame symbol coincides with the discrete valuation v from $K_1^M(E) = E^*$ to $K_0^M(k_v) = \mathbb{Z}$, since it takes $\ell(u\pi^i)$ to i.

Tate obtained a description of $K_2^M(F(x))$ like that of $K_2^M(\mathbb{Q})$, where $F(x)$ is the field of rational functions over a field F (see Milnor [71, p. 106]). We now present Milnor's generalization (Milnor [70, Theorem 2.3]) of this to $K_n^M(F(x))$ using the extended tame symbols. If p is a monic irreducible polynomial in $F[x]$, we use the abbreviated notation

$$A_p = \frac{F[x]}{pF[x]} \ .$$

If v is the p-adic valuation on $F[x]$, recall from (14.52) that inclusion $F[x] \subseteq \mathcal{O}_v$ induces an isomorphism of fields $A_p \cong k_v$ taking $f + pF[x]$ to $f + \mathcal{P}_v$.

(14.59) Theorem. (Milnor, Tate) *For each field F, the extended tame symbols τ_v, for p-adic valuations $v = v_p$ on $F(x)$, combine to form a map τ in a $\mathbb{Z}$-linear split exact sequence*

$$0 \longrightarrow K_n^M(F) \overset{j}{\longrightarrow} K_n^M(F(x)) \overset{\tau}{\longrightarrow} \bigoplus_p K_{n-1}^M(A_p) \longrightarrow 0 \ ,$$

where p runs through the monic irreducibles in $F[x]$.

Proof. Following the extended tame symbol $\tau_v : K_n^M(F(x)) \to K_{n-1}^M(k_v)$ by K_{n-1}^M of the inverse to the isomorphism $A_p \cong k_v$, taking $\overline{q(x)}$ to $\overline{q(x)}$, we get a $\mathbb{Z}$-linear surjection

$$\tau_p : K_n^M(F(x)) \to K_{n-1}^M(A_p)$$

carrying

$$\ell(u_1) \cdots \ell(u_{n-1})\ell(a) \ \text{ to } \ \ell(\overline{u_1}) \cdots \ell(\overline{u_{n-1}})v_p(a)$$

whenever $u_1, \ldots, u_{n-1}$ are in $F[x]$ but not in $pF[x]$, and $a \in F(x)^*$. If I is the set of monic irreducibles in $F[x]$, define the $\mathbb{Z}$-linear map

$$\tau : K_n^M(F(x)) \to \bigoplus_{p \in I} K_{n-1}^M(A_p)$$

so the p-coordinate of $\tau(\alpha)$ is $\tau_p(\alpha)$ for each $p \in I$.

For each integer $d \geq 0$, let L_d denote the subgroup of $K_n^M(F(x))$ generated by all $\ell(u_1) \cdots \ell(u_n)$ for $u_1, \ldots, u_n \in F[x]$ of degree at most d. By multilinearity, if $v_1, \cdots, v_n \in F(x)^*$ and d is the highest degree irreducible polynomial appearing in the unique factorization of $v_1 \cdots v_n$, then $\ell(v_1) \cdots \ell(v_n) \in L_d$. So $K_n^M(F(x))$ is the union of the nested subgroups $L_0 \subseteq L_1 \subseteq L_2 \subseteq \ldots$.

For each $d > 0$, let $I(d)$ denote the set of degree d monic irreducibles in $F[x]$.

(14.60) Lemma. *The composite of τ and projection to the $I(d)$ coordinates induces an isomorphism*

$$t_d : \frac{L_d}{L_{d-1}} \cong \bigoplus_{p \in I(d)} K^M_{n-1}(A_p) ,$$

which is the map $\overline{\tau_p}$ induced by τ_p for each $p \in I(d)$.

Proof. Suppose $d > 0$ and $U(d)$ denotes the set of polynomials in $F[x]$ having degree less than d. For $p \in I(d)$, polynomials in $U(d)$ are not in $pF[x]$; so $\tau_p(L_{d-1}) = \{0\}$, and t_d is a well-defined homomorphism.

Suppose $p \in I(d)$ and $u, v, w, u_1, \ldots, u_{n-2} \in U(d)$, with $\overline{u}\,\overline{v} = \overline{w}$ in A_p. Since uv has degree less than $2d$, $uv = w + rp$, where $r \in F[x]$ is 0 or lies in $U(d)$. If $r = 0$, $\ell(w)\ell(p) = \ell(uv)\ell(p)$. If $r \neq 0$,

$$\ell(w)\ell(p) \; - \; \ell(uv)\ell(p) \; = \; - \, \ell(\frac{w}{uv})\ell(\frac{r}{uv}) ,$$

as in the proof of (14.55). Either way,

$$\ell(u_1) \ldots \ell(u_{n-2})\ell(w)\ell(p) \; - \; \ell(u_1) \ldots \ell(u_{n-2})\ell(uv)\ell(p)$$

lies in L_{d-1}. So the function

$$A^*_p \times \cdots \times A^*_p \; \rightarrow \; \frac{L_d}{L_{d-1}} ,$$

taking $(\overline{u_1}, \ldots, \overline{u_{n-1}})$ to the coset of $\ell(u_1) \ldots \ell(u_{n-1})\ell(p)$ when each $u_i \in U(d)$, is linear in the $(n-1)$-coordinate. By anticommutativity in $K^M_*(F(x))$, it is linear in every coordinate. If $\overline{u_i} + \overline{u_{i+1}} = 1$ in A_p, then $u_i + u_{i+1} = 1$ in $F(x)$. So there is an induced homomorphism

$$\phi_p : K^M_{n-1}(A_p) \; \rightarrow \; \frac{L_d}{L_{d-1}}$$

taking $\ell(\overline{u_1}) \cdots \ell(\overline{u_{n-1}})$ to the coset of $\ell(u_1) \ldots \ell(u_{n-1})\ell(p)$ for $u_1, \ldots, u_{n-1}$ in $U(d)$. Define

$$f_d : \bigoplus_{p \in I(d)} K^M_{n-1}(A_p) \; \rightarrow \; \frac{L_d}{L_{d-1}}$$

to be the sum of the values of ϕ_p on the p-coordinates. Then $t_d \circ f_d$ is the identity map, since it is the identity on each element with p-coordinate $\ell(\overline{u_1}) \ldots \ell(\overline{u_{n-1}})$ and other coordinates 0.

It remains only to show f_d is surjective. Equivalently, we need only show L_d is generated by L_{d-1} and elements $\ell(u_1)\ldots\ell(u_{n-1})\ell(p)$ for $u_1,\ldots,u_{n-1} \in U(d)$ and $p \in I(d)$. Suppose $a, b \in F[x]$ have degree d. Then $a = bq + r$ for $q \in F^*$ and either $r = 0$ or $r \in U(d)$. If $r = 0$,

$$\ell(a)\ell(b) \;=\; \ell(bq)\ell(b) \;=\; \ell(-1)\ell(b) \;+\; \ell(q)\ell(b) \;.$$

Or, if $r \neq 0$, $(a/r) + (-bq/r) = 1$ in $F(x)$; so

$$\begin{aligned}
0 &= \ell(\tfrac{a}{r})\ell(\tfrac{-bq}{r}) \\
&= (\ell(a) - \ell(r))(\ell(b) + \ell(-q) - \ell(r)) \;.
\end{aligned}$$

Either way, $\ell(a)\ell(b)$ is a $\mathbb{Z}$-linear combination of terms $\ell(u)\ell(v)$ with $u \in U(d)$ and degree$(v) \leq d$. Applying this repeatedly, each element of L_d can be written as an element of L_{d-1} plus a $\mathbb{Z}$-linear combination of elements $\ell(u_1)\ldots\ell(u_{n-1})\ell(p)$ with $u_1,\ldots,u_{n-1} \in U(d)$ and $p \in F[x]$ of degree d. Using multilinearity we can assume each $p \in I(d)$. ∎

Returning to the proof of the theorem, for each $d > 0$ let M_d denote the subgroup of the direct sum $\oplus_p K_{n-1}^M(A_p)$, over all monic irreducibles $p \in F[x]$, consisting of those elements with p-coordinate 0 for all p of degree exceeding d. Every element β of the full direct sum belongs to M_d for some d, since only finitely many coordinates of an element are nonzero.

If the degree of p is greater than d, $\tau_p(L_d) = \{0\}$; so $\tau(L_d) \subseteq M_d$. No irreducible polynomial in $F[x]$ has degree 0; so $M_0 = \{0\}$ and $\tau(L_0) = M_0$. Suppose $\tau(L_{d-1}) = M_{d-1}$ and $\beta \in M_d$. By the lemma, there exists $\gamma \in L_d$ so that $\tau(\gamma)$ has the same $I(d)$-coordinates as β. Then $\beta - \tau(\gamma) \in M_{d-1} = \tau(L_{d-1}) \subseteq \tau(L_d)$. So $\beta \in \tau(L_d)$, and $\tau(L_d) = M_d$ for all d. This proves τ is surjective.

Suppose $\alpha \in L_d$ and $\tau(\alpha) = 0$. By the lemma, $\alpha \in L_{d-1}$. This will hold for all $d > 0$; so $\alpha \in L_0$. Since $\tau_p(L_0) = \{0\}$ for all monic irreducibles $p \in F[x]$, L_0 is the kernel of τ. Of course L_0 is the image of the homomorphism

$$j : K_n^M(F) \to K_n^M(F(x)) \;.$$
$$\ell(u_1)\ldots\ell(u_n) \mapsto \ell(u_1)\ldots\ell(u_n)$$

This has left inverse the composite

$$K_n^M(F(x)) \xrightarrow{\;\sigma_{(1/x)}\;} K_n^M(k_{v_\infty}) \cong K_n^M(F) \;,$$

where $\sigma_{(1/x)}$ is the map σ_π defined in (14.58) for the prime $\pi = 1/x$ associated to the discrete valuation v_∞. ∎

14F. Exercises

1. If v is a discrete valuation on a field F, prove the discrete valuation ring $\mathcal{O}_v$ is a euclidean ring.

2. Suppose R is a commutative principal ideal domain, and T is a set of irreducible elements, one generating each maximal ideal of R. If $a, b \in R - \{0\}$, prove a divides b in R if and only if $v_p(a) \leq v_p(b)$ for all $p \in T$. Use this to prove that a^2 divides b^2 if and only a divides b. (The same argument works with 2 replaced by any positive integer n.)

3. Suppose R is a Dedekind domain with field of fractions F. If P is a maximal ideal of R, define $v_P : F^* \to \mathbb{Z}$ so that $v_P(a)$ is the exponent of P in the maximal ideal factorization of aR. (This generalizes the principal ideal domain p-adic valuations.) Show the image of $K_2^M(R) \to K_2^M(F)$ lies in the kernel of all tame symbols $\tau_v : K_2^M(F) \to k_v^*$ for P-adic valuations $v = v_P$.

4. If F is a field and $F(x)$ is the field of rational functions over F, use Lemma (14.60) to prove each element of $K_2^M(F(x))$ has a unique expression

$$\alpha = \beta + \sum_p \ell(n_p)\ell(p) \,,$$

where β comes from $K_2^M(F)$, p runs through the monic irreducibles in $F[x]$, and for each p, n_p is a nonzero polynomial in $F[x]$ of smaller degree than p. *Hint*: Consider the first part of the proof of Theorem (14.56).

5. Imitating the proof of (14.59), prove that, for each $n > 0$, there is a $\mathbb{Z}$-linear split exact sequence

$$0 \longrightarrow K_n^M(\mathbb{Z}) \xrightarrow{\ j\ } K_n^M(\mathbb{Q}) \xrightarrow{\ \tau\ } \bigoplus_p K_{n-1}^M(\mathbb{Z}/p\mathbb{Z}) \longrightarrow 0 \,,$$

where p runs through the positive primes in $\mathbb{Z}$, j is induced by inclusion of $\mathbb{Z}$ in $\mathbb{Q}$, and τ is the composite

$$\tau_p : K_n^M(\mathbb{Q}) \xrightarrow{\ \tau_v\ } K_{n-1}^M(k_v) \cong K_{n-1}^M(\mathbb{Z}/p\mathbb{Z})$$

in the p-coordinate, where $v = v_p$.

6. From the sequence in Exercise 5, deduce that $K_n^M(\mathbb{Q}) \cong \mathbb{Z}/2\mathbb{Z}$ (generated by $\ell(-1)^n$) for each $n > 2$.

7. For an arbitrary discrete valuation v on a field E, calculate $\tau_v(\ell(a)\ell(b)\ell(c))$ for $a, b, c \in E^*$ in a form that does not depend on the choice of prime π in $\mathcal{O}_v$.

8. Verify the ring axioms involving multiplication for the ring $K_*^M(k)\langle\xi\rangle$, and verify the identities $\xi^2 = \ell(-1)\xi$ and $\xi\ell(a) = -\ell(a)\xi$ for each $a \in k^*$.

14G. Norms on Milnor K-Theory

We begin this section with a general discussion of the norm map from a finite-dimensional algebra to a field. Norms are used to route the values of various tame symbols into one field, where they can be compared. This is done in (14.62) to produce Weil's reciprocity law for function fields, and is done again in §§15C and D for quadratic and higher reciprocity laws in number fields. Using Milnor's decomposition of $K_{n+1}^M(F(x))$ in terms of tame symbols (from §14F) Bass and Tate defined a norm between K_n^M groups. We present this definition in (14.63) and describe its properties in the rest of the section. Of special importance for computations is the projection formula (14.67).

If F is a field and A is an F-algebra of finite dimension $n = [A : F]$ as an F-vector space, there are group homomorphisms

$$A^* \xrightarrow{\ \lambda\ } \text{Aut}_F(A) \xrightarrow{\ \rho\ } GL_n(F) \xrightarrow{\ det\ } F^*,$$

where $\lambda(b)$ is left multiplication by b, ρ is matrix representation over a selected basis, and det is the determinant map. The composite is the **algebra norm**

$$N_{A/F} : A^* \to F^*.$$

Since similar matrices have equal determinants, the algebra norm is independent of the choice of F-basis selected to define ρ.

If $f : A \to B$ is an F-algebra isomorphism and $v_1, \ldots, v_n$ is an F-basis of A, then $f(v_1), \ldots, f(v_n)$ is an F-basis of B. If $av_j = \Sigma_i a_{ij} v_i$ with $a_{ij} \in F$, then

$$f(a)f(v_j) = \sum_i a_{ij} f(v_i) .$$

So multiplication by a on A and multiplication by $f(a)$ on B are represented by the same matrices over F. Taking determinants, we see that the triangle

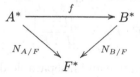

commutes. We refer to this property of norms as **isomorphism invariance**.

If $A \neq 0$, there is a standard embedding $F \to A$, $u \mapsto u \cdot 1_A$, of F as a subfield $F \cdot 1_A$ of the center of A. The composite

$$F^* \xrightarrow{\ \subseteq\ } A^* \xrightarrow{\ N_{A/F}\ } F^*$$

is the nth power map, where $n = [A : F]$, since for $u \in F^*$, $N_{A/F}(u \cdot 1_A) = det(uI_n) = u^n$.

If $b \in A$ has minimal polynomial $p(x) = \Sigma_i a_i x^i$ of degree d over F, then $1, b, b^2, \ldots, b^{d-1}$ is an F-basis of the subalgebra $F[b] \subseteq A$. Over this basis, $\lambda(b)$ is represented by the companion matrix

$$C = \begin{bmatrix} 0 & 0 & \ldots & 0 & -a_0 \\ 1 & 0 & \ldots & 0 & -a_1 \\ 0 & 1 & \ldots & 0 & -a_2 \\ \vdots & \vdots & \ddots & \vdots & \vdots \\ 0 & 0 & \ldots & 1 & -a_{d-1} \end{bmatrix}$$

of $p(x)$, and

$$N_{F[b]/F}(b) = det(C) = (-1)^d a_0 .$$

Suppose $F \subseteq E$ is a finite-degree field extension. Then E is an F-algebra via multiplication in E. In this case, $N_{E/F} : E^* \to F^*$ is called the **field norm**. If A is a finite-dimensional E-algebra, then

$$N_{A/F} = N_{E/F} \circ N_{A/E} .$$

(For a proof, see Jacobson [85, §7.4] or Bourbaki [74, Chapter III, §9.3]). This is known as **transitivity** or **functoriality** of the norm.

If $b \in E^*$ has minimal polynomial over F that factors as

$$(x - b_1) \ldots (x - b_d) = \sum_{i=0}^{d} a_i x^i$$

over an algebraic closure $\overline{F}$ of F, then

$$N_{F(b)/F}(b) = (-1)^d a_0 = b_1 \ldots b_d .$$

Recall from field theory that the **separable degree** $[E : F]_s$ of E over F is the number of extensions of each ring homomorphism $F \to \overline{F}$ to a ring homomorphism $E \to \overline{F}$, and the **inseparable degree**

$$[E : F]_i = \frac{[E : F]}{[E : F]_s}$$

is an integer as well. Both separable and inseparable degrees are multiplicative in a tower of field extensions. Suppose $s_1, \ldots, s_\ell$ are the F-algebra homomorphisms of $F(b)$ into $\overline{F}$, and $\sigma_1, \ldots, \sigma_m$ are the F-algebra homomorphisms of E into $\overline{F}$; so $\ell = [F(b) : F]_s$ and $m = [E : F]_s$. Each s_i extends to exactly m/ℓ of the σ_j. Then

$$N_{E/F}(b) = N_{F(b)/F}(N_{E/F(b)}(b))$$
$$= N_{F(b)/F}(b^{[E:F(b)]})$$
$$= N_{F(b)/F}(b)^{[E:F(b)]} = (b_1 \ldots b_d)^{[E:F(b)]}.$$

But each b_j occurs, as a root of the minimal polynomial of b over F, with the same multiplicity $[F(b) : F]_i$, and $s_1(b), \ldots, s_\ell(b)$ are the distinct members of the list $b_1, \ldots, b_d$. So

$$(b_1 \ldots b_d)^{[E : F(b)]} = \prod_{j=1}^{\ell} s_j(b)^{[F(b) : F]_i [E : F(b)]}$$

$$= \prod_{j=1}^{\ell} s_j(b)^{[E : F(b)]_s [E : F]_i} = \prod_{j=1}^{m} \sigma_j(b)^{[E : F]_i}.$$

This proves

(14.61)
$$N_{E/F}\,(b) = \prod_{j=1}^{[E : F]_s} \sigma_j(b)^{[E : F]_i}.$$

In particular, if $F \subseteq E$ is a finite-degree Galois field extension with Galois group $G = \mathrm{Aut}(E/F)$, then for each $b \in E^*$,

$$N_{E/F}\,(b) = \prod_{\sigma \in G} \sigma(b)\,.$$

Field norms intervene in a connection among all tame symbols on a field of rational functions $F(x)$. Recall from (14.52) that for $v = v_\infty$ (with prime element $1/x$), inclusion $F \subseteq \mathcal{O}_v$ induces an isomorphism $F \cong k_v$. Define

$$\tau_\infty : F(x)^* \times F(x)^* \to F^*$$

to be the tame symbol τ_v followed by the inverse isomorphism $k_v^* \cong F^*$.

(14.62) Theorem. (*Due to Weil* – see Bass and Tate [73, Theorem (5.6)′]) *Suppose F is a field, and for each p-adic valuation $v = v_p$ on $F(x)$ associated with a monic irreducible $p \in F[x]$, τ_v is the tame symbol from $F(x)^* \times F(x)^*$ to k_v^*, and n_v is the field norm from k_v^* to F^*. Then, for all $f, g \in F(x)^*$,*

$$\tau_\infty(f,g)^{-1} = \prod_{p-adic\ v} n_v \circ \tau_v\,(f,g)\,.$$

Proof. Let I denote the set of monic irreducibles in $F[x]$. Since these tame symbols are well-defined on $K_2^M(F(x))$, they are multiplicative in both f and g, and have the same effect on (f, f) as on $(-1, f)$. So it is sufficient to verify the equation in three cases:

Case 1: $f \in F^*$ and $g \in F^*$.

Case 2: One of f, g is in I, the other in F^*.

Case 3: $f \in I$, $g \in I$, and $f \neq g$.

In Case 1, both sides of the equation equal 1. Suppose we are in Case 2 or Case 3. Since f and g are relatively prime polynomials and $\tau_v(f, g) = 1$ whenever $v(f) = v(g) = 0$, the right side of the equation is

$$\left[\prod_{v(g)>0} n_v \circ \tau_v \, (f, g) \right] \left[\prod_{v(f)>0} n_v \circ \tau_v \, (f, g) \right] .$$

By antisymmetry of tame symbols, if the first factor is denoted by $\left(\frac{f}{g}\right)$, the second equals $\left(\frac{g}{f}\right)^{-1}$.

Suppose $\overline{F}$ is an algebraic closure of F and

$$f \; = \; a(x - \alpha_1) \cdots (x - \alpha_m) \; , \quad g \; = \; b(x - \beta_1) \cdots (x - \beta_n)$$

in $\overline{F}[x]$. If $g \in I$ and $v = v_g$, then

$$\left(\frac{f}{g}\right) \; = \; n_v \circ \tau_v \, (f, g) \; = \; n_v(f + \mathcal{P}_v) \; = \; n_v(f(\alpha)) \; ,$$

where $\alpha = x + \mathcal{P}_v \in k_v$. Since $F[x] \subseteq \mathcal{O}_v$ induces an F-algebra isomorphism from $F[x]/gF[x]$ to k_v, the minimal polynomial of α over F is g. If the F-algebra homomorphisms from $k = k_v$ into $\overline{F}$ are denoted by σ_j, then, by (14.61),

$$n_v(f(\alpha)) \; = \; \prod_{j=1}^{[k\,:\,F]_s} \sigma_j(f(\alpha))^{[k\,:\,F]_i} \; = \; \prod_{j=1}^{[k\,:\,F]_s} f(\sigma_j(\alpha))^{[k\,:\,F]_i}$$

$$= \; \prod_{i=1}^{n} f(\beta_i) \; = \; a^n \prod_{i=1}^{n} \prod_{j=1}^{m} (\beta_i - \alpha_j) \; .$$

So

$$\left(\frac{f}{g}\right) \; = \; \begin{cases} a^n \prod\limits_{i,j} (\beta_i - \alpha_j) & \text{if } g \in I \\ 1 & \text{if } g \in F^*; \text{ so } n = 0 \, . \end{cases}$$

Likewise

$$\left(\frac{g}{f}\right) \; = \; \begin{cases} b^m \prod\limits_{i,j} (\alpha_j - \beta_i) & \text{if } f \in I \\ 1 & \text{if } f \in F^*; \text{ so } m = 0 \, . \end{cases}$$

In both Cases 2 and 3, therefore,

$$\left(\frac{f}{g}\right)\left(\frac{g}{f}\right)^{-1} = (-1)^{mn}\frac{a^n}{b^m} .$$

On the other hand, if $v = v_\infty$,

$$
\begin{aligned}
\tau_v(f, g) &= (-1)^{(-m)(-n)}\frac{f^{-n}}{g^{-m}} + \mathcal{P}_v \\
&= (-1)^{mn}\frac{(fx^{-m})^{-n}}{(gx^{-n})^{-m}} + \mathcal{P}_v \\
&= (-1)^{mn}\frac{(fx^{-m} + \mathcal{P}_v)^{-n}}{(gx^{-n} + \mathcal{P}_v)^{-m}} \\
&= (-1)^{mn}\frac{(a + \mathcal{P}_v)^{-n}}{(b + \mathcal{P}_v)^{-m}} = (-1)^{mn}\frac{a^{-n}}{b^{-m}} + \mathcal{P}_v .
\end{aligned}
$$

So

$$\tau_\infty(f, g)^{-1} = (-1)^{mn}\frac{a^n}{b^m} . \qquad \blacksquare$$

Since $E^* \cong K_1^M(E)$ and $F^* \cong K_1^M(F)$, it is natural to ask if there is a norm homomorphism from $K_n^M(E)$ to $K_n^M(F)$ for each $n \geq 0$ having properties like those of the field norm $N_{E/F}$. If $b \in E$ is algebraic over F with minimal polynomial $p \in F[x]$, evaluation at b induces an isomorphism of fields

$$A_p = \frac{F[x]}{pF[x]} \cong F(b) .$$

Using the connection among extended tame symbols given by Milnor's exact sequence

$$0 \longrightarrow K_{n+1}^M(F) \xrightarrow{\ j\ } K_{n+1}^M(F(x)) \xrightarrow{\ \tau\ } \bigoplus_p K_n^M(A_p) \longrightarrow 0 ,$$

described in the last section, Bass and Tate [73] defined norms N_p mapping $K_n^M(A_p)$ to $K_n^M(F)$. Here is their definition in detail:

Recall that j is induced by inclusion $F \subseteq F(x)$ and τ is the p-adic tame symbol τ_p in each coordinate, where p ranges through the set I of monic irreducibles in $F[x]$. The remaining tame symbol on $K_{n+1}^M(F(x))$ is τ_v where $v = v_\infty$ is the discrete valuation on $F(x)$ with $v(a/b) = \text{degree}(b) - \text{degree}(a)$ for nonzero polynomials $a, b \in F[x]$. From (14.52), the composite $F \subseteq \mathcal{O}_v \to k_v$ is an isomorphism of fields, taking u to $u + \mathcal{P}_v$. As in the case $n = 1$, define

$$\tau_\infty : K_{n+1}^M(F(x)) \to K_n^M(F)$$

to be τ_v followed by K_n^M of the inverse isomorphism $k_v \cong F$.

Now $\tau_\infty \circ j = 0$ because elements $u \in F^*$ have $v_\infty(u) = 0$. So if

$$\psi : \bigoplus_p K_n^M(A_p) \to K_{n+1}^M(F(x))$$

is any function with $\tau \circ \psi = 1$, the composite

$$\tau_\infty \circ \psi : \bigoplus_p K_n^M(A_p) \to K_n^M(F)$$

is independent of the choice of ψ. Following τ_∞ by inversion in $K_n^M(F)$ (i.e., by scalar multiplication by -1), we obtain another homomorphism $-\tau_\infty$ from $K_{n+1}^M(F(x))$ to $K_n^M(F)$, and

$$-\tau_\infty \circ \psi : \bigoplus_p K_n^M(A_p) \to K_n^M(F)$$

is also independent of the choice of right inverse ψ of τ.

(14.63) Definition. For each $n \geq 0$ and each monic irreducible polynomial $p \in F[x]$, the **norm**
$$N_p : K_n^M(A_p) \to K_n^M(F)$$
is the homomorphism given by insertion in the p-coordinate, followed by the composite $-\tau_\infty \circ \psi$.

(14.64) Proposition. *For each $n \geq 0$, the composite*

$$K_n^M(F) \longrightarrow K_n^M(A_p) \xrightarrow{N_p} K_n^M(F) \ ,$$

with first map induced by $F \subseteq A_p$, is multiplication by the dimension $[A_p : F]$.

Proof. Suppose $u_1, \ldots, u_n \in F^*$. Then τ sends $\ell(u_1) \cdots \ell(u_n)\ell(p)$ to the insertion of $\ell(\overline{u_1}) \ldots \ell(\overline{u_n})$ in the p-coordinate; so

$$\begin{aligned}
N_p(\ell(\overline{u_1}) \ldots \ell(\overline{u_n})) &= -\tau_\infty(\ell(u_1) \ldots \ell(u_n)\ell(p)) \\
&= -\ell(\overline{u_1}) \ldots \ell(\overline{u_n})v_\infty(p) \\
&= \ell(\overline{u_1}) \ldots \ell(\overline{u_n}) \ degree(p) \ . \qquad \blacksquare
\end{aligned}$$

(14.65) Proposition. *The norms N_p are the unique homomorphisms from $K_n^M(A_p)$ to $K_n^M(F)$ for which*

$$-\tau_\infty = \sum_p N_p \circ \tau_p \ ,$$

where p ranges over the monic irreducibles in $F[x]$.

Proof. Let i_p denote insertion in the p-coordinate of $\oplus_p K_n^M(A_p)$. Since Milnor's exact sequence is split, we can choose a $\mathbb{Z}$-linear right inverse ψ of τ. For $\beta \in K_{n+1}^M(F(x))$, $\beta - \psi(\tau(\beta))$ is in $\ker \tau \subseteq \ker \tau_\infty$. So

$$
\begin{aligned}
-\tau_\infty(\beta) &= -\tau_\infty \circ \psi \circ \tau\ (\beta) \\
&= -\tau_\infty \circ \psi \circ \left(\sum_p i_p \circ \tau_p\ (\beta) \right) \\
&= \sum_p (-\tau_\infty \circ \psi \circ i_p) \circ \tau_p\ (\beta) \\
&= \sum_p N_p \circ \tau_p\ (\beta) \ .
\end{aligned}
$$

Suppose also that N_p' are homomorphisms with

$$
-\tau_\infty = \sum_p N_p' \circ \tau_p \ .
$$

Since τ is surjective, we can choose β so $\tau_q(\beta)$ has any preset value in $K_n^M(A_q)$ and $\tau_p(\beta) = 0$ for each $p \neq q$. Then

$$
N_q(\tau_q(\beta)) = -\tau_\infty(\beta) = N_q'(\tau_q(\beta)) \ ,
$$

proving $N_q = N_q'$ for each monic irreducible $q \in F[x]$. ∎

Note: Taking N_∞ to be the identity map on $K_n^M(F)$, this equation can be rewritten as

$$
\sum_q N_q \circ \tau_q = 0 \ ,
$$

where q ranges through the set consisting of ∞ and the monic irreducibles $p \in F[x]$.

(14.66) Corollary. (Bass, Tate [73]) *For each monic irreducible $p \in F[x]$, the map $N_p : A_p^* \to F^*$ coincides with the field norm $A_p^* \to F^*$.*

Proof. In multiplicative notation (A_p^*, F^* in place of $K_1^M(A_p), K_1^M(F)$), the map $\tau_p : K_2^M(F(x)) \to A_p^*$ is the regular tame symbol τ_v for $v = v_p$, followed by the isomorphism $h : k_v^* \to A_p^*$ restricting the F-algebra isomorphism $k_v \cong A_p$ taking $\overline{x}$ to $\overline{x}$. The norms N_p are the unique homomorphisms $A_p^* \to F^*$ for which

$$
\tau_\infty(-)^{-1} = \prod_p N_p \circ \tau_p\ (-) \ .
$$

By isomorphism invariance of algebra norms, the triangle

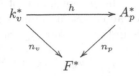

commutes, where n_v and n_p are the field norms. Then

$$\prod_p n_p \circ \tau_p \,(-) \;=\; \prod_p n_p \circ h \circ \tau_v \,(-)$$

$$=\; \prod_p n_v \circ \tau_v \,(-) \;=\; \tau_\infty(-)^{-1}$$

by Weil's Theorem (14.62); so $n_p = N_p$ for all p. ∎

If we simultaneously apply N_p in all degrees, we get a $\mathbb{Z}$-linear map between Milnor rings

$$N_p : K_*^M(A_p) \to K_*^M(F) \ .$$

Likewise, inclusions $F \subseteq F(x)$ and $F \subseteq A_p$ induce ring homomorphisms

$$j : K_*^M(F) \to K_*^M(F(x)) \quad , \quad j_p : K_*^M(F) \to K_*^M(A_p) \ .$$

From Bass and Tate [73, §5] we have the useful **projection formula**:

(14.67) Proposition. *If p is a monic irreducible in $F[x]$, $\alpha \in K_*^M(F)$ and $\beta \in K_*^M(A_p)$, then*

$$N_p(j_p(\alpha)\beta) \;=\; \alpha N_p(\beta) \ .$$

Proof. By means of the ring homomorphisms j_p and j, the rings $K_*^M(A_p)$ and $K_*^M(F(x))$ are $K_*^M(F)$-modules. For each monic irreducible $p \in F[x]$, the extended tame symbol

$$\tau_p : K_*^M(F(x)) \to K_*^M(A_p)$$

is $K_*^M(F)$-linear, since

$$\tau_p(\ell(u_1) \cdots \ell(u_r) \cdot \ell(u_1') \cdots \ell(u_s')\ell(a))$$
$$= \ell(\overline{u_1}) \cdots \ell(\overline{u_r})\ell(\overline{u_1'}) \cdots \ell(\overline{u_s'})v_p(a)$$
$$= \ell(u_1) \cdots \ell(u_r) \cdot \tau_p(\ell(u_1') \ldots \ell(u_s')\ell(a))$$

whenever $u_1, \ldots, u_r \in F^*, u_1', \ldots, u_s' \in \mathcal{O}_v^*$ and $a \in F(x)^*$. Regarding the direct sum $\oplus_p K_*^M(A_p)$ as a $K_*^M(F)$-module by coordinatewise scalar multiplication, it follows that τ is $K_*^M(F)$-linear. So the kernel $j(K_*^M(F))$ of τ is a $K_*^M(F)$-submodule of $K_*^M(F(x))$, and the induced isomorphism

$$\overline{\tau} : \frac{K_*^M(F(x))}{j(K_*^M(F))} \to \bigoplus_p K_*^M(A_p)$$

is $K_*^M(F)$-linear. The calculation involving τ_p above works as well to show τ_∞ is $K_*^M(F)$-linear.

So N_p is a composite of $K_*^M(F)$-linear maps

$$K_*^M(A_p) \xrightarrow{i_p} \bigoplus_q K_*^M(A_q)$$

$$\xrightarrow{\overline{\tau}^{-1}} \frac{K_*^M(F(x))}{j(K_*^M(F))} \xrightarrow{-\tau_\infty} K_*^M(F) \,,$$

and $K_*^M(F)$-linearity of N_p implies

$$N_p(j_p(\alpha)\beta) = N_p(\alpha \cdot \beta) = \alpha \cdot N_p(\beta) = \alpha N_p(\beta) \,. \qquad \blacksquare$$

(14.68) Note. Suppose $E = F(\gamma)$, where $\gamma \in \overline{F}$ has minimal polynomial p over F. Then $A_p \cong E$ as F-algebras. Choose any F-algebra isomorphism $\theta : E \to A_p$, and define N_θ to make the triangle

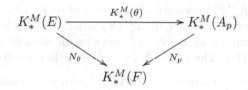

commute. Then N_θ is a group homomorphism. If $a_1, \ldots, a_{n-1} \in F^*$ and $b \in E^*$, then the projection formula implies

$$N_\theta(\ell(a_1)\ldots\ell(a_{n-1})\ell(b)) = N_p(\ell(a_1)\ldots\ell(a_{n-1})\ell(\theta(b)))$$
$$= \ell(a_1)\ldots\ell(a_{n-1})N_p(\ell(\theta(b)))$$
$$= \ell(a_1)\ldots\ell(a_{n-1})\ell(N_{E/F}(b)) \,.$$

So, at least on the $K_*^M(F)$-linear span of $K_1^M(E)$, N_θ is independent of the choice of θ. In Kersten [90, Theorem 20.14] the norm N_θ is shown to be completely independent of θ, isomorphism invariant, and functorial, although we shall make no use of these facts.

14G. Exercises

1. If F is a field, A is an F-algebra, and $[A : F] = n < \infty$, the norm $N_{A/F}$ can be defined as the composite of multiplicative monoid homomorphisms

$$A \xrightarrow{\ \lambda\ } \operatorname{End}_F(A) \xrightarrow{\ \rho\ } M_n(F) \xrightarrow{\ \det\ } F$$

with λ, ρ, and $\det$ as described at the beginning of the section. With this definition, prove that $a \in A$ is a unit of A if and only if $N_{A/F}(a)$ is nonzero.

2. Prove, directly from the definition (14.63) of the norm on Milnor K-theory, that for each monic irreducible $p \in F[x]$ with $p \neq x$,

$$\ell(N_{A_p/F}(\overline{x})) = N_p(\ell(\overline{x})) .$$

Hint: Show $\beta = \ell(x)\ell(p(0)^{-1}p)$ has $\tau_p(\beta) = \ell(\overline{x}), \tau_x(\beta) = \ell(\overline{1}), \tau_q(\beta) = \ell(\overline{1})$ for all monic irreducibles q other than x and p, and $\tau_\infty(\beta) = -\ell((-1)^n p(0))$.

3. Derive (14.64) directly from the projection formula (14.67). Essentially what does (14.64) say when p has degree 1?

4. (Alperin and Dennis [79]) For the field $\mathbb{R}$ of real numbers, show the only nonzero elements of finite additive order in $K_*^M(\mathbb{R})$ are the elements $\ell(-1) \ldots \ell(-1)$ of order 2. So in Corollary (14.47), $K_*^M(\mathbb{R}) = A \overset{\bullet}{\oplus} B$, where $A = \{0, \ell(-1)^n\}$ and B is *uniquely* divisible. *Hint*: Use (14.64) to show the kernel of $K_n^M(\mathbb{R}) \to K_n^M(\mathbb{C})$ consists of elements of order 1 or 2, and (14.44) to show all elements of finite order in $K_n^M(\mathbb{R})$ belong to this kernel. Given x in this kernel, show either x or $x + \ell(-1)^n$ is twice an element y (here use the decomposition in (14.47)). Then $4y = 0$; $2y = 0$ and either $x = 0$ or $x = \ell(-1)^n$.

5. (Kolster) If R is a commutative ring and -1_R is a square in R, prove $\ell(-1_R)\ell(-1_R) = 0$ in $K_2^M(R)$. Show the converse fails by proving -1 is not a square in the field $F = \mathbb{Q}(\sqrt{-2})$, but $\ell(-1)\ell(-1) = 0$ in $K_2^M(F)$ anyway. *Hint*: If $\zeta_8 = e^{2\pi i/8} = (1 + i)/\sqrt{2}$, then $\zeta_8^2 = i$ and $\zeta_8 - \zeta_8^{-1} = \sqrt{-2}$. Let E denote the field $F(\zeta_8) = \mathbb{Q}(\zeta_8)$. Show $i \notin F$, so $[E : F] > 1$. Show

$$(x - \zeta_8)(x + \zeta_8^{-1}) = x^2 - \sqrt{-2}x - 1 ;$$

so $\operatorname{Aut}(E/F) = \{1, \sigma\}$, where $\sigma(\zeta_8) = -\zeta_8^{-1}$. Taking $p = x^2 - \sqrt{-2}x - 1$, there is an F-algebra isomorphism $E \cong A_p$. If N is any of the norms $N_\theta : K_*^M(E) \to K_*^M(F)$ in (14.68), show

$$\ell(-1)\ell(-1) = N(\ell(-1)\ell(\zeta_8)) = N(0) = 0$$

in $K_2^M(F)$.

14H. Matsumoto's Theorem

Having norms on Milnor K-theory, we are at last in a position to prove Matsumoto's Theorem, that $K_2^M(F) \cong K_2(F)$ when F is a field, thereby providing a simple presentation of $K_2(F)$ by generators and relations. Matsumoto's proof of this theorem in his 1969 thesis (Matsumoto [69]), also presented in Milnor's book [71], was group-theoretic, constructing a central extension of $E(F) = SL(F)$ with kernel $K_2^M(F)$. Six years later, Keune [75] found an alternate proof using a dimension shift from K_2 to relative SK_1 and involving the Bass-Kubota presentation of relative SK_1 via Mennicke symbols (see our (11.27)). It is Keune's proof we present below. A third proof due to Hutchinson [90] relies on group homology; it is presented in Rosenberg's book [94] as his Theorem (4.3.15).

The relations (12.31) among Steinberg symbols:

> **M1 :** $\{ab, c\} \;\; = \;\; \{a, c\} \, \{b, c\} \,,$
> $\{a, bc\} \;\; = \;\; \{a, b\} \, \{a, c\} \,,$
>
> **M2 :** $\{a, b\} \;\; = \;\; 1 \quad \text{if} \quad a + b \;\; = \;\; 1 \,,$

show there is, for each field F, a homomorphism of abelian groups

$$\psi : K_2^M(F) \to K_2(F)$$
$$\ell(a)\ell(b) \mapsto \{a, b\} \,.$$

By (12.30), $K_2(F)$ is generated by Steinberg symbols; so ψ is surjective.

(14.69) Matsumoto's Theorem. *For each field F, the map ψ is a group isomorphism. So $K_2(F)$ is an abelian group generated by Steinberg symbols $\{a, b\}$ for $a, b \in F^*$, subject only to the relations M1 and M2.*

Proof. We need only construct a left inverse to ψ. This can be done in two steps: First a homomorphism

$$d : K_2(F) \to SK_1(F[t], (t^2 - t)F[t])$$

is defined by using the connecting homomorphism in the relative sequence. Second, the Mennicke symbol presentation of the relative SK_1 and a property of norms on K_2^M are used to define a homomorphism

$$\overline{\rho} : SK_1(F[t] \,, \; (t^2 - t)F[t]) \; \to \; K_2^M(F)$$

with $\overline{\rho} \circ d \circ \psi = 1$.

Notation: In this proof we are using t rather than x for an indeterminate to avoid confusion with generators $x_{ij}(-)$ of Steinberg groups. We also use the abbreviated notation $R = F[t]$ and $J = (t^2 - t)F[t]$.

The connecting homomorphism ∂_1 from $K_2(R/J)$ to $SK_1(R, J)$ in the relative sequence (13.20) and (13.22) comes from an application of the Snake Lemma to the commutative diagram

$$K_2(R/J)$$

$$\downarrow \cap_I$$

$$\begin{array}{ccccccccc}
1 & \longrightarrow & St'(R, J) & \overset{\subseteq}{\longrightarrow} & St(R) & \longrightarrow & St(R/J) & \longrightarrow & 1 \\
& & \downarrow{\phi} & & \downarrow{\phi} & & \downarrow{\phi} & & \\
\textbf{(14.70)} \quad 1 & \longrightarrow & GL(R, J) & \overset{\subseteq}{\longrightarrow} & GL(R) & \longrightarrow & GL(R/J) & & \\
& & \downarrow & & & & & & \\
& & K_1(R, J) & & & & & &
\end{array}$$

with exact rows. The group $K_2(R/J)$ at the beginning of this "snake" is related to $K_2(F)$ by a commutative diagram

$$\textbf{(14.71)} \quad \begin{array}{ccccc}
K_2(R/J) & \overset{\cong}{\longrightarrow} & K_2(F \times F) & \overset{\cong}{\longrightarrow} & K_2(F) \times K_2(F) \\
\downarrow{\cap_I} & & \downarrow{\cap_I} & & \downarrow{\cap_I} \\
St(R/J) & \overset{\cong}{\longrightarrow} & St(F \times F) & \overset{\cong}{\longrightarrow} & St(F) \times St(F)
\end{array}$$

in which the left isomorphisms are induced by the ring isomorphisms

$$\frac{R}{J} \cong \frac{R}{tR} \times \frac{R}{(t-1)R} \cong F \times F$$

of the Chinese Remainder Theorem and evaluation at 0 and 1, and the right isomorphisms come from (12.8). The bottom row takes

$$x_{ij}(\overline{p(t)}) \mapsto x_{ij}((p(0), p(1))) \mapsto (x_{ij}(p(0)), x_{ij}(p(1))) .$$

Define

$$d : K_2(F) \to SK_1(R, J)$$

to be insertion in the second coordinate $K_2(F) \to K_2(F) \times K_2(F)$, followed by the isomorphism $K_2(F) \times K_2(F) \cong K_2(R/J)$ inverse to the top row of (14.71), followed by the connecting homomorphism $\partial_1 : K_2(R/J) \to SK_1(R, J)$.

For $a, b \in F^*$, Keune computed $d(\{a, b\})$ as follows. In $St(R)$, he defined elements:

$$\begin{aligned}
\alpha(t) &= x_{12}(a)x_{21}(-a^{-1})x_{12}((a-1)t)x_{21}(a^{-1})x_{12}(-a) , \\
\beta(t) &= x_{21}(-a^{-1})x_{12}(a)x_{21}((1-a^{-1})t)x_{12}(-a)x_{21}(a^{-1}) , \\
\gamma(t) &= \alpha(t)x_{21}(a^{-1})\beta(t)x_{21}(-a^{-1})x_{12}((a-1)t) , \\
\delta(t) &= [\gamma(t), h_{13}(b)] .
\end{aligned}$$

The composite

$$St(R) \quad \rightarrow \quad St(R/J) \; \cong \; St(F) \times St(F)$$

takes $\delta(t)$ to $(\delta(0), \delta(1))$. Using the standard facts $x_{ij}(r+s) = x_{ij}(r)x_{ij}(s)$ for $r, s \in F$, and $w_{ij}(u)w_{ij}(-u) = 1$ for $u \in F^*$, yields $\alpha(0) = \beta(0) = 1$,

$$\alpha(1) \; = \; w_{12}(a)x_{12}(-1)x_{21}(a^{-1})x_{12}(-a), \text{ and}$$
$$\beta(1) \; = \; w_{21}(-a^{-1})x_{21}(1)x_{12}(-a)x_{21}(a^{-1}) \; ;$$

so $\gamma(0) = 1$ and $\gamma(1) = w_{12}(a)w_{12}(-1) = h_{12}(a)$. Therefore $\delta(0) = 1$ and $\delta(1) = [h_{12}(a), h_{13}(b)] = \{a, b\}$. So insertion in the second coordinate of $K_2(F) \times K_2(F)$ takes $\{a, b\}$ to $(1, \{a, b\}) = (\delta(0), \delta(1))$, and the isomorphism to $K_2(R/J)$ takes this back to the image of $\delta(t)$ under $St(R) \to St(R/J)$. Following the "snake" ∂_1 through diagram (14.70) shows $\phi(\delta(t))$ is a matrix representing $d(\{a, b\})$.

From direct matrix multiplications,

$$\phi(\alpha(t)) \; = \; \begin{bmatrix} 1 & 0 \\ a^{-2}(1-a)t & 1 \end{bmatrix},$$

$$\phi(\beta(t)) \; = \; \begin{bmatrix} 1 & a(1-a)t \\ 0 & 1 \end{bmatrix},$$

and

$$\phi(\gamma(t)) \; = \; \begin{bmatrix} r & s \\ u & v \end{bmatrix},$$

where

$$r \; = \; (a-1)t \; + \; 1 \, ,$$
$$s \; = \; (a-1)^2(t^2 - t) \; , \quad \text{and}$$
$$u \; = \; - \, (a^{-1} - 1)^2(t^2 - t) \, .$$

Since $\phi(\gamma(t)) \in E(R)$, it has determinant 1. So its inverse is

$$\begin{bmatrix} v & -s \\ -u & r \end{bmatrix},$$

and

$$\phi(\delta(t)) \; = \; [\phi(\gamma(t)), \phi(h_{13}(b))]$$
$$= \; \begin{bmatrix} r & s \\ u & v \end{bmatrix} \begin{bmatrix} b & 0 & 0 \\ 0 & 1 & 0 \\ 0 & 0 & b^{-1} \end{bmatrix} \begin{bmatrix} v & -s \\ -u & r \end{bmatrix} \begin{bmatrix} b^{-1} & 0 & 0 \\ 0 & 1 & 0 \\ 0 & 0 & b \end{bmatrix}$$
$$= \; \begin{bmatrix} 1 + (1 - b^{-1})su & (1-b)rs \\ * & * \end{bmatrix} = \begin{bmatrix} y & z \\ * & * \end{bmatrix},$$

where

$$y = 1 - \frac{b-1}{b} \cdot \frac{(a-1)^4}{a^2}(t^2-t)^2, \quad \text{and}$$

$$z = (1-b)(a-1)^3(t+\frac{1}{a-1})(t^2-t) .$$

Then $d(\{a,b\})$ is the Mennicke symbol $[y,z]_J$. If

$$w = -\frac{a-1}{a^2b}(t-\frac{a}{a-1}) ,$$

then $[y,z]_J = [y+wz,z]_J$

$$= [1 - \frac{b-1}{b}\frac{(a-1)^2}{a}(t^2-t), (1-b)(a-1)^3(t+\frac{1}{a-1})(t^2-t)]_J .$$

Since $\{a,b\} = \{b,a\}^{-1} = \{b,a^{-1}\}$, replacing a by b and b by a^{-1} provides a third expression: $d(\{a,b\}) =$

(14.72) $[1+(a-1)\frac{(b-1)^2}{b}(t^2-t) , \frac{a-1}{a}(b-1)^3(t+\frac{1}{b-1})(t^2-t)]_J .$

Next we consider the construction of $\bar{p}$ from $SK_1(R,J)$ to $K_2^M(F)$.

(14.73) Lemma. (Keune [75]) *Suppose F is a field, and (a,b) is a unimodular pair in $F[t]$ with $a-1$ and b in $(t^2-t)F[t]$. Suppose $g = b/(t^2-t) \neq 0$. Let p represent monic irreducibles in $F[t]$. Then*

$$\sum_{p\nmid a(t^2-t)} v_p(g)N_p(\ell(\bar{a})\ell(\overline{t-1/t}))$$

$$= \sum_{p\nmid g(t^2-t)} v_p(a)N_p(\ell(\bar{g})\ell(\overline{t-1/t})) .$$

Proof. Let z denote $\ell(g)\ell(1-(1/t))\ell(a)$ in $K_3^M(F(t))$. We determine $\tau_v(z)$ for each discrete valuation v on $F(t)$ and apply the summation formula (14.65) for norms of tame symbols. Now

$$\ell(1-\frac{1}{t})\ell(1-t) = 0 = \ell(t)\ell(1-\frac{1}{t}) .$$

So if the prime π in $\mathcal{O}_v$ is t, $t-1$, or $1/t$, then

$$z = \ell(g')\ell(1-\frac{1}{t})\ell(a')$$

with $v(g') = v(a') = 0$. And then

$$\tau_v(z) = -\ell(\overline{g'})\ell(\overline{a'})v(1 - \frac{1}{t}) \ .$$

If $v = v_\infty$ (so $\pi = 1/t$), then $v(1 - (1/t)) = 0$; so $\tau_v(z) = 0$. If, instead, $v = v_t$ or v_{t-1}, the choice of a in $1 + (t^2 - t)F[t]$ implies $v(a) = 0$; so we can take $a' = a$ and $\overline{a'} = \overline{a} = \overline{1}$ in k_v^*. So $\ell(\overline{a'}) = 0$ and again $\tau_v(z) = 0$.

Suppose $v = v_p$, where $p \nmid (t^2 - t)$. If $p \nmid ag$, then $v_p(g) = v_p((t-1)/t) = v_p(a) = 0$; so $\tau_v(z) = 0$. On the other hand, if $p|a$, then $v_p(g) = v_p((t-1)/t) = 0$; so

$$\tau_v(z) \;=\; v_p(a)\ell(\overline{g})\ell(\overline{1 - \frac{1}{t}}) \ ,$$

and

$$\tau_p(z) \;=\; v_p(a)\ell(\overline{g})\ell(\overline{t - 1/t}) \ .$$

Likewise, if $p|g$, then $v_p(a) = v_p((t-1)/t) = 0$ and

$$\tau_p(z) \;=\; v_p(g)\ell(\overline{t - 1/t})\ell(\overline{a}) \;=\; -\, v_p(g)\ell(\overline{a})\ell(\overline{t - 1/t}) \ .$$

The desired equation now follows from

$$- \, v_\infty(z) \;=\; \sum_p N_p \circ \tau_p \, (z)$$

and the $\mathbb{Z}$-linearity of N_p. ∎

To define a homomorphism from $SK_1(R, J)$ to $K_2^M(F)$, one can use the presentation (11.27) for $SK_1(R, J)$ by Mennicke symbols, which applies to a class of rings including $R = F[t]$. Recall that $SK_1(R, J)$ has one generator

$$[a, b]_J \;=\; \begin{bmatrix} a & b \\ c & d \end{bmatrix} E(R)$$

for each pair (a, b) in the set

$$W_J \;=\; \{(a, b) \in R \times R : aR + bR = R, \ a - 1, b \in J\},$$

and the defining relations among these generators are

MS1 : $[a, b]_J \;=\; [a, b + ra]_J$ if $r \in J$,

           $[a, b]_J \;=\; [a + rb, b]_J$ if $r \in R$,

MS2 : $[a_1, b]_J [a_2, b]_J \;=\; [a_1 a_2, b]_J$,

           $[a, b_1]_J \, [a, b_2]_J \;=\; [a, b_1 b_2]_J$.

Following the treatment by Bass [68, Chapter 6, §6] of "q-reciprocities," we prove:

(14.74) Lemma. (Keune [75]) *Suppose F is a field, R is $F[t]$, J is the ideal $(t^2 - t)F[t]$, and $\rho : W_J \to K_2^M(F)$ is defined by $\rho((a, 0)) = 0$ and*

$$\rho((a, (t^2 - t)g)) \;=\; \sum_{p \nmid a(t^2 - t)} v_p(g) N_p(\ell(\overline{a})\ell(\overline{t - 1/t}))$$

if $g \neq 0$. Then ρ induces a homomorphism

$$\overline{\rho} : SK_1(R, J) \to K_2^M(F) .$$
$$[a, b]_J \mapsto \rho((a, b))$$

Proof. We show $\rho((a, b))$ obeys the defining relations MS1 and MS2 given above for $[a, b]_J$. First suppose $r \in J$. If $b \neq 0$ but $a \in R^*$, then $\rho((a, b))$ is a sum of zeros on the right side of equation (14.73). So if $b = 0$ or $b + ra = 0$, then $a \in R^*$ and $\rho((a, b)) = 0 = \rho((a, b + ra))$. On the other hand, if $b = (t^2 - t)g \neq 0, r = (t^2 - t)h$, and $b + ra = (t^2 - t)(g + ha) \neq 0$, then, on the right side of (14.73), $\overline{g} = \overline{g + ha}$ in A_p whenever $v_p(a) \neq 0$; so $\rho((a, b)) = \rho((a, b + ra))$ again.

Now suppose r is in R, but not necessarily in J. If $b = 0$, then $\rho((a, b)) = \rho((a + rb, b))$. Suppose $b \neq 0$. If $p \nmid a(t^2 - t)$ and $v_p(g) \neq 0$, then $p | b$; so $\overline{a} = \overline{a + rb}$ in A_p, and $\rho((a, b)) = \rho((a + rb, b))$ again.

If $b = 0, \rho((a_1, b)) + \rho((a_2, b)) = 0 + 0 = 0 = \rho((a_1 a_2, b))$. Suppose $b \neq 0$. Using the right side of (14.73), $\rho((a_1, b)) + \rho((a_2, b)) = \rho((a_1 a_2, b))$ because $v_p(a_1) + v_p(a_2) = v_p(a_1 a_2)$.

If $b_1 \neq 0$ and $b_2 \neq 0$, say $b_1 = (t^2 - t)g_1$ and $b_2 = (t^2 - t)g_2$; so $b_1 b_2 = (t^2 - t)g_3$ with $g_3 = (t^2 - t)g_1 g_2$. So for $p \nmid (t^2 - t), v_p(g_1) + v_p(g_2) = v_p(g_3)$ and $\rho((a, b_1)) + \rho((a, b_2)) = \rho((a, b_1 b_2))$.

Of course, if $b_1 = b_2 = 0$, then $\rho((a, b_1)) + \rho((a, b_2)) = 0 + 0 = 0 = \rho((a, b_1 b_2))$. Suppose $b_1 \neq 0$ and $b_2 = 0$. Then $a \in R^*$, and the right side of (14.73) is a sum of zeros. So $\rho((a, b_1)) + \rho((a, 0)) = 0 + 0 = 0 = \rho((a, b_1 0))$. The case in which $b_1 = 0$ and $b_2 \neq 0$ is the same. ∎

To complete the proof of Matsumoto's Theorem, we need only check that $\overline{\rho} \circ d \circ \psi = 1$ on $K_2^M(F)$. These maps are group homomorphisms, so it suffices to check this equation on each generator $\ell(a)\ell(b)$ with $a, b \in F^*$. Now $\psi(\ell(a)\ell(b)) = \{a, b\}$. By (14.72),

$$d(\{a, b\}) \;=\; [w, (t^2 - t)g]_J ,$$

where

$$w \;=\; 1 + (a - 1)\frac{(b - 1)^2}{b}(t^2 - t) , \text{ and}$$
$$g \;=\; \frac{a - 1}{a}(b - 1)^3 (t + \frac{1}{b - 1}) .$$

Write q for $t + 1/(b-1)$. If $p \neq q$, $v_p(g) = 0$. And $v_q(g) = 1$. So

$$\overline{\rho}(d(\{a,b\})) \;=\; N_q(\ell(\overline{w})\ell(\overline{t-1/t})) \;.$$

By the Remainder Theorem, each element $\overline{f(t)}$ in A_q equals $\overline{f(1/(1-b))}$, since $1/(1-b)$ is the root of q. So in A_q,

$$\overline{w} \;=\; \overline{1 + (a-1)\frac{(b-1)^2}{b}\left(\frac{1}{(1-b)^2} - \frac{1}{1-b}\right)}$$

$$=\; \overline{1 + \frac{a-1}{b} + \frac{(a-1)(b-1)}{b}}$$

$$=\; \overline{1 + (a-1)} \;=\; \overline{a}\,, \qquad \text{and}$$

$$\overline{t - 1/t} \;=\; \overline{\frac{1}{1-b} - 1} \Big/ \overline{\frac{1}{1-b}}$$

$$=\; \overline{1 - (1-b)} \;=\; \overline{b}\,.$$

Then $\overline{\rho}(d(\{a,b\})) = N_q(\ell(\overline{a})\ell(\overline{b}))$, and this equals $\ell(a)\ell(b)$ because

$$K_2^M(F) \longrightarrow K_2^M(A_q) \xrightarrow{\;N_q\;} K_2^M(F)$$

is multiplication by $\text{degree}(q) = 1$. $\blacksquare$

(14.75) Note and Notation. For each ring homomorphism $f : F \to E$ between fields, the left square

$$
\begin{array}{ccc}
K_2^M(F) & \xrightarrow{\;\psi\;} & K_2(F) \\
\downarrow & & \downarrow \\
K_2^M(E) & \xrightarrow{\;\psi\;} & K_2(E)
\end{array}
\,,
\qquad
\begin{array}{ccc}
\ell(a)\ell(b) & \longrightarrow & \{a,b\} \\
\downarrow & & \downarrow \\
\ell(f(a))\ell(f(b)) & \longrightarrow & \{f(a),f(b)\}
\end{array}
$$

commutes, where the right square shows the effect of the left one on generators. So the isomorphism ψ is a natural isomorphism between functors K_2^M and K_2 from fields to abelian groups. The distinction between $K_2^M(F)$ and $K_2(F)$ becomes only one of notation. Extending this parallel, the groups $K_n^M(F)$ for $n \geq 1$ are often written in the literature as multiplicative abelian groups, with $\ell(a_1), \ldots, \ell(a_n)$ denoted instead by $\{a_1, \ldots, a_n\}$, which is understood to equal $\{a_1\} \cdots \{a_n\}$.

As a direct consequence of the presentation in Matsumoto's Theorem, we obtain a universal property for K_2 of a field:

(14.76) Definition. Suppose F is a field. A **symbol map** on F is any function

$$F^* \times F^* \to G , \quad (a,b) \mapsto f(a,b) ,$$

from $F^* \times F^*$ into an abelian group G, satisfying

$$
\begin{aligned}
f(ab,c) &= f(a,c)f(b,c) , \\
f(a,bc) &= f(a,b)f(a,c) ,
\end{aligned}
$$

and

$$f(a,b) = 1_G \quad \text{if } a+b = 1_F .$$

(14.77) Corollary.. *The Steinberg symbol map* $F^* \times F^* \to K_2(F), (a,b) \mapsto \{a,b\}$, *is initial among all symbol maps on* F. *That is, for each symbol map* $F^* \times F^* \to G$, $(a,b) \mapsto f(a,b)$, *there is a unique homomorphism of abelian groups* $\overline{f} : K_2(F) \to G$ *making the diagram*

$$
\begin{array}{ccc}
F^* \times F^* & \xrightarrow{\ \{\,,\,\}\ } & K_2(F) \\
& \searrow{\scriptstyle f} & \downarrow{\scriptstyle \overline{f}} \\
& & G
\end{array}
$$

commute.   ∎

In Chapters 15 and 16, we investigate symbol maps that are connected to deep number theory and algebra.

14H. Exercises

1. If $f : R \to F$ is any ring homomorphism into an ordered field F, prove $\{-1,-1\}$ is a nontrivial element (of order 2) in $K_2(R)$. Use this to prove $K_2(\mathbb{Z}G) \neq 1$ for all groups G.

2. Using Exercise 1 and §12B, Exercise 6, prove $K_2(\mathbb{Z})$ is cyclic of order 2, generated by $\{-1,-1\}$. Use this to rewrite the exact sequence in the description (14.56) of $K_2^M(\mathbb{Q})$ to be an exact sequence relating $K_2(\mathbb{Z})$ and $K_2(\mathbb{Q})$.

3. In the proof of Matsumoto's Theorem, show the maps $\overline{p}$ and d are isomorphisms.

4. If v is a discrete valuation on a field F, show Matsumoto's relations M1 and M2 are equivalent to the relations

$$\{ab, c\} = \{a, c\}\{b, c\} ,$$
$$\{a, b\} = \{b, a\}^{-1} ,$$
$$\{a, -a\} = 1 , \text{ and}$$
$$\{a, 1 - a\} = 1 \text{ if } v(a) \geq 0 \text{ and } v(1 - a) = 0 .$$

Note: If $R = \mathcal{O}_v$ is the associated discrete valuation ring, it is proved in Dennis and Stein [75] that $K_2(R)$ is presented as an abelian group by the Steinberg symbols $\{a, b\}$ with $a, b \in R^*$, subject only to the above relations together with three more relations, two of which were later proved redundant in Kolster [85].

5. Suppose A is any ring, and $a, b \in A$ with $1 + ab \in A^*$. Define

$$H_{12}(a, b) = x_{21}(-b(1 + ab)^{-1})x_{12}(a)x_{21}(b)x_{12}(-a(1 + ba)^{-1}) .$$

If $\phi : St(A) \to E(A)$ is the standard homomorphism taking $x_{ij}(a)$ to $e_{ij}(a)$, show $\phi(H_{12}(a, b)) = \phi(h_{12}(1 + ab)) = \mathrm{diag}(1 + ab, (1 + ba)^{-1}, 1, 1, ...)$. In particular, this shows

$$\langle a, b \rangle = H_{12}(a, b)h_{12}(1 + ab)^{-1}$$

belongs to $K_2(A)$. This element $\langle a, b \rangle$ is known as a **Dennis-Stein symbol** on A. Note that, when A is the ring $R = \mathcal{O}_v$ in Exercise 4, $K_2(A)$ is presented as an abelian group by the generators $\langle a, b \rangle$ subject only to the relations

$$\langle a, b \rangle = \langle -b, -a \rangle^{-1} ,$$
$$\langle a, b \rangle \langle a, c \rangle = \langle a, b + c + abc \rangle ,$$
$$\langle a, bc \rangle = \langle ab, c \rangle \langle ac, b \rangle ,$$

as shown by Maazen and Stienstra [78].

PART V

Sources of K_2

In §14H, K_2 of a field F is identified as the initial target of all symbol maps on $F^* \times F^*$. So the historical origin of K_2 can be traced to the origin of symbol maps. The first of these is the quadratic norm residue symbol, defined by Hilbert in 1897 in order to extend Gauss' law of quadratic reciprocity. We develop norm residue symbols in Chapter 15, culminating with a concrete version of the localization sequence in the K-theory of a Dedekind domain.

The Brauer group $Br(F)$ of a field F is the set of isomorphism classes of finite-dimensional division F-algebras, multiplied by the tensor product. In Chapter 16 we develop the Brauer group of F and establish the symbol maps assigning to each pair of units in F an analog to the division ring of quaternions. Chapter 16 concludes with a summary of the Tate-Merkurjev-Suslin theorems, which show the strength of this connection between $K_2(F)$ and $Br(F)$.

15

Symbols in Arithmetic

The real symbol is expressed in §15A in terms of norms from quadratic extensions, and this is generalized to a "Hilbert symbol" on any field sufficiently like $\mathbb{R}$. The metric completion of $\mathbb{Q}$ to $\mathbb{R}$ is generalized in §15B. In §15C, each prime number p provides a complete p-adic number field, which is an alternative to $\mathbb{R}$ as a context for $\mathbb{Q}$. Each $\mathbb{Q}_p$ is enough like $\mathbb{R}$ to have a Hilbert symbol. On $\mathbb{Q}^* \times \mathbb{Q}^*$, the product of the real and p-adic Hilbert symbols is 1, and this theorem (15.32) is Hilbert's generalization of the law of quadratic reciprocity. Gauss called this law the "gem of higher arithmetic," provided its first complete proof, and eventually published five more proofs. A lemma of Gauss' is incorporated into the proof we provide, which is due to Tate. Hilbert extended quadratic reciprocity to number fields and anticipated its extension (as "Hilbert's ninth problem") to higher power reciprocity over global fields. In §15D we present this extension, using local class field theory to define the "nth power norm residue symbol" on local fields, with the Hilbert and tame symbols as special cases. We conclude with K-theory exact sequences, (15.44) and (15.45), which neatly summarize the higher reciprocity laws for number fields.

15A. Hilbert Symbols

The first field norm ever used was probably the norm $N_{\mathbb{C}/\mathbb{R}}$ from the complex numbers to the real numbers: $N_{\mathbb{C}/\mathbb{R}}(a+bi) = (a+bi)(a-bi) = a^2 + b^2$; it gives the square of the length of a vector in the complex plane. Its values are all nonnegative. The real symbol $(a, b)_\infty$ can be described in terms of this norm.

(15.1) Lemma. *For $a, b \in \mathbb{R}^*$, $(a, b)_\infty = 1$ if and only if b is a norm from $\mathbb{R}(\sqrt{a})$.*

Proof. If $a > 0$, $\mathbb{R}(\sqrt{a}) = \mathbb{R}$ and $b = N_{\mathbb{R}/\mathbb{R}}(b)$. If $a < 0$, $\mathbb{R}(\sqrt{a}) = \mathbb{C}$ and b is a norm from $\mathbb{C}$ if and only if $b > 0$. ∎

This norm condition can be expressed in terms of quadratic forms.

(15.2) Lemma. *Suppose F is a field and $a, b \in F^*$. Then b is a norm from $F(\sqrt{a})$ if and only if $ax^2 + by^2 = z^2$ has a solution $(x, y, z) \neq (0, 0, 0)$ in F^3.*

Proof. Choose a root $\sqrt{a}$ of $x^2 - a$ in an algebraic closure of F, and let E denote $F(\sqrt{a})$. If $a = c^2$ with $c \in F$, then $\sqrt{a} = \pm c$ and $E = F$. In that case, $b = N_{E/F}(b)$ and $a(c^{-1})^2 + b0^2 = 1^2$.

Suppose, rather, that a is not a square in F. If $b = N_{E/F}(c + d\sqrt{a}) = c^2 - ad^2$ for $c, d \in F$, then $ad^2 + b1^2 = c^2$. For the converse, suppose $ax^2 + by^2 = z^2$ for $x, y, z \in F$ not all zero. Since a is not a square, $y \neq 0$. Then b can be expressed as $N_{E/F}((z/y) + (x/y)\sqrt{a})$. ∎

Note: The quadratic form condition in this lemma makes no reference to $\sqrt{a}$, so the norm condition must be independent of the choice of $\sqrt{a}$.

In 1897, Hilbert [98, §64] generalized the real symbol to other fields F (ordered or not) by defining a map from $F^* \times F^*$ to the multiplicative group $\mathbb{Z}^* = \{\pm 1\}$, taking (a, b) to

$$(a, b)_F \;=\; \begin{cases} 1 & \text{if } b \in N_{F(\sqrt{a})/F}(F(\sqrt{a})^*) \\ -1 & \text{if not .} \end{cases}$$

Taking $x = y = z = 1$ in (15.2), it is evident that $(a, b)_F = 1$ whenever $a + b = 1_F$. Also from (15.2), $(a, b)_F = (b, a)_F$; so $(a, b)_F$ is multiplicative in a if and only if it is multiplicative in b. If $N = N_{F(\sqrt{a})/F}(F(\sqrt{a})^*)$ has index 1 or 2 in F^*, then

$$(a, b_1 b_2)_F \;=\; (a, b_1)_F (a, b_2)_F$$

for all $b_1, b_2 \in F^*$, since $b_1 b_2 \in N$ if and only if b_1 and b_2 are both in N or both outside N. On the other hand, if N has index exceeding 2, there exist cosets $b_1 N \neq N$ and $b_2 N \neq N$, $b_1^{-1} N$, so that $(a, b_1 b_2)_F = (a, b_1)_F = (a, b_2)_F = -1$. We have proved:

(15.3) Proposition. *Suppose F is a field. The following are equivalent:*

   (i) *For each field $E = F(\sqrt{a})$ with $a \in F$, the subgroup $N = N_{E/F}(E^*)$ has index 1 or 2 in F^*,*

   (ii) *The map $(-, -)_F : F^* \times F^* \to \{\pm 1\}$ is multiplicative in each coordinate,*

   (iii) *The map $(-, -)_F : F^* \times F^* \to \{\pm 1\}$ is a symbol map, inducing a homomorphism $K_2(F) \to \{\pm 1\}$ taking $\{a, b\}$ to $(a, b)_F$.* ∎

Whenever $(-, -)_F$ is a symbol map, it is called the **Hilbert symbol** on F. The fact that there is a Hilbert symbol $(-, -)_\infty$ on $\mathbb{R}$ hinges on the fact that $N_{\mathbb{C}/\mathbb{R}}(\mathbb{C}^*)$ is the group of positive real numbers, which has index 2 in $\mathbb{R}^*$. There is a Hilbert symbol on each algebraically closed field F, since $F(\sqrt{a}) = F$ and $N_{F/F}(F^*)$ has index 1 in F^*; but in this case $(a, b)_F = 1$ for all a, b and the induced homomorphism on $K_2(F)$ is trivial.

(15.4) Example. There is no Hilbert symbol on $\mathbb{Q}$, because the norm group $N_{\mathbb{Q}(i)/\mathbb{Q}}(\mathbb{Q}(i)^*)$ has infinite index in $\mathbb{Q}^*$: To see why, note that Kummer's Theorem (7.47) shows the primes $p \in \mathbb{Z}$ that remain irreducible in the factorial ring $\mathbb{Z}[i]$ are those congruent to 3 (mod 4). For any such prime p and any integers a and b, $v_p(a + bi) = v_p(a - bi)$; so

$$v_p(a^2 + b^2) = v_p(N_{\mathbb{Q}(i)/\mathbb{Q}}(a + bi))$$

must be even. If q is an integer not divisible by p, there can be no integers a, b, and c with

$$\frac{p}{q} = \frac{a^2 + b^2}{c^2} = N_{\mathbb{Q}(i)/\mathbb{Q}}(\frac{a}{c} + \frac{b}{c}i) ,$$

since $v_p(pc^2)$ is odd and $v_p(q(a^2 + b^2))$ is even. Thus no two different primes in $3 + 4\mathbb{Z}$ can be congruent modulo norms from $\mathbb{Q}(i)^*$. And there are infinitely many primes in $3 + 4\mathbb{Z}$ (for otherwise their product plus 2 or 4 would be in $3 + 4\mathbb{Z}$, and hence divisible by one of those primes, while relatively prime to their product).

To find fields on which there is a nontrivial Hilbert symbol, perhaps one should look for fields resembling the real numbers. We consider these in the next two sections.

15A. Exercises

1. Suppose F is a field of characteristic other than 2. For $a, b \in F^*$, prove $ax^2 + by^2 = z^2$ has a solution $(x, y, z) \neq (0, 0, 0)$ in F^3 if and only if $aX^2 + bY^2 = 1$ has a solution (X, Y) in F^2. *Hint:* If $z = 0$, try

$$X = \frac{1 - b}{2b(y/x)} \quad \text{and} \quad Y = \frac{1 + b}{2b} .$$

2. If there is a Hilbert symbol on a field F, prove $\{-1, -1\} = 1$ in $K_2(F)$ implies -1 is a sum of two squares in F.

3. Suppose F is a finite field with q elements. Show there is a Hilbert symbol on F if and only if, for each nonsquare a in F^*, there are exactly $q + 1$ solutions (c, d) in $F \times F$ to the equation $c^2 - ad^2 = 1$.

4. Does every finite field have a Hilbert symbol?

15B. Metric Completion of Fields

Aside from Hilbert symbols, there are compelling reasons to study any field resembling $\mathbb{R}$, since the mathematics involving such a field may be as rich as that involving $\mathbb{R}$. A significant property of $\mathbb{R}$ is "completeness": Every Cauchy sequence in $\mathbb{R}$ converges in $\mathbb{R}$. In this section we discuss absolute values on a field, use them to define Cauchy and convergent sequences, and describe how to enlarge each field to a complete field, much as $\mathbb{Q}$ is enlarged to $\mathbb{R}$.

(15.5) Definitions. Let $\mathbb{R}^+$ denote the set of positive real numbers. Suppose F is a field. An **absolute value** on F is any function $F \to \mathbb{R}^+ \cup \{0\}$, $x \mapsto \|x\|$, satisfying, for all $x, y \in F$,

   (i) $\|x\| = 0$ if and only if $x = 0$,

   (ii) $\|xy\| = \|x\|\,\|y\|$,   and

   (iii) $\|x + y\| \le \|x\| + \|y\|$.

A **valued field** $(F, \|\ \|)$ is a field F together with an absolute value $\|\ \|$ on F. By (i) and (ii), $\|\ \| : F^* \to \mathbb{R}^+$ is a homomorphism of multiplicative groups. The image $\|F^*\|$ is the **value group** of $\|\ \|$, a subgroup of $(\mathbb{R}^+, \cdot)$. Since $\|-1\|^2 = \|1\| = 1$ and $\|-1\| \in \mathbb{R}^+$, $\|-1\| = 1$. So $\|-x\| = \|-1\|\,\|x\| = \|x\|$ for all $x \in F$. The axioms (i), (ii), (iii), and this last property imply F is a metric space with metric

$$d(x, y) \quad = \quad \|x - y\| \ .$$

Axiom (iii) is called the **triangle inequality**, since it amounts to $d(x, y) \le d(x, z) + d(z, y)$.

(15.6) Examples.

   (i) The standard absolute value on $\mathbb{C}$ is $|a + bi| = \sqrt{a^2 + b^2}$ for $a, b \in \mathbb{R}$. It extends the standard absolute value $|r| = \max\{r, -r\}$ on $\mathbb{R}$. Given any ring homomorphism σ from a field F into $\mathbb{C}$, there is an absolute value $|\ |_\sigma$ on F defined by $|x|_\sigma = |\sigma(x)|$. Since $\sigma(F)$ contains $\mathbb{Q}$, the value group of $|\ |_\sigma$ contains $\mathbb{Q}^+$, the set of positive rational numbers.

   (ii) If v is a discrete valuation (see (7.25)) on a field F, and $a \in \mathbb{R}$ with $0 < a < 1$, there is an absolute value $|\ |_{v,a}$ on F defined by

$$|x|_{v,a} \quad = \quad \begin{cases} 0 & \text{if } x = 0 \\ a^{v(x)} & \text{if } x \neq 0 \ . \end{cases}$$

Since $v(x+y) \ge \min\{v(x), v(y)\}$ for $x, y \in F^*$ and the function a^x is decreasing on $\mathbb{R}$, this absolute value satisfies a condition

$$|x + y|_{v,a} \quad \le \quad \max\{|x|_{v,a}, \ |y|_{v,a}\}$$

stronger than the triangle inequality. The value group of $|\ |_{v,a}$ is the infinite cyclic subgroup of $(\mathbb{R}^+, \cdot)$ generated by a.

(iii) On each field F there is a **trivial** absolute value, with $\|0\| = 0$ and $\|x\| = 1$ if $x \neq 0$. Its value group is the trivial group $\{1\}$.

(15.7) Definitions. An absolute value $\|\ \|$ on a field F is **ultrametric** if

$$\|x + y\| \ \leq \ \max\{\|x\|,\ \|y\|\}$$

for all $x, y \in F$, and is **archimedean** if it is not ultrametric.

Note that the absolute values $|\ |_\sigma$ in (15.6) (i) are archimedean, since $|1+1|_\sigma = 2$ exceeds $\max\{|1|_\sigma, |1|_\sigma\}$. But the absolute values $|\ |_{v,a}$ and the trivial absolute value are ultrametric. The ultrametric versus archimedean nature of an absolute value is determined by its effect on the prime field:

(15.8) Proposition. *An absolute value $\|\ \|$ on a field F is ultrametric if and only if $\|n1_F\| \leq 1$ for all positive integers n.*

Proof. If $\|\ \|$ is ultrametric, $\|n1_F\| \leq \|1_F\| = 1$ by induction on n. Conversely, if $\|n1_F\| \leq 1$ for all integers $n > 0$, then for each integer $m > 0$ and all $x, y \in F$,

$$\|x + y\|^m \ \leq \ \sum_{r=0}^{m} \left\| \binom{m}{r} 1_F \right\| \ \|x\|^r \ \|y\|^{m-r}$$
$$\leq \ (m+1) \max\{\|x\|,\ \|y\|\}^m \ .$$

Now take mth roots and let $m \to \infty$. ∎

(15.9) Lemma. *If $\|\ \|$ is an ultrametric absolute value on a field F, then the ultrametric inequality $\|x+y\| \leq \max\{\|x\|,\ \|y\|\}$ becomes equality if $\|x\| \neq \|y\|$.*

Proof. Suppose $\|x\| < \|y\|$. Then $\|x + y\| \leq \|y\|$. But $\|y\| = \|x + y - x\| \leq \max\{\|x+y\|,\ \|-x\|\}$ and $\|-x\| = \|x\| < \|y\|$; so $\|x+y\| \geq \|y\|$. ∎

This provides a rather strange geometry on an ultrametric space (F, d): Every triangle is isosceles, and every interior point of a circle is a center of the circle (see Exercise 2).

Suppose $(F, \|\ \|)$ is a valued field. A sequence $\{x_i\}$ in F is **Cauchy** if, for each real $\varepsilon > 0$, $\|x_i - x_j\| < \varepsilon$ for all but finitely many pairs i, j. A sequence $\{x_i\}$ in F **converges** to x in F (or x is a **limit** of $\{x_i\}$) if, for each real $\varepsilon > 0$, $\|x_i - x\| < \varepsilon$ for all but finitely many i. By the triangle inequality, every convergent sequence is Cauchy, and no sequence can converge to two different limits. The latter justifies the notation $\lim x_i = x$ for the assertion that $\{x_i\}$ converges to x.

(15.10) Lemma. *If $(F, \| \ \|)$ is a valued field, then each pair $x, y \in F$ satisfy the inequality $| \ \|x\| - \|y\| \ | \leq \|x - y\|$.*

Proof. By the triangle inequality

$$
\begin{aligned}
\|x\| &\leq \|x - y\| + \|y\| \quad \text{and} \\
\|y\| &\leq \|y - x\| + \|x\| .
\end{aligned}
$$

Solve for $\|x - y\| = \|y - x\|$. ∎

So the absolute value is a continuous map of metric spaces from $(F, \| \ \|)$ to $(\mathbb{R}^+ \cup \{0\}, | \ |)$, for it preserves limits: If $\lim x_i = x$ in $(F, \| \ \|)$, then $\lim \|x_i - x\| = 0$ in $\mathbb{R}$; by (15.10), $\lim \|x_i\| = \|x\|$ in $\mathbb{R}$. Another consequence of (15.10) is that if $\{x_i\}$ is a Cauchy sequence in $(F, \| \ \|)$, then $\{\|x_i\|\}$ is Cauchy in $(\mathbb{R}, | \ |)$. But all Cauchy sequences in $\mathbb{R}$ converge. So we have:

(15.11) Lemma. *If $\{x_i\}$ is a Cauchy sequence in a valued field $(F, \| \ \|)$, then $\lim \|x_i\| = x$ for some $x \in \mathbb{R}^+ \cup \{0\}$.* ∎

(15.12) Definition. A valued field $(F, \| \ \|)$ is **complete** if every Cauchy sequence in F converges in F.

As we just remarked, the field $\mathbb{R}$, with its standard absolute value, is complete. Likewise, $\mathbb{C}$, with its standard absolute value, is complete: If $\{a_n + b_n i\}$ is Cauchy in $\mathbb{C}$, then $\{a_n\}$ and $\{b_n\}$ are Cauchy in $\mathbb{R}$; so $\lim a_n = a$ and $\lim b_n = b$ in $\mathbb{R}$, and $\lim(a_n + b_n i) = a + bi$ in $\mathbb{C}$. The field $\mathbb{Q}$ is not complete under the standard absolute value from $\mathbb{R}$, since every irrational number is the limit of its decimal expansion. For instance, π is the limit of a sequence 3, 3.1, 3.14, 3.141, and so on.

The relationship between $\mathbb{Q}$ and $\mathbb{R}$ is instructive. To make $\mathbb{Q}$ complete by enlarging it within $\mathbb{R}$, we must throw in every real number. So $\mathbb{R}$ is a "completion" of $\mathbb{Q}$. The field $\mathbb{Q}$ is a subfield of the complete field $\mathbb{R}$, and the absolute values on $\mathbb{Q}$ and $\mathbb{R}$ agree on $\mathbb{Q}$.

(15.13) Definitions. An **isometric embedding** from a valued field $(F, \| \ \|_F)$ to a valued field $(E, \| \ \|_E)$ is a ring homomorphism $\sigma : F \to E$ with $\|\sigma(x)\|_E = \|x\|_F$ for all $x \in F$. There is a category $\mathcal{VF}$ with objects the valued fields and arrows the isometric embeddings.

Isometric embeddings are continuous and preserve Cauchy sequences since

$$
\begin{aligned}
\|\sigma(x_i) - \sigma(x)\|_E &= \|x_i - x\|_F , \quad \text{and} \\
\|\sigma(x_i) - \sigma(x_j)\|_E &= \|x_i - x_j\|_F .
\end{aligned}
$$

An isomorphism (invertible arrow) in $\mathcal{VF}$ is an **isometric isomorphism**. Each surjective isometric embedding is an isometric isomorphism — for σ is then an isomorphism in $\mathfrak{Ring}$ and $\|\sigma^{-1}(y)\|_F = \|\sigma(\sigma^{-1}(y))\|_E = \|y\|_E$ for all y in E. Every isometric embedding $\sigma : F \to E$ corresponds to an isometric isomorphism $F \cong \sigma(F)$, where $\sigma(F)$ is a subfield of E and the absolute value on $\sigma(F)$ restricts that on E.

(15.14) More Definitions. Suppose F and E are valued fields. If $S \subseteq E$, its **closure** $\overline{S}$ in E is the set of limits in E of sequences in S. Then S is **dense** in E if $\overline{S} = E$. An isometric embedding $\sigma : F \to E$ is a **metric completion** of $(F, \|\ \|_F)$ if $(E, \|\ \|_E)$ is complete and $\sigma(F)$ is dense in E. If there is such a σ, we also say $(E, \|\ \|_E)$ is a **metric completion** of $(F, \|\ \|_F)$. A sequence $\{x_i\}$ in F is **null** if $\lim x_i = 0$, or equivalently, if $\lim \|x_i\|_F = 0$.

Next we establish the existence of a metric completion for each valued field.

(15.15) Theorem. *In a valued field* $(F, \|\ \|)$, *the Cauchy sequences form a commutative ring* R *under*

$$\{x_i\} + \{y_i\} \ = \ \{x_i + y_i\} \quad and \quad \{x_i\}\{y_i\} \ = \ \{x_iy_i\} \ .$$

The null sequences form an ideal J *of* R. *The quotient* $\widehat{F} = R/J$ *is a field, complete with respect to the absolute value* $\|\ \|\widehat{\ }$ *with*

$$\|\{x_i\} + J\|\widehat{\ } \ = \ \lim \|x_i\| \ .$$

The map taking $x \in F$ *to the coset of the constant sequence* $\{x\}$ *is an isometric embedding of* $(F, \|\ \|)$ *into* $(\widehat{F}, \|\ \|\widehat{\ })$ *with dense image. So* $(\widehat{F}, \|\ \|\widehat{\ })$ *is a metric completion of* $(F, \|\ \|)$.

Proof. If $\{x_i\}$ and $\{y_i\}$ are Cauchy sequences in F, then $\{x_i + y_i\}$ is Cauchy, since for each $\varepsilon > 0$ and all but finitely many i, j,

$$\|(x_i + y_i) - (x_j + y_j)\| \ \le \ \|x_i - x_j\| \ + \ \|y_i - y_j\| \ < \ \frac{\varepsilon}{2} + \frac{\varepsilon}{2} \ = \ \varepsilon \ .$$

Also, $\{x_iy_i\}$ is Cauchy, since

$$\begin{aligned}
\|x_iy_i - x_jy_j\| \ &= \ \|x_iy_i - x_iy_j + x_iy_j - x_jy_j\| \\
&\le \ \|x_i\|\,\|y_i - y_j\| \ + \ \|x_i - x_j\|\,\|y_j\| \ ,
\end{aligned}$$

and the outer two factors are eventually less than their limits plus 1, while the inner two can be made as small as we please. For each $a \in F$, the constant sequence $\{a\}$ is Cauchy. So if $\{x_i\}$ is Cauchy, then $\{-x_i\} = \{-1\}\{x_i\}$ is also

Cauchy. Thus the Cauchy sequences in F form a commutative ring R with identity elements $\{0\}$ and $\{1\}$.

Since every convergent sequence is Cauchy, the null sequences in F belong to R. If $\{x_i\}$ and $\{y_i\}$ are null, then $\|x_i + y_i\| \leq \|x_i\| + \|y_i\|$ implies their sum is null. If $\{x_i\}$ is Cauchy with $\lim \|x_i\| = x \in \mathbb{R}$ and $\{y_i\}$ is null, then $\lim \|x_i y_i\| = x \cdot 0 = 0$; so $\{x_i\}\{y_i\}$ is null. Since $\{0\}$ is null, the set J of null sequences in F is not empty. Thus J is an ideal of R.

The quotient ring $\widehat{F} = R/J$ is a field: Since $\{1\}$ is not null, $\{1\}+J \neq \{0\}+J$. And if $\{x_i\} + J \neq \{0\} + J$, then $\{x_i\}$ is not null, $\lim \|x_i\| = x > 0$, and $x_i \neq 0$ for all but finitely many i. So we can create a sequence $\{y_i\}$ in F with $x_i y_i = 1$ for all but finitely many i. Past those i,

$$\|y_i - y_j\| \;=\; \|\frac{1}{x_i} - \frac{1}{x_j}\| \;=\; \frac{\|x_j - x_i\|}{\|x_i\|\,\|x_j\|} \, ,$$

and the denominator converges to $x^2 > 0$ as $i, j \to \infty$; so it is eventually bounded away from 0. Since $\{x_i\}$ is Cauchy, this shows $\{y_i\}$ is Cauchy too. Since $\{x_i y_i - 1\} = \{x_i\}\{y_i\} - \{1\}$ is null, $\{x_i\} + J$ has multiplicative inverse $\{y_i\} + J$ in $\widehat{F}$.

Suppose $\{x_i\}, \{y_i\} \in R$, so $\lim \|x_i\| = x$ and $\lim \|y_i\| = y$ in $\mathbb{R}$. If $\{x_i\}+J = \{y_i\} + J$, then $\{x_i - y_i\}$ is null and $\lim \|x_i - y_i\| = 0$. From (15.10) it follows that $\lim(\|x_i\| - \|y_i\|) = 0$; so $x = y$. Thus there is an unambiguous map

$$\| \ \|\hat{} : \widehat{F} \to \mathbb{R}^+ \cup \{0\}$$

defined by

$$\|\{x_i\} + J\|\hat{} \;=\; \lim \|x_i\| \,.$$

Since $\{x_i\}$ is null if and only if $\lim \|x_i\| = 0$, the map $\| \ \|\hat{}$ has property (i) of an absolute value. Since $\lim \|x_i y_i\| = \lim \|x_i\| \lim \|y_i\|$, it also has property (ii). And $\lim \|x_i + y_i\| \leq \lim \|x_i\| + \lim \|y_i\|$ shows property (iii) is also true, and $\| \ \|\hat{}$ is an absolute value on $\widehat{F}$.

For each $x \in F$, let $\widehat{x}$ denote the coset $\{x\} + J$ of the constant sequence $\{x\}$. Taking x to $\widehat{x}$ defines a ring homomorphism $\sigma : F \to \widehat{F}$, which is an isometric embedding since $\|\widehat{x}\| = \lim \|x\| = \|x\|$.

Next we show $\sigma(F)$ is dense in $\widehat{F}$ by showing each $\{x_i\} + J = \lim \widehat{x}_i$ in $\widehat{F}$: If $\varepsilon > 0$, then for some n, $\|x_i - x_j\| < \varepsilon$ for all $i, j > n$. So for each $j > n$,

$$\|\widehat{x}_j \;-\; (\{x_i\} + J)\|\hat{} \;=\; \lim_i \|x_j - x_i\| \;<\; \varepsilon \,,$$

as required.

Finally we show $(\widehat{F}, \| \ \|\hat{})$ is complete. Suppose $\{a_i\}$ is a Cauchy sequence in $\widehat{F}$. For each n choose $x_n \in F$ with $\|\widehat{x}_n - a_n\|\hat{} < 1/n$. Suppose $\varepsilon > 0$. For all but finitely many i and j, $\|\widehat{x}_i - a_i\|\hat{}, \|a_i - a_j\|\hat{}$ and $\|a_j - \widehat{x}_j\|\hat{}$ are less than $\varepsilon/3$. By the triangle inequality, $\|x_i - x_j\| = \|\widehat{x}_i - \widehat{x}_j\|\hat{} < \varepsilon$. So $\{x_i\}$ is

Cauchy in $(F, \| \ \|)$. By the choice of x_i, the sequence $\{a_i - \widehat{x}_i\}$ is null in $\widehat{F}$. Since $\{\widehat{x}_i\}$ converges to $\{x_i\} + J$, the sum $\{a_i\} = \{a_i - \widehat{x}_i\} + \{\widehat{x}_i\}$ also converges to $\{x_i\} + J$.   ■

Toward uniqueness of completions we have:

(15.16) Theorem. *Suppose* $\sigma : F \to E$ *and* $\tau : F \to K$ *are isometric embeddings of a valued field* $(F, \| \ \|)$ *into complete valued fields* $(E, \| \ \|_E)$ *and* $(K, \| \ \|_K)$, *and* $\sigma(F)$ *is dense in* E. *Then there is one and only one isometric embedding* $\rho : E \to K$ *with* $\rho \circ \sigma = \tau$.

Proof. The restrictions $s : F \to \sigma(F)$ and $t : F \to \tau(F)$ of σ and τ are isometric isomorphisms. If $\{x_i\}$ and $\{y_i\}$ are sequences in F, applying the isometric embedding $t \circ s^{-1}$ shows $\lim \sigma(x_i) = \lim \sigma(y_i)$ implies $\lim \tau(x_i) = \lim \tau(y_i)$. Each element of E is $\lim \sigma(x_i)$ for some sequence $\{x_i\}$ in F. So there is a function $\rho : E \to K$ taking u to v whenever there is a sequence $\{x_i\}$ in F with $\lim \sigma(x_i) = u$ and $\lim \tau(x_i) = v$.

Since limits preserve sums and products, and ρ takes $\lim \sigma(1_F)$ to $\lim \tau(1_F)$, ρ is a ring homomorphism. Suppose $\lim \sigma(x_i) = u$ and $\lim \tau(x_i) = v$. By (15.10), $\lim \|\sigma(x_i)\|_E = \|u\|_E$ and $\lim \|\tau(x_i)\|_K = \|v\|_K$. Then

$$\|\rho(u)\|_K \ = \ \|v\|_K \ = \ \lim \ \|t \circ s^{-1} \circ \sigma \ (x_i)\|_K$$
$$= \ \lim \ \|\sigma(x_i)\|_E \ = \ \|u\|_E \ ,$$

proving ρ is an isometric embedding. Since $\sigma(F)$ is dense in E and isometric embeddings preserve limits, there is only one isometric embedding $\rho : E \to K$ with $\rho \circ \sigma = \tau$.   ■

(15.17) Corollary. *If* $\sigma : F \to E$ *and* $\tau : F \to K$ *are metric completions of a valued field* $(F, \| \ \|)$, *there is exactly one isometric isomorphism* $\rho : E \to K$ *with* $\rho \circ \sigma = \tau$.   ■

(15.18) Corollary. *If* $\tau : F \to K$ *is an isometric embedding of valued fields and* K *is complete, then the closure* $\overline{\tau(F)}$ *of* $\tau(F)$ *in* K *is a metric completion of* F.

Proof. By (15.15) there is a metric completion $\sigma : F \to E$ of F. By (15.16) there is an isometric embedding $\rho : E \to K$ with $\rho \circ \sigma = \tau$. So $x \mapsto \rho(x)$ defines an isometric isomorphism from E to $\rho(E)$, and $\rho(E)$ is a complete valued field under the absolute value from K. Since $\sigma(F)$ is dense in E and ρ preserves limits, $\rho(\sigma(F)) = \tau(F)$ is dense in $\rho(E)$.   ■

Let's apply these ideas to describe metric completions of the valued fields F in Example (15.6). Suppose first that $\tau : F \to \mathbb{C}$ is a ring homomorphism.

Then τ is an isometric embedding of $(F, | \ |_\tau)$ into $\mathbb{C}$. Since $\mathbb{C}$ is complete, the closure $\overline{\tau(F)}$ of $\tau(F)$ in $\mathbb{C}$ is a metric completion of $(F, | \ |_\tau)$. Being a subfield of $\mathbb{C}$, $\tau(F)$ contains $\mathbb{Q}$; so $\overline{\tau(F)}$ contains $\mathbb{R}$. Then the field $\overline{\tau(F)}$ is either $\mathbb{R}$ or $\mathbb{C}$. If τ is a real embedding $(\tau(F) \subseteq \mathbb{R})$, then $\overline{\tau(F)} = \mathbb{R}$. If instead τ is an imaginary embedding $(\tau(F) \not\subseteq \mathbb{R})$, then $\overline{\tau(F)} = \mathbb{C}$.

Now suppose v is a discrete valuation on F, $0 < a < 1$, and $0 < b < 1$. The function a^x takes $\mathbb{R}^+$ onto the interval $(0, 1)$; so $a^c = b$ for some $c > 0$. Then $| \ |_{v,b} = | \ |_{v,a}^c$. The two absolute values $| \ |_{v,a}$ and $| \ |_{v,b}$ yield the same Cauchy, convergent, and null sequences in F, and the same metric topology. So the field $\widehat{F}$ constructed in (15.15) is independent of the choice of exponential base a in $| \ |_{v,a}$, and we use the notation $\widehat{\mathbf{F}}_v$ for this field. The absolute values $| \ |_{\widehat{v},a}$ on $\widehat{F}_v$ constructed in (15.15) do depend on the choice of a, in the same way as their restrictions : $| \ |_{\widehat{v},b} = | \ |_{\widehat{v},a}^c$. So $\widehat{F}_v$ has the same Cauchy sequences, limits, and metric topology under every choice of a.

(15.19) Proposition. *The discrete valuation v on F extends to a discrete valuation $\widehat{v}$ on $\widehat{F}_v$, so that $| \ |_{\widehat{v},a} = | \ |_{\widehat{v},a}$ for each a with $0 < a < 1$.*

Proof. Fix a and write $| \ |_v$ for $| \ |_{v,a}$ and $| \ |_{\widehat{v}}$ for $| \ |_{\widehat{v},a}$. Recall that the value group

$$|F^*|_v = \langle a \rangle = \{a^m : m \in \mathbb{Z}\} .$$

Except for 0, every real number is the center of an open interval containing at most one of the values a^m. So if $\{x_i\}$ is a Cauchy sequence in $(F, | \ |_v)$ that is not null, the convergent sequence $\{|x_i|_v\}$ is eventually constant:

$$|x_n|_v = |x_{n+1}|_v = |x_{n+2}|_v = \cdots = a^m$$

for some m and n. Then $|\{x_i\} + J|_{\widehat{v}} = \lim |x_i|_v = a^m$, proving $(\widehat{F}_v, | \ |_{\widehat{v}})$ also has value group $\langle a \rangle$.

Of course there is an isometric embedding from $(F, | \ |_v)$ to $(\widehat{F}_v, | \ |_{\widehat{v}})$. By (15.8), $| \ |_v$ ultrametric implies $| \ |_{\widehat{v}}$ is ultrametric. So

$$\widehat{v} = \log_a | \ |_{\widehat{v}}$$

is a discrete valuation on $\widehat{F}$ extending v on F. In fact, $\widehat{v}(\{x_i\} + J) = v(x_n)$, the eventually constant value of the sequence $\{v(x_i)\}$. Applying a^x to the equation defining $\widehat{v}$ yields $| \ |_{\widehat{v},a} = | \ |_{\widehat{v},a}$. ∎

15B. Exercises

1. If the value group of an absolute value $\| \ \|$ on a field F is bounded on the prime field of F, prove it is ultrametric. Use this to show a field with an archimedean absolute value must have characteristic 0.

2. If $\|\ \|$ is an ultrametric absolute value on a field F, prove that for all $x, y, z \in F$, at least two of $\|x - y\|$, $\|x - z\|$, $\|y - z\|$ must be equal. If $r \in \mathbb{R}^+$ and $C = \{y \in F : \|x - y\| = r\}$, we can think of C as a circle with radius r. If z is interior to this circle (meaning $\|x - z\| < r$), show $\|z - y\| = r$ for all $y \in C$. So z is a "center" of the circle.

3. Suppose $(F, \|\ \|)$ is a valued field. Find a category in which the initial objects are the metric completions of $(F, \|\ \|)$. Use this to give an alternate proof of (15.17).

4. If $\|\ \|_1$ and $\|\ \|_2$ are nontrivial absolute values on a field F, prove the following are equivalent:

    (i) F has the same metric topology under $\|\ \|_1$ and $\|\ \|_2$.

    (ii) F has the same open unit ball under $\|\ \|_1$ and $\|\ \|_2$; that is, $\|x\|_1 < 1$ if and only if $\|x\|_2 < 1$.

    (iii) F has the same closed unit ball under $\|\ \|_1$ and $\|\ \|_2$; that is, $\|x\|_1 \leq 1$ if and only if $\|x\|_2 \leq 1$.

    (iv) There is a positive real number c with $\|\ \|_2 = \|\ \|_1^c$.

Hint: For (i) implies (ii), $\|x\| < 1$ is equivalent to $\lim \|x\|^n = 0$. For (ii) implies (iii), if $\|y\|_1 = 1$ and $\|x\|_1 < 1$, show $\|y\|_2 < \|x\|_2^{-1/n}$ for all positive integers n; now take $n \to \infty$ to get $\|y\|_2 \leq 1$. For (iii) implies (iv), show

$$c = \frac{\ln\|x\|_2}{\ln\|x\|_1}$$

is independent of $x \in F^*$ by proving

$$\frac{\ln\|y\|_1}{\ln\|x\|_1} = \frac{\ln\|y\|_2}{\ln\|x\|_2}$$

for $x, y \in F^*$; and this is shown by proving the same rational numbers exceed both. To prove (iv) implies (i), just note that x^c maps $\mathbb{R}^+$ onto $\mathbb{R}^+$.

5. Among the nontrivial absolute values on a field F, say $\|\ \|_1 \sim \|\ \|_2$ if any (hence all) of the assertions (i)–(iv) in Exercise 4 are true. Show $\sim$ is an equivalence relation. If $\|\ \|_1$ is ultrametric, show every $\|\ \|_2$ equivalent to $\|\ \|_1$ is also ultrametric. An equivalence class of nontrivial absolute values on F is sometimes called a **place** of F. Note that each discrete valuation v on F determines a single ultrametric place $\{|\ |_{v,a} : 0 < a < 1\}$ of F and a single ultrametric place $\{|\ |_{\widehat{v,a}}\} : 0 < a < 1\}$ of the completion $\widehat{F}_v$.

6. If $\|\ \|$ is an ultrametric absolute value on a field F and the value group $\|F^*\|$ is discrete — meaning every member is the only member in some open interval, prove $\|F^*\|$ is infinite cyclic and $\|\ \| = |\ |_{v,a}$ for some discrete valuation v on F and real number a with $0 < a < 1$; and show both v and a are uniquely determined by the value group $\|F^*\|$.

7. If a valued field $(F, \| \ \|_F)$ is already complete, prove any completion $(F, \| \ \|_F) \to (E, \| \ \|_E)$ is an isometric isomorphism.

8. Show the only absolute value on a finite field is the trivial one ($\|F^*\| = \{1\}$). Is F complete with respect to this absolute value?

15C. The p-Adic Numbers and Quadratic Reciprocity

A common perception is that the field $\mathbb{Q}$ of rational numbers is just a part of the field $\mathbb{R}$ of real numbers, but that is not the only place to put it. For each prime number p there is a complete valued field $\mathbb{Q}_p$ with $\mathbb{Q}$ as a dense subfield. The p-adic discrete valuation on $\mathbb{Q}^*$ extends to a discrete valuation on $\mathbb{Q}_p^*$. Using this extension we get a tame symbol from $\mathbb{Q}_p^* \times \mathbb{Q}_p^*$ to $(\mathbb{Z}/p\mathbb{Z})^*$; following it by the $(p-1)/2$ power map, we produce a Hilbert symbol on $\mathbb{Q}_p$ if p is odd. On $\mathbb{Q}_2$ we construct a Hilbert symbol directly. On $\mathbb{Q}^* \times \mathbb{Q}^*$, the product of the Hilbert symbols over all primes p turns out to be the real symbol. This product formula (15.32) is Hilbert's extension of the law of quadratic reciprocity. Our proof of it is due to Tate, combining the first proof by Gauss of quadratic reciprocity with the computation of $K_2(\mathbb{Q})$ via tame symbols (14.56).

We continue the notation of the preceding section, and we use freely the basic facts about discrete valuations v, discrete valuation rings $\mathcal{O}_v$, and residue fields k_v, as laid out in (7.25)–(7.26) and the first two pages of §14F.

For each prime number p (> 0) there is a p-adic discrete valuation $v = v_p :$ $\mathbb{Q}^* \to \mathbb{Z}$, where $v(a)$ is the exponent of p in the prime factorization of a. The standard absolute value on $\mathbb{Q}$ associated with this v is $| \ |_p = | \ |_{v,1/p}$ defined by

$$|x|_p = \begin{cases} 0 & \text{if } x = 0 \\ p^{-v(x)} & \text{if } x \neq 0 \ . \end{cases}$$

The metric completion of $(\mathbb{Q}, | \ |_p)$, constructed in (15.15), will be denoted by $(\mathbb{Q}_p, | \ \widehat{|_p})$; here $\mathbb{Q}_p = \widehat{\mathbb{Q}}_v$ is the **field of p-adic numbers**. Recall from (15.15) that $\mathbb{Q}_p = R/J$, where R is the commutative ring of Cauchy sequences in $(\mathbb{Q}, | \ |_p)$ and J is the ideal consisting of the null sequences. So each **p-adic number** is a coset $\overline{\{x_i\}} = \{x_i\} + J$ of a sequence $\{x_i\}$ in $\mathbb{Q}$ that is Cauchy with respect to $| \ |_p$. The absolute value $| \ \widehat{|_p}$ from (15.15) is defined by

$$|\overline{\{x_i\}}| \widehat{|_p} = \lim |x_i|_p \ ,$$

and the sequence $\{|x_i|_p\}$ of real numbers is eventually constant if $\overline{\{x_i\}} \neq 0$.

We shall identify each $x \in \mathbb{Q}$ with the coset $\{x\} + J$ of the constant sequence $\{x\}$, so that $\mathbb{Q}$ becomes a subfield of $\mathbb{Q}_p$. Then $| \ \widehat{|_p}$ extends $| \ |_p$, and there is a discrete valuation $\widehat{v} : \mathbb{Q}_p^* \to \mathbb{Z}$ extending $v : \mathbb{Q}^* \to \mathbb{Z}$, with

$$\widehat{v}(y) = \log_{1/p} |y| \widehat{|_p} \quad \text{and} \quad |y| \widehat{|_p} = p^{-\widehat{v}(y)}$$

for all $y \in \mathbb{Q}_p^*$. If $y = \overline{\{x_i\}}$ for a Cauchy sequence $\{x_i\}$, it follows that $\hat{v}(y)$ is the eventually constant value of $v(x_i)$.

For a useful description of p-adic numbers, recall that $\mathbb{Q}_p$ is the field of fractions of its discrete valuation ring $\mathcal{O}_{\hat{v}}$, which we denote by $\hat{\mathbb{Z}}_p$. A member of $\hat{\mathbb{Z}}_p$ is a **p-adic integer**; it is just a p-adic number y with $y = 0$ or $\hat{v}(y) \geq 0$. So $\hat{\mathbb{Z}}_p$ is the unit ball:

$$\hat{\mathbb{Z}}_p = \{y \in \mathbb{Q}_p : |y|\hat{_p} \leq 1\},$$

and $\hat{\mathbb{Z}}_p^* = \hat{v}^{-1}(0)$ is the unit sphere:

$$\hat{\mathbb{Z}}_p^* = \{y \in \mathbb{Q}_p : |y|\hat{_p} = 1\}.$$

Like every discrete valuation ring, $\hat{\mathbb{Z}}_p$ is a local principal ideal domain. Since $\hat{v}(p) = v(p) = 1$, p is irreducible in $\hat{\mathbb{Z}}_p$, and each $a \in \mathbb{Q}_p^*$ has a unique expression $a = up^n$ with $u \in \hat{\mathbb{Z}}_p^*$ and $n \in \mathbb{Z}$. Then $\hat{v}(a) = n$. Notice that $\mathbb{Q} \cap \hat{\mathbb{Z}}_p$ is the discrete valuation ring $\mathcal{O}_v$, which is the localization $\mathbb{Z}_p = (\mathbb{Z} - p\mathbb{Z})^{-1}\mathbb{Z}$ of $\mathbb{Z}$ and consists of the fractions a/b with $a, b \in \mathbb{Z}$ and $b \notin p\mathbb{Z}$.

(15.20) Proposition.

    (i)   With respect to $|\ |_p$, $\mathbb{Z}$ is dense in $\mathbb{Z}_p$.

    (ii) The p-adic integers are the cosets $\overline{\{x_i\}}$ of Cauchy sequences $\{x_i\}$ in $\mathbb{Z}$ under $|\ |_p$.

Proof. For (i) it suffices to prove that if $x \in \mathbb{Z}_p$ and $\varepsilon > 0$, there is an integer m with $|x - m|_p < \varepsilon$. Write x as a/b, where $a, b \in \mathbb{Z}$ and $b \notin p\mathbb{Z}$. Choose a positive integer n with $p^{-n} < \varepsilon$. Since b and p^n are relatively prime, the Chinese Remainder Theorem implies the cosets $a + b\mathbb{Z}$ and $0 + p^n\mathbb{Z}$ overlap. Choose $m \in \mathbb{Z}$ so $a - bm$ lies in the intersection. Then

$$\left|\frac{a}{b} - m\right|_p = \left|\frac{a - bm}{b}\right|_p \leq \left(\frac{1}{p}\right)^n < \varepsilon.$$

For (ii), if $\{x_i\}$ is a Cauchy sequence in $\mathbb{Z}$ with respect to $|\ |_p$, either $\overline{\{x_i\}} = 0$ or $\hat{v}(\overline{\{x_i\}})$ is the eventually constant $v(x_i) \geq 0$. So $\overline{\{x_i\}} \in \hat{\mathbb{Z}}_p$. Conversely, suppose $\{x_i\}$ is a Cauchy sequence in $(\mathbb{Q}, |\ |_p)$ and $\overline{\{x_i\}} \in \hat{\mathbb{Z}}_p$. If $\overline{\{x_i\}} \neq \overline{\{0\}}$, then $\hat{v}(\overline{\{x_i\}}) = q \geq 0$. So for all but finitely many positive integers i, $v(x_i) = q$ and $x_i \in \mathbb{Z}_p$. For these i, choose $m_i \in \mathbb{Z}$ with $|m_i - x_i|_p < 1/i$. For all other i, take $m_i = 0$. Then $\{m_i\}$ is the sum of the Cauchy sequence $\{x_i\}$ and the null sequence $\{m_i - x_i\}$; so $\{m_i\}$ is Cauchy and $\overline{\{x_i\}} = \overline{\{m_i\}}$. ∎

The description of p-adic integers as cosets of sequences is still a bit messy. But there is a neater description. Suppose $\{x_i\}$ is a Cauchy sequence in $\mathbb{Z}$

with respect to $|\ |_p$. Then for each positive integer n, one and only one coset $\overline{a_n} \in \mathbb{Z}/p^n\mathbb{Z}$ contains x_i for all but finitely many i. In this way $\{x_i\}$ determines a sequence $(\overline{a_1}, \overline{a_2}, ...)$ with $a_n \in \mathbb{Z}$ and $\overline{a_n} \in \mathbb{Z}/p^n\mathbb{Z}$ for each n. If $m < n$, since $x_i \in \overline{a_n}$ for all but finitely many i, the coset in $\mathbb{Z}/p^m\mathbb{Z}$ containing $\overline{a_n} = a_n + p^n\mathbb{Z}$ must be $\overline{a_m} = a_m + p^m\mathbb{Z}$. So the canonical ring homomorphism

$$f : \mathbb{Z}/p^n\mathbb{Z} \ \rightarrow \ \mathbb{Z}/p^m\mathbb{Z}\,, \quad a + p^n\mathbb{Z} \ \mapsto \ a + p^m\mathbb{Z}\,,$$

takes $\overline{a_n}$ to $\overline{a_m}$.

(15.21) Definitions. If $m < n$, $x \in \mathbb{Z}/p^n\mathbb{Z}$, $y \in \mathbb{Z}/p^m\mathbb{Z}$, and $f(x) = y$, say x **goes to** y, or y **comes from** x. Denote by **Seq**(p) the commutative ring whose elements are the sequences $(c_1, c_2, ...)$ with $c_n \in \mathbb{Z}/p^n\mathbb{Z}$ for each n, and whose addition and multiplication are calculated entrywise. Then

$$\varprojlim \ \mathbb{Z}/p^n\mathbb{Z}$$

is the subring of Seq(p) consisting of those sequences $(c_1, c_2, ...)$ in which c_n goes to c_m whenever $m < n$.

The ring $\varprojlim \mathbb{Z}/p^n\mathbb{Z}$ is an "inverse limit." For a general definition of inverse limits, see Exercise 1.

(15.22) Proposition. *There is a ring isomorphism* $\widehat{\mathbb{Z}}_p \cong \varprojlim \ \mathbb{Z}/p^n\mathbb{Z}$ *taking* $\overline{\{x_i\}}$ *to* $(\overline{a_1}, \overline{a_2}, ...)$ *where, for each* n, $\overline{a_n} \in \mathbb{Z}/p^n\mathbb{Z}$ *contains* x_i *for all but finitely many* i.

Proof. The map $R \rightarrow \varprojlim \ \mathbb{Z}/p^n\mathbb{Z}$ taking $\{x_i\}$ to $(\overline{a_1}, \overline{a_2}, ...)$ is a ring homomorphism, with kernel the ideal J of null sequences, and is surjective because it takes $\{a_i\}$ to $(\overline{a_1}, \overline{a_2}, ...)$. ∎

We shall tend to denote each p-adic integer by its associated sequence of cosets $(\overline{a_1}, \overline{a_2}, ...)$. Each ordinary integer $a \in \mathbb{Z}$ then becomes $(\overline{a}, \overline{a}, ...)$. Since $\widehat{\mathbb{Z}}_p$ is a local Dedekind domain with maximal ideal $p\widehat{\mathbb{Z}}_p$, every ideal except $\widehat{\mathbb{Z}}_p$ itself has the form $(p\widehat{\mathbb{Z}}_p)^m = p^m\widehat{\mathbb{Z}}_p$ for a positive integer m. And

$$p^m\widehat{\mathbb{Z}}_p \ = \ \{(\overline{a_1}, \overline{a_2}, ...) \ \in \ \widehat{\mathbb{Z}}_p \ : \ \overline{a_n} = \overline{0} \ \text{for} \ n \le m\}\,,$$

since the latter is an ideal of $\widehat{\mathbb{Z}}_p$ containing p^m but not p^{m-1}. Projection to the nth coordinate $\widehat{\mathbb{Z}}_p \rightarrow \mathbb{Z}/p^n\mathbb{Z}$ is a surjective ring homomorphism extending the canonical map $\mathbb{Z} \rightarrow \mathbb{Z}/p^n\mathbb{Z}$ and may be thought of as reduction mod p^n, since its kernel is $p^n\widehat{\mathbb{Z}}_p$. So two p-adic integers are considered **congruent modulo**

p^n if and only if they have the same nth coordinate (and hence the same first n coordinates).

Projection to the first coordinate $\widehat{\mathbb{Z}}_p \to \mathbb{Z}/p\mathbb{Z}$ induces an isomorphism of rings

$$k_{\widehat{v}} = \frac{\widehat{\mathbb{Z}}_p}{p\widehat{\mathbb{Z}}_p} \cong \frac{\mathbb{Z}}{p\mathbb{Z}},$$

where $k_{\widehat{v}}$ is the residue field associated to the discrete valuation $\widehat{v}$ on $\mathbb{Q}_p$. So the cosets comprising $k_{\widehat{v}}$ are represented by the integers $0, 1, \ldots, p-1$.

Now we turn to the task of showing there is a Hilbert symbol on each p-adic field $\mathbb{Q}_p$. Recall from §15A that there is a Hilbert symbol $(-,-)_F$ on a field F if and only if the map $(-,-)_F : F^* \times F^* \to \mathbb{Z}$, defined by

$$(a,b)_F = \begin{cases} \cdot\, 1 & \text{if } ax^2 + by^2 = z^2 \text{ has a solution} \\ & (x,y,z) \neq (0,0,0) \text{ in } F^3 \\ \\ -1 & \text{if not}, \end{cases}$$

is multiplicative in both a and b. Whether $(-,-)_F$ is bimultiplicative or not, note that $(a,b)_F = (ac^2, bd^2)_F$ whenever $c, d \in F^*$, since $ax^2 + by^2 = z^2$ is equivalent to $ac^2(x/c)^2 + bd^2(y/d)^2 = z^2$.

(15.23) Definition and Notation. If A is a ring and G is a subgroup of $(A^*, \cdot)$, let G^2 denote the set $\{g^2 : g \in G\}$ of squares in G. Since G is abelian, G^2 is a normal subgroup of G; its cosets are the **square classes** in G.

So $(-,-)_F$ is determined by its values on representatives of the square classes in F^*. The same is true of every bimultiplicative function $F^* \times F^* \to \mathbb{Z}^*$.

Recall from a first course in algebra that if F is a field, a finite subgroup of $(F^*, \cdot)$ is always cyclic. In particular, if p is prime, $(\mathbb{Z}/p\mathbb{Z})^*$ is cyclic, generated by some $\overline{r} = r + p\mathbb{Z}$ with $r \in \mathbb{Z}$. Such an integer r is known as a "primitive root mod p."

(15.24) Lemma. *Suppose p is an odd prime and r is a primitive root mod p.*

(i) *For each positive integer n, $(\mathbb{Z}/p^n\mathbb{Z})^*$ has two square classes, represented by 1 and r.*

(ii) *A p-adic integer $(\overline{a_1}, \overline{a_2}, \ldots)$ is a square in $\widehat{\mathbb{Z}}_p^*$ if and only if $\overline{a_1}$ is a square in $(\mathbb{Z}/p\mathbb{Z})^*$.*

(iii) *There are exactly four square classes in $\mathbb{Q}_p^*$, represented by $1, r, p,$ and rp.*

Proof. Since $(\mathbb{Z}/p^n\mathbb{Z})^*$ is abelian, the squaring map is a group endomorphism of $(\mathbb{Z}/p^n\mathbb{Z})^*$ with kernel the set of $\overline{b} = b + p^n\mathbb{Z}$ with p^n dividing $b^2 - 1 = (b-1)(b+1)$. Since the latter factors differ by $2 < p$, p^n divides $b-1$ or $b+1$. So the kernel is $\{\overline{1}, -\overline{1}\}$. The image is the group of squares; so these squares occupy half of $(\mathbb{Z}/p^n\mathbb{Z})^*$ and there are two square classes, represented by $\overline{1}$ and any nonsquare. Since $\overline{r} \in (\mathbb{Z}/p\mathbb{Z})^*$ has even order $p-1$ and generates $(\mathbb{Z}/p\mathbb{Z})^*$, it is not a square there. So $\overline{r} = r + p^n\mathbb{Z}$ is not a square in $(\mathbb{Z}/p^n\mathbb{Z})^*$, proving (i).

For (ii), note first that a p-adic integer $(\overline{a_1}, \overline{a_2}, ...)$ is in $\widehat{\mathbb{Z}}_p^*$, if and only if it is not in the maximal ideal $p\widehat{\mathbb{Z}}_p$, that is, if and only if $a_1, a_2, ...$ are not divisible by p. The first coordinate of a square in $\widehat{\mathbb{Z}}_p^*$ is therefore a square in $(\mathbb{Z}/p\mathbb{Z})^*$. For the converse, suppose $(\overline{a_1}, \overline{a_2}, ...)$ is a p-adic integer and $\overline{a_1} = \overline{b_1}^2$ in $(\mathbb{Z}/p\mathbb{Z})^*$. We construct the coordinates of a square root $(\overline{b_1}, \overline{b_2}, ...)$ of $(\overline{a_1}, \overline{a_2}, ...)$ in $\widehat{\mathbb{Z}}_p^*$ iteratively: Suppose b_n is an integer not divisible by p and $b_n^2 \equiv a_n \pmod{p^n}$. Since $a_{n+1} \equiv a_n \pmod{p^n}$ there are integers c and d with

$$a_{n+1} = a_n + cp^n \quad \text{and} \quad b_n^2 = a_n + dp^n .$$

Since p does not divide $2b_n$, the coset $\overline{2b_n} = 2b_n + p\mathbb{Z}$ is a unit in $\mathbb{Z}/p\mathbb{Z}$. Choose an integer e with $\overline{e} = \overline{2b_n}^{-1}\overline{(c-d)}$ in $\mathbb{Z}/p\mathbb{Z}$. Let

$$b_{n+1} = b_n + ep^n .$$

Then $b_{n+1} \notin p\mathbb{Z}$, $b_{n+1} \equiv b_n \pmod{p^n}$, and modulo p^{n+1},

$$b_{n+1}^2 = (b_n + ep^n)^2 \equiv b_n^2 + 2b_n ep^n$$
$$\equiv b_n^2 + (c-d)p^n = a_{n+1} .$$

So we can generate a p-adic integer $(\overline{b_1}, \overline{b_2}, ...)$ in $\widehat{\mathbb{Z}}_p^*$ whose square is $(\overline{a_1}, \overline{a_2}, ...)$.

Every element of $\mathbb{Q}_p^*$ has a unique expression up^n with $u \in \widehat{\mathbb{Z}}_p^*$ and $n \in \mathbb{Z}$. So every square in $\mathbb{Q}_p^*$ has the form $u^2 p^{2n}$. Since $\widehat{v}_p(r) = v_p(r) = 0$ and r (written as $(\overline{r}, \overline{r}, ...)$) is not a square in $\widehat{\mathbb{Z}}_p^*$, r and 1 are not in the same square class of $\mathbb{Q}_p^*$; neither are p and rp. Comparing p-adic values, 1 and r represent different square classes than p and rp. So $1, r, p$, and rp represent four different square classes in $\mathbb{Q}_p^*$.

It remains to show every element of $\mathbb{Q}_p^*$ belongs to one of these classes. Suppose $u = (\overline{u_1}, \overline{u_2}, ...)$ is in $\widehat{\mathbb{Z}}_p^*$. By (i), for each positive integer n, $\overline{u_n}$ is in the square class of $\overline{1}$ or $\overline{r}$ in $(\mathbb{Z}/p^n\mathbb{Z})^*$. Under each canonical map $\mathbb{Z}/p^n\mathbb{Z} \to \mathbb{Z}/p^m\mathbb{Z}$, units go to units, squares go to squares, $\overline{r}$ goes to $\overline{r}$, and $\overline{1}$ goes to $\overline{1}$. So for some $b \in \text{Seq}(p)^*$, $ub^2 \in \{1, r\}$. So $b^2 \in \{u^{-1}, u^{-1}r\} \subset \widehat{\mathbb{Z}}_p^*$. By (ii), b^2 is a square in $\widehat{\mathbb{Z}}_p^* \subset \mathbb{Q}_p^*$. So u is in the square class of 1 or r in $\mathbb{Q}_p^*$. And for each integer n, up^n is in the square class of $1, r, p$, or rp. ∎

Continue the assumption that p is an odd prime and r is a primitive root mod p. Recall the Legendre symbol, defined for each integer $b \notin p\mathbb{Z}$:

$$\left(\frac{b}{p}\right) = \left(\frac{b}{p}\right)_2 = \begin{cases} 1 & \text{if } \overline{b} \text{ is a square in } \mathbb{Z}/p\mathbb{Z} \\ -1 & \text{if not .} \end{cases}$$

The squares in $(\mathbb{Z}/p\mathbb{Z})^*$ are the even powers of $\bar{r}$, and $\bar{r}$ has order $p-1$. So

$$\left(\frac{b}{p}\right) \equiv b^{(p-1)/2} \pmod{p} .$$

In particular

$$\left(\frac{-1}{p}\right) = (-1)^{(p-1)/2} ,$$

since $1 \not\equiv -1 \pmod{p}$.

Now we evaluate the map

$$(-,-)_p = (-,-)_{\mathbb{Q}_p}$$

on square classes:

(15.25) Lemma. *Suppose p is an odd prime, r is a primitive root mod p, and*

$$\alpha = \left(\frac{-1}{p}\right) = (-1)^{(p-1)/2} .$$

The values of $(-,-)_p$ on square class representatives are

	1	r	p	rp
1	1	1	1	1
r	1	1	-1	-1
p	1	-1	α	$-\alpha$
rp	1	-1	$-\alpha$	α

Proof. If $a = 1$, $ax^2 + by^2 = z^2$ has solution $(1, 0, 1)$; so $(a, b)_p = (b, a)_p = 1$. In $\mathbb{Z}/p\mathbb{Z}$ there are $(p+1)/2$ squares and $(p-1)/2$ nonsquares. So for some integer s, $\bar{r} - \bar{s}^2$ is a square $\bar{t}^2$ in $\mathbb{Z}/p\mathbb{Z}$, with $t \in \mathbb{Z}$. Then $rs^2 + rt^2 \equiv r^2 \pmod{p}$. So $rs^2 + rt^2$ is a square in $\widehat{\mathbb{Z}}_p^*$ and $(r, r)_p = 1$.

If $a, b \in \mathbb{Q}_p^*$ and $ax^2 + by^2 = z^2$ has a solution $(x, y, z) \neq (0, 0, 0)$ in $\mathbb{Q}_p^3$, clearing denominators and dividing out common factors of p yields a solution in $\widehat{\mathbb{Z}}_p^3$ with x, y, z not all in $p\widehat{\mathbb{Z}}_p$. Call this a **primitive** solution.

If $rx^2 + py^2 = z^2$ has a primitive solution, then x and z are not in $p\widehat{\mathbb{Z}}_p$. Considering first coordinates, $\bar{r}$ would be a square in $(\mathbb{Z}/p\mathbb{Z})^*$; so $(r, p)_p = (p, r)_p = -1$. By the same reasoning, $(r, rp)_p = (rp, r)_p = -1$.

If $px^2 + py^2 = z^2$ has a primitive solution (x, y, z), then x and y are not in $p\widehat{\mathbb{Z}}_p$. Dividing the equation by p and taking first coordinates yields $\bar{s}^2 + \bar{t}^2 = \bar{0}$ for $\bar{s}, \bar{t} \in (\mathbb{Z}/p\mathbb{Z})^*$. So $-\bar{1} = (\bar{s}\bar{t}^{-1})^2$ and $\alpha = 1$. Conversely, if $\alpha = 1$, then $-\bar{1}$ is a square in $(\mathbb{Z}/p\mathbb{Z})^*$; so -1 is a square c^2 with $c \in \widehat{\mathbb{Z}}_p^*$, and $(c, 1, 0)$ is a solution. Thus $(p, p)_p = \alpha$. The same argument shows $(rp, rp)_p = \alpha$.

By the same reasoning, if $px^2 + rpy^2 = z^2$ for primitive (x, y, z), then $-\bar{r}$ is a square in $(\mathbb{Z}/p\mathbb{Z})^*$. Conversely, if $-\bar{r}$ is a square in $(\mathbb{Z}/p\mathbb{Z})^*$, $-r = c^2$ for $c \in \widehat{\mathbb{Z}}_p^*$, and $(c, 1, 0)$ is a solution. Since $-\bar{r}$ is a square in $(\mathbb{Z}/p\mathbb{Z})^*$ if and only if $-\bar{1}$ is not, $(p, rp)_p = (rp, p)_p = -\alpha$. ∎

(15.26) Theorem. *For each odd prime p, there is a Hilbert symbol $(-,-)_p$ on $\mathbb{Q}_p$. Identifying $\{\pm\bar{1}\}$ in the residue field $k_{\widehat{v}} = \widehat{\mathbb{Z}}_p/p\widehat{\mathbb{Z}}_p \cong \mathbb{Z}/p\mathbb{Z}$ with $\{\pm 1\} = \mathbb{Z}^*$,*

$$(a,b)_p \;=\; \tau_{\widehat{v}}(a,b)^{(p-1)/2} \,,$$

for all $a, b \in \mathbb{Q}_p^*$, where

$$\tau_{\widehat{v}} : \mathbb{Q}_p^* \times \mathbb{Q}_p^* \;\to\; k_{\widehat{v}}^*$$

is the tame symbol derived from the p-adic discrete valuation $\widehat{v}$ on $\mathbb{Q}_p$.

Proof. From the tame symbol definition (14.53),

$$\tau_{\widehat{v}}(a,b) \;=\; (-1)^{\widehat{v}(a)\widehat{v}(b)} a^{\widehat{v}(b)} b^{-\widehat{v}(a)} \;+\; p\widehat{\mathbb{Z}}_p \,.$$

So on the square class representatives, $\tau_{\widehat{v}}$ has values

	1	r	p	rp
1	$\bar{1}$	$\bar{1}$	$\bar{1}$	$\bar{1}$
r	$\bar{1}$	$\bar{1}$	$\bar{r}$	$\bar{r}$
p	$\bar{1}$	$\bar{r}^{-1}$	$-\bar{1}$	$-\bar{r}^{-1}$
rp	$\bar{1}$	$\bar{r}^{-1}$	$-\bar{r}$	$-\bar{1}$.

Following $\tau_{\widehat{v}}$ by the $(p-1)/2$ power map and the isomorphism from $\{\pm\bar{1}\}$ in $k_{\widehat{v}}^*$ to $\{\pm 1\} = \mathbb{Z}^*$ defines a bimultiplicative map $\mathbb{Q}_p^* \times \mathbb{Q}_p^* \to \{\pm 1\}$, which is therefore constant on square classes. This composite agrees with $(-,-)_p$ on the square class representatives, and hence everywhere. So $(-,-)_p$ is bimultiplicative, and hence a Hilbert symbol. ∎

To establish a Hilbert symbol on $\mathbb{Q}_2$ we must take a different route: The tame symbol in this case takes its values in $k_{\widehat{v}}^* \cong (\mathbb{Z}/2\mathbb{Z})^* = \{\bar{1}\}$; but as we are about to see, $(a,b)_2$ is not always 1.

(15.27) Lemma.

    (i)   *For each integer $n \geq 3$, $(\mathbb{Z}/2^n\mathbb{Z})^*$ has four square classes, represented by $\pm\bar{1}$ and $\pm\bar{5}$.*

    (ii) *A 2-adic integer $(\overline{a_1}, \overline{a_2}, ...)$ is a square in $\widehat{\mathbb{Z}}_2^*$ if and only if $\overline{a_3} = \bar{1}$ in $\mathbb{Z}/8\mathbb{Z}$; that is, $(\widehat{\mathbb{Z}}_2^*)^2 = 1 + 8\widehat{\mathbb{Z}}_2$.*

    (iii) *There are exactly eight square classes in $\mathbb{Q}_2^*$, represented by $\pm 1, \pm 2, \pm 5$, and ± 10.*

Proof. The kernel of the squaring map on $(\mathbb{Z}/2^n\mathbb{Z})^*$ is the set of $\bar{b} \in (\mathbb{Z}/2^n\mathbb{Z})^*$ with 2^n dividing $(b-1)(b+1)$. These factors differ by 2, so they cannot both be divisible by 4. Since b is odd, they are both divisible by 2. So $\bar{b}$ is in the kernel if and only if 2^{n-1} divides $b-1$ or $b+1$. For $n \geq 3$, the kernel has four elements

$\{\pm\overline{1}, \overline{2}^{n-1} \pm \overline{1}\}$. The image is the group of squares. So the number of squares is one-quarter of the order of $(\mathbb{Z}/2^n\mathbb{Z})^*$, and there are exactly four square classes. Since $\pm\overline{1}, \pm\overline{5}$ represent different square classes in $(\mathbb{Z}/8\mathbb{Z})^*$ (where all squares equal $\overline{1}$), they represent different square classes in $(\mathbb{Z}/2^n\mathbb{Z})^*$ for all $n \geq 3$.

For (ii) we first establish a replacement for the iterative process computing square roots in $\widehat{\mathbb{Z}}_p^*$ when p is odd.

Claim. Suppose $(s_1, s_2, ...) \in \widehat{\mathbb{Z}}_2^*$ and $n > 1$. If a square root of s_n in $\mathbb{Z}/2^n\mathbb{Z}$ comes from a square root of s_{n+1} in $\mathbb{Z}/2^{n+1}\mathbb{Z}$, it also comes from a square root of s_{n+2} in $\mathbb{Z}/2^{n+2}\mathbb{Z}$.

Proof of Claim. Say x is a square root of s_n in $\mathbb{Z}/2^n\mathbb{Z}$, $a \in \mathbb{Z}$, $\overline{a}^2 = s_{n+1}$ in $\mathbb{Z}/2^{n+1}\mathbb{Z}$, and x comes from $\overline{a} \in \mathbb{Z}/2^{n+1}\mathbb{Z}$. Since s_{n+1} is a unit, a is odd. Say $a = 2m + 1$. Now $s_{n+2} = \overline{a}^2$ or $\overline{a}^2 + \overline{2}^{n+1}$ in $\mathbb{Z}/2^{n+2}\mathbb{Z}$. In the first case, x comes from the square root $\overline{a}$ of s_{n+2}. In the second case, x comes from $\overline{a} + \overline{2}^n \in \mathbb{Z}/2^{n+2}\mathbb{Z}$; and in that ring, $(\overline{a} + \overline{2}^n)^2 = \overline{a}^2 + \overline{2}^{n+1}\overline{a} = \overline{a}^2 + \overline{2}^{n+1}(\overline{2m+1}) = \overline{a}^2 + \overline{2}^{n+1} = s_{n+2}$. ∎

Any square $(s_1, s_2, ...)$ in $\widehat{\mathbb{Z}}_2^*$ has third coordinate s_3 a square in $(\mathbb{Z}/8\mathbb{Z})^*$, and the only square in that group is 1. For the converse, suppose $(s_1, s_2, ...) \in \widehat{\mathbb{Z}}_2^*$ and $s_3 = \overline{1}$; so $s_2 = \overline{1}$ and $s_1 = \overline{1}$ as well. We construct a square root $(\overline{b_1}, \overline{b_2}, ...) \in \widehat{\mathbb{Z}}_2^*$ iteratively as follows. Take $b_1 = b_2 = 1$. Then $\overline{b_2}$ is a square root of s_2 coming from a square root $\overline{1}$ of s_3. In general, suppose $n \geq 2$, $b_n \in \mathbb{Z}$, and $\overline{b_n}$ is a square root of s_n coming from a square root of s_{n+1}. By the claim, it also comes from a square root $\overline{b_{n+1}}$ of s_{n+2}; so $s_{n+1} = \overline{b_{n+1}}^2$ in $\mathbb{Z}/2^{n+1}\mathbb{Z}$. And $\overline{b_n}$ also comes from $\overline{b_{n+1}} \in \mathbb{Z}/2^{n+1}\mathbb{Z}$. Thus $(\overline{b_1}, \overline{b_2}, ...)$ is constructed in $\widehat{\mathbb{Z}}_2$ with square $(s_1, s_2, ...)$, proving (ii).

Every element of $\mathbb{Q}_2^*$ has a unique expression $u2^n$ with $u \in \widehat{\mathbb{Z}}_2^*$ and $n \in \mathbb{Z}$. So $a = u2^n \in \mathbb{Q}_2^*$ is a square in $\mathbb{Q}_2^*$ if and only if $n = \widehat{v}_2(a)$ is even and u is a square in $\widehat{\mathbb{Z}}_2^*$. Taking third coordinates, $\pm 1, \pm 5$ represent four different square classes; so $\pm 2, \pm 10$ also represent four square classes, as we see when we multiply by 2. Comparing 2-adic values, $\pm 1, \pm 5, \pm 2, \pm 10$ represent eight different square classes.

Suppose $u = (\overline{u_1}, \overline{u_2}, ...) \in \widehat{\mathbb{Z}}_2^*$. By (i), for each positive integer $n, \overline{u_n}$ is in the square class of $\overline{1}, -\overline{1}, \overline{5}$, or $-\overline{5}$ in $(\mathbb{Z}/2^n\mathbb{Z})^*$. For each odd integer a, the square class of $\overline{a}$ in $(\mathbb{Z}/2^n\mathbb{Z})^*$ goes to the square class of $\overline{a}$ in $(\mathbb{Z}/2^m\mathbb{Z})^*$ whenever $m < n$. So for some $b \in \text{Seq}(2)^*$, $ub^2 \in \{\pm 1, \pm 5\}$. Then $b^2 \in \widehat{\mathbb{Z}}_2^*$, with $\mathbb{Z}/8\mathbb{Z}$ coordinate $\overline{1}$. By (ii), b^2 is a square in $\widehat{\mathbb{Z}}_2 \subset \mathbb{Q}_2^*$. So u is in the square class of $1, -1, 5$, or -5. And for each integer n, $u2^n$ is in a square class represented by one of $\pm 1, \pm 5, \pm 2, \pm 10$. ∎

Since $(-, -)_2$ is constant on each pair of square classes, we can compute it by finding its values on the square class representatives.

(15.28) Lemma. *The values of* $(-,-)_2$ *on square class representatives are*

	1	−1	2	−2	5	−5	10	−10
1	1	1	1	1	1	1	1	1
−1	1	−1	1	−1	1	−1	1	−1
2	1	1	1	1	−1	−1	−1	−1
−2	1	−1	1	−1	−1	1	−1	1
5	1	1	−1	−1	1	1	−1	−1
−5	1	−1	−1	1	1	−1	−1	1
10	1	1	−1	−1	−1	−1	1	1
−10	1	−1	−1	1	−1	1	1	−1

Proof. The table will be symmetric since $(a,b)_2 = (b,a)_2$. Note that by (15.27) (ii), every integer in $1+8\mathbb{Z}$ is a square in $\widehat{\mathbb{Z}}_2$. The 1's in the table are accounted for by the following solutions of $ax^2 + by^2 = z^2$:

a	b	(x,y,z)
1	b	$(1,0,1)$
a	$-a$	$(1,1,0)$
−1	2	$(1,1,1)$
−1	5	$(1,1,2)$
−1	10	$(1,1,3)$
2	2	$(1,1,2)$
−2	−5	$(1,1,\sqrt{-7})$
−2	−10	$(3,1,2\sqrt{-7})$
5	5	$(1,2,5)$
−5	−10	$(1,1,\sqrt{-15})$
10	10	$(1,3,10)$

For the −1's, notice that in a primitive solution (x,y,z) of $ax^2 + by^2 = z^2$ for $a,b \in \mathbb{Z}$, either x or y is not divisible by 2; so the fourth coordinates provide a solution $(\overline{u},\overline{v},\overline{w})$ over $\mathbb{Z}/16\mathbb{Z}$, with $u,v,w \in \mathbb{Z}$ and either u or v odd. The squares in $\mathbb{Z}/16\mathbb{Z}$ are $\overline{0},\overline{1},\overline{4}$, and $\overline{9}$. By inspection, such solutions do not exist for the (a,b) with −1 in the table. ∎

Organizing elements of $\widehat{\mathbb{Z}}_2$ according to their square class, every element $a \in \mathbb{Q}_2^*$ has a unique expression

$$a = 2^i(-1)^j 5^k s,$$

where $s \in 1 + 8\widehat{\mathbb{Z}}_2 = (\widehat{\mathbb{Z}}_2^*)^2$, $i \in \mathbb{Z}$, and $j,k \in \{0,1\}$. Suppose $b \in \mathbb{Q}_2^*$ has the corresponding expression

$$b = 2^I(-1)^J 5^K t.$$

If $(-,-)_2$ turns out to be bimultiplicative, the table in (15.28) would imply

$$(a,b)_2 \;=\; (-1)^{iK+jJ+kI} \;.$$

(15.29) Theorem. *With the above notation,* $(-,-)_2$ *is a Hilbert symbol on* $\mathbb{Q}_2$ *and*

$$(a,b)_2 \;=\; (-1)^{iK+jJ+kI}$$

for all $a,b \in \mathbb{Q}_2^*$.

Proof. Define $f(a,b) = (-1)^{iK+jJ+kI}$ so f is a function $\mathbb{Q}_2^* \times \mathbb{Q}_2^* \to \mathbb{Z}^*$. Direct inspection shows $f(a,b) = (a,b)_2$ for all $a,b \in \{\pm1, \pm2, \pm5, \pm10\}$; this inspection is shortened a little by noting that $f(a,b) = f(b,a)$ for all $a,b \in \mathbb{Q}_2^*$. From its definition it is evident that $f(ac^2, bd^2) = f(a,b)$ whenever $c,d \in \mathbb{Q}_2^*$. So $f(a,b) = (a,b)_2$ for all $a,b \in \mathbb{Q}_2^*$.

Now f is bimultiplicative: Suppose

$$a \;=\; 2^i(-1)^j 5^k s \quad \text{and} \quad a' \;=\; 2^{i'}(-1)^{j'} 5^{k'} s'$$

for $i,i' \in \mathbb{Z}$, $j,j',k,k' \in \{0,1\}$, and $s,s' \in (\widehat{\mathbb{Z}}_2^*)^2$. Say j'',k'' are the remainders upon division by 2 of $j+j', k+k'$, respectively. Then

$$aa' \;=\; 2^{i+i'}(-1)^{j''} 5^{k''} s''$$

with $s'' \in (\widehat{\mathbb{Z}}_2^*)^2$. Taking $b = 2^I(-1)^J 5^K t$ with $I \in \mathbb{Z}$, $J,K \in \{0,1\}$, and $t \in (\widehat{\mathbb{Z}}_2^*)^2$,

$$\begin{aligned}
f(aa',b) &= (-1)^{(i+i')K + j''J + k''I} \\
&= (-1)^{iK+jJ+kI}(-1)^{i'K+j'J+k'I} \\
&= f(a,b)f(a',b) \;.
\end{aligned}$$

∎

By Example (15.4), there is no Hilbert symbol $(-,-)_\mathbb{Q}$. But for each prime p, the p-adic Hilbert symbol $(-,-)_p$ restricts to a function

$$(-,-)_p : \mathbb{Q}^* \times \mathbb{Q}^* \to \mathbb{Z}^*,$$

which is bimultiplicative and satisfies $(a,b)_p = 1$ if $a+b=1$ in $\mathbb{Q}$. So the p-adic Hilbert symbols induce group homomorphisms

$$h_p : K_2(\mathbb{Q}) \to \mathbb{Z}^* \;, \quad \{a,b\} \mapsto (a,b)_p,$$

one for each prime p.

When p is odd, $(-,-)_p$ can be evaluated on $\mathbb{Q}^* \times \mathbb{Q}^*$, without reference to p-adic numbers, as

$$(a,b)_p = \tau_v(a,b)^{(p-1)/2},$$

where $v = v_p$ is the p-adic discrete valuation on $\mathbb{Q}$.

The 2-adic Hilbert symbol $(-,-)_2$ can also be evaluated on $\mathbb{Q}^* \times \mathbb{Q}^*$ without direct reference to 2-adic numbers: Each $a \in \mathbb{Q}^*$ has the form $2^i(u/v)$, where u, v are odd integers. And, modulo 8, $uv \equiv (-1)^j 5^k \in \{\pm 1, \pm 5\}$. If w denotes $(-1)^j 5^k$, then $uvw \in 1 + 8\mathbb{Z}$; so $uvw/v^2 w^2 \in (\widehat{\mathbb{Z}}_2^*)^2$ and

$$a = 2^i(-1)^j 5^k \left(\frac{uvw}{v^2 w^2}\right).$$

So these i, j, k are the exponents referred to in (15.29).

The 2-adic Hilbert symbol is used in Tate's proof of quadratic reciprocity, as described in Milnor's book [71, §11]. To prepare for our presentation of this proof, we ask the reader to review the proof of (14.56) in §14F and the subsequent note.

Let $\mathcal{P}$ denote the set of odd (positive) prime numbers. Tate's exact sequence

$$1 \longrightarrow \{\pm 1\} \overset{i}{\longrightarrow} K_2(\mathbb{Q}) \overset{\tau}{\longrightarrow} \bigoplus_{p \in \mathcal{P}} (\mathbb{Z}/p\mathbb{Z})^* \longrightarrow 1$$

$$-1 \longmapsto \{-1, -1\}$$

is split by the left inverse to i induced by the real symbol $(-,-)_\infty$. But $(-1,-1)_2 = -1$ as well, and the 2-adic Hilbert symbol $(-,-)_2$ also induces a left inverse h_2 to i. Tate's original description of $K_2(\mathbb{Q})$ in Milnor [71] uses the resulting isomorphism

(15.30)
$$K_2(\mathbb{Q}) \cong \{\pm 1\} \oplus \bigoplus_{p \in \mathcal{P}} (\mathbb{Z}/p\mathbb{Z})^*,$$

$$\{a, b\} \longmapsto ((a,b)_2, \tau_3(a,b), \tau_5(a,b), \dots)$$

involving the 2-adic Hilbert symbol and the tame symbols at odd primes. Then, how does the real symbol fit into this description?

For the moment, use additive notation for the abelian group $K_2(\mathbb{Q})$ and hence for $\mathrm{Hom}_\mathbb{Z}(M, K_2(\mathbb{Q}))$ whenever M is an abelian group. Composing the isomorphism (15.30) with projection and insertion maps, we get homomorphisms

$$K_2(\mathbb{Q}) \overset{h_2}{\underset{i_2}{\rightleftarrows}} \{\pm 1\}, \quad K_2(\mathbb{Q}) \overset{\tau_p}{\underset{i_p}{\rightleftarrows}} (\mathbb{Z}/p\mathbb{Z})^*$$

with $i_2 \circ h_2 + \sum i_p \circ \tau_p$ the identity on $K_2(\mathbb{Q})$. If $r : K_2(\mathbb{Q}) \to \{\pm 1\}$ is induced by the real symbol, then

$$r = r \circ \left(i_2 \circ h_2 + \sum i_p \circ \tau_p\right) = (r \circ i_2) \circ h_2 + \sum (r \circ i_p) \circ \tau_p.$$

Now $r(i_2(-1)) = r(\{-1, -1\}) = -1$; so $r \circ i_2 : \{\pm 1\} \to \{\pm 1\}$ is the identity map. Each $r \circ i_p : (\mathbb{Z}/p\mathbb{Z})^* \to \{\pm 1\}$ is a homomorphism from $(\mathbb{Z}/p\mathbb{Z})^*$ to its subgroup $\{\pm \bar{1}\}$, followed by the isomorphism $f : \{\pm \bar{1}\} \cong \{\pm 1\}$. The former is determined by its effect on a generator of $(\mathbb{Z}/p\mathbb{Z})^*$, so it must be the $n(p)$ power map $x \mapsto x^{n(p)}$, where $n(p) = (p-1)/2$ or $p - 1$. If we return to multiplicative notation, it follows that for all $a, b \in \mathbb{Q}^*$,

$$
\begin{aligned}
(a,b)_\infty &= (a,b)_2 \prod_{p \in \mathcal{P}} f(\tau_p(a,b)^{n(p)}) \\
&= (a,b)_2 \prod_{p \in \mathcal{P}} (a,b)_p^{m(p)},
\end{aligned}
$$

(15.31)

where $m(p) = 1$ or 2.

Hilbert introduced what we now call Hilbert symbols $(-,-)_F$ in order to generalize Gauss' law of quadratic reciprocity. If p is an odd prime and $v_p(a) = v_p(b) = 0$, then $\tau_p(a,b) = \bar{1}$ and $(a,b)_p = 1$. If p and q are odd primes with $p \equiv (-1)^j 5^k \in \{\pm 1, \pm 5\}$ and $q \equiv (-1)^J 5^K \in \{\pm 1, \pm 5\} \pmod 8$, then

$$
(p,q)_2 = (-1)^{jJ} = \begin{cases} 1 & \text{if } p \text{ or } q \equiv 1 \pmod 4 \\ -1 & \text{if } p \equiv q \equiv 3 \pmod 4 \end{cases}
$$
$$
= (-1)^{(p-1)(q-1)/4}.
$$

If $p \neq q$, then $\tau_p(q,p) = \bar{q} \in (\mathbb{Z}/p\mathbb{Z})^*$, and

$$
(p,q)_p = (q,p)_p = f(\bar{q}^{(p-1)/2}) = \left(\frac{q}{p}\right).
$$

So for $p \neq q$ in $\mathcal{P}$, (15.31) becomes

$$
1 = (-1)^{(p-1)(q-1)/4} \left(\frac{q}{p}\right)^{m(p)} \left(\frac{p}{q}\right)^{m(q)}.
$$

(15.32) Quadratic Reciprocity. *For all $a, b \in \mathbb{Q}^*$,*

$$
(a,b)_\infty = (a,b)_2 \prod_{p \in \mathcal{P}} (a,b)_p \; ;
$$

so for all odd primes $p \neq q$,

$$
1 = (-1)^{(p-1)(q-1)/4} \left(\frac{q}{p}\right) \left(\frac{p}{q}\right).
$$

Proof. (Tate) We need only show that, in (15.31), each $m(p) = 1$ rather than 2. So we evaluate (15.31) at special pairs (a, b).

Suppose $p \in \mathcal{P}$ and $p \equiv -1$ or $-5 \pmod 8$. Then $(p,p)_2 = -1$; so when $a = b = p$, (15.31) becomes $1 = (-1)(p,p)_p^{m(p)}$, and $m(p) \neq 2$.

Or, suppose $p \in \mathcal{P}$ with $p \equiv 5 \pmod 8$. Then $(2,p)_2 = -1$; and when $a = 2$ and $b = p$, (15.31) becomes $1 = (-1)(2,p)_p^{m(p)}$. So again $m(p) \neq 2$.

Suppose the theorem fails, and p is the smallest odd prime with $m(p) = 2$. As we have just seen, p must be in $1 + 8\mathbb{Z}$. So p is a square in $\mathbb{Q}_2^*$ and $(a,p)_2 = 1$ for all $a \in \mathbb{Q}^*$. For each odd prime $q < p$, $m(q) = 1$. So from (15.31),

$$ 1 = (q,p)_q^1 (q,p)_p^2 = (q,p)_q = \left(\frac{p}{q}\right). $$

That is, p is a square modulo each smaller prime.

The needed contradiction is supplied by a modification by Tate of a theorem due to Gauss [86, Theorem 129]:

(15.33) Lemma. *If p is a prime in $1 + 8\mathbb{Z}$, there is a prime $q < \sqrt{p}$ for which p is not a square modulo q.*

Proof. (Milnor [71, pp. 105 106]) Let m denote the largest odd integer less than $\sqrt{p}$, so $m^2 < p < (m+2)^2$. For $i = 1, 3, 5, \ldots, m$,

$$ 0 < \frac{p - i^2}{4} < \frac{(m+2)^2 - i^2}{4} = \frac{m+2+i}{2} \cdot \frac{m+2-i}{2}. $$

Taking the product for all i, $0 < N < (m+1)!$, where

$$ N = \left(\frac{p - 1^2}{4}\right) \left(\frac{p - 3^2}{4}\right) \cdots \left(\frac{p - m^2}{4}\right). $$

Suppose the lemma fails, and p is a square modulo every prime $\ell < \sqrt{p}$. We show this forces N to be a multiple of $(m+1)!$, a contradiction.

If $a_1, \ldots, a_k$ are positive integers and q is prime,

$$ v_q(a_1 \ldots a_k) = \mu(1) + \mu(2) + \cdots $$

where $\mu(s)$ is the number of terms a_i divisible by p^s. Here $\mu(s) = 0$ for $s > \max v_q(a_i)$.

To show $(m+1)!$ divides N, it is enough to show $v_q((m+1)!) \leq v_q(N)$ for each prime q dividing $(m+1)!$. So it is enough to prove that, for each positive integer s with $q^s \leq m+1$, the number of multiples of q^s among $1, 2, \ldots, m+1$ is no more than the number of terms $(p - i^2)/4$ divisible by q^s. These multiples are

$$ q^s, 2q^s, 3q^s, \ldots, rq^s , $$

where $r = [(m+1)/q^s]$ is the greatest integer $\leq (m+1)/q^s$. So we need only find at least r solutions $i \in \{1, 3, 5, \ldots, m\}$ to the congruence

$$ p \equiv i^2 \pmod{4q^s} . $$

First we show there is a solution $i \in \mathbb{Z}$. (Since p is odd, such a solution i will be odd.) By (15.27), $p \in 1 + 8\mathbb{Z}$ implies p is a square modulo every power of 2; so there is a solution if $q = 2$. Suppose instead that q is odd. Then $q^s \neq m + 1$ and $q \leq m < \sqrt{p}$. By hypothesis, p is a square mod q. By (15.24), p is a square mod q^s as well. Under the isomorphism,

$$\frac{\mathbb{Z}}{4q^s\mathbb{Z}} \;\cong\; \frac{\mathbb{Z}}{4\mathbb{Z}} \;\times\; \frac{\mathbb{Z}}{q^s\mathbb{Z}} \,,$$

$\bar{p}$ goes to a square; so it is a square.

If $i \in \mathbb{Z}$ is a solution, then $i \notin q^s\mathbb{Z}$. So i lies between, but not midway between, two consecutive multiples of $2q^s$. So for some integer n,

$$i_0 \;=\; |i - 2q^s n| \;<\; q^s \,.$$

And i_0 is a solution in the interval $(0, q^s)$, while $-i_0$ is a solution in $(-q^s, 0)$. Adding multiples of $2q^s$, we get a solution between each consecutive pair of multiples

$$0, q^s, 2q^s, \ldots, rq^s$$

of q^s that are $\leq m + 1$, and those solutions are r odd integers $\leq m$.   ■ ■

15C. Exercises

1. Suppose $\mathcal{C}$ is a category and

$$\cdots \xrightarrow{f_2} A_2 \xrightarrow{f_1} A_1 \xrightarrow{f_0} A_0$$

is a sequence of arrows in $\mathcal{C}$. A **cone** $(B, \{g_i\})$ **to** this sequence is a commutative diagram in $\mathcal{C}$:

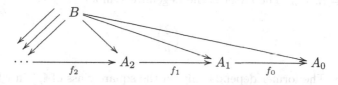

where the arrows coming down are $g_i : B \to A_i$. These cones are the objects of a category $\mathcal{D}$ in which an arrow from $(B, \{g_i\})$ to $(B', \{g_i'\})$ is an arrow in $\mathcal{C}$, $h : B \to B'$, with $g_i' \circ h = g_i$ for each i. A terminal object in $\mathcal{D}$ is called an **inverse limit diagram** for the given sequence, and its apex B is called the **inverse limit**: $B = \varprojlim A_n$. (Compare §4A, Exercise 1.) If $\mathcal{C}$ is a concrete category (as defined in §0), show there is an inverse limit $\varprojlim A_n$ consisting of the sequences $(a_0, a_1, ...)$ with $a_i \in A_i$ for each i, and for which a_n goes to a_m whenever $m < n$.

2. Suppose p is a prime number and $a_0, a_1, ...$ are integers taken from the set $\{0, 1, ..., p-1\}$. Prove the series

$$\sum_{i=0}^{\infty} a_i p^i = a_0 + a_1 p + a_2 p^2 + \cdots$$

converges to a p-adic integer (under $| \ |_p$). Then show each p-adic integer has one and only one expression as such a series. Dividing by powers of p, each p-adic number has an expression

$$\sum_{i=n}^{\infty} a_i p^i = a_n p^n + a_{n+1} p^{n+1} + \cdots$$

for some integer n. This is analogous to the decimal expansion of real numbers — except that negative signs are not needed. Calculate -2 in series form in the ring $\widehat{\mathbb{Z}}_3$.

3. How are the terms of a p-adic integer $(\overline{a_1}, \overline{a_2}, ...)$ related to its series expansion in Exercise 2? How is the p-adic value of a p-adic number determined by its series expansion? Describe the ideals in $\widehat{\mathbb{Z}}_p$ using series.

4. If p is an odd prime number, prove -1 is a sum of two squares in $\mathbb{Q}_p$. *Hint:* Evaluate the p-adic tame symbol on $(-1, -1)$. Can you show -1 is a sum of two squares in $\widehat{\mathbb{Z}}_p$?

5. Show -1 is not a sum of two squares in $\mathbb{Q}_2$, but is a sum of four squares in $\widehat{\mathbb{Z}}_2$, with all four squares belonging to $\mathbb{Z}$.

6. Find the first six coordinates of $\sqrt{2}$ in $\widehat{\mathbb{Z}}_7$. Find the first six coordinates of $\sqrt{-7}$ in $\widehat{\mathbb{Z}}_2$.

7. Use Hilbert's product formula (15.32) to show that, for each odd prime p, $(2, p)_2 = (2, p)_p$. The latter is the Legendre symbol

$$\left(\frac{2}{p} \right)$$

by (15.26). The former depends only on the square class of $\mathbb{Q}_2^*$ in which p lies. Show that $(2, p)_2$ therefore only depends on the congruence class of $p \pmod 8$. Then use the table (15.28) to show

$$\left(\frac{2}{p} \right) = \begin{cases} 1 & \text{if } p \equiv \pm 1 \pmod 8 \\ -1 & \text{if } p \equiv \pm 3 \pmod 8 \end{cases}$$
$$= (-1)^{(p^2 - 8)/2}.$$

8. Use the Legendre symbol identities (for odd primes $p \neq q$ and integers $a, b \neq p\mathbb{Z}$):

(i) $\left(\dfrac{ab}{p}\right) = \left(\dfrac{a}{p}\right)\left(\dfrac{b}{p}\right)$,

(ii) $\left(\dfrac{a}{p}\right) = \left(\dfrac{b}{p}\right)$ if $a \equiv b \pmod{p}$,

(iii) $\left(\dfrac{p}{q}\right) \neq \left(\dfrac{q}{p}\right)$ if and only if $p \equiv q \equiv 3 \pmod 4$,

(iv) $\left(\dfrac{-1}{p}\right) = 1$ if and only if $p \equiv 1 \pmod 4$,

(v) $\left(\dfrac{2}{p}\right) = 1$ if and only if $p \equiv \pm 1 \pmod 8$,

to calculate $\left(\frac{79}{97}\right)$. Verify these identities.

9. Prove the product formula

$$|x| \prod_p |x|_p = 1$$

for all $x \in \mathbb{Q}^*$, where p runs through all prime numbers, and the first absolute value is the standard one on $\mathbb{R}$. Of course, this holds only because $1/p$ was chosen for the exponential base in the definition of $|\ |_p$.

10. (Silvester) In Stein [73, Theorems 2.5 and 2.13], it is proved that, if R is a commutative local ring and J is a radical ideal of R, the kernel of K_2 of the canonical map $R \rightarrow R/J$ is generated by the Steinberg symbols $\{u, 1+q\}$ with $u \in R^*$ and $q \in J$. Use this and the 2-adic symbol to prove $\{-1, -1\} \neq 1$ in $K_2(\mathbb{Z}/4\mathbb{Z})$. *Hint:* First note $\mathbb{Z}/4\mathbb{Z} \cong \widehat{\mathbb{Z}}_2/4\widehat{\mathbb{Z}}_2$ as rings. Then show the 2-adic symbol takes $(-1, -1)$ to -1 but takes each $(u, 1+q)$ to 1 where $u \in \widehat{\mathbb{Z}}_2^*$ and $q \in 4\widehat{\mathbb{Z}}_2$.

15D. Local Fields and Norm Residue Symbols

Hilbert's formulation of quadratic reciprocity as a product formula among Hilbert symbols enabled him to extend reciprocity from $\mathbb{Q}$ to number fields. We describe this extension in (15.34) below.

In the 19th century, cubic and quartic reciprocity laws were developed by Gauss, Jacobi, and Eisenstein (see Lemmermeyer [00]). These take place in $\mathbb{Q}(\zeta_3)$ and $\mathbb{Q}(\zeta_4)$, respectively, where $\zeta_n = e^{2\pi i/n}$ has multiplicative order n. The generalized Legendre symbol in these laws takes its values in the cyclic groups generated by ζ_3 and ζ_4 ($= i$), respectively. In his famous problem list (Browder [76]) partially given at the 1900 International Congress of Mathematicians in

Paris, Hilbert listed as his 9th problem a "Proof of the Most General Law of Reciprocity in Any Number Field." The solution by Artin [27] uses the "class field theory," which sprang from Hilbert's foundational work. Hasse expressed Artin's reciprocity in terms of higher power "norm residue symbols," whose existence and properties were anticipated by Hilbert.

Due to the depth of class field theory, we treat the existence of the reciprocity map (15.35) as essentially an axiom needed to define norm residue symbols. Space limitation forces us to rely on citations for several proofs in this section. The works cited make very rewarding perusal, and we recommend them to the reader.

At the end of this section we report the exact sequence of Moore establishing uniqueness of the reciprocity laws, and a long exact sequence of Quillen, which fits the relations among tame symbols into a K-theoretic context.

We freely use the basic facts and terminology about number fields found in Chapter 7. Suppose F is a number field with ring of algebraic integers R. For each maximal ideal P of R, there is a P-adic discrete valuation $v = v_P : F^* \to \mathbb{Z}$, where $v(a)$ is the exponent of P in the maximal ideal factorization of aR. By (7.27), R/P is a finite field; the size of this field is $q = p^f$, where its characteristic p is the unique prime number $p \in \mathbb{Z}$ with $p \in P$ and f is a positive integer.

The standard absolute value on F associated with this v is $| \ |_P = | \ |_{v,1/q}$, defined by

$$|x|_P = \begin{cases} 0 & \text{if } x = 0 \\ q^{-v(x)} & \text{if } x \neq 0 \,. \end{cases}$$

The metric completion of $(F, | \ |_P)$ constructed in (15.15) is denoted by $(F_P, | \ |_{\widehat{p}})$, and there is a discrete valuation $\widehat{v} : F_P^* \to \mathbb{Z}$ extending v, with

$$|x|_{\widehat{P}} = \begin{cases} 0 & \text{if } x = 0 \\ q^{-\widehat{v}(x)} & \text{if } x \neq 0 \,. \end{cases}$$

Now F_P is the field of fractions of its discrete valuation ring $\mathcal{O}_{\widehat{v}}$, which we denote by $\widehat{R}_P$. Then $F \cap \widehat{R}_P = \mathcal{O}_v$, which by Exercise 1 is the localization $R_P = (R - P)^{-1}R$. Just as in §15C (where R was $\mathbb{Z}$), R is dense in R_P with respect to $| \ |_P$, and $\widehat{R}_P$ consists of the cosets (mod null sequences) of Cauchy sequences in R with respect to $| \ |_P$. The ring $\widehat{R}_P$ can be identified with $\varprojlim R/P^n$, the set of sequences $(\overline{a_1}, \overline{a_2}, ...)$ where, for all n, $a_n \in R$, $\overline{a_n} = a_n + P^n \in R/P^n$, and $\overline{a_{n+1}}$ goes to $\overline{a_n}$ under the canonical ring homomorphism $R/P^{n+1} \to R/P^n$. This identification becomes a ring isomorphism

$$\widehat{R}_P \ \cong \ \varprojlim R/P^n$$

when these sequences are added and multiplied coordinatewise.

If we choose $\pi \in P$ with $\pi \notin P^2$, then $\widehat{v}(\pi) = v(\pi) = 1$ and the nonzero ideals of $\widehat{R}_P$ are

$$\pi^m \widehat{R}_P = \{x \in F_P : \widehat{v}(x) \geq m\}$$

$$= \{(\overline{a_1}, \overline{a_2}, ...) \in \widehat{R}_P : \overline{a_n} = \overline{0} \text{ for } n \leq m\} \,.$$

In particular, projection to the first coordinate $\widehat{R}_P \to R/P$ is a surjective ring homomorphism with kernel $\pi\widehat{R}_P$, inducing an isomorphism of residue fields

$$k_{\widehat{v}} = \frac{\widehat{R}_P}{\pi\widehat{R}_P} \cong \frac{R}{P}.$$

Recall that R/P is a finite field with $q = p^f$ elements and prime characteristic $p \in P$.

Say P is **odd** if p is an odd prime, and P is **even** if $p = 2$. If P is odd and $a \in R$ but $a \notin P$, we can define a Legendre symbol

$$\left(\frac{a}{P}\right) = \begin{cases} 1 & \text{if } \overline{a} \text{ is a square in } R/P \\ -1 & \text{if not .} \end{cases}$$

Then in R/P,

$$\overline{\left(\frac{a}{P}\right)} = \overline{a}^{(q-1)/2},$$

and this determines $\left(\frac{a}{P}\right)$ because $\overline{1} \neq -\overline{1}$ in R/P. By (15.3), for each maximal ideal P of R, even or odd, the map

$$(-,-)_P = (-,-)_{F_P}$$

is a (Hilbert) symbol map on F_P if and only if it is bimultiplicative. Recall the real symbol $(-,-)_{\mathbb{R}}$ is denoted by $(-,-)_{\infty}$.

(15.34) Theorem. (Hilbert)

(i) *For each maximal ideal P of R, the map $(-,-)_P$ is bimultiplicative.*

(ii) *If P is an odd prime, $a \in R$, and $a \notin P$, then*

$$(a,b)_P = \left(\frac{a}{P}\right)^{v_P(b)}.$$

(iii) *If $\mathcal{P}$ is the set of maximal ideals in R and $\mathcal{R}$ is the set of real embeddings $\sigma : F \to \mathbb{R}$ (= ring homomorphisms from F to $\mathbb{R}$), then for each pair $a, b \in F^*$,*

$$\prod_{\sigma \in \mathcal{R}} (\sigma(a), \sigma(b))_{\infty} = \prod_{P \in \mathcal{P}} (a,b)_P.$$

Proof. Specialize (15.43) below to the case $n = 2$. ∎

Part (ii) is just what one would expect if $(-,-)_P$ is the $(q-1)/2$ power of the tame symbol $\tau_{\widehat{v}}$ on F, where $\widehat{v} = \widehat{v}_P$. That $(-,-)_P$ has the latter description can be verified on square class representatives in F_P^* for odd P, just as was done in §15C for $\mathbb{Q}_p$. (See Exercises 2 and 3.)

Part (iii) has a more elegant expression. Say nontrivial absolute values on F are equivalent if they yield the same topology on F. Each equivalence class is called a **place** of F (or a **prime** or **prime spot** of F). Suppose S is the set of places of F. Each place $\mathfrak{p} \in S$ determines an isomorphism class of fields $F_\mathfrak{p}$ that are metric completions of F with respect to an absolute value in $\mathfrak{p}$, and each of these fields has a Hilbert symbol. Following the completion $F \to F_\mathfrak{p}$ by this Hilbert symbol defines a symbol map $(-,-)_\mathfrak{p} : F^* \times F^* \to \{\pm 1\}$, which is independent of the choice of absolute value in $\mathfrak{p}$ and of the choice of completion with respect to that absolute value (see (15.17) and §15B, Exercises 4 and 5). The ultrametric places are represented by the standard P-adic absolute values $|\ |_P$, no two of which are equivalent (by §15B, Exercise 4 (ii)). The archimedean places of F are represented by the absolute values $|\ |_\sigma = |\sigma(\ )|_\mathbb{C}$, where σ is a ring homomorphism, embedding F into $\mathbb{C}$, and no two real embeddings σ yield equivalent absolute values $|\ |_\sigma$ (see Exercise 4). For these absolute values, $\sigma : F \to \mathbb{C}$ is a metric completion when σ is an imaginary embedding and $\sigma : F \to \mathbb{R}$ is a metric completion when σ is a real embedding. The Hilbert symbol $(-,-)_\mathbb{C}$ is trivial, while $(-,-)_\mathbb{R}$ is the real symbol $(-,-)_\infty$. So part (iii) becomes

$$\prod_{\mathfrak{p} \in S} (a,b)_\mathfrak{p} = 1$$

for all $a, b \in F^*$.

To consider the extension of this to higher power reciprocity laws, we quote a result from local class field theory. First we widen the scope of our discussion to include positive characteristic analogs of number fields. Suppose k is a finite field. The rings $\mathbb{Z}$ and $k[x]$ are both euclidean domains in which every nonzero ideal has finite index. Their fields of fractions $\mathbb{Q}$ and $k(x)$ have similar discrete valuations (see (14.51) and (14.52)). A **global field** is a finite-degree field extension of $\mathbb{Q}$ or $k(x)$ for a finite field k. A **local field** is a field F_v that is complete with respect to a discrete valuation v with finite residue field k_v. The ultrametric completions of global fields are local fields.

A field extension $F \subseteq E$ is **abelian** if it is Galois (i.e., normal and separable) with abelian Galois group $\mathrm{Aut}(E/F)$. If F is a local field with algebraic closure $\overline{F}$, let $\mathcal{A}$ denote the set of subfields E of $\overline{F}$ with $F \subseteq E$ abelian. The subfield of $\overline{F}$ generated by all $E \in \mathcal{A}$ is a field $F^{ab} \in \mathcal{A}$, called the **maximal abelian extension** of F.

The following is a major result of local class field theory.

(15.35) Theorem. *Suppose F is a local field. There is a group homomorphism*

$$\theta_F : F \quad \to \quad \mathrm{Aut}(F^{ab}/F) ,$$

so that for each $E \in \mathcal{A}$, θ_F followed by restriction of automorphisms to E

induces a group isomorphism

$$w_{E/F} : \frac{F^*}{N_{E/F}(E^*)} \cong \text{Aut}(E/F) .$$

Proof. See either Serre [79, Chapter XIII, §4] or Cassels and Frohlich [86, Chapter VI, §2]. ∎

The maps θ_F and $w_{E/F}$ are known as **reciprocity maps**.

If $E = F(\sqrt{a})$ for some $a \in F$, an immediate consequence of the reciprocity isomorphism $w_{E/F}$ is $(F^* : N_{E/F}(E^*)) = [E : F] = 1$ or 2; so every local field of characteristic $\neq 2$ has a Hilbert symbol. If F^* has an element of order n, the extensions $F \subseteq F(\sqrt[n]{a})$ are also abelian:

(15.36) Proposition. *Suppose F is any field and F^* has an element ζ of finite order n.*

(i) *If α belongs to a field extension of F, $\alpha^n \in F$, and d is the least positive integer with $\alpha^d \in F$, then d divides n, the minimal polynomial of α over F is $x^d - \alpha^d$, and $F \subseteq F(\alpha)$ is Galois with cyclic Galois group generated by an automorphism τ with $\tau(\alpha) = \zeta^{n/d}\alpha$.*

(ii) *Suppose $F \subseteq E$ is a Galois field extension with cyclic Galois group of order d dividing n. Then $E = F(\alpha)$, where $\alpha^n \in F$.*

Proof. This basic fact from Galois Theory can be found in Lang [93, Chapter VI, §6] or in Jacobson [85, §§4.7 and 4.15]. ∎

(15.37) Definition. Suppose $F = F_v$ is a local field with algebraic closure $\overline{F}$ and F^* has an element ζ of finite order n. For each $a \in F^*$ choose $\alpha \in \overline{F}$ with $\alpha^n = a$. Let E denote $F(\alpha)$ and

$$w_a : \frac{F^*}{N_{E/F}(E^*)} \cong \text{Aut}(E/F)$$

denote the reciprocity isomorphism. For each $b \in F^*$, $w_a(\overline{b})$ takes α to $\zeta^i\alpha$ for some integer i. Define

$$\left(\frac{a, b}{v}\right)_n = \zeta^i = \frac{w_a(\overline{b})(\alpha)}{\alpha} .$$

The map

$$\left(\frac{-, -}{v}\right)_n : F^* \times F^* \to \langle \zeta \rangle$$

is the **nth power norm residue symbol** on F.

The value of this symbol is independent of the choice of $\alpha = \sqrt[n]{a}$ in $\overline{F}$, since $w_a(\overline{b})(\alpha) = \zeta^i\alpha$ implies $w_a(\overline{b})(\beta) = \zeta^i\beta$ for all nth roots $\beta = \zeta^j\alpha$ of a.

(15.38) Proposition. *If $F = F_v$ is a local field and F^* has an element ζ of finite order n, and if $a, a', b, b' \in F^*$,*

(i) $\left(\dfrac{aa', b}{v}\right)_n = \left(\dfrac{a, b}{v}\right)_n \left(\dfrac{a', b}{v}\right)_n$,

(ii) $\left(\dfrac{a, bb'}{v}\right)_n = \left(\dfrac{a, b}{v}\right)_n \left(\dfrac{a, b'}{v}\right)_n$,

(iii) $\left(\dfrac{a, b}{v}\right)_n = 1$ *if $a + b = 1$* ,

(iv) $\left(\dfrac{a, b}{v}\right)_n = 1$ *if and only if b is a norm from $F(\sqrt[n]{a})$* ,

(v) $\left(\dfrac{a, b}{v}\right)_n = 1$ *for all $b \in F^*$ if and only if $a = c^n$ for some $c \in F^*$* ,

(vi) *If d divides n,* $\left(\dfrac{a, b}{v}\right)_n^{n/d} = \left(\dfrac{a, b}{v}\right)_d$.

Proof. If $\alpha, \alpha' \in \overline{F}$ with $\alpha^n = a$ and $\alpha'^n = a'$, then $(\alpha\alpha')^n = aa'$. Since $\theta_F(b)$ is a homomorphism,

$$
\left(\frac{aa', b}{v}\right)_n = \frac{\theta_F(b)(\alpha\alpha')}{\alpha\alpha'} = \frac{\theta_F(b)(\alpha)}{\alpha} \frac{\theta_F(b)(\alpha')}{\alpha'}
$$

$$
= \left(\frac{a, b}{v}\right)_n \left(\frac{a', b}{v}\right)_n ,
$$

proving (i). Since w_a is a homomorphism, if

$$
\left(\frac{a, b}{v}\right)_n = \zeta^i \quad \text{and} \quad \left(\frac{a, b'}{v}\right)_n = \zeta^j ,
$$

then

$$
\left(\frac{a, bb'}{v}\right)_n = \frac{w_a(\overline{bb'})(\alpha)}{\alpha} = \frac{w_a(\overline{b})(w_a(\overline{b'})(\alpha))}{\alpha}
$$

$$
= \frac{w_a(\overline{b})(\zeta^j \alpha)}{\alpha} = \frac{\zeta^j w_a(\overline{b})(\alpha)}{\alpha}
$$

$$
= \zeta^j \zeta^i = \zeta^i \zeta^j ,
$$

proving (ii).

Part (iv) amounts to the fact that $w_a(\overline{b}) = 1$ if and only if $\overline{b} = \overline{1}$ in $F^*/N_{F(\alpha)/F}(F(\alpha)^*)$. We use (iv) to prove (iii): Suppose $a + b = 1$. If

$d = [F(\alpha) : F]$, then $\text{Aut}(F(\alpha)/F)$ is generated by σ with $\sigma(\alpha) = \zeta^{n/d}\alpha$.
So

$$b = 1 - a = \prod_{i=0}^{n-1} (1 - \zeta^i \alpha)$$

$$= \prod_{j=0}^{(n/d)-1} N_{F(\alpha)/F} (1 - \zeta^j \alpha)$$

is a norm from $F(\alpha)$, and $(\frac{a,b}{v})_n = 1$.

For (v), if $w_a(\bar{b})$ fixes α for all $\bar{b}$, then α lies in the fixed field F of $\text{Aut}(F(\alpha)/F)$; so a is an nth power of $c = \alpha \in F^*$. The converse comes from (i) and the equation $\zeta^n = 1$.

Finally, if we fix $a \in F^*$ and denote by $w_{(m)}$ the reciprocity isomorphism $w_{E/F}$ with $E = F(\sqrt[m]{a})$, then for each $b \in F^*$ and each divisor d of n, $w_{(d)}(\bar{b})$ is the restriction to $F(\sqrt[d]{a})$ of $w_{(n)}(\bar{b})$. So

$$\left(\frac{a,b}{v}\right)_d = \frac{w_{(d)}(\bar{b})(\sqrt[d]{a})}{\sqrt[d]{a}} = \frac{w_{(n)}(\bar{b})\left((\sqrt[n]{a})^{n/d}\right)}{(\sqrt[n]{a})^{n/d}}$$

$$= \left(\frac{w_{(n)}(\bar{b})(\sqrt[n]{a})}{\sqrt[n]{a}}\right)^{n/d} = \left(\frac{a,b}{v}\right)_n^{n/d} .$$

$\blacksquare$

Note: If F is a local field of characteristic $\neq 2$, then -1 has order 2 in F^*, and by part (iv) of the preceding proposition, $(\frac{-,-}{v})_2$ is the Hilbert symbol on F. So norm residue symbols are a direct generalization of Hilbert symbols.

For each field K, the elements of finite order in $(K^*, \cdot)$ form a subgroup $\mu(K)$. If $\mu(K)$ is finite, it is cyclic and its order is denoted by $m(K)$. As we see below, $\mu(F)$ is finite for each local field F. There will be an nth power norm residue symbol on F exactly when F^* has an element ζ of order n, that is, exactly when n divides $m(F)$. So what is the number $m(F)$ of roots of unity in a local field F?

To answer this question we employ a generally useful lemma, whose author was the discoverer of p-adic numbers.

(15.39) Hensel's Lemma. *Suppose F is a field that is complete with respect to a discrete valuation v, and $f \in \mathcal{O}_v[x]$ has a coefficient in $\mathcal{O}_v^*$. Reducing coefficients modulo $\mathcal{P}_v$, suppose $\bar{f} = \alpha\beta$ for relatively prime $\alpha, \beta \in k_v[x]$. Then $f = ab$ for some $a, b \in \mathcal{O}_v[x]$ with $\bar{a} = \alpha$, $\bar{b} = \beta$, and $\deg(a) = \deg(\alpha)$.*

Proof. (Neukirch [99]) Suppose $\pi \in F$ with $v(\pi) = 1$, $\deg(f) = D$, and $\deg(\alpha) = d$. So $\deg(\beta) = \deg(\bar{f}) - \deg(\alpha) \leq D - d$. For some t and u

in $\mathcal{O}_v[x]$, $\bar{t}\alpha + \bar{u}\beta = 1$. Lift α, β to $a_0, b_0 \in \mathcal{O}_v[x]$ of the same degree. So $ta_0 + ub_0 - 1$ and $f - a_0b_0$ lie in $\pi\mathcal{O}_v[x]$. If $f = a_0b_0$, we are done.

Suppose $f \neq a_0b_0$. Among the coefficients of $ta_0 - ub_0 - 1$ and $f - a_0b_0$, choose one nonzero coefficient c with the least value $v(c)$. Then $v(c) \geq 1$ and $f \equiv a_0b_0 \pmod{c}$.

We construct a and b as the limits of partial sums in convergent power series; we produce the terms of the series recursively. For $n \geq 1$, suppose there exist $r_i, s_i \in \mathcal{O}_v[x]$ with $\deg(r_i) < d$, $\deg(s_i) \leq D - d$, and for

$$a_{n-1} = a_0 + r_1c + \cdots + r_{n-1}c^{n-1} \, ,$$
$$b_{n-1} = b_0 + s_1c + \cdots + s_{n-1}c^{n-1} \, ,$$

we have $f \equiv a_{n-1}b_{n-1} \pmod{c^n}$. Then

$$f_n = c^{-n}(f - a_{n-1}b_{n-1}) \in \mathcal{O}_v[x] \, ,$$

and $uf_n = qa_0 + r_n$ for some $q, r_n \in F[x]$ with $\deg(r_n) < \deg(a_0) = d$. Since $\deg(a_0) = \deg(\overline{a_0})$, a_0 has leading coefficient in $\mathcal{O}_v^*$; so $q, r_n \in \mathcal{O}_v[x]$. Choose $s_n \in \mathcal{O}_v[x]$ to be what remains when the terms with coefficients in $c\mathcal{O}_v$ are deleted from $tf_n + qb_0$.

Since $ta_0 + ub_0 \equiv 1 \pmod{c}$, we also have

$$f_n \equiv tf_na_0 + uf_nb_0 = tf_na_0 + qa_0b_0 + r_nb_0$$
$$\equiv s_na_0 + r_nb_0 \pmod{c} \, .$$

Now $\deg(r_nb_0) < D$, $\deg(a_0) = d$, $\deg(f_n) \leq D$, and the (unit) leading coefficient of a_0 times the leading coefficient of s_n is not in $c\mathcal{O}_v$. So $\deg(s_n) \leq D - d$.

Modulo c,

$$a_{n-1}s_n + b_{n-1}r_n \equiv a_0s_n + b_0r_n \equiv f_n \, .$$

So if we define

$$a_n = a_0 + r_1c + \cdots + r_nc^n = a_{n-1} + r_nc^n \, ,$$
$$b_n = b_0 + s_1c + \cdots + s_nc^n = b_{n-1} + s_nc^n \, ,$$

then modulo c^{n+1},

$$a_nb_n \equiv a_{n-1}b_{n-1} + (a_{n-1}s_n + b_{n-1}r_n)c^n$$
$$\equiv a_{n-1}b_{n-1} + f_nc^n = f \, .$$

Thus there are infinite series

$$a_0 + r_1c + r_2c^2 + \cdots$$
$$b_0 + s_1c + s_2c^2 + \cdots$$

in $\mathcal{O}_v[x]$ with partial sums a_n of degrees d and b_n of degree at most $D - d$, and these partial sums converge coefficientwise to polynomials $a, b \in F[x]$. Since the values $v(-)$ of a non-null Cauchy sequence in F are eventually constant, $\mathcal{O}_v$ (and $\pi\mathcal{O}_v$) are closed sets. So $a, b \in \mathcal{O}_v[x]$. In the sequence $\{a_n\}$, the coefficients of x^d are constant; so $\deg(a) = \deg(a_0) = d$.

Suppose N is any positive integer. For all but finitely many n, $a_n - a$ and $b_n - b$ belong to $\pi^N\mathcal{O}_v[x]$; so

$$a_n b_n - ab = a_n(b_n - b) + (a_n - a)b \in \pi^N\mathcal{O}_v[x].$$

Thus $\{a_n b_n\}$ converges coefficientwise to ab. Since $f \equiv a_n b_n \pmod{c^{n+1}}$, the sequence $\{a_n b_n\}$ also converges coefficientwise to f. So $f = ab$.

Each $a_n - a_0$ and $b_n - b_0$ lies in $\pi\mathcal{O}_v[x]$, and these converge coefficientwise to $a - a_0$, $b - b_0 \in \pi\mathcal{O}_v[x]$, respectively. So $\bar{a} = \bar{a}_0 = \alpha$ and $\bar{b} = \bar{b}_0 = \beta$. ∎

(15.40) Proposition. *Suppose $F = F_v$ is a local field whose residue field k_v has characteristic p and order $q = p^r$. Let $\mu'(F)$ denote the subgroup of F^* consisting of elements of finite order not divisible by p. Then the canonical map $\mathcal{O}_v \to k_v$ restricts to a group isomorphism $\mu'(F) \cong k_v^*$.*

Proof. By Hensel's Lemma, the polynomial $x^q - x$ splits into q distinct linear factors in $\mathcal{O}_v[x]$; and by Gauss' Lemma, these factors can be chosen monic. So $x^{q-1} - 1$ has $q - 1$ distinct roots in $\mathcal{O}_v$, all of which belong to $\mu'(F)$. Since every element of $\mu'(F)$ is integral over $\mathcal{O}_v$, $\mu'(F) \subseteq \mathcal{O}_v$; so $\mu'(F) \subseteq \mathcal{O}_v^*$. Then $\mu'(F)$ has no element in the zero coset $\bar{0} \in k_v$. Thus it only remains to prove $\mu'(F)$ has at most one element in each of the $q - 1$ remaining cosets.

Suppose $\zeta, \zeta' \in \mu'(F)$ and $\zeta \equiv \zeta' \pmod{\pi}$, where $v(\pi) = 1$. For each positive integer n, there exist $z, z' \in \mu'(F)$ with

$$z^{p^n} = \zeta \quad \text{and} \quad z'^{p^n} = \zeta'.$$

In k_v, $\overline{z - z'}^{p^n} = \bar{\zeta} - \bar{\zeta'} = \bar{0}$. So the irreducible element π divides $z - z'$. Since π divides p, repeated use of the binomial theorem shows

$$\pi^n \text{ divides } z^{p^n} - z'^{p^n} = \zeta - \zeta'.$$

This is true for all $n > 0$, so $\zeta = \zeta'$. ∎

Note: If a local field F_v has positive characteristic ℓ, and k_v has characteristic p, then p, ℓ are two primes that vanish in k_v; so $p = \ell$. Then any root of $x^p - 1 = (x - 1)^p \in F[x]$ equals 1 and $\mu(F)$ has no element of order p. So $\mu(F) = \mu'(F)$ and $m(F) = q - 1$.

On the other hand, in any reference about local fields (such as Jacobson [89, §9.12]), the reader can find a proof that a local field F_v of characteristic 0 is a finite-degree field extension of a copy of the field $\mathbb{Q}_p$ of p-adic numbers (same p as characteristic (k_v)). Then an element ζ of F_v^* of order p^s is a root of a cyclotomic polynomial

$$\Phi(x) \;=\; x^{(p-1)p^{s-1}} + \cdots + x^{p^{s-1}} + 1 \,,$$

which is irreducible in $\mathbb{Q}_p[x]$ because $\Phi(x+1)$ is an Eisenstein polynomial. So $[F_v : \mathbb{Q}_p]$ is a multiple of $\deg(\Phi(x))$, putting a ceiling on the exponent s. Thus for local fields F of characteristic zero, $m(F) = p^s(q-1)$ for some integer $s \geq 0$. Since each global field F is embedded in its metric completions, $m(F)$ is finite for global fields too.

The norm residue symbols include not only the Hilbert symbols, but also the tame symbols:

(15.41) Proposition. *If F_v is a local field and q is the order of its residue field k_v, the tame symbol*

$$\tau_v : F_v^* \times F_v^* \;\to\; k_v^*$$

is the $(q-1)$ power norm residue symbol

$$\left(\frac{-,-}{v}\right)_{q-1} : F_v^* \times F_v^* \;\to\; \mu'(F_v)$$

followed by the isomorphism $\mu'(F_v) \cong k_v^*$, $z \mapsto z + P_v$.

Proof. See Serre [79, Chapter XIV, Proposition 8]. ∎

By (15.38) (vi), each norm residue symbol on a local field F_v is a power of the m_v power norm residue symbol, where $m_v = m(F_v)$ is the number of roots of unity in F_v. So the latter carries the most information. This information is enough to distinguish between elements of finite order in $K_2(F_v)$:

For each abelian group A, let $t(A)$ denote the subgroup of all elements of A with finite order. These elements are known as **torsion** elements and $t(A)$ as the **torsion subgroup** of A.

(15.42) Theorem. *If F_v is a local field, the m_v power norm residue symbol defines a split surjective homomorphism*

$$\lambda : K_2(F_v) \;\to\; \mu(F_v) \,, \quad \{a,b\} \;\mapsto\; \left(\frac{a,b}{v}\right)_{m_v} .$$

The kernel of λ is the group $m_v K_2(F_v)$ of m_v powers of elements in $K_2(F_v)$. This kernel is uniquely divisible; so λ restricts to an isomorphism

$$t(K_2(F_v)) \;\cong\; \mu(F_v) \;=\; t(F_v^*) \,.$$

Proof. Moore [68] proved λ is split surjective with divisible kernel $m_v K_2(F_v)$. Unique divisibility of this kernel was proved when F_v has positive characteristic by Tate [77, Theorem 5.5], and when the characteristic is 0 by Merkurjev [85]. ∎

Note: Since $F_v^* \cong K_1(F_v)$ via $u \mapsto \overline{u \oplus I_\infty}$, this theorem implies $t(K_2(F_v)) \cong t(K_1(F_v))$ for all local fields F_v.

The computation of the full m_v power norm residue map relies on knowing the value of the reciprocity isomorphism $w_{E/F}$ from (15.35), where $F = F_v$ and $E = F_v(a^{1/m_v})$. Much has been written about the computation of $w_{E/F}$ — to find such papers and books in the literature, look for the term **explicit reciprocity**.

The nth power reciprocity laws are generalized by the following product formula, which comes from Artin's Reciprocity Theorem. Suppose F is a global field and S is the set of places of F. Each ultrametric place arises from a discrete valuation v on F, so v is also used to denote that place. Each archimedean place is also denoted by v, with the understanding that if $F_v \cong \mathbb{R}$ and the completion at v corresponds to the embedding $\sigma : F \to \mathbb{R}$, then

$$\left(\frac{a,b}{v}\right)_2 \;=\; (\sigma(a),\sigma(b))_\infty \; ;$$

and if $F_v \cong \mathbb{C}$, every

$$\left(\frac{a,b}{v}\right)_n \;=\; 1 \,.$$

(15.43) Theorem. *If F is a global field and F^* has an element of finite order n, then*

$$\prod_{v \in S} \left(\frac{a,b}{v}\right)_n \;=\; 1$$

for all $a, b \in F^*$.

Proof. See Neukirch [99, Chapter VI, Theorem 8.1]. ∎

Note: If F^* has an element ζ of finite order $n > 2$, then F has no real embeddings; so this product may as well be calculated only over the ultrametric places.

To multiply the values of norm residue symbols, those values must lie in one group — typically in $\mu(F)$, which is a subgroup of each of the cyclic groups $\mu(F_v)$. If $m = m(F)$ and $m_v = m(F_v)$, each homomorphism $\mu(F_v) \to \mu(F)$ is a d_v power map, where d_v is a multiple of m_v/m. The above product formula can be written

$$\prod_{v \in S} \left(\frac{a,b}{v}\right)_{m_v}^{m_v/n} = 1,$$

where n divides m. So it is just the m/n power of the product formula

$$\prod_{v \in S} \left(\frac{a,b}{v}\right)_{m_v}^{m_v/m} = \prod_{v \in S} \left(\frac{a,b}{v}\right)_{m} = 1.$$

(15.44) Theorem. (Moore [68]) *For each global field F, the sequence*

$$K_2(F) \xrightarrow{\ \lambda\ } \bigoplus_{v \in T} \mu(F_v) \xrightarrow{\ \pi\ } \mu(F) \longrightarrow 1$$

is exact, where T is the set of ultrametric or real archimedean places of F, λ is the m_v power norm residue symbol in the v-coordinate, and π is the product for all v of the m_v/m power of the v-coordinate. ∎

The composite $\pi \circ \lambda$ is trivial by the reciprocity law (15.43) with $n = m$. The exactness at $\oplus \mu(F_v)$ also says every product formula

$$\prod_{v \in S} \left(\frac{a,b}{v}\right)_{n(v)} = 1$$

is a power of the one with all $n(v) = m$; so it must have $n(v)$ constant. That is, Theorem (15.43) describes all relations among norm residue symbols on a global field F.

The kernel of λ in (15.44) is known as the **wild kernel** of the global field F. Its elements vanish under the application of every norm residue symbol. Using his "continuous cochain cohomology," Tate found a way to detect some elements of the wild kernel. For example, he showed $\{-1, -1\}$ is a nonvanishing element of the wild kernel of $\mathbb{Q}(\sqrt{-35})$; for details, see the Appendix to Bass and Tate [73].

In (14.56) we used tame symbols to detect all elements of $K_2(\mathbb{Q})$ except $\{-1, -1\}$. The latter goes to 1 under all tame symbols on $\mathbb{Q}$. As it turns out (see §14H, Exercise 2), $K_2(\mathbb{Z})$ is cyclic of order 2, generated by $\{-1, -1\}$, and the exact sequence (14.56) can be rewritten as

$$1 \longrightarrow K_2(\mathbb{Z}) \xrightarrow{\ K_2(f)\ } K_2(\mathbb{Q}) \xrightarrow{\ \tau\ } \oplus_p K_1(\mathbb{Z}/p\mathbb{Z}) \longrightarrow 1,$$

where $f : \mathbb{Z} \to \mathbb{Q}$ is the localization map $n \mapsto n/1$ and, for each prime p, the p-coordinate of τ is the tame symbol associated with $v = v_p$ followed by the isomorphisms

$$k_v^* = (\mathbb{Z}_p/p\mathbb{Z}_p)^* \cong (\mathbb{Z}/p\mathbb{Z})^* \cong K_1(\mathbb{Z}/p\mathbb{Z}),$$

the last by thinking of each unit as the matrix $u \oplus I_\infty$.

More generally, suppose F is the field of fractions of a Dedekind domain R and $\mathcal{P}$ is the set of maximal ideals in R. If $P \in \mathcal{P}$ and $v = v_P$ is the P-adic discrete valuation on F, following the tame symbol defined by v by the standard isomorphisms

$$k_v^* \;=\; (R_P/PR_P)^* \;\cong\; (R/P)^* \;\cong\; K_1(R/P)\,,$$

and applying this composite in the P-coordinate for each $P \in \mathcal{P}$, we define a homomorphism

$$K_2(F) \xrightarrow{\ \tau\ } \bigoplus_{P \in \mathcal{P}} K_1(R/P)$$

whose kernel consists of those elements of $K_2(F)$ not detected by tame symbols. For instance, if $a, b \in R^*$, then $\tau(\{a, b\}) = 1$, since $v_P(a) = v_P(b) = 0$ for all $P \in \mathcal{P}$.

This tame symbol map τ is the beginning of an exact sequence explicitly described in Milnor [71, §13], which we present here without proof of its exactness:

(15.45) Theorem. (Bass, Tate) *If R is a Dedekind domain with field of fractions F and $\mathcal{P}$ is the set of maximal ideals of R, there is an exact sequence*

$$K_2(F) \xrightarrow{\ \tau\ } \bigoplus_{P \in \mathcal{P}} K_1(R/P) \xrightarrow{\ \sigma\ } K_1(R) \xrightarrow{\ K_1(f)\ } K_1(F)$$

$$\xrightarrow{\ V\ } \bigoplus_{P \in \mathcal{P}} K_0(R/P) \xrightarrow{\ \rho\ } K_0(R) \xrightarrow{\ K_0(f)\ } K_0(F) \longrightarrow 0$$

of groups, where $f : R \to F$ is the localization map $a \mapsto a/1$ and τ is the tame symbol map described above.

Proof. The K_0 part of this sequence is the G_0 exact sequence of Swan (6.53) once we identify $K_0(-)$ with $G_0(-)$ for the regular rings R/P, R, and F, through the isomorphism induced by inclusion $\mathcal{P}(-) \subseteq \mathcal{M}(-)$ in one direction and given by the Euler characteristic in the other. The map V takes $x \in F^* \cong K_1(F)$ to $v_P(x)[R/P]$ in the P-coordinate for each $P \in \mathcal{P}$. The homomorphism σ is the product of the maps $K_1(R/P) \to K_1(R)$ taking $x \in (R/P)^*$ to the Mennicke symbol $[a, b] \in SK_1(R)$, where $a, b \in R$ are chosen as follows: The element x is a coset $r + P$ with $r \in R$. As in §7C, Exercise 8, there is an ideal J of R with $P + J = R$ and $PJ = aR$ for some $a \in R$. By the Chinese Remainder Theorem, there exists $b \in (r + P) \cap (1 + J)$. Then $\sigma(x) = [a, b]$. Exactness is proved in Milnor [71, §13]. ∎

The long exact localization sequence (13.39) in Quillen K-theory specializes (see Quillen [73]) to a long exact sequence

$$\cdots \longrightarrow \bigoplus_{P\in\mathcal{P}} K_{n+1}(R/P) \longrightarrow K_{n+1}(R) \longrightarrow K_{n+1}(F)$$

$$\longrightarrow \bigoplus_{P\in\mathcal{P}} K_n(R/P) \longrightarrow K_n(R) \longrightarrow K_n(F) \longrightarrow \cdots$$

when R is a Dedekind domain with field of fractions F. Grayson [79, Corollary 7.13] shows the map from $K_2(F)$ to $\oplus_P K_1(R/P)$ in Quillen's sequence is just the tame symbol map τ; so Quillen's sequence extends (15.45) to the left.

Now suppose R is the ring of algebraic integers in a number field F. Then the residue fields R/P are finite, and each $K_2(R/P) = 1$. Also, by the Bass-Milnor-Serre Theorem (11.33), the kernel $SK_1(R)$ of $K_1(f)$ vanishes. So the localization sequence in this case restricts to a short exact sequence

$$1 \longrightarrow K_2(R) \longrightarrow K_2(F) \stackrel{\tau}{\longrightarrow} \bigoplus_{P\in\mathcal{P}} K_1(R/P) \longrightarrow 1.$$

For this reason, when F is a number field and $R = \text{alg. int.}(F)$, the group $K_2(R)$ is known as the **tame kernel** of F. Of course this tame kernel contains the wild kernel of F, since tame symbols are powers of m_v power norm residue symbols.

15D. Exercises

1. Suppose R is a Dedekind domain with field of fractions F, and P is a maximal ideal of R. If $v = v_p$ is the p-adic discrete valuation on F, prove $\mathcal{O}_v = R_P$, the ring of fractions a/b with $a, b \in R$ and $b \notin P$. *Hint:* If $x \in \mathcal{O}_v$ and $x \notin R_P$, then, since R_P is a discrete valuation ring by (7.22) and (7.26), $x^{-1} \in R_P \subseteq \mathcal{O}_v$. Show $x^{-1} \in R_P^*$, and derive a contradiction.

2. Continuing the notation above, assume P is odd ($2 \notin P$). Use the identification of $\widehat{R}_P$ with $\underleftarrow{\text{Lim}}\, R/P^n$ to show F_P^* has exactly four square classes, and they are represented by $1, r, \pi,$ and $r\pi$, where r is any nonsquare in in $\widehat{R}_P^*$ and π is any element of P not in P^2.

3. Verify that, for odd P, the Hilbert symbol on F_P is the $(q-1)/2$ power of the tame symbol $\tau_{\widehat{v}}$ on F_P, where $\widehat{v} = \widehat{v}_P$ and q is the size of R/P. *Hint:* Evaluate both on square class representatives.

4. If σ, τ are real embeddings of a number field F and $\sigma \neq \tau$, prove $|\ |_\sigma$ and $|\ |_\tau$ yield different topologies on F. *Hint:* If $|a|_\sigma < |a|_\tau$, find $b \in F$ with $|b|_\sigma < 1 < |b|_\tau$.

5. If F is a local field, prove $\{-1, -1\} = 1$ in $K_2(F)$ if and only if -1 is a sum of two squares in F. *Hint:* Say m is the order of $t(F^*)$. If -1 is not a square in F, $m/2$ is odd. Since $\{-1, -1\}^2 = 1, \{-1, -1\} = \{-1, -1\}^{m/2}$. Now use (15.42) and (15.38) (iv) and (vi).

6. The Hasse-Minkowski Theorem (see Lam [80, Chapter 6, (3.2)]) implies -1 is a sum of two squares in a number field F if and only if -1 is a sum of two squares in each metric completion of F. Use this and Exercise 5 to prove $\{-1, -1\}$ is in the wild kernel of a number field F if and only if -1 is a sum of two squares in F. *Note:* By Lam [80, Chapter 11, (2.11)], in a number field F, -1 is a sum of two squares if and only if F has no real embeddings and, at each even prime P of alg. int. (F), the local degree $[F_P : \mathbb{Q}_2]$ is even. For instance, if d is a square-free positive integer, $-1 = x^2 + y^2$ in $\mathbb{Q}(\sqrt{-d})$ if and only if $d \notin 7 + 8\mathbb{Z}$.

16

Brauer Groups

One can trace the historical origin of noncommutative ring theory to generalizations in arbitrary finite dimensions, and later over arbitrary fields F, of the multiplication of points in the complex plane. In a search for an arithmetic of higher dimensional space, Hamilton was forced into four dimensions and noncommutativity in order to produce a division ring. Among the "hypercomplex" number systems discovered in the late 19th and early 20th centuries, the finite-dimensional division algebras played a special role, partly because division by nonzero elements was allowed, but ultimately because of their role in the Wedderburn-Artin structure theory of semisimple rings. The isomorphism classes of finite-dimensional division F-algebras, sharing a common center F, form a group $Br(F)$ developed by Brauer in the early 1930s. This Brauer group has a construction analogous to that of the projective class group $\widetilde{K}_0(R)$, but with algebras in place of modules, tensor products in place of direct sums, and matrix rings in place of free modules. We present this construction of the Brauer group in §16A. In §16B we consider the Brauer group as a functor on posets of subfields and discuss scalar extensions (to "splitting fields") in which division algebras are embedded in matrix rings over fields. The generalizations of quaternions to "cyclic algebras" are discussed briefly in §16C. Finally, in §16D, we display the connection of $K_2(F)$ to $Br(F)$, which is sufficiently strong so that its proof in the 1980s by Merkurjev and Suslin also affirmed the classical conjecture that, if F contains an nth root of unity, the n-torsion part of the Brauer group is generated by cyclic algebras.

16A. The Brauer Group of a Field

Using algebras over a field F, multiplied by tensor product, in place of modules over a ring R, added by direct sum, we describe the analog to f.g. projective R-modules — namely, the finite-dimensional "central simple" F-algebras. Then we develop the "Brauer group" of F, in the same way we constructed the projective class group of R in §4A.

The tensor product of algebras is again an algebra:

(16.1) Proposition. *Suppose R is a commutative ring and A and B are R-algebras. Then the R-module $A \otimes_R B$ is an R-algebra under a multiplication defined on little tensors by $(a_1 \otimes b_1)(a_2 \otimes b_2) = a_1 a_2 \otimes b_1 b_2$.*

Proof. Associativity and commutativity isomorphisms compose in an isomorphism

$$t : (A \otimes_R B) \otimes_R (A \otimes_R B) \;\cong\; (A \otimes_R A) \otimes_R (B \otimes_R B)$$

taking $(a_1 \otimes b_1) \otimes (a_2 \otimes b_2)$ to $(a_1 \otimes a_2) \otimes (b_1 \otimes b_2)$. Multiplications in A and B distribute over addition and satisfy $(xr)y = (x(r1))y = x((r1)y) = x(ry)$ for $r \in R$; so they induce R-linear maps

$$m_A : A \otimes_R A \to A \quad \text{and} \quad m_B : B \otimes_R B \to B .$$

Following the canonical map

$$(A \otimes_R B) \times (A \otimes_R B) \;\to\; (A \otimes_R B) \otimes_R (A \otimes_R B)$$

by t, and then by $m_A \otimes m_B$, defines a multiplication on $A \otimes_R B$ that is balanced over R, so it distributes over addition; and it has the desired effect on little tensors. Evidently it is associative on little tensors, and by distributivity is associative in general. The element $1_A \otimes 1_B$ acts as an identity. The algebra axiom $r(xy) = (rx)y = x(ry)$ works for little tensors x and y since it works for A; and distributivity extends it to all $x, y \in A \otimes_R B$. ∎

For a commutative ring R, suppose A, B, and C are R-algebras and $f : A \otimes_R B \to C$ is R-linear. By the distributive law in the algebra $A \otimes_R B$ and the additivity of f, the map f is an R-algebra homomorphism if and only if

(i) $f((a_1 \otimes b_1)(a_2 \otimes b_2)) = f(a_1 \otimes b_1)f(a_2 \otimes b_2)$ for all $a_1, a_2 \in A$ and $b_1, b_2 \in B$,

and

(ii) $f(1_A \otimes 1_B) = 1_C .$

As an application, we compute $M_n(R) \otimes_R B$:

(16.2) Proposition. *Suppose B is an R-algebra and n is a positive integer. Then $M_n(R) \otimes_R B \cong M_n(B)$ as R-algebras. If also m is a positive integer, $M_n(R) \otimes_R M_m(R) \cong M_{nm}(R)$ as R-algebras.*

Proof. The map $M_n(R) \times B \to M_n(B)$, taking $((r_{ij}), b)$ to $(r_{ij}b)$, is balanced over R, inducing an R-linear (and right B-linear) map α from $M_n(R) \otimes_R B$ to $M_n(B)$. Matrix units ϵ_{ij} form a right B-basis of $M_n(B)$. Let β denote the B-linear map from $M_n(B)$ to $M_n(R) \otimes_R B$ taking ϵ_{ij} to $\epsilon_{ij} \otimes 1_B$ for $1 \leq i, j \leq n$. Since $\alpha \circ \beta$ is the identity on matrix units, β is injective. Since $M_n(R) \otimes_R B$ is spanned as a right B-module by the $\epsilon_{ij} \otimes 1_B$, β is surjective. So β has a two-sided inverse that must be α, and α is an R-linear isomorphism. For matrices $M, M' \in M_n(R)$ and elements $b, b' \in B$,

$$\alpha((M \otimes b)(M' \otimes b')) \;=\; \alpha(MM' \otimes bb') \;=\; MM'bb'$$
$$=\; MbM'b' \;=\; \alpha(M \otimes b)\alpha(M' \otimes b') \,.$$

And $\alpha(I_n \otimes 1_B) = I_n$. So α is an R-algebra isomorphism.

In particular, if m is also a positive integer, we have R-algebra isomorphisms

$$M_n(R) \otimes_R M_m(R) \;\cong\; M_n(M_m(R)) \;\cong\; M_{nm}(R) \,,$$

the second isomorphism being the erasure of internal brackets (see (1.33)). ■

As a further application of criteria (i) and (ii) above, we show $\otimes_R$ is an associative, commutative binary operation, with identity object R, on any full subcategory $\mathcal{C}$ of R-$\mathcal{A}\mathfrak{lg}$ with Obj $\mathcal{C}$ closed under $\otimes_R$ and including R. First suppose $g : A \to A'$ and $h : B \to B'$ are R-algebra homomorphisms. Then

$$g \otimes h : A \otimes_R B \;\to\; A' \otimes_R B'$$
$$a \otimes b \;\mapsto\; g(a) \otimes h(b)$$

is R-linear, multiplicative on little tensors, and takes $1_A \otimes 1_B$ to $1_{A'} \otimes 1_{B'}$. So $g \otimes h$ is an R-algebra homomorphism. If g and h are isomorphisms, $g \otimes h$ is an isomorphism (with inverse $g^{-1} \otimes h^{-1}$). So in $\mathcal{C}$, $A \cong A'$ and $B \cong B'$ imply $A \otimes_R B \cong A' \otimes_R B'$.

The R-linear isomorphisms from §5C,

$$A \otimes_R (B \otimes_R C) \;\cong\; (A \otimes_R B) \otimes_R C \,,$$
$$a \otimes (b \otimes c) \;\mapsto\; (a \otimes b) \otimes c$$

$$A \otimes_R B \;\cong\; B \otimes_R A \,,$$
$$a \otimes b \;\mapsto\; b \otimes a$$

$$R \otimes_R A \;\cong\; A \,,$$
$$r \otimes a \;\mapsto\; ra$$

are multiplicative on little tensors and take 1 to 1 when $A, B, C \in \mathcal{C}$. So, if the category $\mathcal{C}$ is modest, its set $\mathcal{I}(\mathcal{C})$ of isomorphism classes is an abelian monoid under $c(A) \cdot c(B) = c(A \otimes_R B)$.

Suppose F is a field and V is an F-vector space. The F-**dimension** of V is the free rank of V, and will be denoted by $[V : F]$. A **finite-dimensional** (or **f.d.**) F-algebra is an F-algebra A with finite F-dimension $[A : F]$ as a vector space.

If V is an F-vector space (finite- or infinite-dimensional), each subspace V' of V is a direct summand of V, since an F-basis of V' is part of an F-basis of V and the other part spans a complement V'', so $V' \overset{\bullet}{\oplus} V'' = V$. The F-linear inclusion $i : V' \to V$ and the projection $\pi : V \to V'$ along V'' satisfy $\pi \circ i = i_{V'}$. If we also have an F-vector space W with a subspace W', then

$$i \otimes i : V' \otimes_F W' \quad \to \quad V \otimes_F W$$
$$v' \otimes w' \quad \mapsto \quad v' \otimes w'$$

is an injective F-linear map, since $(\pi \otimes \pi) \circ (i \otimes i)$ is the identity map on $V' \otimes_F W'$. So we can safely identify $V' \otimes_F W'$ with the sums of little tensors $v \otimes w$ in $V \otimes_F W$ with $v \in V'$ and $w \in W'$, and we shall do so.

A **subalgebra** of an algebra A is just a submodule of A that is also a subring of A (sharing the same 1). As above, the tensor product of subalgebras of A and B is a subalgebra of $A \otimes_F B$, and the tensor product of ideals in A and B is an ideal of $A \otimes_F B$.

For instance, $A \otimes_F B$ has subalgebras

$$A \otimes_F F1_B \quad \cong \quad A \otimes_F F \quad \cong \quad A \quad \text{and}$$
$$F1_A \otimes_F B \quad \cong \quad F \otimes_F B \quad \cong \quad B \,.$$

These subalgebras (on the left) equal the sets

$$\boldsymbol{A} \otimes \boldsymbol{1} \quad = \quad \{a \otimes 1_B : a \in A\} \quad \text{and}$$
$$\boldsymbol{1} \otimes \boldsymbol{B} \quad = \quad \{1_A \otimes b : b \in B\} \,,$$

respectively, and inverses of the above isomorphisms embed A and B as subalgebras of $A \otimes_F B$:

$$A \quad \cong \quad A \otimes 1 \, , \quad B \quad \cong \quad 1 \otimes B \,.$$
$$a \quad \mapsto \quad a \otimes 1_B \quad b \quad \mapsto \quad 1_A \otimes b$$

The elements $a \otimes 1_B$ commute with the $1_A \otimes b$, and the F-dimensions of $A \otimes 1$ and $1 \otimes B$ multiply up to the F-dimension of $A \otimes_F B$.

For subalgebras A and B of an F-algebra C, let AB denote the set of finite length sums of products ab with $a \in A$ and $b \in B$.

(16.3) Definition. Say a f.d. F-algebra C is the **internal tensor product** $A \overset{\bullet}{\otimes}_F B$ of subalgebras A and B if

(i) $ab = ba$ for all $a \in A$ and $b \in B$,
(ii) $AB = C$, and
(iii) $[A : F][B : F] = [C : F]$.

Above we saw that $A \otimes_F B$ is the internal tensor product $(A \otimes 1) \overset{\bullet}{\otimes} (1 \otimes B)$.

(16.4) Proposition. *Suppose A', B', and C are f.d. F-algebras. There is an F-algebra isomorphism $\mu : A' \otimes_F B' \cong C$ if and only if $C = A \overset{\bullet}{\otimes}_F B$ for some subalgebras $A \cong A'$ and $B \cong B'$.*

Proof. If there is such an isomorphism μ, then $A = \mu(A' \otimes 1)$ and $B = \mu(1 \otimes B')$ are subalgebras of C, and properties (i), (ii), and (iii) for A and B follow from the same properties of $A' \otimes 1$ and $1 \otimes B'$. So $C = A \overset{\bullet}{\otimes}_F B$.

Conversely, suppose $C = A \overset{\bullet}{\otimes}_F B$ for subalgebras A and B. Since multiplication $A \times B \to C$ is balanced over F, there is an F-linear map

$$\mu : A \otimes_F B \quad \to \quad C \ , \quad \sum a_i \otimes b_i \quad \to \quad \sum a_i b_i \ .$$

From (i), μ is multiplicative on little tensors; and $\mu(1 \otimes 1) = 1$. So μ is an F-algebra homomorphism. And μ is surjective by (ii), and therefore injective by (iii). So $A \otimes_F B \cong C$. ∎

An advantage of working over a field F is the canonical expression of elements in $V \otimes_F W$ in terms of bases of V and W:

(16.5) Lemma. *Suppose V and W are vector spaces over a field F, with F-bases X and Y, respectively. For each $c \in V \otimes_F W$, there are unique functions*

$$s : Y \to V \ , \ t : X \to W \ ,$$

$$and \quad f : X \times Y \to F \ ,$$

with finite support, for which

$$c \ = \ \sum_{y \in Y} s(y) \otimes y \ = \ \sum_{x \in X} x \otimes t(x) \ = \ \sum_{x,y} f(x,y)(x \otimes y) \ .$$

Proof. The existence of s, t, and f is immediate from the relations among little tensors. For uniqueness of s, let $\oplus_Y V$ denote the set of functions $Y \to V$ with finite support, made an F-module by pointwise addition and scalar multiplication:

$$
\begin{aligned}
(g_1 + g_2)(y) &= g_1(y) + g_2(y) \,, \\
(rg)(y) &= r(g(y)) \,,
\end{aligned}
$$

for $g, g_1, g_2 \in \oplus_Y V$, $r \in F$, and $y \in Y$. The function

$$
\begin{aligned}
V \times (\oplus_Y F) &\rightarrow \oplus_Y V \,, \\
(v, h) &\mapsto h(-)v
\end{aligned}
$$

where $h(-)v$ is the function taking each y to $h(y)v$, is balanced over F, inducing an F-linear map from $V \otimes_F (\oplus_Y F)$ to $\oplus_Y V$. The F-linear composite

$$
V \otimes_F W \;\cong\; V \otimes_F (\oplus_Y F) \;\rightarrow\; \oplus_Y V
$$

$$
v \otimes \sum_y h(y)y \;\mapsto\; v \otimes h \;\mapsto\; h(-)v
$$

takes $\sum_{y \in Y} s(y) \otimes y$ to s. So s is uniquely determined in this expression. A similar composite

$$
V \otimes_F W \;\cong\; (\oplus_X F) \otimes_F W \;\rightarrow\; \oplus_X W
$$

shows t is unique.

The function

$$
\begin{aligned}
(\oplus_X F) \times (\oplus_Y F) &\rightarrow \oplus_{X \times Y} F \,, \\
(g, h) &\mapsto g \cdot h
\end{aligned}
$$

where $g \cdot h$ takes (x, y) to $g(x)h(y)$, is balanced over F. The resulting F-linear composite

$$
V \otimes_F W \;\cong\; (\oplus_X F) \otimes_F (\oplus_Y F) \;\rightarrow\; \oplus_{X \times Y} F
$$

$$
\left(\sum_x g(x)x \right) \otimes \left(\sum_y h(y)y \right) \;\mapsto\; g \otimes h \;\mapsto\; g \cdot h
$$

takes $\sum_{x,y} f(x,y)(x \otimes y)$ to f, proving f is unique. ∎

Note: The uniqueness of f amounts to the fact that the elements $x \otimes y$ for $x \in X$, $y \in Y$ are distinct and form an F-basis of $V \otimes_F W$.

The **center** $Z(A)$ of an F-algebra A is the set of elements $a \in A$ with $ab = ba$ for all $b \in A$. It is a commutative subring of A containing $F1_A$, so it is a commutative subalgebra of A.

(16.6) Proposition. *If F is a field and A and B are F-algebras, then*
$$Z(A \otimes_F B) = Z(A) \otimes_F Z(B) .$$

Proof. The right side is contained in the left, since each little tensor on the right commutes with $a \otimes b$ for all $a \in A$ and $b \in B$. Suppose X and Y are F-bases of A and B, extending F-bases of $Z(A)$ and $Z(B)$, respectively. Assume

$$c = \sum_{x,y} f(x,y)(x \otimes y)$$

belongs to $Z(A \otimes_F B)$, where $f(x,y) \in F$ and is 0 for all but finitely many $(x,y) \in X \times Y$. Then

$$c = \sum_y ((\sum_x f(x,y)x)) \otimes y$$
$$= \sum_x (x \otimes (\sum_y f(x,y)y)) .$$

For all $a \in A$ and $b \in B$, c commutes with $a \otimes 1$ and $1 \otimes b$. The uniqueness of s and t in the lemma shows $f(x,y) = 0$ whenever $x \notin Z(A)$ or $y \notin Z(B)$. So $c \in Z(A) \otimes_F Z(B)$. ∎

(16.7) Definition. A **central F-algebra** is an F-algebra A whose center $Z(A)$ is $F1_A$.

Now we can draw a parallel between f.g. projective modules over a ring and certain f.d. F-algebras. The matrix rings $M_1(F), M_2(F), M_3(F), \dots$ can serve as F-algebra analogs of f.g. free modules $0, R, R^2, \dots$. The f.g. projective modules over a ring R are the direct summands of f.g. free R-modules. Analogous to them are the F-algebras A that have a complementary F-algebra B, under tensor product.

Assume A and B are F-algebras and $A \otimes_F B \cong M_n(F)$ in F-$\mathcal{A}\mathfrak{lg}$ for some positive integer n. If A and B have finite-dimensional F-subspaces V and W, then $V \otimes_F W$ is isomorphic to a direct summand of $A \otimes_F B$. So

$$[V \otimes_F W : F] = [V : F][W : F] \leq n^2 .$$

Thus A and B must be f.d. F-algebras. If A has a (two-sided) ideal J, then $J \otimes_F B$ is an ideal of $A \otimes_F B$. Since $M_n(F)$ is a simple ring, $J \otimes_F B$ is either 0 or $A \otimes_F B$. Since $[A : F][B : F] = n^2$, $[B : F] \neq 0$. So $[J : F]$ is either 0 or $[A : F]$, forcing $J = 0$ or $J = A$. *This proves A is a simple ring*, as is B, since $B \otimes_F A \cong M_n(F)$ too. And (16.6) shows

$$[Z(A) : F][Z(B) : F] = [Z(A \otimes_F B) : F] = [Z(M_n(F)) : F] .$$

By (8.51), $Z(M_n(F)) = F \cdot I_n$, so the latter dimension is 1. But then $Z(A) = F \cdot 1_A$ and $Z(B) = F \cdot 1_B$.

We have proved:

(16.8) Theorem. *If F is a field, A and B are F-algebras, and $A \otimes_F B \cong M_n(F)$ as F-algebras for some positive integer n, then A and B are nonzero f.d. central simple F-algebras.* ∎

Next we prove the converse by finding a complement under $\otimes_F$ to each nonzero f.d. central simple F-algebra A. For this (and an application in the next section) we need:

(16.9) Lemma. *If A is a simple F-algebra and B is a f.d. central simple F-algebra, then $A \otimes_F B$ is simple.*

Proof. Suppose J is a nonzero ideal of $A \otimes_F B$. Choose an F-basis $w_1, \dots, w_n$ of B. From (16.5), each element of $A \otimes_F B$ has a unique expression $\sum_i a_i \otimes w_i$ with $a_i \in A$; say the **length** of this element is the number of i with $a_i \neq 0$. Choose a nonzero element $\alpha = \sum_i a_i \otimes w_i$ in J of smallest length. Renumber the basis, if necessary, so $a_1 \neq 0$.

Since A is simple, $A a_1 A = A$; so $(A \otimes 1)\alpha(A \otimes 1)$ includes an element $\alpha' = (1 \otimes w_1) + (a_2 \otimes w_2) + \cdots + (a_n \otimes w_n)$ of no greater length than α. For each $a \in A$, $(a \otimes 1)\alpha' - \alpha'(a \otimes 1)$ is an element of J of shorter length than α'; so it equals 0. By uniqueness of the expression $\sum_i s(w_i) \otimes w_i$ in (16.5), each $aa_i - a_i a = 0$ for $2 \leq i \leq n$. So $a_2, \dots, a_n \in Z(A) = F$, and

$$\alpha' = 1 \otimes (w_1 + a_2 w_2 + \cdots + a_n w_n) .$$

Since $w_1 + a_2 w_2 + \cdots + a_n w_n \neq 0$ in the simple ring B, $(1 \otimes B)\alpha'(1 \otimes B) = 1 \otimes B$. So $A \otimes_F B = (A \otimes 1)(1 \otimes B)$ is contained in J. ∎

For each F-algebra A, the additive endomorphisms of A form an F-algebra $\mathrm{End}_{\mathbb{Z}}(A)$, where $(\phi + \psi)(a) = \phi(a) + \psi(a)$, $(r\phi)(a) = r\phi(a) = \phi(a)r$, and $\phi\psi = \phi \circ \psi$, for all $\phi, \psi \in \mathrm{End}_{\mathbb{Z}}(A)$, $r \in F$, and $a \in A$. Among the subalgebras is $\mathrm{End}_F(A)$. Each choice of F-basis of A determines a matrix representation

$$\mathrm{End}_F(A) \cong M_n(F) , \quad n = [A : F] ,$$

which is an F-algebra isomorphism. The opposite ring A^{op} of A has the same center, ideals, and F-dimension as A.

(16.10) Theorem. *If F is a field and A is a f.d. central simple F-algebra, there is an F-algebra isomorphism*

$$A \otimes_F A^{op} \cong \mathrm{End}_F(A) , \quad \sum a_i \otimes a_i' \mapsto \sum a_i(-)a_i' .$$

So $A \otimes_F A^{op} \cong M_n(F)$ as F-algebras, where $n = [A : F]$.

Proof. If $a, a' \in A$, there is an F-linear map $a(-)a' : A \to A$ taking x to axa'. The function

$$A \times A^{op} \to \mathrm{End}_F(A)$$
$$(a, a') \mapsto a(-)a'$$

is balanced over F, inducing an F-linear map

$$\phi : A \otimes_F A^{op} \;\to\; \mathrm{End}_F(A)$$

taking each $a \otimes a'$ to $a(-)a'$. Denoting the multiplication in A^{op} by $\#$,

$$\phi((a_1 \otimes a_1')(a_2 \otimes a_2')) \;=\; \phi(a_1 a_2 \otimes a_1' \# a_2') \;=\; a_1 a_2 (-) a_2' a_1'$$
$$= \; \phi(a_1 \otimes a_1') \circ \phi(a_2 \otimes a_2') \ .$$

Since $\phi(1 \otimes 1) = i_A$ as well, ϕ is a (nonzero) F-algebra homomorphism. Therefore $\ker \phi \neq A \otimes_F A^{op}$. By (16.9), $A \otimes_F A^{op}$ is simple; so $\ker \phi = 0$ and ϕ is injective. Since

$$[\mathrm{End}_F(A) : F] \;=\; [M_n(F) : F] \;=\; n^2$$
$$= \; [A : F]\,[A^{op} : F] \;=\; [A \otimes_F A^{op} : F] \ ,$$

ϕ must also be surjective.   ∎

(16.11) Definition. If R is a nonzero commutative ring, an **Azumaya R-algebra** is an R-algebra A whose center is $R1_A$, with $ann_R(A) = 0$, and for which the R-algebra map

$$A \otimes_R A^{op} \;\to\; \mathrm{End}_R(A) \ ,$$

taking each $a \otimes a'$ to $a(-)a'$, is an isomorphism.

As we see from (16.8) and (16.10), for F a field, the Azumaya F-algebras are the nonzero f.d. central simple F-algebras, and are exactly the F-algebras A for which $A \otimes_F B \cong M_n(F)$ in F-$\mathcal{A}\mathfrak{lg}$ for some $B \in F$-$\mathcal{A}\mathfrak{lg}$ and some positive integer n. Let

$$\mathcal{A}_{\mathfrak{z}}(F)$$

denote the full subcategory of F-$\mathcal{A}\mathfrak{lg}$ whose objects are the Azumaya F-algebras.

For each field F, the category $\mathcal{A}_{\mathfrak{z}}(F)$ is modest: Each object A of $\mathcal{A}_{\mathfrak{z}}(F)$ is F-linearly isomorphic to some F^n, and hence isomorphic in $\mathcal{A}_{\mathfrak{z}}(F)$ to F^n with some multiplication determined by choosing the n coordinates of each product $e_i e_j$. Of course $F \in \mathcal{A}_{\mathfrak{z}}(F)$. And if $A, B \in \mathcal{A}_{\mathfrak{z}}(F)$, there exist positive integers m and n with $A \otimes_F A^{op} \cong M_m(F)$ and $B \otimes_F B^{op} \cong M_n(F)$. Then

$$(A \otimes_F B) \otimes_F (A^{op} \otimes_F B^{op})$$
$$\cong \; (A \otimes_F A^{op}) \otimes_F (B \otimes_F B^{op})$$
$$\cong \; M_m(F) \otimes_F M_n(F) \;\cong\; M_{mn}(F) \ ;$$

so $A \otimes_F B \in \mathcal{A}_{\mathfrak{z}}(F)$. Therefore the set $\mathfrak{I}(\mathcal{A}_{\mathfrak{z}}(F))$ of isomorphism classes is an abelian monoid under the multiplication

$$c(A) \cdot c(B) \;=\; c(A \otimes_F B) \ .$$

(16.12) Definition. For each field F, the **Brauer-Grothendieck group** of F is the group completion

$$BrG(F) \;=\; K_0(\mathcal{A}_\mathfrak{z}(F), \otimes_F)$$

of $(\mathfrak{I}(\mathcal{A}_\mathfrak{z}(F)), \otimes_F)$, written in multiplicative notation. As in (3.20)–(3.22), $BrG(F)$ is generated by elements $[A]$ for $A \in \mathcal{A}_\mathfrak{z}(F)$, and $[A][B] = [A \otimes_F B]$. A typical element is $[A][B]^{-1}$, which can also be written

$$[A \otimes_F B^{op}] \, [B \otimes_F B^{op}]^{-1} \;=\; [A'] \, [M_n(F)]^{-1} \,,$$

where $A' \in \mathcal{A}_\mathfrak{z}(F)$. And $[A] = [B]$ if and only if

$$A \otimes_F C \;\cong\; B \otimes_F C$$

as F-algebras for some $C \in \mathcal{A}_\mathfrak{z}(F)$. Tensoring with C^{op}, we find $[A] = [B]$ if and only if

$$M_n(A) \;\cong\; M_n(B)$$

as F-algebras, for some positive integer n. The identity element of $BrG(F)$ is the class $[F]$.

Continuing the parallel between f.g. projective modules under $\oplus$ and Azumaya F-algebras under $\otimes_F$, $BrG(F)$ is like K_0 of a ring. Analogous to the projective class group $\widetilde{K}_0$ of a ring is the following useful quotient:

(16.13) Definition. For each field F, the **Brauer group** of F is the quotient group

$$Br(F) \;=\; \frac{BrG(F)}{\mathcal{F}} \,,$$

where $\mathcal{F}$ is the subgroup generated by all $[M_n(F)]$ for positive integers n. It is an abelian group with typical element a coset $\overline{[A]}$ of $[A]$, for $A \in \mathcal{A}_\mathfrak{z}(F)$. The operation is

$$\overline{[A]} \; \overline{[B]} \;=\; \overline{[A \otimes_F B]} \,,$$

the identity is $\overline{[F]}$, and

$$\overline{[A]}^{-1} \;=\; \overline{[A^{op}]} \,.$$

Since $M_n(F) \otimes_F M_m(F) \cong M_{mn}(F)$, the elements of $\mathcal{F}$ each have the form $[M_s(F)][M_r(F)]^{-1}$. Thus $\overline{[A]} = \overline{[B]}$ in $Br(F)$ if and only if

$$[A] \, [B]^{-1} \;=\; [M_s(F)] \, [M_r(F)]^{-1} \,, \quad \text{so}$$

$$[M_r(A)] \;=\; [A] \, [M_r(F)] \;=\; [B] \, [M_s(F)] \;=\; [M_s(B)],$$

for positive integers r, s. Combining this with the criterion for equality in $BrG(F)$, we find $\overline{[A]} = \overline{[B]}$ in $Br(F)$ if and only if

$$M_u(A) \;\cong\; M_v(B)$$

as F-algebras, for some positive integers u and v. When the latter condition holds, we say A is **similar** to B, or write $A \sim B$. Evidently similarity is an equivalence relation among the objects of $\mathcal{A}_3(F)$.

Notation: We will focus on $Br(F)$ and make no further reference to $BrG(F)$; so to simplify notation we will write $\overline{[A]}$ as just $[A]$, omitting the bar, to denote the element of $Br(F)$ associated with $A \in \mathcal{A}_3(F)$. It is useful to think of $[A] \in Br(F)$ as the **similarity class** (= equivalence class under similarity) of A in Obj $\mathcal{A}_3(F)$.

Like ideals in $\widetilde{K}_0$ of a Dedekind domain, or anisotropic forms in the Witt ring of a field, division rings act as special representatives of similarity classes making up the Brauer group of a field F. Finite-dimensional F-algebras are left artinian, since a descending chain of left ideals would have descending F-dimensions. So each $A \in \mathcal{A}_3(F)$ is a simple left artinian F-algebra. By the Wedderburn-Artin Theorems (8.28)–(8.33),

$$A \;\cong\; M_n(D) \;\cong\; D \otimes_F M_n(F)$$

in $\mathcal{A}_3(F)$ for a unique $n \geq 1$ and for a division ring $D \in \mathcal{A}_3(F)$, uniquely determined up to isomorphism in $\mathcal{A}_3(F)$. We call D an **underlying division algebra** of A. Put another way, if $\mathbf{Div}(F)$ is the set of isomorphism classes of division rings in $\mathcal{A}_3(F)$, there is a bijection

$$\mathrm{Div}(F) \;\rightarrow\; Br(F) \;,\quad c(D) \;\mapsto\; [D] \;.$$

So $A, B \in \mathcal{A}_3(F)$ are similar if and only if they have the same underlying division algebras. The group operation in $Br(F)$ imposes, through this bijection, a group structure on $\mathrm{Div}(F)$, with $c(D_1) \cdot c(D_2) = c(D_3)$ whenever

$$D_1 \otimes_F D_2 \;\cong\; M_n(D_3) \;\cong\; D_3 \otimes_F M_n(F)$$

as F-algebras. This is the description of the Brauer group in its first appearance in Brauer [32]. From this point of view, calculation of $Br(F)$ amounts to the classification of all f.d. central division F-algebras up to F-algebra isomorphism.

(16.14) Examples.

(i) Suppose F is an algebraically closed field. Every f.d. division F-algebra equals F, since each element is algebraic over F (see (8.44) for details). So $Br(F)$ is the trivial group $\{[F]\}$.

(ii) By a famous theorem of Frobenius, for $\mathbb{R}$ the field of real numbers, every f.d. division $\mathbb{R}$-algebra is $\mathbb{R}$, $\mathbb{C}$, or $\mathbb{H}$ (= the ring of Hamilton's quaternions). For an elementary proof, see Herstein [75, §7.3]. Recall that $\mathbb{H}$ has $\mathbb{R}$-basis $1, i, j, ij$, where $ji = -ij$ and $i^2 = j^2 = (ij)^2 = -1$. Now $\mathbb{C} = \mathbb{R}1 + \mathbb{R}i$ is commutative, hence it is not a central $\mathbb{R}$-algebra. But i does not commute with any quaternion having nonzero j-coefficient; so $\mathbb{C}$ is a maximal commutative subring of $\mathbb{H}$. Any quaternion commuting with both i and j has zero for its i-, j-, and ij-coefficients. So $Z(\mathbb{H}) = \mathbb{R}$. Thus $Br(\mathbb{R})$ is the cyclic group $\{[\mathbb{R}], [\mathbb{H}]\}$ of order 2.

Since $\epsilon_{11} M_n(D) \epsilon_{11} = D\epsilon_{11} \cong D$, the underlying division algebra of any $A \in \mathcal{A}_{\mathfrak{z}}(F)$ is isomorphic to eAe for some idempotent $e \in A$. More generally we have:

(16.15) Proposition. *If F is a field, $A \in \mathcal{A}_{\mathfrak{z}}(F)$ and e is a nonzero idempotent in A, then $eAe \in \mathcal{A}_{\mathfrak{z}}(F)$ and A is similar to eAe.*

Proof. By (8.20), (8.23), and (8.33), A has a simple left ideal L and the f.g. A-modules Ae and A are isomorphic to L^m and L^n for some positive integers m and n, respectively. The map $x \mapsto (-) \cdot x$ defines an F-algebra isomorphism $eAe \cong [\mathrm{End}_A(Ae)]^{op}$, so that, when $e = 1$, $A \cong [\mathrm{End}_A(A)]^{op}$. As in the proof of (8.27), if D is the division F-algebra $\mathrm{End}_A(L)$, then

$$[\mathrm{End}_A(Ae)]^{op} \cong [\mathrm{End}_A(L^m)]^{op} \cong M_m(D)^{op} \cong M_m(D^{op}),$$
$$[\mathrm{End}_A(A)]^{op} \cong [\mathrm{End}_A(L^n)]^{op} \cong M_n(D)^{op} \cong M_n(D^{op}),$$

where the last map in each sequence is the transpose. So eAe and A share the same underlying division algebra D^{op}. ∎

16A. Exercises

1. If R is a commutative ring and A is an R-algebra, prove $A[x] \cong A \otimes_R R[x]$ as R-algebras.

2. If R is a subring of a commutative ring S and G is a group, prove there is an isomorphism $SG \cong S \otimes_R RG$ of S-algebras.

3. If A and B are R-algebras and n is a positive integer, prove

$$M_n(A) \otimes_R B \cong M_n(A \otimes_R B) \cong A \otimes_R M_n(B)$$

as R-algebras.

4. In $M_2(\mathbb{C})$ consider the $\mathbb{R}$-subalgebras

$$A = \left\{ \begin{bmatrix} z & 0 \\ 0 & z \end{bmatrix} : z \in \mathbb{C} \right\}, \quad B = \left\{ \begin{bmatrix} u & v \\ -\overline{v} & \overline{u} \end{bmatrix} : u, v \in \mathbb{C} \right\},$$

where $\overline{w}$ denotes the complex conjugate of w. Prove $M_2(\mathbb{C}) \cong \mathbb{C} \otimes_{\mathbb{R}} \mathbb{H}$ as $\mathbb{R}$-algebras by showing $\mathbb{C} \cong A$, $\mathbb{H} \cong B$, and $M_2(\mathbb{C}) = A \overset{\bullet}{\otimes}_{\mathbb{R}} B$. *Hint.* To show $AB = M_2(\mathbb{C})$, note that A includes iI_2 and B includes

$$\begin{bmatrix} 1 & 0 \\ 0 & 1 \end{bmatrix}, \begin{bmatrix} i & 0 \\ 0 & -i \end{bmatrix}, \begin{bmatrix} 0 & 1 \\ -1 & 0 \end{bmatrix}, \begin{bmatrix} 0 & i \\ i & 0 \end{bmatrix},$$

so it must include ϵ_{st} and $i\epsilon_{st}$ for all $s, t \in \{1, 2\}$.

5. If F is a field, A is an F-algebra, and S is a subset of A, the **centralizer** (or **commutant**) of S in A is

$$Z_A(S) \;=\; \{a \in A : as = sa \quad \text{for all} \quad s \in S\}\,.$$

Show $Z_A(S)$ is a subalgebra of A. If A, B are F-algebras with subalgebras A', B', prove

$$Z_{A \otimes_F B}(A' \otimes_F B') \;=\; Z_A(A') \otimes_F Z_B(B')\,.$$

6. If F is a field, A is an F-algebra, and B is a f.d. central simple F-algebra, prove there is a containment preserving bijection from the set of ideals of A to the set of ideals of $A \otimes_F B$ given by $J \mapsto J \otimes_F B$. *Hint.* For the inverse, take $I \lhd A \otimes_F B$ to

$$I^* \;=\; \{a \in A : a \otimes 1 \in I\}\,.$$

To show $I = I^* \otimes_F B$, first note

$$I \supseteq (I^* \otimes 1)(1 \otimes B) \;=\; I^* \otimes_F B\,.$$

And if $w_1, \ldots, w_n$ is an F-basis of B and $c = \sum a_i \otimes w_i \in I$, use surjectivity of $B \otimes_F B^{op} \to \mathrm{End}_F(B)$ to find $f_j(-) = \sum b_i(-)b_i'$ taking w_i to 1_B if $i = j$, and to 0_B if $i \neq j$. Then $a_j \otimes 1 = (1 \otimes f_j)(c)$ belongs to I. This holds for all j, so $c \in I^* \otimes_F B$.

7. Suppose F is a field. In $\mathcal{A}_{\mathfrak{z}}(F)$, prove the cancellation law $A \otimes_F M_n(F) \cong B \otimes_F M_n(F)$ implies $A \cong B$.

8. Prove $\mathbb{H} \cong \mathbb{H}^{op}$ as $\mathbb{R}$-algebras, by using the final conclusion in Example (16.14) (ii). Can you find the isomorphism?

9. If F is a finite field, prove $Br(F)$ is the trivial group $\{[F]\}$. *Hint.* Review §12B, Exercise 9.

16B. Splitting Fields

If D is a division ring with center F, the noncommutativity in $M_n(D) \cong M_n(F) \otimes_F D$ can be thought of as having two sources — noncommutativity of matrix multiplication, and the noncommutativity in the division ring D. If D has a finite F-basis $v_1, \ldots, v_m$, this distinction is illusory: For $D = \overset{\bullet}{\oplus} F v_i$ with an F-algebra multiplication determined by the v_i-coefficients of each product $v_j v_k$, if $F \subseteq E$ is a field extension, D is contained in an E-algebra $ED = \overset{\bullet}{\oplus} E v_i$ with the same products $v_j v_k$. In this section we see that $E \otimes_F D \cong ED \cong M_n(E)$ for some field extension $F \subseteq E$. So the noncommutativity in D is also matrix-theoretic.

(16.16) Proposition. *There is a functor **Br** from the category of fields and inclusions of subfields to the category of abelian groups, where $Br(F)$ is the Brauer group of a field F; and if $i_{E/F} : F \to E$ is an inclusion of a subfield, then*

$$Br(i_{E/F}) : Br(F) \quad \to \quad Br(E)$$

takes $[A]$ to $[E \otimes_F A]$ for each Azumaya F-algebra A.

Proof. Suppose $A \in \mathcal{A}_{\mathfrak{z}}(F)$ with $[A : \Gamma] = n$, and $F \subseteq E$ is a field extension. As E-vector spaces,

$$E \otimes_F A \quad \cong \quad E \otimes_F F^n \quad \cong \quad (E \otimes_F F)^n \quad \cong \quad E^n \ .$$

By (16.6), $Z(E \otimes_F A) = E \otimes_F F 1_A = E(1_E \otimes 1_A)$. And, by (16.9), $E \otimes_A A$ is a simple ring. So $E \otimes_F A \in \mathcal{A}_{\mathfrak{z}}(E)$.

If D is a division ring in $\mathcal{A}_{\mathfrak{z}}(F)$ and $A, B \in \mathcal{A}_{\mathfrak{z}}(F)$ have the same underlying division ring D, then both of the algebras $E \otimes_F A \cong E \otimes_F D \otimes_F M_m(F)$ and $E \otimes_F B \cong E \otimes_F D \otimes_F M_n(F)$ have the same underlying division ring as $E \otimes_F D$. So $E \otimes_F (-)$ preserves similarity, and there is a function $Br(i_{E/F}) : Br(F) \to Br(E)$ taking $[A]$ to $[E \otimes_F A]$ for all $A \in \mathcal{A}_{\mathfrak{z}}(F)$. It is a group homomorphism, since, for $A, B \in \mathcal{A}_{\mathfrak{z}}(F)$,

$$
\begin{aligned}
E \otimes_F (A \otimes_F B) \quad &\cong \quad (E \otimes_F A) \otimes_F B \quad \cong \quad (A \otimes_F E) \otimes_F B \\
&\cong \quad ((A \otimes_F E) \otimes_E E) \otimes_F B \\
&\cong \quad (E \otimes_F A) \otimes_E (E \otimes_F B)
\end{aligned}
$$

as E-algebras. That Br preserves identity arrows and composites follows from the isomorphisms

$$F \otimes_F A \quad \cong \quad A \ , \quad L \otimes_E (E \otimes_F A) \quad \cong \quad L \otimes_F A \ . \qquad \blacksquare$$

(16.17) Definitions. Suppose $F \subseteq E$ is a field extension and $A \in \mathcal{A}_3(F)$. We say E is a **splitting field** for A, or E **splits** A, if

$$E \otimes_F A \cong M_n(E)$$

as E-algebras, for some positive integer n.

(16.18) Proposition. *Suppose $F \subseteq E$ is a field extension and $A \in \mathcal{A}_3(F)$.*

 (i) *The field E splits A if and only if $[A]$ is in the kernel of $Br(i_{E/F})$.*
 (ii) *If E splits A, every extension field of E splits A.*
 (iii) *The algebraic closure $\overline{F}$ of F splits A.*
 (iv) *If $A \cong M_n(D)$ for a division ring D in $\mathcal{A}_3(F)$, then A and D have the same splitting fields.*

Proof. If $A \cong M_n(D)$ is similar to D' for division rings $D, D' \in \mathcal{A}_3(F)$, then D is similar to, and hence isomorphic to, D'; so $A \cong M_n(D) \cong M_n(D')$. That is, the division rings similar to A are the underlying division rings of A. Applying this to $E \otimes_F A$ in $\mathcal{A}_3(E)$, the field E is an underlying division ring for $E \otimes_F A$ if and only if $E \otimes_F A$ is similar to E, proving (i). Properties (ii), (iii), and (iv) are direct consequences of (i) and triviality of $Br(\overline{F})$. ∎

For each field F and each $A \in \mathcal{A}_3(F)$, $\overline{F} \otimes_F A \cong M_n(\overline{F})$ as F-algebras, where $\overline{F}$ is an algebraic closure of F and n is a positive integer. If $A \cong F^m$ as F-vector spaces, then $\overline{F} \otimes_F A \cong \overline{F} \otimes_F F^m \cong \overline{F}^m$ as $\overline{F}$-vector spaces; so $[A : F] = [\overline{F} \otimes_F A : \overline{F}] = [M_n(\overline{F}) : \overline{F}] = n^2$. So the F-dimension of each Azumaya F-algebra A is a square n^2; the integer $n = \sqrt{[A : F]}$ is known as the **degree** of A. The **index** of $A \cong M_n(D)$ is the degree of its underlying division algebra D. So index (D) = degree $(D) = \sqrt{[D : F]}$.

(16.19) Theorem. *Suppose F is a field, D is a f.d. central division F-algebra of index d, and E is a maximal commutative subring of D. Then E is a splitting field for D, and $[D : E] = [E : F] = d$.*

Proof. Each commutative subring of D is an integral domain B, and its field of fractions

$$\{ab^{-1} : a, b \in B , \ b \neq 0\}$$

in D is also a commutative subring of D. Since E is maximal, it is already a field. Since EF is a commutative subring of D, $F \subseteq E$.

Now $A = \mathrm{End}_{\mathbb{Z}}(D)$ is an F-algebra, and there are F-algebra homomorphisms

$$\lambda : E \ \to \ A \quad , \quad \rho : D \ \to \ A$$
$$x \ \mapsto \ x \cdot (-) \qquad x \ \mapsto \ (-) \cdot x$$

with images the subalgebras $\lambda(E)$ and $\rho(D)$ of A. Then $\lambda(E)\rho(D)$ is a subalgebra of A contained in $\mathrm{End}_E({}_E D)$. The E-linearity of these elements of $\lambda(E)\rho(D)$ follows from the fact that elements of $\lambda(E)$ and $\rho(D)$ commute with each other; and this also tells us there is a surjective F-algebra homomorphism

$$\mu : E \otimes_F D \;\;\to\;\; \lambda(E)\rho(D) \;.$$
$$x \otimes y \;\;\mapsto\;\; x(-)y$$

Since $E \otimes_F D$ is simple and $\mu \neq 0$, μ is also injective. So

$$E \otimes_F D \;\;\cong\;\; \lambda(E)\rho(D) \;\;\subseteq\;\; \mathrm{End}_E({}_E D) \;.$$

To complete the proof, we need only show the latter containment is equality; for then $[E : F][D : F] = [D : E]^2[E : F] = [D : E][D : F]$, which implies $[D : E] = [E : F] = \sqrt{[D : F]} = d$ and $E \otimes_F D \cong M_d(E)$.

Let R denote $\lambda(E)\rho(D)$. It will suffice to prove that, for each list $d_1, \ldots, d_t$ of E-linearly independent elements in D, and each list $d_1', \ldots, d_t'$ in D, there exists $r \in R$ with $r(d_i) = d_i'$ for each i. (In the literature this is what it means to say R is a "dense ring of E-linear transformations on D.") We prove this by induction on t.

For $t = 1$, an E-linearly independent element d_1 is necessarily nonzero; then for each $d_1' \in D$,
$$d_1' \;=\; d_1 d_1^{-1} d_1' \;=\; \rho(d_1^{-1} d_1')(d_1) \;,$$
and $\rho(d_1^{-1} d_1') \in R$. Now assume elements of R suffice to send each list of $t-1$ E-linearly independent elements to every list of $t-1$ elements in D.

Claim. *If $d_1, \ldots, d_t$ are E-linearly independent elements of D, there exists $r \in R$ with $r(d_i) = 0$ for $1 \leq i \leq t-1$ and $r(d_t) \neq 0$.*

Proof of Claim. Suppose not. If $r, s \in R$ with $r(d_i) = s(d_i)$ for $1 \leq i \leq t-1$, then $(r-s)(d_t) = 0$ and $r(d_t) = s(d_t)$. So we can define a function $f : D^{t-1} \to D$ by

$$f(r(d_1), \ldots, r(d_{t-1})) \;=\; r(d_t)$$

whenever $r \in R$. Now D is an R-module via evaluation, and hence D^{t-1} is an R-module as well. With these actions, f is R-linear:

$$f(r \cdot (d_1, \ldots, d_{t-1}) + s \cdot (d_1, \ldots, d_{t-1})) \;=\; f((r+s) \cdot (d_1, \ldots, d_{t-1}))$$
$$=\; (r+s) \cdot d_t \;=\; r \cdot d_t + s \cdot d_t$$
$$=\; f(r \cdot (d_1, \ldots, d_{t-1})) + f(s \cdot (d_1, \ldots, d_{t-1})) \;, \text{ and}$$

$$f(s \cdot (r \cdot (d_1, \ldots, d_{t-1}))) \;=\; f((s \circ r) \cdot (d_1, \ldots, d_{t-1}))$$
$$=\; (s \circ r) \cdot d_t \;=\; s \cdot (r \cdot d_t) \;=\; s \cdot f(r \cdot (d_1, \ldots, d_{t-1})) \;.$$

As in (1.31), f must be a matrix multiplication:

$$f(x_1, \ldots, x_{t-1}) = [\phi_1 \cdots \phi_{t-1}] \begin{bmatrix} x_1 \\ \vdots \\ x_{t-1} \end{bmatrix} = \sum_{i=1}^{t-1} \phi_i(x_i) \,,$$

where each $\phi_i \in \text{End}_R(D)$. Now an R-linear map from D to D is right D-linear, so it is left multiplication by some $y \in D$. It is also left E-linear, so $ye = ye1_D = ey1_D = ey$ for all $e \in E$. Then $E[y]$ is a commutative subring of D. Maximality of E forces $y \in E$. So each ϕ_i is $y_i \cdot (-)$ for some $y_i \in E$. Then

$$d_t = f(d_1, \ldots, d_{t-1}) = \sum_{i=1}^{t-1} y_i d_i \,,$$

violating the E-linear independence of $d_1, \ldots, d_t$. This proves the claim.   ∎

Now choose $r_1, \ldots, r_t \in R$ with

$$r_i(d_j) = \begin{cases} z_i \neq 0 & \text{if } j = i \\ 0 & \text{if } j \neq i \,. \end{cases}$$

By the case $t = 1$, there exist r_i' with $r_i'(z_i) = d_i'$; then

$$r = \sum_{i=1}^{t} r_i' \circ r_i \in R$$

takes d_i to d_i' for each i, as required.   ∎

We close this section with an explanation of the term "splitting field." Suppose F is a field, $A \in \mathcal{A}_3(F)$, $[A : F] = n^2$, and $A \cong M_m(D)$ for a division ring $D \in \mathcal{A}_3(F)$. The matrix units times an F-basis of D form an F-basis of $M_m(D)$; so $[D : F] = n^2/m^2$ and $m \leq n$.

Since A is a simple artinian ring, it has only one isomorphism class $c(P)$ of simple modules, by (8.33). Since A is semisimple, every f.g. A-module M is semisimple, so it is isomorphic to P^r for some integer r determined by $[M : F]$. In particular, there are A-linear isomorphisms

$$A \cong M_m(D)\epsilon_{11} \overset{\bullet}{\oplus} \cdots \overset{\bullet}{\oplus} M_m(D)\epsilon_{mm} \cong P^m \,,$$

and A contains a list of m nonzero mutually orthogonal idempotents e_i (corresponding to ϵ_{ii}). If $A \cong M_1 \oplus \cdots \oplus M_s$ for nonzero A-modules M_i, each M_i is a quotient of A; so it is f.g. and isomorphic to P^{r_i} for some $r_i \geq 1$. So $r_1 + \cdots + r_s = m$ and $s \leq m$. If A contains nonzero mutually orthogonal idempotents $e_1, \ldots, e_t$ with sum e, then

$$A = Ae_1 \overset{\bullet}{\oplus} \cdots \overset{\bullet}{\oplus} Ae_t \overset{\bullet}{\oplus} A(1 - e) \,,$$

and hence $t \leq m$. Thus m is the maximum number of nonzero A-modules in a direct sum decomposition of A. So we say A is **split** (or **maximally split**) if $m = n$.

(16.20) Lemma. *If $A \in \mathcal{A}_3(F)$ and $[A : F] = n^2$, the following are equivalent:*

(i) *A is split,*

(ii) *$A \cong M_n(F)$ in $\mathcal{A}_3(F)$,*

(iii) *A is isomorphic to a direct sum of n nonzero A-modules,*

(iv) *A contains n nonzero mutually orthogonal idempotents,*

(v) *Either F is finite or some $a \in A$ has minimal polynomial over F that splits into n distinct linear factors in $F[x]$.*

Proof. The equivalence of (i) through (iv) follows from the preceding discussion. If $A \cong M_n(F)$ and F is not finite, some $a \in A$ corresponds to a diagonal matrix $\text{diag}(c_1, \ldots, c_n)$, where $c_1, \ldots, c_n$ are n different elements of F. Then the minimal polynomial of a over F is $\Pi_i(x - c_i)$.

Conversely, if F is finite, the only division ring in $Az(F)$ is F (according to Wedderburn's Theorem, §12B, Exercise 9); so $A \cong M_n(F)$. Or, if $a \in A$ has minimal polynomial $p(x) = \Pi_i(x - c_i)$, where $c_1, \ldots, c_n$ are n different elements of F, then A has a subalgebra

$$F[a] \;\cong\; \frac{F[x]}{p(x)F[x]} \;\cong\; \bigoplus_i \frac{F[x]}{(x - c_i)F[x]} \;\cong\; F^n$$

by the Chinese Remainder Theorem. So A contains n nonzero mutually orthogonal idempotents. ∎

Note: For a field extension $F \subseteq E$, E is a splitting field of $A \in \mathcal{A}_3(F)$ if and only if the Azumaya E-algebra $E \otimes_F A$ is split.

16B. Exercises

1. Show there is a functor Br from the category $\mathfrak{Field}$ of fields and all ring homomorphisms between them to $\mathcal{Ab}$, defined for each arrow $\sigma : F \to E$ by

$$Br(\sigma) : Br(F) \;\to\; Br(E) \,,$$
$$[A] \;\mapsto\; [E \otimes_F A]$$

where F acts on E through σ.

2. Show there is a "restriction of scalars" functor Br from the category of fields and ring isomorphisms between them to $\mathcal{Ab}$, defined for each arrow $\sigma : F' \cong F$ by

$$Br(\sigma) : Br(F) \;\to\; Br(F') \,,$$
$$[A] \;\mapsto\; [A_\sigma]$$

where A_σ is A with action of F' through σ.

3. If $\tau : F \to E$ is a ring homomorphism between fields, each Azumaya F-algebra is also an Azumaya $\tau(F)$-algebra, with scalars acting through the inverse isomorphism $\tau(F) \cong F$. Say E **splits** $A \in \mathcal{A}_\mathfrak{z}(F)$ **over** τ if E is a splitting field of A as a $\tau(F)$-algebra. Suppose F is a field, A is a f.d. simple F-algebra with center K, $F \subseteq E$ is a finite degree field extension, and $\sigma, \tau : K \to E$ are two ring homomorphisms fixing F. Prove E splits A over σ if and only if E splits A over τ.

4. Suppose $F \subseteq E$ is a field extension and $p(x)$ is an irreducible polynomial in $F[x]$. Show there is an E-algebra isomorphism

$$E \otimes_F \frac{F[x]}{p(x)F[x]} \cong \frac{E[x]}{p(x)E[x]} .$$

5. Suppose G is a finite group and F is a field of characteristic zero. So FG is isomorphic as an F-algebra to

$$M_{n(1)}(D_1) \oplus \cdots \oplus M_{n(r)}(D_r)$$

for f.d. division F-algebras D_i with centers K_i. If $F \subseteq E$ is a field extension, prove E is a splitting field for G (as defined in (8.46)) if and only if, for each i, every embedding of K_i into $\overline{E}$ (an algebraic closure of E) takes K_i into E, and E splits D_i over each such embedding. *Hint:* Using Exercise 2 of §16A and Exercise 4 above, show

$$EG \cong \bigoplus_i M_{n(i)} \left(\frac{E[x]}{p_i(x)E[x]} \otimes_{K_i} D_i \right) ,$$

where $K_i = F(\alpha_i)$ and α_i has minimal polynomial $p_i(x)$ over F. The action of K_i on the left argument is via $a(\alpha_i) \mapsto \overline{a(x)}$ for $a(x) \in F[x]$. Now use the Chinese Remainder Theorem, remembering that $p_i(x)$ must be separable since F has characteristic zero. So

$$EG \cong \bigoplus_{i,j} M_{n(i)}(E(\alpha_{ij}) \otimes_{K_i} D_i) ,$$

where for each i there is one root $\alpha_{ij} \in \overline{E}$ chosen from the roots of each irreducible factor of $p_i(x)$ in $E[x]$, and the action of K_i is through the embedding $a(\alpha_i) \mapsto a(\alpha_{ij})$ for each $a(x) \in F[x]$. Comparing centers shows E is a splitting field of G if and only if each $\alpha_{ij} \in E$, and $E \otimes_{K_i} D_i$ is a matrix ring over E for each embedding $K_i \to E$ over F.

Note: This argument works equally well in positive characteristics if FG is a **separable F-algebra**, meaning a semisimple ring with each field F_i separable over F.

6. If A is a ring, an A-module M is **faithful** if $aM = 0$ for some $a \in A$ implies $a = 0$. Prove the following theorem by using the approach in the proof of (16.19).

Jacobson Density Theorem. *If A is a ring with a faithful simple module M, the map*

$$\rho : A \;\to\; \mathrm{End}_{\mathbb{Z}}(M) \,, \; a \;\mapsto\; a \cdot (-) \,,$$

is an injective ring homomorphism with image a dense ring of D-linear transformations on M, where the division ring $D = \mathrm{End}_A(M)$ acts on M by evaluation.

Note: If A is left artinian, M will be a finite-dimensional D-vector space, and $\rho(A)$ will equal $\mathrm{End}_D(M)$, the centralizer of the centralizer of $\rho(A)$ in $\mathrm{End}_{\mathbb{Z}}(M)$. So the Jacobson Density Theorem generalizes the double centralizer property proved by another method in §8C, Exercise 11, for simple artinian rings A. Often the Density Theorem is used as a cornerstone of the theory of algebras, leading to the proofs of many results, including the Wedderburn-Artin Theorem and (16.19).

7. Use (16.19) to prove every f.d. central division $\mathbb{R}$-algebra A has index 1 or 2, and has a copy inside $M_2(\mathbb{C})$. This is a first step toward proving $A \cong \mathbb{R}$ or $\mathbb{H}$.

8. If D is a division ring in $\mathcal{A}_{\mathfrak{z}}(F)$ and n is a positive integer, show $A = M_n(D)$ has a maximal commutative F-subalgebra B, with $[B : F]^2 = [A : F]$.

9. Prove:

Skolem-Noether Theorem. *If F is a field, $A \in \mathcal{A}_{\mathfrak{z}}(F)$, and $\phi : B \to B'$ is an F-algebra isomorphism between subalgebras of A, then $\phi = u(-)u^{-1}$ for some $u \in A^*$.*

Hint: If M is a simple A-module and $D = \mathrm{End}_A(M)$, then M is a $D \otimes_F B$-module via $(d \otimes b)m = d(bm)$, and is also a $D \otimes_F B$-module M' via $(d \otimes b) * m = d(\phi(b)m)$. Since $D \otimes_F B$ is a simple artinian ring and $[M : F] = [M' : F]$, there is a $D \otimes_F B$-linear isomorphism $\theta : M \to M'$. Then θ is F-linear and

$$\theta(d(bm)) \;\; = \;\; (d \otimes b) * \theta(m) \;\; = \;\; d(\phi(b)\theta(m)) \,.$$

Take $b = 1$ and use the double centralizer property to show $\theta = u \cdot (-)$ for some $u \in A^*$. Now take $d = 1$ and use the fact that M is a faithful A-module.

16C. Twisted Group Rings

An ample source of f.d. simple F-algebras is the set of simple components of semisimple group rings FG. When G has a normal subgroup H with coset representatives $g_1, \ldots, g_m$ in G, the group ring FG is $\overset{\bullet}{\oplus} FHg_i$, and the multiplication depends on that in FH, the automorphisms $g_i(-)g_i^{-1}$ of FH, and the representative of the coset containing each g_ig_j. Often a simple component of FG arises by replacing FH by a homomorphic image that is a field (see Exercises 1 and 2).

There is a general construction of such algebras. Suppose R is a commutative ring, $\text{Aut}(R)$ is the group of ring automorphisms of R under composition, G is a group, $\theta : G \to \text{Aut}(R)$ is a group homomorphism, and $f : G \times G \to R^*$ is a function. The free R-module A based on a set $\{u_g : g \in G\}$, indexed by G, has an R-bilinear multiplication determined by

$$(ru_g)(su_h) \;=\; [r \cdot \theta(g)(s) \cdot f(g,h)]u_{gh}$$

whenever $r, s \in R$ and $g, h \in G$. This multiplication is associative if and only if

$$(16.21) \qquad\qquad \theta(g)(f(h,k)) \;=\; \frac{f(gh,k)}{f(g,hk)}\, f(g,h)$$

for all g, h, and k in G. In that case, A is a ring with multiplicative identity

$$1_A \;=\; f(1,1)^{-1}u_1 \;.$$

Then 1_A is R-linearly independent, and $R \to R1_A$, $r \mapsto r1_A$, is an injective ring homomorphism; so we can (and will) identify r with $r1_A$, making R a subring of A and making scalar multiplication the restriction to $R \times A$ of ring multiplication in A. To multiply in A one only needs the ring axioms and the relations

$$u_g u_h \;=\; f(g,h)u_{gh} \quad \text{and} \quad u_g r = \theta(g)(r)u_g$$

for $g, h \in G$ and $r \in R$. Assuming (16.21), A is called the **twisted group ring**

$$R \circ {}^{\theta}_{f} G$$

with **twist** θ and **factor set** f.

(16.22) Examples.

(i) If the factor set is **trivial** (meaning $f(g,h) = 1$ for all $f, g \in G$), and the twist is the trivial homomorphism ($\theta(g) = i_R$ for all $g \in G$), and if $U = G$ (with $u_g = g$), then $R \circ {}^{\theta}_{f} G$ is just the group ring RG.

(ii) If R is a field L, G is a finite subgroup of $\text{Aut}(L)$, and $\theta : G \to \text{Aut}(L)$ is inclusion, then $R \circ {}^{\theta}_{f} G$ is the **crossed product**

$$(L/F \,;\, f) \,,$$

where F is the fixed field L^G. By Artin's Theorem in Galois theory, $G = \text{Aut}(L/F)$.

(iii) Under the hypotheses of (ii), if $G = \langle g \rangle$ is cyclic of order n, $c \in F^*$, and

$$f(g^i, g^j) \;=\; \begin{cases} 1 & \text{if } \ i+j < n \\ c & \text{if } \ i+j \geq n \end{cases}$$

for $i, j \in \{0, 1, \ldots, n-1\}$, then $(L/F; f)$ is the **cyclic algebra**

$$(\mathbf{L/F, g, c}) \ .$$

If $u = u_g$, this algebra has L-basis $1, u, \ldots, u^{n-1}$, with $u^n = c$, and $u\ell = g(\ell)u$ for each $\ell \in L$.

(iv) Suppose F is a field and the multiplicative group F^* has an element ζ of order n. Choosing $a \in F^*$, let R denote the quotient ring $F[x]/(x^n - a)F[x]$. If $p(x) \in F[x]$, let $\overline{p(x)}$ denote the coset of $p(x)$ in R. By uniqueness of remainders in the division algorithm for $F[x]$, R has F-basis $1, \overline{x}, \overline{x}^2, \ldots, \overline{x}^{n-1}$. The F-algebra homomorphism $F[x] \to F[x]$ taking x to ζx fixes $x^n - a$; so it induces an F-algebra homomorphism $\sigma : R \to R$, $\overline{p(x)} \mapsto \overline{p(\sigma(x))}$. Since $\sigma^n = i_R$, σ is an automorphism of R. Since $\overline{x}, \zeta\overline{x}, \ldots, \zeta^{n-1}\overline{x}$ are distinct in R, σ generates a cyclic subgroup $G = \langle \sigma \rangle$ of $\mathrm{Aut}(R)$. Say $\theta : \langle \sigma \rangle \to \mathrm{Aut}(R)$ is inclusion and $f : \langle \sigma \rangle \times \langle \sigma \rangle \to F^*$ is defined by

$$f(\sigma^i, \sigma^j) = \begin{cases} 1 & \text{if} \quad i + j < n \\ b & \text{if} \quad i + j \geq n \end{cases}$$

for some $b \in F^*$ and all $i, j \in \{1, \ldots, n-1\}$. Then $R \circ_f^\theta G$ is the nth **power norm residue algebra**

$$(\mathbf{a, b, \zeta})_{\mathbf{F}} \ .$$

As in cyclic algebras we write u for u_σ; so $(a, b, \zeta)_F$ has R-basis $1, u, u^2, \ldots, u^{n-1}$ and $u^n = b$, while $ur = \sigma(r)u$ for each $r \in R$. And $(a, b, \zeta)_F$ is an F-algebra (since σ fixes F) with F-basis

$$\{\overline{x}^i u^j : 0 \leq i, j \leq n - 1\}$$

and multiplication determined by the relations

$$\overline{x}^n = a \, , \quad u^n = b \, , \quad \text{and} \quad u\overline{x} = \zeta\overline{x}u \ .$$

Note: The examples (ii), (iii), and (iv) generalize Hamilton's quaternions: If $\sigma \in \mathrm{Aut}(\mathbb{C})$ is complex conjugation and $f : \langle \sigma \rangle \times \langle \sigma \rangle \to \mathbb{R}^*$ is defined by $f(1, 1) = f(1, \sigma) = f(\sigma, 1) = 1$ and $f(\sigma, \sigma) = -1$, then

$$\mathbb{H} = \mathbb{C} \circ_f^\theta \langle \sigma \rangle = (\mathbb{C}/\mathbb{R}; f) = (\mathbb{C}/\mathbb{R}, \sigma, -1) = (-1, -1, -1)_{\mathbb{R}} \ .$$

However, it is the 2nd power norm residue algebras that are usually called quaternion algebras.

(16.23) Proposition. *Each crossed product $A = (L/F; f)$ is an Azumaya F-algebra with L as a maximal commutative subring.*

Proof. Since $F = L^G$, the extension $F \subseteq L$ is Galois and $G = \operatorname{Aut}(L/F)$. So the order of G equals both $[A : L]$ and $[L : F]$, and hence $[A : F]$ is the square of the order of G, so it is finite.

Suppose $z = \sum \ell_g u_g$ in A commutes with every element of L. If $h \in G$ and $h \neq i_L$, there exists $\ell \in L$ with $h(\ell) \neq \ell$. Then

$$0 = z\ell - \ell z = \sum (\ell_g g(\ell) - \ell \ell_g) u_g$$

forces $\ell_h = 0$. So $z = \ell_1 u_1 = \ell_1 f(1, 1) \in L$. Thus L is a maximal commutative subring of A. This forces $Z(A) \subseteq L$. And ℓ belongs to $Z(A)$ if and only if, for all $g \in G$, $\ell u_g = u_g \ell = g(\ell) u_g$. So $Z(A) = L^G = F$.

Suppose J is an ideal of A with a nonzero element $y = \sum \ell_g u_g$. Choose such a y with the fewest nonzero coefficients ℓ_h. If there are two nonzero coefficients ℓ_h and ℓ_k, where $h \neq k$ in G, then $h(\ell) \neq k(\ell)$ for some $\ell \in L^*$. Then J contains

$$w = y - h(\ell)^{-1} y \ell = \sum \ell_g (1 - h(\ell)^{-1} g(\ell)) u_g = \sum \ell'_g u_g ,$$

where $\ell'_g = 0$ whenever $\ell_g = 0$, $\ell'_h = 0$, and $\ell'_k \neq 0$. So $w \neq 0$, $w \in J$, and w has fewer nonzero coefficients than y, which is a contradiction. Therefore y has only one term $\ell_g u_g$. Taking $\gamma = g^{-1}$, J also contains

$$f(\gamma, g)^{-1} f(1, 1)^{-1} u_\gamma \ell_g^{-1} y = 1_A .$$

So $J = A$, proving A is simple. ∎

(16.24) Proposition. *Suppose F is a field and the group F^* has an element ζ of finite order n. For each pair $a, b \in F^*$, the nth power norm residue algebra $A = (a, b, \zeta)_F$ is an Azumaya F-algebra with $R = F[x]/(x^n - a)F[x]$ as a maximal commutative subring.*

Proof. The proof is similar to that for crossed products. First, note that the dimension $[A : F] = [A : R][R : F] = n^2$ is finite. Suppose $z = \sum r_i u^i$ commutes with $\overline{x}$. Comparing coefficients of u^i in the equation $z\overline{x} = \overline{x} z$, we find $r_i \zeta^i \overline{x} = \overline{x} r_i$. Since $\overline{x}^n = a \in F^*$, $\overline{x} \in R^*$. So $r_i \zeta^i = r_i$. If $0 < i < n$, this means $r_i = 0$. So $z = r_0 1 = r_0 \in R$. This proves R is a maximal commutative subring of A, and $Z(A) \subseteq R$.

Now suppose $r \in Z(A)$. Then $ru = ur = \sigma(r)u$; so $r = \sigma(r)$. If $r = \sum c_i \overline{x}^i$ with $c_i \in F$, then $\sigma(r) = \sum c_i \zeta^i \overline{x}^i$. So $c_i = c_i \zeta^i$ for each i. Since ζ has order n, $c_i = 0$ for $0 < i < n$, and $r = c_0 \overline{x}^0 = c_0 \in F$. Since A is an F-algebra, $F \subseteq Z(A)$. So $Z(A) = F$.

Suppose J is an ideal of A with a nonzero element $y = \sum r_i u^i$. Choose such a y with the fewest nonzero coefficients r_i. As in the proof of (16.23), but with $\overline{x}$ in place of ℓ, it follows that $y = ru^i$ for some $r \in R$ and $i \in \mathbb{Z}$. Since $u^n = b \in F^*$, $u \in A^*$. Then J contains the nonzero element $yu^{-i} = r \in R$. Choose a nonzero $r = \sum c_j \overline{x}^j$ in $R \cap J$ with the fewest nonzero coefficients. If c_s and c_t are nonzero with $0 \le s < t < n$, then $R \cap J$ contains

$$uru^{-1} - \zeta^s r \;=\; \sum_{j=0}^{n-1} c_j(\zeta^j - \zeta^s)\overline{x}^j \,,$$

which has fewer nonzero coefficients than r but is still nonzero since the coefficient $c_t(\zeta^t - \zeta^s) \neq 0$. This contradiction proves $r = c\overline{x}^j$ for some $c \in F^*$ and $j \in \mathbb{Z}$. So $r \in A^*$, $J = A$, and A is simple. ∎

16C. Exercises

1. Suppose $F \subseteq E$ is a finite-degree Galois field extension, $G = \mathrm{Aut}(E/F)$, and $1 : G \times G \to E^*$ is the constant map to 1_E. Prove $(E/F; 1) \cong M_n(F)$, where $n = [E : F]$. *Hint:* Show E is an $(E/F; 1)$-module with

$$\Big(\sum r_g u_g\Big) \cdot e \;=\; \sum r_g g(e) \,.$$

So there is an F-algebra homomorphism

$$\rho : (E/F; 1) \;\to\; \mathrm{End}_F(E)$$

taking x to $x \cdot (-)$. Compare F-dimensions and recall that $(E/F; 1)$ is simple to show ρ is an isomorphism.

2. If D_n is the dihedral group with generators a, b and relations $a^n - 1$, $b^2 = 1$, $bab^{-1} = a^{-1}$, show for each factor $d > 2$ of n there is a surjective $\mathbb{Q}$-algebra homomorphism

$$\rho_d : \mathbb{Q}D_n \;\to\; \Big(\frac{\mathbb{Q}(\zeta_d)}{\mathbb{Q}(\zeta_d + \zeta_d^{-1})} ; 1\Big)$$

taking a to $\zeta_d = e^{2\pi i/d}$ and taking b to complex conjugation σ on $\mathbb{Q}(\zeta_d)$. Use a $\mathbb{Q}(\zeta_d + \zeta_d^{-1})$-basis 1, ζ_d of $\mathbb{Q}(\zeta_d)$, to obtain, as in Exercise 1, a representation of D_n in $M_2(\mathbb{Q}(\zeta_d + \zeta_d^{-1}))$. Show this is a full (hence absolutely irreducible) representation over $\mathbb{Q}(\zeta_d + \zeta_d^{-1})$, and hence over $\mathbb{R}$. What happens if we try this for $d = 1$ or 2?

3. Suppose G is a group, R is a commutative ring, and $\theta : G \to \mathrm{Aut}(R)$ is a group homomorphism. Denote $\theta(g)(r)$ by $\,^g r$. Let G^n denote $G \times \cdots \times G$ (n factors) and $[G^n, R^*]$ the set of functions from G^n to R^*. Then $[G^n, R^*]$ is an abelian group under pointwise multiplication. Define group homomorphisms

$$[G, R^*] \xrightarrow{\;\delta_1\;} [G^2, R^*] \xrightarrow{\;\delta_2\;} [G^3, R^*]$$

so that

$$\delta_1(c)(g,h) \;=\; \frac{c(g)\,{}^g c(h)}{c(gh)}\,, \quad \delta_2(f)(g,h,k) \;=\; \frac{{}^g f(h,k) f(g,hk)}{f(gh,k) f(g,h)}\,.$$

By (16.21), the factor sets f for twisted group rings $R \circ_f^\theta G$ with twist θ comprise the kernel of δ_2. Prove $\mathrm{im}(\delta_1) \subseteq \ker(\delta_2)$; and if f, f' belong to the same coset in $\ker(\delta_2)/\mathrm{im}(\delta_1)$, then $R \circ_f^\theta G \cong R \circ_{f'}^\theta G$ by an R-linear ring homomorphism. *Hint:* If $c \in [G, R^*]$, describe $R \circ_f^\theta G$ in terms of the altered R-basis $\{v_g : g \in G\}$ with $v_g = c(g) u_g$.

4. If $F \subseteq E$ is a finite-degree Galois field extension with $G = \mathrm{Aut}(E/F)$, and δ_1, δ_2 are defined as in the preceding exercise, for the twist θ given by the identity on G, then elements of $\ker \delta_2$ are **2-cocycles**, elements of im δ_1 are **2-coboundaries**, and the quotient group $\ker \delta_2/\mathrm{im}\, \delta_1$ is the **Galois cohomology group $H^2(G, E^*)$**. Prove $f \mapsto [(E/F; f)]$ defines an injection

$$\gamma : H^2(G, E^*) \;\to\; Br(F)\,.$$

Hint: Similar Azumaya F-algebras of equal F-dimension are isomorphic. So, by Exercise 3, it is enough to show an isomorphism ϕ from $(E/F; f)$ to $(E/F; f')$ implies f/f' lies in $\mathrm{im}\, \delta_1$. By the Skolem-Noether Theorem (§16B, Exercise 9), there is an automorphism ψ of $(E/F; f')$ carrying $\phi(e)$ to e for each $e \in E$. Then $\psi \circ \phi$ is an E-linear ring isomorphism. For each $g \in G$, take $(\psi \circ \phi)(u_g) = w_g$. Since E is a maximal commutative subring of $(E/F; f')$ and $w_g u_g^{-1}$ centralizes E, it belongs to E^*, defining $c : G \to E^*$. Show $f = \delta_1(c) f'$.

5. Continuing the assumptions in Exercise 4, suppose f and g are two elements of $\ker \delta_2$. Show

$$(E/F; f) \otimes_F (E/F; g) \;\sim\; (E/F; fg)\,,$$

so γ is a group homomorphism. *Hints:* Say $E = F(\alpha)$ and $p(x)$ is the minimal polynomial of α over F. In the expression

$$e \;=\; \prod_{\rho \in G - \{1\}} \frac{\alpha \otimes 1 - 1 \otimes \rho(\alpha)}{(\alpha - \rho(\alpha)) \otimes 1}$$

show the denominator is a unit in $E \otimes_F E$, the numerator is nonzero, and hence e is a nonzero element of the commutative subalgebra $E \otimes_F E$. Under the coefficientwise ring homomorphism $E[x] \cong (1 \otimes E)[x]$, show $p(x)$ goes to a polynomial with $\alpha \otimes 1$ as a root, proving $(\alpha \otimes 1)e = (1 \otimes \alpha)e$. Extend this to $(x \otimes 1)e = (1 \otimes x)e$ for all $x \in E$. Conclude that $e^2 = e$. Taking A to be $(E/F; f) \otimes_F (E/F; g)$, we know $A \sim eAe$ by (16.15). Now

$$eAe \;=\; \sum_{\sigma, \tau \in G} e(E \otimes E) e e(u_\sigma \otimes u_\tau) e\,.$$

Show $e(E \otimes E)e = e(E \otimes 1)e$ is a field E' isomorphic to E over F. Show $e(u_\sigma \otimes u_\tau)e = 0$ if $\sigma \neq \tau$, and $e(u_\sigma \otimes u_\sigma)e = e(u_\sigma \otimes u_\sigma)$, and call the latter element v_σ. Show $v_\sigma e(x \otimes 1)e = e(\sigma(x) \otimes 1)ev_\sigma$ and $v_\sigma v_\tau = e(f(\sigma, \tau)g(\sigma, \tau) \otimes 1)ev_{\sigma\tau}$. Now construct an isomorphism $eAe \cong (E/F; fg)$.

6. Suppose a crossed product $A = (E/F; f)$ has index d. Prove $[A]^d = [F]$ in $Br(F)$. *Hints:* Since $A \cong M_r(D)$ for a division ring $D \in \mathcal{A}_3(F)$, we also have $A \cong \text{End}_{D^{op}}(V)$, where $V = M_r(D)\epsilon_{11}$. Since $E \subseteq A$, V is also an E-vector space. Then $[V : F] = [V : E][E : F]$ and $[V : F] = [V : D^{op}][D^{op} : F] = rd^2 = [E : F]d$; so $[V : E] = d$. For each $g \in G = \text{Aut}(E/F)$, $u_g x = g(x)u_g$ for all $x \in E$. Suppose $v_1, \dots, v_d$ is an E-basis of V, and

$$u_g v_j = \sum_{i=1}^{n} e_{ij} v_i \quad (e_{ij} \in E).$$

Let $M(g)$ denote the matrix $(e_{ij}) \in M_n(E)$. Prove the equation $u_g u_h = f(g, h)u_{gh}$ yields the matrix equation

$$M(g)\, g(M(h)) = f(g, h)\, M(gh).$$

Now take determinants to get

$$f(g, h)^d \det M(gh) = g(\det M(h)) \det M(g).$$

Show $\det M(g) \neq 0$ for all $g \in G$. Conclude that

$$f(g, h)^d = \frac{{}^g c(h)c(g)}{c(gh)} = \delta_1(c),$$

where $c(-) = \det M(-) : G \to E^*$. Now use Exercises 1, 4, and 5.

7. Prove each crossed product $(E/F : f)$, with $G = \text{Aut}(E/F)$ a finite cyclic group, is a cyclic algebra. *Hint:* Say $[E : F] = n$ and G is generated by g. Show $1 = u_g^0, u_g^1, \dots, u_g^{n-1}$ are multiples of the basis $\{u_\sigma : \sigma \in G\}$ by scalars in E^*. Take $u = u_g$ and show G consists of the conjugations $u^i(-)u^{-i}$ restricted to E. Show the unit u^n lies in the center of $(E/F; f)$.

8. Suppose $F \subseteq E$ is a Galois field extension of finite degree with cyclic Galois group $G = \text{Aut}(E/F)$ generated by g. Prove there is an injective group homomorphism

$$\gamma' : \frac{F^*}{N_{E/F}(E^*)} \to Br(F)$$

taking the coset of $a \in F^*$ to the similarity class $[(E/F, g, a)]$. *Hint:* If $a \in F^*$, let f_a denote the factor set

$$f_a(g^i, g^j) = \begin{cases} 1 & \text{if } i + j < n \\ a & \text{if } i + j \geq n \end{cases}$$

for $i, j \in \{0, \ldots, n-1\}$. Then $(E/F, g, a) = (E/F; f_a)$. Since $f_a f_b = f_{ab}$, there is a homomorphism from F^* into $Br(F)$ taking a to $[(E/F, g, a)]$, by Exercise 5. Show $f_a \in \operatorname{im} \delta_1$ if and only if $a \in N_{E/F}(E^*)$ — the c in $[G, E^*]$ will be defined by $c(1) = 1$ and

$$c(g^i) = bg(b)g^2(b) \cdots g^{i-1}(b)$$

for $1 \leq i \leq n-1$, where $b = c(g) \in E^*$ and $N_{E/F}(b) = a$.

Note: Under the hypotheses in Exercise 8, the homomorphism γ' factors as an isomorphism

$$\frac{F^*}{N_{E/F}(E^*)} \cong H^2(G, E^*)$$

$$\bar{a} \longmapsto f_a$$

followed by γ; the surjectivity of this isomorphism comes from Exercise 7.

9. Prove the following theorem of Wedderburn: If the order of $\bar{a}$ in the quotient $F^*/N_{E/F}(E^*)$ is $n = [E : F]$, the cyclic algebra $(E/F, g, a)$ is a division ring. *Hint:* Use Exercises 6 and 8.

10. Use Exercise 9 to construct two division rings D_1, D_2 in $\mathcal{A}_3(\mathbb{Q})$ of index 3, with $D_1 \not\cong D_2$ as $\mathbb{Q}$-algebras. *Hint:* Say $\zeta = e^{2\pi i/7}$ and $E = \mathbb{Q}(\zeta + \zeta^{-1})$. Show $\mathbb{Q} \subseteq E$ is a degree 3 Galois field extension with cyclic Galois group G generated by the automorphism taking $\zeta + \zeta^{-1}$ to $\zeta^2 + \zeta^{-2}$. By Washington [97, Proposition 2.16], $\mathbb{Z}[\zeta + \zeta^{-1}]$ is the ring of algebraic integers in E. So by Kummer's Theorem (7.47), $2\mathbb{Z}[\zeta + \zeta^{-1}]$ factors into maximal ideals the same way the minimal polynomial $x^3 + x^2 - 2x - 1$ of $\zeta + \zeta^{-1}$ over $\mathbb{Q}$ factors in $(\mathbb{Z}/2\mathbb{Z})[x]$. Show it remains irreducible. Then use the 2-adic valuation on E to prove 2 is not in $N_{E/\mathbb{Q}}(E^*)$.

16D. The K_2 Connection

Suppose F is a field and F^* has an element ζ of (finite) order n. Closely following Milnor [71, §15], we prove the similarity class in $Br(F)$ of the nth power norm residue algebra $(a, b, \zeta)_F$, defined above, is multiplicative in a and b, and $(a, b, \zeta)_F$ is split when $a + b = 1$ or a is an nth power in F. According to Matsumoto's presentation (14.69) and (14.77) of $K_2(F)$, this results in a group homomorphism $\overline{\{a, b\}} \mapsto [(a, b, \zeta)_F]$ from $K_2(F)/nK_2(F)$ to the group $_nBr(F)$ of elements in $Br(F)$ whose nth power is 1. We discuss, without proofs, the Tate-Merkurjev-Suslin Theorem that this is an isomorphism, and the consequences for $Br(F)$ and $K_2(F)$.

(16.25) Theorem. *If a field F has an element ζ of finite multiplicative order n, there is a group homomorphism*

$$R_{n,F} : \frac{K_2(F)}{nK_2(F)} \ \to \ {}_nBr(F)$$

taking the coset $\overline{\{a,b\}}$ to the class $[(a,b,\zeta)_F]$ for each pair $a,b \in F^*$.

Proof. If $a = \alpha^n$ for some $\alpha \in F^*$, the minimal polynomial $t^n - a$ of $\overline{x} \in (a,b,\zeta)_F$ splits into n different linear factors

$$(t - \alpha)(t - \zeta\alpha) \cdots (t - \zeta^{n-1}\alpha)$$

in $F[t]$. By (16.20), $(\alpha^n, b, \zeta)_F \cong M_n(F)$ as F-algebras.

Now suppose a, b_1, b_2 are any units in F. The Azumaya F-algebra

$$A \ = \ (a, b_1, \zeta)_F \otimes_F (a, b_2, \zeta)_F$$

has an F-basis consisting of all $X_1^i U_1^j X_2^k U_2^\ell$ for i,j,k,ℓ in $\{0, 1, \ldots, n-1\}$, where

$$\begin{aligned} X_1 &= \overline{x} \otimes 1, & U_1 &= u \otimes 1, \\ X_2 &= 1 \otimes \overline{x}, & U_2 &= 1 \otimes u. \end{aligned}$$

Multiplication in A satisfies

$$\begin{aligned} U_1 X_1 &= \zeta X_1 U_1, & U_2 X_2 &= \zeta X_2 U_2, \\ X_1^n &= X_2^n = a, & U_1^n &= b_1, & U_2^n &= b_2, \end{aligned}$$

and the elements X_1, U_1 commute with the elements X_2, U_2. Now A can be divided another way as an internal tensor product: Consider the subalgebras

$$\begin{aligned} B &= F[X_1, U_1 U_2] \ \cong \ (a, b_1 b_2, \zeta)_F, \\ C &= F[X_1^{-1} X_2, U_2] \ \cong \ (1, b_2, \zeta)_F. \end{aligned}$$

The generators $X_1, U_1 U_2$ of B commute with the generators $X_1^{-1} X_2, U_2$, the subalgebra BC is all of A, and $[B : F][C : F] = n^4 = [A : F]$. So, by (16.4),

$$A \ \cong \ (a, b_1 b_2, \zeta)_F \otimes_F (1, b_2, \zeta)_F.$$

Now 1 is an nth power in F, so the second factor is isomorphic to $M_n(F)$, and A is similar to $(a, b_1 b_2, \zeta)_F$. Thus

$$[(a, b_1, \zeta)_F][(a, b_2, \zeta)_F] \ = \ [(a, b_1 b_2, \zeta)_F]$$

in the Brauer group $Br(F)$.

There is an F-algebra isomorphism from $(a, b, \zeta)_F$ to $(b, a, \zeta^{-1})_F$ switching $\overline{x}$ and u. Applying this, we also have

$$[(a_1, b, \zeta)_F] \, [(a_2, b, \zeta)_F] \quad = \quad [(a_1 a_2, b, \zeta)_F]$$

for units a_1, a_2, b of F.

To establish that $(a, b, \zeta)_F$ splits when $a + b = 1$, we use the nearly commutative multiplication in $(a, b, \zeta)_F$ to get a noncommutative Binomial Theorem: For $r \geq 0$, define polynomials $p_r(t) \in F[t]$ by

$$
\begin{aligned}
p_0(t) &= 1 \\
p_1(t) &= t - 1 \\
p_2(t) &= (t - 1)(t^2 - 1) \\
&\vdots \\
p_m(t) &= (t - 1)(t^2 - 1) \dots (t^m - 1) \\
&\vdots
\end{aligned}
$$

For $0 \leq i \leq r$, define the **noncommutative binomial coefficient** $b_i^r(t) \in F(t)$ by

$$b_i^r(t) \quad = \quad \frac{p_r(t)}{p_i(t) p_{r-i}(t)} \ .$$

(16.26) Lemma. *For $0 < i \leq r$,*

$$b_i^r(t) \quad = \quad t^i b_i^{r-1}(t) \quad + \quad b_{i-1}^{r-1}(t) \ .$$

So for $0 \leq i \leq r$, $b_i^r(t) \in F[t]$.

Proof.

$$
\begin{aligned}
b_i^r(t) &= \frac{p_r(t)}{p_i(t) p_{r-i}(t)} \\[2mm]
&= \frac{p_{r-1}(t)}{p_{i-1}(t) p_{r-i-1}(t)} \frac{t^r - 1}{(t^i - 1)(t^{r-i} - 1)} \\[2mm]
&= \frac{p_{r-1}(t)}{p_{i-1}(t) p_{r-i-1}(t)} \left[t^i \left(\frac{1}{t^i - 1} \right) + \frac{1}{t^{r-i} - 1} \right] \\[2mm]
&= t^i b_i^{r-1}(t) \quad + \quad b_{i-1}^{r-1}(t) \ .
\end{aligned}
$$

Since $b_0^r(t) = b_r^r(t) = 1$ for all $r \geq 1$, the preceding formula proves $b_i^r(t)$ is a polynomial for $0 \leq i \leq r$ by induction on r. ∎

(16.27) Lemma. *In the nth power norm residue algebra $(a, b, \zeta)_F$,*

$$(\overline{x} + u)^r \; = \; \sum_{i=0}^{r} b_i^r(\zeta) \overline{x}^i u^{r-i}$$

for all $r \geq 1$. In particular, $(\overline{x} + u)^n = a + b$.

Proof. When $r = 1$, the first equation holds because $b_0^1(t) = b_1^1(t) = 1$. Assume $r > 1$ and the first equation is true with $r - 1$ in place of r. Then

$$
\begin{aligned}
(\overline{x} + u)^r \; &= \; (\overline{x} + u) \sum_{i=0}^{r-1} b_i^{r-1}(\zeta) \overline{x}^i u^{r-1-i} \\[1mm]
&= \; \sum_{i=0}^{r-1} b_i^{r-1}(\zeta) \overline{x}^{i+1} u^{r-1-i} \; + \; \sum_{i=0}^{r-1} \zeta^i b_i^{r-1}(\zeta) \overline{x}^i u^{r-i} \\[1mm]
&= \; \overline{x}^r \; + \; u^r \; + \; \sum_{i=1}^{r-1} [\zeta^i b_i^{r-1}(\zeta) + b_{i-1}^{r-1}(\zeta)] \overline{x}^i u^{r-i} \\[1mm]
&= \; \sum_{i=0}^{r} b_i^r(\zeta) \overline{x}^i u^{r-i}
\end{aligned}
$$

by the preceding lemma.

Now take $r = n$. Since ζ has order n, $p_n(\zeta) = 0$ but $p_i(\zeta) \neq 0$ for $0 < i < n$. So $t - \zeta$ is a factor of the numerator but not the denominator of

$$b_i^n(t) \; = \; \frac{p_n(t)}{p_i(t) p_{n-i}(t)}$$

if $0 < i < n$. So the polynomial $b_i^n(t) \in F[t]$ has ζ as a root. Then

$$
\begin{aligned}
(\overline{x} + u)^n \; &= \; b_n^n(\zeta) \overline{x}^n \; + \; b_0^n(\zeta) u^n \\
&= \; \overline{x}^n \; + \; u^n \; = \; a \; + \; b.
\end{aligned}
$$
∎

Returning to the proof of the theorem, no monic polynomial of degree $r < n$ in $F[t]$ can have $\overline{x} + u$ as a root, since $1, u, \ldots, u^{n-1}$ are R-linearly independent and the coefficient in $(\overline{x}+u)^r$ of u^r is 1. So $t^n - (a+b)$ is the minimal polynomial over F of $\overline{x} + u$. If $a + b = 1$, this minimal polynomial splits into n different linear factors $\Pi(t - \zeta^i)$ in $F[t]$; so by (16.20), $(a, b, \zeta)_F$ splits. That is,

$$[(a, b, \zeta)_F] \; = \; 1$$

in $Br(F)$ if $a + b = 1$.

By Matsumoto's Theorem, there is a homomorphism of abelian groups

$$K_2(R) \quad \to \quad Br(F) \ .$$
$$\{a, b\} \quad \mapsto \quad [(a, b, \zeta)_F]$$

Since $[(a, b, \zeta)_F]^n = [(a^n, b, \zeta)_F] = 1$, this map takes its values in $_nBr(F)$ and its kernel contains $nK_2(F)$. So it induces the required homomorphism $R_{n,F}$. ∎

To understand the image of $R_{n,F}$, consider the relationship between norm residue and cyclic algebras, which one might expect from the similarity of their definitions.

(16.28) Proposition. *If F is a field and ζ has order n ($< \infty$) in F^*, every n^2-dimensional cyclic algebra $C = (E/F, g, b)$ is an nth power norm residue algebra $(a, b, \zeta)_F$; and every nth power norm residue algebra $A = (a, b, \zeta)_F$ is similar to a d^2-dimensional cyclic algebra $(E/F, g, b)$, where d divides n and $E = F(\alpha)$ for some α with $\alpha^n = a$.*

Proof. By (15.36), the field E in the cyclic algebra C is a $F(\beta)$ for some β with $\beta^n \in F^*$, and the cyclic group $\langle g \rangle = \mathrm{Aut}(E/F)$ is generated by an automorphism h with $h(\beta) = \zeta\beta$. Then $g = h^\ell$ and $h = g^m$, where $\ell m \equiv 1 \pmod{n}$. Taking $\alpha = \beta^m$, we also have $\alpha^n \in F$, $E = F(\alpha)$ and $g(\alpha) = g(\beta)^m = (\zeta^\ell \beta)^m = \zeta\alpha$. Say $\alpha^n = a$. Since $[C : F] = n^2$, $[E : F] = n$. So E has F-basis $1, \alpha, \ldots, \alpha^{n-1}$. Then C has an F-basis consisting of all $\alpha^i u^j$ with $i, j \in \{0, \ldots, n-1\}$, and the multiplication in C obeys the relations $\alpha^n = a, u^n = b$, and $u\alpha = \zeta\alpha u$. So, replacing $\overline{x}$ by α defines an F-algebra isomorphism from $(a, b, \zeta)_F$ to C.

For the second assertion, suppose a and b are any units of F and $A = (a, b, \zeta)_F$. By (16.15), eAe is similar to A for each nonzero idempotent $e \in A$. We obtain such an idempotent from a Chinese Remainder Theorem decomposition of $R = F[x]/(x^n - a)F[x]$ that is invariant under conjugation by a power of u, and show eAe is a cyclic algebra of the required form.

In an algebraic closure of F, choose a root α of $x^n - a$. Say d is the least positive integer with $\alpha^d \in F$ and write $\alpha^d = c$, $F(\alpha) = E$. By (15.36), $n = md$ for some integer m, $[E : F] = d$, and $x^d - c$ is the minimal polynomial of α over F. Further, (15.36) tells us $F \subseteq E$ is Galois with cyclic Galois group generated by an automorphism g with $g(\alpha) = \zeta^m\alpha$. If we define

$$N(x) \quad = \quad \frac{x^m - 1}{x - 1} \quad = \quad 1 + x + \cdots + x^{m-1} \ ,$$

then, in $F[x]$,

$$
\begin{aligned}
x^n - a \quad &= \quad x^{dm} - c^m \quad = \quad c^m((x^d/c)^m - 1) \\
&= \quad c^m(x^d/c - 1)N(x^d/c) \\
&= \quad c^{m-1}(x^d - c)N(x^d/c) \ .
\end{aligned}
$$

Since $x^n - a$ has n roots $\zeta^i \alpha$ in E, it is a separable polynomial; so $x^d - c$ and $N(x^d/c)$ are relatively prime in $F[x]$. Under the Chinese Remainder Theorem isomorphism of F-algebras

$$R = \frac{F[x]}{(x^n - a)F[x]} \longrightarrow \frac{F[x]}{(x^d - c)F[x]} \times \frac{F[x]}{N(x^d/c)F[x]} ,$$
$$\overline{p(x)} \longmapsto (\overline{p(x)}, \overline{p(x)})$$

the element $e = N(\overline{x}^d/c)/m$ goes to $(\overline{1}, \overline{0})$. So e is idempotent, and there is a composite F-algebra isomorphism ψ

$$Re \cong \frac{F[x]}{(x^d - c)F[x]} \times \{\overline{0}\} \cong E$$

taking $\overline{x}e$ to α.

Let σ denote the automorphism $u(-)u^{-1}$ of A, taking $\overline{x}$ to $\zeta\overline{x}$. Then $\sigma^m(\overline{x}^d/c) = \zeta^{md}\overline{x}^d/c = \overline{x}^d/c$; so $\sigma^m(e) = e$. Now

$$\begin{aligned}
eAe &= e(R + Ru + \cdots + Ru^{n-1})e \\
&= Re + Re\sigma(e)u + \cdots + Re\sigma^{n-1}(e)u^{n-1} .
\end{aligned}$$

We compute each $e\sigma^i(e)$. For brevity, write $y = \overline{x}^d/c$; so $c = N(y)/m$ and $\sigma(y) = \zeta^d y$. Since $y^m = \overline{x}^n/\alpha^n = a/a = 1$, it follows that $(y - 1)N(y) = 0$; so $yN(y) = N(y)$ and $ye = e$. Then

$$e \, \sigma^i(e) = e \frac{N(\sigma^i(y))}{m} = e \frac{N(\zeta^{di}y)}{m} = \frac{N(\zeta^{di}y)e}{m}$$

$$= \frac{N(\zeta^{di})}{m} e = \begin{cases} e & \text{if } i \in m\mathbb{Z} \\ 0 & \text{if } i \notin m\mathbb{Z} , \end{cases}$$

the latter because $N(\zeta^{dmj}) = N(1) = m$, while if $i \notin m\mathbb{Z}$, ζ^{di} is a root of $x^m - 1$ but not of $x - 1$. Therefore

$$eAe = Re + Reu^m + \cdots + Reu^{m(d-1)} .$$

So it has Re-basis $1, u^m, \ldots, u^{m(d-1)}$, with $(u^m)^d = u^n = b$, and

$$u^m \overline{x}e = \sigma^m(\overline{x}e)u^m = \zeta^m \overline{x}eu^m .$$

So the F-linear isomorphism

$$eAe \cong (E/F, g, b) ,$$

taking $(\overline{x}e)^i u^{mj}$ to $\alpha^i u^j$ for $i, j \in \{0, \ldots, d - 1\}$, is an isomorphism of F-algebras. $\blacksquare$

For an arbitrary field F, we now summarize some standard facts about Azumaya F-algebras, drawn from Reiner [75], Jacobson [89], and Kersten [90]. A subfield K of $B \in \mathcal{A}_{\mathfrak{z}}(F)$ is called **self-centralizing** if $Z_B(K) = K$. If $F \subseteq E$ is a finite-degree field extension, then E splits $A \in \mathcal{A}_{\mathfrak{z}}(F)$ if and only if E is a self-centralizing subfield of some $B \in \mathcal{A}_{\mathfrak{z}}(F)$ similar to A. If $F \subseteq E$ happens to be Galois, B is a crossed product $(E/F; f)$. Every $A \in \mathcal{A}_{\mathfrak{z}}(F)$ is split by E for some finite-degree Galois field extension $F \subseteq E$; so crossed products generate $Br(F)$. In light of §16C, Exercise 6, it follows that $Br(F)$ is a torsion group – every element has finite order.

If, as above, $F \subseteq E$ is not only Galois but cyclic (meaning $\mathrm{Aut}(E/F)$ is cyclic), then B is a cyclic algebra $(E/F, g, a)$. For F a local field, every $A \in \mathcal{A}_{\mathfrak{z}}(F)$ is a cyclic algebra $(E/F, g, a)$, where E is an unramified extension of F, and the index of A equals the order of $[A]$ in $Br(F)$.

Now suppose F is a global field. The work of Albert, Brauer, Hasse, and Noether, in the first half of the 20th century, showed $A, B \in Az(F)$ are similar if and only if $F_p \otimes_F A$, $F_p \otimes_F B$ are similar for each completion F_p of F. In combination with the Grunwald-Wang Theorem in global class field theory, this shows each $A \in \mathcal{A}_{\mathfrak{z}}(F)$ is a cyclic algebra with index equal to the order of $[A]$ in $Br(F)$.

So for local and global fields F containing an element ζ of multiplicative order n, the homomorphism

$$R_{n,F} : \frac{K_2(F)}{nK_2(F)} \quad \to \quad {}_n Br(F)$$
$$\{a, b\} \quad \mapsto \quad [(a, b, \zeta)_F]$$

is surjective by (16.28). In 1976, Tate [76] proved $R_{n,F}$ is injective for local and global fields. The K-theory used in Tate's proof is limited to the projection formula for the norm on K_2.

In the early 1980s, Merkurjev and Suslin extended this remarkable isomorphism to arbitrary fields:

(16.29) Merkurjev-Suslin Theorem. *Suppose F is a field and F^* has an element ζ of finite order n. The homomorphism $R_{n,F}$ is an isomorphism.* ∎

The proof is well beyond the scope of this book, relying on the K-theory of schemes, etale cohomology, and the localization sequence for Quillen K-theory. For an exposition of this proof, see Section (8.2) in Srinivas [96]. For the case $n = 2$, Merkurjev found a more elementary proof, given a beautiful and informative exposition by Wadsworth in [86]. The algebraic consequences of the Merkurjev-Suslin Theorem include a highly accessible description of Azumaya F-algebras:

(16.30) Corollary. *If F is a field and F^* has an element of order n $(< \infty)$, every $A \in \mathcal{A}_{\mathfrak{z}}(F)$ of index n is similar to a tensor product $C_1 \otimes_F \cdots \otimes_F C_m$ of cyclic algebras $C_i \in \mathcal{A}_{\mathfrak{z}}(F)$ of F-dimension d_i^2, where each d_i divides n.*

Proof. By §16C, Exercise 6, $[A]^n = 1$ in $Br(F)$. So $[A] \in {}_nBr(F)$ and

$$[A] = R_{n,F}(\{\overline{a_1, b_1}\} \cdots \{\overline{a_m, b_m}\})$$
$$= [(a_1, b_1, \zeta)_F] \cdots [(a_m, b_m, \zeta)_F] .$$

By (16.28), each $(a_i, b_i, \zeta)_F$ is similar to a cyclic algebra C_i of F-dimension d_i^2, where d_i divides n. So

$$[A] = [C_1] \cdots [C_m] = [C_1 \otimes_F \cdots \otimes_F C_m] . \qquad \blacksquare$$

As we mentioned above, every Azumaya algebra over a global field is a cyclic algebra – not just similar to a tensor product of cyclic algebras. Using Tate's proof of the injectivity of $R_{n,F}$ for global fields, Lenstra [76] obtained a parallel result for K_2:

(16.31) Theorem. *If F is a global field, each element of $K_2(F)$ is a single Steinberg symbol $\{a, b\}$.* $\qquad \blacksquare$

16D. Exercises

1. Suppose F is a field and ζ has order n $(< \infty)$ in F^*. Suppose $a, b \in F^*$. Prove the following are equivalent:

  (i) $(a, b, \zeta)_F \cong M_n(F)$ in $\mathcal{A}_{\mathfrak{z}}(F)$;
  (ii) $b \in N_{E/F}(E^*)$, where $E = F(\sqrt[n]{a})$;
  (iii) $a \in N_{K/F}(K^*)$, where $K = F(\sqrt[n]{b})$.

Hint: By (16.28), $A = (a, b, \zeta)_F$ is similar to a cyclic algebra $(E/F, g, b)$, where $E = F(\alpha)$ for a root α of $x^n - a$. Apply §16C, Exercise 8, to prove (i) is equivalent to (ii). In $K_2(F)$, $\{a, b\}^{-1} = \{b, a\}$; so $(b, a, \zeta)_F$ is similar to the opposite ring of $(a, b, \zeta)_F$.

Note: This equivalence explains the name "norm residue algebra" for $(a, b, \zeta)_F$. The map $R_{n,F}$ is also called the **nth power norm residue homomorphism** for the same reason. When F is a local field, ${}_nBr(F)$ is cyclic of order n and there is an isomorphism ${}_nBr(F) \cong \langle \zeta \rangle$ relating $R_{n,F}$ to the nth power norm residue symbol (15.37). For details, see Milnor [71, §15.9].

2. Suppose F is a field of characteristic $\neq 2$, so that -1 has order 2 in F^*. If $a, b \in F^*$, prove the following are equivalent:

(i) $\{a, b\} \in 2K_2(F)$,
(ii) $(a, b, -1)_F \cong M_2(F)$ in $\mathcal{A}_3(F)$,
(iii) $ax^2 + by^2 = 1$ for some $x, y \in F$.

Hint: The Merkurjev-Suslin Theorem is not needed for this. Assertions (ii) and (iii) are equivalent by the preceding exercise, (15.2), and §15A, Exercise 1. Since $R_{2,F}$ is well-defined, (i) implies (ii). For the reverse implication, assume b is a norm from $F(\sqrt{a})$. If a is a square in F, then $\{a, b\} \in 2K_2(F)$. If a is not a square in F, $b = u^2 - av^2$ with $u, v \in F$. If u or v is 0, show $\{a, b\} \in 2K_2(F)$. If u and v are nonzero, expand

$$1 \;\; = \;\; \{a\frac{v^2}{u^2} \,,\, 1 - a\frac{v^2}{u^2}\}$$

to get $\{a, b\} \in 2K_2(F)$.

Note: For a generalization to $\{a, b\} \in nK_2(F)$, see Milnor [71, Theorem 15.2].

3. Why doesn't the preceding exercise prove injectivity of $R_{2,F}$ when $1 \neq -1$ in F?

4. If F is a field and F^* has an element of order n ($< \infty$), prove every $A \in \mathcal{A}_3(F)$ with $[A] \in {}_nBr(F)$ has a splitting field E with $F \subseteq E$ Galois with finite abelian Galois group.

5. If F is a field and F^* has an element of order n ($< \infty$), give a presentation of the abelian group ${}_nBr(F)$ by generators and relations, based on Matsumoto's presentation of $K_2(F)$.

6. If F is a field of characteristic $\neq 2$ and the 2-primary part of the torsion subgroup of $K_2(F)$ is cyclic, prove there is a Hilbert symbol on F. *Hint:* In this case, $K_2(F)/2K_2(F) \cong \{\pm 1\}$. Use Exercise 2.

APPENDIX

A. Sets, Classes, Functions

The ambition of putting a structure on the collection of all groups, all rings, etc., must be tempered by concerns in set theory brought about by certain paradoxes. One of the most famous is the paradox discovered by Bertrand Russell: If there is a set u of all sets, one should be able to form the subset $s = \{x \in u : x \notin x\}$ consisting of all sets that are not members of themselves; but then $s \in s$ if and only if $s \notin s$.

To avoid such difficulties, one must be circumspect about the formation of sets. Not every imaginable collection can be called a set. The widely accepted foundation for modern mathematics is the list of Zermelo-Fraenkel axioms for the formation of sets, which we outline below. Along with these axioms we shall assume there is a collection U called the **universe**, that U is the collection of all sets, and that $x \in a \in U$ implies $x \in U$. This has the odd effect that every element of a set is also a set.

When one first hears of sets, they are described as collections of objects, and not all these objects are, themselves, collections. So restricting our attention to sets of sets may seem unnatural. However, this approach is sufficient for the foundations of mathematics; numbers of all kinds can be constructed as sets. And, since collections of objects that are not sets do not play a role in the foundations, we follow a version of Occam's razor: One should not make unnecessary assumptions.

Within the context of our universe U, the Zermelo-Fraenkel axioms can be summarized as follows: Sets with the same members are equal. There is a set $\emptyset$ with no members, called the **empty set**. If x and y are sets, there is a set $\{x, y\}$ whose members are x and y. For each set a, there is a set $\mathcal{P}ower(a)$ whose members are the subsets of a. If a is a set, there is a "union" set $\cup a$ whose members are the members of the members of a. (When $a = \{x, y\}$, $\cup a$ is written $x \cup y$.) The axiom of **replacement** says that if $\sigma(x, y)$ is a sentence for which, for each $x \in a$, $\sigma(x, y)$ and $\sigma(x, z)$ imply $y = z$, then there is a set b whose members are those y for which $\sigma(x, y)$ for some $x \in a$. One important consequence of this axiom is that a collection of sets a_i, indexed by the members i of a set I, is also a set $\{a_i : i \in I\}$. A second consequence is that, if a is a set and $p(x)$ is a sentence, there is a set $\{x \in a : p(x)\}$ whose members are those x

in a for which $p(x)$ is true. The latter assertion is used to define the intersection of sets.

The axiom of **regularity** says every nonempty set a has a member b with $a \cap b = \emptyset$. A consequence is that, among members of a set, there are no infinite chains $\cdots \in a_3 \in a_2 \in a_1$, since $\{a_i : i \geq 1\}$ would violate this condition. In particular, no set can be a member of itself. If a is a set, so are $\{a, a\} = \{a\}$ and $a^+ = a \cup \{a\}$. Call a^+ the **successor** of a. A **successor set** is any set a with $\emptyset \in a$ and with $x^+ \in a$ whenever $x \in a$. The axiom of **infinity** says there is a successor set. (Within any successor set, the intersection of all successor subsets is a successor set $\{\emptyset, \emptyset^+, \emptyset^{++}, \dots\}$, which can be used as a model for the nonnegative integers $\{0, 1, 2, \dots\}$.)

If x and y are sets, define the **ordered pair** (x, y) to be $\{x, \{x, y\}\}$. Using regularity, it is easy to show $(x, y) = (x', y')$ if and only if $x = x'$ and $y = y'$. If a and b are sets, there is a **cartesian product** set $a \times b$ whose members are the ordered pairs (x, y) with $x \in a$ and $y \in b$. We define a **function** $f : a \to b$, from a set a to a set b, to be an ordered pair $((a, b),\ G_f)$, where the **graph** G_f of f is a subset of $a \times b$ for which

(i) $\forall\ x \in a,\ \exists\ y \in b$ with $(x, y) \in G_f$, and

(ii) $(x, y) \in G_f$ and $(x, z) \in G_f$ implies $y = z$.

If $x \in a$, we write $f(x)$ to mean the $y \in b$ with $(x, y) \in G_f$.

The axiom of **choice** says that, if I is a set, and for each $i \in I$, a_i is a nonempty set, then there is a "choice" function

$$f : I \to \cup \{a_i : i \in I\}$$

with $f(i) \in a_i$ for each i. More crudely, this axiom says that one can simultaneously choose one element from each set a_i. This concludes our axioms for set theory.

Russell's paradox is avoided in this system of assumptions: The axiom of replacement provides, for each set a, a set $\{x \in a : x \notin x\}$. But it does not create a set $\{x \in U : x \notin x\}$ because, if U were a set, we would have $U \in U$, violating regularity. Subcollections of U are called **classes**. Every set is a class, because $x \in a \in U$ implies $x \in U$. But some classes (such as U) are too large to be sets. There are collections that are not even classes. For instance, $\{U\}$ is a collection with one member U; since U is not a set, $\{U\}$ is not a class!

A group is a pair $(G, \cdot)$, where $\cdot$ is a binary operation on G — that is, a function $f : G \times G \to G$. So $(G, \cdot)$ is a set and we can form the class of all groups. Similarly there is a class of all rings and, for each ring R, a class of all R-modules. Such classes of objects are used in the categories discussed in Chapter 0.

In the construction of a free R-module in (1.8) we use the assertion:

(A.1) Proposition. *No function is a member of its own domain.*

Proof. If $f : a \to b$ is a function and $f \in a$, then

$$a \in (a, b) \in ((a, b), G_f) = f \in a,$$

which violates the axiom of regularity. ∎

B. Chain Conditions, Composition Series

In a first course in linear algebra, we learn that a subspace of a finite-dimensional vector space is also finite-dimensional. Is every submodule of a f.g. R-module also a f.g. R-module? If so, every left ideal of R (being an R-submodule of $R \cdot 1$) would be finitely generated, and this need not be true:

(B.1) Examples.

(i) In $M_2(\mathbb{Q})$ there is a subring

$$R = \begin{bmatrix} \mathbb{Z} & \mathbb{Q} \\ 0 & \mathbb{Z} \end{bmatrix} = \left\{ \begin{bmatrix} a & b \\ 0 & c \end{bmatrix} : a, c \in \mathbb{Z}, \ b \in \mathbb{Q} \right\} .$$

This ring R has a left ideal

$$J = \begin{bmatrix} 0 & \mathbb{Q} \\ 0 & 0 \end{bmatrix} = \left\{ \begin{bmatrix} 0 & q \\ 0 & 0 \end{bmatrix} : q \in \mathbb{Q} \right\} ,$$

which is not finitely generated: For

$$\begin{bmatrix} a & b \\ 0 & c \end{bmatrix} \begin{bmatrix} 0 & q \\ 0 & 0 \end{bmatrix} = \begin{bmatrix} 0 & aq \\ 0 & 0 \end{bmatrix} ;$$

so if $q_1, \cdots, q_n \in \mathbb{Q}$,

$$R \begin{bmatrix} 0 & q_1 \\ 0 & 0 \end{bmatrix} + \cdots + R \begin{bmatrix} 0 & q_n \\ 0 & 0 \end{bmatrix}$$

$$= \begin{bmatrix} 0 & \mathbb{Z}q_1 + \cdots + \mathbb{Z}q_n \\ 0 & 0 \end{bmatrix} \neq \begin{bmatrix} 0 & \mathbb{Q} \\ 0 & 0 \end{bmatrix}$$

because the elements of $\mathbb{Z}q_1 + \cdots + \mathbb{Z}q_n$ have a common denominator, while the elements of $\mathbb{Q}$ do not.

(ii) If F is a field, the ring $R = F[x_1, x_2, ...]$ of polynomials in the infinite list of indeterminates $x_1, x_2, ...$ has an ideal $J = \langle x_1, x_2, ... \rangle$ that is not finitely generated: For if $p_1, \cdots, p_n \in J$, there is a finite set T of positive integers with

$$\langle p_1, \cdots, p_n \rangle \subseteq \langle \{x_i : i \in T\} \rangle .$$

Then there is a positive integer $j \notin T$, and

$$x_j \notin \langle \{x_i : i \in T\} \rangle .$$

In each of these examples, the ring R contains an infinite chain of left ideals $J_1 \subsetneq J_2 \subsetneq \cdots$; in (i) take

$$J_n = \begin{bmatrix} 0 & \frac{1}{2^n}\mathbb{Z} \\ 0 & 0 \end{bmatrix} ,$$

and in (ii) take

$$J_n = \langle x_1, \cdots, x_n \rangle .$$

So we are led to consider conditions on chains in a poset:

Suppose S is a set with a partial order $\prec$ (so $\prec$ is reflexive, antisymmetric, and transitive). If $T \subseteq S$, a **maximal element** of T is any element $t \in T$ for which there is no strictly larger element $t' \in T$ (i.e., no $t' \in T$ with $t \prec t'$ and $t \neq t'$). An **ascending chain** in S is any sequence

$$\{x_i\}_{i=1}^{\infty} = (x_1, x_2, ...)$$

of elements $x_i \in S$ with $x_i \prec x_{i+1}$ for each i. An ascending chain $\{x_i\}_{i=1}^{\infty}$ is called **strict** if $x_i \neq x_{i+1}$ for each i, or is called **stationary** if there is a positive integer n for which $x_i = x_{i+1}$ for all $i \geq n$.

(B.2) Definition. A poset S is said to have the **ascending chain condition** (or **ACC**) if every ascending chain in S is stationary.

(B.3) Proposition. *Suppose S is a poset. The following are equivalent:*

(i)   *S has ACC.*
(ii) *There is no strict ascending chain in S.*
(iii) *Every nonempty subset of S has a maximal element.*

Proof. A strict chain would not be stationary; so (i) implies (ii). Suppose there is a nonempty subset T of S with no maximal element. For each $t \in T$ there is a nonempty set

$$L_t = \{t' \in T : t \prec t' , t \neq t'\}$$

of elements strictly larger than t. By the axiom of choice, there is a function $f : T \to T$ with $f(t) \in L_t$ for each $t \in T$. Then, beginning with any $t \in T$,

$$t , \; f(t) , \; f(f(t)) , \ldots$$

is a strict chain in S. So (ii) implies (iii). If we assume (iii), every ascending chain $\{x_i\}_{i=0}^{\infty}$ has a term x_n that is maximal in the set $\{x_i : i \geq 1\}$. Then $x_n = x_{n+1} = \ldots$. So (iii) implies (i). ∎

For each partial order $\prec$ on a set S, there is a "dual" partial order $\prec_d$ on S, defined so that $x \prec_d y$ if and only if $y \prec x$. Each definition and assertion about posets applies as well to $\prec_d$, and so has a dual version in which $\prec$ is reversed. Reversing $\prec$ in the preceding discussion, maximal elements become **minimal elements**, ascending chains become **descending chains**, and ACC becomes **DCC**, the **descending chain condition**. The terms "strict" and "stationary" are unchanged. In this language, the (equally valid) dual of (B.3) becomes:

(B.4) Corollary. *Suppose S is a poset. The following are equivalent:*

   (i)   *S has DCC.*
  (ii) *There is no strict descending chain in S.*
 (iii) *Every nonempty subset of S has a minimal element.* ∎

Of course, it is possible for an individual poset $(S, \prec)$ to have ACC but not DCC or DCC but not ACC, for a partial order and its dual can have different properties.

(B.5) Definitions. An R-module M is called **noetherian** (resp. **artinian**) if the set of R-submodules of M, partially ordered by $\subseteq$, has ACC (resp. DCC).

(B.6) Definitions. A ring R is called **left noetherian** (resp. **left artinian**) if R is noetherian (resp. artinian) as a left R-module — or, equivalently, if R has ACC (resp. DCC) on its set of left ideals.

(B.7) Correspondence Theorem. *If an R-linear map $f : M \to N$ has kernel K and image $f(M) = I$, there is an isomorphism of categories $F : \mathcal{C} \to \mathcal{D}$ from the poset $\mathcal{C}$ of R-submodules of M that contain K to the poset $\mathcal{D}$ of R-submodules of I (both posets ordered by containment). If $L \in \mathrm{Obj}(\mathcal{C})$,*

$$F(L) = f(L) = \{f(x) : x \in L\} .$$

The inverse $G : \mathcal{D} \to \mathcal{C}$ of F takes each object $J \in \mathrm{Obj}(\mathcal{D})$ to

$$G(J) = f^{-1}(J) = \{x \in M : f(x) \in J\} .$$

If $L_1 \subseteq L_2$ are objects of $\mathcal{C}$, there is an R-module isomorphism:

$$L_2/L_1 \cong f(L_2)/f(L_1) .$$

Proof. It is straightforward to verify that $f(L)$ is an R-submodule of $f(M) = I$, $f(L_1) \subseteq f(L_2)$, $f^{-1}(J)$ is an R-submodule of M that contains K, and, if $J_1 \subseteq J_2$ are objects of $\mathcal{D}$, then $f^{-1}(J_1) \subseteq f^{-1}(J_2)$. Thus the functors F and G are defined. Since $K \subseteq L$, $f^{-1}(f(L)) = L$; since $J \subseteq I$, $f(f^{-1}(J)) = J$; so G is inverse to F. The composite of R-linear maps:

$$L_2 \to f(L_2) \to f(L_2)/f(L_1)$$
$$x \mapsto f(x) \mapsto f(x) + f(L_1)$$

is surjective with kernel $f^{-1}(f(L_1)) = L_1$; so the induced R-linear map

$$L_2/L_1 \to f(L_2)/f(L_1)$$
$$x + L_1 \mapsto f(x) + f(L_1)$$

is an isomorphism. ∎

(B.8) Proposition. *For each exact sequence*

$$0 \longrightarrow N \xrightarrow{\ f\ } M \xrightarrow{\ g\ } Q \longrightarrow 0$$

in R-$\mathfrak{Mod}$, M is noetherian (resp. artinian) if and only if both N and Q are noetherian (resp. artinian).

Proof. By the Correspondence Theorem (B.7), any strict ascending or descending chain in N or Q would induce one in M, because f is injective and g is surjective. To see that a strict ascending or descending chain in M would induce such chains in N or Q, we only need to check that, for R-submodules $M_1 \subseteq M_2$ of M, if $f^{-1}(M_1) = f^{-1}(M_2)$ and $g(M_1) = g(M_2)$, then $M_1 = M_2$.

To verify this, we chase elements: If $x \in M_2$, $g(x) \in g(M_2) = g(M_1)$; so $g(x) = g(y)$ for some $y \in M_1$. Then $g(x - y) = 0$. By exactness at M, $x - y = f(z)$ for some $z \in N$. Since $x - y \in M_2$, $z \in f^{-1}(M_2) = f^{-1}(M_1)$. So $x - y = f(z) \in M_1$. But $y \in M_1$; so $x = y + f(z) \in M_1$. Thus $M_2 = M_1$. ∎

In particular, if N is an R-submodule of M, then M noetherian (resp. artinian) implies the same for both N and M/N, since the inclusion and canonical map make an exact sequence:

$$0 \longrightarrow N \xrightarrow{\ i\ } M \xrightarrow{\ c\ } M/N \longrightarrow 0 .$$

(B.9) Corollary. *If R is a left noetherian (resp. left artinian) ring, then every f.g. left R-module M is noetherian (resp. artinian).*

Proof. Induct on the number n of generators $s_1, \cdots, s_n$ of M. If $n = 1$, the result follows from the surjective R-linear map $R \to Rs_1 = M$, $r \mapsto rs_1$. If $n > 1$, use the exact sequence

$$ 0 \longrightarrow N \overset{\subseteq}{\longrightarrow} M \longrightarrow M/N \longrightarrow 0 \, , $$

where $N = Rs_1 + \cdots + Rs_{n-1}$ and

$$ \frac{M}{N} = \frac{N + Rs_n}{N} \cong \frac{N}{N \cap Rs_n} $$

by the Diamond Isomorphism Theorem. ∎

Up to now we have treated "noetherian" and "artinian" in a symmetric way. But it is the noetherian condition that addresses the opening question of this section:

(B.10) Proposition. *An R-module M is noetherian if and only if every R-submodule of M is finitely generated.*

Proof. Suppose the R-module M is noetherian. By (B.3) (i $\Rightarrow$ iii), the set of finitely generated R-submodules of M has a maximal member N. If $M \neq N$, there is an element $m \in M - N$. But then N is properly contained in the f.g. R-submodule $N + Rm$ of M, which is contrary to the choice of N. So $M = N$ and M is finitely generated. By (B.8), every R-submodule of M is also noetherian and hence finitely generated.

Conversely, suppose every R-submodule of M is finitely generated. Assume $\{M_i\}_{i=1}^{\infty}$ is an ascending chain of R-submodules of M. Then $\cup_{i=1}^{\infty} M_i$ is also an R-submodule of M, so it has a finite spanning set $s_1, \cdots, s_n$. Each s_i belongs to some M_j; so some M_m contains $\{s_1, \cdots, s_n\}$. Then $p \geq m$ implies

$$ M_m \subseteq M_p \subseteq \bigcup_{i=1}^{\infty} M_i \subseteq M_m \, ; $$

so the chain is stationary. ∎

(B.11) Corollary. *For a ring R, the following are equivalent:*

   (i)   *Every R-submodule of each finitely generated R-module is finitely generated.*

   (ii) *Every left ideal of R is finitely generated.*

   (iii) *R is left noetherian.*

Proof. Since $R = R \cdot 1$ is a f.g. R-module, (i) implies (ii). If (ii) holds, by (B.10) R is a noetherian left R-module, and hence a left noetherian ring, proving (iii). If (iii) holds, (B.9) says every f.g. R-module is noetherian; so (i) follows from (B.10). ∎

In elementary linear algebra, a linear transformation $T : V \to V$, from a finite-dimensional vector space V to itself, is injective if and only if it is surjective; for the rank of T plus the nullity of T is the dimension of V. For modules over a left noetherian ring this remains partially true:

(B.12) Proposition. *Suppose R is a left noetherian ring and M is a f.g. left R-module. If an R-linear map $f : M \to M$ is surjective, it is also injective.*

Proof. Since $f(0_M) = 0_M$, there is an ascending chain $\{ker(f^n)\}$ of R-submodules of M. By (B.9), M is noetherian. So for some n, $ker(f^{n+1}) = ker(f^n)$. If $x \in M$, then $x = f^n(y)$ for some $y \in M$ because f is surjective. Then $f(x) = 0_M$ implies $y \in ker(f^{n+1}) = ker(f^n)$; so $x = f^n(y) = 0$. ∎

The converse of (B.12) fails even in some simple cases:

(B.13) Example. Since every ideal of $\mathbb{Z}$ is principal, $\mathbb{Z}$ is a left noetherian ring. But multiplication by 2 is a $\mathbb{Z}$-linear map $\mathbb{Z} \to \mathbb{Z}$ that is injective but not surjective.

Note that the definitions and assertions of the preceding section have an equally valid version in which "R-module" is replaced by "right R-module," R-$\mathcal{M}o\eth$ is replaced by $\mathcal{M}o\eth$-R, and scalars are written on the right. However, it is possible for a ring to be left noetherian, or left artinian, but not right noetherian, or right artinian, respectively.

From a first course in group theory, the reader may recall the concept of a composition series for a finite group and the Jordan-Hölder Theorem, which asserts the uniqueness of the simple composition factors. Here we present the similar theory of composition series for modules, with simple groups replaced by simple modules (i.e., modules $M \neq 0$, with no submodules except 0 and M).

(B.14) Definitions. Suppose S is a set with a partial order $\prec$. A **finite chain** in S is any list $x_0, \cdots, x_n$ of elements of S, where n is a positive integer and

$$x_0 \succ x_1 \succ \cdots \succ x_n .$$

The number n of steps in this chain is called its **length**. If $\mathcal{C}$ and $\mathcal{D}$ are finite chains in S, and every element listed in $\mathcal{D}$ is also listed in $\mathcal{C}$, then $\mathcal{D}$ is called a **subchain** of $\mathcal{C}$, and $\mathcal{C}$ is called a **refinement** of $\mathcal{D}$.

For the rest of this section, suppose R is a ring, M is a left R-module, and $S(M)$ is the set of all R-submodules of M, partially ordered by containment.

(B.15) Definition. Two finite chains in $S(M)$,

$$M_0 \supseteq \cdots \supseteq M_m \quad \text{and} \quad N_0 \supseteq \cdots \supseteq N_n ,$$

are called **equivalent** if $m = n$ and there is a permutation σ of $\{0, \cdots, m-1\}$ for which there are R-linear isomorphisms

$$\frac{M_i}{M_{i+1}} \cong \frac{N_{\sigma(i)}}{N_{\sigma(i)+1}}$$

for each $i = 0, \cdots, m-1$.

(B.16) Theorem (O. Schreier). *In $S(M)$ any two finite chains*

$$M_0 \supseteq \cdots \supseteq M_m \quad \text{and} \quad N_0 \supseteq \cdots \supseteq N_n ,$$

with $M_0 = N_0$ and $M_m = N_n$, have equivalent refinements.

Proof. For $0 \le i < m$ and $0 \le j \le n$, let

$$M_{ij} = M_{i+1} + (M_i \cap N_j) .$$

For $0 \le i \le m$ and $0 \le j < n$, define

$$N_{ij} = N_{j+1} + (M_i \cap N_j) .$$

Then

$$M_i = M_{i0} \supseteq M_{i1} \supseteq \cdots \supseteq M_{in} = M_{i+1} , \quad \text{and}$$

$$N_j = N_{0j} \supseteq N_{1j} \supseteq \cdots \supseteq N_{mj} = N_{j+1} .$$

So the first chain has a refinement consisting of the M_{ij} with $0 \le i < m$ and $0 \le j < n$, followed by M_m at the end; and the second chain has a refinement consisting of the N_{ij} with $0 \le i < m$ and $0 \le j < n$, followed by N_n at the end. It only remains to prove that, for $0 \le i < m$ and $0 \le j < n$, there is an R-linear isomorphism:

$$\frac{M_{ij}}{M_{i,j+1}} \cong \frac{N_{ij}}{N_{i+1,j}} .$$

But this follows from the Diamond Isomorphism Theorem, since

$$\begin{aligned}
M_{i,j+1} + (M_i \cap N_j) &= M_{ij} , \\
N_{i+1,j} + (M_i \cap N_j) &= N_{ij} , \quad \text{and} \\
M_{i,j+1} \cap (M_i \cap N_j) &= (M_{i+1} \cap N_j) + (M_i \cap N_{j+1}) \\
&= N_{i+1,j} \cap (M_i \cap N_j) ;
\end{aligned}$$

so both quotients are R-linearly isomorphic to

$$\frac{M_i \cap N_j}{(M_{i+1} \cap N_j) + (M_i \cap N_{j+1})} \quad . \qquad \blacksquare$$

As with ascending and descending chains, call a finite chain $M_0 \supseteq \cdots \supseteq M_m$ in $S(M)$ **strict** if $M_i \neq M_{i+1}$ for each $i = 0, \cdots, m-1$. By the Correspondence Theorem (B.7), if $M_{i+1} \subsetneqq M_i$ are R-submodules of M, there is no intermediate R-submodule (i.e., no $N \in S(M)$ with $M_{i+1} \subsetneqq N \subsetneqq M_i$) if and only if the quotient M_i/M_{i+1} is simple.

(B.17) Definitions. A strict finite chain $M_0 \supsetneqq \cdots \supsetneqq M_m$ in $S(M)$ is called a **composition series** for M if it has no other strict refinement – that is, if $M_0 = M$, $M_m = 0$, and each M_i/M_{i+1} is simple. In that case, the simple quotients M_i/M_{i+1} ($0 \leq i < m$) are called the **composition factors** of the composition series.

The composition factors are almost uniquely determined by M:

(B.18) Corollary. (Jordan-Hölder) *Any two composition series for M are equivalent.*

Proof. By Schreier's Theorem the two composition series have equivalent refinements. Deleting repetitions of modules, they also have equivalent strict refinements. $\qquad \blacksquare$

(B.19) Corollary. *An R-module M has a composition series if and only if M is both noetherian and artinian.*

Proof. Suppose M has a composition series of length m. By Schreier's Theorem (B.16), and after deleting repetitions of submodules, every strict finite chain $M_0 \supsetneqq \cdots \supsetneqq M_r$ in $S(M)$ has a strict refinement equivalent to the given composition series; so $r \leq m$. So M has no strict ascending or strict descending chain of submodules.

Conversely, suppose $S(M)$ has no strict ascending or descending chains. By (B.3) (ii $\Rightarrow$ iii), since M has at least one proper submodule 0, M has a maximal proper submodule M_1. If $M_1 \neq 0$, we can repeat the argument (since $S(M_1) \subset S(M)$) to show M_1 has a maximal proper submodule M_2. Continuing, we create a chain $M \supsetneqq M_1 \supsetneqq M_2 \supsetneqq \cdots$, which must be finite, since $S(M)$ has no strict descending chains. For this chain to stop, it must reach $M_m = 0$. By the choice of each M_i, this chain has no other strict refinement, so it is a composition series for M. $\qquad \blacksquare$

Combining this with (B.9) proves:

(B.20) Corollary. *If R is a ring that is both left noetherian and left artinian, every f.g. left R-module M has a composition series.* ■

(B.21) Note. Left artinian rings are automatically left noetherian by the Hopkins-Levitzki Theorem. For a clear exposition of this theorem, see Lam [91, Theorem (4.15)]. But artinian modules need not be noetherian.

As with chain conditions, the definitions and assertions about composition series remain valid if "left" is replaced by "right" and "R-module" is replaced by "right R-module."

SPECIAL SYMBOLS

The page number locates the definition.

REFERENCES

Boldface page numbers in parentheses locate citations of each entry in this book.

Almkvist, G.

[73] Endomorphisms of finitely generated projective modules over a commutative ring, *Ark. Mat.* **11** (1973), 263–301. **(94)**

Alperin, R. C. and Dennis, R. K.

[79] K_2 of quaternion algebras, *J. Algebra* **56** (1979), 262–273. **(556)**

Anderson, D. F.

[78] Projective modules over subrings of $k[X, Y]$ generated by monomials, *Pacific J. Math.* **79** (1978), 5–17. **(515)**

Artin, E.

[27] Beweis des allgemeinen Reziprozitätsgesetzes, *Hamb. Abh.* **5** (1927), 353–363. **(596)**

Baeza, R.

[78] *Quadratic Forms over Semilocal Rings*, Lecture Notes in Math. 655, Springer-Verlag, Berlin - New York 1978. **(159)**

Bak, A.

[69] On modules with quadratic forms, in *Algebraic K-Theory and its Geometric Applications*, Lecture Notes in Math. 108, Springer-Verlag, Berlin 1969, 55–66. **(82)**

Bass, H.

[64A] K-theory and stable algebra, *Inst. Hautes Études Sci. Publ. Math.* **22** (1964), 5–60. **(131, 329, 344, 358)**

[64B] The stable structure of quite general linear groups, *Bull. Amer. Math. Soc.* **70** (1964), 429–433. **(342)**

[68] *Algebraic K-Theory* W.A. Benjamin, Inc., New York - Amsterdam 1968. **(132, 185, 199, 312, 378, 484, 485, 486, 515, 561)**

[74] Introduction to some methods of algebraic K-theory. Expository Lectures from the CBMS Regional Conference held at Colorado State University, Ft. Collins, Colo., August 24-28, 1973, *C.B.M.S. Reg. Conf. Ser.* No. 20, Amer. Math. Soc., Providence, R.I. 1974. **(333, 483)**

Bass, H., Heller, A., and Swan, R. G.

[64] The Whitehead group of a polynomial extension, *Inst. Hautes Etudes Sci. Publ. Math.* **22** (1964), 61–79. **(484)**

Bass, H., Lazard, M. and Serre, J. - P.

[64] Sous-groupes d'indice fini dans $SL(n, \mathbb{Z})$, *Bull. Amer. Math. Soc.* **70** (1964), 385–392. **(370)**

Bass, H., Milnor, J. and Serre, J. - P.
  [67] Solution of the congruence subgroup problem for $SL_n(n \geq 3)$ and $Sp_{2n}(n \geq$
        2), *Inst. Hautes Études Sci. Publ. Math.* **33** (1967), 59–137.   (**370, 374,**
        392, 395–396)
Bass, H. and Schanuel, S.
  [62] The homotopy theory of projective modules, *Bull. Amer. Math. Soc.* **68**
        (1962), 425–428.   (**329**)
Bass, H. and Tate, J.
  [73] The Milnor ring of a global field, in *Algebraic K-Theory, II: "Classical"*
        Algebraic K-Theory and Connections with Arithmetic (Proc. Conf., Seattle,
        Wash., Battelle Memorial Inst., 1972), Lecture Notes in Math. 342, Springer-
        Verlag, Berlin 1973, 349–446.   (**525, 533, 549, 551, 553, 554, 606**)
Bergman, G. M.
  [74] Coproducts and some universal ring constructions, *Trans. Amer. Math.*
        Soc. **200** (1974), 33–38.   (**113**)
Borel, A. and Serre, J. - P.
  [58] Le théorème de Riemann-Roch, *Bull. Soc. Math. France* **86** (1958), 97–136.
        (**83**)
Borevich, Z. I. and Vavilov, N. A.
  [85] Location of subgroups in the general linear group over a commutative ring,
        Proc. Steklov Inst. Math. **165** (1985), 27–46; translation from *Trudy Mat.*
        Inst. Steklov **165** (1984), 24–42.   (**368**)
Bourbaki, N.
  [74] *Elements of Mathematics. Algebra, Part I*, Hermann, Paris; Addison-Wesley
        Publ. Co., Reading, Mass. 1974, translation from *Elements de Mathema-*
        tique, Algebre, Hermann, Paris 1971, Chapters 1–3.   (**548**)
Brauer, R.
  [32] Über die algebraische Struktur von Schiefkörpern, *J. Reine Angew. Math.*
        166 (1932), 241–252; also in *Richard Brauer: Collected Papers*, Volume 1
        (Ed. by P. Fong and W.J. Wong), M.I.T. Press, Cambridge, Mass. 1980,
        103–114.   (**620**)
Brenner, J. L.
  [60] The linear homogeneous group III, *Ann. of Math.* **71** (1960), 210–223.
        (**370**)
Brouwer, L. E. J.
  [12] Über Abbildungen vom Mannigfaltigkeiten, *Math. Ann.* **71** (1911/ 1912),
        97–115.   (**112**)
Browder, F. E. (Editor)
  [76] *Mathematical Developments Arising from Hilbert Problems*, Proceedings
        of Symposia in Pure Mathematics, Vol. XXVIII, Amer. Math. Soc., Provi-
        dence, R.I. 1976.   (**595**)
Burnside, W.
  [55] *Theory of Groups of Finite Order*, 2nd ed., Dover Publications, Inc., New
        York 1955.   (**137**)
Cartan, H. and Eilenberg, S.
  [99] *Homological Algebra, with an Appendix by David A. Buchsbaum*, Reprint
        of the 1956 original, Princeton Univ. Press, Princeton, N.J. 1999.   (**471**)

Cassels, J. W. S. and Frohlich, A. (Editors)

[86] *Algebraic Number Theory*, Proceedings of the instructional conference held at the University of Sussex, Brighton, September 1–17, 1965, Reprint of the 1967 original, Academic Press, Inc. (Harcourt Brace Jovanovich, Publishers), London 1986. **(599)**

Chen, H. Y.

[94] K_0-groups of connected rings, *Nanjing Daxue Xuebao Shuxue Bannian Kan* **11** (1994), 106–108. **(251)**

Cohn, P. M.

[66] Some remarks on the invariant basis property, *Topology* **5** (1966), 215–228. **(112, 113)**

[94] Rings with a single projective, *Algebra Colloq.* **1** (1994), 121–127. **(113)**

Curtis, C. W. and Reiner, I.

[62] *Representation Theory of Finite Groups and Associative Algebras*, Pure and Applied Mathematics, Vol. XI, John Wiley & Sons, New York - London 1962. **(272)**

[87] *Methods of Representation Theory, Vol. II, With Applications to Finite Groups and Orders*, Pure and Applied Mathematics, John Wiley & Sons, New York 1987. **(364, 434, 483)**

Dedekind, R.

[93] Eleventh supplement to: Dirichelt, P. G. Lejeune, in *Vorlesungen über Zahlentheorie*, Reprint of the 1893 4th edition, Chelsea Publishing Co., New York 1968. **(84, 210, 226)**

Dennis, R. K. and Geller, S. C.

[76] K_i of upper triangular matrix rings, *Proc. Amer. Math. Soc.* **56** (1976), 73–78. **(185)**

Dennis, R. K. and Stein, M. R.

[75] K_2 of discrete valuation rings. *Advances in Math.* **18** (1975), 182–238. **(455, 565)**

Dieudonné, J.

[43] Les déterminants sur un corps non commutatif, *Bull. Soc. Math. France* **71** (1943), 27–45. **(329)**

Eilenberg, S. and Steenrod, N.

[52] *Foundations of Algebraic Topology*, Princeton Univ. Press, Princeton, N.J. 1952. **(431, 456)**

Elman, R. and Lam, T. Y.

[72] Pfister forms and K-theory of fields, *J. Algebra* **23** (1972), 181–213. **(529)**

Estes, D. and Ohm, J.

[67] Stable range in commutative rings, *J. Algebra* **7** (1967), 343–362. **(130)**

Fröhlich, A.

[83] *Galois Module Structure of Algebraic Integers*, Ergebnisse der Mathematik und ihrer Grezgebiete (3), Springer-Verlag, Berlin 1983. **(264)**

Gauss, C. F.

[86] *Disquisitiones Arithmeticae*, translated and with a preface by Arthur A. Clarke, Revised by William C. Waterhouse, Cornelius Greither and A. W. Grootendorst and with a preface by Waterhouse, Springer-Verlag, New York 1986. **(537, 592)**

Grayson, D.

[76] Higher algebraic K-theory. II (after Daniel Quillen), *Algebraic K-Theory (Proc. Conf. Northwestern Univ., Evanston, Ill., 1976)*, Lecture Notes in Math. 551, Springer-Verlag, Berlin 1976, 217–240. **(486)**

[79] Localization for flat modules in algebraic K-theory, *J. Algebra* **61** (1979), 463–496. **(608)**

Hahn, A. J. and O'Meara, O. T.

[89] *The Classical Groups and K-Theory*, with a foreword by J. Dieudonné, Grundlehren der Mathematischen Wissenschaften 291, Springer-Verlag, Berlin 1989. **(360)**

Herstein, I. N.

[75] *Topics in Algebra*, 2nd ed., Xerox College Publishing, Lexington, Mass. - Toronto, Ont. 1975. **(428, 620)**

Hilbert, D.

[98] *The Theory of Algebraic Number Fields*, translated from the German and with a preface by Iain T. Adamson, with an introduction by Franz Lemmermeyer and Norbert Schappacher, Springer-Verlag, Berlin 1998. **(570)**

Hutchinson, K.

[90] A new approach to Matsumoto's theorem, *K-Theory* **4** (1990), 181–200. **(557)**

Jacobson, N.

[85] *Basic Algebra. I*, 2nd ed., W.H. Freeman and Company, New York 1985. **(94, 320, 548, 599)**

[89] *Basic Algebra II*, 2nd ed., W.H. Freeman and Company, New York 1989. **(217, 223, 603, 642)**

Jones, G. A.

[86] Congruence and noncongruence subgroups of the modular groups; a survey, *Proceedings of Groups - St. Andrews 1985*, London Math. Soc. Lecture Note Ser. 121, Cambridge Univ. Press, Cambridge 1986, 223–234. **(370)**

Kelley, J. L. and Spanier, E. H.

[68] Euler characteristics, *Pacific J. Math.* **26** (1968), 317–339. **(94)**

Kersten, I.

[90] *Brauergruppen von Körpern* Aspects of Mathematics, D6, Friedr. Vieweg & Sohn, Braunschweig 1990. **(555, 642)**

Keune, F.

[75] $(t^2 - t)$-reciprocities on the affine line and Matsumoto's theorem, *Invent. Math.* **28** (1975), 185–192. **(557, 560, 562)**

[78] The relativization of K_2, *J. Algebra* **54** (1978), 159–177. **(447)**

Klein, F. and Fricke, R.

[92] *Vorlesungen über die Theorie der elliptischen Modulfunktionen*, Vol. 1, Teubner, Leipzig 1890-92. **(370)**

Kolster, M.

[85] K_2 of noncommutative local rings, *J. Algebra* **95** (1985), 173–200. **(565)**

Kubota, T.

[66] Ein arithmetischer Satz über eine Matrizengruppe, *J. Reine Angew. Math.* **222** (1966), 55–57. **(374, 378, 394)**

Lam, T. - Y.

[68A] Artin exponent of finite groups, *J. Algebra* **9** (1968), 94–119. **(312)**

[68B] Induction theorems for Grothendieck groups and Whitehead groups of finite groups, *Ann. Sci. École Norm. Sup.* **1** (1968), 91–148.   **(312)**

[76] Series summation of stably free modules, *Quart. J. Math. Oxford Ser. (2)* **27** (1976), 37–46.   **(113)**

[78] *Serre's Conjecture*, Lecture Notes in Math. 635, Springer-Verlag, Berlin - New York 1978.   **(119)**

[80] *The Algebraic Theory of Quadratic Forms*, Revised second printing, Mathematics Lecture Note Series, Benjamin/Cummings Publishing Co., Inc., Advanced Book Program, Reading, Mass. 1980.   **(81, 609)**

[91] *A First Course in Noncommutative Rings*, Graduate Texts in Math. 131, Springer-Verlag, New York 1991.   **(655)**

Lang, S.

[93] *Algebra*, 3rd ed., Addison-Wesley Publishing Co., Reading, Mass. 1993.   **(258, 320, 599)**

Lemmermeyer, F.

[00] *Reciprocity Laws. From Euler to Eisenstein*, Springer Monographs in Mathematics, Springer-Verlag, Berlin 2000.   **(595)**

Lenstra, H. W., Jr.

[76] K_2 of a global field consists of symbols, *Algebraic K-Theory (Proc. Conf., Northwestern Univ., Evanston, Ill., 1976)*, Lecture Notes in Math. 551, Springer-Verlag, Berlin 1976, 69–73.   **(643)**

Maazen, H. and Stienstra, J.

[78] A presentation for K_2 of split radical pairs, *J. Pure Appl. Algebra* **10** (1977/78), 271–294.   **(565)**

MacLane, S. and Birkhoff, G.

[88] *Algebra*, 3rd ed., Chelsea Publishing Co., New York 1988.   **(320)**

Magnus, W.

[35] Über n-dimensionale Gittertransformationen, *Acta Math.* **64** (1935), 353–367.   **(402)**

Magurn, B. A., van der Kallen, W. and Vaserstein, L. N.

[88] Absolute stable rank and Witt cancellation for noncommutative rings, *Invent. Math.* **91** (1988), 525–542.   **(82, 131)**

Masley, J. M. and Montgomery, H. L.

[76] Cyclotomic fields with unique factorization, *J. Reine Angew. Math.* **286/287** (1976), 248–256.   **(210)**

Matsumoto, H.

[69] Sur les sous-groupes arithmétiques des groupes semi-simples déployés, *Ann. Sci. École Norm. Sup.* **2** (1969), 1–62.   **(557)**

Matsumura, H.

[89] *Commutative Ring Theory*, translated from the Japanese by M. Reid, 2nd ed., Cambridge Studies in Advanced Mathematics 8, Cambridge Univ. Press, Cambridge 1989.   **(128)**

McConnell, J. C. and Robson, J. C.

[87] *Noncommutative Noetherian Rings*, with the cooperation of L. W. Small, Pure and Applied Mathematics, John Wiley & Sons, Ltd., Chichester 1987.   **(110, 132)**

Mennicke, J. L.

[65] Finite factor groups of the unimodular group, *Ann. of Math.* **81** (1965), 31–37.   **(370, 374)**

Merkurjev, A. S.
  [85] The group K_2 for a local field, *Proc. Steklov Inst. Math.* **165** (1985),
      125–129, translated from *Trudy Mat. Inst. Steklov.* **165** (1984), 115–118.
      (605)
Milnor, J.
  [70] Algebraic K-theory and quadratic forms, *Invent. Math.* **9** (1969/1970),
      318–344. **(521, 529, 540, 543)**
  [71] *Introduction to Algebraic K-Theory*, Annals of Mathematical Studies, No.
      72, Princeton Univ. Press, Princeton, N.J. 1971. **(402, 403, 413, 537,
      539, 543, 557, 590, 592, 607, 636, 643, 644)**
  [78] Analytic proofs of the "hairy ball theorem" and the Brouwer fixed-point
      theorem, *Amer. Math. Monthly* **85** (1978), 521–524. **(112)**
Moore, C. C.
  [68] Group extensions of p-adic and adelic linear groups, *Inst. Hautes Études
      Sci. Publ. Math.* **35** (1968), 157–222. **(605, 606)**
Nesterenko, Yu. P. and Suslin, A. A.
  [90] Homology of the general linear group over a local ring, and Milnor's K-
      theory, *Math. USSR-Izv.* **34** (1990), 121-145, translated from *Izv. Akad.
      Nauk SSSR Ser. Mat.* **53** (1989), 121–146. **(533)**
Neukirch, J.
  [99] *Algebraic Number Theory*, translated from the 1992 German original and
      with a note by Norbert Schappacher, with a foreward by G. Harder, Grundlehren
      der Mathematischen Wissenschaften 322, Springer-Verlag, Berlin 1999. **(601,
      605)**
Nielsen, J.
  [24] Die Gruppe der dreidimensionalen Gittertransformationen, *Det. Kgl. Danske
      Videnskabernes Selskab. Math-fysiske Meddelelser V, 12, Kopenhagen* (1924),
      1–29. **(402)**
Ojanguren, M. and Sridharan, R.
  [71] Cancellation of Azumaya algebras, *J. Algebra* **18** (1971), 501–505. **(109)**
Quillen, D.
  [72] On the cohomology and K-theory of the general linear groups over a finite
      field, *Ann. of Math.* **96** (1972), 552–586. **(524)**
  [73] Higher algebraic K-theory, I, in *Algebraic K-Theory, I: Higher K-Theories
      (Proc. Conf. Battelle Memorial Inst., Seattle, Wash. 1972*, Lecture Notes in
      Math. 341, Springer-Verlag, Berlin 1973, 85–147. **(608)**
  [76] Projective modules over polynomial rings, *Invent. Math.* **36** (1976), 167–
      171. **(110)**
Reiner, I.
  [75] *Maximal Orders*, London Mathematical Society Monographs, No. 5, Aca-
      demic Press (A subsidiary of Harcourt Brace Jovanovich, Publishers), London
      - New York 1975. **(642)**
Rentschler, R. and Gabriel, P.
  [67] Sur la dimension des anneaux et ensembles ordonnés, *C. R. Acad. Sci.
      Paris Sér. A-B* **265** (1967), A712–A715. **(131)**
Rim, D. S.
  [59] Modules over finite groups, *Ann. of Math.* **69** (1959), 700–712. **(84,
      437, 471)**

Rosenberg, J.
  [94] *Algebraic K-Theory and its Applications*, Graduate Texts in Math. 147, Springer-Verlag, New York 1994. **(61, 557)**
Rowen, L. H.
  [91] *Ring Theory*. Student edition, Academic Press, Inc., Boston, Mass. 1991. **(132)**
Schur, J.
  [04] Über die Darstellung der endlichen Gruppen dutch gebrochene lineare Substitutionem, *J. Reine Angew. Math.* **127** (1904), 20–40. **(412)**
Serre, J. - P.
  [58] Modules projectifs et espaces fibrés à fibre vectorielle, in *Séminaire P. Dubreil, M. - L. Dubreil - Jacotin et C. Pisot, 1957/58, Fasc. 2, Exposé 23, Secrétariat mathématique, Paris.* **(84, 110, 517)**
  [73] *A Course in Arithmetic*, translated from the French, Graduate Texts in Math. 7, Springer-Verlag, New York - Heidelberg 1973. **(395)**
  [77] *Linear Representations of Finite Groups*, translated from the 2nd French edition by Leonard L. Scott, Graduate Texts in Math. 42, Springer-Verlag, New York - Heidelberg 1977. **(312)**
  [79] *Local Fields*, translated from the French by Marvin Jay Greenberg, Graduate Texts in Math. 67, Springer-Verlag, New York - Berlin 1979. **(599, 604)**
Silvester, J. R.
  [81] *Introduction to Algebraic K-Theory*, Chapman & Hall, London - New York 1981. **(455)**
Springer, T. A.
  [72] A remark on the Milnor ring, *Nederl. Akad. Wetensch. Proc. Ser.* **A 75** = *Indag. Math.* **34** (1972), 100–105. **(533)**
Srinivas, V.
  [96] *Algebraic K-Theory*, 2nd ed., Progress in Math. 90, Birkhäuser Boston, Inc., Boston, Mass. 1996. **(642)**
Stafford, J. T.
  [90] Absolute stable rank and quadratic forms over noncommutative rings, *K-Theory* **4** (1990), 121–130. **(82, 131)**
Stein, M. R.
  [71] Relativizing functors on rings and algebraic K-theory, *J. Algebra* **19** (1971), 140–152. **(432)**
  [73] Surjective stability in dimension 0 for K_2 and related functors, *Trans. Amer. Math. Soc.* **178** (1973), 165–191. **(595)**
  [78] Stability theorems for K_1, K_2 and related functors modeled on Chevalley groups, *Japan J. Math.* **4** (1978), 77–108. **(130)**
Steinberg, R.
  [62] Générateurs, relations et revêtements de groupes algébriques. *Colloq. Théorie des Groupes Algébriques (Bruxelles, 1962), Librairie Universitaire Louvain*; Gauthier-Villars, Paris (1962), 113–127. **(403, 413)**
Steinitz, E.
  [11] Rechteckige Systeme und Moduln in algebraischen Zahlenkörpern I, *Math. Ann.* **71** (1911), 328–354. **(243)**

668   *References*

[12] Rechtekige Systeme und Moduln in algebraischen Zahlenkörpern II, *Math.*
      Ann. **72** (1912), 297–345.   **(243)**
Stewart, I. and Tall, D.
[87] *Algebraic Number Theory*, 2nd ed., Chapman & Hall, London 1987.   **(210,**
      239, 455)
Suslin, A. A.
[76] Projective modules over polynomial rings are free, *Soviet Math. Dokl.* **17**
      (1976), 1160–1164, translated from *Dokl. Akad. Nauk SSSR* **229** (1976),
      1063–1066.   **(110)**
[77] On the structure of the special linear group over polynomial rings, *Math.*
      USSR Izv. **11** (1977), 221–238, translated from *Izv. Akad. Nauk SSSR Ser.*
      Mat. **41** (1977), 235–252.   **(346)**
[79] The cancellation problem for projective modules and related topics, in *Ring*
      Theory (Proc. Conf., Univ. Waterloo, Waterloo, 1978), Lecture Notes in
      Math 734, Springer-Verlag, Berlin 1979, 323–338.   **(112)**
[82] Mennicke symbols and their applications in the *K*-theory of fields, in *Al-*
      gebraic K-Theory. Part I (Oberwolfach, 1980), Lecture Notes in Math. 966,
      Springer-Verlag, Berlin - New York 1982, 334–356.   **(521)**
Suslin, A. A. and Yarosh, V. A.
[91] Milnor's K_3 of a discrete valuation ring, *Algebraic K-Theory*, Adv. Soviet
      Math. 4, Amer. Math. Soc., Providence, RI 1991, 155–170.   **(533)**
Swan, R. G.
[59] Projective modules over finite groups, *Bull. Amer. Math. Soc.* **65** (1959),
      365–367.   **(84)**
[60] Induced representations and projective modules, *Ann. of Math.* **71** (1960),
      552–578.   **(203, 312)**
[68] *Algebraic K-Theory*, Lecture Notes in Math. 76, Springer-Verlag, Berlin -
      New York 1968.   **(86, 115, 195, 197, 251, 515)**
[70] *K-Theory of Finite Groups and Orders*, Notes taken by E. Graham Evans,
      Lecture Notes in Math. 149, Springer- Verlag, Berlin - New York 1970.
      (312)
[71] Excision in algebraic *K*-theory, *J. Pure Appl. Algebra* **1** (1971), 221–252.
      (446, 447, 467)
[74] A Cancellation theorem for projective modules in the metastable range,
      Invent. Math. **27** (1974), 23–43.   **(132)**
Tate, J.
[76] Relations between K_2 and Galois cohomology, *Invent. Math.* **36** (1976),
      257–274.   **(642)**
[77] On the torsion in K_2 of fields, in *Algebraic Number Theory (Kyoto Internat.*
      Sympos. Res. Inst. Math. Sci., Univ. Kyoto, 1976, Japan Soc. Promotion
      Sci., Tokyo 1977, 243–261.   **(605)**
Taylor, M. J.
[81] On Fröhlich's conjecture for rings of integers of tame extensions, *Invent.*
      Math. **63** (1981), 41–79.   **(264)**
Tulenbaev, M. S.
[81] The Schur multiplier of the group of elementary matrices of finite order,
      J. Soviet Math. **17** (1981), 2062–2067, translated from *Zap. Nauchn. Sem.*
      Leningrad. Otdel. Mat. Inst. Steklov. (LOMI) **86** (1979), 162–169.   **(348)**

Uchida, K.
  [71] Class numbers of imaginary abelian number fields. III, *Tôhoku Math. J.*
        23 (1971), 573–580. **(210)**
van der Waerden, B. L.
  [91] *Algebra. Vol. I, II*, Based in part on lectures by E. Artin and E. Noether,
        translated from the 7th German edition by Fred Blum and John R. Schulen-
        berger, Springer-Verlag, New York 1991. **(473)**
Vaserstein, L. N.
  [69] K_1-theory and the congruence subgroup problem, *Math. Notes* **5** (1969),
        141–148, translated from *Mat. Zametki* **5** (1969), 233–244. **(346, 348)**
  [71] The stable range of rings and the dimension of topological spaces, *Funct.
        Anal. Appl.* **5** (1971), 102–110, translated from *Funk. Anal. i Prilozen.* **5**
        (1971), 17–27. **(122)**
  [81] On the normal subgroups of GL_n over a ring, in *Algebraic K-Theory,
        Evanston 1980 (Proc. Conf., Northwestern Univ., Evanston, Ill., 1980)*, Lec-
        ture Notes in Math. 854, Springer-Verlag, Berlin - New York 1981, 456–465.
        (368)
Voevodsky, V.
  [99] Voevodsky's Seattle lectures: *K*-theory and motivic cohomology, in *Alge-
        braic K-Theory (Seattle, WA, 1997)*, Proc. Sympos. Pure Math. 67, Amer.
        Math. Soc., Providence, R.I. 1999, 283-303 **(531)**
Wadsworth, A. R.
  [86] Merkurjev's elementary proof of Merkurjev's theorem, in *Applications of
        Algebraic K-Theory to Algebraic Geometry and Number Theory, Parts I, II
        (Boulder, Colo., 1983)*, Contemp. Math. 55, Amer. Math. Soc., Providence,
        R.I. 1986, 741–776. **(642)**
Washington, L. C.
  [97] *Introduction to Cyclotomic Fields*, 2nd ed., Graduate Texts in Math. 83,
        Springer-Verlag, New York 1997. **(210, 472, 475, 476, 636)**
Whitehead, J. H. C.
  [41] On incidence matrices, nuclei and homotopoy types, *Ann. of Math.* **42**
        (1941), 1197–1239. **(329)**
  [49] Combinatorial homotopy I, II, *Bull. Amer. Math. Soc.* **55** (1949), 213–245,
        453–496. **(329)**
  [50] Simple homotopy types, *Amer. J. Math.* **72** (1950), 1–57. **(329)**
Wilson, J. S.
  [72] The normal and subnormal structure of general linear groups, *Proc. Cam-
        bridge Philos. Soc.* **71** (1972), 163–177. **(368)**
Witt, E.
  [37] Theorie der quadratischen Formen in beliebigen Korpern, *J. Reine Angew.
        Math.* **176** (1937), 31–44. **(79, 82, 158)**

INDEX